Proceedings of the Second ISAAC Congress

International Society for Analysis, Applications and Computation

Volume 7

The titles published in this series are listed at the end of this volume.

Proceedings of the Second ISAAC Congress

Volume 1

This project has been executed with Grant No. 11–56 from the Commemorative Association for

the Japan World Exposition (1970)

Edited by

Heinrich G.W. Begehr
Freie Universität, Berlin, Germany

Robert P. Gilbert
University of Delaware, Newark, Delaware, U.S.A.

and

Joji Kajiwara
Kyushu University, Fukuoka, Japan

KLUWER ACADEMIC PUBLISHERS
DORDRECHT / BOSTON / LONDON

A C.I.P. Catalogue record for this book is available from the Library of Congress.

ISBN 0-7923-6597-6
ISBN 0-7923-6754-5 (Set)

Published by Kluwer Academic Publishers,
P.O. Box 17, 3300 AA Dordrecht, The Netherlands.

Sold and distributed in North, Central and South America
by Kluwer Academic Publishers,
101 Philip Drive, Norwell, MA 02061, U.S.A.

In all other countries, sold and distributed
by Kluwer Academic Publishers,
P.O. Box 322, 3300 AH Dordrecht, The Netherlands.

Printed on acid-free paper

Printed in the Netherlands.

Preface

This book is the Proceedings of the Second ISAAC Congress. **ISAAC** is the acronym of

the International Society for Analysis,
its Applications and Computation.

The president of ISAAC is Professor Robert P. Gilbert, the second named editor of this book, e-mail: gilbert@math.udel.edu .

The Congress is world-wide valued so highly that an application for a grant has been selected and this project has been executed with Grant No. 11-56 from

✿**the Commemorative Association for**
the Japan World Exposition (1970).

The finance of the publication of this book is exclusively the said Grant No. 11-56 from ✿. Thus, a pair of each one copy of two volumes of this book will be sent to all contributors, who registered at the Second ISAAC Congress in Fukuoka, free of charge by the Kluwer Academic Publishers.

Analysis is understood here in the broad sense of the word, including differential equations, integral equations, functional analysis, and function theory. It is the purpose of **ISAAC** to promote analysis, its applications, and its interaction with computation. With this objective, ISAAC organizes international Congresses for the presentation and discussion of research on analysis. **ISAAC** welcomes new members and those interested in joining **ISAAC** are encouraged to look at the web-site

http://www.math.udel.edu/ gilbert/isaac/index.html

and

http://www.math.fu-berlin.de/ rd/ag/isaac/newton/index.html.

The First **ISAAC** Congress was held at the University of Delaware, Newark, Delaware, U. S. A. during June 1997. The Second **ISAAC** Congress was held during August 1999 at Fukuoka Institute of Technology, Fukuoka, Japan.

As a novelty at the opening ceremony of the second congress ten ISAAC Awards were handed over to those younger mathematicians who were elected by an Award Committee. The awardees were

Daniele Andreucci (Roma, Italy; pdes, free boundary problems),
Dao–Qing Dai (Guangzhou, China; complex methods in pdes, wavelet analysis),
Lê Hai Khôi (Hanoi, Vietnam; several complex variables, pdes),
P. Alexandrovich Krutitskii (Moscow, Russia; pdes, applications),
Pei–Chu Hu (Jinan, China; complex analysis),
Song–Ying Li (Irvine, Ca, USA; complex analysis of one and several variables),
Stefano Marmi (Firence, Italy; dynamical systems, global analysis),
Sergiy A. Plaksa (Kiev, Ukraine; complex analysis, hypercomplex analysis),
Takashi Shioya (Fukuoka, Japan; differential geometry, global analysis),
Dachun Yang (Beijing, China; harmonic analysis, pdes).

The price was US $ 500 for those awardees who were attending the congress. The fund for these awards was raised by the local organizers Professor J. Kajiwara and Professor M. Nishihara, the representative and the secretary-general of the said Grant No. 11-56 from respectively, with which this project has been executed. The award committee consisted of H. Begehr, A. Bourgeat, A.K. Brüning, A. Fasano, C.H. Fitzgerald, A. Jeffrey, J. Kajiwara, P. Tamrazov, C.C. Yang (Chairman).
The ISAAC society will continue in encouraging young researchers in analysis, its applications and computation by awarding prizes for high quality research at its international congresses.

The proceedings of the First ISAAC congress were published in 8 volumes. Only 4 of them have appeared in this series. The 8 volumes are

Partial Differential and Integral Equations, eds. H.G.W. Begehr, R.P. Gilbert, G.–C. Wen, this series 2 (1999).

Reproducing Kernels and their Applications, eds. S. Saitoh, R. Alpay, J.A. Ball, T. Ohsawa, this series 3 (1999).

Recent Developments in Complex Analysis and Computer Algebra, eds. R.P. Gilbert, J. Kajiwara, Y.S. Xu, this series 4 (1999).

Direct and Inverse Problems of Mathematical Physics, eds. R.P. Gilbert, J. Kajiwara, Y.S. Xu, this series 5 (2000).

Inverse Problems, Tomography, and Image Processing. Proc. of sessions from 1. int. congress of ISAAC. Ed. A.G. Ramm. Plenum Press, New York, (1998).

Spectral and Scattering Theory. Proc. of sessions from 1. int. congress of ISAAC. Ed. A.G. Ramm. Plenum Press, New York, (1998).

Dirac Operators in Analysis. Eds. J. Ryan, D. Stuppa. Pitman Research Notes No. 394. Addison Wesley Longman, Harlow, (1998).

Orthogonal Polynomials and Computer Algebra. Eds. R.A. Askey, W. Koepf, T.H. Koornwinder. Special issue of Journal of Symbolic Computation. J. Symb. Comput. 28, (6) (1999).

Herein are the Proceedings of the Second Congress. From the Table of Contents it is evident that Complex Analysis remains the dominant research area of the attendees of the Congress. After that there is a concentration in Real Analysis, Numerical and Computational Mathematics, and also Applied Mathematics. In the area of Complex Analysis there are many sub-areas such as Hardy Spaces, Riemann-Hilbert problems, hyperanalytic function theory, conformal mappings, Nevanlinna theory, meromorphic functions, complex dynamics, Riemann surfaces, entire functions, quasianalytic functions, and Bergman spaces, to name a few. There is, of course, some overlap of topics. For instance, in the area of differential equations which normally belongs to Real Analysis there is also the Function Theoretic school; moreover, we will find numerous Applied papers. Also in the area of Numerical Analysis and Computation we will find papers on numerical conformal mappings.

In Real Analysis we find papers on the central limit theorem, oscillatory integrals, wavelets, Brownian intersection measures, Carathéodory convergence, iterative function systems, distribution theory, Minkowski inequalities, operator algebras, Fourier hyperfunctions, Stieltjes moments, Whitney jets, Fatou sets, fractional powers of operators, ultradistributions, etc..

In the Applications and Differential Equations we have Hamiltonian theory, the Fredholm determinant, wavelet analysis, Schwarzschild space-time, singular integral operators, eigenvalue problems, quadrature formulae for hypersingular integrals, remote sensing, plasticity, sound synthesis, Navier Stokes equations, nonlinear parabolic equations, Schrödinger equations, Sobolev spaces, wave propagation, gauge-invariant equations.

The subarea of Function Theoretic methods in differential equations was also represented with papers on Clifford analysis, Cauchy singular integrals, hyperanalytic boundary value problems, monogenic functions, oblique derivative problems, etc.

The main responsibility of collecting and evaluating the contributions to these volumes was taken by the session organizers. They had organized the following sessions:

Value Distribution Theory and Complex Dynamics (P.C. Hu, X.H. Hua, C.C. Yang),
Nevanlinna Theory and Related Topics (S. Mori),
Riemann Surfaces (Y. Imayoshi, H. Sato, H. Shiga),
$\overline{\partial}$–Equations on Complex Manifolds (T. Ohsawa),
Function Space Theory in Complex Ball and Related Topics (E.G. Kwon),
Infinite Dimensional Holomorphy (S. Dineen),
Complex Dynamics of Higher Dimension (M. Suzuki),
Complex and Funtional Analytic Methods in Partial Differential Equations (H. Begehr),
Integral Equations and Boundary Value Problems (G.–C. Wen),
Analytic Extension Formulas and their Applications (S. Saitoh),
Complex and Numerical Analysis (J. Kajiwara),
Analytic Functionals, Hyperfunctions, and Generalized Functions (M. Morimoto),
Constructive Methods in Applied Problems (P. Krutitskii),
Special Session (M. Reissig),
Potential Theory (P. Tamrazov),
Functional Analysis (A. Inoue),

Spectral Theory of Differential Operators and Applications (B. Vainberg, S. Molchanov),
Function Spaces and Applications (V. Burenkov, S. Samko),
Ocean Acoustics (R.P. Gilbert, G. Makrakis),
Inverse Problem in Acoustics (R.P. Gilbert, Y. Xu),
Direct and Inverse Scattering in Extended Inhomogeneous Environments (L. Fishman),
Computational Biology (K. Harimoto),
Curve Theory and its Applications to Coding Theory (T. Kato).

Contributions in the same session adjoin each other in order of reception in this book .

The main work of composing the material to these two volumes were achieved by the last named editor. It was also he who has managed to get financial support for printing these volumes. We are grateful to ✿ the Commemorative Association for their Grant No. 11–56:

The Mathematical Society of Japan consented to support the Second ISAAC Congress 7th April 1998. The Fukuoka Institute of Technology acknowledged to hold the congress at FIT and consented to support the congress 21st October 1998. Consequently, we could freely use conference rooms, lecture rooms equipped with airconditioners and school bus between FIT and Sasaguri, where participants from emerging countries, i. e. from present and former socialist countries, countries under the support of IMF and developing countries, stayed at the hotel said later.

The moral support of MSJ and the substantial support of FIT made way for the local organizing committee to acquire the support of the said Grant No. 11–56 from ✿ 17th June 1999. The Grant No. 11–56 from ✿ the Commemorative Association supported partially air fares and fully the expenses necessary for the hotel at Sasaguri said later, of all participants from the emerging countries.

The congress is also supported by Fukuoka Prefecture, Educational Committee of Fukuoka Prefecture, Convention Bureau of Fukuoka City and the Association for the Japan National Authorized Mathematics Certification. By the support of the Educational Committee of Fukuoka Prefecture, participants from the emerging countries got lodgings at the

Fukuoka Prefectural Socioeducation Center at Sasaguri. In order to restore our mathematics, which contributes progress and harmony of humankind, for citizen, the local organizing committee held a dialog forum with citizen associated to the congress 22nd August 1999 by a support of the Yokatopia Foundation, that is, the Commemorative Association for the Asian and Pacific Exposition. The companies, Fujitsu Kyushu Digital Technology Ltd., Gordon & Breach Publishers, Kinokuniya Company Ltd., Nippon Culture Service Ltd. and NTT Software Corporation Ltd. donated to the congress. A lot of Japanese colleagues invited their foreign acquaintances to the congress by their own Grants-in-Aid from the Ministry of Education, Culture and Sports of Japan, or Funds from the Japan Society for the Promotion of Science.

We would like to express our gratitude again to the aforementioned donations, furtherance and supports, by which we, mathematicians, can continue contributions to the progress and harmony of humankind since Sir **ISAAC** Newton.

Thanks are due to Dr. Miki Tsuji-Aoyagi, who established the pseudo-Kluwer style of this book which matches tex files typeset not only by LATeX, but also Amsart and AmsTeX, and who helped the executive editor in compilation of tex files which were difficult to compile for him, to Professor Yukio Fukushima, who helped him to get dviout of tex files with figures, e.g. ✿, written in postscript and Mrs. Barbara G. Wengel for helping adjusting the manuscripts typeset in every kind of TeX to the pure Kluwer style.

In view of this fine collection of papers the Second ISAAC Congress may be seen as a great success. ISAAC's mandate, that is nourishing the growth of Analysis, its Applications, and Computation, is well documented in these Proceedings. We look forward to the Third ISAAC Congress which is to be held in Berlin, Germany by the first named editor, e-mail: begehr@math.fu-berlin.de , during August 2001 which we hope will carry on the fine tradition of the ISAAC Congresses.

Heinrich G. W. Begehr
Robert P. Gilbert
Joji Kajiwara

CONTENTS

Volume 1

Volume 2

1 A CENTRAL LIMIT THEOREM FOR THE SIMPLE RANDOM WALK ON A CRYSTAL LATTICE

Motoko KOTANI
Toshikazu SUNADA

Mathematical Institute
Graduate School of Science, Tohoku University
Aoba, Sendai, 980-8578, Japan

kotani@math.tohoku.ac.jp
sunada@math.tohoku.ac.jp

1. INTRODUCTION

Consider a *lattice graph* X realized in a k-dimensional vector space V (we shall use the same symbol X also for the set of vertices by abuse of notations). What we should have here in mind as a lattice graph is a generalization of classical lattices graphs such as the hyper-cubic lattice in $\mathbb{R}^k$, the triangular lattice and the hexagonal lattice in $\mathbb{R}^2$. We shall show, by using our previous result [1], that, as the mesh of X goes to zero, the simple (isotropic) random walk on X "converges" to the Brownian motion on V with a suitable Euclidean structure. More precisely, we shall prove

Theorem 1.1. *Let $p(n,x,y)$ $(x,\ y \in X)$ be the transition probability at the n-th step for the simple random walk on X. There exist a positive number D and an inner product $\langle \cdot, \cdot \rangle$ on V such that, with a fixed $t > 0$*

H.G.W. Begehr et al. (eds.), Proceedings of the Second ISAAC Congress, Vol. 1, 1–6.

and $n\delta^2 = 2Dt$, *we have*

$$(1) \qquad \lim_{n\uparrow\infty} \sum_{y\in X} p(n,x,y) f(\delta y) = (4\pi t)^{-k/2} \int_V \exp\left(-\frac{\|\mathbf{y}\|^2}{4t}\right) f(\mathbf{y}) d\mathbf{y}$$

for every continuous function f *on* V *with compact support, where* $\|\mathbf{y}\|^2 = \langle \mathbf{y}, \mathbf{y}\rangle$ *and* $d\mathbf{y}$ *denotes the Lebesgue measure on* V *associated with the Euclidean structure.*

See the last section for an explicit description of the Euclidean structure on V.

The result above generalizes the well-known convergence theorem for the random walk on the hyper-cubic lattice which can be proved by means of the classical central limit theorem for the sum of independent random variables. In [1], we have proved the theorem for the *standard realization* of a lattice graph. This note is a supplement to [1].

Acknowledgement The authors are grateful to Dr. T. Shirai for his suggestion of validity of the theorem for general realizations of lattice graphs.

2. Generalized lattice graphs

A special feature of the lattice graph X realized in V is that there exists a lattice subgroup Γ of V (a discrete subgroup of maximal rank) which preserves X as Γ acts on V by translations, and that the quotient of X by the Γ-action is finite. The inclusion $\iota : \Gamma \to V$ yields an isomorphism $\iota_{\mathbb{R}} : \Gamma \otimes \mathbb{R} \to V$. Thus we may identify V with $\Gamma \otimes \mathbb{R}$. From this view, we are led to the following definition of generalized lattice graph.

Let X be a locally finite, connected graph which admits a free action of an abelian group Γ of infinite order with a finite quotient graph X_0. By a *realization* of X, we mean a (piecewise linear) map Φ of X into the vector space $\Gamma \otimes \mathbb{R}$ satisfying

$$(2) \qquad \Phi(\sigma x) = \Phi(x) + \sigma \otimes 1 \qquad (\sigma \in \Gamma,\ x \in X).$$

We shall introduce an inner product on $\Gamma \otimes \mathbb{R}$ (and hence on V) in the following way. Since X is an abelian covering space of X_0 with the covering transformation group Γ, one has a surjective homomorphism $\rho : \mathrm{H}_1(X_0, \mathbb{Z}) \to \Gamma$. Let $\rho_{\mathbb{R}} : \mathrm{H}_1(X_0, \mathbb{R}) \to \Gamma \otimes \mathbb{R}$ be the linear extension of ρ. The vector space $\mathrm{H}_1(X_0, \mathbb{R})$ carries the inner product obtained by restricting the one on the 1-chain group for X_0 which is defined by

$$(3)\qquad \langle e, e'\rangle = \begin{cases} 1 & (e = e') \\ -1 & (e = \overline{e'}) \\ 0 & (e \neq e, \overline{e'}), \end{cases}$$

where e, e' denote oriented edges in X_0, and $\overline{e'}$ denotes the inverse edge of e'. We now introduce an inner product on $\Gamma \otimes \mathbb{R}$ in such a way that $\rho_{\mathbb{R}} : (\text{Ker } \rho_{\mathbb{R}})^{\perp} \to \Gamma \otimes \mathbb{R}$ is an isometry.

Let D be the number of unoriented edges in X_0. Note that

$$(4)\qquad 2D = \sum_{x \in X} \deg x.$$

With the Euclidean structure on $\Gamma \otimes \mathbb{R}$ defined above, we have

Theorem 2.1. *For a fixed $t > 0$ and $n\delta^2 = 2Dt$,*

$$(5)\qquad \lim_{n\uparrow\infty} \sum_{y\in X} p(n,x,y) f(\delta\Phi(y)) = (4\pi t)^{-k/2} \int_{\Gamma\otimes\mathbb{R}} \exp\left(-\frac{\|\mathbf{y}\|^2}{4t}\right) f(\mathbf{y}) d\mathbf{y}$$

for every continuous function f on $\Gamma \otimes \mathbb{R}$ with compact support.

Proof. The homomorphism $\rho : \mathrm{H}_1(X_0, \mathbb{Z}) \to \Gamma$ induces an injection of $\mathrm{Hom}(\Gamma, \mathbb{R})$ into $\mathrm{H}^1(X_0, \mathbb{R})$. We shall identify $\mathrm{Hom}(\Gamma, \mathbb{R})$ with its image in $\mathrm{H}^1(X_0, \mathbb{R})$, and also identify $\mathrm{H}^1(X_0, \mathbb{R})$ with the space of "harmonic 1-forms" (the discrete Hodge-Kodaira theorem). Let Φ_0 be the standard realization of X in $\Gamma \otimes \mathbb{R}$; namely

$$(6)\qquad \langle \Phi(x), \omega \rangle = \int_{x_0}^{x} \tilde{\omega},$$

where $\omega \in \mathrm{Hom}(\Gamma, \mathbb{R})$ and $\tilde{\omega}$ is the lift of ω to X (see [3]). Note that

$$(7)\qquad \Phi_0(\sigma x) = \Phi_0(x) + \sigma \otimes 1 \qquad (\sigma \in \Gamma,\ x \in X).$$

In the previous paper [1], we have proved that the claim in the theorem is true for the standard realization.

Put $F = \Phi - \Phi_0$. We then observe that

$$(8)\qquad F(\sigma x) = F(x) \qquad (\sigma \in \Gamma, x \in X).$$

In particular, F is bounded, and hence since f is uniformly continuous, we find that, for a given $\epsilon > 0$, there exists a positive δ_0 with

$$(9)\qquad |f(\delta\Phi(y)) - f(\delta\Phi_0(y))| < \epsilon \qquad (y \in X)$$

for every δ with $0 < \delta < \delta_0$. We thus have

$$(10) \qquad |\sum_{y\in X} p(n,x,y)f(\delta\Phi(y)) - \sum_{y\in X} p(n,x,y)f(\delta\Phi_0(y))| < \epsilon$$

and

(11)

$$|\lim_{n\uparrow\infty} \sum_{y\in X} p(n,x,y)f(\delta\Phi(y)) - (4\pi t)^{-k/2}\int_{\Gamma\otimes\mathbb{R}} \exp(-\frac{\|\mathbf{y}\|^2}{4t})f(\mathbf{y})d\mathbf{y}| \le \epsilon.$$

Since ϵ is arbitrarily small, this completes the proof. □

Remark. The claim for the standard realization Φ_0 is a consequence of the following *local central limit theorem*:

Put $d_0(x,y) = \|\Phi_0(x) - \Phi_0(y)\|$ and $C(X) = (2D)^{k/2-1}\mathrm{vol}(\Gamma\otimes\mathbb{R}/\Gamma)$.

1. If X is a non-bipartite graph, then

$$\lim_{n\uparrow\infty}\left[(4\pi n)^{k/2}p(n,x,y)(\deg y)^{-1} - C(X)\exp(-\frac{D}{2n}d_0(x,y)^2)\right] = 0$$

uniformly for all $x, y \in X$.

2. If X is a bipartite graph with a bipartition $V = A \amalg B$, and
   (a) if $x, y \in A$ or $x, y \in B$, then $p(n,x,y) = 0$ for odd n and

$$\lim_{n\uparrow\infty}\left[(4\pi n)^{r/2}p(n,x,y)(\deg y)^{-1} - 2C(X)\exp(-\frac{D}{2n}d_0(x,y)^2)\right] = 0$$

uniformly for $x, y \in A$ or for $x, y \in B$, where n runs over even numbers,

   (b) if $x \in A$, $y \in B$ or $x \in B$, $y \in A$, then $p(n,x,y) = 0$ for even n and

$$\lim_{n\uparrow\infty}\left[(4\pi n)^{r/2}p(n,x,y)(\deg y)^{-1} - 2C(X)\exp(-\frac{D}{2n}d_0(x,y)^2)\right] = 0$$

uniformly for $x \in A, y \in B$ or for $x \in B, y \in A$, where n runs over odd numbers.

3. COMPUTATION OF THE INNER PRODUCT

We shall now explicitly describe the inner product on $\Gamma\otimes\mathbb{R}$ defined in the previous section. Let $\sigma_1, \dots, \sigma_k$ be elements in Γ such that $\sigma_1\otimes 1, \dots, \sigma_k\otimes 1$ form a $\mathbb{Z}$-basis of $\Gamma\otimes\mathbb{Z}$. Since the inner product is determined by the matrix $S = (\langle\sigma_i\otimes 1, \sigma_j\otimes 1\rangle)$, it suffices to give an algorithm for the computation of S. Take closed paths $c_1, \dots, c_n$ in X_0 such that the homology classies $[c_1], \dots, [c_n]$ form a $\mathbb{Z}$-basis of $\mathrm{H}_1(X_0, \mathbb{Z})$. For instance, we may choose $c_1, \dots, c_n$ in the following way.

Take a maximal (spanning) tree T in X_0. The tree T contains all vertices of X_0, and the number of unoriented edges in T is $N-1$ where N is the number of vertices in X_0. The number of unoriented edges not contained in T is equal to n, the rank of $\mathrm{H}_1(X_0, \mathbb{Z})$. Give an orientation $\{e_1, \ldots, e_D\}$ to the graph X_0 such that

$$e_1, \ldots, e_n \notin T, \tag{12}$$

$$e_{n+1}, \ldots, e_D \in T, \tag{13}$$

where we should note that $D = N+n-1$. For each i with $1 \leq i \leq n$, take a minimal path c_i' in T with $o(c_i') = t(e_i)$ and $t(c_i') = o(e_i)$ where $o(c)$ (resp. $t(c)$) denotes the origin (resp. terminus) of a path c. Put $c_i = e_i c_i'$. We observe that each c_i is a circuit, i.e. it has no self-intersection, and that $[c_1], \ldots, [c_n]$ form a basis of $\mathrm{H}_1(X_0, \mathbb{Z})$.

We now take a lift of c_i to X, which we denote by $\tilde{c}_i$. Then there exists $\gamma_i \in \Gamma$ such that $\gamma_i o(\tilde{c}_i) = t(\tilde{c}_i)$. Note that $\rho([c_i]) = \gamma_i$. Write

$$\gamma_i \otimes 1 = \sum_{j=1}^{k} r_{ij} \sigma_j \otimes 1. \tag{14}$$

Let $u_1, \ldots, u_n \in \mathrm{H}^1(X_0, \mathbb{R})$ be the dual basis of $[c_1], \ldots, [c_n]$, and $v_1, \ldots, v_k \in \mathrm{Hom}(\Gamma, \mathbb{R})$ be the dual basis of $\sigma_1 \otimes 1, \ldots, \sigma_k \otimes 1$. Then

$$v_i = \sum_{j=1}^{n} r_{ji} u_j. \tag{15}$$

Thus with respect to the dual inner product on

$$\mathrm{H}^1(X_0, \mathbb{R}) = \mathrm{Hom}(\mathrm{H}_1(X_0, \mathbb{R}), \mathbb{R}), \tag{16}$$

we have

$$\langle v_i, v_j \rangle = \sum_{h,k=1}^{n} r_{hi} r_{kj} \langle u_h, u_k \rangle. \tag{17}$$

Here we should note that the dual inner product on $\mathrm{Hom}(\Gamma, \mathbb{R})$ $(= \mathrm{Hom}(\Gamma \otimes \mathbb{R}), \mathbb{R}))$ coincides with the restriction of the dual inner product on $\mathrm{H}^1(X_0, \mathbb{R})$. If we put $V = (\langle v_i, v_j \rangle)$, $R = (r_{ij})$, $U = (\langle u_h, u_k \rangle)$, we obtain

$$V = {}^t RUR. \tag{18}$$

Note that $S = V^{-1}$, $(\langle [c_i], [c_j] \rangle) = U^{-1}$. Therefore we have

$$S = ({}^t R I^{-1} R)^{-1}, \tag{19}$$

where $I = (\langle [c_i], [c_j] \rangle)$. The *intersection matrix* I is easily computed by looking at how closed paths c_i and c_j intersect.

References

[1] M.Kotani and T.Sunada, *Albanese maps and diagonal long time asymptotics for the heat kernel*, *preprint* 1998.
[2] M.Kotani and T.Sunada, *The Jacobian torus associated with a finite graph*, *preprint* 1998.
[3] M.Kotani and T.Sunada, *Standard realizations of crystal lattices via harmonic maps*, *preprint* 1998.

2 LEVEL STATISTICS FOR QUANTUM HAMILTONIANS

Some Preliminary Ideas toward Mathematical Justification of the Theory of Berry and Tabor

Nariyuki MINAMI

Institute of Mathematics, University of Tsukuba
Tsukuba-shi, Ibaraki, 305-8571 Japan

minami@sakura.cc.tsukuba.ac.jp

Abstract

According to Berry and Tabor, the spectrum of a classically integrable quantum Hamiltonian (the regular spectrum) should generically have the same statistical property of the Poisson point process. In particular, the spacings between energy levels obey the exponential distribution, which is the phenomena called "level clustering ". Strict justification of this assertion seems to be a very difficult mathematical problem. In this note, the author shall introduce a weakened notion of level clustering and give a sufficient condition for this in terms of factorial moments of the number of levels in intervals of fixed length. A preliminary result is also stated, which is closely related to the justification of this sufficient condition in the case of regular spectra. All proofs are omitted, and will be published elsewhere.

H.G.W. Begehr et al. (eds.), Proceedings of the Second ISAAC Congress, Vol. 1, 7–14.

1 An outline of Berry-Tabor's theory and the definition of clustering and repulsion of energy levels.

Consider a bounded physical system of $d-$degrees of freedom which is described by a classical Hamiltonian $H(p,q)$, $(p,q) \in \mathbf{R^d} \times \mathbf{R^d}$. Let $H(\hbar)$ be its quantization, and $\{E_n(\hbar)\}_{n\geq 1}$ be the energy levels (eigenvalues) of $H(\hbar)$. Here $\hbar$ is the Planck's constant, and we suppose that the spectrum of $H(\hbar)$ is purely discrete.

Now the assertion of Berry and Tabor ([2]) is the following: Suppose the classical dynamical system associated to $H(p,q)$ is completely integrable. Then the spectrum $\{E_n(\hbar)\}_{n\geq 1}$ of $H(\hbar)$ resembles, in the limit of $\hbar \searrow 0$, the Poisson point process. In particular, the level spacing distribution is $e^{-t}dt$ under suitable normalization. This phenomena was called "level clustering "by Berry and Tabor because the large part of spacing distribution is concentrated near zero, which means the statistical tendency of the appearance of closely lying levels. On the other hand, when the classical dynamical system is chaotic, the level spacing distribution $p(t)dt$ should satisfy $p(t) \sim Ct^\gamma$, $t \searrow 0$ with $\gamma > 0$, which is the phenomena called level repulsion.

In short, Berry and Tabor conjectured that the regular or chaotic properties of a classical dynamical system manifest themselves in the level statistics of its quantization.

In order to give a precise meaning to the level spacing distribution, we need to somehow normalize (unfold) the spectrum of $H(\hbar)$, so that it looks like a typical realization of a stationary point process. In doing this, we follow the idea of Berry and Tabor, thereby accomplishing simultaneously this normalization and the limiting procedure $\hbar \searrow 0$. Let us assume that the specrum $\{E_n(\hbar)\}_{n\geq 1}$ satisfies the following conditions:

(A1) All levels are non-degenerate and $E_n(\hbar) \geq 0$.

(A2) $E_n(\hbar) \searrow 0$ monotonically as $\hbar \searrow 0$.

(A3) For each fixed $E > 0$, there is a constant $\nu(E) > 0$ such that

$$N_\hbar(E) \equiv \sharp\{n \geq 1 \mid E_n(\hbar) \leq E\} \sim \nu(E)\hbar^{-d} \ (\hbar \searrow 0) \ . \quad (1)$$

Let us take $E > 0$, which we shall fix throughout. By (A1) and (A2), we can define $\hbar_n = \hbar_n(E)$ as the unique solution of the equation

$E_n(\hbar) = E$. Let us define the unfolded level λ_n by $\lambda_n = \nu(E)\hbar_n(E)^{-d}$. Then from (A3), it is easily seen that

$$\sharp\{n \geq 1 \mid \lambda_n \leq L\} \sim L \quad (L \to \infty) . \tag{2}$$

Thus the unfolded levels $\{\lambda\}_n$ are asymptotically uniformly distributed.

We introduce the following notation:

$$N(t) = N(t;c) = \sharp\{n \geq 1 \mid \lambda_n \in (t, t+c]\} ; \tag{3}$$

$$\pi_k(c;L) = \frac{1}{L}\int_0^L 1_{\{N(t;c)=k\}} dt ; \tag{4}$$

$$\mu_k(c;L) = \frac{1}{k!}\frac{1}{L}\int_0^L N(t)\{N(t)-1\}\cdots\{N(t)-k+1\}dt ; \tag{5}$$

$$\rho(c) = \frac{\sharp\{n \geq 1 \mid \lambda_n \leq L \ , \ \lambda_{n+1} - \lambda_n \leq c\}}{\sharp\{n \geq 1 \mid \lambda_n \leq L\}} . \tag{6}$$

Here $c > 0$ and $k = 0, 1, 2, \ldots$.

Definition 1 *We shall say that one has the strict sense level clustering if*

$$\pi_k(c) \equiv \lim_{L\to\infty} \pi_k(c;L) = e^{-c}\frac{c^k}{k!} \tag{7}$$

holds for each $c > 0$ and $k \geq 0$.

Obviously, the strict sense level clustering is equivalent to the weak convergence as $L \to \infty$ of the probability distribution $\{\pi_k(c;L)\}_{k=0}^{\infty}$ on $\mathbf{Z}_+$ to the Poisson distribution with mean c . As is well known in probability theory, a sufficient condition for this weak convergence is that the moments of all orders of $\{\pi_k(c;L)\}_{k=0}^{\infty}$ converges to those of the Poisson distribution. Namely we have

Proposition 1 *If one has*

$$\mu_k(c) \equiv \lim_{L\to\infty} \mu_k(c;L) = \frac{c^k}{k!} \tag{8}$$

for each $c > 0$ and $k = 1, 2, \ldots$, then one has the strict sense level clustering.

If one has the strict sense level clustering, then the limiting level spacing distribution exists and is the exponential distribution $e^{-c}dc$. This is a direct consequence of the following more general proposition:

Proposition 2 *If* $\pi_0(c) = \lim_{L\to\infty} \pi_0(c;L)$ *exists and is differentiable with respect to* c *, then* $\rho(c) = \lim_{L\to\infty} \rho(c;L)$ *also exists, and one has*

$$\rho(c) = 1 + \frac{d}{dc}\pi_0(c) \ . \tag{9}$$

It seems to be a difficult problem to prove the strict sense level clustering for any concrete Hamiltonian. Moreover, as was shown in [5] , there is an example of one-dimensional Hamiltonian for which the level statistics, in a slightly different formulation, can be rigorously performed, but the obtained level spacing distribution is different from the exponential distribution. In that case, we have still $\rho'(c) > 0$, so that we should say that the level clustering is taking place, and if we take the high disorder limit in the system, then the data $\rho(c)$ converges to the distribution function of $e^{-c}dc$. This circumstance has led the present author to the opinion that we should difine the notion of level clustering in much weaker sense, nevertheless retaining physical significance. Thus taking the broadest statistical sense of the clustering and repulsion, we make the following definition:

Definition 2 *We shall say that one has the wide sense level clustering [resp. repulsion] if*

$$\liminf_{c\searrow 0} \frac{\underline{\rho}(c)}{c} > 0 \quad \left[resp. \ \limsup_{c\searrow 0} \frac{\bar{\rho}(c)}{c} = 0 \right] \tag{10}$$

holds. Here we have set

$$\underline{\rho}(c) = \liminf_{L\to\infty} \rho(c;L) \quad ; \quad \bar{\rho}(c) = \limsup_{L\to\infty} \rho(c;L) \ . \tag{11}$$

We can prove the following

Proposition 3 *(i) Suppose one has* $\mu_1(c) = c$ *for any* $c > 0$ *and*

$$\bar{\mu}_2(c) \equiv \limsup_{L\to\infty} \mu_2(c;L) = o(c^2) \quad (c \searrow 0) \ . \tag{12}$$

Then one has the wide sense level repulsion.
(ii) Suppose one has $\mu_1(c) = c$ *,* $\mu_2(c) = \frac{1}{2}c^2$ *for any* $c > 0$ *and*

$$\bar{\mu}_3(c) \equiv \limsup_{L\to\infty} \mu_3(c;L) = o(c^2) \quad (c \searrow 0) \ . \tag{13}$$

Then one has the wide sense level clustering.

2 Level statistics for regular spectra

Suppose that the classical Hamiltonian system associated to $H(p,q)$ is completely integrable. Then one can transform the variable (p,q) into the action–angle variable $(I,\varphi)=(I_1,\ldots,I_d;\varphi_1,\ldots,\varphi_d)$, and $H(p,q)=H(I)$ depends only on the action variable (see [1]).

Now following Percival ([7]) , Berry and Tabor ([2]), we shall say that $H(\hbar)$ has regular spectrum if its energy levels are (approximately) given by quantizing the action variables $I_1,\ldots,I_d$ appearing in $H(I)$. Namely, we suppose that the energy levels of $H(\hbar)$ are

$$E_n(\hbar)=H(\hbar(n+\frac{1}{4}\alpha)) \quad ; \quad n=(n_1,\ldots,n_d)\in \mathbf{Z}_+^{\mathbf{d}} , \tag{14}$$

where $\alpha\in \mathbf{Z}_+^{\mathbf{d}}$ is the Maslov index. This procedure is called the EBK quantization after the names of Einstein, Brillouin and Keller, and it gives the exact spectrum in such concrete examples as the rectangular billiards and the harmonic oscillators.

Under mild conditions on $H(I)$ (e.g. $H(I)$ being convex, positive with $H(0)=0$) , regular spectra satisfy the conditions (A1)–(A3) stated in §1 with

$$\nu(E)=\int\cdots\int_{\mathbf{R}_+^{\mathbf{d}}} 1_{\{H(I)\le E\}}dI_1\cdots dI_d \ . \tag{15}$$

Let us apply the level statistics formulated in §1 to our regular spectrum. Note that the suffix n distinguishing the levels is now d-dimensional.

For $x\in \mathbf{R}_+^{\mathbf{d}}\setminus\{\mathbf{0}\}$ define $\hbar(x)=\hbar_E(x)\ge 0$ by the equation $H(\hbar(x)\cdot x)=E$, where $E>0$ will be fixed throughout, and let $\lambda(x)=\nu(E)\hbar_E(x)^{-d}$. Our unfolded levels are then given by $\lambda(n+\frac14\alpha)$, $n\in\mathbf{Z}_+^{\mathbf{d}}$. Since $\lambda(\beta x)=\beta^d\lambda(x)$ for $\beta>0$, we have the equivalence

$$\lambda(n+\frac{1}{4}\alpha)\in(t,t+c] \ \Leftrightarrow n+\frac{1}{4}\alpha\in\Pi_c(t) , \tag{16}$$

where we have defined

$$\Pi_c(t)=\{x\in\mathbf{R}_+^{\mathbf{d}} \mid \mathbf{t^{1/d}}\lambda(\varphi(\mathbf{x}))^{-\mathbf{1/d}}<|\mathbf{x}|<(\mathbf{t}+\mathbf{c})^{\mathbf{1/d}}\lambda(\varphi(\mathbf{x}))^{-\mathbf{1/d}}\} , \tag{17}$$

with

$$\varphi(x)=\frac{x}{|x|} \ . \tag{18}$$

It is easy to see $|\Pi_c(t)|=c$, where $|\Pi_c(t)|$ is the volume of $\Pi_c(t)$.

Hence the number $N(t) = N(t;c)$ of unfolded levels is equal to the number of lattice points (which are shifted by $\frac{1}{4}\alpha$) in the domain $\Pi_c(t)$, namely we have

$$N(t) = \sharp\{\Pi_c(t) \cap (\mathbf{Z}_+^{\mathbf{d}} + \frac{\mathbf{1}}{\mathbf{4}}\alpha)\} \ . \tag{19}$$

¿From this geometric consideration and $|\Pi_c(t)| = c$, it is obvious that we always have $\mu_1(c) = c$. It is, however, a non-trivial question to prove or disprove $\mu_k(c) = c^k/k!$ for $k \geq 2$. Sarnak ([8]) proved $\mu_2(c) = c^2/2$ by a number theoretic method for the quantized uniform motion on a two dimensional flat torus. (Actually he proved that the pair correlation function is proportional to c , which turns out to be equivalent to $\mu_2(c) = c^2/2$ after a suitable normalization.) On the other hand, Sinai [9] and Major [4] (see also [6]) considered the case in which $d = 2$, $\alpha = 0$ and the curve (written in polar coordinate) $r = f(\varphi) = \lambda(\varphi)^{-1/d}$ which defines the boundary of $\Pi_c(t)$ is very random. Especially, Major proved that $\mu_k(c) = c^k/k!$, $k \geq 1$ holds for almost all realization of the random curve $r = f(\varphi)$.

But the randomness they assume is so strong that the curve $r = f(\varphi)$ consisting the boundary of $\Pi_c(t)$ cannot be smooth, which violates the natural connection between level statistics and the lattice points counting. On the other hand, their proof suggests us that it would be possible to prove $\mu_k(c) = c^k/k!$ for $k = 1,\ldots,K$ and for any finite K , if the boundary of $\Pi_c(t)$ has "n-dimensional randomness "with n sufficiently large. (Presumably $n \geq 2K$ would suffice.)

For example, consider the d-dimensional billiard of a particle of mass 1 in the rectangle $\Lambda = \Pi_{j=1}^d[0,a_j]$. Then its classical Hamiltonian is $H(I) = \frac{\pi^2}{2}\sum_{j=1}^d (I_j/a_j)^2$, if expressed in terms of action variables, and its quantization is $H(\hbar) = -\frac{\hbar^2}{2}\Delta$ with the Dirichlet boundary condition on $\partial\Lambda$. Its energy levels (eigenvalues) are exactly given by

$$E_n(\hbar) = H(\hbar n) = \frac{\pi^2\hbar^2}{2}\sum_{j=1}^{d}(\frac{n_j}{a_j})^2 \quad (n = (n_1,\ldots,n_d)\ ,\ n_j \geq 1)\ . \tag{20}$$

We can then conjecture that if $(a_1\ldots,a_d)$ is a d-dimensional random variable with smooth density, then for sufficiently large d , one has $\mu_k(c) = c^k/k!$ for $k = 2$ and 3 for almost all $(a_1\ldots,a_d)$, yielding the wide sense level clustering according to Proposition 3.

3 A preliminary result

Let $d = 2$ in the rectangular billiard of §2. In this case we have

$$\Pi_c(t) = \{x \in \mathbf{R}_+^2 \mid \mathbf{t} < \frac{\pi}{4}\mathbf{a_1 a_2} \sum_{\mathbf{j=1}}^{\mathbf{2}} (\frac{\mathbf{x_j}}{\mathbf{a_j}})^{\mathbf{2}} < \mathbf{t} + \mathbf{c}\} \; . \tag{21}$$

Since this depends only on the ratio a_1/a_2 , we consider

$$\tilde{\Pi}_c(t) = \{x \in \mathbf{R}_+^2 \mid \mathbf{t} < \sum_{\mathbf{j=1}}^{\mathbf{2}} (\frac{\mathbf{x_j}}{\mathbf{a_j}})^{\mathbf{2}} < \mathbf{t} + \mathbf{c}\} \tag{22}$$

instead of $\Pi_c(t)$ in order to have two-dimensional randomness.

Let $\xi(t) = \xi(t; a_1, a_2)$ be the number of lattice points in $\tilde{\Pi}_c(t)$, namely $\xi(t) = \sharp(\tilde{\Pi}_c(t) \cap \mathbf{Z}_+^2)$. Finally let (a_1, a_2) be a two-dimensional random variable whose distribution has a density $p(a_1, a_2)$ which is C^1 and has a compact support in $(0, \infty)^2$. Then by a direct computation, we can prove the following

Theorem *We have*

$$\lim_{t\to\infty} E[\xi(t)] = E[\lambda_c] \; ; \tag{23}$$

and

$$\lim_{t\to\infty} E[\xi(t)(\xi(t) - 1)] = E[\lambda_c^2] \; , \tag{24}$$

where $E[\cdot]$ *means the integration with respect to* $p(a_1, a_2) da_1 da_2$ *, and*

$$\lambda_c \equiv \frac{\pi c}{4} a_1 a_2 = |\tilde{\Pi}_c(t)| \; , \tag{25}$$

which is the area of $\tilde{\Pi}_c(t)$ *independently of* t .

This result is consistent with that of Cheng, Lebowitz and Major ([3]) , in which they computed the second moment of the number of lattice points in the domain which has a shape similar to our $\Pi_c(t)$ with nonrandom boundary but has randomly shited center.

References

[1] Arnold, V.I.: Mathematical methods of classical mechanics. Springer (1978)

[2] Berry, M.V., Tabor, M.: Level clustering in the regular spectrum. Proc. R. Soc. Lond. A. vol.356 (1977) 375–394

[3] Cheng, Z., Lebowitz, J.L., Major, P.: On the number of lattice points between two enlarged and randomly shifted copies of an oval. Probab. Th. Rel. Fields vol.100 (1994) 253–268

[4] Major, P.: Poisson law for the number of lattice points in a random strip with finite area. Probab. Th. Rel. Fields vol.92 (1992) 423–464

[5] Minami, N.: Level clustering in a finite system. Progr. Theor. Phys. Supplement No.116 (1994) 359–368

[6] Minami, N.: On the Poisson limit theorems of Sinai and Major. preprint

[7] Percival, I.C.: Regular and irregular spectra. J. Phys. B. vol.6 (1973) L229–L232

[8] Sarnak, P.: Values at integers of binary quadratic forms. *Harmonic analysis and number thory* CMS Conf. Proc. vol.21, AMS (1997) 181–203

[9] Sinai, Ya.G.: Poisson distribution in a geometric problem. Adv. Sov. Math. (AMS) vol.3 (1991) 199–214

3 FERMION PROCESS AND FREDHOLM DETERMINANT

Tomoyuki SHIRAI

Department of Mathematics, Tokyo Institute of Technology
Oh-okayama, Meguro-ku, Tokyo 152-8551, Japan

shirai@math.titech.ac.jp

Yoichiro TAKAHASHI

Research Institute for Mathematical Sciences, Kyoto University
Sakyo-ku, Kyoto 606-8502, Japan

takahasi@kurims.kyoto-u.ac.jp

Abstract

We construct a family of probability measures on a configuration space associated with Fredholm determinants for symmetric integral operators. For these probability measures we show basic limit theorems (law of large numbers, central limit theorem and large deviation.) Also we deal with shift dynamical systems naturally corresponding to these probability measures on $\mathbf{Z}^1$ and investigate their ergodic properties.

1. Introduction

Let $\mathcal{H}_N$ be the space of $N \times N$-Hermitian matrices, and $P(dX)$ the Gaussian probability measure on $\mathcal{H}_N$ given by

$$(1.1) \qquad P(dX) = Z_N^{-1} \exp(-\mathrm{Tr}(X^2))dX,$$

where Z_N^{-1} is the normalized constant and dX is the Lebesgue measure on $\mathcal{H}_N$ identified with $\mathbf{R}^{N^2}$. The pair $(\mathcal{H}_N, P(dX))$ is called Gaussian Unitary Ensemble (GUE). GUE has been investigated by many authors. One of the main concerns about GUE is to study the joint distributions of

H.G.W. Begehr et al. (eds.), Proceedings of the Second ISAAC Congress, Vol. 1, 15–23.

N eigenvalues and its limiting behavior as $N \to \infty$. The joint eigenvalue distribution is well known [8] and given by

$$\mu_N(dx_1, dx_2, \ldots, dx_N) = \tag{1.2}$$

$$C_N^{-1} \prod_{1 \le i < j \le N} (x_i - x_j)^2 \exp\left(-\sum_{i=1}^N x_i^2\right) dx_1, dx_2, \ldots, dx_N.$$

In order to obtain the limiting behavior as $N \to \infty$, we need to take a scaling limit. Here we show two typical scalings. First we consider the empirical measure

$$\xi_N = \frac{1}{N} \sum_{i=1}^N \delta_{x_i/\sqrt{N}}, \tag{1.3}$$

where $\{x_i\}_{i=1}^N$ are the eigenvalues of GUE. Then as $N \to \infty$, one can obtain the so-called Wigner's semi-circle law, that is,

$$\xi_N \to \frac{1}{\pi}\sqrt{2 - x^2}dx \tag{1.4}$$

almost surely.

The second one is the following: set $\lambda_N = \pi/\sqrt{2N}$. Then,

$$\begin{aligned} \lim_{N\to\infty} \lambda_N^n \rho_n^{(N)}(\lambda_N x_1, \ldots, \lambda_N x_n) &= \det\left(K(x_i, x_j)\right)_{i,j=1}^n \\ &:= \rho_n(x_1, \ldots, x_n), \end{aligned} \tag{1.5}$$

where

$$K(x, y) = \frac{\sin \pi(x - y)}{\pi(x - y)} \tag{1.6}$$

and $\rho_n^{(N)}(x_1, x_2, \ldots, x_n)$ is the density correlation function defined by

$$\rho_n^{(N)}(x_1, x_2, \ldots, x_n) = \tag{1.7}$$

$$\frac{N!}{(N-n)!} \int_{\mathbf{R}^{N-n}} \mu_N(x_1, x_2, \ldots, x_N) dx_{n+1} dx_{n+2} \ldots dx_N.$$

In particular, $\rho_1^{(N)}(x)/N$ is the eigenvalue density distribution.

It is natural to ask that does there exist a probability measure on a configuration space which has $\{\rho_n\}_{n=1}^\infty$ as correlation functions? Moreover does there exist a probability measure on a configuration space which has

$$\rho_n(x_1, x_2, \ldots, x_n) = \det\left(K(x_i, x_j)\right)_{i,j=1}^n \tag{1.8}$$

as correlation functions for a kernel $K(x, y)$? The answer is a yes under some conditions for $K(x, y)$.

Let R be a locally compact Hausdorff space with countable basis, $\lambda(dx)$ be a nonnegative Radon measure on R and Q be the space of nonnegative integer-valued Radon measures on R. An element ξ of Q may be expressed as

$$\xi = \sum_i \delta_{x_i} \quad (x_i \in R) \tag{1.9}$$

and then we write

$$\langle \xi, f \rangle = \sum_i f(x_i) \tag{1.10}$$

for a function f on R so far as the right hand side of (1.10) makes sense. It is well known that a measure on Q is characterized by its Laplace transform

$$\int_Q \mu(d\xi) \exp(-\langle \xi, f \rangle). \tag{1.11}$$

Now we define the correlation function ρ_n in this setting. For $\xi \in Q$ and any bounded measurable function f_n on R^n with compact support, denote

$$\langle \xi_n, f_n \rangle = \sum_{x_1, x_2, \ldots, x_n \in \xi}^{*} f_n(x_1, x_2, \ldots, x_n) \tag{1.12}$$

where $\sum^*$ denotes the sum over all mutually distinct points $x_1, x_2, \ldots, x_n$. If $\langle \xi_n, f_n \rangle$ is μ-integrable for any bounded continuous function f_n with compact support on R^n, then the formula

$$\int_Q \langle \xi_n, f_n \rangle \mu(d\xi) = \int_{R^n} f_n(x_1, \ldots, x_n) \lambda_n(dx_1 \ldots dx_n) \tag{1.13}$$

defines a measure λ_n on R^n which is called the n-th correlation measure of μ. Furthermore if λ_n is absolutely continuous with respect to $\lambda^{\otimes n}$, its density $\rho_n(x_1, \cdots, x_n)$ is called the n-th correlation function.

The following is a basic result for the question above.

Theorem 1.1. *Let K be a bounded symmetric integral operator on $L^2(R, \lambda(dx))$ whose spectrum Spec(K) is contained in the unit interval $[0, 1]$. Assume that, for each compact subset Λ of R, the restriction $K_\Lambda = 1_\Lambda K 1_\Lambda$ of K to the subspace $L^2(R, \lambda(dx))$ is of trace class(1_Λ being the indicator of the set Λ) and that the spectrum Spec(K_Λ) is contained in the half open interval $[0, 1)$. Then there exists a unique probability measure μ on the configuration space Q such that*

$$\int_Q \mu(d\xi) \exp(-\langle \xi, f \rangle) = \det(I - K_\varphi) \tag{1.14}$$

for each nonnegative measurable function f on R with compact support, where K_φ stands for the trace class operator defined as

$$K_\varphi(x,y) = \sqrt{\varphi(x)}K(x,y)\sqrt{\varphi(y)} \tag{1.15}$$

and

$$\varphi(x) = 1 - \exp(-f(x)). \tag{1.16}$$

Furthermore, the correlation functions of the probability measure μ exist and are given by

$$\rho_n(x_1, \dots, x_n) = \det\Big(K(x_i, x_j)\Big)_{i,j=1}^n \quad (n \geq 1) \tag{1.17}$$

and

$$\rho_0 = 1.$$

The Palm measure is the basic concept in point process theory as well as the correlation function (cf. [3]). If μ has mean λ_1, there exists a probability measure μ^x on Q for λ_1-a.e.x satisfying the following equation

$$\int_Q \mu(d\xi) \int_R \xi(dx)u(\xi, x) = \int_R \lambda_1(dx) \int_Q \mu^x(d\xi)u(\xi + \delta_x, x) \tag{1.18}$$

for any bounded measurable function $u(\xi, x)$ on $Q \times R$ with compact support in x. The probability measure μ^x on Q is called the Palm measure.

The point processes associated with Fredholm determinant is closed under the operation to take the Palm measure $\mu \mapsto \mu^x$.

Theorem 1.2. *The Palm measure μ^x of μ at x is also given by the formula (1.14) with $K(x_1, x_2)$ replaced by*

$$K^x(x_1, x_2) = K(x_1, x_2) - K(x, x_1)K(x, x_2)/K(x, x).$$

The idea of the proof of Theorem 1.1 is simple. Note that $Q(\Lambda) \cong \bigcup_{n=0}^\infty \Lambda^n/\sim$ for each compact set Λ. We define the density function σ_Λ on $\bigcup_{n=0}^\infty \Lambda^n/\sim$ by

$$\sigma_\Lambda(x_1, \dots, x_n) = \det(I - K_\Lambda)\det(J[\Lambda](x_i, x_j))_{i,j=1}^n \text{ on } \Lambda^n, \tag{1.19}$$

where $J[\Lambda] = K_\Lambda(I - K_\Lambda)^{-1}$ and for $n = 0$

$$\sigma_\Lambda(\emptyset) = \det(I - K_\Lambda) \text{ on } \Lambda^0. \tag{1.20}$$

This defines a probability measure μ_Λ on $Q(\Lambda)$. It suffices to check Kolmogorov's consistency condition for the family of probability measures $\{\mu_\Lambda\}_{\Lambda \subset R}$. For the correlation functions we use the principle of inclusion and exclusion.

Remark. Macchi [6, 7] was the first to study such a point process and called it a fermion process. The fermion processes, especially, those associated with the sine kernel and some other special kernels, have been studied in detail by [1, 11, 12]

Example 1.3. Let $R = \mathbf{R}^d$ the real line, dx be the Lebesgue measure and K be the convolution operator with a continuous kernel $k(x)$ defined by

$$k(x) = \left(\frac{1}{2\pi}\right)^d \int_{\mathbf{R}^d} \widehat{k}(t) e^{-itx} dt. \tag{1.21}$$

for an even function $\widehat{k} \in L^1(\mathbf{R}^d)$ taking values in $[0,1]$. Then one can check the required condition in Theorem 1.1, and so K defines a translation invariant probability measure on the configuration space $Q = Q(\mathbf{R}^d)$. Especially, in the case of $d = 1$, the sine kernel $k(x)$ is the Fourier transform of the indicator function of the interval $[-\pi, \pi]$ so that $\mathrm{Spec}(K) = \{0\} \cup \{1\}$ and it is easy to check that $\mathrm{Spec}(K_\Lambda) \subset (0,1)$ whenever Λ is a bounded interval.

2. Convolution on $\mathbf{R}^d$

In this section, we always assume the kernel $k(x)$ satisfies the conditions given in above Example 1.3. In this case, the corresponding measure μ is translation invariant as mentioned above.

Lemma 2.1. *Let f be a bounded measurable function with compact support and set $f_N = f(\cdot/N)$. Then, as $N \to \infty$,*

$$\begin{aligned} &\int_Q \langle \xi, f_N \rangle^2 \mu(d\xi) - \left(\int_Q \langle \xi, f_N \rangle \mu(d\xi) \right)^2 \\ &\sim N^d \int_{\mathbf{R}^d} f(x)^2 dx \times \left(\frac{1}{2\pi}\right)^d \int_{\mathbf{R}^d} \widehat{k}(t)(1 - \widehat{k}(t)) dt. \end{aligned} \tag{2.1}$$

Remark. The quantity above for an indicator function f is called the number variance in random matrix theory. Unless $\widehat{k}$ takes values only 0 and 1 almost everywhere, the number variance does not vanish. ¿From this point of view, the point field associated with the sine kernel is a typical degenerate case because the Fourier transform of the sine kernel $\sin \pi x / \pi x$ is an indicator function of the interval $[-\pi, \pi]$

The following are basic limit theorems.

Proposition 2.2. *(Law of large numbers) Let f be a bounded measurable function with compact support. Then*

$$\langle \xi, \frac{f_N}{N^d} \rangle \to \int_{\mathbf{R}^d} f(x)k(0)dx \quad \mu\text{-a.e.}\xi \text{ and in } L^1(Q,\mu), \tag{2.2}$$

where $f_N(\cdot) = f(\cdot/N)$.

Proposition 2.3. *(Central limit theorem) Let f be a bounded measurable function with compact support and assume $\int_{\mathbf{R}^d} f(x)dx = 0$. Then*

$$\lim_{N\to\infty} \int_Q \mu(d\xi) \exp\left(i\langle \xi, \frac{f_N}{N^{d/2}} \rangle \right) = \exp\left(-\frac{1}{2}\sigma^2 \|f\|_2^2 \right), \tag{2.3}$$

where

$$\sigma^2 = \left(\frac{1}{2\pi}\right)^d \int_{\mathbf{R}^d} \widehat{k}(t)(1-\widehat{k}(t))dt \tag{2.4}$$

and $f_N(\cdot) = f(\cdot/N)$.

Proposition 2.4. *(Large deviation) Let f be a nonnegative function with compact support and set $\varphi = 1 - \exp(-f(x))$. Then,*

$$\lim_{N\to\infty} \frac{1}{N^d} \log \int_Q \mu(d\xi) \exp\left(-\langle \xi, f_N \rangle\right) = \tag{2.5}$$

$$\left(\frac{1}{2\pi}\right)^d \int_{\mathbf{R}^d} dt \int_{\mathbf{R}^d} dx \log\left(1 - \widehat{k}(t)\varphi(x)\right),$$

where $f_N(\cdot) = f(\cdot/N)$.

3. Shift

The following is the discrete version of Theorem 1.1.

Theorem 3.1. *Let R be a countable set, λ be a measure on R and K be a bounded operator on $\ell^2(R,\lambda)$ with Spec$(K) \subset [0,1]$. Then there exists a unique probability measure μ on the configuration space $Q = Q(R)$ which satisfies the following property: for any disjoint finite subsets Λ_0 and Λ_1 with union $\Lambda = \Lambda_0 \cup \Lambda_1$,*

$$\mu(0^{\Lambda_0}1^{\Lambda_1}) = \det(P_{\Lambda_0}(I_\Lambda - K_\Lambda) + P_{\Lambda_1}K_\Lambda), \tag{3.1}$$

where P_{Λ_i} are the projection operators onto $\ell^2(\Lambda_i,\lambda)$. Here we abbreviated notation as

$$\mu(0^{\Lambda_0}1^{\Lambda_1}) = \mu\left(\xi \in Q \,;\, \xi\{i\} = 0 \ (i \in \Lambda_0),\ \xi\{j\} = 1 \ (j \in \Lambda_1)\right). \tag{3.2}$$

Remark. In the continuous case, Poisson point process is not contained in our class, however, in the discrete case, Bernoulli measure is the fermion point process associated with the operator αI for $0 < \alpha < 1$. The difference comes from the fact that the identity operator I restricted on a compact set on a discrete space is a compact operator while that on a continuum space is not.

In the discrete case, we can check Kolmogorov's consistency condition directly. The following simple identity is crucial.

Lemma 3.2. *Let A be an $N \times N$-matrix and P_Λ an $N \times N$-diagonal matrix such that*

$$P_\Lambda(i,i) = \begin{cases} 1 & \text{if } i \in \Lambda^c \\ 0 & \text{if } i \in \Lambda, \end{cases} \tag{3.3}$$

for each subset $\Lambda \subset \{1, 2, \ldots, n\}$. Then

$$\sum_{\Lambda \subset \{1,2,\ldots,n\}} (-1)^{|\Lambda|} \det(A - P_\Lambda) = 1. \tag{3.4}$$

Moreover, if A is a symmetric matrix such that $0 \leq A \leq I$, then

$$(-1)^{|\Lambda|} \det(A - P_\Lambda) \geq 0. \tag{3.5}$$

Example 3.3. Let $R = \mathbf{Z}^d$, λ be the Haar measure on $\mathbf{Z}^d$ such that $\lambda(\{x\}) = 1$ for any $x \in \mathbf{Z}^d$, and K be a convolution operator with kernel k defined by

$$k(n) = \left(\frac{1}{2\pi}\right)^d \int_{\mathbf{T}^d} \widehat{k}(\theta) e^{in\theta} d\theta, \tag{3.6}$$

for an even function $\widehat{k} \in L^1(\mathbf{T}^d)$ taking values in $[0, 1]$. Then K satisfies the conditions required in Theorem 3.1 and the corresponding probability measure μ is translation invariant.

Remark. The discrete versions of Proposition 2.2, 2.3 and 2.4 can be proved in the same way. Especially, the discrete version of Proposition 2.4 is a generalization of Szegö's theorem for Toeplitz operators [10].

¿From now on we consider the case where $R = \mathbf{Z}^1$ and K is a convolution operator. Since μ has no multiple points, we can identify $Q(\mathbf{Z}^1)$ with $\{0,1\}^{\mathbf{Z}^1}$, and so we obtain a shift dynamical system $(\{0,1\}^{\mathbf{Z}^1}, \mu, \sigma)$ where σ stands for the translation to the left or the shift transformation: $(\sigma\xi)\{n\} = \xi\{n+1\}$. This provides us a new family of shift dynamical systems. We show ergodic properties of the shifts.

Theorem 3.4. *Let K be a convolution operator with kernel k given by (3.6). Then the shift dynamical system $(\{0,1\}^{\mathbf{Z}}, \mu, \sigma)$ is mixing. Moreover, if $\|K\| < 1$ and*

$$\sum_{n\in\mathbf{Z}} |n||k(n)|^2 < \infty, \tag{3.7}$$

the shift is weak Bernoulli, i.e., $(\{0,1\}^{\mathbf{Z}}, \mu, \sigma)$ is isomorphic to the Bernoulli system.

Proposition 3.5. *Let K be a convolution operator with kernel k given by (3.6). Then the shift dynamical system $(\{0,1\}^{\mathbf{Z}}, \mu, \sigma)$ has positive entropy:*

$$h_\mu(\sigma) = - \lim_{\Lambda\to\mathbf{Z}^1} \frac{1}{|\Lambda|} \sum_{\Lambda_0\subset\Lambda} \mu(0^{\Lambda_0}1^{\Lambda_1}) \log \mu(0^{\Lambda_0}1^{\Lambda_1}) > 0, \tag{3.8}$$

where $\Lambda = \Lambda_0 \cup \Lambda_1$.

Definition 3.6. *For the potential $\{\Phi_n\}_{n\geq 1}$; $\Phi_n : R^n \to \mathbf{R}$, we define the total energy U by*

$$U(x|\xi) = \Phi_1(x) + \sum_{y\in\xi} \Phi_2(x,y) + \sum_{y,z\in\xi} \Phi_3(x,y,z) + \cdots . \tag{3.9}$$

The probability measure μ is called a Gibbs measure if

$$\int_Q \mu(d\xi)F(\xi) = \int_Q \mu(d\xi)e^{-U(x|\xi_{\Lambda^c})}F(\delta_x + \xi_{\Lambda^c}) \tag{3.10}$$

for any compact Λ and bounded Borel measurable function F, where ξ_{Λ^c} is the restriction of ξ to Λ.

Theorem 3.7. *Let K be the convolution operator on $\ell^2(\mathbf{Z}^d)$ and assume the Fourier transform of the convolution kernel k takes the values in the unit open interval: $0 < \widehat{k}(\xi) < 1$. Let j be the convolution kernel of $J = (I-K)^{-1}K$ and set $J_\xi = (j(x-y))_{x,y\in\xi}$ and $j_\xi = (j(x))_{x\in\xi}$ for $\xi \subset \mathbf{Z}^d$. Then, the measure μ is Gibbs and*

$$U(0|\xi) = -\log(j(0) - \langle J_\xi^{-1} j_\xi, j_\xi\rangle). \tag{3.11}$$

Furthermore, the measure μ is the unique Gibbs measure with potential (3.11).

References

[1] A. Borodin and G. Olshanski, *Point processes and the infinite symmetric group part III: fermion point processes*, math.RT/9804088.

[2] I. P. Cornfeld, S. V. Fomin and Ya. G. Sinai, *Introduction to Ergodic Theory*, Springer Verlag, 1982.

[3] D. J. Daley and D. Veres-Jones, *An Introduction to the Theory of Point Processes*, Springer Verlag, 1988.

[4] M. Kac, *Toeplitz matrices, translation Kernel and a related problem in probability theory*, Duke Math. J. 21(1954), 501-509.

[5] N. M. Katz and P. Sarnak, *Random Matrices, Frobenius Eigenvalues, and Monodromy*, AMS Colloquium Publications, 1999.

[6] O. Macchi, *The coincidence approach to stochastic point processes*, Adv. Appl. Prob. 7. 83-122(1975).

[7] O. Macchi, *The fermion process – a model of stochastic point process with repulsive points*, Transactions of the Seventh Prague Conference on Information Theory, Statistical Decision Functions, Random Processes and of the Eighth European Meeting of Statisticians (Tech. Univ. Prague, Prague, 1974), Vol. A, pp. 391–398.

[8] M. L. Mehta, *Random Matrices*, 2nd edition, Academic Press, 1991.

[9] T. Shirai and Y. Takahashi, *Random point field associated with certain Fredholm determinant I : fermion, poisson, boson point processes*, preprint.

[10] T. Shirai and Y. Takahashi, *Random point field associated with certain Fredholm determinant II : fermion shift and its ergodic properties*, preprint.

[11] A. Soshnikov, *Gaussian fluctuation for the number of particles in Airy, Bessel, Sine and other determinantal random point fields*, math-ph/9907012.

[12] H. Spohn, *Interaction Brownian Particles : a study of Dyson's model*, Hydrodynamic Behavior and Interacting Particle Systems, ed. G. Papanicalaou, 1987.

4 STRONG TYPE ESTIMATION FROM WEAK TYPE ESTIMATES FOR SOME INTEGRAL OPERATORS

Nobuhiko FUJII

Department of Mathematics, Tokai University
3-20-1 Orido, Shimizu Shizuoka, 424-8610 Japan

nfujii@scc.u-tokai.ac.jp

Abstract

We show that if an integral operator T is of weak type (p, p) for some $p \in [1, \infty)$ then T is of 'strong' type (q, q), namely T is L^q-bounded, for all $q \in (p, \infty)$ under some additional assumptions for the operator T and the measure μ.

1 Fundamental Lemma

Let μ be a locally finite, nonnegative Borel measure on the n-dimensional Euclidean space $\mathbf{R}^n$ or an absolutely continuous measure with respect to the Lebesgue measure dx on $\mathbf{R}^n$, namely,

$$d\mu = w(x)dx, \qquad w(x) \geq 0.$$

We assume that μ satisfies the doubling condition that there exists a positive constant C_0 for cubes I and $\tilde{I}$ such that if $I \subset \tilde{I}, |\tilde{I}| = 2^n|I|$, then

$$\mu(\tilde{I}) \leq C_0\mu(I),$$

H.G.W. Begehr et al. (eds.), Proceedings of the Second ISAAC Congress, Vol. 1, 25–30.

where $|E|$ denotes the Lebesque measure of a measurable set E and a cube I denotes a measurable subset $\prod_{j=1}^{n}[a_j, a_j + l)$, $a_j, l \in \mathbf{R}, l > 0$, of $\mathbf{R}^n$.

Let $f(x)$ be a finitely real-valued (Borel or Lebesgue) measurable function defined on $\mathbf{R}^n$. We define a real number $m_f(I)$ for $f(x)$ and a finite cube I as follows:

$$m_f(I) = \max\left\{\lambda \in \mathbf{R};\ \mu\{x \in I; f(x) < \lambda\} \leq \frac{1}{2}\mu(I)\right\}.$$

Also we define for a locally integrable function $f(x)$ and a cube I

$$f_I = \frac{1}{\mu(I)} \int_I f(x) d\mu$$

It is said that $m_f(I)$ is the median over the cube I and f_I is the mean value over I.

It is obvious from the definition that

$$\mu\{x \in I; f(x) \geq m_f(I)\} \geq \frac{1}{2}\mu(I).$$

So we have

$$m_f(I) \leq 2f_I.$$

We also define that for any measurable function $f(x)$ and $A > 1$

$$M^{\sharp}_{A,f}(I) = \min\left\{\lambda \geq 0; \mu\{x \in I; |f(x) - m_f(I)| > \lambda\} \leq \frac{\mu(I)}{A}\right\}.$$

Lemma 1. *Let $f(x)$ be a finitely real-valued measurable function. For a cube Q and positive numbers $A > 2C_0$ and $B > 0$ there exists a measurable function $g(x)$, families $\mathcal{F}_|$ $(| \geq \infty)$ of dyadic subcubes of Q and sequences $\{a_I^{(j)}\}$, $I \in \mathcal{F}_|, | \geq \infty$, of numbers such that*

(i) *if I and J in $\mathcal{F}_j$ and $I \neq J$ then $I \cap J = \phi$,*

(ii)

$$\bigcup_{I \in \mathcal{F}_{j+1}} I \subset \bigcup_{I \in \mathcal{F}_j} I,$$

(iii)

$$\sum_{J \in \mathcal{F}_{j+k}, J \subset I} \mu(J) \leq \left(\frac{2}{A}\right)^k \mu(I)$$

for all $I \in \mathcal{F}_|$, $| \geq \prime$, $\| \geq \infty$, *where* $\mathcal{F}_0 = \{Q\}$,

(iv)

$$f(x) = g(x) + \sum_{j=1}^{\infty} \sum_{I \in \mathcal{F}_j} a_I^{(j)} \chi_I(x) \qquad \textit{for} \quad a.e. x \in Q,$$

where $\chi_I(x)$ *is the characteristic funtion of* I.

(v)

$$|g(x)| \leq |m_f(Q)| + 2 \sup_{I \subset J} M^{\sharp}_{A,f}(J) + B$$

$$\textit{for } a.e. x \in \bigcup_{I \in \mathcal{F}_j} I \setminus \left(\bigcup_{J \in \mathcal{F}_{j+1}} J \right),$$

(vi)

$$|a_I^{(j)}| \leq 2 \sup_{I \subset J} M^{\sharp}_{A,f}(J) + B$$

This lemma shows that a function $f(x)$ is majorized by the sharp maximal functions

$$M^{\sharp}_{A,f}(x) = \sup_{I \ni x} M^{\sharp}_{A,f}(I)$$

nearly pointwise, but not pointwise. This decomposition lemma is essentially due to Carleson [C] and its proof may be found in [F2]. We have the following well-known result by Fefferman and Stein [FS] from this lemma:

Theorem 2. *Suppose that* μ *have* A_∞ *property, that is, there exist a positive constants* C_1 *and* δ *such that for a cube* I *and any measurable subset* E *of* I

$$\mu(E) \leq C_1 \left(\frac{|E|}{|I|} \right)^{\delta} \mu(I).$$

And let $\Psi(t)$ *be a nonnegative function on* $[0, \infty)$ *which satisfies that*

$$\Psi(2t) \leq C_2 \Psi(t) \qquad \textit{for} \quad \textit{any} \quad t > 0.$$

Then there exists positive constants A *and* C_3, *independent of* $f(x)$, *such that if* $\lim_{|Q| \to \infty} m_f(Q) = 0$, *where* $Q \ni 0$,

$$\int_{\mathbf{R}^n} \Psi(|f(x)|) d\mu \leq C_3 \int_{\mathbf{R}^n} \Psi(M^{\sharp}_{A,f}(x)) d\mu.$$

This inequality and its proof was stated in [F1].

2 Maximal Function Estimates

Let $K_\alpha(x)$ be a function defined outside of the origin such that

$$|K_\alpha(x)| \le \frac{C_4}{|x|^{n-\alpha}} \qquad (0 \le \alpha < n)$$

We may define an integral transformation by $K_\alpha(x)$ formally as follows:

$$T_\alpha f(x) = K_\alpha * f(x) = \int_{\mathbf{R}^n} K_\alpha(x-y) f(y) dy$$

for suitable functions $f(x)$.

If $\alpha = 0$, then $T_0 f$ is a singular integral operator and if $0 < \alpha < n, T_\alpha f$ is a fractional integral or the Riesz potential of f. If we assume that $T_\alpha f$ is of weak type, then we may obtain the following estimation due to the arguments of Strömberg, Jawerth and Torchinsky.

Lemma 3. *Suppose that T_0 is of weak type (p,p) for some p, $1 \le p < \infty$:*

$$\mu\{x; |T_0 f(x)| > \lambda\} \le \frac{C_5}{\lambda^p} \int_{\mathbf{R}^n} |f(x)|^p d\mu \qquad \text{for all} \quad \lambda > 0 \qquad (1)$$

And if $K_0(x)$ satisfies the smoothness condition:

$$|K_0(x-y) - K_0(-y)| \le C_6 \frac{|x|^\varepsilon}{|y|^{n+\varepsilon}} \qquad for \quad |y| > 2|x|, \qquad (2)$$

where $\varepsilon > 0$ is independent of x and y. Then there exists a positive constant C_7 for $A > C_0$ such that

$$M^\sharp_{A,T_0 f}(I) \le C_7 \left\{ \left(\frac{A}{\mu(I)} \int_I |f(x)|^p d\mu \right)^{\frac{1}{p}} + \sup_{J \supset I} \frac{1}{|J|} \int_J |f(x)| dx \right\}$$

for any smooth function $f(x)$ with compact support and a cube I.

If $0 < \alpha < n$, then $K_\alpha(x)$ is locally integrable with respect to the Lebesgue measure. So we may consider the following situation.

Let $\varphi(x)$ is a bounded function with compact support. We set

$$T_\alpha^* f(x) = \sup_{t>0} |\varphi_t * K_\alpha * f(x)|$$

where $\varphi_t(x) = \frac{1}{t^n}\varphi\left(\frac{x}{t}\right)$. Then by using the maximal theorem we have for a cube I and any $\lambda > 0$

$$|\{x \in I; \sup_{t>0} |\varphi_t * (K_\alpha * (f\chi_I))\chi_I(x)| > \lambda\}| \leq \frac{C_8|I|^{\alpha/n}}{\lambda} \int_I |f(x)|dx\,.$$

Thus we have the following lemma.

Lemma 4. *If $K_\alpha(x) = \dfrac{1}{|x|^{n-\alpha}}$ and $d\mu = dx$, then for a cube I and $A > 1$*

$$M^\sharp_{A,T_\alpha^* f}(I) \leq C_9 A \sup_{J \supset I} \frac{1}{|J|^{1-\alpha/n}} \int_J |f(x)|dx.$$

3 Results

From Lemma 3 we have the well-known result by the routine argument, however which is a weaker result than usual.

Theorem 5. *Let $1 \leq p < \infty$ and let $d\mu = w(x)dx$ satisfy Muckenhoupt's A_p condition :*

$$\sup_I \frac{1}{|I|} \int_I w(x)dx \left(\frac{1}{|I|}\int_I w(x)^{1-p'}dx\right)^{p-1} < \infty,$$

where $(1-p)(1-p') = 1$.

Suppose T_0 satisfies (1) and (2) in Lemma 3 and $p < q < \infty$. Then T_0 is bounded on $L^q(w(x)dx, \mathbf{R}^n)$;

$$\int_{\mathbf{R}^n} |T_0 f(x)|^q w(x)dx \leq C_9 \int_{\mathbf{R}^n} |f(x)|^q w(x)dx$$

where the constant C_9 is independent of f.

From our point of view we can argue a two-weight inequality for the opertors $T_0 f$ and $T^*_\alpha f$. We mention a result only for $T^*_\alpha f$ defined by $K_\alpha(x) = \dfrac{1}{|x|^{n-\alpha}}$.

Theorem 6. *Let $w(x)$ and $v(x)$ be nonnegative functions on $\mathbf{R}^n$, and let $1 < p \leq q < \infty$. If there exist positive constants $\delta, \varepsilon, 0 < \varepsilon < 1$, and C_{10} such that*

$$\int_E w(x)dx \left(\frac{1}{|I|^{1-\alpha/n}} \int_{c(n,\epsilon)I} v^{1-p'}(x)dx \right)^q \leq$$

$$C_{10} \left(\frac{|E|}{|I|} \right)^\delta \left(\int_F v^{1-p'}(x)dx \right)^{q/p} < \infty$$

for any cube I and measurable subsets E and F of I with

$\mathrm{E} \cap F = \phi$ *and*

$|F| \geq \varepsilon |I|$.

Then there exists a positive constant C_{11} such that

$$\left(\int_{\mathbf{R}^n} |T^*_\alpha f(x)|^q w(x)dx \right)^{1/q} \leq$$

$$C_{11} \left(\int_{\mathbf{R}^n} |f|^p v(x)dx \right)^{1/p}$$

*for all measurble functions*f(x).

References

[C] L. Carleson, *Two remarks on H^1 and BMO*, Adv. in Math. **22** (1976), 269-277.

[FS] C. Fefferman and E. M. Stein, *H^p spaces of several variables*, Acta Math.**129**(1972), 137-193.

[F1] N. Fujii, *A proof of the Fefferman-Stein-Strömberg inequality for the sharp maximal functions*, Proc. Amer. Math. Soc. **106** (1989), 371-377.

[F2] ——, *A condition for a two-weight norm inequality for singular integral operators*, Studia Math. **98**(1991), 175-190.

5 CONJUGATE FOURIER SERIES AND INTEGRALS OF SEVERAL VARIABLES IN THE L-1 SENSE

Zhongkai LI[1]

Department of Mathematics,
Capital Normal University,
Beijing 100037, China

lizk@mail.cnu.edu.cn

MOS subject classification Categories: 42B20, 42B08

Keywords: Fourier series of several variables, Cesáro means, conjugate series.

Abstract

The conjugate Fourier series and integrals of several variables in the $\ell-1$ sense are introduced. Their Cesàro and Riesz summability is studied by defining the appropriate conjugate functions.

1. Introduction

In the theory of Fourier series and integrals of several variables, the spherical sums and the rectangular sums have been intensively studied. The former was initiated by the classical paper of Bochner in 1936, and stimulated by some papers of Stein, Weiss and Shapiro (cf. [11-13]). For the later one, some basic results can be found in Zygmund's book [15] and the more developments in the survey papers of Golubov [8] and

[1]The author is supported by National Natural Science Foundation of China, Project 19601025.

H.G.W. Begehr et al. (eds.), Proceedings of the Second ISAAC Congress, Vol. 1, 31–39.

D'yachenko [7]. By the $l-p$ modulus of the summation multiindex, spherical sums and the rectangular sums are often called $l-2$ sums and $l-\infty$ sums respectively. Correspondingly, the $l-1$ sums have less been considered. Only in recent years, some basic results are obtained by Berens, Xu and the author (cf. [1-4]). What is interesting is that Xu first ran into the problem on the $l-1$ sum of multiple Fourier series when he studied the general multivariate orthogonal polynomial expansions.

To be precise, let us introduce the necessary notations. Let $\mathbb{R}^d$ denote the Euclidean space of dimension d and let $\mathbb{T}^d = \mathbb{R}^d (\text{mod } 2\pi\mathbb{Z}^d)$ denote the d-dimentional torus. For a vector x in $\mathbb{R}^d$, $|x|_p$ defines its l_p-norm, $1 \le p \le \infty$. To the multiindex $m = (m_1, \dots, m_d)$ in $\mathbb{Z}^d$, we associate the $l-p$ modulus to be the l_p-norm of m considered as a point in $\mathbb{R}^d$.

For a function $f \in L(\mathbb{T}^d)$, the nth $l-p$ partial sum of its Fourier series is defined by

$$(1.1)\qquad S_{n,d,p}(f;\theta) = \sum_{|m|_p \le n} \hat{f}(m) e^{i<m,\theta>}, \qquad \theta \in \mathbb{T}^d \quad \text{and} \quad n \in \mathbb{N}_0,$$

where

$$(1.2)\qquad \hat{f}(m) = \frac{1}{(2\pi)^d} \int_{\mathbb{T}^d} f(\theta) e^{-i<m,\theta>} d\theta$$

is the Fourier coefficient of f w.r.t. the multiindex $m = (m_1, \dots, m_d) \in \mathbb{Z}^d$, and $< m, \theta >$ defines the inner product of m and $\theta = (\theta_1, \dots, \theta_d)$.

When $p = 2$ and $p = \infty$, (1.1) corresponds to the spherical sums and the rectangular sums respectively.

In the case of $p = 1$, as mentioned above, only a few of results on the Cesàro summability of (1.1) are known. As usual, the Cesàro means of order δ of (1.1) for $p = 1$ is defined by

$$(1.3)\qquad S^{\delta}_{n,d,1}(f;\theta) = \binom{n+\delta}{n}^{-1} \sum_{k=0}^{n} \binom{n-k+\delta-1}{n-k} S_{n,d,1}(f;\theta).$$

Berens and Xu proved in [1-3] that (i) for $\delta > 0$ and $f \in L^p(\mathbb{T}^d)$, $1 \le p \le \infty$, $\lim_{n\to\infty} \|S^{\delta}_{n,d,1}(f) - f\|_p = 0$; and (ii) the $(C, 2d-1)$ means define positive operators and the order $2d-1$ is best posible. Also, together with the author, they proved in [4] that (iii) for $\delta > 0$ and $f \in L^p(\mathbb{T}^d)$, $1 \le p \le \infty$, $S^{\delta}_{n,d,1}(f;\theta)$ converges to $f(\theta)$ almost everywhere. Actually, (i) and (iii) are proved only for the Fourier integrals, but by a general result of Stein (cf. [13]), they also hold for Fourier series. It should be pointed out that part (i) has to be attributed to Podkorytov [10], who verified the norm convergence for a general class of summability method w.r.t. an arbitrary polyhedral modulus of the multiindex m, which includes the Riesz or Cesàro means in the $l-1$ sense. We also like to introduce a result of Bochner, that the partial sums of Fourier integrals based on any convex polyhedron in $\mathbb{R}^d$ having a finite number of sides converges in L^p norm for $1 < p \leq 2$, of which a proof appeared in Herz [9].

An attractive feature on the study of the $l-1$ sums is that a closed form for the Dirichlet kernel can be given as a divided difference of a univariate function, the arguments in the difference being the independent variables (cf. [1-3]), which makes developing a rich theory in the $l-1$ sense possible.

In the present paper, we will introduce and study the conjugate Fourier series and integrals and the conjugate functions of several variables in the $l-1$ sense. To this end, we first recall the conjugacy in the $l-2$ sense.

In the $l-2$ case, let $K(x) = \frac{\omega(x/|x|_2)}{|x|_2^d}$ be a Calderón-Zygmund kernel, where $\omega(\xi)$ satisfies the following two properties: (i) $\omega(\xi)$ is in Lipα on S^{d-1}, $\alpha > 0$; and (ii) $\int_{S^{d-1}} \omega(\xi)d\sigma(\xi) = 0$, $d\sigma(\xi)$ being the natural volume element on S^{d-1}. Define the Calderón-Zygmund transform (singular integral) of a function $g \in L(\mathbb{R}^d)$ by

$$\tilde{g}(x) = \text{p.v.} g * K(x) = \lim_{\varepsilon \to 0} \int_{\mathbb{R}^d \setminus B(0,\varepsilon)} g(x-y)K(y)dy. \tag{1.4}$$

It is shown in [5] that $\tilde{g}(x)$ exists almost everywhere and the principle value Fourier transform $\hat{K}(y)$ of $K(x)$ exists for every y, and moreover, if $\tilde{g} \in L(\mathbb{R}^d)$, $\hat{\tilde{g}}(y) = \hat{f}(y)\hat{K}(y)$.

Next, let $K^*(x)$ be the periodic analogue of $K(x)$ as $K^*(x) = K(x) + \sum_{m \neq 0}[K(x+2\pi m) - K(2\pi m)]$ for $x \neq 2\pi m$. The series is convergent absolutely and uniformly for $x \in \mathbb{T}^d$ and $K^*(x+2\pi m) = K^*(x)$. If $f(x) \in L(\mathbb{T}^d)$ and is extended by periodicity to $\mathbb{R}^d$, it then follows that

$$\tilde{f}(x) = \lim_{\varepsilon \to 0} \frac{1}{(2\pi)^d} \int_{\mathbb{T}^d \setminus B(0,\varepsilon)} f(x-y)K^*(y)dy \tag{1.5}$$

exists and is finite for a.e. $x \in \mathbb{T}^d$. It is shown that if $\tilde{f} \in L(\mathbb{T}^d)$, then its mth Fourier coefficient $\hat{\tilde{f}}(m) = \hat{f}(m)\hat{K}^*(m) = \hat{f}(m)\hat{K}(m)$, $\hat{K}^*(m)$ being the principle value Fourier coefficient of $K^*(x)$ (cf. [6]). Naturally, the conjugate Fourier series of f with the Fourier series $\sum \hat{f}(m)e^{i<m,\theta>}$

should be

$$\sum_{m\neq 0}\hat{K}(m)\hat{f}(m)e^{i<m,x>}. \tag{1.6}$$

If $K(x)=k_j(x)=x_j/|x|_2^{d+1}$, $j=1,\dots,d$, (1.4) reduces to the Riesz transform

$$R_jg(x)=\lim_{\varepsilon\to 0}\int_{\mathbb{R}^d\setminus B(0,\varepsilon)}\frac{y_j}{|y|_2^{d+1}}g(x-y)dy, \tag{1.7}$$

and for $y\neq 0$, $\hat{k}_j(y)=-iy_j/|y|_2 2^d\Gamma(\frac{d+1}{2})\pi^{\frac{d-1}{2}}$. Thus for this case, if $R_jg\in L(\mathbb{R}^d)$,

$$\widehat{R_jg}(y)=-\frac{iy_j}{|y|_2 2^d\Gamma(\frac{d+1}{2})\pi^{\frac{d-1}{2}}}\hat{f}(y), \tag{1.8}$$

and if $f(x)\in L(\mathbb{T}^d)$, its conjugate series (1.6) reduces to

$$-i\sum_{m\neq 0}\frac{m_j}{|m|_2}\hat{f}(m)e^{i<m,\theta>},\qquad j=1,\dots,d, \tag{1.9}$$

after dropping a constant.

Now we introduce the conjugate Fourier series and integrals and the conjugate functions in the $l-1$ sense. We first consider the conjugate Fourier integrals.

Let g be a function in $L(\mathbb{R}^d)$ and let $\hat{g}(\alpha)=\frac{1}{(2\pi)^d}\int_{\mathbb{R}^d}g(x)e^{-i<\alpha,x>}dx$, $\alpha\in\mathbb{R}^d$, be its Fourier transform. The inverse Fourier transform, i.e. the Fourier integral, is then formally defined by

$$g(x)=\int_{\mathbb{R}^d}\hat{g}(\alpha)e^{i<\alpha,x>}d\alpha,\qquad x\in\mathbb{R}^d. \tag{1.10}$$

Motivated by (1.8), the conjugate Fourier integrals $\tilde{g}_j(x)$ in the $l-1$ sense are defined such that if $\tilde{g}_j(x)\in L(\mathbb{R}^d)$, $\hat{\tilde{g}}_j(\alpha)=\frac{-i\alpha_j}{|\alpha|_1}\hat{g}(\alpha)$, namely, in formal,

$$\tilde{g}_j(x)=\int_{\mathbb{R}^d}\frac{-i\alpha_j}{|\alpha|_1}\hat{g}(\alpha)e^{i<\alpha,x>}d\alpha,\qquad x\in\mathbb{R}^d, \tag{1.11}$$

for $j=1,2,\dots,d$. Their partial sums (integrals) are defined by

$$\tilde{S}_{R,j}(g,x)=\int_{0<|\alpha|_1<R}\frac{-i\alpha_j}{|\alpha|_1}\hat{g}(\alpha)e^{i<\alpha,x>}d\alpha,\qquad\text{for}\quad j=1,\dots,d. \tag{1.12}$$

For a function $f \in L(\mathbb{T}^d)$ with the Fourier series $\sum \hat{f}(m)e^{i<m,\theta>}$, motivated by (1.9), its conjugate series in the $l-1$ sense are defined by

$$\sum_{m\neq 0} \frac{-im_j}{|m|_1}\hat{f}(m)e^{i<m,\theta>}, \qquad \text{for} \quad j=1,\dots,d. \tag{1.13}$$

The purpose of this paper is to study, in the $l-1$ sense, the existence of the conjugate Fourier integrals on $\mathbb{R}^d$ and the conjugate functions on $\mathbb{T}^d$, and the Riesz or Cesáro summability of the conjugate integrals (1.11) and the conjugate series (1.13).

In contrast to the univariate case and the multivariate spherical case, the singularities of the conjugate functions in the $l-1$ sense, which may also be called Riesz or Hilbert transforms, are not isolated, but are some hyperplanes, and the order of singularity increases when two or more hyperplanes cross (see (2.5)-(2.8) below).

2. The Conjugate Fourier Integrals in the $l-1$ Sense

For a function $g \in L(\mathbb{R}^d)$, its conjugate Fourier integrals are formally defined by (1.11), and the partial sums(integrals) $\tilde{S}_{R,j}(g,x)$ given by (1.12) can be written as

$$\tilde{S}_{R,j}(g;x) = g * \tilde{D}_{R,j}(x), \tag{2.1}$$

for $j=1,\dots,d$, where,

$$\tilde{D}_{R,j}(x) = \int_{0<|\alpha|_1<R} \frac{-i\alpha_j}{|\alpha|_1} e^{i<\alpha,x>} d\alpha \tag{2.2}$$

are called the conjugate Dirichlet kernels.

First we have

Lemma 2.1. *If d is even,*

$$\tilde{D}_{R,j}(x) = 2^d(-1)^{\frac{d}{2}} \frac{\partial}{\partial x_j}[x_1^2,\dots,x_d^2]u^{\frac{d-1}{2}} \int_0^{R\sqrt{u}} \frac{\sin t}{t} dt, \tag{2.3}$$

and if d is odd,

$$\tilde{D}_{R,j}(x) = 2^d(-1)^{\frac{d-1}{2}} \frac{\partial}{\partial x_j}[x_1^2,\dots,x_d^2]u^{\frac{d-1}{2}} \int_0^{R\sqrt{u}} \frac{1-\cos t}{t} dt, \tag{2.4}$$

where $[t_1,\dots,t_d]f(u)$ *denotes the* $(d-1)$*th divided difference of* f *at the arguments* $t_1,\dots,t_d$*,* u *being the divided difference variable.*

Now, for $j=1,\dots,d$, we introduce the functions $G_j(x)$: if d is even,

$$G_j(x)=2^d(-1)^{\frac{d}{2}}\frac{\partial}{\partial x_j}[x_1^2,\dots,x_d^2]u^{\frac{d-1}{2}}\int_0^\infty\frac{\sin t}{t}dt, \tag{2.5}$$

and if d is odd,

(2.6)

$$G_j(x)=2^d(-1)^{\frac{d+1}{2}}\{\int_0^1\frac{1}{t}\frac{\partial}{\partial x_j}[x_1^2,\dots,x_d^2]u^{\frac{d-1}{2}}\cos\sqrt{u}tdt + \frac{\partial}{\partial x_j}[x_1^2,\dots,x_d^2]u^{\frac{d-1}{2}}\int_1^\infty\frac{\cos\sqrt{u}t}{t}dt\}.$$

The conjugate functions in the $l-1$ sense are defined by

$$\tilde{g}_j(x)=\lim_{\varepsilon\to 0}\tilde{g}_{\varepsilon,j}(x), \tag{2.7}$$

if the limit exists in some sense (a.e. or in norm), where,

$$\tilde{g}_{\varepsilon,j}(x)=\int_{y\in\mathbb{R}^d,|y|_1>\varepsilon}g(x-y)G_j(y)dy, \tag{2.8}$$

for $j=1,\dots,d$, are called the truncated conjugate functions.

For $d=1$, (2.7) coincides with the Hilbert transform of single variable, since from (2.6), $G(x)=\frac{2}{x}$ and there is only one such $G(x)$ in this case. For $d=2$, if follows from (2.5) that $G_j(x)=\frac{2^2\text{sign}x_j}{(|x_1|+|x_2|)^2}\int_0^\infty\frac{\sin t}{t}dt$, for $j=1,2$.

The main theorems of this section are stated as follows.

Theorem 2.1. *For* $g\in L(\mathbb{R}^d)$ *and* $j=1,\dots,d$*, the limit in (2.7) exists almost everywhere in* $\mathbb{R}^d$*, and the conjugate mappings* $T_jg(x)=\tilde{g}_j(x)$ *are of* $L^p(\mathbb{R}^d)$ *boundedness for* $1<p<\infty$ *and of weak* $L(\mathbb{R}^d)$ *boundedness.*

Theorem 2.2. *Let* $\delta>0$*. For* $g\in L(\mathbb{R}^d)$ *and* $j=1,\dots,d$*, we have*

$$\lim_{R\to\infty}\{\tilde{S}^\delta_{R,j}(g;x)-\tilde{g}_{1/R,j}(x)\}=0 \tag{2.9}$$

almost everywhere, and also in L*-norm, where* $\tilde{S}^\delta_{R,j}(g;x)$ *are the Riesz means of order* δ *of the conjugate Fourier integrals (1.11), namely,*

$$\tilde{S}^\delta_{R,j}(g;x)=\int_{0<|\alpha|_1<R}(1-\frac{|\alpha|_1}{R})^\delta\frac{-i\alpha_j}{|\alpha|_1}\hat{g}(\alpha)e^{i<\alpha,x>}d\alpha. \tag{2.10}$$

3. The Conjugate Fourier Series in the $l-1$ Sense

For a function $f \in L(\mathbb{T}^d)$ with the Fourier series $\sum \hat{f}(m)e^{i<m,\theta>}$, its conjugate Fourier series in the $l-1$ sense are defined by (1.13), and their partial sums are

$$\tilde{S}_{n,j}(f;\theta) = \sum_{0<|m|_1\leq n} \frac{-im_j}{|m|_1}\hat{f}(m)e^{i<m,\theta>} \tag{3.1}$$
$$= f * \tilde{D}_{n,j}(\theta),$$

for $j = 1, \dots, d$, where,

$$\tilde{D}_{n,j}(\theta) = \sum_{0<|m|_1\leq n} \frac{-im_j}{|m|_1}e^{i<m,\theta>} \tag{3.2}$$

are the conjugate Dirichlet kernels.

First we have

Lemma 3.1. *For $j = 1, \dots, d$,*

(3.3)
$$\tilde{D}_{n,j}(\theta) = 2\frac{\partial}{\partial\theta_j}[\cos\theta_1, \dots, \cos\theta_d]\sum_{m=1}^{n}\frac{1}{m}\cos(\frac{d+1}{2}\pi + m\varphi)\sin^{d-1}\varphi,$$

where the divided difference variable is $u = \cos\varphi$.

For $j = 1, \dots, d$, we introduce the functions $G_j^*(\theta)$ as

(3.4)
$$G_j^*(\theta) = 2\frac{\partial}{\partial\theta_j}[\cos\theta_1, \dots, \cos\theta_d]\sum_{m=1}^{\infty}\frac{1}{m}\cos(\frac{d+1}{2}\pi + m\varphi)\sin^{d-1}\varphi.$$

Using [15, Vol.I, (2.8)], we have

Lemma 3.2. *For $j = 1, \dots, d$, if d is even,*

$$G_j^*(\theta) = (-1)^{\frac{d}{2}+1}\frac{\partial}{\partial\theta_1}[\cos\theta_1, \dots, \cos\theta_d](\pi - \varphi)\sin^{d-1}\varphi, \tag{3.5}$$

and if d is odd,

$$G_j^*(\theta) = 2(-1)^{\frac{d+1}{2}}\frac{\partial}{\partial\theta_j}[\cos\theta_1, \dots, \cos\theta_d]\sin^{d-1}\varphi\log\frac{1}{2\sin\frac{1}{2}\varphi}, \tag{3.6}$$

where, the divided difference variable is $u = \cos\varphi$.

The conjugate functions related to the conjugate Fourier series (1.13) in the $l-1$ sense are defined by

$$\tilde{f}_j(\theta) = \lim_{\varepsilon\to 0} \tilde{f}_{\varepsilon,j}(\theta), \tag{3.7}$$

if the limit exists in some sense (a.e. or in norm), where,

$$\tilde{f}_{\varepsilon,j}(\theta) = \int_{\varphi\in\mathbb{T}^d, |\varphi|_1>\varepsilon} f(\theta-\varphi)G_j^*(\varphi)d\varphi, \tag{3.8}$$

for $j = 1, \ldots, d$, are the truncated conjugate functions.

For $d = 1$, (3.7) coincides with the conjugate function of single variable, since from (3.6), $G_j(\theta) = \cot\frac{1}{2}\theta$ and there is only one such $G(\theta)$ in this case.

The main theorems of this section are stated as follows.

Theorem 3.1. *For* $f \in L(\mathbb{T}^d)$ *and* $j = 1, \ldots, d$, *the limit in (3.7) exists almost everywhere in* $\mathbb{T}^d$, *and the conjugate mappings* $T_j f(\theta) = \tilde{f}_j(\theta)$ *are of* $L^p(\mathbb{T}^d)$ *boundedness for* $1 < p < \infty$ *and of weak* $L(\mathbb{T}^d)$ *boundedness.*

Theorem 3.2. *Let* $\delta > 0$. *For* $f \in L(\mathbb{T}^d)$ *and* $j = 1, \ldots, d$, *we have*

$$\lim_{n\to\infty} \{\tilde{S}^{\delta}_{n,j}(f;\theta) - \tilde{f}_{1/n,j}(\theta)\} = 0 \tag{3.9}$$

almost everywhere in $\mathbb{T}^d$, *and also in* L*-norm, where* $\tilde{S}^{\delta}_{n,j}(f;\theta)$ *are the Cesàro means of order* δ *of the conjugate Fourier series (1.13), namely for* $j = 1, \ldots, d$,

$$\tilde{S}^{\delta}_{n,j}(f;\theta) = \binom{n+\delta}{n}^{-1} \sum_{k=0}^{n} \binom{n-k+\delta-1}{n-k} \tilde{S}_{k,j}(f;\theta), \tag{3.10}$$

and $\tilde{S}_{n,j}(f;\theta)$ *are the partial sums (3.1) of (1.13).*

References

1. H. Berens and Y. Xu, *Fejér means for multivariate Fourier series*, Math. Z. **221**(1996), 449-465.
2. H. Berens and Y. Xu, *$l-1$ summability for multiple Fourier integral*, in "Approximation Theory VIII", C. Chui and L. Schumaker eds., World Sci. Publ. Co., Singapore, 1995, 55-62.
3. H. Berens and Y. Xu, *$l-1$ summability of multiple Fourier integrals and positivity*, Math. Proc. Camb. Phil. Soc. **122**(1997), 149-172.

4. H. Berens, Zh.-K. Li and Y. Xu, *On $l-1$ Riesz summability of the inverse Fourier integral*, preprint.
5. A. P. Calderon and A. Zygmund, *On the existence of certain singular integrals*, Acta Math. **88**(1952), 85-139.
6. A. P. Calderon and A. Zygmund, *Singular integrals and periodic functions*, Studia Math. **14**(1954), 249-271.
7. M. I. D'yachenko, *Some problems in the theory of multiple trigonometric series*, Russian Math. Surveys **47**(5)(1992), 103-171.
8. B. I. Golubov, *Multiple Fourier series and Fourier integrals*, J. Soviet. Math. **24**(1984), 639-673.
9. C. S. Herz, *On the mean inversion of Fourier and Hankel transforms*, Proc. Nat. Acad. Sci. USA **40**(1954), 996-999.
10. A. N. Podkorytov, *Summation of multiple Fourier series over polyhedra*, Vestnik Leningrad Univ. Math. **13**(1981), 69-77.
11. V. L. Shapiro, *Fourier series in several variables*, Bull. Amer. Math. Soc. **70** (1964), 48-93.
12. V. L. Shapiro, , *Topics in Fourier and geometric analysis*, Mem. Amer. Math. Soc., 1964.
13. E. M. Stein and G. Weiss, "Introduction to Fourier Analysis on Euclidean Spaces", Princeton Univ. Press, Princeton, 1971.
14. Y. Xu, *Christoffel functions and Fourier series for multivariate orthogonal polynomials*, J. Approx. Theory **82**(1995), 205-239.
15. A. Zygmund, "Trigonometric Series", 2nd Edition, Cambridge Univ. Press, Cambridge, 1968.

6 ADMISSIBLE WAVELETS AND SIEGEL DOMAINS

Heping LIU[1]

School of Mathematical Sciences, Peking University
Beijing 100871, The People's Republic of China

hpliu@pku.edu.cn

1. INTRODUCTION

To decompose Hardy functions Grossmann and Morlet ([2])introduced the admissible wavelet of one-dimension. Let us recall some facts of one-dimensional wavelet analysis.

All automorphisms of the upper half-plane $\mathbf{U} = \{z \in \mathbf{C} : \mathrm{Im}z > 0\}$ are given by $\mathbf{SL}(2, \mathbf{R})$ according to the linear fractional transforms

$$z \longmapsto gz = \frac{az+b}{cz+d}, \quad z \in \mathbf{U}, g \begin{pmatrix} a & b \\ c & d \end{pmatrix} \in \mathbf{SL}(2, \mathbf{R}).$$

Let $\mathbf{SL}(2, \mathbf{R}) = \mathbf{NAK}$ be the Iwasawa decomposition, where

$$\mathbf{N} = \left\{ \begin{pmatrix} 1 & b \\ 0 & 1 \end{pmatrix} : \quad b \in \mathbf{R} \right\},$$

$$\mathbf{A} = \left\{ \begin{pmatrix} \sqrt{a} & 0 \\ 0 & \frac{1}{\sqrt{a}} \end{pmatrix} : \quad a \in \mathbf{R}^+ \right\},$$

[1]The Project Supported by the National Natural Science Foundation of P. R. China and the Excellent Young Teachers Fund of the Ministry of Education of P. R. China.

H.G.W. Begehr et al. (eds.), Proceedings of the Second ISAAC Congress, Vol. 1, 41–49.

$$\mathbf{K}=\left\{\begin{pmatrix}\cos\theta & \sin\theta\\ -\sin\theta & \cos\theta\end{pmatrix}:\quad \theta\in[0,2\pi)\right\}.$$

Then the affine automorphism group of the upper half-plane turns out to be the Iwasawa subgroup $\mathbf{P}=\mathbf{NA}$ of $\mathbf{SL}(2,\mathbf{R})$. $\mathbf{P}$ acts also on the real line, the boundary of the upper half-plane, and coincides with the affine group "$ax+b$". The affine group $\mathbf{P}$ is a nonunimodular group with the left Haar measure $d\mu_l(b,a)=\frac{dbda}{a^2}$.

Let $H^2(\mathbf{R})$ denote the Hardy space, which consists of the boundary values of the Hardy space of holomorphic functions on the upper half-plane. The affine group $\mathbf{P}$ has a natural unitary representation U on $H^2(\mathbf{R})$ defined by

$$U_{(b,a)}f(x)=\frac{1}{\sqrt{a}}f(\frac{x-b}{a}),\quad f\in H^2(\mathbf{R}). \tag{1}$$

U is irreducible and square-integrable in the sense of following: there exists in $H^2(\mathbf{R})$ at least one function ϕ, not identically zero, such that

$$\frac{1}{\|\phi\|^2_{L^2(\mathbf{R})}}\int_{\mathbf{P}}\left|\langle\phi,U_{(b,a)}\phi\rangle_{L^2(\mathbf{R})}\right|^2\frac{dbda}{a^2}=C_\phi<+\infty. \tag{2}$$

The condition (2) is called the admissibility condition. If $\phi\in H^2(\mathbf{R})$ satisfies the admissibility condition (2), we say that ϕ is an admissible wavelet and define the wavelet transform of a function $f\in H^2(\mathbf{R})$ with respect to ϕ by

$$W_\phi f(b,a)=C_\phi^{-\frac{1}{2}}\langle f,U_{(b,a)}\phi\rangle_{L^2(\mathbf{R})}. \tag{3}$$

The wavelet transform $W_\phi f$ is an isometric operator from $H^2(\mathbf{R})$ into $L^2(\mathbf{P},d\mu_l)$. Any function $f\in H^2(\mathbf{R})$ can be recovered from its wavelet transform through the formula

$$f(x)=C_\phi^{-\frac{1}{2}}\int_{\mathbf{P}}W_\phi f(b,a)U_{(b,a)}\phi(x)\,\frac{dbda}{a^2}. \tag{4}$$

In essence, the above result is, as pointed out by Grossmann and Morlet, a special case of the general theory of square integrable representation of nonunimodular groups, developed by Duflo and Moore ([1]). It is more important that Grossmann and Morlet characterize the admissibility condition by

$$C_\phi=\int_0^{+\infty}\frac{|\widehat{\phi}(\xi)|^2}{\xi}\,d\xi<+\infty. \tag{5}$$

The condition (5) is easier to verify and not involved with the affine group and its representation.

Remark. We can identify the affine group with the upper-half plane by setting $z = b + ia$. Therefore the wavelet transfrom can be used to establish the relation between the various function spaces on the real line and the upper half-plane, which is useful for studying the Toeplitz-Hankel type operators (cf. [4]).

Now we want to extend the wavelet analysis to the higher dimensions. As we have seen above, it is natural to consider the objects such as the generalized upper half-plane of several complex variables, i.e., the Siegel domain, and its (distinguished) boundary, its affine automorphism group or the Iwasawa subgroup of its automorphism group. To be more illustrative, here we state the results in two typical special cases. The same results hold in more generalized cases such as the affine homogenous Siegel domains of type II and the symmetric tube domains (cf. [6], [7]) .

2. THE CASE ONE

First we consider the generalized upper half-plane $\mathbf{U}^{n+1}$ in $\mathbf{C}^{n+1}$ defined by $\mathbf{U}^{n+1} = \{(z, z_{n+1}) \in \mathbf{C}^{n+1} : \mathrm{Im} z_{n+1} > |z|^2\}$, where $z = (z_1, \cdots, z_n) \in \mathbf{C}^n$. $\mathbf{U}^{n+1}$ is a Siegel domain of type II and is holomorphically equivalent to the unit ball $\mathbf{B}^{n+1}$ in $\mathbf{C}^{n+1}$. If $n = 0$, $\mathbf{U}^{n+1}$ degenerates into the the upper half-plane $\mathbf{U}$. The Cayley transform Γ from $\mathbf{B}^{n+1}$ onto $\mathbf{U}^{n+1}$ is given by

$$\begin{cases} z_j &= \frac{iw_j}{1-w_{n+1}}, \quad j = 1, \cdots, n, \\ z_{n+1} &= i\frac{1+w_{n+1}}{1-w_{n+1}}, \end{cases}$$

where $w = (w_1, \cdots, w_n, w_{n+1}) \in \mathbf{B}^{n+1}$.

All automorphisms of $\mathbf{B}^{n+1}$ are given by $\mathbf{SU}(n+1, 1)$ through the linear fractional transforms

$$\widetilde{T}_g : w \longmapsto gw = \left(\frac{Aw^t + B}{Cw^t + D}\right)^t, \quad w \in \mathbf{B}^{n+1},$$

where $g = \begin{pmatrix} A & B \\ C & D \end{pmatrix} \in \mathbf{SU}(n+1, 1)$, w^t denotes the transpose of w. $\mathbf{SU}(n+1,1)$ acts also on $\mathbf{U}^{n+1}$ by means of maps $T_g = \Gamma\widetilde{T}_g\Gamma^{-1}$, and is regarded as the automorphism group of $\mathbf{U}^{n+1}$. Let $\mathbf{SU}(n+1, 1) = \mathbf{NAK}$ be the Iwasawa decomposition, where

$$n(z,t) = \begin{pmatrix} I_n & iz^t & -iz^t \\ i\overline{z} & 1 - \frac{|z|^2 - it}{2} & \frac{|z|^2 - it}{2} \\ i\overline{z} & -\frac{|z|^2 - it}{2} & 1 + \frac{|z|^2 - it}{2} \end{pmatrix},$$

$$a(\zeta) = \begin{pmatrix} I_n & 0 & 0 \\ 0 & \cosh\zeta & \sinh\zeta \\ 0 & \sinh\zeta & \cosh\zeta \end{pmatrix},$$

$$\boldsymbol{N} = \{n(z,t), \quad z \in \boldsymbol{C}^n, \;\; t \in \boldsymbol{R} \quad \}, \qquad \boldsymbol{A} = \{a(\zeta) : \quad \zeta \in \boldsymbol{R}\},$$

$$\boldsymbol{K} = \left\{ \begin{pmatrix} A & 0 \\ 0 & D \end{pmatrix} : \quad A \in \boldsymbol{U}(n+1), \det A \cdot D = 1 \right\}.$$

The Iwasawa subgroup $\mathbf{P}$ is the semidirect product $\mathbf{NA}$. For $(z, z_{n+1}) \in \mathbf{C}^{n+1}$, we set $t = \mathrm{Re} z_{n+1}, \rho = \mathrm{Im} z_{n+1} - |z|^2$, and use the coordinates $z \in \mathbf{C}^n, t \in \mathbf{R}, \rho \in \mathbf{R}$. We can identify $\mathbf{N}$ with $\partial\mathbf{U}^{n+1}$ and $\mathbf{P}$ with $\mathbf{U}^{n+1}$ by taking $\rho = e^{2\zeta}$.

The group law of $\mathbf{N}$ is given by

$$(z,t)(z',t') = (z+z', t+t'+2\mathrm{Im} z\overline{z'}), \tag{6}$$

where $z\overline{z'} = \sum_{j=1}^n z_j \overline{z'_j}$. So $\mathbf{N}$ (or $\partial\mathbf{U}^{n+1}$) is nothing but the Heisenberg group, denoted by $\mathbf{H}^n$. The dilations of $\mathbf{H}^n$ are defined by $\rho(z,t) = (\sqrt{\rho}z, \rho t), \rho > 0$, which are consistent with the dilations of $\mathbf{U}^{n+1}$ defined by $\rho(z, z_{n+1}) = (\sqrt{\rho}z, \rho z_{n+1})$. The group law of $\mathbf{P}$ is given by

$$(z,t,\rho)(z',t',\rho') = (z + \sqrt{\rho}z', t + \rho t' + 2\sqrt{\rho}\mathrm{Im} z\overline{z'}, \rho\rho'). \tag{7}$$

$\mathbf{P}$ is a nonunimodular group with the left Haar measure $d\mu_l(z,t,\rho) = \frac{dz dt d\rho}{\rho^{n+2}}$, where dz denotes the Lebesgue measure of $\mathbf{C}^n$.

$\mathbf{N}$ acts on $\mathbf{U}^{n+1}$ by the translations with respect to $\partial\mathbf{U}^{n+1} = \mathbf{H}^n$

$$T_{(z,t)} : (z',t',\rho') \longmapsto (z+z', t+t'+2\mathrm{Im} z\overline{z'}, \rho'),$$

$$(z',t',\rho') \in \mathbf{U}^{n+1}, (z,t) \in \mathbf{N}.$$

$\mathbf{A}$ acts on $\mathbf{U}^{n+1}$ by the dilations

$$T_{a(\zeta)} : (z',t',\rho') \longmapsto \rho(z',t',\rho') = (\sqrt{\rho}z', \rho t', \rho\rho'),$$

$$(z',t',\rho') \in \mathbf{U}^{n+1}, a(\zeta) \in \mathbf{A},$$

where $\rho = e^{2\zeta}$. So the action of $\mathbf{P}$ on $\mathbf{U}^{n+1}$ coincides with the multiplication of $\mathbf{P}$ under the identification of $\mathbf{P}$ and $\mathbf{U}^{n+1}$. Obviously $\mathbf{P}$ act on $\partial\mathbf{U}^{n+1}$ also, and the action of $\mathbf{P}$ on $\partial\mathbf{U}^{n+1}$ (or $\mathbf{H}^n$) is similar to that of the affine group "$ax + b$" on $\partial\mathbf{U}$ (or $\mathbf{R}$).

Let $H^2(\mathbf{H}^n)$ denote the Hardy space on the Heisenberg group, which consists of the boundary values of the Hardy space of holomorphic functions on $\mathbf{U}^{n+1}$. We consider the unitary representation U of $\mathbf{P}$ on

$H^2(\mathbf{H}^n)$ defined by

$$
\begin{aligned}
U_{(z,t,\rho)}f(z',t') &= \rho^{-\frac{n+1}{2}} f(T^{-1}_{(z,t,\rho)}(z',t')) \\
&= \rho^{-\frac{n+1}{2}} f\Big(\frac{z'-z}{\sqrt{\rho}}, \frac{t'-t-2\mathrm{Im} z\overline{z'}}{\rho}\Big), \qquad (8) \\
&\qquad f \in H^2(\mathbf{H}^n), (z,t,\rho) \in \mathbf{P}.
\end{aligned}
$$

U is irreducible and square-integrable. If a function $\phi \in H^2(\mathbf{H}^n)$, not identically zero, satisfies the admissibility condition

$$
\frac{1}{\|\phi\|^2_{L^2(\mathbf{H}^n)}} \int_{\mathbf{P}} \left|\langle \phi, U_{(z,t,\rho)}\phi\rangle_{L^2(\mathbf{H}^n)}\right|^2 \frac{dzdtd\rho}{\rho^{n+2}} = C_\phi < +\infty, \tag{9}
$$

we say that ϕ is an admissible wavelet and define the wavelet transform of a function $f \in H^2(\mathbf{H}^n)$ with respect to ϕ by

$$
W_\phi f(z,t,\rho) = C_\phi^{-\frac{1}{2}} \langle f, U_{(z,t,\rho)}\phi\rangle_{L^2(\mathbf{H}^n)}. \tag{10}
$$

We give the characterization of the admissibility condition similar to (5) (cf. [9]).

Theorem 1 Let $\phi \in H^2(\mathbf{H}^n)$, not identically zero. Then ϕ is an admissible wavelet if and only if

$$
C_\phi = \int_0^{+\infty} \|\widehat{\phi}(\lambda)\|^2_{H-S} \frac{d\lambda}{\lambda} < \infty, \tag{11}
$$

where $\|\cdot\|_{H-S}$ denotes the Hilbert-Schmidt norm of operators.

As consequence, the wavelet transform $W_\phi f$ is an isometric operator from $H^2(\mathbf{H}^n)$ into $L^2(\mathbf{P}, d\mu_l)$ and any function $f \in H^2(\mathbf{H}^n)$ can be recovered from its wavelet transform. The wavelet transform can be used to study the various function spaces on $\mathbf{U}^{n+1}$ (cf. [9]).

We note that the Iwasawa subgroup $\mathbf{P}$ is not the affine automorphism group of $\mathbf{U}^{n+1}$. In fact, let $\mathbf{M}$ be the centralizer of $\mathbf{A}$ in $\mathbf{K}$, *i.e.*,

$$
\mathbf{M} = \left\{ m(u,\theta) = \begin{pmatrix} e^{i\theta}u & 0 & 0 \\ 0 & e^{i\theta} & 0 \\ 0 & 0 & e^{i\theta} \end{pmatrix} : \begin{array}{l} u \in \mathbf{U}(n), \\ \det u = e^{-i(n+2)\theta}, \\ 0 \le \theta < 2\pi, \end{array} \right\}.
$$

The group $\mathbf{PM}$ is the minimal parabolic subgroup of $\mathbf{SU}(n+1,1)$ and gives all affine automorphisms of $\mathbf{U}^{n+1}$. The action of $\mathbf{M}$ on $\mathbf{U}^{n+1}$ is given by

$$
T_u : (z',t',\rho') \longmapsto (uz',t',\rho'), \quad (z',t'\rho') \in \mathbf{U}^{n+1}, u \in \mathbf{U}(n).
$$

We can also consider the unitary representation U^1 of the affine automorphism group $\mathbf{PM}$ on $H^2(\mathbf{H}^n)$ defined by

$$
\begin{aligned}
U^1_{(z,t,\rho,u)}f(z',t') &= \rho^{-\frac{n+1}{2}} f(T_u^{-1}T^{-1}_{(z,t,\rho,u)}(z',t')) \\
&= \rho^{-\frac{n+1}{2}} f\Big(\frac{u^{-1}(z'-z)}{\sqrt{\rho}}, \frac{t'-t-2\mathrm{Im} z\overline{z'}}{\rho}\Big), \qquad (12)\\
&\qquad f \in H^2(\mathbf{H}^2), (z,t,\rho,u) \in \mathbf{P}.
\end{aligned}
$$

U^1 is still square-integrable. If a function $\phi \in H^2(\mathbf{H}^2)$, not identically zero, satisfies the following admissibility condition

$$
\frac{1}{\|\phi\|^2_{L^2(\mathbf{H}^n)}} \int_{\mathbf{PM}} \left| \langle \phi, U^1_{(z,t,\rho,u)}\phi \rangle_{L^2(\mathbf{H}^n)} \right|^2 \frac{dzdtd\rho du}{\rho^{n+2}} = C_\phi < +\infty. \qquad (13)
$$

we say that ϕ is an admissible wavelet under U^1. The admissibility condition (13) is also equivalent to (11) (cf. [3]).

3. THE CASE TWO

Next we consider the Siegel upper half-plane $\mathbf{T}_\Omega$ defined as follows. Let $V = \mathrm{Sym}(n,\mathbf{R})$ be the space of all $n \times n$ symmetric real matrices. $\Omega \in V$ consists of all positive definite matrices. Then $\mathbf{T}_\Omega = \{z = x+iy : x \in V, y \in \Omega\}$. $\mathbf{T}_\Omega$ is a tube domain (Siegel domain of type I) and is holomorphically equivalent to the Cartan domain $\mathbf{D}$ of class II defined by $\mathbf{D} = \{z = x+iy : x, y \in V, I_n - z^*z \in \Omega\}$, where z^* denotes the conjugate transpose of z. When $n = 1$, $\mathbf{T}_\Omega$ is the upper half-plane $\mathbf{U}$. We mainly concern the case $n > 1$. All automorphisms of $\mathbf{T}_\Omega$ are given by $\mathbf{Sp}(n,\mathbf{R})$ through the linear fractional transforms

$$
z \longmapsto gz = (az+b)(cz+d)^{-1}, \quad z \in \mathbf{T}_\Omega, g = \begin{pmatrix} a & b \\ c & d \end{pmatrix} \in \mathbf{Sp}(n,\mathbf{R}).
$$

Let T denote the set of $n \times n$ upper triangular matrices with positive diagonal entrices. $\mathbf{Sp}(n,\mathbf{R})$ has the following (partial Iwasawa) decomposition $\mathbf{Sp}(n,\mathbf{R}) = \mathbf{NAK}$, where

$$
\begin{aligned}
\mathbf{N} &= \left\{ x = \begin{pmatrix} I_n & x \\ 0 & I_n \end{pmatrix} : \quad x \in V \right\}, \\
\mathbf{A} &= \left\{ a = \begin{pmatrix} a^t & 0 \\ 0 & a^{-1} \end{pmatrix} : \quad a \in T \right\}, \\
\mathbf{K} &= \left\{ \begin{pmatrix} a & b \\ -b & a \end{pmatrix} : \quad a + ib \in \mathbf{U}(n) \right\}.
\end{aligned}
$$

The Iwasawa subgroup $\mathbf{P} = \mathbf{NA}$ can be identified with $\mathbf{T}_\Omega$ by identifying $(x, a) \in \mathbf{P}$ with $x + ia^t a \in \mathbf{T}_\Omega$. The group law of $\mathbf{P}$ is given by

$$(x,a)(x',a') = (x + a^t x' a, a'a). \tag{14}$$

The action of $\mathbf{P}$ on $\mathbf{T}_\Omega$ coincides with the multiplication of $\mathbf{P}$ under the identification of $\mathbf{P}$ and $\mathbf{T}_\Omega$. $\mathbf{P}$ is a nonunimodular group with the left Haar measure $d\mu_l(x,a) = 2^n \prod_{j=1}^n a_{jj}^{-(r+j+1)} dx\, da$, where dx and da are separately the Lebesgue measures on V and T. $\mathbf{P}$ also acts on V, the distinguished boundary of $\mathbf{T}_\Omega$. The action of $\mathbf{P}$ on V is similar to that of the affine group "$ax + b$" on $\mathbf{R}$.

Let $H^2(V)$ denote the Hardy space, which consists of the boundary values of the Hardy space of holomorphic functions on $\mathbf{T}_\Omega$. We consider the unitary representation U of $\mathbf{P}$ on $H^2(V)$ defined by

(15)

$$U_{(x,a)} f(x') = |a|^{-\frac{n+1}{2}} f\big((a^t)^{-1}(x' - x)a^{-1}\big), \quad f \in H^2(V), (x,a) \in \mathbf{P}.$$

U is square-integrable. If a function $\phi \in H^2(V)$, not identically zero, satisfies the admissibility condition

$$\frac{1}{\|\phi\|^2_{L^2(V)}} \int_P |\langle \phi, U_{(x,a)}\phi\rangle_{L^2(V)}|^2 d\mu_l(x,a) C_\phi < \infty, \tag{16}$$

we say that ϕ is an admissible wavelet and define the wavelet transform of a function $f \in H^2(V)$ with respect to ϕ by

$$W_\phi f(x,a) = C_\phi^{-\frac{1}{2}} \langle f, U_{(x,a)}\phi\rangle_{L^2(V)}. \tag{17}$$

Jiang and Peng ([5]) gives a characterization of the admissibility condition in terms of the left Haar measure of $\mathbf{P}$. Here we give the characterization of the admissibility condition, which is equivalent to their result but not involved with the group $\mathbf{P}$ (cf. [6]). Let $\Delta_1(x), \cdots, \Delta_n(x)$ denote the inverted principal minors of $x \in V$.

Theorem 2 Suppose $\phi \in H^2(V)$, not identically zero. Then ϕ is an admissible wavelet if and only if

$$C_\phi = \int_\Omega |\widehat{\phi}(x)|^2 \prod_{j=i}^n |\Delta_j(x)|^{-1}\, dx < \infty. \tag{18}$$

As same as in the case one, the wavelet transform $W_\phi f$ is an isometric operator from $H^2(V)$ into $L^2(\mathbf{P}, d\mu_l)$ and any function $f \in H^2(V)$ can be recovered from its wavelet transform. The wavelet transform can also be used to study the various function spaces on $\mathbf{T}_\Omega$ (cf. [5], [6]).

But we do have a great difference when considering the affine automorphism group of $\mathbf{T}_\Omega$. Let $\mathbf{M}$ be the subgroup of $\mathbf{Sp}(n,\mathbf{R})$ defined by

$$\mathbf{M} = \left\{ m = \begin{pmatrix} m & o \\ 0 & m \end{pmatrix} : \quad m \in \mathbf{SO}(n) \right\}.$$

The group $\mathbf{PM}$ is a parabolic subgroup of $\mathbf{Sp}(n,\mathbf{R})$ and gives all affine automorphisms of $\mathbf{T}_\Omega$.

We can also consider the unitary representation U^1 of the affine automorphism group $\mathbf{PM}$ of $\mathbf{T}_\Omega$ on $H^2(V)$ defined by

$$\begin{aligned} U^1_{(x,a,m)} f(x') = |a|^{-\frac{n+1}{2}} f\big(m^{-1}(a^t)^{-1}(x'-x)a^{-1}m\big), \\ f \in H^2(V), (x,a,m) \in \mathbf{PM}. \end{aligned} \tag{19}$$

U^1 is still square-integrable. A function $\phi \in H^2(V)$, not identically zero, is called an admissible wavelet under U^1 if ϕ satiafies the admissibility condition

$$\frac{1}{\|\phi\|^2_{L^2(V)}} \int_{\mathbf{PM}} |\langle \phi, U_{(x,a,m)}\phi\rangle_{L^2(V)}|^2 d\mu_l(x,a,m) = C_\phi < \infty, \tag{20}$$

where $d\mu_l(x,a,m)$ is the left Haar measure of $\mathbf{PM}$. The characterization of the admissibility condition is follows (cf. [8]).

Theorem 3 Suppose $\phi \in H^2(V)$, not identically zero. Then ϕ is an admissible wavelet under U^1 if and only if

$$C_\phi = \int_\Omega |\widehat{\phi}(x)|^2 |x|^{-\frac{n+1}{2}}\, dx < \infty, \tag{21}$$

where $|x| = \Delta_n(x)$ is the determinant of x.

The condition (21) is different from (18). This is due to that the rank of $\mathbf{Sp}(n,\mathbf{R})$ is larger than one. As the generalization of the admissibility condition (5), the condition (21) seems more simple and natural than (18).

References

[1] M. Duflo and C. C. Moore, On the regular representation of a nonunimodular locally compact group, *J. Funct. Anal.* **21** (1976), 209-243.

[2] A. Grossmann and J. Morlet, Decomposition of Hardy functions into square integrable wavelets of constant shape, *SIAM J. Math. Anal.* **15** (1984), 723-736.

[3] J. He and H. Liu, Admissible wavelets associated with the affine automorphism group of the Siegel upper half-plane, *J. Math. Anal. Appl.* **208** (1997), 58-70.

[4] Q. Jiang and L. Peng, Wavelet transform and Toeplitz-Hankel type operators, *Math. Scand.* **70** (1992), 247-264.

[5] Q. Jiang and L. Peng, Admissible wavelets on the Siegel domain of type one, *Sci. in China, Series A* **41** (1998), No. 9, 897-909.
[6] H. Liu, Wavelet transforms and symmetric tube domains, *J. Lie Th.* **8** (1998), 351-366.
[7] H. Liu, Admissible wavelets associated with the affine homogeneous Siegel domain of type II, Preprint.
[8] H. Liu, Admissibility condition and the affine automorphism group, Preprint.
[9] H. Liu and L. Peng, Admissible wavelets associated with the Heisenberg group, *Pacific J. Math.* **180** (1997), No.1, 101-123.

7 SOME RESULTS ON A CLASS OF OSCILLATORY INTEGRALS

Shanzhen LU [1]

Department of Mathemathematics
Beijing Normal University
Beijing 100875
The People's Republic of China

lusz@bnu.edu.cn

1. Background

Many operators on harmonis analysis are related to some versions of oscillatory integrals. In fact, Fourier transform $\hat{f}(x)$ is just a simple oscillatory integral:

$$\hat{f}(x) = \int_{\mathbb{R}^n} e^{2\pi i x \cdot y} f(y)\, dy.$$

Let us now consider the Bochner-Riesz multiplier defined by

$$(\widehat{Sf})(\xi) = (1 - |\xi|^2)_+^\alpha \hat{f}(\xi), \; 0 < \alpha \leq \frac{n-1}{2}.$$

By a simple arguments, L^p-boundedness of S can be reduced to one of the following oscillatory integral operator T:

$$Tf(x) = \int_{\mathbb{R}^n} e^{i|x-y|} |x-y|^{-(\frac{n-1}{2}+\alpha)} f(y)\, dy.$$

Precisely speaking, S is bounded on $L^p(\mathbb{R}^n)$ if and only if T is bounded on $L^p(\mathbb{R}^n)$, $1 < p < \infty$. As above, the study of some operators related to

[1]Supported by the National Natural Science Foundation of China (Grant No. 19131080).

H.G.W. Begehr et al. (eds.), Proceedings of the Second ISAAC Congress, Vol. 1, 51–57.

Precisely speaking, S is bounded on $L^p(\mathbb{R}^n)$ if and only if T is bounded on $L^p(\mathbb{R}^n)$, $1 < p < \infty$. As above, the study of some operators related to curvature can also be reduced to the study of certain oscillatory integral. For example, let us consider the Hilbert transform associated with a homogeneous curve γ:

$$Kf(x) = \int_{-\infty}^{\infty} f(x - \gamma(t))\frac{dt}{t}.$$

It is easy to check that L^2-boundedness of K is equivalent with L^2-boundedness of the following multiplier

$$m(\xi) = \int_{-\infty}^{\infty} e^{i\xi\cdot\gamma(t)}\frac{dt}{t},$$

where m is an oscillatory integral. The oscillatory singular integrals in my talk here is taken the following form

$$Tf(x) = \int_{\mathbb{R}^n} e^{i\,P(x,y)}K(x-y)f(y)\,dy,$$

where K is a Calderón-Zygmund kernel, $P(x,y)$ is a real polynomial defined on $\mathbb{R}^n \times \mathbb{R}^n$. Such oscillatory integrals are closely connected with singular Radon transform.

2. L^p-boundedness

(1) Let us begin with a Phong-Stein's result when $P(x,y) = (Bx, y)$: bilinear form. Precisely,

$$Tf(x) = p.v. \int_{\mathbb{R}^n} e^{i(Bx,y)}K(x-y)\,dy,$$

where K is a CZK. For simplicity, we first restricted to homogeneous CZK. That means K satisfies (a) $K \in C^1(\mathbb{R}^n\backslash\{0\})$; (b) $K(x) = \Omega(x')/|x|^n$ with Ω homogeneous of degree 0 on S^{n-1}; (c) $\int_{S^{n-1}} \Omega\,d\sigma(x') = 0$.

Theorem 1 ([PS]) *If $1 < p < \infty$, then T is bounded on $L^p(\mathbb{R}^n)$.*

The idea of proof

$$T\begin{cases} H_E^1 \to L^1 \\ L^\infty \to BMO_E \end{cases} \overset{\text{interpolation}}{\Longrightarrow} L^p \to L^p,$$

where H_E^1 is a variant of H^1:

$$H_E^1 = \{f : f = \sum \lambda_k a_k,\ \text{each } a_k \text{ is a } H_E^1 \text{ atom},\ \sum|\lambda_k| < \infty.\},$$

$$a(x) \text{ is a } H_E^1 \text{ atom} \Longleftrightarrow \begin{cases} \operatorname{supp} a \subset Q, \\ \|a\|_\infty \le 1/|Q|, \\ \int a(y)e^{i(Bx,y)}dy = 0. \end{cases}$$

$$BMO_E = (H_E^1)'.$$

Question 1. T: $H^1 \to L^1$?

(2) For general case of $P(x,y)$, $P(x,y)$ is a polynomial defined on $\mathbb{R}^n \times \mathbb{R}^n$, Ricci and Stein got a result.

Theorem 2 ([RS]) *Let*

$$Tf(x) = p.v. \int_{\mathbb{R}^n} e^{iP(x,y)} K(x-y) f(y)\, dy,$$

where K is a CZK. If $1 < p < \infty$, then T is bounded on $L^p(\mathbb{R}^n)$. The proof of Theorem 2 does not employ interpolation. It is based on a series of Von der corput lemmas. latter on, Pan give another proof by a method similar to one of Theorem 1.

Question 2. Is the smooth assumption on K necessary ?

(3) For the case of $p = 1$, Chanillo and Christ give an answer.

Theorem 3 ([CC]) *Let T be as in Theorem 2. Then T: $L^1(\mathbb{R}^n) \to WL^1(\mathbb{R}^n)$.*

(4) Part answer for Question 1 when $P(x,y) = P(x-y)$. In this case, T is a convolution operator. Hu-Pan give the following result.

Theorem 4 ([HP]) *Let*

$$Tf(x) = p.v. \int_{\mathbb{R}^n} e^{iP(x-y)} K(x-y) f(y)\, dy,$$

where K is a CZK, $P(u)$ is a polynomial defined on $\mathbb{R}^n$. Then T : $H^1(\mathbb{R}^n) \to H^1(\mathbb{R}^n)$.

(5) Part answer for Question 2 when K is certain rough kernel. We found that the smooth condition (a) on kernel K is not necessary. K is called a rough kernel if K satisfies (b), (c) and (d) where (d) $\Omega \in L^q(S^{n-1})$, $1 < q \le \infty$. In this case, Lu and Zhang got

Theorem 5 ([LZ]) *Let*

$$Tf(x) = p.v. \int_{\mathbb{R}^n} e^{iP(x,y)} K(x-y) f(y)\, dy,$$

where $P(x,y)$ is a polynomial on $\mathbb{R}^n \times \mathbb{R}^n$, K is a rough kernel with $\Omega \in L^q(S^{n-1})$ $(1 < q \le \infty)$. Then $T : L^p(\mathbb{R}^n) \to L^p(\mathbb{R}^n)$. It should be pointed out that we use the rotation method due to Calderón in the proof of Theorem 5.

Remark: *The condition of $\Omega \in L^q(S^{n-1})$ in Theorem 5 is first weaken into $\Omega \in L\log^+ L(S^{n-1})$ by Jiang and Lu* ([JL]). *Recently, it is weaken into $\Omega \in H^1(S^{n-1})$ by Fan and Pan* ([FP]) *when $P(x,y) = P(x-y)$. For general $P(x,y)$, it is still an open problem.*

3. Criterion on L^p-boundedness of Oscillatory integral operators T

Since T is not a Calderón-Zygmund operator, we can not use $T1$ theorem due to David and Journe to judge whether T is L^2 or L^p bounded. A natural conjucture is as follows. Let

$$Tf(x) = p.v. \int_{\mathbb{R}^n} e^{iP(x,y)} K(x-y) f(y)\, dy$$

and

$$\tilde{T}f(x) = p.v. \int_{\mathbb{R}^n} K(x-y) f(y)\, dy.$$

Note that when $P(x,y) = P_0(x) + P_1(y)$,

$$T: L^p \to L^p \Leftrightarrow \tilde{T}: L^p \to L^p \ (1 < p < \infty).$$

However, for nontrivial polynomial $P(x,y)$, that means, $P(x,y)$ does not take the form of $P_0(x) + P_1(y)$, the above conclusion is false. By use of rotation method, we get a convenient criterion to judge whether T is bounded on L^p.

Theorem 6 (**[LZ]**) *Let*

$$Tf(x) = p.v. \int_{\mathbb{R}^n} e^{iP(x,y)} K(x-y) f(y)\, dy$$

and

$$\tilde{T}_0 f(x) = p.v. \int_{|x-y|<1} K(x-y) f(y)\, dy,$$

where $P(x,y)$ is a nontrivial polynomial, K is a rough kernel with $\Omega \in L^q(S^{n-1})$ $(1 < q \le \infty)$. Then

$$T: L^p \to L^p \Longleftrightarrow \tilde{T}_0: L^p \to L^p \ (1 < p < \infty).$$

4. Generalization to multilinear operators

Let us consider a simple oscillatory integrals of 1-dimension

$$Sf(x) = \int_{-\infty}^{\infty} e^{iP(x,y)} \frac{1}{|x-y|} \log\left(\frac{1+|x|}{1+|y|}\right) f(y)\, dy.$$

If we denote $A_1(x) = \log(1+|x|)$, $A_2(x) = x$, and $\Omega(x) =$Sgn(x), then S can be rewritten as

$$Sf(x) = \int_{-\infty}^{\infty} e^{iP(x,y)} \frac{\Omega(x-y)}{|x-y|^2} (A_1(x) - A_1(y))(A_2(x) - A_2(y)) f(y)\, dy.$$

It is easy to see that S is a multilinear oscillatory integrals. The singular integral corresponding to S is a multiplinear singular integral

$$\tilde{S}f(x) = \int_{-\infty}^{\infty} \frac{\Omega(x-y)}{|x-y|^2}(A_1(x) - A_1(y))(A_2(x) - A_2(y))f(y)\,dy.$$

Since we have known $\tilde{S}$ more than S, it is our desire to judge properties of S by use of $\tilde{S}$. More general, let us consider the following multilinear oscillatory integral

$$T_{A_1,A_2}f(x) = \int_{\mathbb{R}^n} e^{iP(x,y)} \frac{\Omega(x-y)}{|x-y|^{n+M-1}} \prod_{j=1}^{2} R_{m_j}(A_j;\, x,y) f(y)\,dy,$$

where $M = m_1 + m_2$, Ω is homogeneous of degree zero, $R_m(A;\, x,y)$ denotes the m-th order Taylor series remainder of A at x expanded about y, more precisely,

$$R_m(A;\, x,y) = A(x) - \sum_{|\alpha|<m} \frac{1}{\alpha!} D^\alpha A(y)(x-y)^\alpha.$$

Suppose that A_1 has derivatives of order $m_1 - 1$ in $\mathrm{BMO}(\mathbb{R}^n)$, A_2 has derivatives of order m_2 in $L^{r_0}(\mathbb{R}^n)$, $1 < r_0 \le \infty$. Then the following generalized version of Theorem 6 was obtained by Chen, Hu and Lu.

Theorem 7 ([CHL]) *Let Ω be homogeneous of degree zero and belong to $L^q(S^{n-1})$ for some $q > 1$. If A_1 and A_2 are defined as above, then for $1/r = 1/p + 1/r_0$, $1 < p, r < \infty$, the following two facts are equivalent:*
(i) *If $P(x,y)$ is a nontrivial real-valued polynomial, then*

$$T_{A_1,A_2} : L^p \to L^r.$$

(ii) *The truncated operator*

$$S_{A_1,A_2} : L^p \to L^r,$$

where

$$S_{A_1,A_2}f(x) = \int_{|x-y|<1} \frac{\Omega(x-y)}{|x-y|^{n+M-1}} \prod_{j=1}^{2} R_{m_j}(A_j;\, x,y) f(y)\,dy.$$

Remark. *In fact, CZK discussed above as a homogeneous Calderón-Zygmund kernel. For the standard Calderón-Zygmund kernel $K(x,y)$, K only satisfies* (i) $|K(x,y)| \le C/|x-y|^n$, $x \ne y$, (ii) $|\nabla_x K(x,y)| + |\nabla_y K(x,y)| \le C/|x-y|^{n+1}$, *and K does not have homogeneous relation.* However, if we replace the homogeneous CZK by the standard CZK, then similar results corresponding to Theorem 5 and Theorem 7 can be also established respectively by Zhang-Liu and Lu using different method (see

[ZL] and [L]). Precisely, let us consider a multilinear oscillatory singular integral operator

$$T_{A_1,A_2}f(x) = p.v. \int_{\mathbb{R}^n} e^{iP(x,y)} \frac{K(x,y)}{|x-y|^{M-1}} \prod_{j=1}^{2} R_{m_j}(A_j;\, x,y) f(y)\, dy,$$

where K is a Calderón-Zygmund standard kernel.

Theorem 8 ([L]) *If A_1 and A_2 are defined as above. Then for $1/r = 1/p + 1/r_0$, $1 < p, r < \infty$, the following two facts are equivalent:* (i) *If $P(x,y)$ is a nontrivial real-valued polynomial, then*

$$T_{A_1,A_2} : L^p \to L^r$$

(ii) *The truncated operator*

$$S_{A_1,A_2} : L^p \to L^r$$

where

$$S_{A_1,A_2}f(x) = p.v. \int_{|x-y|<1} \frac{K(x,y)}{|x-y|^{M-1}} \prod_{j=1}^{2} R_{m_j}(A_j;\, x,y) f(y)\, dy.$$

Since there is no any homogeneous for the Calderón-Zygmund standard kernel, the proving method of Theorem 7 does not work. We partly employ the technique of [RS] and some estimates on multilinear operators due to Cohen and Gosselin ([CG]).

Acknowledgement The author would like to thank Professor Joji Kajiwara and Professor Masaru Nishihara for inviting me to organize the session on Harmonic Analysis and its Applications during the period of the second congress of ISAAC.

References

[CC] S. Chanillo and M. Christ, Weak (1,1) bound for oscillatory singular integral, Duke Math. J., 55 (1987), 141-155.

[CG] J. Cohen and J. Gosselin, A BMO estimate for multilinear singular integrals, Illinois J. Math., 30 (1985), 445-464.

[CHL] W. G. Chen, G. E. Hu and S. Z. Lu, Criterion of (L^p, L^r) boundedness for a class of multilinear oscillatory singular integrals, Nagoya Math. J., 149 (1998), 1-19.

[FP] D. S. Fan and Y. B. Pan, Singular integral operators with rough kernel supported by subvarieties, Amer. J. Math., 119 (1997), 799-839.

[HP] Y. Hu and Y. B. Pan, Boundedness of oscillatory singular integrals on Hardy spaces, Ark. Mat., 30 (1992), 311-320.

[JL] Y. S. Jiang and S. Z. Lu, Oscillatory singular integrals with rough kernel, Math. and its Appl., Harmonic Analysis in China, Kluwer Academic Publishers, 135-145.

[L] S. Z. Lu, Multilinear oscillatory integrals with Calderón-Zygmund kernel, Sci. in China (Ser. A), 42:10 (1999), 1039-1046.

[LZ] S. Z. Lu and Y. Zhang, Criterion on L^p-boundedness for a class of oscillatory singular integral with rough kernel, Rev. Mat. Iberoamericana, 8 (1992), 201-219.

[P] Y. B. Pan, Hardy spaces and oscillatory singular integrals, Rev. Mat. Iberoamericana, 7 (1991), 55-64.

[PS] D. H. Phong and E. M. Stein, Hilbert integrals, singular integrals, and Radon transforms I, Acta Math., 157 (1986), 99-157.

[RS] F. Ricci and E. M. Stein, Harmonic analysis on nilpotent groups and singular integrals I: Oscillatory integrals, J. Funct. Anal., 73 (1987), 179-194.

[ZL] Y. Zhang and H. P. Liu, Criterion of L^2-Boundedness for a class of Oscillatory integrals, Approx. Theory & its Appl., 7:4 (1991), 1-5.

8 WEIGHTED HARDY SPACES ON A DOMAIN

Akihiko MIYACHI

Department of Mathematics, Tokyo Woman's Christian University
Zempukuji, Suginami-ku, Tokyo, 167-8585 JAPAN

miyachi@twcu.ac.jp

§ Introduction

Weighted Hardy spaces were first studied by Gacía-Cuerva [GC] for the case of weight functions in the A_∞ class. Generalization to the case of weight functions satisfying only the doubling condition was given by Strömberg and Torchinsky [ST]. Both of these works are concerned with the weighted Hardy spaces on $\mathbb{R}^n$. If we try to generalize the arguments of [GC] and [ST] so that we can deal with weighted Hardy spaces on a domain of $\mathbb{R}^n$, we meet with considerable technical difficulties since in those papers the convolution and the Fourier transform on $\mathbb{R}^n$ are used as basic tools and the use of testing functions in the Schwartz class $\mathcal{S}$ seems to be unavoidable.

The purpose of the present note is to consider weighted Hardy spaces on a domain of $\mathbb{R}^n$. The key idea goes back to the works of Uchiyama [U1] and [U2]. Here we give only the statements of the main results. The details will be given elsewhere.

Throughout this note Ω denotes an arbitrary open subset of $\mathbb{R}^n$.

We use the following notations. We write $\mathcal{D}'(\Omega)$ to denote the set of distributions on Ω. The symbol $B(x,r)$ ($x \in \mathbb{R}^n$, $\infty > r > 0$) denotes the open ball with the center x and with radius r. If $B = B(x,r)$ and $0 < a < \infty$, then $aB = B(x, ar)$.

§ Maximal functions and doubling measures

H.G.W. Begehr et al. (eds.), Proceedings of the Second ISAAC Congress, Vol. 1, 59–64.

First we shall introduce two kinds of maximal functions for distributions on Ω.

Definition 1. *We take a function* $\phi \in C_0^\infty(\mathbb{R}^n)$ *such that* $\operatorname{supp}\phi \subset B(0,1)$ *and* $\int \phi(x)dx = 1$. *We set* $\phi_t(x) = t^{-n}\phi(t^{-1}x)$ *for* $0 < t < \infty$. *For* $f \in \mathcal{D}'(\Omega)$ *and* $0 < \epsilon < 1$, *we define the radial maximal function* $f^{+,\epsilon}$ *by*

$$f^{+,\epsilon}(x) = \sup\{|\langle f, \phi_t(x - \cdot)\rangle|; 0 < t < \epsilon \operatorname{dis}(x, \Omega^c)\} \qquad (x \in \Omega).$$

Definition 2. *For* $f \in \mathcal{D}'(\Omega)$, $1 < a < \infty$, *and* $0 < s < \infty$, *we define the grand maximal function* $f_s^{*,a}$ *by*

$$f_s^{*,a}(x) = \sup\{|\langle f, \psi\rangle|\} \qquad (x \in \Omega)$$

with the sup taken over all those $\psi \in C_0^\infty(\mathbb{R}^n)$ *for which there exists a ball* $B = B_\psi$ *such that* $aB \subset \Omega$, $B \ni x$, *and* $\|\psi\|_{\Lambda(s)} \leq$ *(radius of* $B)^{-n-s}$.

In the above, we used the following notation:

$$\|\psi\|_{\Lambda(s)} = \sum_{|\nu|=[s]} \sup\left\{\frac{|f^{(\nu)}(x) - f^{(\nu)}(y)|}{|x-y|^{s-[s]}}\right\} \quad \text{for} \quad s \notin \mathbb{N}$$

and

$$\|\psi\|_{\Lambda(s)} = \sum_{|\nu|=s-1} \sup\left\{\frac{|f^{(\nu)}(x) - 2f^{(\nu)}((x+y)/2) + f^{(\nu)}(y)|}{|x-y|}\right\}$$

for $s \in \mathbb{N}$ (here $f^{(\nu)} = \partial^\nu f$).

Next we introduce the class of doubling measures on Ω.

Definition 3. *We write* $\operatorname{Doubl}(\Omega, \sigma)$, $0 < \sigma < \infty$, *to denote the set of all those Borel measures* λ *on* Ω *which satisfy this condition: There exist* $T \in (1, \infty)$ *and* $K \in [1, \infty)$ *such that for all those balls* B *satisfying* $TB \subset \Omega$ *and for all* $t \geq 1$ *we have*

$$(1) \qquad 0 < \lambda(B) \leq K t^\sigma \lambda(t^{-1}B) < \infty.$$

We set $\operatorname{Doubl}(\Omega) = \bigcup_{0<\sigma<\infty} \operatorname{Doubl}(\Omega, \sigma)$.

It can be shown that if $\lambda \in \operatorname{Doubl}(\Omega, \sigma)$ then for *every* $T \in (1, \infty)$ there exists a $K = K_T \in [1, \infty)$ such that (1) holds for all those balls B satisfying $TB \subset \Omega$ and for all $t \geq 1$.

§ Basic inequalities and definition of Hardy spaces

It is obvious that

$$(2) \qquad f^{+,\epsilon}(x) \leq c f_s^{*,1/\epsilon}(x) \qquad (\forall x \in \Omega).$$

The fact that we also have the estimate which is, in a sense, reverse to the above one is the key to our study of Hardy spaces. In order to give the reverse estimate, we use the following maximal function.

Let $\lambda \in \mathrm{Doubl}(\Omega)$. For nonnegative measurable functions g on Ω and for $1 < a < \infty$ and $0 < r < \infty$, we define

$$M_{\lambda,r}^{\Omega,a}(g)(x) = \sup \left(\lambda(B)^{-1} \int_B g(y)^r d\lambda(y) \right)^{1/r} \qquad (x \in \Omega)$$

with the sup taken over all those balls B such that $aB \subset \Omega$ and $B \ni x$.

By slightly modifying the well known arguments for the Hardy-Littlewood maximal function, we can prove that, for $\lambda \in \mathrm{Doubl}(\Omega)$, the inequality

$$\|M_{\lambda,r}^{\Omega,a}(g)\|_{L^p(\Omega,\lambda)} \leq c\|g\|_{L^p(\Omega,\lambda)} \tag{3}$$

holds if $3 < a < \infty$ and $r < p \leq \infty$.

The following theorem gives the inequality reverse to (2).

Theorem 4. *Let $\lambda \in \mathrm{Doubl}(\Omega, \sigma)$ with $0 < \sigma < \infty$. Suppose $0 < s < \infty$, $\sigma \leq n + s$, $3 < a < \infty$, and $0 < \epsilon < 1$. Then the following inequality holds for all $f \in \mathcal{D}'(\Omega)$ and for all $x \in \Omega$:*

$$f_s^{*,a}(x) \leq cM_{\lambda,\sigma/(n+s)}^{\Omega,a}(f^{+,\epsilon})(x).$$

This theorem can be proved by using the argument given by Uchiyama [U1] and [U2] (cf. also [M1]).

The following corollary immedately follows from Theorem 4 and the inequalities (2) and (3).

Corollary 5. *Let $\lambda \in \mathrm{Doubl}(\Omega, \sigma)$ with $0 < \sigma < \infty$. Suppose $0 < s < \infty$, $\sigma \leq n + s$, $3 < a < \infty$, $0 < \epsilon < 1/3$, and $\sigma/(n+s) < p \leq \infty$. Then for all $f \in \mathcal{D}'(\Omega)$ we have*

$$\|f^{+,\epsilon}\|_{L^p(\Omega,\lambda)} \approx \|f_s^{*,a}(x)\|_{L^p(\Omega,\lambda)}.$$

Now we define the Hardy space $H^p(\Omega, \lambda)$.

Definition 6. *Let $\lambda \in \mathrm{Doubl}(\Omega)$ and $0 < p < \infty$. Take an ϵ such that $0 < \epsilon < 1/3$. For $f \in \mathcal{D}'(\Omega)$, we set*

$$\|f\|_{H^p(\Omega,\lambda)} = \|f^{+,\epsilon}\|_{L^p(\Omega,\lambda)}.$$

We define $H^p(\Omega, \lambda)$ as the set of all those $f \in \mathcal{D}'(\Omega)$ such that $\|f\|_{H^p(\Omega,\lambda)} < \infty$.

¿From Corollary 5, we see the following.

(1) $\|f\|_{H^p(\Omega,\lambda)}$ does not depend, up to the equivalence of quasinorms, on the choice of $\epsilon \in (0,1/3)$ and ϕ which are used in the definition of $f^{+,\epsilon}$.

(2) $H^p(\Omega,\lambda)$ does not depend on the choice of ϵ and ϕ.

(3) A distribution f on Ω belongs to $H^p(\Omega,\lambda)$ if and only if $f_s^{*,a}$ belongs to $L^p(\Omega,\lambda)$, where λ, σ, p, a, and s satisfy the assumptions of Corollary 5.

§ Atomic decomposition

As in the case of usual H^p spaces, from the characterization of $H^p(\Omega,\lambda)$ by the grand maximal function (as given in (3) above) we can deduce the characterization of $H^p(\Omega,\lambda)$ in terms of the atomic decomposition. Before we give the atomic decomposition, we fix a notation.

We write L^∞ and $\|\cdot\|_\infty$ to denote the L^∞ class and the L^∞ norm on $\mathbb{R}^n$ with respect to the Lebesgue measure. If $f \in L^\infty$ and if $\operatorname{supp} f \subset \Omega$, then we regard f as an element of $\mathcal{D}'(\Omega)$ in the usual way, i.e., we define $f \in \mathcal{D}'(\Omega)$ by

$$\langle f, \psi \rangle = \int_\Omega f(x)\psi(x)dx \qquad (\psi \in C_0^\infty(\Omega)).$$

Notice that these do not concern with the measure λ.

The following theorem gives the atomic characterization of $H^p(\Omega,\lambda)$. This theorem can be proved by the method similar to the one given in [M2].

Theorem 7. *Let* $\lambda \in \mathrm{Doubl}(\Omega)$ *and* $0 < p < \infty$. *Let* u, a, *and* b *be real numbers such that* $20 < u < \infty$ *and* $20 < a < b < a^2/20$. *Let* m *be a nonnegative integer such that* $n + m + 1 > \max\{\sigma/p, \sigma\}$.

(1) *Suppose* $\{g_i\}$ *is a sequence of* L^∞ *functions and* $\{B_i\}$ *is a sequence of balls. Also suppose* $uB_i \subset \Omega$, $\operatorname{supp} g_i \subset B_i$, $\int g_i(x)x^\nu dx = 0$ *for* $|\nu| \leq \ m$, *and* $\sum_i \|g_i\|_\infty \chi_{B_i} \in L^p(\Omega,\lambda)$. *Then the series* $\sum_i g_i$ *converges unconditionally in* $H^p(\Omega,\lambda)$ *and*

$$\left\|\sum_i g_i\right\|_{H^p(\Omega,\lambda)} \leq c\left\|\sum_i \|g_i\|_\infty \chi_{B_i}\right\|_{L^p(\Omega,\lambda)}$$

(2) *Suppose* $\{h_j\}$ *is a sequence of* L^∞ *functions and* $\{Q_j\}$ *is a sequence of balls. Also suppoe* $aQ_j \subset \Omega$, $bQ_j \cap \Omega^c \neq \emptyset$, $\operatorname{supp} h_j \subset Q_j$,

and $\sum_j \|h_j\|_\infty \chi_{Q_j} \in L^p(\Omega, \lambda)$. *Then the series* $\sum_j h_j$ *converges unconditionally in* $H^p(\Omega, \lambda)$ *and*

$$\left\| \sum_j h_j \right\|_{H^p(\Omega,\lambda)} \leq c \left\| \sum_j \|h_j\|_\infty \chi_{Q_j} \right\|_{L^p(\Omega,\lambda)}$$

(3) *Suppose* $f \in H^p(\Omega, \lambda)$. *Then there exist* $\{g_i\}$ *and* $\{B_i\}$ *satisfying the conditions of* (1) *and there exist* $\{h_j\}$ *and* $\{Q_j\}$ *satisfying the conditions of* (2) *such that*

$$f = \sum_i g_i + \sum_j h_j.$$

Moreover $\{g_i\}$, $\{B_i\}$, $\{h_j\}$, *and* $\{Q_j\}$ *can be taken so that they satisfy*

$$\left\| \left(\sum_i \|g_i\|_\infty^r \chi_{B_i}\right)^{1/r} \right\|_{L^p(\Omega,\lambda)} + \left\| \left(\sum_j \|h_j\|_\infty^r \chi_{Q_j}\right)^{1/r} \right\|_{L^p(\Omega,\lambda)}$$
$$\leq \quad c_r \|f\|_{H^p(\Omega,\lambda)}$$

for every $0 < r < \infty$.

§ The case of A_p weight

It is easy to see that the theory of A_p weight works on Ω, an arbitrary open subset of $\mathbb{R}^n$, and that $H^p(\Omega, \lambda)$, $1 < p < \infty$, coincides with $L^p(\Omega, \lambda)$ if and only if $d\lambda(x) = w(x)dx$ with w an A_p weight. Precise statement reads as follows.

Definition 8. *Let* $1 < p < \infty$ *and let* w *be a nonnegative measurable function on* Ω. *We say* w *is of class* $A_p(\Omega)$, $1 < p < \infty$, *if there exist* $3 < S < \infty$ *and* $1 \leq G < \infty$ *such that the inequalities*

$$0 < \int_B w(x)dx < \infty$$

and

$$\left(|B|^{-1} \int_B w(x)dx\right)^{1/p} \left(|B|^{-1} \int_B w(x)^{-p'/p}\right)^{1/p'} \leq G \tag{4}$$

hold for all those balls B *satisfying* $SB \subset \Omega$.

It can be shown that if $w \in A_p(\Omega)$, $1 < p < \infty$, then for *every* $S \in (3, \infty)$ there exists a $G = G_S \in [1, \infty)$ such that (4) holds for all those balls B satisfying $SB \subset \Omega$.

We have the following theorem. In this theorem we write $L^1_{\text{loc}}(\Omega, dx)$ to denote the set of all functions on Ω which are integrable with respect to the Lebesgue measure on every compact subset of Ω .

Theorem 9. (1) *Suppose* $1 < p < \infty$, $w \in A_p(\Omega)$, *and* λ *is the measure defined by* $d\lambda(x) = w(x)dx$. *Then* $\lambda \in \mathrm{Doubl}(\Omega)$, *both* $H^p(\Omega, \lambda)$ *and* $L^p(\Omega, \lambda)$ *are subsets of* $L^1_{\mathrm{loc}}(\Omega, dx)$, *and* $H^p(\Omega, \lambda) = L^p(\Omega, \lambda)$. *Moreover we have*

$$\|f\|_{H^p(\Omega,\lambda)} \approx \|f\|_{L^p(\Omega,\lambda)}$$

for all the elements of $H^p(\Omega, \lambda) = L^p(\Omega, \lambda)$.

(2) *Let* $\lambda \in \mathrm{Doubl}(\Omega)$. *If there exists a positive constant* A *for which the inequality*

$$\|f\|_{H^p(\Omega,\lambda)} \leq A\|f\|_{L^p(\Omega,\lambda)}$$

holds for all $f \in C_0^\infty(\Omega)$, *then* λ *is absolutely continuous with respect to the Lebesgue measure and the Radon-Nikodym derivative* $w = d\lambda/dx$ *is of class* $A_p(\Omega)$.

References

[GC] García-Cuerva, *Weighted* H^p *Spaces*, Dissert. Mathematicae 162, Warszawa, 1979.

[M1] A. Miyachi, Maximal functions for distributions on open sets, Hitotsubashi J. Arts Sci. **28**(1987), 45–58.

[M2] A. Miyachi, H^p spaces over open subsets of $\mathbb{R}^n$, Studia Math. **44**(1990), 205–228.

[ST] J.-O. Strömberg and A. Torchinsky *Weighted Hardy Spaces*, Lecture Notes in Math. 1381, Springer-Verlag, 1989.

[U1] A. Uchiyama, A maximal function characterization of H^p on the space of homogeneous type, Trans. Amer. Math. Soc. **262**(1980), 579–592.

[U2] A. Uchiyama, On the radial maximal function of distributions, Pacific J. Math. **121**(1986), 467–483.

9 COMMUTATORS OF SINGULAR INTEGRAL OPERATORS ON MORREY SPACES WITH SOME GROWTH FUNCTIONS

Takahiro MIZUHARA

Department of Mathematical Sciences
Faculty of Science, Yamagata University
Yamagata 990-8560, Japan

mizuhara@sci.kj.yamagata-u.ac.jp

Abstract. The talk will be concerned with the boundedness of commutators $[b, S]$, between some singular integral operator S and multiplication operator by a locally integrable function b, on Morrey spaces $L^{p,\Phi}(R^n)$ with some growth function Φ. Our results (Theorems 2.1, 2.2 and 3.1) generalize partly the classical results due to Di Fazio and Ragusa [DiFRag]. Also we obtain two new results (Theorems 2.3 and 3.2).

1. Introduction.

Let $\Phi = \Phi(x, r)$, be a growth function on $R_+^{n+1} = R^n \times R_+$, that is, a positive and non-decreasing function with repect to $r > 0$. We say that the growth function $\Phi(x, r)$ satisfies the Δ_2-condition (or doubling condition) for $r > 0$ if there exists constant $D = D(\Phi) \geq 1$, independent of (x, r), such that

$$\Phi(x, 2r) \leq D\Phi(x, r), \quad (x, r) \in R_+^{n+1}, \tag{1.1}$$

H.G.W. Begehr et al. (eds.), Proceedings of the Second ISAAC Congress, Vol. 1, 65–73.

or equivalently,

$$\Phi(x,2r)/D \le \Phi(x,r) \le \Phi(x,2r), \quad (x,r) \in R_+^{n+1}.$$

In this case, we write $\Phi \in \Delta_2$ simply.

Remark. Nakai [Nak] assumed a slightly weak condition on $\Phi(x,r)$ replacing (1.1) ; there exists a constant $C > 0$ such that, for all $(x,r) \in R_+^{n+1}$,

$$r \le t \le 2r \Longrightarrow C^{-1} \le \Phi(x,t)/\Phi(x,r) \le C. \tag{1.2}$$

However, for simplicity, we describe the results under the assumption (1.1). Of course our results are also valid under the condition (1.2).

Function Spaces. Let R^n be the n-dimensional Euclidean space and let $B = B(x,r)$ be the ball centered at $x \in R^n$ and with radius $r > 0$. Let $Q = Q(x,r)$ be the cube centered at $x \in R^n$ and with sides of length $r > 0$, where the cube will always mean a compact cube with sides parallel to the axes and nonempty interior. $|B|$ and $|Q|$ stand for the Lebesgue measures of ball B and cube Q, respectively. Let $0 < p < \infty$.

Definition 1.1 (Morrey spaces). (Mizuhara [Miz$_2$]). *We denote by $L^{p,\Phi} = L^{p,\Phi}(R^n)$ the space of locally integrable functions f, defined on R^n, for which there exists a constant C, independent of balls $B = B(x,r)$, such that*

$$\int_{B(x,r)} |f(y)|^p dy \le C^p \Phi(x,r) \tag{1.3}$$

for all balls $B = B(x,r)$. Let $\|f\|_{p,\Phi}$ be the smallest constant C satisfying (1.3). Then the space $L^{p,\Phi}$ becomes a quasi-Banach space with quasi-norm $\|\cdot\|_{p,\Phi}$ in the sense of Triebel [Tri]. *In particular, if $1 \le p < \infty$, then the space $L^{p,\Phi}$ becomes a Banach space with norm $\|\cdot\|_{p,\Phi}$. The balls $B = B(x,r)$ in (1.3) can be replaced by cubes $Q = Q(x,r)$.*

These Morrey spaces $L^{p,\Phi}(R^n)$ are natural extensions of those ones due to the author [Miz$_1$].

When $\Phi(x,r) = r^\lambda$, $\lambda \ge 0$, then L^{p,r^λ} is the classical Morrey space denoted by $L^{p,\lambda}$ simply. The classical Morrey spaces $L^{p,\lambda}$, $0 < \lambda < n$, were originally introduced by Morrey [Mor] in 1938 and used by himself and the others in the problems related to the calculus of variations and the theory of elliptic PDE's. We refer to Campanato [Cam], Giaguinta [Gia], Kufner-John-Fučik [KufJohFuc] and Peetre [Pee$_1$, Pee$_2$].

Let $BMO(R^n)$ be the John-Nirenberg space of all functions of bounded mean oscillation, that is, $BMO(R^n)$ is a Banach space, modulo constants, with norm $\|\cdot\|_*$ defined by

$$\|b\|_* = \sup_B |B|^{-1} \int_B |b(y) - b_B| dy, \quad \text{where} \quad b_B = |B|^{-1} \int_B b(y) dy.$$

The space $BMO(R^n)$ is identified with the dual space of the Hardy space $H^1(R^n)$ in the sense of Fefferman-Stein ([FefSte$_2$]).

Let $\Lambda_\alpha(R^n)$, $0 < \alpha < n$, be the space of all Lipschitz continuous functions of order α on R^n. The space $\Lambda_\alpha(R^n)$ is homogeneous in the sense of dilations. The dual space of $H^p(R^n)$ can be identified with the Lipschitz space $\Lambda_\alpha(R^n)$, $\alpha = n(1/p - 1)$.

Classical operators. Let f be a locally integrable function on R^n. The Hardy-Littlewood maximal operator M is defined by

$$Mf(x) = \sup_{x \in B} |B|^{-1} \int_B |f(y)| dy$$

where the supremum is taken over all balls B containing x and $|B|$ is the volume of the ball B. We introduce two variants of M. Let $0 < q < \infty$ and

$$M_q f(x) = \{(M|f|^q)(x)\}^{1/q}.$$

The sharp maximal function $f^\sharp(x)$ is defined by

$$f^\sharp(x) = \sup_{x \in B} |B|^{-1} \int_B |f(y) - f_B| dy, \quad \text{where} \quad f_B = |B|^{-1} \int_B f(y) dy.$$

Let T be a Calderon-Zygmund singular integral operator $Tf = k * f$ defined by the kernel k satisfying the conditions;

$$\|\hat{k}\|_\infty \le C, \quad |k(x)| \le C|x|^{-n} \quad \text{for } 0 \ne x \in R^n,$$
$$|k(x) - k(x-y)| \le C|y|/|x|^{n+1} \quad \text{for } |y| \le |x|/2.$$

Let I_α, $0 < \alpha < n$, be the fractional integral operator (or Riesz potential operator) of order α defined by

$$(I_\alpha f)(x) = \int_{R^n} \frac{f(y)}{|x-y|^{n-\alpha}} dy$$

for a suitable function f. Related to $I_\alpha f$, the fractional maximal function $M^*_{\alpha/n} f(x)$, which appeared in [MucWhe] as $f^*_\alpha(x)$, is defined by

$$M^*_{\alpha/n} f(x) = f^*_\alpha(x) = \sup_{x \in Q} \frac{1}{|Q|^{1-\alpha/n}} \int_Q |f(y)| dy.$$

Let b a locally integrable function and S be an operator. We define the commutator $[b, S]$ between the operator S and the multiplication operator by b, by $[b, S] = bS - Sb$.

In this note we show the boundednesss of the commutator $[b, S]$, for $b \in BMO(R^n)$ or $b \in \Lambda_\alpha(R^n)$, on Morrey spaces $L^{p,\Phi}(R^n)$ with some growth function Φ. Our results partly generalize the recent results due to Di Fazio and Ragusa [DiFRag] in the case of the classical Morrey spaces $L^{p,\lambda}(R^n), 0 < \lambda < n, 1 < p < \infty$.

The letters C's will denote positive constants, which may have different values in each line.

Acknowledgement. I would like to thank Professor Lu, Shan Zhen for his invitation to this Congress.

2. Commutators between Calderon-Zygmund singular integral operators and multiplication operator by a function $b \in BMO(R^n) \cup \Lambda_\alpha(R^n)$.

G_a-condition. We consider the following condition on growth function $\Phi(x, r)$; $\Phi(x, r)/r^{a+1}$ is non-increasing with respect to $r > 0$ and $\Phi(x, t)/t^{a+1}$ is integrable on $[r, \infty)$ for all $r > 0$ and any $x \in R^n$, and in addition there exists a constant C, independent of $(x, r) \in R^{n+1}_+$, such that

$$\int_r^\infty \frac{\Phi(x,t)}{t^{a+1}} dt \leq C \frac{\Phi(x,r)}{r^a}, \tag{2.1}$$

for some $a > 0$. In this case, we write $\Phi \in G_a$ simply.

We first observe the following properties of G_a, $a > 0$;

Lemma 2.1. (i) *If* $0 < a' < a < \infty$, *then* $G_{a'} \subset G_a \subset \Delta_2$.

(ii) *If* $\Phi \in \Delta_2$ *with a doubling constant* D, $1 \leq D < 2^n$, *then* $\Phi \in G_n$.

(iii) *Let* C *be a constant satisfying* (2.1). *If* $a > 0$ *and* $aC > 1$, *then* $G_a \subset G_{a\gamma}$ *for some* γ, $0 < \gamma < 1$. *More precisely, if* $\Phi \in G_a$, $a > 0$ *and* $aC > 1$, *then there exist constants* $\gamma = \gamma(C, a)$, $0 < \gamma < 1$, *and* $C' = C'(C, a, \gamma) > 0$ *such that for any* $(x, r) \in R^{n+1}_+$

$$\int_r^\infty \frac{\Phi(x,t)}{t^{a\gamma+1}} dt \leq C' \frac{\Phi(x,r)}{r^{a\gamma}}.$$

In particular, when $a = n$, we have $G_n \subset G_{n\gamma}$ for some γ, $0 < \gamma < 1$.

Using this Lemma and the weighted maximal inequalitiy due to Fefferman-Stein [FefSte$_1$] (also see [GarFra]), we can show the following first result ;

Theorem 2.1. *Let $0 < p < \infty$. We assume that $\Phi \in G_n$. Then there exists a constant $C > 0$, independent of f, such that*

$$\|Mf\|_{p,\Phi} \leq C\|f^{\sharp}\|_{p,\Phi}$$

for all $f \in L^{p,\Phi} \cap L_c^{\infty}(R^n)$.

Remark. Let $L_c^{\infty}(R^n)$ be the set of all essentially bounded functions on R^n with compact support. If there exists a constant C, independent of $(x,r) \in R_+^{n+1}$, $0 < r \leq 1$, such that

$$\frac{r^n}{\Phi(x,r)} \leq C,$$

and if there exists a constant C', independent of $(x,r) \in R_+^{n+1}$, $1 < r$, such that

$$0 < C' \leq \Phi(x,r),$$

then the inclusion $L_c^{\infty}(R^n) \subset L^{p,\Phi}(R^n)$ is valid. Applying this criterion to classical Morrey spaces, we have the inclusion $L_c^{\infty}(R^n) \subset L^{p,\lambda}(R^n)$. But we don't know whether $L_c^{\infty}(R^n) \subset L^{p,\Phi}(R^n)$ is valid or not in general.

Our second aim is to show the following ;

Theorem 2.2. *Let $1 < p < \infty$, $b \in BMO(R^n)$ and T be a Calderon-Zygmund singular integral operator. We assume that $\Phi \in G_n$. Then the commutator $[b,T]$ is bounded in $L^{p,\Phi}$. More precisely, there exists a constant C, independent of b and f, such that*

$$\|[b,T](f)\|_{p,\Phi} \leq C\|b\|_*\|f\|_{p,\Phi}$$

for all $b \in BMO(R^n)$ and $f \in L^{p,\Phi} \cap L_c^{\infty}(R^n)$.

This theorem follows from Theorem 2.1 and the following three lemmas ;

Lemma 2.2. *Let $1 < q,\ s < \infty$, $b \in BMO(R^n)$ and T be a Calderon-Zygmund singular integral operator. Then there exists constant C independent of b and f such that*

$$\{[b,T](f)\}^{\sharp}(x) \leq C\|b\|_*\{M_q(Tf)(x) + (M_s f)(x)\}$$

for almost all $x \in R^n$ *and all* $f \in L_c^\infty(R^n)$

Lemma 2.3. *Let* $0 < q < p < \infty$. *We assume that* $\Phi \in G_n$. *Then the maximal operator* M_q *is a bounded cperator in* $L^{p,\Phi}(R^n)$ *and*

$$\|M_q f\|_{p,\Phi} \leq C\|f\|_{p,\Phi}$$

for some constant C *independent of* $f \in L^{p,\Phi}(R^n)$.

Lemma 2.4. *Let* $1 < p < \infty$. *We assume that* $\Phi \in G_n$. *Then the Calderon-Zygmund singular integral operator* T *is a bounded operator in* $L^{p,\Phi}(R^n)$ *and*

$$\|Tf\|_{p,\Phi} \leq C\|f\|_{p,\Phi}$$

for some constant C *independent of* $f \in L^{p,\Phi}(R^n)$.

In the case $b \in \Lambda_\alpha(R^n)$, $0 < \alpha < n$, we obtain the following :

Theorem 2.3. *Let* $1 < p < q < \infty$, $0 < \alpha = n(1/p - 1/q) < n$. *We assume that* $\Phi \in G_{n-p\alpha}$. *If* $b \in \Lambda_\alpha(R^n)$, *then the commutator* $[b, T]$ *is a bounded operator from* $L^{p,\Phi}(R^n)$ *into* $L^{q,\Phi^{q/p}}(R^n)$ *and*

$$\|[b,T]f\|_{q,\Phi^{q/p}} \leq C\|b\|_{\Lambda_\alpha(R^n)}\|f\|_{p,\Phi}$$

for some constant C *independent of* $b \in \Lambda_\alpha(R^n)$ *and* $f \in L^{p,\Phi} \cap L_c^\infty(R^n)$.

This theorem follows from the following lemma ;

Lemma 2.5. *Let* $1 < p < q < \infty$, $0 < \alpha = n(1/p - 1/q) < n$. *We assume that* $\Phi \in G_{n-p\alpha}$. *Then the fractional integral operator* I_α *is a bounded operator from* $L^{p,\Phi}(R^n)$ *into* $L^{q,\Phi^{q/p}}(R^n)$ *and*

$$\|I_\alpha f\|_{q,\Phi^{q/p}} \leq C\|f\|_{p,\Phi}$$

for some constant C *independent of* $f \in L^{p,\Phi}$.

3. Commutators between the fractional integral operator and multiplication operator by a function $b \in BMO(R^n) \cup \Lambda_\alpha(R^n)$.

We show the following in the case $b \in BMO(R^n)$;

Theorem 3.1. *Let* $1 < p < q < \infty$, $0 < \alpha = n(1/p - 1/q) < n$. *We assume that* $\Phi \in G_{n-p\alpha}$ *and* $\Phi^{q/p} \in G_n$. *If* $b \in BMO(R^n)$, *then the* *commutator* $[b, I_\alpha]$ *is a bounded operator from* $L^{p,\Phi}(R^n)$ *into* $L^{q,\Phi^{q/p}}(R^n)$ *and*

$$\|[b, I_\alpha]f\|_{q,\Phi^{q/p}} \leq C\|b\|_*\|f\|_{p,\Phi}$$

for some constant C *independent of* $b \in BMO(R^n)$ *and* $f \in L^{p,\Phi} \cap L_c^\infty(R^n)$.

This theorem follows from the following pointwise estimate and the bounbedness of the fractional maximal operator ;

Lemma 3.1. *Let* $0 < \alpha < n,\ 1 < u,\ t < n/\alpha$ *and* $b \in BMO(R^n)$. *Then there exists constant* C *independent of* b *and* f *such that*

$$\{[b, I_\alpha](f)\}^\sharp(x) \leq C\|b\|_*\{M_u(I_\alpha f)(x) + (M^*_{\alpha t/n}|f|^t)^{1/t}(x)\}$$

for almost all $x \in R^n$ *and all* $f \in L_c^\infty(R^n)$

Lemma 3.2. *Let* $1 < p < q < \infty,\ 0 < \alpha = n(1/p - 1/q) < n$. *We assume that* $\Phi \in G_{n-p\alpha}$. *Then the fractional maximal operator* $M^*_{\alpha/n}$ *is a bounded operator from* $L^{p,\Phi}(R^n)$ *into* $L^{q,\Phi^{q/p}}(R^n)$ *and*

$$\|M^*_{\alpha/n}f\|_{q,\Phi^{q/p}} \leq C\|f\|_{p,\Phi}$$

for some constant C *independent of* $f \in L^{p,\Phi}$.

Lastly we establish the following result ;

Theorem 3.2. *Let* $1 < p < q < \infty,\ 0 < \alpha, \beta,\ \alpha+\beta = n(1/p-1/q) < n,\ 1 < p < n/(\alpha+\beta)$. *We assume that* $\Phi \in G_{n-p(\alpha+\beta)}$. *If* $b \in \Lambda_\alpha(R^n)$, *then the commutator* $[b, I_\beta]$ *is a bounded operator from* $L^{p,\Phi}(R^n)$ *into* $L^{q,\Phi^{q/p}}(R^n)$ *and*

$$\|[b, I_\beta]f\|_{q,\Phi^{q/p}} \leq C\|b\|_{\Lambda_\alpha}\|f\|_{p,\Phi}$$

for some constant C *independent of* $b \in \Lambda_\alpha(R^n)$ *and* $f \in L^{p,\Phi} \cap L_c^\infty(R^n)$.

This theorem follows from Lemma 2.5 as Theorem 2.3.

References.

[Cam] Campanato,S. Proprieta di una famiglia di spazi funzionali, Ann. Scuola Norm. Sup. Pisa, 18(1964), 137-160.

[ChiFra] Chiarenza,F. and Frasca,M. Morrey spaces and Hardy-Littlewood maximal function, Rend. Mat.(7), 7(1987), 273-279.

[DifRag] Di Fazio,G. and Ragusa,M.A. Commutators and Morrey spaces, Boll.U.M.I.(7), 5-A(1991), 323-332.

[FefSte$_1$] Fefferman,C. and Stein,E.M. Some maximal inequalities, Amer. J. Math. 93(1971), 107-115.

[FefSte$_2$] Fefferman,C. and Stein,E.M. H^p spaces of several variables, Acta Math. 129(1972), 127-193.

[GarRub] Garcia-Cuerva,J. and Rubio de Francia,J.L. Weighted norm inequalities and related topics, North-Holland, 1985.

[Gia] Giaquinta,M. Multiple integrals in the calculus of variations and nonlinear elliptic systems, Annals of Mathematics Studies, No.105, Princeton Univ. Press, Princeton, New Jersey, 1983.

[KufJohFuc] Kufner,A., John,O. & Fučik,S. Function spaces, Academia, Praha, 1977.

[Miz$_1$] Mizuhara,T. Boundedness of some classical operators on generalized Morrey spaces, Harmonic Analysis (S.Igari, editor), ICM-90 Satellite Proceedings, Springer-Verlag, Tokyo (1991), 183-189.

[Miz$_2$] Mizuhara,T. Commutators of singular integral operators on Morrey spaces with general growth functions, RIMS 1102, June 1999, Harmonic Analysis and Nonlinear Partial Diferential Equations, Kyoto University, Kyoto(1999), 49-63.

[Mor] Morrey,C.B.Jr. On the solutions of quasi-linear elliptic partial differential equations, Trans. A.M.S. 43(1938), 126-166.

[MucWhe] Muckenhoupt,B. & Wheeden,R., Weighted norm inequalities for fractional integrals, Trans. A.M.S. 192(1974), 261-274.

[Nak] Nakai,E. Hardy-Littlewood maximal operator, singular integral operators and the Riesz potentials on generalized Morrey spaces, Math. Nachr. 166(1994), 95-103.

[Pee$_1$] Peetre,J. On convolution operators leaving $L^{p,\lambda}$ spaces invariant, Ann. Mat. Pura Appl.(4), 72(1966), 295-304.

[Pee$_2$] Peetre,J. On the theory of $L_{p,\lambda}$ spaces, J. Funct. Anal. 4(1969), 71-87.

[Ste] Stein,E.S. Singular integrals and differentiability properties of functions, Princeton University Press, Princeton, New Jersey, 1970.

[Tor] Torchinsky,A. Real-Variable Methods in Harmonic Analysis, Academic Press, Inc. 1986.
[Tri] Triebel,H. Theory of Function Spaces, Monographs in Math.Vol.78, Birkhäuser, 1983.

10 ON GENERALIZED FRACTIONAL INTEGRALS IN THE ORLICZ SPACES

Eiichi NAKAI

Department of Mathematics, Osaka Kyoiku University
Kashiwara, Osaka 582-8582, Japan

enakai@cc.osaka-kyoiku.ac.jp

1991 Mathematics Subject Classification: 26A33, 46E30
Keywords and phrases: fractional integtal, Riesz potential, Orlicz space, space of homogeneous type

1. Definitions

Let $X = (X, d, \mu)$ be a space of homogeneous type, i.e. X is a topological space endowed with a quasi-distance d and a positive measure μ such that

$$d(x,y) \geq 0 \quad \text{and} \quad d(x,y) = 0 \text{ if and only if } x = y,$$
$$d(x,y) = d(y,x),$$
$$d(x,y) \leq K_1\,(d(x,z) + d(z,y)),$$

the balls $B(x,r) = \{y \in X : d(x,y) < r\}$, $r > 0$, form a basis of neighborhoods of the point x, μ is defined on a σ-algebra of subsets of X which contains the balls, and

$$0 < \mu(B(x,2r)) \leq K_2\,\mu(B(x,r)) < \infty,$$

where $K_i \geq 1$ ($i = 1,2$) are constants independent of $x, y, z \in X$ and $r > 0$.

H.G.W. Begehr et al. (eds.), Proceedings of the Second ISAAC Congress, Vol. 1, 75–81.

We assume that $X = (X, d, \mu)$ is Q-homogeneous ($Q > 0$), i.e.

$$K_3{}^{-1} r^Q \le \mu(B(x,r)) \le K_3 r^Q, \tag{1.1}$$

where $K_3 \ge 1$ is a constant independent of $x \in X$ and $r > 0$. The n-dimensional Euclidean space $\mathbb{R}^n$ is n-homogeneous.

For a function $\phi : (0, +\infty) \to (0, +\infty)$, let

$$I_\phi f(x) = \int_X f(y) \frac{\phi(d(x,y))}{d(x,y)^Q} \, d\mu(y).$$

We consider the following conditions on ϕ:

$$\frac{1}{A_1} \le \frac{\phi(s)}{\phi(r)} \le A_1 \quad \text{for} \quad \frac{1}{2} \le \frac{s}{r} \le 2, \tag{1.2}$$

$$\frac{\phi(r)}{r^Q} \le A_2 \frac{\phi(s)}{s^Q} \quad \text{for} \quad s \le r, \tag{1.3}$$

$$\int_0^1 \frac{\phi(t)}{t} \, dt < \infty, \tag{1.4}$$

where $A_i > 0$ ($i = 1, 2$) are independent of $r, s > 0$.

If $\phi(r) = r^\alpha$, $0 < \alpha < Q$, then I_ϕ is the fractional integral or the Riesz potential denoted by I_α.

A function $\Phi : [0, +\infty) \to [0, +\infty]$ is called a Young function if Φ is convex, $\lim_{r \to +0} \Phi(r) = \Phi(0) = 0$ and $\lim_{r \to +\infty} \Phi(r) = +\infty$. Any Young function is increasing.

For a Young function Φ, the complementary function is defined by

$$\Psi(r) = \sup\{rs - \Phi(s) : s \ge 0\}, \quad r \ge 0.$$

Then Ψ is also a Young function. For example, if $\Phi(r) = r^p/p$, $1 < p < \infty$, then $\Psi(r) = r^{p'}/p'$, $1/p + 1/p' = 1$. If $\Phi(r) = r$, then $\Psi(r) = 0 (0 \le r \le 1), = +\infty (r > 1)$.

If a Young function Φ satisfies

$$0 < \Phi(r) < +\infty \quad \text{for} \quad 0 < r < +\infty. \tag{1.5}$$

then Φ is continuous and bijective from $[0, +\infty)$ to itself. The inverse function Φ^{-1} is also increasing and continuous.

For a Young function Φ, let

$$L^\Phi(X) = \left\{ f \in L^1_{\text{loc}}(X) : \int_X \Phi(\epsilon |f(x)|)\, d\mu(x) < +\infty \text{ for some } \epsilon > 0 \right\},$$

$$\|f\|_\Phi = \inf \left\{ \lambda > 0 : \int_X \Phi\left(\frac{|f(x)|}{\lambda} \right) d\mu(x) \le 1 \right\},$$

$$L^\Phi_{weak}(X) = \left\{ f \in L^1_{\text{loc}}(X) : \sup_{r>0} \Phi(r)\, m(r, \epsilon f) < +\infty \text{ for some } \epsilon > 0 \right\},$$

$$\|f\|_{\Phi,weak} = \inf \left\{ \lambda > 0 : \sup_{r>0} \Phi(r)\, m\left(r, \frac{f}{\lambda} \right) \le 1 \right\},$$

where $m(r, f) = \mu(\{x \in X : |f(x)| > r\})$.

A function Φ said to satisfy the ∇_2-condition, denoted $\Phi \in \nabla_2$, if

$$\Phi(r) \le \frac{1}{2k} \Phi(kr), \quad r \ge 0,$$

for some $k > 1$.

Let $Mf(x)$ be the maximal function, i.e.

$$Mf(x) = \sup_{B \ni x} \frac{1}{|B|} \int_B |f(y)|\, d\mu(y),$$

where the supremum be taken over all balls B containing x.

We assume that Φ satisfies (1.5). Then M is bounded from $L^\Phi(X)$ to $L^\Phi_{weak}(X)$ and

$$\|Mf\|_{\Phi,weak} \le C_0 \|f\|_\Phi. \tag{1.6}$$

If $\Phi \in \nabla_2$, then M is bounded on $L^\Phi(X)$ and

$$\|Mf\|_\Phi \le C_0 \|f\|_\Phi. \tag{1.7}$$

2. Main results

Our main results are as follows.

Theorem 2.1. *Let ϕ satisfy (1.2)~(1.4). Let Φ_i $(i = 1, 2)$ be Young functions with (1.5). Assume that there exist constants $A, A', A'' > 0$ such that, for all $r > 0$,*

$$\int_r^\infty \Psi_1 \left(\frac{\phi(t)}{A \int_0^r (\phi(s)/s)\, ds\, {\Phi_1}^{-1}(1/r^Q) t^Q} \right) t^{Q-1}\, dt \le A', \tag{2.1}$$

$$\int_0^r \frac{\phi(t)}{t}\, dt\, {\Phi_1}^{-1}\left(\frac{1}{r^Q} \right) \le A''\, {\Phi_2}^{-1}\left(\frac{1}{r^Q} \right), \tag{2.2}$$

where Ψ_1 is the complementary function with respect to Φ_1. Then, for any $C_0 > 0$, there exists a constant $C_1 > 0$ such that, for $f \in L^{\Phi_1}(X)$,

$$\Phi_2\left(\frac{|I_\phi f(x)|}{C_1\|f\|_{\Phi_1}}\right) \le \Phi_1\left(\frac{Mf(x)}{C_0\|f\|_{\Phi_1}}\right). \tag{2.3}$$

Therefore I_ϕ is bounded from $L^{\Phi_1}(X)$ to $L^{\Phi_2}_{weak}(X)$. Moreover, if $\Phi_1 \in \nabla_2$, then I_ϕ is bounded from $L^{\Phi_1}(X)$ to $L^{\Phi_2}(X)$.

Remark 2.1. In the case $\Phi_1(r) = r$, (2.1) is equivalent to

$$\frac{\phi(t)}{t^Q} \le A\frac{\int_0^r (\phi(s)/s)\,ds}{r^Q}, \quad 0 < r \le t.$$

This inequality follows from (1.2) and (1.3).

For functions $\theta, \kappa : (0, +\infty) \to (0, +\infty)$, we denote

$$\theta(r) \sim \kappa(r)$$

if there exists a constant $C > 0$ such that

$$C^{-1}\theta(r) \le \kappa(r) \le C\theta(r), \quad r > 0.$$

A function $\theta : (0, +\infty) \to (0, +\infty)$ is said to be almost increasing (almost decreasing) if there exists a constant $C > 0$ such that

$$\theta(r) \le C\theta(s) \quad (\theta(r) \ge C\theta(s))$$

for $r \le s$.

Corollary 2.2. *Let ϕ satisfy* (1.2)$\sim$(1.4). *Let Φ_i $(i = 1, 2)$ be Young functions with* (1.5). *Assume that*

$$\int_0^r \frac{\phi(t)}{t}\,dt\ {\Phi_1}^{-1}\left(\frac{1}{r^Q}\right)$$

is almost decreasing and that there exist constants $A, A' > 0$ such that, for all $r > 0$,

$$\int_r^\infty \frac{\phi(t)}{t}{\Phi_1}^{-1}\left(\frac{1}{t^Q}\right)dt \le A\int_0^r \frac{\phi(t)}{t}\,dt\ {\Phi_1}^{-1}\left(\frac{1}{r^Q}\right) \tag{2.4}$$

$$\le A'\ {\Phi_2}^{-1}\left(\frac{1}{r^Q}\right). \tag{2.5}$$

Then (2.3) *holds. Therefore I_ϕ is bounded from $L^{\Phi_1}(X)$ to $L^{\Phi_2}_{weak}(X)$. Moreover, if $\Phi_1 \in \nabla_2$, then I_ϕ is bounded from $L^{\Phi_1}(X)$ to $L^{\Phi_2}(X)$.*

Remark 2.2. If $\phi(r)/r^\varepsilon$ is almost increasing for some $\varepsilon > 0$ and $\phi(t)/t^Q$ is almost decreasing, then ϕ satisfies (1.2)$\sim$(1.4) and $\int_0^r (\phi(t)/t)\,dt \sim \phi(r)$. If $r^\varepsilon\phi(r){\Phi_1}^{-1}(1/r^Q)$ is almost decreasing for some $\varepsilon > 0$, then (2.4) holds.

Corollary 2.3. *Let $\phi(r) = r^\alpha$ with $0 < \alpha < Q$. Let Φ_i $(i = 1, 2)$ be Young functions with* (1.5). *Assume that there exist constants $A, A' > 0$ such that, for all $r > 0$,*

$$\int_r^\infty t^{\alpha-1}\Phi_1{}^{-1}\left(\frac{1}{t^Q}\right) dt \le A r^\alpha \Phi_1{}^{-1}\left(\frac{1}{r^Q}\right), \tag{2.6}$$

$$\le A'\Phi_2{}^{-1}\left(\frac{1}{r^Q}\right). \tag{2.7}$$

Then (2.3) *holds. Therefore I_α is bounded from $L^{\Phi_1}(X)$ to $L^{\Phi_2}_{weak}(X)$. Moreover, if $\Phi_1 \in \nabla_2$, then I_α is bounded from $L^{\Phi_1}(X)$ to $L^{\Phi_2}(X)$.*

Remark 2.3. Results similar to this corollary are in [5] and [9]. Kokilashvili and Krbec [5] considered the boundedness of I_α with weights, and gave a necessary and sufficient condition on the weights so that weighted inequalities hold. Torchinsky [9] treated sublinear operators with weak type (p_i, q_i) $(i = 1, 2)$ and used interpolation.

Corollary 2.4 (Hardy-Littlewood-Sobolev). *Let $\phi(r) = r^\alpha$, $\Phi_1(r) = r^p$ and $\Phi_2(r) = r^q$ with $0 < \alpha < Q$, $1 \le p < Q/\alpha$ and $Q/q = Q/p - \alpha$. Then* (2.3) *holds. Therefore I_α is bounded from $L^1(X)$ to $L^q_{weak}(X)$ for $p = 1$ and from $L^p(X)$ to $L^q(X)$ for $1 < p < Q/\alpha$.*

3. Examples

Example 3.1. Let ϕ satisfy (1.2) and

$$\phi(r) = \begin{cases} r^{\alpha_1} & \text{for small } r, \\ r^{\alpha_2} & \text{for large } r, \end{cases}$$

where $0 < \alpha_1 < Q$ and $-\infty < \alpha_2 < Q$. Let Φ_1 and Φ_2 be convex and

$$\Phi_1(\xi) = \begin{cases} \xi^{p_2}, \\ \xi^{p_1}, \end{cases} \quad \Phi_2(\xi) = \begin{cases} \xi^{q_2}, \\ \xi^{q_1}, \end{cases} \quad \text{for} \begin{cases} \text{small } \xi, \\ \text{large } \xi, \end{cases}$$

where

$$1 < p_1 < Q/\alpha_1, \; Q/q_1 \ge Q/p_1 - \alpha_1,$$

$$\begin{cases} 1 < p_2 < Q/\alpha_2, \; Q/q_2 \le Q/p_2 - \alpha_2, & \text{when } 0 < \alpha_2 < Q, \\ 1 < p_2 < q_2 < \infty, & \text{when } \alpha_2 = 0, \\ 1 < p_2 \le q_2 < \infty, & \text{when } \alpha_2 < 0. \end{cases}$$

Then (2.3) holds and I_ϕ is bounded from $L^{\Phi_1}(X)$ to $L^{\Phi_2}(X)$.

Example 3.2. Let ϕ satisfy (1.2) and

$$\phi(r)=\begin{cases}(\log(1/r))^{-\alpha_1} & \text{for small } r,\\ (\log r)^{\alpha_2} & \text{for large } r,\end{cases}$$

where $1<\alpha_1<\infty$ and $-\infty<\alpha_2<\infty$. Let Φ_1 and Φ_2 be convex and

$$\Phi_1(\xi)=\begin{cases}1/\exp(1/\xi^{p_2}),\\ \exp(\xi^{p_1}),\end{cases}\quad \Phi_2(\xi)=\begin{cases}1/\exp(1/\xi^{q_2}),\\ \exp(\xi^{q_1}),\end{cases}\quad \text{for}\begin{cases}\text{small } \xi,\\ \text{large } \xi,\end{cases}$$

where

$$0<p_1<1/(\alpha_1-1),\ 1/q_1\geq 1/p_1-(\alpha_1-1), \tag{3.1}$$

(3.2)

$$\begin{cases}0<p_2<1/(\alpha_2+1),\ 1/q_2\leq 1/p_2-(\alpha_2+1), & \text{when } -1<\alpha_2<Q,\\ 0<p_2<q_2<\infty, & \text{when } \alpha_2=-1,\\ 0<p_2\leq q_2<\infty, & \text{when } \alpha_2<-1.\end{cases}$$

Then (2.3) holds and I_ϕ is bounded from $L^{\Phi_1}(X)$ to $L^{\Phi_2}(X)$.

Example 3.3. Let ϕ satisfy (1.2) and

$$\phi(r)=\begin{cases}(\log(1/r))^{-1}(\log\log(1/r))^{-\alpha_1} & \text{for small } r,\\ (\log r)^{-1}(\log\log r)^{\alpha_2} & \text{for large } r,\end{cases}$$

where $1<\alpha_1<\infty$ and $-\infty<\alpha_2<\infty$. Let Φ_1 and Φ_2 be convex and

$$\Phi_1(\xi)=\begin{cases}1/\exp\exp(1/\xi^{p_2}),\\ \exp\exp(\xi^{p_1}),\end{cases}\quad \Phi_2(\xi)=\begin{cases}1/\exp\exp(1/\xi^{q_2}),\\ \exp\exp(\xi^{q_1}),\end{cases}\quad \text{for}\begin{cases}\text{small } \xi,\\ \text{large } \xi,\end{cases}$$

where p_1, p_2, q_1 and q_2 satisfy (3.1) and (3.2). Then (2.3) holds and I_ϕ is bounded from $L^{\Phi_1}(X)$ to $L^{\Phi_2}(X)$.

Example 3.4. Let ϕ satisfy (1.2) and

$$\phi(r)=\begin{cases}(\log(1/r))^{-\alpha_1} & \text{for small } r,\\ (\log r)^{\alpha_2} & \text{for large } r,\end{cases}$$

where $\alpha_1>1$ and $\alpha_2>-1$. Let Φ_1 and Φ_2 be convex and

$$\Phi_1(\xi)=\xi^p,\quad \Phi_2(\xi)=\begin{cases}\xi^p(\log(1/\xi))^{-p(\alpha_2+1)} & \text{for small } \xi,\\ \xi^p(\log\xi)^{p(\alpha_1-1)} & \text{for large } \xi,\end{cases}$$

where $1\leq p<\infty$. Then (2.3) holds and I_ϕ is bounded from $L^1(X)$ to $L^{\Phi_2}_{weak}(X)$ for $p=1$ and from $L^p(X)$ to $L^{\Phi_2}(X)$ for $1<p<\infty$.

Example 3.5. Let ϕ satisfy (1.2) and

$$\phi(r) = \begin{cases} r^Q(\log(1/r))^{\alpha_1} & \text{for small } r, \\ r^Q(\log r)^{-\alpha_2} & \text{for large } r, \end{cases}$$

where $\alpha_i > 0$ $(i = 1, 2)$. Let Φ_1 and Φ_2 be convex and

$$\Phi_1(\xi) = \begin{cases} \xi(\log(1/\xi))^{-\beta_2}, \\ \xi(\log \xi)^{\beta_1}, \end{cases} \quad \Phi_2(\xi) = \begin{cases} 1/\exp((1/\xi)^{1/(\alpha_2-\beta_2)}), \\ \exp(\xi^{1/(\alpha_1-\beta_1)}), \end{cases} \quad \text{for} \begin{cases} \text{small } \xi, \\ \text{large } \xi, \end{cases}$$

where $0 \leq \beta_i < \alpha_i$ $(i = 1, 2)$. Then (2.3) holds and I_ϕ is bounded from $L^{\Phi_1}(X)$ to $L^{\Phi_2}_{weak}(X)$.

References

[1] P. Cifuentes, J. R. Dorronsoro and J. Sueiro, *Boundary tangential convergence on spaces of homogeneous type*, Trans. Amer. Math. Soc. 332 (1992), 331–350

[2] R. R. Coifman and G. Weiss, *Analyse harmonique non-commutative sur certains espaces homogenes*, Lecture Notes in Math., vol.242, Springer-Verlag, Berlin and New York, 1971.

[3] ______ *Extensions of Hardy spaces and their use in analysis*, Bull. Amer. Math. Soc. **83** (1977), 569–645.

[4] A. E. Gatto and S. Vági, *Fractional integrals on spaces of homogeneous type*, in Analysis and Partial Differential Equations, edited by Cora Sadosky, Marcel Dekker, New York, 1990, 171–216.

[5] V. Kokilashvili and M. Krbec, *Weighted inequalities in Lorentz and Orlicz spaces*, World Scientific, Singapore, New Jersey, London and Hong Kong, 1991.

[6] E. Nakai and H. Sumitomo, *On generalized Riesz potentials and spaces of some smooth functions*, preprint.

[7] M. M. Rao and Z. D. Ren, *Theory of Orlicz Spaces*, Marcel Dekker, Inc., New York, Basel and Hong Kong, 1991.

[8] E. M. Stein, *Harmonic Analysis, real-variable methods, orthogonality, and oscillatory integrals*, Princeton University Press, Princeton, NJ, 1993.

[9] A. Torchinsky, *Interpolation of operations and Orlicz classes*, Studia Math. **59** (1976), 177–207.

11 WEAK (1,1) ESTIMATES FOR LITTLEWOOD-PALEY FUNCTIONS WITH ROUGH KERNELS

Shuichi SATO

Department of Mathematics
Faculty of Education, Kanazawa University
Kanazawa, 920-1192 Japan

shuichi@kenroku.kanazawa-u.ac.jp

1991 Mathematics Subject Classification: Primary 42B25, 42B20
Keywords and phrases: Littlewood-Paley functions, Marcinkiewicz integrals, singular integrals

Abstract

We consider Marcinkiewicz integrals on $\boldsymbol{R}^n$ and $\boldsymbol{T}^n$ arising from rough kernels satisfying the $L \log L$-condition on S^{n-1} and prove the weak type $(1,1)$ estimates. We also prove the weighted weak type $(1,1)$ estimates on $\boldsymbol{R}^n$ with A_1-weights. In this case the $L \log L$-condition is replaced by the L^q-condition with $q > 1$.

This is a report of a joint work [4] with D. Fan (Wisconsin-Milwaukee). Let $\psi \in L^1(\boldsymbol{R}^n)$, for which we always assume

$$\int_{\boldsymbol{R}^n} \psi(x)\, dx = 0.$$

H.G.W. Begehr et al. (eds.), Proceedings of the Second ISAAC Congress, Vol. 1, 83–87.

Define a Littlewood-Paley square function $g_\psi(f)$ on $\boldsymbol{R}^n$ by

$$g_\psi(f)(x) = \left(\int_0^\infty |\psi_t * f(x)|^2 \frac{dt}{t}\right)^{1/2},$$

initially for $f \in \mathcal{S}(\boldsymbol{R}^n)$ (the Schwartz space), where $\psi_t(x) = t^{-n}\psi(t^{-1}x)$.

For the L^2-boundedness of g_ψ, the following is known (see [4]).

Theorem 1 *Suppose*

(1) $\int_{\boldsymbol{R}^n} |\psi(x)| \log(2 + |x|)\, dx < \infty$;

(2) $\int_{\boldsymbol{R}^n} |\psi(x)| \log(2 + |\psi(x)|)\, dx < \infty$.

Then g_ψ is bounded on L^2.

Let $\Omega \in L^1(S^{n-1})$. (We denote by S^{n-1} the unit sphere.) We say $\Omega \in L \log L$ if

$$\|\Omega\|_{L\log L} = \int_{S^{n-1}} |\Omega(\theta)| \log(2 + |\Omega(\theta)|)\, d\sigma(\theta) < \infty,$$

where σ denotes the Lebesgue surface measure on S^{n-1}.

Theorem 1 easily implies the following.

Corollary 1 *Let $\psi(x) = |x|^{-n}h(|x|)\Omega(x')$, where $x' = x/|x|$. Then g_ψ is bounded on L^2 if*

(1) $\Omega \in L \log L$;

(2) $\int_0^\infty |h(r)| \log(2 + |h(r)|)\, dr/r < \infty$;

(3) $\int_0^\infty |h(r)| \log(2 + r)\, dr/r < \infty$;

(4) $\int_0^\infty |h(r)| \log(2 + r^{-1})\, dr/r < \infty$.

Let

$$\psi(x) = |x|^{-n+\rho}\Omega(x')\chi_{(0,1]}(|x|) \qquad (\rho > 0).$$

We assume $\Omega \in L^1$ and $\int \Omega(\theta)\, d\sigma(\theta) = 0$. Define the Marcinkiewicz integral $\mu_\rho(f)$ on $\boldsymbol{R}^n$ by $\mu_\rho(f) = g_\psi(f)$.

When Ω satisfies a Lipschitz condition on S^{n-1}, Stein [8] proved that μ_1 is of weak type $(1,1)$. In [5] Hörmander proved that μ_ρ is of weak type $(1,1)$ under a Dini condition on Ω. Recently, Ding-Fan-Pan [1], [2], [3] and Sato [6] have studied the L^p-boundedness ($p > 1$) of the Littlewood-Paley functions with rough kernels.

In [4] we proved the following.

Theorem 2 *If $\Omega \in L\log L$, then μ_ρ is of weak type $(1,1)$, that is,*

$$\sup_{\lambda>0} \lambda\,|\{x \in \boldsymbol{R}^n : \mu_\rho(f)(x) > \lambda\}| \le C\|f\|_1,$$

where C is a constant independent of f, and $|E|$ denotes the Lebesgue measure of a set E.

To prove Theorem 2, we proved the following.

Theorem 3 *Let $\psi(x) = |x|^{-n}h(|x|)\Omega(x')$. Suppose*

(1) $\Omega \in L\log L$;

(2) $\mathrm{supp}(h) \subset [1,2]$;

(3) $\int_0^\infty |h(r)|^2\,dr < \infty$;

(4) $\int_0^1 \omega(t)\,dt/t < \infty$, *where*

$$\omega(t) = \sup_{|u|<t/2}\left(\int_0^\infty |h(r-u) - h(r)|^2\,dr\right)^{1/2}$$

Then g_ψ is of weak type $(1,1)$.

We can show that Theorem 3 implies Theorem 2 (see [4]). We also proved the weighted versions of Theorems 2 and 3. Recall that
the weight class A_1 of Muckenhoupt is the collection of weights w such that

$$M(w)(x) \le cw(x) \quad \text{a.e.} \quad \text{for some constant } c,$$

where M denotes the Hardy-Littlewood maximal operator. For a weight w on $\boldsymbol{R}^n$ we write $\|f\|_{p,w} = \|fw^{1/p}\|_p$ and $w(E) = \int_E w$.

Theorem 4 *Suppose $\Omega \in L^q$ for some $q > 1$ and $w^{q'} \in A_1$, $q' = q/(q-1)$. Then*

$$\sup_{\lambda>0} \lambda w\left(\{x \in \boldsymbol{R}^n : \mu_\rho(f)(x) > \lambda\}\right) \le C_w\|f\|_{1,w}.$$

Theorem 5 *Let $q > 1$ and $w^{q'} \in A_1$. For $\psi(x) = |x|^{-n}h(|x|)\Omega(x')$ suppose*

(1) $\Omega \in L^q$;

(2) $\operatorname{supp}(h) \subset [1,2]$;

(3) $\int_0^\infty |h(r)|^2\, dr < \infty$;

(4) $\omega(t) \le ct^\alpha$ *for some* $\alpha > 0$ ($\omega(t)$ *is as in Theorem* 3).

Then

$$\sup_{\lambda>0} \lambda w\left(\{x \in \boldsymbol{R}^n : g_\psi(f)(x) > \lambda\}\right) \le C_w \|f\|_{1,w}.$$

We can see that Theorem 5 implies Theorem 4 as Theorem 3 implies Theorem 2. To prove Theorems 3 and 5 we proved vector-valued versions of results of Seeger [7]. To prove Theorem 5 we also need to apply a modified version of the interpolation argument of Vargas [9].

The results of Seeger [7] and the method which proves Theorem 5 can be used to prove the following result.

Theorem 6 *Let* $K(x) = |x|^{-n}\Omega(x')$, *where* $\Omega \in L^\infty$ *and* $\int \Omega(\theta)\, d\sigma(\theta) = 0$. *Define a singular integral operator* T *by* $T(f) = \text{p.v.}\, K * f$ $(f \in \mathcal{S}(\boldsymbol{R}^n))$. *Then we have*

$$\sup_{\lambda>0} \lambda w\left(\{x \in \boldsymbol{R}^n : |T(f)(x)| > \lambda\}\right) \le C_w \|f\|_{1,w}$$

for $w \in A_1$.

See Vargas [9] for the 2-dimensional case.

Finally, we proved an analog of Theorem 2 on the n-torus. Let $\boldsymbol{T}^n$ be the n-dimensional torus. $\boldsymbol{T}^n$ can be identified with $\boldsymbol{R}^n/\Lambda$, where Λ is the unit lattice which is the additive group of points in $\boldsymbol{R}^n$ having integral coordinates. For $\tilde{f} \in C^\infty(\boldsymbol{T}^n)$, $\tilde{f}$ has the Fourier series

$$\tilde{f}(x) = \sum_{k\in\Lambda} a_k e^{2\pi i\langle k,x\rangle}.$$

Let $\tilde{\mu}_\rho(\tilde{f})$ be the Marcinkiewicz integral on $\boldsymbol{T}^n$ defined by

$$\tilde{\mu}_\rho(\tilde{f})(x) = \left(\int_0^\infty |\tilde{\psi}_t * \tilde{f}(x)|^2 \,\frac{dt}{t}\right)^{1/2},$$

where

$$\tilde{\psi}_t * \tilde{f}(x) = \sum_{k\in\Lambda} \hat{\psi}(tk) a_k e^{2\pi i\langle k,x\rangle} = \sum_{k\ne 0} \hat{\psi}(tk) a_k e^{2\pi i\langle k,x\rangle}$$

since $\hat{\psi}(0) = 0$. (Here $\hat{\psi}(\xi) = \int \psi(x) e^{-2\pi i\langle x,\xi\rangle}\, dx$ is the Fourier transform.)

Then we have the following.

Theorem 7 *Let* $\psi(x) = |x|^{-n+\rho}\Omega(x')\chi_{(0,1]}(|x|)$ *be as in Theorem 2. Then* $\tilde{\mu}_\rho$ *is of weak type* $(1,1)$ *on* $\boldsymbol{T}^n$.

We proved this result by a transference method, assuming Theorem 2.

References

[1] Y. Ding, D. Fan and Y. Pan, *On Littlewood-Paley functions and singular integrals,*, preprint.

[2] Y. Ding, D. Fan and Y. Pan, *L^p-boundedness of Marcinkiewicz integrals with Hardy space function kernel,* Acta Math. Sinica (to appear).

[3] Y. Ding, D. Fan and Y. Pan, *Weighted boundedness for a class of rough Marcinkiewicz integrals,*

Indiana Univ. Math. J. (to appear).

[4] D. Fan and S. Sato, *Weak type* $(1,1)$ *estimates for Marcinkiewicz integrals with rough kernels,* preprint.

[5] L. Hörmander, *Estimates for translation invariant operators in L^p spaces,* Acta Math. 104 (1960), 93–139.

[6] S. Sato, *Remarks on square functions in the Littlewood-Paley theory,* Bull. Austral. Math. Soc. **58** (1998), 199–211.

[7] A. Seeger, *Singular integral operators with rough convolution kernels,* J. Amer. Math. Soc. **9** (1996), 95–105.

[8] E. M. Stein, *On the functions of Littlewood-Paley, Lusin, and Marcinkiewicz,* Trans. Amer. Math. Soc. **88** (1958), 430–466.

[9] A. Vargas, *Weighted weak type* $(1,1)$ *bounds for rough operators,* J. London Math. Soc. (2) **54** (1996), 297–310.

12 A NOTE ON AVERAGE DENSITIES OF BROWNIAN INTERSECTION MEASURES

Narn-Rueih SHIEH

Department of Mathematics, National Taiwan University
Taipei, Taiwan

shiehnr@math.ntu.edu.tw

1991 Mathematics Subject Classification: Primary 42B25, 42B20
Keywords and phrases: Littlewood-Paley functions, Marcinkiewicz integrals, singular integrals

Abstract

The purpose of this note is to complement the paper "Small scale limit theorems for the intersection local times of Brownian motions" (P Mörters and NR Shieh, Electronic J. Probab. vol.4(1999) Paper no.9 Pages 1–23), which is reported at The 2nd ISAAC Session 17. In this note, we use an alternative method to prove one crucial step in the proof of main theorems in that paper. Namely, we shall use a device to prove that: once the average density at a typical point, say $x = 0$, holds, then the same result holds for a generic point x.

H.G.W. Begehr et al. (eds.), Proceedings of the Second ISAAC Congress, Vol. 1, 89–95.

1 Introduction and Main Result

Bedford–Fisher[BF] introduced the notion of *average densities of order n*, henceafter AD_n for short, of fractal sets and measures in $\mathbb{R}^d$. This notion gives a fine local description of fractals and have various applications; see Bedford[B], Falconer[F, Chapter 6]. It is also related to Mandelbrot's concept of fractal lacunarity[M]. In Mörters–Shieh[MS], it is proved complete results on the AD_2 and the AD_3 for an important fractal measure associated with Brownian sample paths. The problem was initiated in Shieh[S]. We mention that a good reference for the mathematical theory of random fractals arising from the sample functions of Brownian motions and Lévy processes is Taylor[T]. In [MS], the proofs of main results are devided into two parts. One part is to prove that the results hold at $x = 0$, and the other part is to prove that the results at a generic x can be ruduced from those at $x = 0$. The measure studied in [MS] is *Brownian intersection measure.* Remark: we will use this shorter term here in this note, instead of the longer one in [MS]. For the simplicity of the presentation, we will present in this note only for the intersection measure of two spatial Brownian paths; the case for intersections of several planar Brownian paths can be treated exactly in the same way.

Let I denote the intersection of two independent Brownian paths X, Y in $\mathbb{R}^3$. It is well known that, with probability one, I is of Hausdorff dimension 1(Note that we have suppressed the randomness sign ω from I). However the exact Hausdorff measure for I is much more complicated. To study this and some related problems, a certain canonical measure on I, denoted by μ in all the followings, was constructed and studied by Dynkin[D], Geman–Horowitz–Rosen[GHR] and Le Gall[L1]. The measure μ is expressed heuristically as

$$\mu(A) = \int_A \int_s \int_t \delta_x(X(s))\delta_x(Y(t))dsdtdx,$$

where δ_x is the "δ–function" at x. The precise definition of μ can be seen in [MS, Section 2]. The following result is proved in [MS], among others.

Theorem 1.1 *Let $X(s), Y(t)$, where $s, t \in \mathbb{R}$, be two independent Brownian paths in $\mathbb{R}^3$, $X(0) = Y(0) = 0$, and let $I = X(\mathbb{R}) \cap Y(\mathbb{R})$. Let μ denote the intersection measure associated with I Then, with probability one, the average density of order two of μ at x, with respect to the gauge* *function $\phi(r) = r$, is given by*

$$AD_2(\mu, x) = \frac{4}{\pi},$$

for μ almost all $x \in I$.

We recall that the above AD_2 is defined by

$$AD_2(\mu, x) = \lim_{R\to\infty} \frac{1}{R} \int_0^R \frac{\mu(B(x, e^{-r}))}{2e^{-r}} dr,$$

whenever the limit exists($B(x, r)$ denotes the closed ball with center x and radius r).

It is interesting to compare Theorem 1.1 with a result of Falconer–Xiao[FX]; they proved at there, among others, the AD_2 for the occupation measure of a single Brownian path in $\mathbb{R}^d$, $d \geq 3$. See also Mörters[M] for the planar case. The proof of Theorem 1.1 consists of two parts: firstly to prove the existence of $AD_2(\mu, 0)$ to evaluate its value, and secondly to prove the existence of $AD_2(\mu, x)$ for μ almost every $x \in I$. For proof of the first part, the method is that one revealed in [BF] and [FX], namely to use the ergodic theorem applied to a certain suitable scaling flow on the path space. For the second part, the proof is more complicated than those arguments in proofs of the above cited papers, since we need more considerable effort on the stopping arguments for the case of Brownian intersections; see also [L1, p119] for such a remark. We do not reproduce the first part here. Now we prove the second part by a defferent device, rather than the method of Palm distributions in [MS]. This device is more probabilistic in formality, though we suspect that it is indeed close related to Palm distributions in essence.

2 The $AD_2(\mu, x)$ at a generic x

To prove that $AD_2(\mu, x) = 4/\pi$ at μ almost every $x \in I$, we need a lemma which provides a certain form of the Markov property, at times which the two paths X, Y have intersection. We note that these times are typically *not* stopping times. This is the difficulty comparing with the results in [BF] and [FX], where standard applications of Markov property is allright. The lemma and its proof are motivated from Le Gall[L2, Theorem IX–1] which concerns with the case of the self–intersections. To begin with, we consider independent Brownian motions in $\mathbb{R}^3$ in the following setting. Let $C_3 = \{f : \mathbb{R}_+ \to \mathbb{R}^3,\ \ f$ is continuous and $f(0) =$

$0\}$, $\mathcal{B}(C_3)$ the σ–algebra generated by cylinder subsets of C_3, and W the standard Wiener measure on $(C_3, \mathcal{B}(C_3))$. Let $\Omega = C_3^{\otimes 2}$, $\mathcal{F} = \mathcal{B}(C_3)^{\otimes 2}$, and $P = W^{\otimes 2}$; so that $w = (w_1, w_2) \in \Omega$ represents two independent spatial Brownian motions. For any continuous–paths process $U(u), u \in \mathbb{R}$, and any $t : 0 < t < \infty$ let two random elements $U_{t,\infty}$ and $U_{-\infty,t}$ in C_3 be defined respectively by

$$\begin{aligned} U_{t,\infty}(t') &= U(t+t') - U(t) \\ U_{-\infty,t}(t') &= U(t-t') - U(t), \quad t' \geq 0. \end{aligned}$$

Lemma 2.1 *Let $W_j(u), u \geq 0, \quad j = 1,2,3,4$, be four independent Brownian motions in $\mathbb{R}^3$, $W_j(0) = 0$. For any Borel A in $(0,\infty)^2$ and any nonnegative measurable functional F on $(C_3)^{\otimes 4}$,*

$$\begin{aligned} & E\Big[\int\int_A \alpha(dsdt) F(X_{-\infty,s}, X_{s,\infty}, Y_{-\infty,t}, Y_{t,\infty})\Big] \\ = \;& E[F(W_1, W_2, W_3, W_4)] \times E[\alpha(A)]. \end{aligned}$$

In the above, the $\alpha(dsdt)$ denotes the local time measure for Brownian intersections, which was defined in [GHR] and [L1]. We postpone the proof of Lemma 2.1 to the end of this section, and proceed the second part of the proof of Theorem 1.1 as follows. Let F be the indicator of the event H which is

$$\begin{aligned} H = \{(w_1, w_2, w_3, w_4) \in (C_3)^{\otimes 4} : \quad & AD_2(\mu_{i,j}, 0) = 1/\pi \\ & \text{for } \; i = 1,2 \;\; j = 3,4\}, \end{aligned}$$

where $\mu_{i,j}$ is the intersection measure associated with $w_i[0,\infty) \cap w_j[0,\infty)$. In view of first part of the proof of Theorem 1.1(see [MS] for this),

$$F(W_1, W_2, W_3, W_4) = 1 \quad a.s.$$

Thus, by Lemma 2.1,

$$E\Big[\int\int_A \alpha(dsdt) F(X_{-\infty,s}, X_{s,\infty}, Y_{-\infty,t}, Y_{t,\infty})\Big] = E[\alpha(A)] \quad \forall A,$$

which implies that, with probability one, for $\alpha(dsdt)$ almost all $(s,t) \in (0,\infty)^2$,

$$\begin{aligned} AD_2(X_{-\infty,s}(0,\infty) \cap Y_{-\infty,t}(0,\infty), 0) &= 1/\pi, \\ AD_2(X_{-\infty,s}(0,\infty) \cap Y_{t,\infty}(0,\infty), 0) &= 1/\pi, \\ AD_2(X_{s,\infty}(0,\infty) \cap Y_{-\infty,t}(0,\infty), 0) &= 1/\pi, \\ AD_2(X_{s,\infty}(0,\infty) \cap Y_{t,\infty}(0,\infty), 0) &= 1/\pi. \end{aligned}$$

In the above, we have used alternative notations for the AD_2 of those intersection measures associated with the indicated paths intersections. Using the facts that $X_{-\infty,s}[0,\infty) = X(-\infty,s]-X(s)$, that $X_{s,\infty}[0,\infty) = X[s,\infty) - X(s)$, and that so are true for Y, we have then that, with probability one, for $\alpha(dsdt)$ almost all $(s,t) \in \text{Supp}(\alpha)$,

$$\begin{aligned} AD_2(X(-\infty,s] \cap Y(-\infty,t], X(s) = Y(t) \in I) &= 1/\pi, \\ AD_2(X(-\infty,s] \cap Y[t,\infty), X(s) = Y(t) \in I) &= 1/\pi, \\ AD_2(X[s,\infty) \cap Y(-\infty,t], X(s) = Y(t) \in I) &= 1/\pi, \\ AD_2(X([s,\infty) \cap Y[t,\infty,), X(s) = Y(t) \in I) &= 1/\pi. \end{aligned}$$

Since any three independent Brownian trails in $\mathbb{R}^3$ cannot have intersections, except possibly at the endpoints, and the four trails $X(-\infty,s], X[s,\infty), Y(-\infty,t], Y[t,\infty)$ are independent, we see that, whenever $X(s) = Y(t) \in I$ and $r > 0$,

$$\begin{aligned} & \mu(X(\mathbb{R}) \cap Y(\mathbb{R}) \cap B(X(s) = Y(t), r)) \\ = \; & \mu(X(-\infty,s] \cap Y(-\infty,t] \cap B(X(s) = Y(t), r)) \\ + \; & \mu(X(-\infty,s] \cap Y[t,\infty) \cap B(X(s) = Y(t), r)) \\ + \; & \mu(X[s,\infty) \cap Y(-\infty,t] \cap B(X(s) = Y(t), r)) \\ + \; & \mu(X[s,\infty) \cap Y[t,-\infty) \cap B(X(s) = Y(t), r)). \end{aligned}$$

Therefore, with probability one, for $\alpha(dsdt)$ almost all $(s,t) \in (0,\infty)^2$,

$$AD_2(\mu, X(s = Y(t) \in I) = 4/\pi,$$

by the four AD_2 densities given in the above. Since μ is simply the image measure of α under $(s,t) \to X(s)$, we conclude that, with probability one,

$$AD_2(\mu, x) = 4/\pi,$$

for μ almost all $x \in I$. Thus, the proof of of Theorem 1.1 will be complete, once we have proceeded

Proof of Lemma 2.1. It suffices to prove the assertion for bounded continuous F and for compact rectangle A. The l.h.s. of the assertion of Lemma 2.1 is the limit

$$\lim_{\epsilon\downarrow 0} E\Big[\int\int_A \alpha^\epsilon(dsdt) F(X_{-\infty,s}, X_{s,\infty}, Y_{-\infty,t}, Y_{t,\infty})\Big],$$

where α^ϵ is the approximating measure of α. The expectation in the above display is

(2.1)

$$\int\int_A \frac{dsdt}{((4/3)\pi\epsilon^3)^2} E\Big[F(X_{-\infty,s}, X_{s,\infty}, Y_{-\infty,t}, Y_{t,\infty})\cdot$$
$$\text{Leb}(B(X(s),\epsilon)\cap B(Y(t),\epsilon))\Big].$$

Note that

$$\text{Leb}\,(B(X(s),\epsilon)\cap B(Y(t),\epsilon)) = \text{Leb}\,(B(0,,\epsilon)\cap B(Y(t)-X(s),\epsilon)).$$

Conditioning w.r.t. $Y(t)-X(s)$ and noting that $Y(t)-X(s)$ is $N(0,t+s)^{\otimes 3}$ distributed, we see that (2.1) is

$$\int\int_A \frac{dsdt}{((4/3)\pi\epsilon^3)^2}\int_{|y|\le 2\epsilon} dy p_{t+s}(y)\text{Leb}\,(B(0,\epsilon)\cap B(y,\epsilon))\times$$
$$E\Big[F(X_{-\infty,s}, X_{s,\infty}, Y_{-\infty,t}, Y_{t,\infty})|Y(t)-X(s)=y\Big],$$

where p_{s+t} denotes the density of $N(0,s+t)^{\otimes 3}$. The assertion of the lemma is obtained by the three observations that

$$\lim_{\epsilon\downarrow 0}\sup_{|y|\le 2\epsilon} E\Big[F(X_{-\infty,s}, X_{s,\infty}, Y_{-\infty,t}, Y_{t,\infty})|Y(t)-X(s)=y\Big] =$$
$$E\Big[F(W_1,W_2,W_3,W_4)\Big],$$

$$\frac{1}{((4/3)\pi\epsilon^3)^2}\int_{|y|\le 2\epsilon} dy\text{Leb}\,(B(0,\epsilon)\cap B(y,\epsilon)) = 1, \quad \text{and}$$

$$E[\alpha(A)] = \int\int_A p_{s+t}(0)dsdt. \quad \Box$$

Acknowledgement. The paper [MS] is reported at The 2nd ISAAC "Harmonic Analysis and its Applications" Session. The author thanks Professor Shan–zhen Lu for invitation to the occasion. This work is supported in part by a grant *NSC 1999/2000 2115–M–002–006.*

References

[B] Bedford,T.(1991). *Applications of dynamical systems theory to fractals.* In: Fractal Geometry and Analysis(ed. J. Belair and S. Dubuc) 1-44, Kluwer, Dordrecht.

[BF] Bedford, T. and Fisher, A.T.(1992). *Analogues of the Lebesgue density theorem for fractal sets of reals and integers* . Proc. London Math. Soc. **64** 95-124.

[D] Dynkin, E.B.(1984). *Additive functionals of several time–reversible Markov processes* . J. Functional Anal. **42** 64-101.

[F] Falconer, K.J.(1997). *Techniques in Fractal Geometry.* Wiley New York 1997.

[FX] Falconer, K.J. and Xiao, Y.M.(1995). *Average densities of the image and zero set of stable processes* . Stoch. Processes Appl. **55** 271-283.

[GHR] Geman, D., Horowitz, J. and Rosen, J.(1984). *A local time analysis of intersections of Brownian paths in the plane* . Ann. Probab. **12** 86-107.

[L1] Le Gall, J.F.(1987). *The exact Hausdorff measure of Brownian Multiple points* . In: Seminar on Stochastic Processes 1986, 103-137, Birkhöuser Boston 1987.

[L2] Le Gall, J.F.(1992). *Some properties of planar Brownian motion.* In: Lecture Notes in Math. vol. 1527, Springer Verlag.

[M] Mandelbrot, B.B.(1995). *Measures of fractal lacunarity: Minkowski content and alternatives* . In: Fractal Geometry and Stochastics, 15-42, Birkhöuser Boston 1995.

[M] Mörters, P.(1998). *The average density of the path of planar Brownian motion* . Stoch. Processes Appl. **74** 133-149.

[MS] Mörters, P. and Shieh, N.R.(1999). *Small scale limit theorems for the intersection local times of Brownian motions.* Electronic J. Probab. **4** Paper no.9 Pages 1–23.

[S] Shieh, N.R.(1997). *A growth condition for Brownain intersection points* . In: Trends in Probability and Related Analysis, Proceedings of SAP'96, 265-272, World Scientific Singapore 1997.

[T] Taylor, S.J.(1986). *The measure theory of random fractals* . Math. Proc. Camb. Phil. Soc. **100** 383-406.

13 PROBLEM OF INTEGRAL GEOMETRY ON PARABOLOIDS WITH PERTURBATION

Akbar H. BEGMATOV[1]

Sobolev Institute of Mathematics, 4 Acad. Koptyug prosp.
Novosibirsk 630090, Russia

bah@math.nsc.ru

1991 Mathematics Subject Classification: 45D05, 53C65, 44A12
Key words: Problems of integral geometry, ill–posed problems, uniqueness theorem, analytic representattion of solution.

1. This paper studies a problem of integral geometry in a three-dimensional layer on a family of paraboloids with a perturbation that is an integral with a weight function over the interiors of the paraboloids. Note that the exposition is based on the author's works [1, 2]. Uniqueness questions of a similar problem on the plane were studied in [3]. Various analytic representations for solutions to problems of integral geometry on parabolas (paraboloids) were given in [4, 5]. Other classes of integral geometry problems on the plane and in the three–dimensional space were considered in [6, 7].

Let us introduce notation: $x \in \mathbf{R}^3$, $\xi \in \mathbf{R}^3$; $\bar{x} = (x_1, x_2)$, $\bar{\xi} = (\xi_1, \xi_2)$; $\lambda \in \mathbf{R}^2$;

[1] This work was partly supported by a grant No. 15/99 from the State Committee of Republic of Uzbekistan on Science and Technique.

H.G.W. Begehr et al. (eds.), Proceedings of the Second ISAAC Congress, Vol. 1, 97–103.

$L = \{x \in \mathbf{R}^3 : 0 < x_3 < h, h < \infty\}$; $\bar{L} = \{x \in \mathbf{R}^3 : 0 \le x_3 \le h\}$; $\mathcal{A} = \{x \in \mathbf{R}^3 : |x_k| < a_k\,(k = 1, 2), a_k < \infty; 0 < x_3 < h\}$. Denote by $\{\mathcal{P}(x)\}$ the family of the paraboloids having as vertices the points x and determined by the relations

$$\mathcal{P}(x) = \{\xi : x_3 - \xi_3 = |\bar{x} - \bar{\xi}|^2;\ 0 \le x_3 \le h\}.$$

By $p(x)$ we denote the projection of $\mathcal{P}(x)$ onto the plane $x_3 = 0$; by $P(x)$ we denote the domain bounded by the surface $\mathcal{P}(x)$ and the plane $x_3 = 0$.

Problem 1. Determine a function $u(x)$ given the sums of the integrals of u over the surfaces $\mathcal{P}(x)$ and the integrals of u with known weight function K over the domains $P(x)$

$$\int\limits_{p(x)} u(\xi)\, d\bar{\xi} + \int\limits_{P(x)} K(x, \xi) u(\xi)\, d\xi = f(x). \tag{1}$$

2. We construct a simple representation for a Fourier image of a solution to equation (1) in the case when the second summand from the left side is identical zero. From this formula it follows that the Problem 1 is strongly unstable, i.e. the instability of the solution to equation (1) is of the same type that in the Cauchy problem for Laplace's equation.

Theorem 1. *Assume that $f(x)$ is known for all $x \in \bar{L}$ and the weight function $K(x, \xi) = 0$ for all x from the layer $\bar{L}$. Then a solution to equaton (1) in the class $C_0^2(L)$ is unique and the representation*

$$v(\lambda, x_3) = \frac{1}{2\pi} \frac{\partial^2}{\partial x_3^2} \int\limits_0^{x_3} I_0\Big(|\lambda|\sqrt{x_3 - \xi_3}\Big)\, F(\lambda, \xi_3)\, d\xi_3$$

is valid.

Proof. Let the function K be identical zero. We arrive at the problem of integral geometry

$$\int\limits_{p(x)} u(\xi)\, d\bar{\xi} = f(x). \tag{2}$$

Let us denote: $< \cdot, \cdot >$ is the scalar product in $\mathbf{R^2}$;

$$v(\lambda, x_3) = \frac{1}{2\pi} \int\limits_{\mathbf{R^2}} e^{i<\lambda, \bar{x}>} u(x)\, d\bar{x}; \quad F(\lambda, x_3) = \frac{1}{2\pi} \int\limits_{\mathbf{R^2}} e^{i<\lambda, \bar{x}>} f(x)\, d\bar{x}.$$

Applying the Fourier transform with respect to (x_1, x_2) to equation (2) and using an integral representation for the Bessel function J_0

$$J_0(z) = \frac{1}{\pi}\int_0^{\pi} \cos(z\cos\theta)\, d\theta,$$

we obtain

$$F(\lambda, x_3) = 2\pi \int_0^{x_3} J_0\Big(|\lambda|\sqrt{x_3 - \xi_3}\,\Big)\, v(\lambda, \xi_3)\, d\xi_3. \tag{3}$$

Applying to equation (3) the Volterra operator with the kernel

$$\frac{ch(|\lambda|\sqrt{x_3 - \xi_3}\,)}{\sqrt{x_3 - \xi_3}},$$

we get

$$\int_0^{x_3} \frac{v(\lambda,\tau)}{\sqrt{x_3 - \tau}}\, d\tau = \frac{1}{2\pi}\frac{\partial}{\partial x_3}\int_0^{x_3} \frac{ch(|\lambda|\sqrt{x_3 - \xi_3}\,)}{\sqrt{x_3 - \xi_3}}\, F(\lambda, \xi_3)\, d\xi_3.$$

The solution to this equation is given by

$$v(\lambda, x_3) = \frac{1}{2\pi^2}\frac{\partial}{\partial x_3}\int_0^{x_3} \frac{\partial}{\partial \xi_3}\Big[\int_0^{\xi_3} \frac{ch(|\lambda|\sqrt{\xi_3 - \tau}\,)}{\sqrt{\xi_3 - \tau}}\, F(\lambda,\tau)\, d\tau\Big]\frac{d\xi_3}{\sqrt{x_3 - \xi_3}}. \tag{4}$$

¿From (4) it is easy to see that

$$v(\lambda, x_3) = \frac{1}{2\pi}\frac{\partial^2}{\partial x_3^2}\int_0^{x_3} I_0\Big(|\lambda|\sqrt{x_3 - \xi_3}\,\Big)\, F(\lambda, \xi_3)\, d\xi_3,$$

where I_0 is the Bessel function of imaginary argument.

3. Here we give the main result of this work: uniqueness theorem for considered problem of integral geometry with perturbation under rather general assumptions on the weihgt function.

Theorem 2. *Assume that $f(x)$ is known for all $x \in \bar{L}$; the weight function K has all continuous derivatives up to the second order inclusively and vanishes together with its derivatives on the surface of the paraboloid $\mathcal{P}(x)$. Then a solution to equation* (1) *in the class $C_0^3(\mathcal{A})$ is unique.*

Proof. Let the right–hand side of (1) be identical zero. Then $f_0(x) = -f_1(x)$, where f_0 and f_1 are the first and the second summand in the left–hand side of (1) respectively.

We consider the function

$$q(x,y) = \frac{\partial}{\partial y} \int\limits_0^{x_3} \int\limits_0^{2\pi} u(x_1 + \sqrt{1-y}\,\rho\cos\varphi,\ x_2 + \sqrt{1-y}\,\rho\sin\varphi, \xi_3)\, d\varphi\, d\xi_3,$$

where $\rho = \sqrt{x_3 - \xi_3}$; $y \in [0;1]$.

It is easy to see that the function q satisfies an equation

$$\frac{\partial}{\partial x_3} q(x,y) = -\frac{1}{4} \int\limits_0^y \Delta_{\bar{x}}\, q(x,\eta)\, d\eta + q_0(x),$$

where

$$\Delta_{\bar{x}} = \frac{\partial^2}{\partial x_1^2} + \frac{\partial^2}{\partial x_2^2}, \quad q_0(x) = -\frac{1}{4} \Delta_{\bar{x}}\, f_0(x).$$

Introduce the notation:

$$Q(\lambda, x_3, y) = \frac{1}{2\pi} \int\limits_{\mathbf{R^2}} e^{i<\lambda,\bar{x}>} q(x,y)\, d\bar{x}; Q_0(\lambda, x_3) = \frac{1}{2\pi} \int\limits_{\mathbf{R^2}} e^{i<\lambda,\bar{x}>} q_0(x)\, d\bar{x};$$

$$q_1(\lambda, y) = \begin{cases} Q(\lambda, h, y), & \text{if } 0 \le y \le 1, \\ 0, & \text{if } y > 1. \end{cases}$$

Lemma 1. *For any fixed λ the function Q can be extended from the domain $\Omega_0 = \{(x_3,y) : 0 \le x_3 \le h, 0 \le y \le 1\}$ to the whole half–strip $\Omega = \{(x_3,y) : 0 \le x_3 \le h, 0 \le y < \infty\}$ as a solution to the Cauchy problem*

$$\frac{\partial}{\partial x_3} Q(\lambda, x_3, y) = \frac{|\lambda|^2}{4} \int\limits_0^y Q(\lambda, x_3, \eta)\, d\eta + Q_0(\lambda, x_3), \tag{5}$$

$$Q(\lambda, h, y) = q_1(\lambda, y). \tag{6}$$

Note that the Cauchy problem (6) for the evolutionary equation (5) is uniformly correct (see [8]).

Take $b > 0$. Introduce the functions

$$G(\lambda, x_3, y; b) = e^{b(x_3-h)^2 - bh^2} Q(\lambda, x_3, y), \tag{7}$$

$$g_0(x;b) = e^{b(x_3-h)^2-bh^2}q_0(x), \tag{8}$$

and

$$g(x,y;b) = \frac{1}{2\pi}\int\limits_{\mathbf{R}^2} e^{-i<\bar{x},\lambda>}G(\lambda,x_3,y;b)\,d\lambda. \tag{9}$$

Lemma 2. *The function G satisfies the equation*

$$\frac{\partial}{\partial x_3}G(\lambda,x_3,y;b) =$$

$$\frac{|\lambda|^2}{4}\int\limits_0^y G(\lambda,x_3,\tau;b)\,d\tau + 2b(x_3-h)\,G(\lambda,x_3,y;b) + G_0(\lambda,x_3;b).$$

For any fixed λ, x_3 and b the function G can be considered as an element of the Hilbert space $\mathcal{H}$ of functions defined on the half–axis $0 \le y < \infty$ with the inner product

$$(h_1,h_2) = \int\limits_0^\infty e^{-ay}h_1(y)h_2(y)\,dy, \qquad a>0.$$

Let I be an integral operator in $\mathcal{H}$: $Ip = \int\limits_0^y p(\tau)\,d\tau$. The adjoint operator is defined by the equality: $I^*g = e^{ay}\int\limits_y^\infty e^{-a\tau}g(\tau)\,d\tau$.

Lemma 3. *The following equality is valid:*

$$(I^*I - II^*)\,G = \frac{1}{a}\int\limits_0^\infty e^{-a\tau}\,G(\tau)\,d\tau.$$

These lemmas imply the main inequality

$$4b\int\limits_0^h\int\limits_0^\infty\int\limits_{\mathbf{R}^2} e^{-ay}\,g^2(x,y;b)\,d\bar{x}\,dy\,dx_3 \le$$

$$\frac{1}{4}\int\limits_0^\infty\int\limits_{\mathbf{R}^2} e^{-ay}(I+I^*)\Delta_{\bar{x}}g(\bar{x},h,y;b)\cdot g(\bar{x},h,y;b)\,d\bar{x}\,dy + \frac{1}{a}\int\limits_0^h\int\limits_{\mathbf{R}^2} g_0^2(x;b)\,d\bar{x}\,dx_3. \tag{10}$$

Assume that $0 \le y \le 1$. It is obvious that

$$g(\bar{x}, h, y; b) = e^{-bh^2} q(\bar{x}, h, y; b). \tag{11}$$

Using (7)–(9) and (11), we obtain

$$\frac{1}{4}\int_0^\infty \int_{\mathbf{R}^2} e^{-ay}(I + I^*)\, \Delta_{\bar{x}} g(\bar{x}, h, y; b) \cdot g(\bar{x}, h, y; b)\, d\bar{x}\, dy \le C_1 e^{-2bh^2}, \tag{12}$$

$$\int_0^h \int_{\mathbf{R}^2} g_0^2(x; b)\, d\bar{x}\, dx_3 \le C_2 \int_0^1 \int_0^h \int_{\mathbf{R}^2} e^{-ay}\, g^2(x, y; b)\, d\bar{x}\, dx_3\, dy, \tag{13}$$

where C_1 and C_2 are some constants.

Assume that the function u is not identically zero. Hence, the function g is not identically zero, too, and there exists a number ε ($0 < \varepsilon < 1$) such that

$$\int_0^\infty \int_0^h \int_{\mathbf{R}^2} e^{-ay}\, g^2(x, y; b)\, d\bar{x}\, dx_3\, dy \ge C_3 e^{-2bh^2 \cdot \varepsilon}, \tag{14}$$

where C_3 is some constant.

It follows from inequalities (10) and (12)–(14) that

$$e^{-2bh^2 \cdot \delta} \ge B,$$

where $\delta = 1 - \varepsilon$, $0 < \delta < 1$; $B = C_4 b - C_5$, C_4 and C_5 are some constants.

It is easy to see that this iqnequality fails for sufficiently large values of the parameter b, which proves the assertion of the theorem.

4. Let $\{\Gamma(x)\}$ be a family of surfaces that smoothly fill the layer L and are parametrized uniquely by means of their vertices $x \in L$.

Problem 2. Determine a function $u(x)$ given by integrals of $u(x)$ over the surfaces $\Gamma(x)$ with a weight function $g(x, \xi)$

$$\int_{\Gamma(x)} g(x, \xi) u(\xi)\, ds = f(x), \tag{15}$$

for all x from the layer L.

Problem of solving equation (15) is an integral geometry problem of Volterra type. We show that the uniqueness question for such problems of integral geometry reduces to the studying uniqueness of solutions of the problem of integral geometry on a family of paraboloids with perturbation considered above.

References

[1] Begmatov Akbar H., *On a problem of integral geometry with perturbation in the three–dimensional space,* Doklady Akademii Nauk (to appear).

[2] Begmatov Akbar H., *Integral geometry problem with perturbation in the three–dimensional space,* Sibirsk. Mat. Zh. (to appear).

[3] Lavrent′ev M. M., *Integral geometry problems with perturbation on the plane,* Sibirsk. Mat. Zh. **37**, No. 4, 1996, 851–857; English transl. in Siberian Math. J. **37**, No. 4, 1996, 747–752.

[4] Bukhgeĭm A. L., *Some problems of integral geometry,* Sibirsk. Mat. Zh. **13**, No. 1, 1972, 34–42; English transl. in Siberian Math. J. **13**, 1972.

[5] Lavrent′ev M. M. and Savel′ev L. Ya., *Linear operators and ill–posed problems,* "Nauka", Moscow, 1991; English transl., Consultants Bureau, New York, 1995.

[6] Begmatov Akbar H., *On a class of problems of integral geometry in the plane,* Doklady Akademii Nauk, **331**, No. 3, 1993, 261–262; English transl. in Doklady Mathematics **48**, No. 1, 1994, 56–58.

[7] Begmatov Akbar H., *Reducing problems of integral geometry in the three–dimensional space to perturbed polysingular integral equations,* Doklady Akademii Nauk, **360**, No. 5, 1998, 583–585; English transl. in Doklady Mathematics **57**, No. 3, 1998, 424–426.

[8] Krein S. G., *Linear differential equations in Banach space,* "Nauka", Moscow, 1967; English transl., Amer. Math. Soc., Providence, R. I., 1971.

14 THE CONNECTION BETWEEN DISCRETE AND CONTINUOUS REALISATIONS OF LEAST SQUARES METHOD

Y. V. CHEBRAKOV and V. V. SHMAGIN

St.-Petersburg Technical University, Apt.11, Bldg.3, Nevsky Ave.
191186, St.-Petersburg, Russia

chebra@phdeg.hop.stu.neva.ru

1991 Mathematics Subject Classification: 65R20, 62J02, 65D10
Key words: Numerical solutions of integral equations, fitting techniques, continuous approach

Abstract

This paper is devoted to investigating numerical solutions of some integral equations with taking into account theoretical statements of discrete and operational approaches to quantitative processing numerical information. We demonstrate, firstly, if fitting techniques are used in the frame of continuous approach, then they are effective for decreasing computation errors in numerical solutions of integral equations, and, secondly, for some fitting problems, discrete solutions may be obtained from continuous ones.

H.G.W. Begehr et al. (eds.), Proceedings of the Second ISAAC Congress, Vol. 1, 105–112.

I Introduction

It is well-known [2, 8, 11], that a set of methods for solving integral and differential equations requires an application of fitting techniques. According to [7], in the beginning of 18-th center researches had 4 different methods to solve fitting problems: the Mayer–Laplace method of averages, the Boscovich–Laplace method of least absolute deviations, Laplace method of minimizing the largest absolute residual, and Legendre method of minimizing the sum of squared residuals. The only single way of choosing between these methods was to compare results of estimates and residuals.

As we have seen from reviews [4, 5] of theoretical and computation difficulties, existing in modern data analysis, the actual position of affairs with fitting techniques does not change practically up today: although we have appreciable quantity of different data analysis algorithms, we have no an effective way of choosing the best of them. The most impressive statement of the subject is contained in Y. P. Adler introduction [1]: "The regression analysis degraded ingloriously to the level of a heuristic method, in which the residual analysis and common sense of interpreter play the leading role". From our point of view [3–5], to bring modern fitting techniques out of "a degraded condition", it is necessary to replace the classical (discrete) approach to quantitative processing numerical information by more progressive operational (continuous) one.

The main goals of this paper are to attract the attention of the researches, occupied in solving integral and differential equations, to (a) problems of certification of using computational methods; (b) connections, existing between discrete and continuous approaches to quantitative processing numerical information; (c) new techniques of solving integral and differential equations, which arise in the frame of continuous approach to quantitative processing data.

II Main Theoretical Statements of the Classical (Discrete) Approach to Data Analysis

As well-known [3–5, 6, 10], for found experimental dependence $\{y_n, \mathbf{X}_n\}$ $(n = 1, 2, \ldots, N)$ and given approximative function $F(\mathbf{A}, \mathbf{X})$, in the measurement (natural science) experiments the main problems of classical regression analysis theory are finding estimates of $\mathbf{A}'$ and y' and variances of $\delta\mathbf{A}'$ and $\delta(y - y')$, where $\mathbf{A}'$ is an estimate of vector parameter $\mathbf{A}$ of the function $F(\mathbf{A}, \mathbf{X})$ and $\{y'_n\} = \{F(\mathbf{A}', \mathbf{X}_n)\}$. In particular, if $F(\mathbf{A},\mathbf{X}) = \sum_{l=1}^{L} a_l h_l$ {$F(\mathbf{A},\mathbf{X})$ is a linear model}, where $h_l(\mathbf{X})$ are some functions on $\mathbf{X}$, then in classical regression analysis theory standard solution of discussed problems has form

$$\mathbf{A}' = (\mathbf{H}^T\mathbf{H})^{-1}\mathbf{H}^T\mathbf{Y}, (\delta\mathbf{A}')^2 = s/(N-L)\mathrm{diag}(\mathbf{H}^T\mathbf{H})^{-1},$$
$$\delta_p(y_n - y'_n) = \pm t_p s\sqrt{1 + \mathbf{H}_n^T(\mathbf{H}^T\mathbf{H})^{-1}\mathbf{H}_n}, \tag{1}$$

where $\mathbf{H}$ is a matrix $L \times N$ in size with n-th row $\mathbf{H}_n = (h_1(\mathbf{X}_n), h_2(\mathbf{X}_n), \ldots, h_L(\mathbf{X}_n))$; $\mathbf{H}^T$ is the transposed matrix $\mathbf{H}$; $\mathbf{Y} = \{y_n\}$; $s = \sum_{n=1}^{N}(y_n - y'_n)^2/(N-L)$; the value of t_p is determined by t-Student distribution table and generally depends on the assigned value of the significance level of p and the value of N–L (the number of freedom degree); at the assigned value of the significance level of p, the notation of $\delta_p(y - y')$ means confidence interval for possible deviations of experimental values of y from computed values $y' = F(\mathbf{A}',\mathbf{X})$.

As it is shown in [4, 5] {see also Section IV}, the solution (1) has a set of defects. The main of these defects is exhibited in a fact, that, although for each contaminated data array a family of analytical approximative functions exists, the only single approximative function may be found by (1). This defect is successfully overcome in the frame of continuous approach to quantitative data processing, which main statements are considered in the next section.

III Main Theoretical Statements of the Operational (Continuous) Approach to Data Analysis

For any measurement (natural science) experiment, according to [3–5], the data analysis model has form

$$y_n = g(F(\mathbf{A}',\mathbf{X}_n) + e_n), \tag{2}$$

where y_n is n-th value of dependent variable ($n = 1, 2, \ldots, N$); $F(\mathbf{A},\mathbf{X})$ is an approximative function; e_n is n-th value of random variable; g is a function, allowing to describe the way of dependent variable measurement. Thus, the continuous approach to quantitative data processing differs from the classical one, discussed in Section II, in the presence of the measurement function g in the data analysis model. Function g is necessary to show that n-th value of dependent variable is measured within an error ϵ_n. For instance, if $\epsilon_n = \epsilon$ for any n, then the measurement function g has the form

$$g_\epsilon(y) = \epsilon[y/\epsilon] + \epsilon \ \text{ if } \ |y - \epsilon[y/\epsilon]| \geq \epsilon/2, \ \text{ else } \ g_\epsilon(y) = \epsilon[y/\epsilon], \tag{3}$$

where record $[b]$ means the integer part from b.

The main computative problems of the continuous approach to quantitative data processing are finding (a) the value of vector parameter $\mathbf{A}$ of the fitting function $F(\mathbf{A}, \mathbf{X})$ within an error $\delta\mathbf{A}$ on the given experimental dependence $\{y_n, \mathbf{X}_n\}$; (b) the general analytical solution, which presents itself a set of equivalent analytical functions $g_\epsilon(F(P_{m_i}(\mathbf{C}_i, \xi), \mathbf{X}))$, where $P_{m_i}(\mathbf{C}_i, \xi)$ are some multinomials of degree m_i with a variable ξ and vector parameter $\mathbf{C}_i$; (c) the value of variance $\delta\mathbf{A}'$.

The methods for solving problems (a)–(c) one may find in [3–5]. The connection, existing between the discrete solution (1) and a set of equivalent analytical functions $g_\epsilon(F(P_{m_i}(\mathbf{C}_i, \xi), \mathbf{X}))$ is pointed in Section IV.

IV Connections Between Discrete and Continuous Approaches to Data Analysis

To reveal connections between discrete and continuous approaches to data analysis, we discuss below the results of regression analyzing the experimental data array given in table. This array is obtained by Russian chemist D. I. Mendeleev in 1881, when he investigated the solvability (y, relative units) of sodium nitrate ($NaNO_3$) on the water temperature (x, °C).

Table. D. I. Mendeleev data array.

n	x_n	y_n	$y_n - y'_n$
1	0	66.7	−0.80
2	4	71.0	0.02
3	10	76.3	0.10
4	15	80.6	0.05
5	21	85.7	−0.07
6	29	92.9	0.17
7	36	99.4	0.58
8	51	113.6	1.73
9	68	125.1	−1.56

By using classical formulae (1) for analyzing the data array $\{y_n, x_n\}$, Y. V. Linnik [9] states that (a) these data, as it was noted by D. I. Mendeleev, are well-fitted by linear model $y' = 67.5 + 0.871x$ $\{\delta\mathbf{A}' = (0.5; 0.2)\}$, although the correspondence between experimental and computed on linear model values of y is slightly getting worse at the beginning and end of investigated temperature region (see the values $\{y_n - y'_n\}$ in table); (b) for the discussed data array the confidence interval of $\delta_p(y - y')$ should be computed at the significance level of $p = 0.9$:

$$\delta_{0.9}(y_n - y'_n) = \pm 0.593\sqrt{1 + (x_n - 26)^2/451.1}\ . \tag{4}$$

As we noted early [4, 5], since the values of $|y_n - y'_n|$ are greater than the values, computed from (4) at $n = 8$ and 9, the standard way (1), offered by Y. V. Linnik for determining the confidence interval $\delta(y - y')$, is out of character with the discussed experimental data array. It follows from results presented in table, if one assumes that $\delta(y - y') \geq \max |y_n - y'_n| = 1.73$, then the broken connections of the confidence interval $\delta(y - y')$ with D. I. Mendeleev data array will be pieced up. But values of $\delta\mathbf{A}'$, calculated from (1), disagree with the values $\delta(y - y') \geq 1.73$, and, consequently, standard values of $\delta\mathbf{A}'$ are out of character with D. I. Mendeleev data array also.

One can easily examine, if values of y' are determined from the formula

$$y'(\xi, x) = g_\epsilon(68.12 + 0.02\xi + (0.85652 - 0.00046\xi)x), \tag{5}$$

found in the frame of the continuous approach to data analysis {see Section III}, then for each n the difference $|y_n - y'_n|$ will keep within the limit of the chosen above value for the confidence interval $\delta(y - y')$, where $\epsilon = 1.9$; $0 \leq \xi \leq 35$; the function g is determined by (3). If one puts extreme values of ξ=0 and 35 in (5), one will be able to determine the exact limit of the variation for the linear model parameters a_0 and a_1: $a_0 = 67.77 \pm 0.35$ and $a_1 = 0.865 \pm 0.008$. By comparing this (continuous) solution with the standard LS-solution, one can conclude that LS-estimations of parameter a_0 and a_1 $\{\mathbf{A}' = (67.5 \pm 0.5; 0.87 \pm 0.2)\}$ are pretty near equal of the mean values of these parameters in the general analytical solution (5). However, values of variances $\delta a'_0$ and $\delta a'_1$, computed by standard method, disagree with exact values determined by (5).

V Investigating Numerical Solutions of Some Integral Equations in the Frame of Discrete and Continuous Approaches to Data Analysis

Effectiveness of numerical solutions of integral equations by quadrature is demonstrated in [11] by the example of solving the following

Volterra integral equation of the first kind

$$\int_0^x (2 + x^2 - s^2) y(s) ds = f(x) \tag{6}$$

with $f(x) = x^2$. In particular, it is mentioned, if $0 \le x \le 3.5$ and step $h = 0.05$, then numerical solutions $\{y_n\}$ of the equation (6) coincide with the analytical solution $y_{\text{a}} = s\ \exp(-s^2/2)$ in the first three significant digits. It should be noted that the dependence $\{\Delta y, x\} = \{y - y_{\text{a}}, x\}$ for the considered case is a fair-curve. From point of view of modern data analysis theory [4, 6, 10], it means that the found decision is inadequate and so additional investigations are required to examine its properties.

1. We note that, in real-life natural science investigations, values of the function $f(x) = x^2$, located in the right part of (6), are given often within an error ϵ_n and, consequently, values of f_n are determined by the correlation (2) with the measurement function g (3). In particular, assigning $\epsilon = 0.0001$ in (3) and $h = \pi/63$, we found that the behaviour of the dependence $\{\Delta y, x\}$ has changed significantly for this case: max $|\Delta y| = 1.1$, where max $|\Delta y|$ is maximum absolute deviation value of numerical solutions $\{y_n\}$ of the equation (6) from the analytical solution y_{a}. Thus, we conclude, that effectiveness of numerical solution of (6) by quadrature is achieved in [11], because the function $f(x)$ values are given with infinite accuracy {within the error $\epsilon << 0.0001$}.

2. Let us change slightly the form of integral equation (4):

$$\int_1^x (2 + x^2 - s^2) y(s) ds = f(x) \tag{7}$$

and assume that $f(x) = g_\epsilon(x^2 - 1)$. The analytical solution of the equation (7) is $s\ \exp(1/2 - s^2/2)$. If $1 \le x \le 3.5$, $\epsilon = 0.001$ and 0.0001, we obtain that in numerical solutions of the equation (7), values of max $|\Delta y|$ increase with decreasing value of ϵ: max $|\Delta y|$=0.41 and 0.90 respectively. This fact means only that the value $\epsilon = 0.001$ is insufficient for obtaining the robust solutions of the equation (7).

VI A Set of Methods for Decreasing Computation Errors in Numerical Solutions of Integral Equations

Let us clear up a question whether any methods exist by which, one may decrease computation errors in numerical solutions obtained in Section V.

1. Since the (experimental) data arrays $\{g_\epsilon(x_n^2), x_n)\}$ and $\{g_\epsilon(x_n^2 - 1), x_n)\}$, defined in Section V, are contaminated ones, it is evident that fitting technique is able to smooth these dependences. Indeed, if these dependences are replaced by ones, obtained from multinomials of the second degree $Q_2(\mathbf{B}, x) = b_0 + b_1 x + b_2 x^2$, in which the value of vector parameter $\mathbf{B}'$ is determined by least square method {see (1)}, then the magnitudes of max $|\Delta y|$ are significantly decreased in numerical solutions of equations (6) and (7): max $|\Delta y| = 0.12$, 0.24 and 0.70 respectively for cases, considered in points 1 and 2 of Section V.

2. To decrease the contamination level of numerical solutions of integral equations (6) and (7) still more than in point 1 of this section, we remind {see Section III and IV} that the general analytical solution of the problem on fitting arrays $\{g_\epsilon(x_n^2), x_n)\}$ and $\{g_\epsilon(x_n^2 - 1), x_n)\}$ by multinomial $Q_2(\mathbf{B}, x)$ has form of a set of equivalent analytical functions:

$$F_1(\xi_1, x) = g_{\epsilon_1}\{-0.62 \cdot 10^{-4} + 0.2 \cdot 10^{-5}\xi_1 +$$

$$(0.783 \cdot 10^{-4} - 0.23 \cdot 10^{-5}\xi_1)x + (0.99998085 + 0.54 \cdot 10^{-6}\xi_1)x^2\},$$

$$F_2(\xi_2, x) = g_{\epsilon_1}\{-1.0023 + 0.1 \cdot 10^{-3}\xi_2 +$$

$$(0.002033 - 0.93 \cdot 10^{-4}\xi_2)x + (0.999569 + 0.20 \cdot 10^{-4}\xi_2)x^2\},$$

$$F_3(\xi_3, x) = g_{\epsilon_2}\{-1.00022 + 0.1 \cdot 10^{-5}\xi_3 +$$

$$(0.1981 \cdot 10^{-3} - 0.93 \cdot 10^{-5}\xi_3)x + +(0.9999584 + 0.20 \cdot 10^{-5}\xi_3)x^2\}, \quad (8)$$

where $F_1(\xi_1, x)$, $F_2(\xi_2, x)$ and $F_3(\xi_3, x)$ are fitting functions respectively for arrays $\{g_{\epsilon_1}(x_n^2), x_n)\}$, $\{g_{\epsilon_1}(x_n^2 - 1), x_n)\}$ and $\{g_{\epsilon_2}(x_n^2 - 1), x_n)\}$; the function g is determined by (3); $0 \le \xi_1 \le 49$; $0 \le \xi_2 \le 36$; $0 \le \xi_3 \le 36$; $\epsilon_1 = 0.001$ and $\epsilon_2 = 0.0001$.

One may decrease computation errors in numerical solutions of integral equations (6) and (7) by formulae (8), if dependences $\{g_\epsilon(x_n^2), x_n)\}$ and $\{g_\epsilon(x_n^2 - 1), x_n)\}$ are replaced not by ones, obtained from multinomials $Q_2(\mathbf{B}, x)$, but by ones, obtained from the set of equivalent analytical functions (8): max $|\Delta y| = 0.00046$ ($\xi_1 = 31$), 0.0055 ($\xi_2 = 15$) and 0.58 ($\xi_3 = 0$) respectively for cases, considered in points 1 and 2 of Section V and point 1 of this section.

Analyzing the presented results, we conclude that if fitting techniques are used with taking into account the operational (continuous) approach to quantitative processing numerical information, then they are effective for decreasing computation errors in numerical solutions of integral equations.

References

[1] Adler Y. P., *The Introduction to Russian edition of the book* [10]. Finansy i statistika, Moscow, 1982.

[2] Barker C. T. H.,*The numerical treatment of integral equations.* Clarendon Press, Oxford, 1977.

[3] Chebrakov Y. V. and Shmagin V. V., *Correct solutions of fit problems in different experimental situations.* In: Advanced Mathematical Tools in Metrology III. World Scientific, Singapore, 1997, 246–249.

[4] —— *Regression data analysis for physicists and chemists.* S.-Petersburg, S.-Petersburg State Univ. Press, 1998.

[5] —— *Computative paradoxes in modern data analysis.* Smarandache Notions J. 1999, **10** (1–2–3), 61–80.

[6] Draper K. R. and Smith H., *Applied regression analysis.* Wiley, New York, 1981.

[7] Hald A., *A history of mathematical statistics: From* 1750 *to* 1930. Wiley, New York, 1998.

[8] Langtangen H. P., *Computational partial differential equations: Numerical methods and diffpack programming.* Springer, Berlin–Tokyo, 1999.

[9] Linnik Y. V., *Least squares method and foundations of mathematical-statistical theory of observations processing* (In Rissian). Moscow, Publ. House of phys.-math. lit., 1958.

[10] Mosteller F. and Tukey J. W., *Data analysis and regression: A second course in statistics.* Reading, Massachusetss, 1978.

[11] Verlan A. F. and Sizikov V. S., *Integral equations: Methods, algorithms, programs* (In Russian). Naukova Dumka, Kiev, 1986.

15 AN EIGENVALUE PROBLEM FOR ANALYTIC FUNCTIONS

Dao Qing DAI

Department of Mathematics, Zhongshan University

Guangzhou 510275, China

stsddq@zsu.edu.cn

Ming Sheng LIU

Department of Mathematics, South China Normal University

Guangzhou 510631, China

liumsh@scnu.edu.cn

1991 Mathematics Subject Classification: 35J35, 30E25, 35J45
Key words: Eigenvalue problem, Analytic functions

Abstract: An eigenvalue problem for analytic functions with Riemann-Hilbert-Poincaré boundary conditions is studied. It is shown that the spectrum are composed of discrete points and continuous curves in the complex plane.
Key Words: Eigenvalue problem, Analytic functions.
AMS Subject Classification: 35J35, 30E25, 35J45.

1. Introduction

Let Ω be the unit disk $\{z : |z| < 1\}$ in the complex plane $\mathbb{C}$. In this paper, we consider the following eigenvalue problem: Find a function $u(z)$, which is analytic in the unit disk Ω and a constant λ such that

H.G.W. Begehr et al. (eds.), Proceedings of the Second ISAAC Congress, Vol. 1, 113–118.

they satisfy the boundary condition

$$\mathrm{Re}\,[\frac{\partial u}{\partial n} + \lambda\alpha(z)u] = 0, z \in \partial\Omega, \tag{1}$$

where $\frac{\partial u}{\partial n}$ denotes the outer normal derivative of u. The objective is to find out those λ such that the equation (1) has non zero solutions.

We consider only the case that

$$\alpha(z) = a\bar{z} + b, \tag{2}$$

in this note, where a and b are complex constants, $a \neq 0$. The results about general $\alpha(z)$ will be reported in a forthcoming publication.

Boundary value problems for analytic functions, e.g. Riemann problem and Riemann-Hilbert problem[1,3,4], have been solved successfully. It seems that few work have been done for the problem (1). In some cases, one usually assumes that λ is sufficiently small, e.g. for second order equation in [2], see also [5]. We are concerned with the eigenvalue λ of the problem (1).

2. The Main Results

By simple calculations, we have for analytic function u

$$\frac{\partial u}{\partial n} = z\frac{du}{dz} \text{ on } \partial\Omega. \tag{3}$$

From (2) and (3), we rewrite (1) as

$$\mathrm{Re}[zu' + \lambda(\frac{a}{z} + b)u] = 0, z \in \partial\Omega. \tag{4}$$

Since the function u is analytic in Ω, the function $zu' + \lambda(a/z + b)u$ is analytic in $\Omega\backslash\{0\}$ and has a pole of order at most one at 0. So it follows that $u(z)$ satisfies (4) if and only if it satisfies [3,4]

$$zu' + \lambda(\frac{a}{z} + b)u = \frac{c}{z} - \bar{c}z + ia_0, z \in \Omega, \tag{5}$$

where c is an arbitrary complex number and a_0 is an arbitrary real number.

Let

$$u(z) = \sum_{k=0}^{\infty} u_k z^k, z \in \Omega, \tag{6}$$

with $\sum_{k=0}^{\infty} |u_k| < \infty$. Then from (5), we get

$$\sum_{k=1}^{\infty} ku_k z^k + \lambda a \sum_{k=0}^{\infty} u_k z^{k-1} + \lambda b \sum_{k=0}^{\infty} u_k z^k = \frac{c}{z} - \bar{c}z + ia_0,$$

i.e.

$$\sum_{k=2}^{\infty}(k-1+b)u_{k-1}z^k + \lambda a\sum_{k=0}^{\infty}u_k z^k + \lambda b u_0 z = c + ia_0 z - \overline{c}z^2, \tag{7}$$

By comparing the coefficients of z^k on both sides of (7), we get the system of algebraic equations:

$$\lambda a u_0 = c, \tag{8}$$

$$\lambda a u_1 + \lambda b u_0 = ia_0, \tag{9}$$

$$\lambda a u_2 + (1+\lambda b)u_1 = -\overline{c}, \tag{10}$$

$$\lambda a u_{k+1} + (k+\lambda b)u_k = 0, k \geq 2. \tag{11}$$

I. We first assume that $\lambda = 0$

From (8)–(11), we get

$$c = a_0 = 0,$$
$$u_k = 0, \text{ for all } k \geq 1.$$

Hence we have

Theorem 1 *$\lambda = 0$ is an eigenvalue of the equation (1), the corresponding linearly independent eigenfunctions are* 1 *and* i.

II. Secondly, we assume that

$$\lambda b + k^* = 0 \text{ for some } k^* \geq 2,$$

then $b \neq 0$ and

$$\lambda = -\frac{k^*}{b} \neq 0.$$

From (11), we get

$$u_{k^*+1} = 0$$

and

$$u_n = 0, \text{ for any } n \geq k^* + 1.$$

From (8)–(11), we get

$$u_0 = \frac{c}{\lambda a}, \tag{12}$$

$$u_1 = \frac{ia_0}{\lambda a} - \frac{bc}{\lambda a^2}, \tag{13}$$

$$u_2 = -\frac{\overline{c}}{\lambda a} - \frac{1+\lambda b}{\lambda a}\left(\frac{ia_0}{\lambda a} - \frac{bc}{\lambda a^2}\right), \tag{14}$$

$$u_{k+1} = -\frac{k+\lambda b}{\lambda a} u_k, \text{ for all } 2 \le k \le k^* - 1. \tag{15}$$

We have

Theorem 2 *If $b \neq 0$, then for any integer $k \geq 2$, $\lambda = -k/b$ is the eigenvalue of the equation (1). The eigenfunction is*

$$u(z) = \sum_{k=0}^{k^*} u_k z^k,$$

which is a polynomial of degree k^* and depends on 3 real constants linearly.

III. Finally, we assume that

$$\lambda \neq 0 \text{ and } k + \lambda b \neq 0 \text{ for all } k \geq 2. \tag{16}$$

Lemma 1 *For $k \geq 2$, we must have*

$$u_k = 0. \tag{17}$$

Proof If $u_2 \neq 0$, for any integer $n \geq 0$, we get from (11) that

$$\begin{aligned} u_{n+1} &= \left(-\frac{n+\lambda b}{\lambda a}\right) u_n \\ &= \cdots = \left(-\frac{n+\lambda b}{\lambda a}\right) \cdots \left(-\frac{2+\lambda b}{\lambda a}\right) u_2 \\ &\to \infty \text{ as } n \to +\infty, \end{aligned}$$

which contradicts to $\sum_{k=0}^{\infty} |u_k| < \infty$. Therefore

$$u_2 = 0.$$

Hence from (11), we have

$$u_k = 0, \text{ for all } k \geq 2.$$

which proves (17).

Lemma 2 *If $1 + \lambda b = 0$, then*

$$c = 0, \text{ and } u_0 = 0,$$

$$u_k = 0 \text{ for all } k \geq 2,$$

$$u_1 = -\tfrac{i a_0 b}{a}.$$

Proof Since $1 + \lambda b = 0$, we get from (14) that

$$u_2 = -\frac{\overline{c}}{\lambda a},$$

so it follows from Lemma 1 that

$$c = 0.$$

Hence lemma 2 follows from (12)–(14) and Lemma 1.

Theorem 3 *If $b \neq 0$, $\lambda = -1/b$ is an eigenvalue of the problem (1) and there is only one eigenfunction*

$$u(z) = -\frac{ib}{a} z.$$

From (14) and (17), we get

$$\lambda(a^2\overline{c} - b^2 c + i a_0 ab) = bc - i a_0 a,$$

If $a^2\overline{c} - b^2 c + i a_0 ab = 0$ we must have

$$bc - i a_0 a = 0.$$

Hence $a^2\overline{c} = 0$ and $c = 0, a_0 = 0$. The equation (1) has only the trivial solution.

We therefore get

$$\lambda = \frac{bc - i a_0 a}{a^2\overline{c} - b^2 c + i a_0 ab}. \tag{18}$$

Remark: It is interesting to find out that λ depends continuously(but not linearly) on c and a_0. Its path in the complex plane forms continuous curve.

We have

Theorem 4 *Let λ be defined by (18) and the arbitrary real constant a_0 and the arbitrary complex constant c be such that*

$$\lambda \neq 0 \ \textit{and} \ k + \lambda b \neq 0, k \geq 1.$$

Then λ is an eigenvalue of the equation (1) and its eigenfunction is

$$u(z) = (a^2\overline{c} - b^2 c + i a_0 ab)\left(\frac{c}{a(bc - i a_0 a)} - \frac{1}{a} z\right).$$

Acknowledgments The first-named author is partly supported by the Alexander von Humboldt Stiftung and the National Science Foundation of China.

References

[1] Begehr H., *Complex analytic methods for partial differential equations,* An introductory text. World Scientific, Singapore etc., 1994.

[2] Begehr H. and Wen G. C., *Oblique derivative problems for elliptic systems of second order equations in infinite domains,* Zeitschrift für Analysis und ihre Aawendungen, 18 (1999), No.1, 193–204.

[3] Gakhov F. D., *Boundary value problems,* Nauka, Moscow, 1977.

[4] Lu J. K., *Boundary value problems for analytic functions,* World Scientific, Singapore, 1993.

[5] Wolfersdorf L. von, *On the theory of nonlinear generalized Poincaré problems for harmonic functions,* In: Continum Mechanics and Related Problems of Analysis(eds.: M. Balavadze et al.). Tbilisi: Metsniereba 1993, 330–337.

16 ON QUADRATURE FORMULAE OF HYPERSINGULAR INTEGRALS[1]

Jin Yuan DU

Ji Cheng Hu

Department of Mathematics, Wuhan University

Wuhan 430072, P. R. China

jydu@whu.edu.cn

1991 Mathematics Subject Classification: 35J35, 30E25, 35J45
Key words: Eigenvalue problem, Analytic functions

Abstract. Some quadrature formulae of hypersingular integrals are established and their estimate of remainder and convergence are also given.

§1 Introduction

In [1] H. R. Kutt presented a numerical method for computing singular integrals

$$\mathbf{K}_{m+1}(f,t) = \int_a^b \frac{\tau - t}{|\tau - t|^{m+2}} f(\tau) d\tau, \quad a < t < b, \ m = 0, 1, \cdots. \tag{1.1}$$

[1] Supported by the NNSF and the SF of SME of China.

H.G.W. Begehr et al. (eds.), Proceedings of the Second ISAAC Congress, Vol. 1, 119–127.

In 1981 [2], D. F. Paget considered the quadrature formulae for the singular integrals

$$\mathbf{I}_2(f,t) = \int_a^b \frac{w(\tau)f(\tau)}{(\tau-t)^2}d\tau, \quad a<t<b. \tag{1.2}$$

In [3,4] Du discussed the quadrature formulae for more general singular integrals

$$\mathbf{I}_{m+1}(f,t) = \int_a^b \frac{w(\tau)f(\tau)}{(\tau-t)^{m+1}}d\tau, \quad a<t<b,\ m=0,1,\cdots. \tag{1.3}$$

We know that singular integrals of the form

$$\mathbf{I}^*_{m+1}(f,t) = \int_a^b \frac{w(\tau)f(\tau)}{|\tau-t|(\tau-t)^{m}}d\tau, \quad a<t<b,\ m=0,1,\cdots, \tag{1.4}$$

or, in more general forms

$$\mathcal{S}_{m+1}(f,t) = \int_a^b \frac{w(\tau)f(\lambda(\tau))}{|\lambda(\tau)-\lambda(t)|^{m+1}}d\tau, \quad a<t<b,\ m=0,1,\cdots, \tag{1.5}$$

$$\mathcal{S}^{\#}_{m+1}(f,t) = \int_a^b \frac{w(\tau)(\lambda(\tau)-\lambda(t))}{|\lambda(\tau)-\lambda(t)|^{m+2}}f(\lambda(\tau))d\tau, \quad a<t<b,\ m=0,1,\cdots, \tag{1.6}$$

also appear frequently in problems of the theories of elasticity. Here $w(\tau)$ is a given non-negative and integrable function, which is known as the weight function, $\lambda(\tau)$ is a given simple arc in the complex plane and $f(z)$ is a function defined on the arc λ. The definitions of (1.3)–(1.6) will be given in §4 and §6 under some assumptions made on w, λ and f. In this paper, we shall discuss their numerical evaluation.

§2 Peano derivatives

Definition 2.1 If f is defined on the set X , t is the cluster point of X, and

$$f(\tau) = \sum_{j=0}^{n} \frac{a_j}{j!}(\tau-t)^j + \mathrm{o}((\tau-t)^n), \tag{2.1}$$

where a_j's are some constants, then a_n is called the generalized derivative of n-th order of f at $x=t$ and symbolized by $a_n = f^{[n]}(t)$.

We denote the polynomial in the right hand side of (2.1) by $T_t^n f$, obviously,

$$(T_t^n f)(\tau) = \sum_{j=0}^{n} \frac{f^{[j]}(t)}{j!}(\tau - t)^j,$$

which is called the generalized Taylor polynomial of n-th order of f at $\tau = t$. a_n is called the n-th Peano derivative of f at $\tau = t$. If $f^{[n]}(t)$ exists, we write $f \in P^{[n]}(t)$. We call

$$(d_t^n f)(\tau) = \frac{f(\tau) - (T_t^n f)(\tau)}{(\tau - t)^n}, \quad \tau \neq t, \tag{2.2}$$

the difference quotient of n-th order of f at $\tau = t$. Obviously $d_t^n f \in P^{[0]}(t)$ and $(d_t^n f)^{[0]}(t) = 0$.

Lemma 2.1 *$f \in P^{[n+m]}(t)$ iff $d_t^n f \in P^{[m]}(t)$ and*

$$(d_t^n f)^{[m]}(t) = \begin{cases} \dfrac{m!}{(n+m)!} f^{[n+m]}(t), & \text{if } m = 1, 2, \cdots, \\ 0, & \text{if } m = 0. \end{cases}$$

Following [5,6], we introduce some subclasses of the class $P^{[n]}(t)$. If $d_t^n f \in H$ (the Hölder class), we then write $f \in H^{[n]}(t)$. If $|(d_t^n f)(\tau)| \leq C|\tau - t|^\mu$, $0 < \mu \leq 1$ (the Hölder condition at $\tau = t$), where C and μ are constants, we then write $f \in h^{[n]}(t)$. Let $\mathcal{K}^{[n]}(t)$ denote one of the classes $P^{[n]}(t)$, $h^{[n]}(t)$ and $H^{[n]}(t)$.

Lemma 2.2 *If $g \in \mathcal{K}^{[n]}(t)$ and $f \in \mathcal{K}^{[n]}(w)$, where $w = g(t)$, then $f \circ g \in \mathcal{K}^{[n]}(t)$.*

Definition 2.2 The function f is defined on the set X of the complex plane, $t \in X$ and is its interior point. Let $z = x + iy$, $t = \alpha + i\beta$, where x, y, α, β are real. If there are some homogeneous polynomials

$$p_j(x, y) = \frac{1}{j!} \sum_{k=0}^{j} C_j^k a_{k,j-k} x^k y^{n-k} \quad (j = 0, \cdots, n) \tag{2.3}$$

(we regard $p_0 = a_{0,0}$ as a constant), such that $f(z) = \sum\limits_{j=0}^{n} p_j(x - \alpha, y - \beta) + o((z-t)^n)$, then p_n is called the (real) generalized derivative of n-th order of f at $z = t$ as well as these constants $a_{k,n-k}$ $(k = 0, \cdots, n)$ are called the (real) Peano partial derivatives of n-th order of f at $z = t$ and symbolized by $a_{k,n-k} = \partial^{[n]} f / \partial x^k \partial y^{n-k}$ $(k = 0, \cdots, n)$. Let $\left(P_t^j f\right)(z) = p_j(x - \alpha, y - \beta)$. If $(P_t^n f)$ exists, we write $f \in RP^{[n]}(t)$. Let

$$(RT_t^n f)(z) = \sum_{j=0}^{n} \frac{1}{j!}\left(P_t^j f\right)(z), \quad (Rd_t^n f)(z) = \frac{f(z) - (RT_t^n f)(z)}{(z-t)^n},$$

$$z \neq t. \tag{2.4}$$

As before, $Rd_t^n f \in RP^{[0]}(t)$ and $(Rd_t^n f)^{[0]}(t) = 0$. If $Rd_t^n f \in H$, we write $f \in RH^{[n]}(t)$. If $|(Rd_t^n f)(z)| \leq C|z-t|^\mu$, where C, μ are constants and $0 < \mu \leq 1$, we then write $f \in Rh^{[n]}(t)$. We agree that $R\mathcal{K}^{[n]}(t)$ denotes one of the classes $RP^{[n]}(t)$, $Rh^{[n]}(t)$ and $RH^{[n]}(t)$.

Lemma 2.3 *If $g \in \mathcal{K}^{[n]}(t)$ and is a real variable function, $f \in R\mathcal{K}^{[n]}(w)$, where $w = g(t)$, then the composite function (if exists) $f \circ g \in \mathcal{K}^{[n]}(t)$.*

We call $(f|_{(t,b]})^{[n]}(t)$ the n-th right sided generalized derivative of f at t, denoted as $f_+^{[n]}(t)$. The meanings of the symbols $T_{t\pm}^n f$, $d_{t\pm}^n f$, $P_\pm^{[n]}(t)$, $\mathcal{K}_\pm^{[n]}(t)$, etc. all are obvious.

Lemma 2.4 *Suppose that f is defined on $[a,b]$ and $t \in [a,b]$. If $f' \in \mathcal{K}^{[n-1]}(t), n \geq 1$, then $f \in \mathcal{K}^{[n]}(t)$. and $(f')^{[k]}(t) = f^{[k+1]}(t)$ $(0 \leq k \leq n-1)$.*

§3 The sectioned Hermite interpolation

Let $\mathcal{F}$ be a function space, E^s the s-dimensional Euclidean space, $\mathbf{r}_s : \mathcal{F} \to E^s$ is a linear operator, then we say it to be an s-dimensional discretization operator on $\mathcal{F}$. In addition, let a subspace $\mathcal{U} \subset \mathcal{F}$ with $\dim \mathcal{U} = s$ be given, such that

$$\text{If } \mathbf{r}_s f = (0, \cdots, 0) \in E^s \text{ and } f \in \mathcal{U}, \text{ then } f = 0. \tag{3.1}$$

Interpolation problem $(\mathcal{F}, \mathbf{r}_s, \mathcal{U})$: *To find an operator Intpol: $\mathcal{F} \to$* $\mathcal{U}$, such that $\mathbf{r}_s \circ Intpol = \mathbf{r}_s$.

This problem is known as the interpolation problem from $\mathcal{F}$ to $\mathcal{U}$ according as the discretization operator $\mathbf{r}_s$. The operator *Intpol* is called the interpolation operator from $\mathcal{F}$ to $\mathcal{U}$ according as the discretization operator $\mathbf{r}_s$, $\mathcal{U}$ is called the interpolating space of this interpolation problem. This problem is uniquely solvable, i.e., $Intpol[f] = \sum_{j=1}^{s} f_j (\mathbf{r}_s|_\mathcal{U})^{-1} e_j$, where $\{e_j,\ j = 1, \cdots, e_s\}$ is the normal basis of E^s and $\{f_1, \cdots, f_s\} = \mathbf{r}_s f$. Obviously, all $(\mathbf{r}_s|_\mathcal{U})^{-1} e_j$ $(j = 1, \cdots, s)$ form the basis of $\mathcal{U}$, each of which is called the fundamental function of the interpolation problem.

Let $a \leq x_1 < x_2 < \cdots < x_n \leq b$, $a < t < b$ and $t \neq x_j$, $\nu_1, \nu_2, \cdots, \nu_n$, $m \in \mathcal{N}_0$ (the set of all nonnegative integers). Write $\vec{x} =$

$(x_1, x_2, \cdots, x_n)$, $\vec{\nu} = (\nu_1, \nu_2, \cdots, \nu_n)$, $(\vec{\nu}, m) = (\nu_1, \nu_2, \cdots, \nu_n, m)$, $|\vec{\nu}| = \sum_{j=1}^n \nu_j$, $|(\vec{\nu}, m)| = |\vec{\nu}|+m$, $s = |\vec{\nu}|+n+m+1$, $D(\tau)=\prod_{j=1}^n (\tau - x_j)^{\nu_j+1}$, and $D_j(\tau)=\prod_{k=1,k\neq j}^n (\tau - x_j)^{\nu_k+1}$.

Example 3.1 Take $\mathcal{F} = P^{[\vec{\nu},m]}(\vec{x}, t^+) := (\bigcap_{j=1}^n P^{[\nu_j]}(x_j)) \bigcap (P_+^{[m]}(t))$, $\mathbf{r}_s = \mathbf{r}\left[\begin{smallmatrix}\vec{\nu},m\\ \vec{x},t^+\end{smallmatrix}\right]$,

$$\mathbf{r}\left[\begin{smallmatrix}\vec{\nu},m\\ \vec{x},t^+\end{smallmatrix}\right] f = \Big(f^{[0]}(x_1), \cdots, f^{[\nu_1]}(x_1), \cdots,$$

$$f^{[0]}(x_n), \cdots, f^{[\nu_n]}(x_n), f_+^{[0]}(t), \cdots, f_+^{[m]}(t)$$

and $\mathcal{U} = \Pi_{s-1}(t) := \operatorname{sgn}(\cdot - t)\Pi_{s-1}$, which is the class of all sectioned polynomials with the knot point t. We may prove that (3.1) holds. So the triple $(P^{[\vec{\nu},m]}(\vec{x}, t^+), \mathbf{r}\left[\begin{smallmatrix}\vec{\nu},m\\ \vec{x},t^+\end{smallmatrix}\right], \Pi_{s-1}(t))$ forms really an interpolation problem. We put up

$$\begin{cases} W_{j,k}(\tau) = \dfrac{\operatorname{sgn}(\tau - t)}{\operatorname{sgn}(x_j - t)} \dfrac{D(\tau)(\tau - t)^{m+1}}{k!\,(\nu_j - k)!} \dfrac{\partial^{\nu_j - k}}{\partial x^{\nu_j - k}} \left[\dfrac{1}{D_j(x)(x-t)^{m+1}(\tau - x)}\right]_{x=x_j}, \\ U_k(\tau) = \operatorname{sgn}(\tau - t) \dfrac{D(\tau)(\tau - t)^{m+1}}{k!(m-k)!} \dfrac{\partial^{m-k}}{\partial x^{m-k}} \left[\dfrac{1}{D(x)(\tau - x)}\right]_{x=t} \end{cases} \tag{3.2}$$

They are just the fundamental functions. Thus the operator *intpol* becomes now

$$\mathbf{H}\left[\begin{smallmatrix}\vec{\nu},\, m\\ \vec{x},t|^+\end{smallmatrix}\right] f(\tau) = \sum_{j=1}^n \sum_{k=0}^{\nu_j} W_{j,k}(\tau) f^{[k]}(x_j) + \sum_{k=0}^m U_k(\tau) f_+^{[k]}(t). \tag{3.3}$$

It is called the sectioned Hermite interpolation polynomial with ν_j order at x_j and (right) m order at the knot t of f in the class $P^{[\vec{\nu},m]}(\vec{x}, t^+)$.

§4 Hypersingular integrals

Let f is defined on $[a, b]$ and $f \in h_+^{[m]}(t)$ $(m \geq 0)$. Define [6]

$$\int_t^b \frac{f(\tau)}{(\tau - t)^{m+1}} d\tau = \sum_{j=0}^{m-1} \frac{-f_+^{[j]}(t)}{j!(m-j)(b-t)^{m-j}} + \frac{f_+^{[m]}(t)}{m!} \ln(b-t) +$$

$$\int_t^b \frac{(d_{t^+}^n f)(\tau)}{\tau - t} d\tau, \; a \leq t < b, \tag{4.1}$$

which is called the (right) one-sided singular integral of order $m+1$ for f at $\tau = t$.

Similarly, when $f \in h_-^{[m]}(t)$, then define

$$\int_a^t \frac{f(\tau)}{(\tau-t)^{m+1}}d\tau=\sum_{j=0}^{m-1}\frac{f_-^{[j]}(t)}{j!(m-j)(a-t)^{m-j}}-\frac{f_-^{[m]}(t)}{m!}\ln(t-a)+$$

$$\int_a^t \frac{\left(d_{t-}^m f\right)(\tau)}{\tau-t}d\tau,\ a<t\le b, \tag{4.2}$$

which is called the (left) one-sided singular integral of order $m+1$ for f at $\tau=t$.

If $f\in h_{\pm}^{[m]}(t)$, by the addition property we define

$$\int_a^b \frac{f(\tau)}{(\tau-t)^{m+1}}d\tau=\int_a^t \frac{f(\tau)}{(\tau-t)^{m+1}}d\tau+\int_t^b \frac{f(\tau)}{(\tau-t)^{m+1}}d\tau,\ a<t<b, \tag{4.3}$$

$$\int_a^b \frac{f(\tau)}{|\tau-t|(\tau-t)^{m}}d\tau=\int_t^b \frac{f(\tau)}{(\tau-t)^{m+1}}d\tau-\int_a^t \frac{f(\tau)}{(\tau-t)^{m+1}}d\tau,\ a<t<b, \tag{4.4}$$

which is called the (absolutely) singular integral of order $m+1$ for f at $\tau=t$.

Lemma 4.1 *Suppose that L is a closed smooth curve. If $f'\in h^{[m-1]}(t)$, then* [7, 8] *$f\in h^{[m]}(t)$ and*

$$\int_L \frac{f(\tau)}{(\tau-t)^{m+1}}d\tau=\frac{1}{m}\int_L \frac{f'(\tau)}{(\tau-t)^{m}}d\tau. \tag{4.5}$$

Consequently, if $f^{(m)}\in H$, then

$$\int_L \frac{f(\tau)}{(\tau-t)^{m+1}}d\tau=\frac{1}{m!}\int_L \frac{f^{(m)}(\tau)}{\tau-t}d\tau. \tag{4.6}$$

Lemma 4.2 *Let $t\in(a,b)$. Suppose that f is an absolutely integrable function on $[a,b]$ and $f^{(m)}\in H$ on certain neighbored of t, then*

$$\frac{d^m}{dt^m}\int_a^b \frac{f(\tau)}{\tau-t}d\tau=m!\int_a^b \frac{f(\tau)}{(\tau-t)^{m+1}}d\tau,\ \ t\in(a,b). \tag{4.7}$$

§5 Singular quadrature formulae

We know that, for the weight function w there is a system of orthogonal polynomials $\{p_n\}_0^\infty$. Let

$$q_n(z)=\int_a^b \frac{w(\tau)p_n(\tau)}{\tau-z}d\tau,\ \ z\in(a,b), \tag{5.1}$$

which is called the associated function of p_n.

Lemma 5.1 *If* $w \in H^{(m)} = \bigcup_{\epsilon>0} H^{(m)}[a+\epsilon, b-\epsilon]$, *then*

$$q_n^{(k)}(z) = k! \int_a^b \frac{w(\tau)p_n(\tau)}{(\tau - z)^{k+1}} d\tau, \quad k = 0, \cdots, m, \tag{5.2}$$

$$q_n^{(m+1)}(x_{n,j}) = (m+1)! \int_a^b \frac{w(\tau)p_{n,j}(\tau)}{(\tau - x_{n,j})^{m+1}} d\tau, \quad j = 1, \cdots, n, \tag{5.3}$$

where $p_{n,j}(\tau) = \dfrac{p_n(\tau)}{\tau - x_{n,j}}$, $x'_{n,j}s$ *are* n *simple zeros of* p_n.

Theorem 5.1 *If* $f \in C^{2n+m+1}(t)$, *then*

$$\mathbf{I}^*_{m+1}(f,t) = \mathbf{Q}_{m+1,n}(f,t) + \mathbf{R}_{m+1,n}(f,t), \tag{5.4}$$

where $\mathbf{Q}_{m+1,n}(f,t) = \mathbf{I}^*_{m+1}(\mathbf{H}\begin{bmatrix} \vec{1}_n, & m \\ \vec{x}, & t|+ \end{bmatrix} f, t)$, *i.e.*,

$$\mathbf{Q}_{m+1,n}(f,t) = \begin{cases} \displaystyle\sum_{j=1}^{n} \frac{H_{n,j} f(x_{n,j})}{|x_{n,j}-t|(x_{n,j}-t)^m} + \frac{1}{m!}\left[\frac{q_n(t)}{p_n(t)} f(t)\right]_+^{(m)}, \\ \qquad\qquad \text{if } t \neq x_{n,j}, \\ \displaystyle\sum_{j=1, j\neq r}^{n} \frac{H_{n,j} f(x_{n,j})}{|x_{n,j}-t|(x_{n,j}-t)^m} + \frac{1}{(m+1)!}\left[\frac{q_n(t)}{p_{n,r}(t)} f(t)\right]_+^{(m+1)}, \\ \qquad\qquad \text{if } t = x_{n,r}, \end{cases}$$

$$H_{n,j} = \frac{q_n(x_{n,j})}{p'_n(x_{n,j})}, \ \mathbf{R}_{m+1,n}(f,t) = \frac{r_n\left[\text{sgn}(\cdot - t)f\right]^{(2n+m+1)}(\xi)}{c_n^2(2n+m+1)!}, a \le \xi \le b.$$

Theorem 5.2 *If* $f \in C^{m+1}(t)$, *then* $\lim_{n\to\infty} \mathbf{Q}_{m+1,n}(f,t) = \mathbf{I}^*_{m+1}(f,t)$, *more precisely,*

$$\left|\mathbf{Q}_{m+1,n}(f,t)\right| \le C\omega\Big((\text{sgn}(\cdot - t)f)^{(m+1)}, 1/n\Big) \quad (C \text{ is a constant}).$$

§6 Applications

We rewrite (1.5) and (1.6) as follows.

$$\mathcal{S}_{m+1}(f,t) = \begin{cases} \displaystyle\int_a^b \frac{w(\tau)f(\lambda(\tau))}{|\tau - t|(\tau-t)^m \left|\bigwedge_t^1 \lambda(\tau)\right|^{m+1}} d\tau, & \text{if } m \text{ is even}, \\ \displaystyle\int_a^b \frac{w(\tau)f(\lambda(\tau))}{(\tau-t)^{m+1} \left|\bigwedge_t^1 \lambda(\tau)\right|^{m+1}} d\tau, & \text{if } m \text{ is odd}, \end{cases} \tag{6.1}$$

$$\mathcal{S}^{\#}_{m+1}(f,t) = \begin{cases} \int_a^b \dfrac{w(\tau)f(\lambda(\tau))}{(\tau-t)^{m+1}\Lambda\left[{}^1_t\right]\overline{\lambda}(\tau)\left|\Lambda\left[{}^1_t\right]\lambda(\tau)\right|^m}d\tau, & \text{if } m \text{ is even},\\ \int_a^b \dfrac{w(\tau)f(\lambda(\tau))}{|\tau-t|(\tau-t)^{m}\Lambda\left[{}^1_t\right]\overline{\lambda}(\tau)\left|\Lambda\left[{}^1_t\right]\lambda(\tau)\right|^m}d\tau, & \text{if } m \text{ is odd}, \end{cases} \tag{6.2}$$

where

$$\Lambda\left[{}^1_t\right]\lambda(\tau) = \begin{cases} \dfrac{\lambda(\tau)-\lambda(t)}{\tau-t}, & \text{if } \tau \neq t,\\ \lambda'(t), & \text{if } \tau = t \end{cases},$$

$$\Lambda\begin{bmatrix} 1 \\ t \end{bmatrix}\overline{\lambda}(\tau) = \begin{cases} \dfrac{\overline{\lambda(\tau)}-\overline{\lambda(t)}}{\tau-t}, & \text{if } \tau \neq t,\\ \overline{\lambda'(t)}, & \text{if } \tau = t. \end{cases}$$

Following §4, for the existence of (6.1) and (6.2), we assume that, λ is a simple smooth arc and $\lambda \in h^{[m+1]}(t)$, f is defined in certain neighbored of λ and $f \in Rh^{[m]}(t)$, w is that in last section. In these assumptions, both (6.1) and (6.2) are well defined. We further assume $\lambda \in h^{[m+2]}(t)$ and $f \in Rh^{[m+1]}(t)$, then

$$\mathcal{S}_{m+1}(f,t) \approx \begin{cases} \displaystyle\sum_{j=1}^{n} \frac{H_{n,j}f(\lambda(x_{n,j}))}{|\lambda(x_{n,j})-\lambda(t)|^{m+1}} + \frac{1}{m!}\left[\frac{q_n(f\circ\lambda)}{p_n\left|\Lambda\left[{}^1_t\right]\lambda\right|^{m+1}}\right]_+^{(m)}(t), \\ \qquad \text{if } t \neq x_{n,j}, \\ \displaystyle\sum_{j=1, j\neq r}^{n} \frac{H_{n,j}f(\lambda(x_{n,j}))}{|\lambda(x_{n,j})-\lambda(t)|^{m+1}} + \\ \displaystyle\frac{1}{(m+1)!}\left[\frac{q_n(f\circ\lambda)}{p_{n,r}\left|\Lambda\left[{}^1_t\right]\lambda\right|^{m+1}}\right]_+^{(m+1)}(t), \\ \qquad \text{if } t = x_{n,r}; \end{cases}$$

$$
\mathcal{S}^{\#}_{m+1}(f,t) \approx \begin{cases} \displaystyle\sum_{j=1}^{n} \frac{H_{n,j} f\left(\lambda\left(x_{n,j}\right)\right)}{\left(\overline{\lambda}\left(x_{n,j}\right)-\overline{\lambda}(t)\right)\left|\lambda\left(x_{n,j}\right)-\lambda(t)\right|^{m}} + \\ \displaystyle\frac{1}{m!}\left[\frac{q_n(f\circ\lambda)}{p_n \bigwedge[_t^1]\overline{\lambda}\left|\bigwedge[_t^1]\lambda\right|^{m}}\right]_{+}^{(m)}(t), & t \neq x_{n,j}, \\ \displaystyle\sum_{j=1,j\neq r}^{n} \frac{H_{n,j} f\left(\lambda\left(x_{n,j}\right)\right)}{\left(\overline{\lambda}\left(x_{n,j}\right)-\overline{\lambda}(t)\right)\left|\lambda\left(x_{n,j}\right)-\lambda(t)\right|^{m}} + \\ \displaystyle\frac{1}{(m+1)!}\left[\frac{q_n(f\circ\lambda)}{p_{n,r} \bigwedge[_t^1]\overline{\lambda}\left|\bigwedge[_t^1]\lambda\right|^{m}}\right]_{+}^{(m+1)}(t), & t = x_{n,r}. \end{cases}
$$

References

[1] Kutt H. R., *The numerical evaluation of principal value integrals by finite—part integration*, Numer. Math., **24**(1975), 205–210.

[2] Paget D. F., *The numerical evaluation of Hadamard finite-part integrals*, Numer. Math., **36**(1981), 447–453.

[3] Du Jinyuan, *The quadrature formulas for singular integrals of higher order*, Chinese Ann. of Math., **6A**(1985), 5, 625–636.

[4] Du Jinyuan, *The quadrature formulas of the closed type and the transformed weight type for singular integrals of higher order*, J. of Math. (PRC), **6**(1986), 4, 439–454.

[5] Du Jinyuan, *A note on Riemann boundary value problem of non-normal type*, J. of Math. (PRC), **10**(1990), 1, 111–120.

[6] Lu Jianke, *Peano derivatives and singular integrals of arbitrary order*, Wuhan University J. of Natural Sciences, **1**(1990), 1, 9–13.

[7] Lu Jian-ke, *Boundary Value Problems for Analytic Functions*, World Scientific., 1993.

[8] Lu Jianke, *Singular integrals of higher order and their applications to the solutions of singular integral equations*, Sci. & Tech. of Wuhan University , **2**(1977), 106–122.

17 THEORETICAL AND NUMERICAL ANALYSIS OF INVERSION OF SATELLITE REMOTE SENSING

S.-X. HUANG

P.O. Box 003, Nanjing 211101, P.R. China

huangsx@public1.ptt.js.cn

Jun LI

CIMSS, University of Wiscosin-Madison

Madison, WI 53706, U.S.A.

huangsx@public1.ptt.js.cn

1 Introduction

The weather satellite makes it possible to acquire meteorological data on a global basis, which are powerful means of weather monitoring /predicting and research of atmospheric sciences. By remote sensing we mean to inverse the space/ time patterns of meteorological elements from the signals of atmospheric sound, light and electromagnetic waves. Additionally, such a satellites is capable of deriving quantitative information on meteorological parameters beyond the limits of nations and orography , and with this means it is likely to get data over oceans covering more than 70% of the earth's surfaces).

H.G.W. Begehr et al. (eds.), Proceedings of the Second ISAAC Congress, Vol. 1, 129–136.

So far, developed for retrieval have been numerous schemes that can be separated into there broad categories: the statistical type has the merit of so fast calculation that it is the first to be accepted in operational practice and also the demerit of the need to treat conventional data statistically. And if such data were lacking, particularly in inaccessible areas, e.g., oceans, deserts, plateaus and polar regions, nothing wound be done about the inversion. On the other hand, many physical properties are ignored in establishing the remote sensing expression, thus damaging the accuracy. Next, the physical technique is based on the ill-posed character of the integral equation of the first kind and contraction of an effective algorithm for directly retrieving atmospheric parameters expect the demerit that fewer passages lead to considerable error when using an interpolation formula to make numerical integration and in particular ,the interpolation points are not always coincident with the curve of realistic temperature distribution, thus exerting great effect on the desire accuracy. In recent years remarkable progress has been achieved in this field, including the simultaneous physical inversion (Smith, W.L., 1985), the relaxation iterative method (Susskind, 1984), the simultaneous minimum variance retrieval technique (Fleming, 1986) and the 3I inversion(Chedin, 1985),i.e., Improved Initialization Inversion. For the statistical-physical inversion coming from the combination of Smith's and Chedin's methods cited above with statistical results, the reader can be referred to relevant contribution by Deepak et al. (1985). If we consider the nonlinear effect of the model, Newton nonlinear iterative method is introduced by Li Jun (1994).

In this paper, we apply a new method called improved discrepancy principle (IDP) to inverse atmospheric parameters on the inverse test of new generation geostationary operational environmental satellite (GOES-8), and show that the result obtained by this method is more accurate than that in use.

2 RTE and some difficulties of atmospheric remote sensing inversion

Infrared and microwave bands are widely used in the atmospheric sounding, the former producing measurements less accurate and being subject to cloud effect but with part of the probing much more evident. For details the reader is referred to the treaties by Zeng(1974); Zhao et al.(1987).

As we know the atmospheric radiation transfer equation(RTE) takes the form

$$R = \epsilon B_s \tau_s - \int_0^{p_s} B d\tau + (1-\epsilon) \int_0^{p_s} B d\tau^* + R', \tag{2.1}$$

where R is spectral radiance, R is Plank radiance, which is a function of temperature, i.e., an implicit function of pressure level p, τ is the total transmittance of the atmospheric above the pressure level, p_s is surface pressure, ϵ is surface emissivity, superscript "s" indicates surface value of physical quantities. $\tau = \tau_c \cdot \tau_w, \tau_c$ is co_2 component transmittance , $\tau_w = \exp(-\int_0^p \frac{kq}{g} dp)$, q is mixing ratio of absorbing constituent, k is absorbing coefficient of absorbing constituent,R' is reflected radiation of sun,

$$R' = c_r {\tau_s}^2 \rho_r, c_r = 2.16 \times 10^{-5} \cos\theta B_r(T_{sun}),$$

where T_{sun} is surface temperature of sun, $T_{sun} = 5600k$. θ is zenith angle of sun.

This model (2.1) (susskind,1984) consider the effects of surface emissivity and solar reflectivity.

Now, we give some difficulties of atmospheric sensing inversion (Huang,1996).

1. It is clear that the inversion is a problem of nonlinear equation of the first kind, ill-posed in nature.

2. Determining initial data.

3. Nonlinearity.

4. Calculation of kernel function (weighting function). Atmospheric radiation transmission is quantity not exactly known, but obtained from numerical calculation that gives rise to severe difficulties in exploring the properties of weighting function.

5. Designing iteration scheme of steady approximate solution and selecting a point for iteration ending are very difficult. In utilizing iteration for steady approximate solutions, it is not true that the more the iterative steps, reach the accuracy. The iteration convergence to the solution, but after stabilization, divergence occurs and the operation should be ceased, when the minimum difference arises between the approximate and exact solutions. Thus leading to the fact that the choice of a point for iteration ending is of particular importance. It is also important that an effective scheme has to be found, responsible for weakening of k's-dependence an auxiliary information, because choice of the number of iterative steps, k, rests on the structure of initial data, attributes of weighting function and other auxiliary messages.

6. The presence of cloud increases the complexity of inversion. In the case that there is cloud, inversion becomes more unstable due to increased inversion parameters and complexity of radiation transfer, and more difficult to obtain the solution due to the lack of atmosphere profiles in presence of cloud and cloud parameters.

3 Improved discrepancy principle (IDP) to inverse atmospheric parameters

Now we employ new method called IDP to inverse atmospheric parameters, e.g., atmospheric temperature ,humidity and etc.. Because this model is nonlinear, at first we discuss linearized model, by use of discrepancy principle(DP) to choose optimal parameter, but linearized model is an approximate for nonlinear, then in the inverse process, if we employ the DP for linearzed model, the accuracy of inversion will be improved greater or less extent, if we continue iterating, the accuracy will be improved unremarkable, so we must consider the nonlinear factor.

Now we employ the following inverse process:

Step 1: Linearization and discretization of model (2.1)

The first order variation of eq.(2.1) yields

$$\delta R = \epsilon\tau_s\delta B_s + \epsilon B_s\delta\tau_s + B_s\tau_s\delta\epsilon - \int_0^{p_s}\delta B d\tau - \int_0^{p_s} B\delta(d\tau) +$$

$$(1-\epsilon)\int_0^{p_s}\delta B d\tau^* + (1-\epsilon)\int_0^{p_s} B d(\delta\tau^*) - \delta\epsilon\int_0^{p_s} B d\tau^* + 2c_r\tau_s\rho_r\delta\tau_s + c_r{\tau_s}^2\delta\rho_r.$$

Set $\delta B = \frac{\partial B}{\partial T}\delta T, \delta R = \frac{\partial R}{\partial T_B}\delta T_B$and $\beta(p) = \left(\frac{\partial B}{\partial T}\right)\cdot\left(\frac{\partial R}{\partial T_B}\right)^{-1}$, then we obtain

$$(\delta T_B) = W_{T_s}\delta T_s + \int_0^{p_s} W_T\delta T dp + \int_0^{p_s} W_q\delta lnq dp + W_\epsilon\delta\epsilon + W_{\rho_r}\delta\rho_r,$$

where

$$W_{T_s} = \beta_s\epsilon\tau_s, \quad W_T = -\beta\frac{\partial\tau}{\partial p} + \beta(1-\epsilon)\frac{\partial\tau^*}{\partial p},$$

$$W_q = \left((T_s - T_a)\epsilon\tau_s\beta_s + \frac{2c_r{\tau_s}^2\rho_r}{\partial R/\partial T_B} - 2(1-\epsilon)\int_0^{p_s}\beta\tau^*\frac{\partial T}{\partial p}dp\right)\frac{\partial ln\tau_w}{\partial p}$$
$$+\left(\int_0^{p_s}\beta[\tau + (1-\epsilon)\tau^*]\frac{\partial T}{\partial p}dp\right)\frac{\partial ln\tau_w}{\partial p},$$

$$W_\epsilon = \beta_s \tau_s T_s - \int_0^{p_s} \beta \frac{\partial \tau^*}{\partial p} dp, \quad W_{\rho_r} = \frac{c_r \tau_s{}^2}{\partial R / \partial T_B}.$$

where T_s is surface skin temperature, T_a is surface air temperature, T_B is the brightness temperature.

Let us consider now a vector X, which contains L (level of temperature) atmospheric temperatures, L mixing ratio of water vapor and one surface temperature and $\epsilon_i, \rho_{r_i} (i = 1, ..., N), Y$ is a vector, which contains N satellite brightness temperature (i.e., channel numbers are N). Then equation (3.1) for jth channel becomes

$$\delta T_B = W_{T_s} \delta T_S + \int_0^{p_s} W_q \delta lnq dp + \sum_{i=1}^{N} W_{\rho_i}^* \delta \rho_{r_i},$$

where

$$W_{\epsilon_i}{}^* = \xi(N_i, j) W_{\epsilon_{N_i}}, W_{\rho_{r_i}}{}^* = \xi(N_i, j) W_{\rho_{r_{N_i}}}, \xi(i^*, j) = \{^{i, i^* = j}_{0, i^* \neq j},$$

$X = X(T_1, ..., T_2, q_1, ..., q_L, \epsilon_1, ..., \epsilon_N, \rho_{r_1}, ..., \rho_{r_N}), Y = (Y_1, ..., Y_N)$. Let X_0 is the initial guess value of atmospheric profile and $Y_0 = F(X_0)$, then nonlinear equation 2.1 linearized at X_0: $F'(X_0)(X - X_0) = Y - Y_0$, where $F'(X_0)$ is $Freche't$ derivative of F at X_0. Set

$$J(X) = (Y^m - F(X))^T \cdot E^{-1} (Y^m - F(X)) + \gamma (X - X_0)^T \cdot (X - X_0) = min, \tag{3.2}$$

where Y^m is measurement data, E is the satellite observational error covariance matrix, it includes the measurement errors of satellite and the model transimittance error. Let Y be accurate data. Then $\|Y^m - Y\| \le \sigma^2$, where σ is $\sigma^2 = \frac{1}{N} \sum_{k=1}^{N} e_k{}^2$, e_k is observational error of kth frequency band including instruments noise and model error of atmospheric transimissivity, i.e., $e_k{}^2 = \epsilon^2 + f_k{}^2$.

By use of DP, we obtain iterative result $(\gamma_1, X_0(\gamma_0))$, where γ_0 is the first guess value (empirical parameters).

Step 2: setting out from (γ_1, X_1), then nonlinear equation linearized at X_1, by use of DP, we obtain iterative result $(\gamma_2, X_1(\gamma_1)) \equiv (\gamma_2, X_2)$, and so on. We call this method by name IDP. This method can choose parameter γ effectively, reasonably and objectively in the meantime, and consider nonlinearity of problem also. We apply this method to inverse of probe data GOES-8, the calculating results indicated that this more accurate than empirical parameter method and general DP.

Enforcing plan is following

$$X_{n+1} = X_n + (\gamma_n I + F'^T(X_n) E^{-1} F'(X_n)^T \cdot F'^T(X_n) E^{-1} (Y^m - F(X_n))),$$

$$\gamma_{n+1} = \gamma_n - \frac{G_n}{G_n{}'}, \quad G_n = (F(X_{n+1}) - Y^m)^T \cdot E^{-1}(F(X_{n+1}) - Y^m) - \sigma^2,$$

$$\begin{aligned} G_n{}' = \; & 2(F(X_{n+1}) - Y^m)^T \cdot E^{-1} F'(X_n) \\ & \left(\gamma_n I + F'^T(X_n) E^{-1} F'(X_n)\right)^{-2} \cdot F'^T(X_n) E^{-1}(F(X_{n+1}) - Y^n). \end{aligned}$$

In the particle iterative process, if $\|X_{n+1} - X_n\| \le \xi$, then the iterative will be stopped, where ξ is very small value. In the particle numerical calculation, we obtain higher accuracy, which only needs iterative process two times.

The table 1 gives instrument noise of GOES-8 for every band.

Table 1. Instrument Noise for GOES-8

$$Unit : mWm^{-2}sr^{-1}((cm)^{-1}))^{-1}$$

Channel No.	Center Wavenumber	Instrument Noise
1	680.71	1.7600
2	695.73	1.2100
3	711.88	0.9800
4	732.50	0.7400
5	747.53	0.6800
6	790.40	0.3200
7	829.07	0.2000
8	906.55	0.1300
9	1029.55	0.1600
10	1339.55	0.0800
11	1422.01	0.0700
12	1535.43	0.1100
13	2184.93	0.0120
14	2207.49	0.0098
15	2247.11	0.0090
16	2421.74	0.0039
17	2509.51	0.0043
18	2666.41	0.0018

4 Numerical experiment

We apply IDP on the inverse test of a group of GOES-8 practical observation data. This data is a part of the atmospheric radiation measurement (ARM) at cloud and radiation tests(CART), CART located middle-north Oklahoma and middle-south Kansa, GOES-8 has 18 HIRS(High-resolution infrared radiation sounds), these infrared observation data includes information of atmospheric temperature, humidity, and etc.. All data of GOES-8 is collected at the case of serene. We take one month data from June 18 to July 17, 1997, and only choose that data, which related to radiosone data, which were 56 specimens. Radiation transfer model (RTM) is employed high-speed model of CIMMS (Cooperative Institute for Meteorological Satellite Studies , University of Wisconsin). We take the result of numerical predictive model of NCEP(National Center Environmental Predictive, U.S.A.) as inverse initial guess value. In the inverse process, we divide atmospheric pressure into 42 layers from 0.1 hpa to 1050 hpa. The observation error consists of instrumental noise and model error of 0.2k. Take two tests: Choosing empirical parameter γ, which obtain optimum parameter $\gamma = 0.1$ with statistical model. We do the first test of minimum information method; For the second test, Choosing $\gamma = 0.1$ as initial guess value γ_0, by use of IDP, we proceed to inverse it, in general, the iterative times are three or four nearly, then we compare the inverse result with radiosone data result, and obtain root mean square errors (RMSE).

Fig.1, which will be sent at a request to huangsx@public1.ptt.js.cn , is RMSE for each layer temperature by use of two methods. Because GOES is difficult to provide additional information of numerical predictive temperature, so we are difficult to improve initial guess value, but by use of IDP, the results show that the accuracy of temperature of level layers is better than that in use. Fig.2 is RMSE of each layer relative humidity by use of two methods, it is very clear that the accuracy of relative humidity is better than that empirical inversion.

References

[1] Chedin. A. et al., J. Climate Appl. Meteorol., 24 (1985), 128–143.

[2] Deepak A. et al., Advance in Remote Sensing Retrieval Methods, A. Deepak publishing, 1985.

[3] Fleming H. E. et al., Preprint Vol. 2th conference on satellite meteorology remote sensing and application, 986.

[4] Huang Sixun., Advances in Atmos. Sci., 13 (1996), 489–504.

[5] Li Jun et al., Adv. in Atmos. Sci., 11 (1994), No.2: 128–138.

[6] Smith W. L., Woolf H. M., Hayden C. M., and Schreiner S. J., The Simultaneous export retrieval package, University of Wisconsin-Madison, 1985, 224–253.

[7] Susskind J. et al., J. G. R., 89 (1984), 4677–4697.

[8] Zeng Q. C., Principles of Atmospheric Infrared Remote Sensing, Science Press, 1974.

[9] Zhao B. L. et al., Principle of Atmospheric Sounding, China Meteo. Press, 1987.

18 OPTIMIZATION OF VECTOR-VALUED INTEGRAL EQUATIONS FOR A CLASS[1]

Chuan Gan HU

Department of Mathematics, Nankai University

Tianjin 300071, China

chuganhu@public1.tpt.tj.cn

Li Xin MA

Department of Mathematics, Dezhou Normal College

Tianjin 300071, China

Dezhou, Shandong 253023, China

Abstract: In this paper, the optimization of integral equations for a class with a parameter is presented in a locally convex space.

[1] Project was supported by the National Natural Science Foundation of China

H.G.W. Begehr et al. (eds.), Proceedings of the Second ISAAC Congress, Vol. 1, 137–143.

1. Introduction

Assume that the dynamics, to be given by the following linear and finite-dimensional system

$$\begin{cases} x(t) = \int_\sigma^t (Ax(t) + Bu(t))dt, \qquad x(\sigma) = 0 \\ y(t) = Cx(t) + Du(t) \end{cases} \tag{1}$$

is the minimal realization of the transfer function matrix

$$G(s) = \begin{bmatrix} G_{11}(s) & G_{12}(s) \\ G_{21}(s) & G_{22}(s) \end{bmatrix} \tag{2}$$

with $\hat{u} = K\hat{y}$, where A, B, C, D, K are real matrices, $\hat{u}$ and $\hat{y}$ are the Laplace transforms of u and y respectively.

The standard problem is to find a real-rational proper K to minimize the H_∞-norm of the transfer matrix from $\hat{w}$ to $\hat{z}$

$$\hat{z} = [G_{11} + G_{12}K(I - G_{22}K)^{-1}G_{21}]\hat{w}.$$

For this purpose the model-matching criterion is

$$\sup\{\|\hat{z}\|_2 : \hat{w} \in H_2, \|\hat{w}\|_2 \le 1\} = \text{minimum},$$

where H_2 denotes the Hardy space H_2.

The above problem has been solved before 1987[3-5]. In 1993, problem (1) was extended from finite dimensional spaces to infinite dimensional Hilbert spaces(see[7]).

In this paper, problem (1) is extended to locally convex spaces contained Hilbert spaces in order to enlarge the scope of solutions of (1) via an indication by a parameter.

2. Basic theory

Assume that $\mathbf{R}$ is the real field and $\mathbf{R}^n$ is the Cartesian product of n copies of $\mathbf{R}$, here n is any positive integer, and $\mathbf{C}$ is a complex plane.

We apply the $G(s)$ in the model matching problem to $G(s, \xi)$, where ξ in $\mathbf{R}^n$ and $G(s, \xi)$ is in $C^\infty(\mathbf{R}^n)$(locally convex space) for each fixed s in $\mathbf{C}$ and in H_∞ for each fixed ξ in $\mathbf{R}^n$.

Definition 1. The locally convex space VH_∞ consists of all complex-valued parameter functions $F(s, \xi)$ of a complex variable s and a parameter ξ, which are analytic and bounded in $\text{Re}(s) > 0$ for any fixed

ξ in $\mathbf{R}^n$, and $F(s,\xi)$ is in $C^\infty(\mathbf{R}^n)$ for any fixed s in the open right half-plane. The VH_∞-norm of $F(s,\xi)$ in VH_∞ is defined by $\|F\|_\infty = \sum_1^\infty g_k/2^k(1+g_k)$, where $g_k = \sup\{\|F(\cdot,\xi)\|_\infty : -k < \xi_i < k, i = 1,\ldots,n \text{ and } \xi = (\xi_1,\ldots,\xi_n)\}$.

Definition 2. The subset of VH_∞ consisting of all real-rational functions of s and ξ will be denoted by VRH_∞.

Definition 3. We call $F(s,\xi)$ to be strong proper if $F(s,\xi) \in VRH_\infty$ with

$$\sup\{|F(\infty,\xi)| : \xi \in \mathbf{R}^n\} < \infty$$

and strictly strong proper if $F(\infty,\xi) = 0$.

Definition 4. We call $F(s,\xi)$ to be stable, if $F(s,\xi) \in VRH_\infty$ and $F(s,\xi)$ has no poles in the closed right half-plane $\mathrm{Re}(s) \geq 0$ for any fixed ξ in $\mathbf{R}^n$.

Theorem 1. *If $F(s,\xi)$ is real-rational about s in $\mathrm{Re}(s) > 0$, then $F(s,\xi) \in VRH_\infty$ if and only if $F(s,\xi)$ is strong proper and stable for any fixed $\xi \in \mathbf{R}^n$.*

Definition 5. A scalar-valued parameter function $T(s,\xi)$ in VRH_∞ is inner if

$$T(-s,\xi)T(s,\xi) = 1$$

and outer if it has no zeros in $\mathrm{Re}(s) > 0$.

Theorem 2. *Every scalar-valued parameter function $T(s,\xi)$ in VRH_∞ has a factorization $T(s,\xi) = T_i(s,\xi)T_o(s,\xi)$, $T_i(s,\xi)$ inner, $T_o(s,\xi)$ outer.*

We define

$$G(s,\xi) = \begin{bmatrix} T_1(s,\xi) & T_2(s,\xi) \\ T_3(s,\xi) & 0 \end{bmatrix}, \quad K(s,\xi) = -Q(s,\xi),$$

then the model-matching problem is $\|T_1 - T_2QT_3\|_\infty$ =minimum, where $T_i(i=1,2,3) \in VRH_\infty$. The constraint that K stabilizes G is equivalent to that $Q \in VRH_\infty$.

3. Infimal model-matching error

Definition 6. The linear time invariant system S_1 defined by

$$x(t,\xi) = \int_\sigma^t (A(\xi)x(t,\xi) + B(\xi)u(t,\xi))dt, \quad x(\sigma,\xi) = 0 \tag{3}$$

$$y(t,\xi) = C(\xi)x(t,\xi), \tag{4}$$

where $A(\xi)$ is $n \times n$, $B(\xi)$ is $n \times m$ and $C(\xi)$ is a constant $r \times n$ matrix depending on ξ, is said to be completely controllable (c.c. for short), if the $n \times nm$ controllability matrix

$$U(\xi) = [B(\xi) \quad A(\xi)B(\xi) \ldots A^{n-1}(\xi)B(\xi)]$$

has rank n(for any fixed ξ in $\mathbf{R}^n$), denoted by $(A(\xi),\ B(\xi))$.

Definition 7. The system S_1 described by (3) and (4) is completely observable (c.o. for short), if the observability matrix

$$V^T(\xi) = [C(\xi) \quad C(\xi)A(\xi) \ldots C(\xi)A^{n-1}(\xi)]^T$$

has rank n, denoted by $(A(\xi), C(\xi))$.

Definition 8. Given an $r \times m$ matrix $G(s,\xi)$ whose elements are rational functions of s (for any fixed ξ in $\mathbf{R}^n$), we wish to find matrices $A(\xi)$, $B(\xi)$ and $C(\xi)$ depending on ξ, having dimensions $n \times n, n \times m$ and $r \times n$ respectively, such that

$$G(s,\xi) = C(\xi)\left(sI_n - A(\xi)\right)^{-1} B(\xi),$$

where I_n is the unit matrix of order n, $[A(\xi),\ B(\xi),\ C(\xi),\ 0]$ is termed a realization of $G(s,\xi)$ of order n, and is not, of course, unique. All such the above realizations will include matrices $G(s,\xi)$ having the least dimensions–be called the minimal realizations.

Suppose $R(s,\xi) = [r_{ij}(s,\xi)]$ is an $p \times m$ strictly proper rational-fraction matrix of s (for any fixed ξ in $\mathbf{R}^n$).

Theorem 3. *A realization* $[A(\xi),\ B(\xi),\ C(\xi),\ 0]$ *of a given transfer matrix* $G(s,\xi)$ *is minimal if* $(A(\xi), B(\xi))$ *is c.c. and* $(A(\xi), C(\xi))$ *c.o.*

The Lyapunov equations are

$$A(\xi)L_c(\xi) + L_c(\xi)A^T(\xi) = B(\xi)B^T(\xi), \tag{5}$$

$$A^T(\xi)L_o(\xi) + L_o(\xi)A(\xi) = C^T(\xi)C(\xi). \tag{6}$$

Define the two controllability and observability gramians:

$$\begin{aligned} L_c(\xi) &= \int_0^\infty e^{-A(\xi)t}B(\xi)B^T(\xi)e^{-A^T(\xi)t}dt, \\ L_o(\xi) &= \int_0^\infty e^{-A^T(\xi)t}C^T(\xi)C(\xi)e^{-A(\xi)t}dt. \end{aligned}$$

Theorem 4. *$L_c(\xi)$ and $L_o(\xi)$ are the unique solutions of* (5) *and* (6) *respectively.*

Define

$$\begin{aligned} f(s,\xi) &= [A(\xi),\, w(\xi),\, C(\xi),\, 0], \\ g(s,\xi) &= [-A^T(\xi),\, \lambda^{-1}(\xi)L_o(\xi)w(\xi),\, B^T(\xi),\, 0], \end{aligned} \tag{7}$$

and

$$X(s,\xi) = R(s,\xi) - \lambda(\xi) f(s,\xi)/g(s,\xi). \tag{8}$$

So

$$f(s,\xi) = C(\xi)(sI - A(\xi))^{-1}w(\xi) \in VRH_2^{\perp},$$

and

$$g(s,\xi) = B^T(\xi)(sI + A^T(\xi))^{-1}\lambda^{-1}(\xi)L_o(\xi)w(\xi) \in VRH_2.$$

Definition 9. Let α denote the infimal model-matching error

$$\alpha = \inf\{\|T_1 - T_2QT_3\|_\infty : Q \in VRH_\infty\}. \tag{9}$$

A matrix Q inVRH_∞ satisfying $\alpha = \|T_1 - T_2QT_3\|_\infty$ will be called optimal, where α is a model-matching error.

When $T_i(s,\xi)$ are scalar-valued, then there is no need for both $T_2(s,\xi)$ and $T_3(s,\xi)$. So we may as well suppose $T_3(s,\xi) = 1$. It is also assumed that $T_2^{-1}(s,\xi) \in VRH_\infty$ to avoid the trivial instance of the problem.

Returning to the model-matching problem, bring in an inner-outer factorization of $T_2(s,\xi)$: $T_2(s,\xi) = T_{2i}(s,\xi)T_{2o}(s,\xi)$, we have

$$|||T_1 - T_2Q|||_\infty = |||R - X|||_\infty.$$

We conclude that

$$\alpha = \inf\{\|R - X\|_\infty : X \in VRH_\infty\} = \mathrm{dist}(R, VRH_\infty).$$

Definition 10. The VL_p space, $1 \le p < \infty$, will be viewed as pth power integrable functions about s and ξ. When $p = \infty$, VL_∞ is the space of essentially bounded functions(for any fixed ξ in $\mathbf{R}^n$).

Definition 11. The VRL_p space, $1 \le p \le \infty$, will be viewed as a subset of VL_p, which consists of all real-rational functions of s and ξ.

Definition 12.

(i) Let $F(s,\xi) \in VL_\infty$ and $g(s,\xi) \in VL_2$. Then the operator

$$\Lambda_{F(s,\xi)} : \Lambda_{F(s,\xi)} g(s,\xi) = F(s,\xi) g(s,\xi)$$

is called the Laurent operator.

(ii) A related operator is $\Lambda_{F(s,\xi)}|VH_2$, the restriction of $\Lambda_{F(s,\xi)}$ to VH_2, which maps VH_2 to VL_2 where $F(s,\xi) \in VL_\infty$.

(iii) For $F(s,\xi)$ in VL_∞, the Hankel operator with symbol $F(s,\xi)$, denoted by $\Gamma_{F(s,\xi)}$, maps VH_2 to $VH_2^\perp$ and is defined as

$$\Gamma_{F(s,\xi)} := \Pi_1 \Lambda_{F(s,\xi)} | VH_2,$$

where $VL_2 = VH_2 \oplus VH_2^\perp$, and Π_1 is the projection from VL_2 onto $VH_2^\perp$.

Using similar to the classical methods we have the following conclusion:

Theorem 5. *There exists a closest VRH_∞-function $X(s,\xi)$ to a given VRL_∞-function $R(s,\xi)$, and $\|R - X\|_\infty = \|\Gamma_R\|$.*

Theorem 6. *The infimal model-matching error α equals $\|\Gamma_R\|$ and the unique optimal X equals $R(s,\xi) - \lambda(\xi) f(s,\xi)/g(s,\xi)$.*

For a single-input and single-output design in the form of parameter valued case, we have similar to ordinary computing method.

Example.

$$P(s,\xi) = [(s-1)(s-2)]/[(s+1)(s^2+s+1+\xi^2)] \in VRH_\infty,$$

$$\omega_1 = 0.01, \epsilon = 0.1.$$

From the above method, we derive

$$K(s,\xi) = 0.615(s+0.4)(s+1)(s^2+s+1+\xi^2)/$$

$$(s^4 + 6.145s^3 + 12.54s^2 + 13.53s + 0.0232).$$

Note $K(s,\xi) \notin RH_\infty$, but $K(s,\xi) \in VRH_\infty$.

Acknowledgment. The authors would like to thank the Organizing Committee of the Second Congress ISAAC 1999 Fukuoka for its support, Prof. S.S. Chern for his guidance, encouragement and help, and Prof. G.D. Hu, Prof. J. Kajiwara and Prof. C.C. Yang for their enduring help and support in this work.

References

[1] Barnett S., *Introduction to Mathematical Control Theory*, Clarennon Press, 1985.

[2] Curtain R. F. & Prichard A. J., *Infinite Dimensional Linear Systems Theory*, Springer-Verlag, Berlin, 1978.

[3] Francis B. A. *A Course in H_∞-Control Theory*, Springer-Verlag, Berlin, Heidelberg, New York, 1987.

[4] Francis B. A. & Zames G., *On H_∞-optimal sensitivity theory for SISO feedback system,* IEEE Trans. Auto. Cont., Vol. AC-29 (1984), 9–16.

[5] Francis B. A. & Doyle J. C., *Linear control theory with an H_∞ optimality criterion,* SIAM J. Control and Optimization, 25(1987), 815–844.

[6] Hu C. G. & Yang C. C., *Vector-Valued Functions and their Applications*, Kluwer Academic Publishers, Dordrecht, 1992.

[7] Keulen B. V., *H_∞-control with measurement-feedback for linear infinite- dimensional systems*, Journal of Mathematical Systems, Estimation and Control, 3(1993), 373–411.

[8] Nehari Z., *On bounded bilinear forms*, Ann. of Math., 65(1957), 153–162.

[9] Petrushev P. P. & Popov V. A., *Rational Approximation of Real Function*, Cambridge University Press, New York, 1987.

[10] Power S. C. *Hankel Operators on a Hibert Space*, Pitman, London, 1982.

[11] Rudin W., *Functional Analysis*, McGraw-Hill, Inc, 1973.

[12] Zames G. & Francis B. A., *Feedback, minimax sensitivity, and optimal robustness,* IEEE Trans. Auto. Cont, Vol.AC-28 (1983), 585–601.

19 NONLINEAR RIEMANN-HILBERT PROBLEMS OF FIRST ORDER QUASILINEAR ELLIPTIC SYSTEM[1]

Ming Zhong LI

College of Sciences, Shanghai University
Shanghai 200072, China

1991 Mathematics Subject Classification: 30G20, 35J55.
Key words: Nonlinear Riemann-Hilbert problem, nonlinear singular integral equation, existence theorem.

Abstract: In this paper we study nonlinear Riemann-Hilbert problems for the first order quasilinear elliptic system with the general form. Because its equivalent form is nonlinear singular integral equation and the boundary condition is strongly nonlinear, so its proof is more difficult. We overcome these difficulties by means of the results which are got by V. S. Vinogradov and me in the papers[1,2] and integral equation theory. We have proved the existence theorem of the solution for the BVP, which generalize the results of nonlinear Riemann-Hilbert problem for the analytic function and first order elliptic system[3–5].

[1]Project supported by the National Natural Science Foundation of China(19671056) and Shanghai Scientific Foundation.

H.G.W. Begehr et al. (eds.), Proceedings of the Second ISAAC Congress, Vol. 1, 145–151.

STATEMENT AND MAIN RESULTS OF THE PROBLEMS

We consider the following nonlinear boundary value problems for the first order quasilinear elliptic system with the general form

$$w_{\bar{z}} - q_1(z,w)w_z - q_2(z,w)\bar{w}_{\bar{z}} = F(z,w), \ \ z \in G, \tag{1}$$

$$\psi(s, u(e^{is}), v(e^{is})) = f(s), \ \ e^{is} = z \in \Gamma, \tag{2}$$

where G is unit disk and $\Gamma = \{z \mid |z| = 1\}$ is its boundary, F, ψ, f, q_i are known functions and q_i satisfy the following uniformly elliptic condition

$$|q_1(z,w)| + |q_2(z,w)| \leq q_0 < 1, \quad z \in \overline{G}, \ w \in E, \tag{3}$$

where E is a complex plane.

Now we introduce the parameter $t \in [0,1]$ and consider the following BVP

$$w_{\bar{z}} - q_1(z,w)w_z - q_2(z,w)\bar{w}_{\bar{z}} = tF(z,w), \ \ z \in G, \tag{4}$$

$$t\psi(s,u,v) + (1-t)(u\psi_u(s,u,v) + v\psi_v(s,u,v)) = tf(s), \ \ e^{is} = z \in \Gamma. \tag{5}$$

Obviously, when $t = 0$, it is Riemann-Hilbert problems for the homogeneous first order quasilinear elliptic system, and when $t = 1$, it is the problem (1)–(2).

Find derivatives of (4),(5) with respect to t, we may obtain the equation

$$\begin{gathered}\dot{w}_{\bar{z}} - q_1(z,w)\dot{w}_z - q_2(z,w)\overline{\dot{w}}_{\bar{z}} \\ +[A_1(z,w) + A_1^*(z,w,w_z)]\dot{w} + [A_2(z,w) + A_2^*(z,w,w_z)]\overline{\dot{w}} = F(z,w)\end{gathered} \tag{6}$$

with the boundary condition

$$\mathrm{Re}[\overline{\lambda(t,w)}\dot{w}(z)] = g(t,\dot{w}(z)) = g(t,\dot{w}(z)), \ \ z \in \Gamma, \ t \in [0,1], \tag{7}$$

where

$$\begin{cases} A_1(z,w) = -tF_w(z,w), \ \ A_2(z,w) = -tF_{\overline{w}}(z,w), \\ A_1^*(z,w,w_z) = q_{1w}(z,w)w_z + q_{2w}(z,w)\overline{w}_{\bar{z}}, \\ A_2^*(z,w,w_z) = q_{1\overline{w}}(z,w)w_z + q_{2\overline{w}}(z,w)\overline{w}_{\bar{z}}, \end{cases} \tag{8}$$

and

$$\begin{cases} \dot{w} = \dfrac{dw}{dt} = \dot{u} + i\dot{v}, & \lambda = a + ib, \ |\lambda| = 1, \\ a = \dfrac{\tilde{a}}{|\tilde{a} + i\tilde{b}|}, & \tilde{a} = \psi_u + (1-t)\psi_{uu}u + (1-t)\psi_{uv}v, \\ b = \dfrac{\tilde{b}}{|\tilde{a} + i\tilde{b}|}, & \tilde{b} = \psi_v + (1-t)\psi_{vu}u + (1-t)\psi_{vv}v, \\ g = \dfrac{\tilde{g}}{|\tilde{a} + i\tilde{b}|}, & \tilde{g} = u\psi_u + v\psi_v - \psi + f. \end{cases} \tag{9}$$

Assumption A:

(i) $q_i(z, w)$ and $F(z, w)$ are L_p-integrable with respect to $z \in \overline{G}$ and continuous with respect to $w \in E$ and their partial derivatives up to first order satisfy Hölder-Lipschitz condition. When $w(z) \in D_{1,p}(\overline{G})$, $q_i(z, w(z))$ and $F(z, w(z))$ are uniformly bounded for $D_{1,p}$ norms.

(ii) $\psi(s, u, v)$ and its derivatives up to second order uniformly satisfy Hölder-Lipschitz condition. The H-L norms of the $\psi(s, u, v)$ and its partial derivatives up to first order are sufficiently small.

(iii) $f(s)$, $s \in [0, 2\pi]$ is 2π-periodic real continuous function and $f'(s) \in L_p(\Gamma)$, $p > 2$.

(iv) For arbitrary fixed w, the argument of λ is a single-valued function, $\lambda(z, w) \neq 0$, and then $\frac{1}{2\pi} \int_\Gamma d \arg \lambda(z, 0) = 0$.

It is not difficult to know that, under Assumption A(i), when $w(z) \in D_{1,p}(\overline{G})$, $A_i(z, w(z))$, $A_i^*(z, w, w_z) \in L_p(\overline{G}), i = 1, 2$ and are uniformly bounded for the norms.

We know that the equation (6) is equivalent to the following integral equation

$$\begin{aligned} \dot{w}(z) &= \phi(z) + T(q_1(z, w(z))\dot{w}_z + q_2(z, w(z))\overline{\dot{w}}_{\bar{z}}) \\ &\quad -T(A_1(z, w(z))\dot{w} + A_2(z, w(z))\overline{\dot{w}}) - T(A_1^*(z, w(z), w_z)\dot{w} \\ &\quad +A_2^*(z, w(z), w_z)\overline{\dot{w}}) + T(F(z, w(z))), \end{aligned} \tag{10}$$

where $Tf = -\frac{1}{\pi} \iint_G \frac{f(\zeta)}{\zeta - z} d\sigma_\zeta$, $\phi(z)$ is an arbitrary holomorphic function in G.

By using of the Green formula, we can transfer the integral equation (10) to the following equivalent formula

$$\begin{aligned}\dot{w}(z) &= \phi_1(z) - \frac{1}{\pi}\iint_G \frac{q_1(\zeta, w(\zeta))\dot{w}(\zeta)}{(\zeta - z)^2} d\sigma_\zeta - q_2(z, w(z))\overline{\dot{w}(z)} \\ &+\frac{1}{\pi}\iint_G \frac{[\frac{\partial q_1(\zeta,w(\zeta))}{\partial \zeta} + A_1(\zeta, w(\zeta)) + A_1^*(\zeta, w(\zeta), w_\zeta)]\dot{w}(\zeta)}{\zeta - z} d\sigma_\zeta \\ &+\frac{1}{\pi}\iint_G \frac{[\frac{\partial q_2(\zeta,w(\zeta))}{\partial \zeta} + A_2(\zeta, w(\zeta)) + A_2^*(\zeta, w(\zeta), w_\zeta)]\overline{\dot{w}(\zeta)}}{\zeta - z} d\sigma_\zeta \\ &-\frac{1}{\pi}\iint_G \frac{F(\zeta, w(\zeta))}{\zeta - z} d\sigma_\zeta,\end{aligned} \tag{11}$$

where $\phi_1(z)$ is also an arbitrary holomorphic function

$$\phi_1(z) = \phi(z) + \frac{1}{2\pi i}\int_\Gamma \frac{q_1(\zeta, w(\zeta))\dot{w}(\zeta)}{\zeta - z} d\bar{\zeta} + \frac{1}{2\pi i}\int_\Gamma \frac{q_2(\zeta, w(\zeta))\overline{\dot{w}(\zeta)}}{\zeta - z} d\zeta.$$

Obviously, when $w(\zeta), \dot{w}(\zeta) \in C_\alpha(\overline{G})$, $\phi_1(z) \in C_\alpha(\overline{G})$.

Now we consider the boundary condition (7) and introduce substitution

$$\dot{w} = w^* e^{ip(z)}, \tag{12}$$

where $p(z)$ is the Schwartz integral

$$p(z) = p(z, 0) = \mathrm{Sch}(\arg \lambda(t, 0)) = \frac{1}{2\pi}\int_0^{2\pi} \arg(\lambda(t, 0))\frac{e^{i\theta} + z}{e^{i\theta} - z} d\theta. \tag{13}$$

Then

$$\dot{w}_{\bar{z}} = e^{ip(z)} {w^*}_{\bar{z}}.$$

Note

$$p(z) = p_1(z) + ip_2(z),$$

where $p_1(z) = \mathrm{Re}[p(z)]$, $p_2(z) = \mathrm{Im}[p(z)]$.

So the boundary condition (7) can be transferred as

$$\mathrm{Re}[w^*(z)] = g^*(z, w^*(z)) = e^{p_2(z)} g(z, w) + \mathrm{Re}[\bar{\lambda}(z, 0) - \bar{\lambda}(z, w))e^{ip_1(z)} w^*]. \tag{14}$$

Substituting (12) into (11), the integral equation (11) is also written as

$$w^*(z) - q_2(z, w(z))e^{-2ip_1(z)}\overline{w^*(z)} - e^{-ip(z)}\Pi(q_1(\zeta, w(\zeta))e^{ip(\zeta)}w^*(\zeta))$$

$$= \phi_*(z) - e^{-ip(z)}T([\frac{\partial q_1(\zeta, w(\zeta))}{\partial \zeta} + A_1(\zeta, w(\zeta)) +$$

$$A_1^*(\zeta, w(\zeta), w_\zeta)]e^{ip(\zeta)}w^*(\zeta))$$

$$-e^{-ip(z)}T([\frac{\partial q_2(\zeta, w(\zeta))}{\partial \bar{\zeta}} + A_2(\zeta, w(z)) + A_2^*(\zeta, w(\zeta), w_\zeta)])e^{\overline{ip(\zeta)}}\overline{w^*(\zeta))}$$

$$+e^{-ip(z)}T(F(\zeta, w(\zeta))), \tag{15}$$

where

$$\Pi f = -\frac{1}{\pi}\iint_G \frac{f(\zeta)}{(\zeta - z)^2} d\sigma_\zeta,$$

$\phi^*(z) = e^{-ip(z)}\phi(z)$ is an arbitrary holomorphic function in G.

Introducing holomorphic function $p^*(z)$,

$$p^*(z) = p^*(z, 0) = -\frac{1}{2\pi i}\int_\Gamma \frac{\arg \lambda(t, 0)(\tau + z)}{(\tau - z)\tau} d\tau. \tag{16}$$

Thus the singular integral equation (15) is transferred as follows

$$w^*(z) - q_2(z, w(z))e^{-2ip_1(z)}\overline{w^*(z)}$$

$$-[e^{-ip(z)}\Pi(q_1(\zeta, w(\zeta))e^{ip(\zeta)}w^*(\zeta)) -$$

$$e^{-ip^*(z)}\Pi^*(q_1(\zeta, w(\zeta))e^{ip(\zeta)}w^*(\zeta))]$$

$$+e^{-ip(z)}T([\frac{\partial q_1(\zeta, w(\zeta))}{\partial \zeta} + A_1(\zeta, w(\zeta)) + A_1^*(\zeta, w(\zeta), w_\zeta)]e^{ip(\zeta)}w^*(\zeta))$$

$$+e^{ip^*(z)}T^*([\frac{\partial q_1(\zeta, w(\zeta))}{\partial \zeta} + A_1(\zeta, w(\zeta)) + A_1^*(\zeta, w_(\zeta), w_\zeta)]e^{ip(\zeta)}w^*(\zeta))$$

$$+e^{-ip(z)}T([\frac{\partial q_2(\zeta, w(\zeta))}{\partial \bar{\zeta}} + A_2(\zeta, w(\zeta)) +$$

$$A_2^*(\zeta, w(\zeta), w_\zeta)]e^{-\overline{ip(\zeta)}}\overline{w^*(\zeta)})$$

$$+e^{ip^*(z)}T^*([\frac{\partial q_2(\zeta, w(\zeta))}{\partial \bar{\zeta}} + A_2(\zeta, w(\zeta)) +$$

$$A_2^*(\zeta, w(\zeta), w_\zeta)]e^{-\overline{ip(\zeta)}}\overline{w^*(\zeta)})$$

$$-e^{-ip(z)}T(F(\zeta, w(\zeta))) + e^{ip^*(\zeta)}T^*\overline{F(\zeta, w(\zeta))} = \phi_0^*(z), \tag{17}$$

where $\phi_0^*(z)$ is a holomorphic function, when $q_2(z, w(z)) = 0$, it can be expressed as

$$\phi_0^*(z) = \frac{1}{2\pi i}\int_\Gamma \frac{g^*(\zeta, w^*(\zeta))(\zeta + z)}{(\zeta - z)\zeta} d\zeta + ic_0$$

and

$$T^* f = -\frac{z}{\pi}\iint_G \frac{\overline{f(\zeta)}}{1 - \bar{\zeta}z} d\sigma_\zeta, \ \Pi^* f = -\frac{z^2}{\pi}\iint_G \frac{\overline{f(\zeta)}}{(1 - \bar{\zeta}z)^2} d\sigma_\zeta.$$

Based on this result we further prove the following important theorems:

Theorem 1. *Under Assumption A, for arbitrary given $\omega(z) \in C_\alpha(\overline{G}) \cap D_{1,p}(\overline{G})$, when the H-L norms of the $\psi(s, u, v)$ and its first order derivatives are sufficiently small, the BVP for the first order elliptic system*

$$\begin{aligned} \dot{w}_{\bar{z}} - q_1(z, \omega(z))\dot{w}_z - q_2(z, \omega(z))\overline{\dot{w}_{\bar{z}}} + [A_1(z, w) + A_1^*(z, w, \omega_z)]\dot{w} \\ + [A_2(z, w) + A_2^*(z, w, \omega_z)]\overline{\dot{w}} = F(z, w) \end{aligned} \tag{18}$$

with boundary condition (7) must be solvable and the solution $\dot{w}(z)$ belongs to $C_{\alpha_0}(\overline{G}) \cap D_{1,p_0}(\overline{G})$, which dependents on arbitrarily read constant c_0.

Theorem 2. *Under Assumption A, when the H-L coefficients of the $\psi(s, u, v)$ and its derivatives up to second order and $M_{q'}, H_{q'}, H_A, A_0 = ||F_w(z, 0)||_{p_0}$ are sufficiently small, then the nonlinear BVP (6),(7) has unique solution $w^*(z)$, which belongs to $C_{\alpha_0}(\overline{G}) \cap D_{1,p_0}(\overline{G})$, where α_0, p_0 are constants, and*

$$\frac{1}{2} < \alpha_0 < 1, \ 2 < p_0 < p.$$

References

[1] Vinogradov V. S., *On bounded solution for linear elliptic systems in the plane,* DAH, 121 (1958), 399-402.

[2] Li Mingzhong, *On a class of nonlinear boundary value problems for the first order elliptic system with the general form,* Fudan Journal (Natural Science), 1 (1987), 85–93.

[3] Wolfersdorf L. V., *A class of nonlinear Riemann-Hilbert problems for holomorphic functions,* Math. Nachr., 116 (1984), 89–107.

[4] Begehr H. & Hsiao G. C., *Nonlinear boundary value problems for a class of elliptic systems,* Lecture Note in Math., Springer-Verlag, Berlin, 1980.

[5] Sigrid Keneht & Ali Seif Mshimba, *On the existence of a solution to the Riemann-Hilbert problems for a partial differential equation,* Complex variables, 6 (1986), 240–263.

20 THE ALGORITHM IMPLEMENTATION OF CAUCHY SINGULAR INTEGRAL IN DAUBECHIES WAVELETS ON THE INTERVAL

W. LIN
Q. LI

Department of Mathematics, Zhongshan University
Guangzhou 510275, P. R. China

stslw@zsu.edu.cn

1991 Mathematics Subject Classification: 30E20, 41A50
Key words: Cauchy singular integral, non-standard form, Daubechies wavelets on the interval

Abstract: Based on Daubechies wavelets on the interval, a fast algorithm for calculating Cauchy singular integral is presented. Before this, the sparseness of the representation of Cauchy singular integral in Daubechies wavelets on the interval are also studied. In the last section, we compare the errors obtained by the three methods: general wavelets, periodic wavelets and wavelets on the interval.

The numerical result shows obviously that the errors obtained by the last method are independent on the location of the variance and the

H.G.W. Begehr et al. (eds.), Proceedings of the Second ISAAC Congress, Vol. 1, 153–161.

choice of the coarsest scale.

I Introduction

Assume L is a set of curve $L_1, \ldots, L_n$ in complex plain, where L_i and L_j are not interest for $i \neq j$. Given a complex function $f(t)$ defined on L. Suppose $t_0 \in L$ and $O(t_0, \eta)$ is a circle, t_0 is its center and η is its radius. $O(t_0, \eta)$ inter sets L in two points t', t''. Denote $L_\eta = t'\hat{t}''$. If

$$\lim_{\eta \to 0} \frac{1}{2\pi i} \int_{L_\eta} \frac{f(t)}{t - t_0} dt \tag{1.1}$$

exists, the limit is called **Cauchy singular integral** of $f(t)$ along L and denote

$$\frac{1}{2\pi i} p.v. \int_L \frac{f(t)}{t - t_0} dt. \tag{1.2}$$

It is well known that many boundary value problems can be converted into the corresponding Cauchy singular integral equations which have been extensively studied in both analytic form and numerical form. Since most of the Cauchy singular integral equations do not have analytic solutions, numerical solutions catch more attentions than analytic one. There were many references, such as [DS, Ge, Go, P], on this problem before the wavelet enters.

In [BCR], Beylkin, Coifman and Rokhlin introduced a wavelet-based algorithm for the rapid numerical application of a class of linear operators to arbitrary vectors. The algorithm is established on the theory of orthonormal wavelets with compact supports firstly constructed by Daubechies in [D].

In [CDV], the authors constructed a kind of orthonormal basis of $L^2[0,1]$ from Daubechies wavelet basis of $L^2(R)$. They inherit the vanishing moment property of the wavelets in $L^2(R)$ and the most successful point is that the numbers of wavelets and scaling functions in each level are the same, which is exceptionally useful in numerical implementation.

Let ϕ be a scaling function and ψ be its corresponding wavelet which has N vanishing moments. Suppose supp$\phi(x)$=supp$\psi(x) = [1-N, N]$. Let ϕ_k^{left} and $\psi_k^{left}, k = 0, \ldots, N-1$, be the left boundary scaling functions and wavelets and ϕ_k^{right} and $\psi_k^{right}, k = -N, \ldots, -1$, be the right boundary scaling functions and wavelets.

The following are the notations frequently used in our further discussion:

$$\psi_{j,k}^{left}(x) \quad = \quad 2^{-j/2}\psi_k^{left}(2^{-j}x), \quad k = 0, \ldots, N-1$$

$$\begin{aligned}\psi_{j,k}(x) &= 2^{-j/2}\psi(2^{-j}x-k), \quad k=N,\ldots,2^j-N-1\\ \psi^{right}_{j,k}(x) &= 2^{-j/2}\psi^{right}_k(2^{-j}x), \quad k=-N,\ldots,-1.\end{aligned}$$

We denote $\phi^{left}_{j,k}, \phi_{j,k}$ and $\phi^{right}_{j,k}$ similarly as $\psi^{left}_{j,k}, \psi_{j,k}$ and $\psi^{right}_{j,k}$. The boundary functions satisfy the following scaling relations:

$$\phi^{left}_{-j,k} = \sum_{l=0}^{N-1} H^{left}_{k,l}\phi^{left}_{-j-1,l} + \sum_{m=N}^{N+2k} h^{left}_{k,m}\phi_{-j-1,m},$$

$k=0,\ldots,N-1$,

$$\psi^{left}_{-j,k} = \sum_{l=0}^{N-1} G^{left}_{k,l}\phi^{left}_{-j-1,l} + \sum_{m=N}^{N+2k} g^{left}_{k,m}\phi_{-j-1,m},$$

$k=0,\ldots,N-1$,

$$\phi^{right}_{-j,k} = \sum_{l=-N}^{-1} H^{right}_{k,l}\phi^{right}_{-j-1,l} + \sum_{m=-N+1+2k}^{-N-1} h^{right}_{k,m}\phi_{-j-1,m},$$

$k=-N,\ldots,-1$

$$\psi^{right}_{-j,k} = \sum_{l=-N}^{-1} G^{right}_{k,l}\phi^{right}_{-j-1,l} + \sum_{m=-N+1+2k}^{-N-1} g^{right}_{k,m}\phi_{-j-1,m},$$

$k=-N,\ldots,-1$.

where $H^{left}_{k,l}, h^{left}_{k,m}, G^{left}_{k,l}, g^{left}_{k,m}, H^{right}_{k,l}, h^{right}_{k,m}, G^{right}_{k,l}, g^{right}_{k,m}, 0\le k,l\le N-1, N\le m\le N+2k$ are the boundary filters which are introduced in [CDV]. The interior functions ϕ and ψ satisfy the scaling relation in common sense.

Denote $\phi^0_{j,k} = \phi^{left}_{j,k}, \psi^0_{j,k} = \psi^{left}_{j,k}$ for $0\le k\le N-1$ and $\phi^1_{j,2^j-k}(x) = \phi^{right}_{j,-k}(x-1)$, $\psi^1_{j,2^j-k}(x) = \psi^{right}_{j,k}(x-1)$ for $1\le k\le N$. Then when $2\ge 2N$, the set

$$\begin{aligned}&\bigcup_{j\ge J}[\{\psi^0_{-j,k}: k=0,\ldots,N-1\}\bigcup\{\psi_{-j,k}: m=N,\ldots,2^j-N-1\}\\ &\qquad\bigcup\{\psi^1_{-j,2^j-N+k}: k=0,\ldots,N-1\}]\\ &\bigcup\{\phi^0_{-J,k}: k=0,\ldots,N-1\}\bigcup\{\phi_{-j,m}: m=N,\ldots,2-N-1\}\\ &\qquad\bigcup\{\phi^1_{-J,2-N+k}: k=0,\ldots,N-1\}\end{aligned}$$

is an orthonormal basis of $L^2[0,1]$. For the convenience of the later discussion, we end this section with a new notation:

$$\phi^{\epsilon}_{-j,k} = \begin{cases}\phi^0_{-j,k} & 0\le k\le N-1,\\ \phi_{-j,k} & N\le k\le 2^j-N-1,\\ \phi^1_{-j,k} & 2^j-N\le k\le 2^j-1.\end{cases}$$

II Sparseness of Cauchy singular integral

A domain $[x, x+a] \times [t, t+b] \subset R^2$ is called **separated from the diagonal** if $a + \max\{a, b\} \le t - x$ or $b + \max\{a, b\} \le x - t$.

Denote

$$\begin{aligned}
\eta_{k,l}^{-j} &= \int_0^1 \int_0^1 \frac{1}{t-x} \phi_{-j,k}^{\epsilon}(t) \phi_{-j,l}^{\epsilon}(x) dt dx, \\
\alpha_{k,l}^{-j} &= \int_0^1 \int_0^1 \frac{1}{t-x} \psi_{-j,k}^{\epsilon}(t) \psi_{-j,l}^{\epsilon}(x) dt dx, \\
\beta_{k,l}^{-j} &= \int_0^1 \int_0^1 \frac{1}{t-x} \phi_{-j,k}^{\epsilon}(t) \psi_{-j,l}^{\epsilon}(x) dt dx, \\
\gamma_{k,l}^{-j} &= \int_0^1 \int_0^1 \frac{1}{t-x} \psi_{-j,k}^{\epsilon}(t) \phi_{-j,l}^{\epsilon}(x) dt dx.
\end{aligned}$$

Let $\alpha^{-j} = (\alpha_{k,l}^{-j}), \beta^{-j} = (\beta_{k,l}^{-j})$ and $\gamma^{-j} = (\gamma_{k,l}^{-j})$. Denote matrix $M_j = \begin{pmatrix} \alpha^{-j} & \beta^{-j} \\ \gamma^{-j} & 0 \end{pmatrix}$ and a compound matrix

$$M = \begin{pmatrix} M_1 & & \\ & M_2 & \\ & & \ddots \end{pmatrix}.$$

We call the matrix M the **non-standard form** of Cauchy singular integral operator.

Denote

$$I_{j,l}^0 = \text{supp}\psi_{j,l}^0 = \text{supp}\phi_{j,l}^0, \; I_{j,l} = \text{supp}\psi_{j,l} = \text{supp}\phi_{j,l},$$

$$I_{j,l}^1 = \text{supp}\psi_{j,l}^1 = \text{supp}\phi_{j,l}^1,$$

and

$$I_{j,k}^{\epsilon} = \begin{cases} I_{j,k}^0 & 0 \le k \le N-1 \\ I_{j,k} & N \le k \le 2^j - N - 1 \\ I_{j,k}^1 & 2^j - N \le k \le 2^j - 1. \end{cases}$$

Lemma 2.1 *If domain $I_{j,k}^{\epsilon} \times I_{j,l}^{\epsilon}$ is separated from the diagonal, then*

$$|\alpha_{k,l}^{-j}|, |\beta_{k,l}^{-j}|, |\gamma_{k,l}^{-j}| \le \frac{1}{2^j 3^{N-1}}. \tag{2.1}$$

Proof. The proof is parallel to the that in [A].

Lemma 2.2 *If $I_{j,k}^{\epsilon} \times I_{j,l}^{\epsilon}$ is not separated from the diagonal, then $|k-l| < 4N-2$.*

Proof. We just need to consider the following cases.

Case 1: $k = 0, \ldots, N-1, l = N, \ldots, 2^j - N - 1$. Note that $I^0_{j,k} = [0, \frac{N+k}{2^j}]$ and $I_{j,l} = [\frac{1-N+l}{2^j}, \frac{N+l}{2^j}]$, i.e., $a_k = \frac{N+k}{2^j}, b_l = \frac{2N-1}{2^j}, a_k < b_l$.

$$\begin{aligned}
& I^0_{j,k} \times I_{j,l} \text{ is not separated from diagonal} \\
\Rightarrow \quad & t_l - x_k = \frac{1-N+l}{2^j} < a_k + \max\{a_k, b_l\}, \text{ and} \\
& x_k - t_l = -\frac{1-N+l}{2^j} < b_l + \max\{a_k, b_l\} \\
\Rightarrow \quad & l - k < 4N - 2 \\
& \text{and } l - k \leq l - (N-1) < 4N - 2. \text{ (Notice that } k \leq N-1)
\end{aligned}$$

Case 2: $k, l = N, \ldots, 2^j - N - 1$. Note that $I_{j,k} = [\frac{1-N+k}{2^j}, \frac{N+k}{2^j}]$ and $I_{j,l} = [\frac{1-N+l}{2^j}, \frac{N+l}{2^j}]$, i.e., $a_k = b_l = \frac{2N-1}{2^j}$.

$$\begin{aligned}
& I_{j,k} \times I_{j,l} \text{ is not separated from the diagonal} \\
\Rightarrow \quad & |x_k - t_l| < 2a_k \\
\Rightarrow \quad & |k - l| < 4N - 2.
\end{aligned}$$

Case 3: $k = 2^j - N, \ldots, 2^j - 1, l = N, \ldots, 2^j - N - 1$. Note that $I^1_{j,k} = [\frac{k-N+1}{2^j}, \frac{N+K}{2^j}]$ and $I_{j,l} = [\frac{1-N+l}{2^j}, \frac{N+l}{2^j}], i.e., a_k = \frac{2^j-k+N-1}{2^j}, b_l = \frac{2N-1}{2^j}, a_k < b_l$.

$$\begin{aligned}
& I^1_{j,k} \times I_{j,l} \text{is not separated from the diagonal} \\
\Rightarrow \quad & x_k - t_l = \frac{k-l}{2^j} < a_k + \max\{a_k, b_l\} = \frac{2^j - k + 3N - 2}{2^j}, \text{ and} \\
& t_l - x_k = -\frac{l-k}{2^j} < b_l + \max\{a_k, b_l\} = \frac{4N-2}{2^j} \\
\Rightarrow \quad & k - l < 2^j - k + 3N - 2 \leq 4N - 2 \text{ (Notice that } 2^j - k \leq N \\
& \text{and } l - k < 4N - 2).
\end{aligned}$$

Theorem 2.1 *The non-standard form of Cauchy singular integral in the two dimensional wavelet basis on interval is sparse.*

Proof. The sparseness of the non-standard form is easy to get from Lemmas 2.1 and 2.2.

III Algorithm and numerical result

When $0 \leq k \leq N-1, k \leq l \leq N-1$.

$$\alpha^{-j}_{k,l} \;=\; \int_0^1 \int_0^1 \frac{1}{t-x} \psi^{\epsilon}_{-j,k}(t) \psi^{\epsilon}_{-j,l}(t) dt dx$$

$$
\begin{aligned}
&= \int_0^1\int_0^1 \frac{1}{t-x}[\sum_{n=0}^{N-1} G_{k,n}^{left}\phi_n^{-j-1}(t)+\sum_{m=N}^{N+2k} g_{k,m}^{left}\phi_m^{-j-1}(t)] \\
&\quad [\sum_{p=0}^{N-1} G_{l,p}^{left}\phi_p^{-j-1}(t)+\sum_{q=N}^{N+2l} g_{l,q}^{left}\phi_q^{-j-1}(t)]dtdx \\
&= \sum_{n,p} G_{k,n}^{left}G_{l,p}^{left}\eta_{n,p}^{-j-1}+\sum_{n,q} G_{k,n}^{left}g_{l,q}^{left}\eta_{n,q}^{-j-1} \\
&+ \sum_{m,p} g_{k,m}^{left}G_{l,p}^{left}\eta_{m,p}^{-j-1}+\sum_{m,q} g_{k,m}^{left}g_{l,q}^{left}\eta_{m,q}^{-j-1}.
\end{aligned}
$$

Denote

$$
\begin{aligned}
G_{k,p}^0 &= \begin{cases} G_{k,p}^{left}, & 0\le p\le N-1, \\ g_{k,p}^{left}, & N\le p\le 2N+k, \end{cases} \\
H_{k,p}^0 &= \begin{cases} H_{k,p}^{left}, & 0\le p\le N-1, \\ h_{k,p}^{left}, & N\le p\le 2N+k, \end{cases} \\
G_{k,p}^1 &= \begin{cases} G_{k,p-2^j}^{right}, & 2^j-N\le p\le 2^j-1, \\ g_{k,p-2^j}^{right}, & 2^j-2N-k-1\le p\le 2^j-N-1, \end{cases} \\
H_{k,p}^1 &= \begin{cases} H_{k,p-2^j}^{right}, & 2^j-N\le p\le 2^j-1, \\ h_{k,p-2^j}^{right}, & 2^j-2N-k-1\le p\le 2^j-N-1. \end{cases}
\end{aligned}
$$

Let $k' = k-2^j, l' = l-2^j$. Then we get

$$
\alpha_{k,l}^{-j} = \sum_{p,q=0}^{N+2k} G_{k,p}^0 G_{l,q}^0 \eta_{p,q}^{-j-1}. \tag{3.1}
$$

Similarly, we get

$$
\alpha_{k,l}^{-j} = \begin{cases}
\sum_{p=0}^{N+2k}\sum_q G_{k,p}^0 g_{q-2l}\eta_{p,q}^{-j-1}, \\
0\le k\le N-1, \qquad N\le l\le 2^j-N-1, \\
\sum_{p=0}^{N+2k}\sum_{q=-1}^{-N+2k'} G_{k,p}^0 G_{k',q}^1 \eta_{p,2^{j+1}+q}^{-j-1}, \\
0\le k\le N-1, \qquad 2^j-N\le l\le 2^j-1, \\
\sum_{p,q} g_{p-2k}g_{q-2l}\eta_{p,q}^{-j-1}, \\
N\le k\le 2^j-N-1, \qquad k\le l\le 2^j-N-1, \\
\sum_p\sum_{q=-1}^{-N+2l} g_{p-2k}G_l^1\eta_{p,2^{j+1}+q}^{-j-1}, \\
N\le k\le 2^j-N-1, \qquad 2^j-N\le l\le 2^j-1, \\
\sum_{p=-1}^{-N+2k'}\sum_{q=-1}^{-N+2l'} G_{k',p}^1 G_{l',q}^1 \eta_{2^{j+1}+p,2^{j+1}+q}^{-j-1}, \\
2^j-N\le k\le 2^j-1, \qquad k\le l\le 2^j-1,
\end{cases} \tag{3.2}
$$

and

$$\beta_{k,l}^{-j}=\begin{cases}\sum_{p,q=0}^{N+2k} G_{k,p}^{0}H_{l,q}^{0}\eta_{p,q}^{-j-1}\\ 0\le k\le N-1, \qquad k\le l\le N-1,\\ \sum_{p=0}^{N+2k}\sum_{q} G_{k,p}^{0}h_{q-2l}\eta_{p,q}^{-j-1},\\ 0\le k\le N-1, \qquad N\le l\le 2^{j}-N-1,\\ \sum_{p=0}^{N+2k}\sum_{q=-1}^{-N+2k'} G_{k,p}^{0}H_{k',q}^{1}\eta_{p,2^{j+1}+q}^{-j-1},\\ 0\le k\le N-1, \qquad 2^{j}-N\le l\le 2^{j}-1,\\ \sum_{p,q} g_{p-2k}h_{q-2l}\eta_{p,q}^{-j-1},\\ N\le k\le 2^{j}-N-1, \qquad k\le l\le 2^{j}-N-1,\\ \sum_{p}\sum_{q=-1}^{-N+2l} g_{p-2k}H_{l}^{1}\eta_{p,2^{j+1}+q}^{-j-1},\\ N\le k\le 2^{j}-N-1, \qquad 2^{j}-N\le l\le 2^{j}-1,\\ \sum_{p=-1}^{-N+2k'}\sum_{q=-1}^{-N+2l'} G_{k',p}^{1}H_{l',q}^{1}\eta_{2^{j+1}+p,2^{j+1}+q}^{-j-1},\\ 2^{j}-N\le k\le 2^{j}-1, \qquad k\le l\le 2^{j}-1.\end{cases} \tag{3.3}$$

In the following, we compare the errors of the numerical calculation of Cauchy singular integral in three methods: general wavelets, periodic wavelets, wavelets on the interval. In all cases, the experiments are performed for $N = 64$. So the finest level is 6. We have three choices of the coarsest level: 5, 4, 3. We calculate the Cauchy singular integral of functions $\sin(x)$. From the following tables of the errors, we will find that the errors of the calculation in wavelets on the interval is the best in two senses.

Example. $f(j) = \sin(j/64)$.

Table 1: The coarsest level=5

	general wavelets	periodic wavelets	wavelets on the interval
0	0.2354015684	1.6298055096	0.0000000042
1	0.0033585840	0.0000000116	0.0000000046
2	0.0034136426	0.0000000104	0.0000000055
3	0.0034705367	0.0000000104	0.0000000178
4	0.0035293595	0.0000000106	0.0000000090
⋮	⋮	⋮	⋮
59	0.0520578868	0.0000000123	0.0000000537
60	0.0694105176	0.0000000144	0.0000000171
61	0.1041157809	0.0000000170	0.0000000124
62	0.2082315727	0.0000000237	0.0000000114
63	0.8064907480	0.4030569518	0.0000000104

Table 2: The coarsest level=4

	general wavelets	periodic wavelets	wavelets on the interval
0	0.4204908703	2.9755180665	0.0000000067
1	0.0585814227	0.3605826124	0.0000000102
2	0.0154712542	0.2081824364	0.0000000097
3	0.0118560977	0.0000000209	0.0000000234
4	0.0120211730	0.0000000219	0.0000000152
⋮	⋮	⋮	⋮
59	0.2937336725	0.0000000248	0.0000000616
60	0.4046616206	0.0000000293	0.0000000257
61	0.3086294307	0.6505531436	0.0000000196
62	0.7270115116	1.1267910771	0.0000000183
63	1.0423279004	0.7049797065	0.0000000168

Firstly, they are independent on the locations of the variants. Secondly, they are also independent on the choice of the coarsest level. Comparing with wavelets on the interval, the errors of the calculation in periodic wavelets have obviously boundary effect. More coarse level chosen, more points are distorted. Among these three methods, the first is the worst one since the accuracy is really by far not enough no matter in the boundary or in the interior.

Table 3: The coarsest level=3

	general wavelets	periodic wavelets	wavelets on the interval
0	1.0423279004	3.7633519745	0.0000000082
1	0.1319450910	0.6935604068	0.0000000145
2	0.0162616030	0.1189613003	0.0000000139
3	0.0272695904	0.0138025283	0.0000000275
4	0.0043331195	0.1545354990	0.0000000192
⋮	⋮	⋮	⋮
59	0.3223202511	1.0194288768	0.0000000657
60	0.4433595805	1.2681867776	0.0000000296
61	0.6789812493	1.2391206658	0.0000000223
62	0.8098405151	1.2844972694	0.0000000222
63	1.1019435933	0.8049614264	0.0000000210

References

[A] B. Alpert, *A class of Bases in L^2 for the sparse representation of integral operators*, SIAM J. Math. Anal. 24 (1993), No. 1, 246–262.

[BCR] G. Beylkin, R. Coifman and V. Rokhlin, *Fast wavelet transforms and numerical algorithms I*, Comm. on Pure and Appl. Math. XLIV (1991), 141–183.

[CDV] A. Cohen, I. Daubechies and P. Vial, *Wavelet on the interval and fast wavelet transforms*, Applied and computational Harmonic Analysis, No. 1 (1993).

[D] I. Daubechies, *Orthonormal bases of compactly supported wavelets*, Comm. on Pure and Appl. Math., XLI (1988), 909–996.

[DS] C. Dagnino and E. Santi, *Spline product quadrature rules for Cauchy singular integrals*, J. of Comp. and Appl. Math. 33 (1990), 133–140.

[Ge] A. Gerasoulis, *Piecewise-polynomial quadratures for Cauchy singular integrals*, SIAM J.Numer. Anal. 23 (1986), 891–902.

[Go] M.A. Goldberg, *Introduction to the numerical solution of Cauchy singular integral equations*, M.A. Goldberg, ed., Numerical solution of integral equations, Mathematical concepts and methods in science and engineering 42, 1990, 183–308.

[P] R. Philip, *Application of approximating splines for the solution of Cauchy singular integral equations*, Applied Numerical Math ematics 15 (1994), 285–297.

21 CLOSED FORM SOLUTION FOR A HYPERSINGULAR INTEGRAL EQUATION OF ORDER N+1

Xing LI[1]

Dept. of Math., Ningxia University
Yinchuan 750021, P. R. China

li_x@nxu.edu.cn

Abstract: We consider the solution in closed form for a hypersingular integral equation of order $n+1$ of the type

$$\phi(x) + \frac{(x-a)^{\frac{1}{2}}(x-b)^{\frac{1}{2}}}{\pi i} \,⨍_a^b \frac{\phi(t)}{(t-x)^{n+1}} dt = g(x), x \in (a,b), n = 1, 2, \cdots .$$

By using a complex function-theoretic method in a straightforward manner, after introducing a new unknown function, this equation is transformed into a differential Riemann boundary value problem on the cut (a,b), which is amenable to its closed form solution.

[1] The author is grateful for support of NSFC and the Teaching and Research Award Fund for Outstanding Young Teachers in Higher Education Institutions of M. O. E., of P. R. China.

H.G.W. Begehr et al. (eds.), Proceedings of the Second ISAAC Congress, Vol. 1, 163–167.

1 INTRODUCTION

The solution of hypersingular integral equations is currently gaining increasing interest due to their extensive applications in mechanics and so on. Because the integral involved in the hypersingular integral equations is not defined in the usual sense, not even as a Cauchy principal value, but as the finite-part, so the numerical methods, for example, the collocation method, the Galerkin method, together with different singular integration techniques are frequently used to solve hypersingular integral equations. For instance, the elliptic wing case of Prandtl's equation, a special hypersingular integral equation of the second kind, as given by

$$\phi(x) - \frac{\alpha(1-x^2)^{\frac{1}{2}}}{\pi} \,\text{⨎}\!\int_{-1}^{1} \frac{\phi(t)dt}{(t-x)^2} = f(x), -1 < x < 1, \tag{1}$$

with $\phi(\pm 1) = 0$, for which $f(x) = (\frac{2\pi k}{\beta})(1-x^2)^{\frac{1}{2}}$, $\alpha > 0$, $\beta > 0$ and k are known constants, has been investigated by Dragos in [2] and [3] using collocation method. Recently, a straightforward analysis involving the complex function-theoretic method is employed by Chakraborti et al. to determine the closed-form solution of the equation (1) in [1]. And the author obtained the solution in closed form of the periodic hypersingular integral equation in the form

$$\phi(x) + \frac{f(x)}{\pi i} \,\text{⨎}\!\int_{-1}^{1} \phi(t)\cot^2(t-x)dt = g(x), -1 < x < 1, \tag{2}$$

where $g(x) \in H$ is a given periodic function with period π, $f(x)$ is a known special periodic fundamental function.

2 FORMULATION AND SOLUTION

In this paper, we consider the solution of a hypersingular integral equation of order $n+1$ of the type

$$\phi(x) + \frac{w(x)}{\pi i} \,\text{⨎}\!\int_{a}^{b} \frac{\phi(t)}{(t-x)^{n+1}} dt = g(x),\ x \in (a,b),\ n = 1, 2, \cdots, \tag{3}$$

with

$$\phi^{(j)}(a) = \phi^{(j)}(b) = 0, j = 0, 1, \cdots, n-1, \tag{4}$$

where we assume that $\phi(x)$ possess derivations up to order n in (a,b) and $\phi^{(n)}(x)$ satisfies a Hölder condition, namely

$$|f^{(n)}(x_1) - f^{(n)}(x_2)| \leq A|x_1 - x_2|^{\mu},$$

whenever x_1 and x_2 both lie in (a,b), A is a constant and $0 < \mu \le 1$, i.e. $\phi(x) \in C^{n,\mu}(a,b)$. $w(x) = (x-a)^{\frac{1}{2}}(x-b)^{\frac{1}{2}}$ is the fundamental function associated with the mixed boundary value problem of a plate with a crack $\gamma = (a,b)$, and $g(x) \in H$ is a given function on (a,b). $\phi(x)$ is to be determined.

The integral occurring on the left of the equation (2) must be interpreted as a Hadamard finite-part integral, defined by (see [4], [6] and [10])

$$\begin{aligned}\, & \! = \!\,\end{aligned}$$

$$\begin{aligned}\int_a^b \frac{\phi(t)}{(t-x)^{n+1}}dt &= \lim_{\epsilon\to 0}\left\{\int_a^{x-\epsilon}\frac{\phi(t)}{(t-x)^{n+1}}dt + \int_{x+\epsilon}^b \frac{\phi(t)}{(t-x)^{n+1}}dt\right\} \\ &- \lim_{\epsilon\to 0}\sum_{j=0}^{n-1}\frac{f^j(x)}{j!}\left\{\frac{1-(-1)^{n-j}}{(n-j)\epsilon^{n-j}}\right\}\frac{1}{n!}\int_a^b\frac{\phi^{(n)}(t)}{(t-x)}dt \\ &+ \sum_{j=0}^{n-1}\frac{1}{n\cdots(n-j)}\left\{\frac{\phi^{(j)}(a)}{(a-x)^{n-j}} - \frac{\phi^{(j)}(b)}{(b-x)^{n-j}}\right\},\ x\in(a,b). \qquad (5)\end{aligned}$$

Because $\phi(t) \in C^{n,\mu}(a,b)$, $0 < \mu \le 1$, then the Cauchy principal-value integral on the right hand side of (5) exists.

By the conditions (4) from equation (5) we have

$$\int_a^b \frac{\phi(t)}{(t-x)^{n+1}}dt = \frac{1}{n!}\int_a^b \frac{\phi^{(n)}(t)}{(t-x)}dt,\ x\in(a,b). \qquad (6)$$

Furthermore, we have a simpler formula[8]

$$\int_a^b \frac{\phi(t)}{(t-x)^{n+1}}dt = \frac{1}{n!}\frac{d^n}{dx^n}\int_a^b \frac{\phi(t)}{(t-x)}dt,\ x\in(a,b). \qquad (7)$$

For solving hypersingular integral equation (3) we introduce a sectionally holomorphic function of the complex variable $z = x + iy$ in the complex plane

$$\Phi(z) = \frac{1}{2\pi i}\int_a^b \frac{\phi(t)}{(t-z)}dt, \qquad (8)$$

then, we have the Plemelj formulae for the integral with Cauchy kernel (see [9] and [7])

$$\Phi^{\pm}(t) = \pm\frac{1}{2}\phi(x) + \frac{1}{2\pi i}\int_a^b \frac{\phi(t)}{(t-x)}dt,\ x\in(a,b), \qquad (9)$$

where $\Phi^{\pm}(x)$ are the limiting values of $\Phi(z)$ on the positive and negative sides of the cut (a, b), which cut the complex plane along the segment (a,b) of the real axis.

By formula (9) the hypersingular integral equation (3) is transferred into a differential Riemann boundary value problem with jump along the slit (a,b) on the x axis immediately,

$$\left[1+\frac{w(x)}{n!}\frac{d^n}{dx^n}\right]\Phi^+(x)-\left[1-\frac{w(x)}{n!}\frac{d^n}{dx^n}\right]\Phi^-(x)=g(x), x\in(a,b). \quad (10)$$

For solving this differential Riemann boundary value problem, we introduce a new function

$$\Psi(z)=\left[1+\frac{w(z)}{n!}\frac{d^n}{dz^n}\right]\Phi(z), \quad (11)$$

where $w(z)=(z-a)^{\frac{1}{2}}(z-b)^{\frac{1}{2}}$.

Thus, we obtain a Riemann jump boundary value problem immediately

$$\Psi^+(x)-\Psi^-(x)=g(x), x\in(a,b). \quad (12)$$

Apparently, the Riemann jump problem (12) is the simplest Riemann boundary value problem. We have its solution in the form (see [7]),

$$\Psi(z)=\frac{1}{2\pi i}\int_a^b\frac{g(t)}{(t-z)}dt. \quad (13)$$

Then, we may use the classical method to solve the classical ordinary differential equation (11) with respect unknown function $\Phi(z)$. After obtaining $\Phi(z)$ we may obtain $\phi(x)$ by Plemelj formula, namely

$$\phi(x)=\Phi^+(x)-\Phi^-(x). \quad (14)$$

References

[1] Chakrabarti A., Mandal B., Basu B. N., and Banerjea S., *Solution of a hypersingular integral equation of the second kind,* ZAMM, 77(1997), No.4, 319–320.

[2] Dragos L., *A collocation method for the integration of Prandtl's equation,* ZAMM, 74(1994), 289–290.

[3] Dragos L., *Integration of Prandtl's equation by the aid of quadrature formulae of Gauss type*, Quart. Appl. Math. 52(1994), 23–29.

[4] Fox C., *A generalization of the Cauchy principal value,* Canadian J. Math., 9(1957), No.1, 110–119.

[5] Li X., *Solution of a class of periodic hypersingular integral equations,* Finite or Infinite Dimensional Complex Analysis: 7th International Colloquium, Marcel Dekker, to appear.

[6] Lu J. K., *On singular integrals with singularities of high fractional order and their applications*, Acta Mathematica Scientia, 2(1982), No.2, 211–218.

[7] J. K. Lu, *Boundary value problems for analytic functions*, World Scientific, Singapore, 1993.

[8] Monegato G., *On the weights of certain quadratures for the numerical evalution of Cauchy principal value integrals and their derivatives,* Numer. Math., 50(1987), 273–281.

[9] Muskhelishvili N. I., *Singular integral equations*, Noordhoff, Groningen, 1953.

[10] Wang C. R., *Hadamard principal value of singular integral* $\int_L \frac{f(\tau)}{(\tau-t)^{n+1}} d\tau$, Chin. Ann. of Math., 3A(1982), No.2, 195–202.

22 CYCLICALLY SYMMETRIC CRACK PROBLEMS OF DIFFERENT MEDIA II[1]

Jianke LU

Department of Mathematics, Wuhan University
Wuhan 430072, P. R. China

jydu@whu.edu.cn

Abstract: The first fundamental problem with cyclic symmetry for the infinite elastic plane consisting of two different isotropic media with cracks on the interface is studied. It is reduced to a singular integral equation on a closed contour with two nodes only, which is rather simple for its solution.

0. Introduction

The plane elastic problems of different isotropic media with n cracks on the interface (incomplete welding problem) were discussed in [1]. They may be reduced to singular integral equations along the interface by complex variable methods. By the general theory (cf. [2] or [3]), there are n arbitrary constants in the general solution of such an equation, which should be determined by a system of algebraic linear equations, whose coefficients are hard to be calculated out. On the otherhand, the cyclically symmetric problems for an infinite elastic plane of a single isotropic medium with cracks were discussed in [4], which shows that

[1] Project supported by NNSF of China (No. 19871064)

H.G.W. Begehr et al. (eds.), Proceedings of the Second ISAAC Congress, Vol. 1, 169–174.

the general solution of the obtained equation contains only one arbitrary constant, which may be determined much easier.

In this paper, we illustrate that the method used there may be also applied to cyclically symmetric crack problems of two different media. We confine ourselves to the case, where the cracks lie on the interface though the method is also effective in general. By the same reason, only the first fundamental problem will be considered.

1. Preliminations

We shall briefly recall certain notations and results in [1] and [4], which will be applied to our cyclically symmetric problems.

Assume the infinite elastic plane consists of two different isotropic media with a cyclically symmetric and sufficiently smooth closed contour L as the interface, oriented counter-clockwisely. There are n cyclically symmetric cracks $l_j = a_j b_j, j = 0, 1, \cdots n-1$ $(n \geq 2)$ on L, $a_j = \omega^j a_0, b_j = \omega^j b_0$, where $\omega = e^{\frac{2\pi i}{n}}$ is called the cyclic constant and n, the order of symmetry. The center of symmetry is assumed to be at the origin. The interior and the exterior regions bounded by L are denoted by S^+ and S^- with elastic constants κ^+, μ^+ and κ^-, μ^- respectively. Denote arc $l'_j = b_j a_{j+1} \subset L$ $(j = 0, 1, \cdots, n-1, a_n = a_0)$ and $l = \sum_{j=0}^{n-1} l_j, l' = \sum_{j=0}^{n-1} l'_j$.

For the first fundamental problem, given the external cyclically symmetric loads $X_n^\pm(t) + iY_n^\pm(t), t \in L$, or in detail, $X_j^\pm(t) + iY_j^\pm(t) \in H$ (class of Hőlder continuous functions), $t \in l_j, j = 0, 1, \cdots, n-1$, $X^\pm(\omega t) = \omega X^\pm(t), Y^\pm(\omega t) = \omega Y^\pm(t)$, and the cyclically symmetric displacement difference $g(t) = [u^+(t) + iv^+(t)] - [u^-(t) + iv^-(t)], t \in l'$ $(g(t) = g_j(t)$ for $t \in l'_j)$ with $g'(t) \in H$, where $u^\pm(t) + iv^\pm(t)$ are the displacements of the positive and the negative sides at $t \in l'$. By assumption, $g(\omega t) = \omega g(t)$ and hence $g'(\omega t) = g'(t)$. Moreover, for simplicity, we assume that there is no rotation nor stress at infinity. We want to determine the elastic equilibrium, i.e., seek for the stresses $\sigma_\rho, \sigma_\theta, \tau_{\rho\theta}$ at any point $z = \rho e^{i\theta}$ in polar coordinates, which may be expressed in terms of Kolosov functions $\Phi(z), \Psi(z)$[2] :

$$\begin{aligned} \sigma_\rho + \sigma_\theta &= 4\mathrm{Re}[\Phi(z)], \\ \sigma_\theta - \sigma_\rho + 2i\tau_{\rho\theta} &= 2[\overline{z}\Phi'(z) + \Psi(z)]e^{2i\theta}, \end{aligned} \tag{1.1}$$

where $z = \rho e^{i\theta}$, and $\Phi(z)$, $\Psi(z)$, in our case, are sectionally holomorphic with L as the jumping curve, satisfying[3]

$$\Phi(\omega z) = \Phi(z), \Psi(\omega z) = \overline{\omega}^2 \Psi(z), \tag{1.2}$$

and $\Phi(\infty) = \Psi(\infty) = 0$.

Our problem is: given $X_n^{\pm}(t) + iY_n^{\pm}(t), t \in l$ as well as $g^{\pm}(t), t \in l'$, find the elastic equilibrium.

Without loss of generality, we may assume

$$X_j + iY_j = \int_{l_j} [X_n^+(t) + iY_n^+(t)]ds + \int_{l_j} [X_n^-(t) + iY_n^-(t)]ds = 0, j = 0, 1, \cdots n-1$$

(s being the arc-length parameter on l_j), and denote

$$f^{\pm}(t) = f_j^{\pm}(t) = \pm i \int_{a_j}^{t} [X_n^{\pm}(t) + iY_n^{\pm}(t)]ds, t \in l_j, j = 0, 1, \cdots, n-1, \tag{1.3}$$

and so $f_j^{\pm}(a_j) = 0, f_j^+(b_j) = f_j^-(b_j)$. Denote

$$F(t) = f^+(t) - f^-(t), \ G(t) = f^+(t) + f^-(t), \quad t \in l. \tag{1.4}$$

Obviously,

$$f^{\pm}(\omega t) = \omega f^{\pm}(t), F(\omega t) = \omega F(t), G(\omega t) = \omega G(t)$$

so that both $F'(t)$ and $G'(t)$ are cyclically symmetric:

$$F'(\omega t) = F'(t), G'(\omega t) = G'(t), \ t \in l. \tag{1.5}$$

2. The Known Results for General Case

We shall briefly sketch some results for the general case (without cyclic symmetry) obtained in [1].

We may express $\Phi(z)$ and $\Psi(z)$ in terms of an unknown function $\Omega(t), t \in L$, which is the unique solution of the singular integral equation

$$\begin{aligned} \mathbf{K}_1\Omega &\equiv \frac{1}{\pi i}\int_L \frac{\Omega(t)}{t-t_0}dt + \frac{1}{2}(\mathbf{K}_0\Omega)(t_0) \\ &= \frac{1}{2}G'(t_0) + \frac{1}{2\pi i}\frac{\partial}{\partial t_0}\int_L \frac{F(t)}{\overline{t}-\overline{t_0}}d\overline{t}, \quad t_0 \in l, \end{aligned} \tag{2.1}$$

$$\begin{aligned} \mathbf{K}_2\Omega &\equiv A\Omega(t_0) + \frac{B}{\pi i}\int_L \frac{\Omega(t)}{t-t_0}dt - (\beta^+ - \beta^-)(\mathbf{K}_0\Omega)(t_0) \\ &= 4g'(t_0) - \frac{\beta^+ - \beta^-}{\pi i}\frac{\partial}{\partial t_0}\int_L \frac{F(t)}{\overline{t}-\overline{t_0}}d\overline{t}, \quad t_0 \in l', \end{aligned} \tag{2.2}$$

in class h_0 (for notation, cf. [2]), i.e., having integrable singularities at a_j,b_j, satisfying the supplementary conditions

$$\int_{l_j}(\mathbf{K}_2\Omega)(t)dt = 4[g(b_j) - g(a_j)] - \frac{\beta^+ - \beta^-}{\pi i}\int_{l_j} F(t)\left(\frac{1}{\overline{t} - \overline{b_j}} - \frac{1}{\overline{t} - \overline{a_j}}\right)d\overline{t}, \quad j = 0, 1, \cdots, n-1, \tag{2.3}$$

where

$$\mathbf{K}_0\Omega \equiv \frac{1}{\pi i}\int_L \frac{\partial}{\partial t_0}\log\frac{t - t_0}{\overline{t} - \overline{t_0}}\Omega(t)dt + \frac{1}{\pi i}\int_L \frac{\partial}{\partial t_0}\frac{t - t_0}{\overline{t} - \overline{t_0}}\overline{\Omega(t)}d\overline{t}, \quad t_0 \in L, \tag{2.4}$$

and

$$A = \alpha^+ + \alpha^- + \beta^+ + \beta^-, \quad B = \alpha^+ - \alpha^- - \beta^+ + \beta^-, \quad \alpha^\pm = \kappa^\pm/\mu^\pm, \quad \beta^\pm = 1/\mu^\pm. \tag{2.5}$$

After $\Omega(t)$ is obtained, then $\Phi(z)$ and $\Psi(z)$ are given by, respectively,

$$\Phi(z) = \frac{1}{2\pi i}\int_L \frac{\Omega(t)}{t - z}dt, \quad z \notin L, \tag{2.6}$$

$$\Psi(z) = -\frac{1}{2\pi i}\int_L \frac{\overline{\Omega(t)}}{t - z}d\overline{t} - \frac{1}{2\pi i}\int_L \frac{\overline{t}\Omega(t)}{(t - z)^2}dt + \frac{1}{2\pi i}\int_l \frac{\overline{F'(t)}}{t - z}d\overline{t}, \quad z \notin L. \tag{2.7}$$

3. Main Results

For our considered cyclically symmetric case, it is easily seen that $\Omega(t)$ is cyclically symmetric: $\Omega(\omega t) = \Omega(t)$ and so does $(\mathbf{K}_0\Omega)(t_0)$, which implies that both (2.1) and (2.2) remain unchanged, if t_0 is replaced by ωt_0. At the same time, conditions in (2.3) would be fulfilled, if one of them does, e.g.,

$$\int_{l_0}(\mathbf{K}_2\Omega)(t)dt = 4[g(b_0) - g(a_0)] - \frac{\beta^+ - \beta^-}{\pi i}\int_{l_0} F(t)\left(\frac{1}{\overline{t} - \overline{b_0}} - \frac{1}{\overline{t} - \overline{a_0}}\right)d\overline{t}. \tag{3.1}$$

As in [4], let $\zeta = z^n$, $\tau = t^n (t \in L)$. In the ζ-plane, L is mapped to Λ with multiple order n, l_0 to an arc $\lambda = \alpha_0\beta_0$ on Λ and l_0' to $\lambda' = \Lambda\backslash\lambda$ ($\alpha_0 = a_0^n$, $\beta_0 = b_0^n$). Denote $\Omega(t) = \Omega^*(\tau)$.

Note that, by the results in [4], $\mathbf{K}_0\Omega$ in (2.4) may be written as

$$\mathbf{K}_0^*\Omega^* \equiv -\frac{1}{\pi i}\int_\Lambda \frac{\Omega^*(\tau)}{\tau - \tau_0}d\tau + \widetilde{\mathbf{K}_0^*}\Omega^*, \tag{3.2}$$

where

$$\widetilde{\mathbf{K}_0^*}\Omega^* \equiv \frac{1}{\pi i}\int_\Lambda \frac{\tau_0^{1-\frac{1}{n}}\overline{\tau}^{\frac{1}{n}}}{\tau^{1-\frac{1}{n}}\overline{\tau_0}^{\frac{1}{n}}}\frac{\Omega^*(\tau)}{\overline{\tau}-\overline{\tau_0}}d\tau\frac{d\overline{\tau_0}}{d\tau_0} + \frac{n\tau_0^{1-\frac{1}{n}}}{\pi i}\int_\Lambda \frac{\partial}{\partial\tau_0}\frac{|\tau|^{\frac{2}{n}}\,\overline{\tau_0} - |\tau_0|^{\frac{2}{n}}\,\overline{\tau}}{\overline{\tau_0}^{\frac{1}{n}}(\overline{\tau}-\overline{\tau_0})}\frac{\overline{\Omega^*(\tau)}}{\overline{\tau}}d\overline{\tau}, \tag{3.3}$$

and so $\mathbf{K}_1\Omega,\ \mathbf{K}_2\Omega$ in (2.1), (2.2) become respectively

$$\mathbf{K}_1^*\Omega^* \equiv \frac{1}{2\pi i}\int_\Lambda \frac{\Omega^*(\tau)}{\tau-\tau_0}d\tau + \frac{1}{2}\widetilde{\mathbf{K}_0^*}\Omega^*(\tau_0), \quad \tau_0\in\lambda, \tag{3.4}$$

$$\mathbf{K}_2^*\Omega^* \equiv A\Omega^*(\tau_0) + \frac{\alpha^+-\alpha^-}{\pi i}\int_\Lambda \frac{\Omega^*(\tau)}{\tau-\tau_0}d\tau - (\beta^+-\beta^-)\widetilde{\mathbf{K}_0^*}\Omega^*(\tau_0), \quad \tau_0\in\lambda'. \tag{3.5}$$

Write $F^*(\tau) = F(t), G^*(\tau) = G(t), g^*(\tau) = g(t)$, so

$$F'(t) = n\tau^{1-\frac{1}{n}}F^{*\prime}(\tau)$$

and so do similar formulas for $G'(t), g'(t)$. We see that

$$\begin{aligned}\frac{1}{2\pi i}\frac{\partial}{\partial t_0}\int_L \frac{F(t)}{\overline{t}-\overline{t_0}}d\overline{t} &= \frac{1}{2\pi i}\int_L \frac{F'(t)}{\overline{t}-\overline{t_0}}d\overline{t}\frac{d\overline{t_0}}{dt_0} \\ &= \frac{1}{2\pi i}\int_\Lambda \frac{n\tau^{1-\frac{1}{n}}F^{*\prime}(\tau)}{\overline{\tau}-\overline{\tau_0}}d\overline{\tau}\left(\frac{\overline{\tau_0}}{\tau_0}\right)^{\frac{1}{n}-1}\frac{d\overline{\tau_0}}{d\tau_0} \quad (\tau_0 = t_0^n),\end{aligned} \tag{3.6}$$

which will be denoted by $H(\tau_0)$. Hence equations (2.2),(2.3) become respectively

$$\mathbf{K}_1^*\Omega^* = \frac{n}{2}\tau_0^{1-\frac{1}{n}}G^{*\prime}(\tau_0) + H(\tau_0), \quad \tau_0\in\lambda, \tag{3.7}$$

$$\mathbf{K}_2^*\Omega^* = 2n\tau_0^{1-\frac{1}{n}}g^{*\prime}(\tau_0) - 2(\beta^+-\beta^-)H(\tau_0), \quad \tau_0\in\lambda', \tag{3.8}$$

while the supplementary condition (3.1) becomes

$$\frac{1}{n}\int_\lambda \mathbf{K}_2^*\Omega^*(\tau)\frac{d\tau}{\tau^{1-\frac{1}{n}}} = 4[g^*(\beta_0) - g^*(\alpha_0)] - \frac{\beta^+-\beta^-}{\pi i}\int_\lambda F^*(\tau)\left(\frac{1}{\overline{\tau}-\overline{\beta_0}} - \frac{1}{\overline{\tau}-\overline{\alpha_0}}\right)d\overline{\tau}. \tag{3.9}$$

Thus, our problem is reduced to solve (3.7)–(3.8) on the closed contour $\Lambda = \lambda+\lambda'$ with additional condition (3.9) in class h_0 (the only two nodes are α_0, β_0). By the general theory[1], there occurs only one arbitrary constant in the general solution of (3.7)–(3.8), which may be uniquely determined by (3.9).

After $\Omega^*(\tau)$ is obtained and so does $\Omega(t) = \Omega^*(t^n)$, the Kolosov functions are given by (2.6), (2.7).

Remark The first term in (3.4) and the first two terms in (3.5) do not constitute the characteristic part of (3.7)–(3.8), which is not important for the method of numerical solution.

4. Some Comments

(i) If there exist stresses (but no rotation) at infinity, we have $\Phi(\infty) = \Gamma$ (real) and $\Psi(\infty) = \Gamma'$ (complex). Then, on the right sides of (2.6), (2.7), constant terms Γ and Γ' should be added respectively and consequently terms $-2\Gamma - \overline{\Gamma}' \frac{d\overline{t_0}}{dt_0}$ and $-2B\Gamma + 2(\beta^+ - \beta^-)\overline{\Gamma}' \frac{d\overline{t_0}}{dt_0}$ should be added in (2.1) and (2.2) respectively as well as appropriate term in (2.3). Consequent changes should be made for corresponding equations in the ζ-plane. By the way, we note that $\Gamma' \neq 0$ occur only in case $n = 2$ by cyclic symmetry, which denotes a tension (or compression) in a single direction.

(ii) If $X_n^+(t) + iY_n^+(t) = -(X_n^-(t) + iY_n^-(t))$, $t \in l$, which follows $F'(t) = 0$ and(or) $g(t) = 0$, which usually occur in practice, then all the terms of the equations involving $F^{*\prime}(\tau)$ and(or) $g^{*\prime}(\tau)$ would disappear, which make them considerably simpler.

(iii) In case $n = 2$, all the equations involved become simpler since $|\tau|^{\frac{2}{n}} = |\tau|$, etc.

(iv) If L is the unit circle $|t| = 1$ and so Λ is also the unit circle $|\tau| = 1$, then all the equations obtained reduce to very simple ones and the solution of our problem may be expressed in closed form (in terms of integrals), which would not be discussed here in detail.

References

[1] Lu Jianke, *Plane elastic problems of different media with cracks on the interface,* Applied Math. J. of Chin. Univ., **8**(1993), No. 1, 15–24.

[2] Muskhelishville N. I., *Some Basic Problems of the Mathematical Theory of Elasticity,* Noordhoff, Groningen, 1953.

[3] Lu Jianke, *Complex Variable Methods in Plane Elasticity,* World Scientific, Singapore, 1995.

[4] Lu Jianke, *On cyclically symmetric crack problems,* Wuhan Univ. J. of Natural Sci., **2**(1997), No. 2, 132–138.

23 LINEAR CONJUGATE BOUNDARY VALUE PROBLEMS FOR FIRST ORDER ORDINARY SYSTEM OF LINEAR DIFFERENTIAL EQUATIONS WITH SINGULAR OR SUPER SINGULAR COEFFICIENTS

Nusrat RAJABOV

Tajik State National University
Dushanbe 734025, Tajikistan
nrajabov@td.silk.org

Let $\Gamma_o = \{0 < x < b\}$ be a set of the point on the real axis, containing the point $x = c$, $0 < c < b$. In $\Gamma = \Gamma_o \backslash \{c\}$ we consider the following first order linear system

$$|x - c|^\alpha Y_j'(x) + \sum_{k=1}^{2} P_{jk}(x) Y_k(x) = f_j(x), \ j = 1, 2, \tag{1}$$

where α=constant> 0, $P_{jk}(x)$, $f_j(x)$ $(j, k = 1, 2)$ are given functions in Γ with first genus singularity at $x = c$. The problem is to find a manifold solution in term of arbitrary constant and investigations of boundary valued problems, for the first order ordinary differential equation and first order ordinary system of differential equations with singular or super-singular points and its application, which has been studied in [1]-[9].

Formerly the authors in [4]-[9] independent from position weak singularity, singular or super-singular point ($c = a$; $c = b$; $a < c < b$),

H.G.W. Begehr et al. (eds.), Proceedings of the Second ISAAC Congress, Vol. 1, 175–183.

power singularity ($\alpha < 1,\ \alpha = 1, \alpha > 1$), and the roots of characteristic equation

$$\lambda^2 P_{21}(c-0) + (P_{11}(c-0) - P_{22}(c-0))\lambda - P_{12}(c-0) = 0 \qquad (2)$$

$$\mu^2 P_{21}(c+0) + (P_{11}(c+0) - P_{22}(c+0)\mu - P_{12}(c+0) = 0 \qquad (3)$$

construction of general solution of system (1) in Γ. For the investigation of behavior of the solution of system (1) in the neighborhood of the singular or super singular point $x = c$, we had already obtained the general solution of the system (1), applied for classification of posing of the Cauchy type boundary value problems with the set condition in the singular or super-singular point.

In this paper, on the basis of formerly obtained integral representation independent of power singularity, the roots of characteristic equation at singular point, posing and investigation of the linear conjugate boundary value problems with the set conditions in the singular or super-singular point.

Problem R_1. Let $a < c < b$. Find the solution of the system (1), when the roots of the characteristic equations (2), (3) are real and different, of class $C'(\Gamma)$, satisfying one of the following conditions:

$$\left.\begin{matrix} Y_1(c-0) + \lambda_j Y_2(c-0) = A_j[Y_1(c+0) + \mu_j Y_2(c+0)] + B_j \\ j = 1,2,\ by\ \alpha < 1; \end{matrix}\right\} \qquad (4)$$

$$\left.\begin{matrix} [(c-x)^{-P_c^{\lambda_j}}(Y_1(x) + \lambda_j Y_2(x))]_{x=c-0} = \\ A_j[(x-c)^{-P_c^{\mu_j}}(Y_1(x) + \mu_j Y_2(x))]_{x=c+0} + B_j, \\ j = 1,2\ by\ \alpha = 1; \end{matrix}\right\} \qquad (5)$$

$$\left.\begin{matrix} \exp(P_c^{\lambda_j}\omega_c^\alpha(x)) \cdot (Y_1(x) + \lambda_j Y_2(x))]_{x=c-0} = A_j\cdot \\ \exp(-P_c^{\mu_j}\omega_c^\alpha(x)) \cdot (Y_1(x) + \mu_j Y_2(x))]_{x=c+0} + B_j, \\ j = 1,2\ by\ \alpha > 1; \end{matrix}\right\} \qquad (6)$$

in point $x = c$, and one of the following conditions in end points of interval Γ_o:

$$\left.\begin{matrix} 1)Y_j(a) = D_j'(j = 1,2); & 2)Y_j(b) = D_j^2(j = 1,2); \\ 3)Y_1(a) = D_1^3, Y_2(b) = D_2^3; & 4)Y_1(b) = D_1^4, Y_2(a) = D_2^4 \end{matrix}\right\}, \qquad (7)$$

where A_j, B_j, D_j^k, $j = 1,2;\ 1 \le k \le 4$ are given constants, $\omega_c^\alpha(x) = [(\alpha-1)|x-c|^{\alpha-1}]^{-1}$, $P_c^{\lambda_1} = A_c^- + B_c^-$, $P_c^{\lambda_2} = A_c^- - B_c^-$, $P_c^{\mu_1} = A_c^+ + B_c^+$,

$P_c^{\mu_2} = A_c^+ - B_c^+$, $A_c^{\pm} = 2^{-1}(P_{11}(c \pm 0) + P_{22}(c \pm 0))$, $B_c^{\pm} = 2^{-1}\sqrt{D_c^{\mp}}$, $D_c^{\pm} = (P_{22}(c \pm 0) - P_{11}(c \pm 0))^2 + 4P_{12}(c \pm 0) \cdot P_{21}(c \pm 0)$.

Problem R_2. Let $a < c < b$. Find the solution of the system (1), when the roots of the characteristic equations (2), (3) are real and equal by class $C'(\Gamma)$, satisfying one of the following conditions:

$$Y_j(c-0) = A_j Y_j(c+0) + B_j,\ j = 1,2,\ by\ \alpha < 1;$$

$$[(c-x)^{-P_c^{\lambda}}(Y_j(x)]_{x=c-0} = A_j[(x-c)^{-P_c^{\mu}} Y_j(x)]_{x=c+0} + B_j,$$
$$j = 1,2\ by\ \alpha = 1;$$

$$[\exp(-P_c^{\lambda}\omega_c^{\alpha}(x)) \cdot (Y_j(x)]_{x=c-0} = A_j \cdot [\exp(P_c^{\mu}\omega_c^{\alpha}(x)) \cdot Y_j(x)]_{x=c+0} + B_j,$$
$$j = 1,2\ by\ \alpha > 1;$$

in point $x = c$, and one of the conditions in (7) in end points of interval Γ_o, where A_j, B_j, D_j^k, $j = 1,2$, $1 \leq k \leq 4$ are given constants,

$$\lambda = 2^{-1}(P_{22}(c-0) - P_{11}(c-0)) \cdot (P_{21}(c-0)))^{-1},$$

$$\mu = 2^{-1}P_{21}(c+0) \cdot (P_{22}(c+0) - P_{11}(c+0)).$$

Problem R_3. Let $a < c < b$. Find the solution of the system (1), when the roots of the characteristic equations (2), (3) are complex-conjugate, by class $C'(\Gamma)$, satisfying one of the following conditions:

$$[\varphi_1^-(x)Y_1(x) + (\lambda_1\varphi_1^-(x) - \lambda_2\varphi_2^-(x)) \cdot Y_2(x)]_{x=c-0} = A_1.$$

$$[\varphi_1^+(x)Y_1(x) + (\mu_1\varphi_1^+(x) + \mu_2\varphi_2^+(x)) \cdot Y_2(x)]_{x=c+0} = B_1,$$

$$[\varphi_2^-(x)Y_1(x) + (\lambda_2\varphi_1^-(x) + \lambda_1\varphi_2^-(x)) \cdot Y_2(x)]_{x=c-0} = A_2.$$

$$[-\varphi_2^+(x)Y_1(x) + (\mu_2\varphi_1(x) - \mu_1\varphi_2(x)) \cdot Y_2(x)]_{x=c+0} = B_2,$$

by $\alpha < 1$;

$$\{(c-x)^{-A_c^-}[\varphi_1^-(x)Y_1(x) + (\lambda_1\varphi_1^-(x) - \lambda_2\varphi_2^-(x)) \cdot Y_2(x)]\}_{x=c-0} =$$

$$= A_1\{(x-c)^{A_c^+}[\varphi_1^+(x)Y_1(x) + (\mu_1\varphi_1^+(x) + \mu_2\varphi_2^+(x)) \cdot Y_2(x)]\}_{x=c+0} + B_1,$$

$$\{(c-x)^{-A_c^-}[\varphi_2^-(x)Y_1(x) + (\lambda_2\varphi_1^-(x) + \lambda_1\varphi_2^-(x)) \cdot Y_2(x)]\}_{x=c-0} =$$

$$= A_2\{(x-c)^{A_c^+}[-\varphi_2^+(x)Y_1(x) + (\mu_2\varphi_1(x) - \mu_1\varphi_2(x)) \cdot Y_2(x)]\}_{x=c+0} + B_2$$

by $\alpha = 1$;

$$\{\exp(A_c^-\omega_c^{\alpha}(x))[\varphi_1^-(x)Y_1(x) + (\lambda_1\varphi_1^-(x) - \lambda_2\varphi_2^-(x)) \cdot Y_2(x)]\}_{x=c-0} =$$

$$= A_1\{\exp(-A_c^+\omega_c^\alpha(x))[\varphi_1^+(x)Y_1(x) + (\mu_1\varphi_1^+(x) + \mu_2\varphi_2^+(x))\cdot Y_2(x)]\}_{x=c+0} + B_1,$$

$$\{\exp(A_c^-\omega_c^\alpha(x))[\varphi_2^-(x)Y_1(x) + (\lambda_2\varphi_1^-(x) + \lambda_1\varphi_2^-(x))\cdot Y_2(x)]\}_{x=c-0} =$$

$$= A_2\{\exp(-A_c^+\omega_c^\alpha(x))[-\varphi_2^+(x)Y_1(x) + (\mu_2\varphi_1^+(x) - \mu_1\varphi_2^+(x))\cdot Y_2(x)]\}_{x=c+0} + B_2,$$

by $\alpha > 1$; where $\varphi_1^\pm(x) = \cos[B_c^\pm\omega_c^\alpha(x)]$, $\varphi_2^\pm(x) = \sin[B_c^\pm\omega_c^\alpha(x)]$, by $\alpha > 1$; $\varphi_1^\pm(x) = \cos[B_c^\pm\ln|x-c|]$, $\varphi_2^\pm(x) = \sin[B_c^\pm\ln|x-c|]$, by $\alpha = 1$; $\varphi_1^\pm(x) = \cos(B_c^\pm x)$, $\varphi_2^\pm(x) = \sin(B_c^\pm x)$, by $\alpha < 1$, in point $x = c$, and one of the conditions (7) in end points of interval Γ_o, where A_j, B_j, D_j^k, $j = 1, 2$, $1 \le k \le 4$ are given constants,

$$\lambda_1 = (2P_{21}(c+0))^{-1}\cdot(P_{22}(c+0) - P_{11}(c+0)),$$

$$\mu_1 = (2P_{21}(c-0))^{-1}\cdot(P_{22}(c-0) - P_{11}(c-0)),$$

$$\lambda_2 = (2P_{21}(c+0))^{-1}\cdot\sqrt{D_c^+},$$

$$\mu_2 = (2P_{21}(c-0))^{-1}\cdot\sqrt{D_c^-},$$

$$D_c^\pm = |(P_{22}(c\pm 0) - P_{11}(c\pm 0))^2 + 4P_{12}(c\pm 0)\cdot P_{21}(c\pm 0)|.$$

Remark 1. Analogous problems can be divided into three cases: 1) The roots of equation (2) are real and different, and roots of equation (3) are real and equal or complex-conjugate; 2) The roots of equation (2) are real and equal, and roots of equation (3) are real and different or complex-conjugate; 3) The roots of the equation (2) are complex-conjugate, and roots of the equation (3) are real and different or real and equal.

For solving Problems $R_j (1 \le j \le 3)$ and problems indicated in Remark 1, we are based on the integral representations in [4]-[6], [8], [9]. For Problem R_1, we use the integral representation

$$y_j(x) = \begin{cases} \Omega_{jc}^{\alpha,+}[c_1, c_2, f_1, f_2], & \text{by } a < x < c, \\ \Omega_{jc}^{\alpha,-}[c_3, c_4, f_1, f_2], & \text{by } c < x < b \end{cases}, \tag{8}$$

obtained in [9], where $\Omega_{jc}^{\alpha,\pm}, j = 1, 2$ are given integral operators independent of the valued $\alpha(\alpha < 1; \alpha = 1; \alpha > 1)$, in which $c_j, 1 \le j \le 4$ are arbitrary constants. Moreover, we use the integral representations [4] and [9]

$$\left.\begin{aligned} y_1 + \mu_j y_2 &= K_c^{\alpha,-}[P_c^{\mu_j}, F_1^- + \mu_j F_2^-, c_{j+1}], \text{ by } a < x < c, \\ y_1 + \lambda_j y_2 &= K_c^{\alpha,+}[P_c^{\lambda_j}, F_1^+ + \lambda_j F_2^+, c_j], \text{ by } c < x < b, \end{aligned}\right\} \tag{9}$$

where $F_j^{\pm}(x) = F_j^{\pm}[f_j, y_1, y_2] = f_j(x) - \sum\limits_{k=1}^{2}(P_{jk}(x) - P_{jk}(c\pm 0))y_k(x), j = 1, 2, K_c^{\alpha,\pm}$ are given integral operators dependent on the valued α. The integral operators $K_c^{\alpha,\pm}$ with $\alpha > 1$ possess the following forms

$$K_c^{\alpha,-}[P, Q, c_1] = \exp[-P(c-0)w_c^{\alpha}(x)][c_1 - \int_x^c \exp(P(c-0)w_c^{\alpha}(t)](c-t)^{-\alpha}Q(t)dt],$$

$$K_c^{\alpha,+}[P, Q, c_2] = \exp[P(c+0)w_c^{\alpha}(x)][c_2 + \int_c^x \exp(-P(c+0)w_c^{\alpha}(t))(t-c)^{-\alpha}Q(t)dt].$$

By using the integral representation (9), condition (4) with $\alpha < 1$, condition (5) with $\alpha = 1$ and condition (6) with $\alpha > 1$, we obtain the following two algebraic systems

$$c_1 = A_1c_3 + B_1,\ c_2 = A_2c_4 + B_2. \tag{10}$$

Applying the linear, integral operators $\Omega_{jc}^{\alpha,\pm}$, the formula (8) can be written the form

$$y_j(x) = \begin{cases} E_j^1(x)c_1 + E_j^2(x)c_2 + E_j^3(x), \ by\ a < x < c, \\ F_j^1(x)c_3 + F_j^2(x)c_4 + F_j^3(x), \ by\ c < x < b, \end{cases} \tag{11}$$

where $E_j^k(x), F_j^k(x), j = 1, 2, 1 \le k \le 3$ are known functions.

If we use the representation (11), system (10) and condition (7), then for Problem R_1, we have the following algebraic systems

$$\left.\begin{array}{l} c_1 = A_1c_3 + B_1 \\ c_2 = A_2c_4 + B_2 \\ E_j^1(a)c_1 + E_j^2(a)c_2 + E_j^3(a) = D_j^1 \\ j = 1, 2 \end{array}\right\}, \quad \left.\begin{array}{l} c_1 = A_1c_3 + B_1 \\ c_2 = A_4c_4 + B_2 \\ F_j^1(b)c_3 + F_j^2(b)c_4 + F_j^3(b) = D_j^2 \\ j = 1, 2 \end{array}\right\},$$

$$\left.\begin{array}{l} c_1 = A_1c_3 + B_1 \\ c_2 = A_2c_4 + B_2 \\ E_1^1(a)c_1 + E_1^2(a)c_2 + E_1^3(a) = D_1^3 \\ F_2^1(b)c_3 + F_2^2(b)c_4 + F_2^3(b) = D_2^3 \end{array}\right\}, \quad \left.\begin{array}{l} c_1 = A_1c_3 + B_1 \\ c_2 = A_2c_4 + B_2 \\ F_1^1(b)c_3 + F_1^2(b)c_4 + F_1^3(b) = D_1^4 \\ E_2^1(a)c_1 + E_2^2(a)c_2 + E_2^3(a) = D_2^4 \end{array}\right\}$$

For solving these algebraic systems, we find $c_j, 1 \le j \le 4$.

$$If\ \Delta_1 \equiv A_1A_2(E_1^1(a)E_2^2(a) - E_1^2(a)E_2^1(a)) \neq 0,$$

$$\Delta_2 \equiv F_1^1(b)F_2^2(b) - F_2^1(b)F_1^2(b) \neq 0,\ \Delta_3 \equiv A_1A_2E_1^2(a)\cdot$$

$$\cdot F_2^1(b) \neq 0,\ \Delta_4 \equiv A_2F_1^1(b)(E_2^1(a) - E_2^1(a)) \neq 0,$$

then the above algebraic systems have a unique solution.

Moreover we need to find the constants $c_j, 1 \leq j \leq 4$. Substituting the values of $c_j (1 \leq j \leq 4)$ in integral representation (8), we can find the solution of Problem R_1.

Finally, we write the above result as:

Theorem 1. *Let in system (1), $\alpha > 1$, $P_{jk}(x)$, $f_j(x) \in C(\Gamma)$ $(j,k = 1,2)$, in point $x = c$, we have first genus singularity and in neighborhood point $x = c$ satisfying the following conditions*

$$P_{1j}(x) - P_{1j}(c-0) + \lambda_1 (P_{2j}(x) - dP_{2j}(c-0)) = O[\exp(-2B_c^- \omega_c^\alpha(x))(c-x)^{\beta_{1j}}],$$

$$\beta_{1j} > \alpha - 1$$

by $x \to c - 0$, $j = 1,2$;

$$P_{1j}(x) - P_{1j}(c+0) + \lambda_2 (P_{2j}(x) - P_{2j}(c+0)) = O[\exp(-2B_c^+ \omega_c^\alpha(x))(x-c)^{\beta_{2j}}],$$

$$\beta_{2j} > \alpha - 1$$

by $x \to c + 0$, $j = 1,2$; $P_c^{\lambda_j} < 0\,(j = 1,2)$, $P_c^{\mu_j} > 0\,(j = 1,2)$. Besides, let $\Delta_j \neq 0, 1 \leq j \leq 4$. Then problem R_1 has a unique solution, which is given with the aid of obvious formula.

Remark 2. In the case, when one of the $\Delta_j = 0, 1 \leq j \leq 4$, then Problem R_1 has the results similar to the linear conjugate boundary value problem (Problem 6) in the theory of analytic functions [10].

For solving Problem R_2, we use the integral representation [9]

$$y_j(x) = \begin{cases} \Omega_{1+j}^{\alpha,-}[c_1, f_1, f_2] \text{ by } a < x < c, \\ \Omega_{1+j}^{\alpha,+}[c_2, f_1, f_2] \text{ by } c < x < b, \end{cases} \tag{12}$$

where $\Omega_{1+j}^{\alpha,\pm}[d, f_1, f_2] = \exp[\pm P_c^\pm w_c^\alpha(x)] K_{1+j,c}^{\alpha,\pm}[d, f_1, f_2]$ by $\alpha > 1$, and $\Omega_{1+j}^{1,\pm}[d, f_1, f_2] = |x - c|^{\mp P_c^\pm} K_{1+j}^{1,\pm}[c_2, f_1, f_2]$, by $\alpha = 1, K_{1+j}^{\alpha,\pm}$ are given integral operators, c_1, c_2 are arbitrary constants, $P_c^\pm = 2^{-1}(P_{11}(c \pm 0) + P_{22}(c \pm 0))$.

The integral representation (12) is obtained based on the assumption, i.e. the solution $y_j(x)$, $j = 1,2$ of the system (1) is satisfied the following conditions

$$\lim_{x \to c \pm 0} [\exp(\mp P_c^\pm w_c^\alpha(x))(y_1(x) + \lambda^\pm y_2(x))] = 0 \tag{13}$$

by $\alpha > 1$, and the following conditions

$$\lim_{x \to c \pm 0} [|x - c|^{\pm P_c^\pm} (y_1(x) + \lambda^\pm y_2(x))] = 0, \tag{14}$$

by $\alpha = 1$, where $\lambda^+ = \mu, \lambda^- = \lambda$, coefficients of the system (1) are satisfied the following conditions: functions $P_{jk}(x), j, k = 1, 2$ exactly that, functions

$$V_{11}^{c\pm 0}(x) = P_{21}(x) - P_{21}(c \pm 0), V_{12}^{c\pm 0}(x) = P_{22}(x) - P_{22}(c \pm 0),$$

$$V_{21}^{c\pm 0}(x) = P_{11}(x) - P_{11}(c \pm 0) + \lambda^{\pm}(P_{21}(x) - P_{21}(c \pm 0)),$$

$$V_{22}^{c\pm 0}(x) = P_{12}(x) - P_{12}(c \pm 0) + \lambda^{\pm}(P_{22}(x) - P_{22}(c \pm 0)),$$

in the neighborhood of point $x = c$, apply to zero, i.e. which is satisfied the conditions:

$$V_{1j}^{c\pm 0}(x) = 0(|x - c|^{\beta_{1j}}), \beta_{1j} > \alpha - 1, j = 1, 2, \tag{15}$$

$$V_{2j}^{c\pm 0}(x) = 0(|x - c|^{\beta_{2j}}), \beta_{2j} > 2(\alpha - 1), j = 1, 2, \tag{16}$$

by $x \to c \pm 0, P_c^{\lambda,+} > 0, P_c^{\lambda,-} < 0, P_c^{\lambda,\pm} = 2^{-1}(P_{11}(c \pm 0) + P_{22}(c \pm 0))$.

We note that in the case, when solution of system (1) is satisfied the condition (13) with $\alpha > 1$ and condition (14) with $\alpha = 1$, then for the solution of problem R_2, sufficiently is set, conditions in singular or super-singular point $x = c$, no need is give conditions (7), as in this case, the general solution of system (1) includes two arbitrary constants.

By using the integral representation (12) and conditions of problem R_2, with $\alpha < 1, \alpha = 1$ and $\alpha > 1$ respectively, we have the following two algebraic systems

$$\left.\begin{aligned} \lambda c_1 &= A_1 c_1 + B_1 \\ c_1 &= A_2 c_2 + B_2 \end{aligned}\right\}$$

¿From this, if $A_1 - \lambda A_2 \neq 0$, then

$$c_1 = \frac{A_1 B_2 - B_1 A_2}{A_1 - \lambda A_2}, \quad c_2 = \frac{\lambda B_2 - B_1}{A_1 - \lambda A_2}. \tag{17}$$

Substituting the values of c_1, c_2 in the integral representation (12), we can find a solution of problem R_2.

Thus, we have

Theorem 2. *Let in system* (1) $P_{jk}(x), f_j(x) \in C(\Gamma)(j, k = 1, 2)$, *in point $x = c$ can have first genus singularity and in neighborhood point $x = c$ by $\alpha > 1$ and $\alpha = 1$ correspondingly satisfying conditions* (15), (16), $P_c^{\lambda,+} > 0, P_c^{\lambda,-} < 0$. *The solution of the system* (1) *at $\alpha > 1$ and $\alpha = 1$ correspondingly by $x \to c \pm 0$ is satisfied conditions* (13), (14). *Besides let $A_1 - \lambda A_2 \neq 0$. Then problem R_2 has a unique solution, which is given by formula* (12), *where c_1, c_2 are given by formula* (17).

Remark 3. In the case, when $A_1 - \lambda A_2 = 0$, then Problem R_2 has the results similar to the linear conjugate boundary value problem in theory of analytic functions [10].

Remark 4. In the case, when the solution of the system (1) does not satisfy the conditions (13) or (14), then Problem R_2 has the results similar to Theorem 1.

Remark 5. The similar assertion in Theorem 1 may be obtained for Problem R_3 and problems indicated in Remark 1.

References

[1] Rajabov N., Integral representation and boundary value problems for some differential equations with singular lines or singular surfaces (Introduction in the theory nonmodel the second order hyperbolic equation with singular lines), part 4 (Russian), Tajik State University Publishers, Dushanbe, 1985, 147 pp.

[2] Rajabov N., Linear hyperbolic equations with two singular lines. Integral equations and boundary value problems. World Scientific, Singapore, New-Jersey, London, Hong-Kong, 1991, 170–175.

[3] Rajabov N., Linear hyperbolic equations of the fourth order with two singular lines. Proceeding of Asian Mathematical Conference, 1990 (Hong-Kong, 1990), World Scientific, 1992, 387–393.

[4] Rajabov N., The system two first order ordinary differential equation with super-singular points. Proceedings of international scientific conference on differential equations with singular coefficients, 1996 (Dushanbe, 1996), 69.

[5] Rajabov N. and Mirzoev A. M., Cauchy type problems for the two first order system ordinary linear differential equation with one super-singular point. Proceedings of international scientific conference on differential equations with singular coefficients, 1996 (Dushanbe, 1996), 71.

[6] Rajabov N. and Chojaev S., To theory of the one class two first order system ordinary differential equation with one super-singular point. Proceedings of international scientific conference on differential equations with singular coefficients, 1996 (Dushanbe, 1996), 74.

[7] Rajabov N., Linear conjugate boundary value problem for the two second order linear ordinary differential equation with super-singular coefficients. ICM 1998, International Congress of Mathematicians (Berlin, August 18-27, 1998, Abstracts of Short Communications and Poster Sessions), 186.

[8] Rajabov N., An introduction to the theory of partial differential equations with super-singular coefficients (forsi), Tehran University Publishers, Tehran, 1996, 230 p.

[9] Rajabov N., Introduction to ordinary differential equation with super-singular coefficients (English), Dushanbe, 1998, 160 p.

[10] Gachov F. D., Boundary value problems, "Nauka", M., 1977, 640 p.

24 INITIAL-MIXED BOUNDARY VALUE PROBLEMS FOR PARABOLIC EQUATIONS OF SECOND ORDER WITH MEASURABLE COEFFICIENTS IN A HIGHER DIMENSIONAL DOMAIN

Guo Chun WEN

Department of Mathematics, Peking University
Beijing 100871, China

wengc@pku.edu.cn

1991 Mathematics Subject Classification: 35K20, 35K55, 35K60
Key words: Initial-mixed problem, parabolic equations, measurable

Abstract. In [1], Yu. A. Alkhutov and I. T. Mamedov discussed the solvability of the Dirichlet problem for a linear nondivergence uniformly parabolic equations with Cordes coefficients. In this paper, we try to discuss the solvability of initial-mixed boundary value problem for nondivergence uniformly parabolic equations of second order with measurable coefficients in a higher dimensional domain, which includes the Dirichlet problem and initial-oblique derivative problem for the equations as special cases.

H.G.W. Begehr et al. (eds.), Proceedings of the Second ISAAC Congress, Vol. 1, 185–192.

I Formulation of initial-mixed problem for parabolic equation

Let Ω be a bounded domain in $\mathbb{R}^n$ and the boundary $\partial\Omega \in \mathbb{C}^2$. Denote $Q = \Omega \times I$, $I = 0 < t \leq T$, $0 < T < \infty$, $S = S_1 \cup S_2$ is the parabolic boundary, where $S_1 = \Omega \times \{t = 0\}$ is the bottom and $S_2 = \partial\Omega \times \bar{I}$ is the lateral boundary. We consider the nonlinear parabolic equation of second order

$$F(x,t,u,D_xu,D_x^2u)-u_t=0, \text{ i.e. } Lu=\sum_{i,j=1}^{n} a_{ij}u_{x_ix_j}+\sum_{i=1}^{n} b_iu_{x_i}+cu-u_t=d \tag{1.1}$$

in Q, where $D_xu = (u_{x_i})$, $D_x^2u = (u_{x_ix_j})$, and

$$a_{ij}=\int_0^1 F_{\tau r_{ij}}(x,t,u,p,\tau r)d\tau, b_i=\int_0^1 F_{\tau p_i}(x,t,u,\tau p,0)d\tau, r=D_x^2u, p=D_xu,$$
$$c=\int_0^1 F_{\tau u}(x,t,\tau u,0,0)d\tau, d=-F(t,x,0,0,0), r_{ij}=\frac{\partial u}{\partial x_i\partial x_j}, p_i=\frac{\partial u}{\partial x_i}. \tag{1.2}$$

Suppose that (1.1) satisfies Condition C, i.e. for arbitrary functions $u_1(x,t)$, $u_2(x,t) \in B^* = \tilde{C}^{1,0}_{\beta,\beta/2}(Q^*) \cap \tilde{W}^{2,1}_2(Q^*)$, $F(x,t,u,D_xu,D_x^2u)$ satisfy the condition

$$F(x,t,u_1,D_xu_1,D_x^2u_1)-F(x,t,u_2,D_xu_2,D_x^2u_2)=$$
$$\sum_{i,j=1}^{n} \tilde{a}_{ij}u_{x_ix_j}+\sum_{i=1}^{n}\tilde{b}_iu_{x_i}+\tilde{c}u, \tag{1.3}$$

where $\tilde{C}^{1,0}_{\beta,\beta/2}(Q^*) = C^{0,0}(\overline{Q}) \cap C^{1,0}_{\beta,\beta/2}(Q^*)$, $\tilde{W}^{2,1}_2(Q^*) = W^{2,0}_2(Q^*) \cap W^{0,1}_2(Q^*)$, $0 < \beta < 1$, D^* is any closed subset in $\bar{D}\backslash E$, $E = \{(x,t)|\, a = 0, (x,t) \in S_2\}$, $u = u_1 - u_2$ and

$$\tilde{a}_{ij}=\int_0^1 F_{u_{x_ix_j}}(x,t,\tilde{u},\tilde{p},\tilde{r})d\tau, \tilde{b}_i=\int_0^1 F_{u_{x_i}}(x,t,\tilde{u},\tilde{p},\tilde{r})d\tau, \tilde{u}=u_2+\tau(u_1-u_2),$$
$$\tilde{c}=\int_0^1 F_u(x,t,\tilde{u},\tilde{p},\tilde{r})d\tau, \tilde{p}=D_x[u_2+\tau(u_1-u_2)], \tilde{r}=D_x^2[u_2+\tau(u_1-u_2)], \tag{1.4}$$

and $\tilde{a}_{ij}, \tilde{b}_i, \tilde{c}, d$ satisfy the conditions

$$q_0\sum_{j=1}^{n}|\xi_j|^2 \leq \sum_{i,j=1}^{n} \tilde{a}_{ij}\xi_i\xi_j \leq q_0^{-1}\sum_{j=1}^{n}|\xi_j|^2,$$

$$\sup \sum_{i,j=1}^{n} \tilde{a}_{ij}^2 / \inf [\sum_{i=1}^{n} \tilde{a}_{ii}]^2 \leq q_1 < \frac{1}{n-1/2}. \tag{1.5}$$

$$|\tilde{a}_{ij}| \leq k_0, c > \inf_Q c_0, L_p[\tilde{b}_i, \overline{Q}] \leq k_0, L_p[\tilde{c}, \overline{Q}] \leq k_0, L_p[d, \overline{Q}] \leq k_1, i, j = 1, ..., n, \tag{1.6}$$

where a is as stated in (1.8) below, $q_0 (0 < q_0 < 1), q_1, k_0, k_1, p(> n+2)$ are non-negative constants and c_0 is a real constant. Moreover, for almost every point $(x,t) \in Q$ and $D_x^2 u$, $\tilde{a}_{ij}(x,t,u,D_x u, D_x^2 u)$, $\tilde{b}_i(x,t,u,D_x u)$, $\tilde{c}(x,t,u)$ are continuous in $u \in I\!R, D_x u \in I\!R^n$.

The so-called initial-mixed boundary value problem (Problem M), i.e. to find a continuously differentiable solution $u = u(x,t) \in B^* = \tilde{C}^{1,0}_{\beta,\beta/2}(Q^*) \cap \tilde{W}^{2,1}_2(Q^*)$ satisfying the initial-boundary conditions:

$$u(x,0) = f(x), \quad x \in \Omega, \tag{1.7}$$

$$lu = a\frac{\partial u}{\partial \nu} + bu = g(x,t), \text{ i.e. } lu = \sum_{j=1}^{n} d_j \frac{\partial u}{\partial x_j} + bu = g(x,t), \ (x,t) \in S_2, \tag{1.8}$$

in which $d_j(x,t) = \sum_{i=1}^n a_{ij} \cos(\nu, x_i)$, $f(x)$, $g(x,t), b(x,t)$ satisfy the conditions

$$C^2_\alpha[f(x), \bar{\Omega}] \leq k_2, C^{2,1}_{\alpha,\alpha/2}[g(x,t), S_2] \leq k_2, C^{1,1}_{\alpha,\alpha/2}[b(x,t), S_2] \leq k_0,$$

$$C^{2,1}_{\alpha,\alpha/2}[d_j(x,t), S_2] \leq k_0 (j = 1, \cdots, n), \cos(\nu, n) \geq d_0 > 0, a + !b \geq 1 \text{ on } S_2, \tag{1.9}$$

where n is the unit outward normal on S_2, $\alpha (0 < \beta \leq \alpha < 1), k_0$, $k_2, d_0\,(0 < d_0 < 1)$ are non-negative constants. In particular, if Problem M with the conditions $a = 0, \nu = s$, $b = 1$ on S_2, then Problem M is the Dirichlet boundary value problem, which will be called Problem D, and if Problem M with the conditions $a > 0$ on S_2, herein s is the unit tangent vector on S_2, then Problem M is the initial-regular derivative boundary value problem, which will be called Problem O.

By the maximin principle of solutions for parabolic equations, we can prove the uniqueness of solutions of Problem M.

Theorem 1.1 *If equation* (1.1) *satisfies Condition C, then the solutions of Problem M is unique.*

II A priori estimates of solutions for initial-mixed problem for (1.1)

Theorem 2.1 *Under the same condition in Theorem* 1.1, *any solution* $u(x,t)$ *of Problem M for* (1.1) *satisfies the estimate*

$$\| u \|_{C^{0,0}(\bar{Q})} \le M_1, \tag{2.1}$$

where M_1 *is a non-negative constant only dependent to* q, α, k, p, Q, *i.e.* $M_1 = M_1(q,\ \alpha, k,\ p, Q)$, $q = (q_0, q_1)$, $k = (k_0, k_1, k_2)$.

Proof Let $u(x,t)$ be any solution of Problem M and substitute it into (1.1). There is no harm in assuming that the coefficient $c \ge 0$ in (1.1) in Q, because the requirement can be realized through a transformation of function $u = v\exp(Bt)$, where B is an appropriately large number, such that $B > \inf_Q c$. Now, we find a solution $\psi(x,t)$ of (1.1) with the initial-boundary condition $\psi(x,t) = 0$ on S. On the basis of the results in [4], we can obtain the estimates of $\psi(x,t)$:

$$C^{1,0}_{\beta,\beta/2}[\psi, \bar{Q}] \le M_2,\ \| \psi \|_{\tilde{W}_2^{2,1}(Q)} \le M_3, \tag{2.2}$$

in which $\beta(0 < \beta \le \alpha)$, $M_j = M_j(q, \alpha, k, p, Q)(j = 2, 3)$ are non-negative constants. Next, we find out the solution $\Psi(x,t)$ of the homogeneous equation

$$Lu = \sum_{i,j=1}^{n} a_{ij} u_{x_i x_j} + \sum_{i=1}^{n} b_i u_{x_i} + cu - u_t = 0 \text{ in } Q, \tag{2.3}$$

with the initial-boundary condition $\Psi(x,t) = 1$ on S. Similarly, we can prove that $\Psi(x,t)$ satisfies the estimates

$$C^{1,0}_{\beta,\beta/2}[\Psi, \bar{Q}] \le M_4,\ \| \Psi \|_{\tilde{W}_2^{2,1}(G)} \le M_5,\ 0 < M_6 \le \Psi \le 1, \tag{2.4}$$

where $M_j = M_j(q, \alpha, k, p, G) \ge 0, j = 4, 5, 6$. According to the method in [3], we see that the function

$$U(x,t) = [u(x,t) - \psi(x,t)]/\Psi(x,t) \tag{2.5}$$

is a solution of the initial-boundary value problem

$$\sum_{i,j=1}^{n} a_{ij} u_{x_i x_j} + \sum_{i=1}^{n} \tilde{b}_i u_{x_i} - u_t = 0,\ \tilde{b}_i = b_i + \sum_{i=1}^{n} a_{ij} (\ln \Psi)_{x_j} \text{ in } Q, \tag{2.6}$$

$$U(x,0) = [g(x) - \psi(x,0)]/\Psi(x,0) = U_0(x) \text{ on } \Omega, \tag{2.7}$$

$$a\frac{\partial U}{\partial\nu}+\tilde{b}U=\tilde{g},\ \tilde{b}=b+a\frac{\partial\ln\Psi}{\partial\nu},\ \tilde{g}=[g-a\frac{\partial\psi}{\partial\nu}-b\psi]/\Psi \text{ on } S_2, \quad (2.8)$$

where $\tilde{b}>0$ on S_2. By means of the maximin principle of solutions of (2.6), we know that $U(x,t)$ attains its maximum and minimum at points $P^*=(x^*,t^*), P_*=(x_*,t_*)\in S$ respectively, we can get

$$C[U,\bar{Q}]=\max[U(x^*,t^*),|U(x_*,t_*)|]\le\max[\max_D|U_0(x)|,\max_{S_2}|\tilde{g}|/\min_{S_2}\tilde{b}]. \quad (2.9)$$

Combining (2.2),(2.4) and (2.9), the estimate (2.1) is derived.

Theorem 2.2 *Under the same condition as in Theorem* 2.1, *any solution* $u(x,t)$ *of Problem* M *for* (1.1) *satisfies the estimates*

$$\tilde{C}^{1,0}_{\beta,\beta/2}[u,Q^*]\le M_7,\ \parallel u\parallel_{\tilde{W}^{2,1}_2(Q^*)}\le M_8, \quad (2.10)$$

where $Q^*=\bar{Q}\cap\{\cap_{(x^*,t^*)\in\partial S_2^*}[|x-x^*|^2+|t-t^*|]\ge\varepsilon>0\}$, ε *is a small positive number and* $S_2^*=\{(x,t)\in S_2, a(x,t)=0, t=constant\}, \partial S_2^*$ *is the set of end points of* $S_2^*, \beta(0<\beta\le\alpha), M_j=M_j(q,\alpha,k,p,Q)(j=7,8)$ *are non-negative constants.*

Proof By using the method as stated in [3], we can prove that any solution $u(x,t)$ of Problem M for (1.1) satisfies the estimates

$$C^{1,0}_{\beta,\beta/2}[u,G]\le M_9,\ \parallel u\parallel_{\tilde{W}^{2,1}_2(G)}\le M_{10}, \quad (2.11)$$

in which $G=Q_m, Q_m=\{(x,t)\in\bar{Q},\text{dist}(x,\partial\Omega)\ge 1/m\}, m$ is a positive integer, $\beta(0<\beta\le\alpha), M_j=M_j(q,\alpha,k,p,\varepsilon,m,Q)(j=9,10)$ are non-negative constants. Next, we give the estimates of the solution $u(x,t)$ near the lateral S_2 of Q. We arbitrarily choose an inner point $P^*=(x^*,t^*)$ of S_2^*, and denote $\tilde{S}_2=\{(x,t)\in S_2^*, |x-x^*|^2+|t-t^*|<\varepsilon^*\}$, where $\varepsilon^*(<\varepsilon)$ is an appropriately small positive number such that $\tilde{S}_2\cap S_2\subset S_2^*\backslash\partial S_2^*$. By using a similar method as stated in [4], we can obtain estimates of $u(x,t)$ in the neighborhood $\hat{Q}_2=\{(x,t)\in Q\cup\tilde{S}_2, \text{dist}((x,t),S_2\backslash\tilde{S}_2)\ge\varepsilon^*>0\}$ of $\tilde{S}_2$, namely

$$C^{1,0}_{\beta,\beta/2}[u,\hat{S}_2]\le M_{11},\parallel u\parallel_{\tilde{W}^{2,1}_2(\hat{Q})}\le M_{12}, \quad (2.12)$$

where $M_j=M_j(q,\alpha,k,p,\varepsilon,\varepsilon^*,Q), j=11,12$. Moreover, we discuss the surface $S_3=\{(x,t)\in S_2, a(x,t)>0,\cos(\nu,n)\ne 0\}$. According to the method in [4], we can also derive the estimates of $u(x,t)$:

$$C^{1,0}_{\beta,\beta/2}[u,\tilde{Q}_3]\le M_{13},\parallel u\parallel_{\tilde{W}^{2,1}_2(\tilde{Q}_3)}\le M_{14}, \quad (2.13)$$

where $\tilde{Q}_3=\{(x,t)\in Q\cup S_3, \text{dist}((x,t),S_2\backslash S_3)\geq\varepsilon^*>0\}$, $M_j=M_j(q,\alpha,k,p,\varepsilon,\varepsilon^*,Q)$, $j=13,14$. Combining (2.11)–(2.13), the estimates in (2.10) are derived.

III The solvability of initial-mixed boundary value problem for (1.1)

Theorem 3.1 *Let* (1.1) *satisfy Condition C. Then Problem M for* (1.1) *is solvable.*

Proof We are free to choose a positive integer m and consider Problem M_m for (1.1) with the initial-boundary conditions

$$u(x,0)=g(x)+g_m(x,0) \text{ on } \Omega, \tag{3.1}$$

$$[a(x,t)+\frac{1}{m}]\frac{\partial u}{\partial\nu}+b(x,t)u=g(x,t) \text{ on } S_2, \tag{3.2}$$

where $g_m(x,t)$ is an appropriate solution of the oblique derivative problem for the homogeneous equation (2.3) with the boundary condition

$$(a+\frac{1}{m})\frac{\partial g_m}{\partial\nu}+bg_m=-\frac{1}{m}\frac{\partial g}{\partial\nu} \text{ on } S_2, \tag{3.3}$$

such that $g_m(x,t)$ satisfies estimates similar to (2.11) with $G=\overline{Q}$, and $g_m(x,t)\to 0$ as $m\to\infty$. According to the result in [4], we know that g_m exists and Problem M_m has the solution $u_m(x,t)$, $m=1,2,...$, which satisfy the estimate (2.10). Hence from $\{u_m(x,t)\}$, we can select a subsequence $\{u_{m_k}(x,t)\}$, which uniformly converges to a solution $u_0(x,t)$ of (1.1) in any closed subset $Q_*\subset\bar{Q}\backslash S_2^*$, and $u_0(x,t)$ satisfies the initial-boundary conditions (1.7) and

$$a(x,t)\frac{\partial u_0}{\partial\nu}+b(x,t)u_0=g(x,t) \text{ on } S_2\backslash S_2^*. \tag{3.4}$$

It remains to prove that $u_0(x,t)$ is continuous in $\bar{Q}$ and satisfies the boundary condition (1.8). We select an arbitrary point $P^*=(x^*,t^*)\in S_2^*$ (it can be replaced by any surface $S\subset S_2^*$), and let G_β denote the point set $\{[|x-x^*|^2+|t-t^*|]<\beta]\cap S_2\}$, where β is a sufficiently small positive number. We construct a real continuous function $f(x,t)$ as follows:

$$f(z,t)=\begin{cases} M_{15}+1, (x,t)\in S_2\backslash G_{\beta/2}, \\ \eta>0 \text{ in } G_{\beta/4}, \end{cases} \eta\leq f(x,t)\leq M_{15}+1 \text{ in } G_{\beta/2}\backslash G_{\beta/4}, \tag{3.5}$$

where M_{15} is an undetermined positive number. The function $f(x,t)$ satisfies the estimate

$$C^{1,0}_{\mu-\varepsilon}[f,S_2] \le \frac{M_{16}}{\beta^{1+\mu-\varepsilon}},\ \frac{1}{2} < \mu-\varepsilon < 1, \tag{3.6}$$

where ε is a sufficiently small positive constant, $M_{16} = M_{16}(M_{15}, S_2)$. Let $\hat{u}_m(x,t)$ be a solution of the homogeneous equation (2.3) satisfying the boundary condition $\hat{u}_m(x,t) = f(x,t)$ on S_2. It is not difficult to see that $\hat{u}_m(x,t)$ satisfies the estimate

$$C^{1,0}[\hat{u}_m(x,t), \bar{Q}] \le M_{17}/\beta^{1+\mu-\varepsilon},$$

where $M_{17} = M_{17}(q,\alpha,k,Q)$. Now, we extend $b(x,t)$ from S_2^* to S_2, such that we have a new function $b^*(x,t) \in C^{1,0}_{\alpha,\alpha/2}(S_2)$, $b^*(x,t) > 0$ and $b^*(x,t) = b(x,t)$ on $G_{\beta/2}$. Moreover, we can find a solution $u^*(x,t)$ of equation (1.1) with the boundary condition $u_m^*(x,t) = g(x,t)/b^*(x,t)$ on S_2. Setting

$$\tilde{u}_\pm(x,t) = \pm\hat{u}_m(x,t) - u_m(x,t) + u_m^*(x,t),$$

it is easily seen that $\tilde{u}_\pm(x,t)$ are solutions of $Lu = 0$. In the following, we verify that

$$\tilde{u}_+(x,t) \ge 0,\ \tilde{u}_-(x,t) \le 0,\ (x,t) \in \bar{Q}.$$

In fact, obviously $\tilde{u}_+(x,t) \ge M_{15} + 1 - M_{15} > 0$ on $S_2 \backslash G_{\beta/2}$, where

$$M_{15} = M + \max_{S_2} \frac{g(z,t)}{b_2^*(x,t)} > 0,\ M = \max_{\bar{Q}} |u_m(x,t)|,$$

herein the constant M is similar to M_1 in (2.1). If $\tilde{u}_+(x,t)$ takes a negative minimum in $\bar{Q}$, then there exists a point $P' = (x',t') \in G_{\beta/2}$, such that $\tilde{u}_+(x',t') \le \min_{\bar{Q}} \tilde{u}_+(x,t)$. However if choose β small enough, and then select n large enough, then we have

$$[a(x,t) + \frac{1}{m}]\frac{\partial \tilde{u}_+}{\partial \nu} + b(x,t)\tilde{u}_+ > 0 \text{ on } G_{\beta/2}.$$

This shows that $\tilde{u}_+(x,t)$ cannot attain a negative minimum on $G_{\beta/2}$. On the basis of the maximin principle of solutions of $Lu_m = 0$, we can obtain

$$\tilde{u}_+(x,t) = \hat{u}_m(x,t) - u_m(x,t) + u_m^*(x,t) \ge 0, \text{ i.e. } u_m(x,t) - u_m^*(x,t) \le \hat{u}_m(x,t)$$

in $\bar{Q}$. By the same reasoning, we have $\tilde{u}_-(x,t) \le 0$, i.e. $u_m(x,t) - u_m^*(x,t) \ge -\hat{u}_m(x,t)$ in $\bar{Q}$. From $|\hat{u}_m(x,t)| \le \eta$ on $G_{\beta/4}$, it follows that $|u_m(x,t) - u_m^*(x,t)| < \eta$ on $G_{\beta/4}$. By the equicontinuity of $\{\hat{u}_m(x,t)\}$ in $\bar{Q}$, it is seen $|u_m(x,t) - u_m^*(x,t)| \le |\hat{u}_m(x,t)| \le 2\eta$ in a neighborhood of (x^*, t^*) in $\bar{Q}$. Denote by $\tilde{u}_0(x,t)$ the limit function of $\{u_m(x,t) - u_m^*(x,t)\}$ in Q, it is clear that $|\tilde{u}_0(x,t)| \le 2\eta$. Noting that η is an arbitrary positive number, we see that $\tilde{u}_0(x,t)$ at $P^* = (x^*, t^*)$ is continuous, and $\tilde{u}_0(x^*, t^*) = 0$. Hence $u_0(x,t) = \tilde{u}_0(x,t) + u_0^*(x,t)$ at (x^*, t^*) is also continuous, where $u_0^*(x,t)$ is a limit function of a subsequence of $\{u_m^*(x,t)\}$. This completes the proof.

References

[1] Alkhutov Yu. A. and Mamedov I. T., *The first boundary value problem for nondivergence second order parabolic equations with discontinuous coefficients.* Math. USSR Sbornik 59 (1988), 471-495.

[2] Ladyzhenskaya O. A., Solonnikov V. A. and Ural'ceva N. N., *Linear and Quasilinear Equations of Parabolic Type.* Trans. Math. Monographs 23, Amer. Math. Soc., Providence, RI, 1968.

[3] Wen G. C. and Begehr H., *Boundary Value Problems for Elliptic Equations and Systems.* Longman Scientific and Technical, Harlow, 1990.

[4] Wen G. C. and Tain Mao-ying, *Initial-oblique derivative problems for nonlinear parabolic equations with measurable coefficients.* Comm. in Nonlinear Sci. & Numer. Simu. 2 (1998), 109–113.

25 STABILITY ESTIMATES IN STATE-ESTIMATION FOR A HEAT PROCESS

Dinghua XU[1]

Department of Basic Sciences, East China Geological Institute
Road 149, Shanghai 200072, P. R. China

dhxujc@ecgi.jx.cn

Masahiro YAMAMOTO[2]
Department of Mathematical Sciences, The University of Tokyo
3-8-1 Komaba, Meguro, Tokyo 153, Japan

myama@ms.u-tokyo.ac.jp

1991 Mathematics Subject Classification: 35R25, 35K05
Key words: State-estimation, heat equation, Lipschitz stability, conditional stability, Carleman estimate

Abstract: This paper deals with a heat equation with the homogeneous Dirichlet boundary condition:

$$\begin{cases} u_t(t,x) = \Delta u(t,x), & 0 < t < T, \quad x \in \Omega, \\ u(t,x) = 0, & 0 < t < T, \quad x \in \partial\Omega, \end{cases}$$

[1] Supported by Jiangxi Provincial Natural Scientific Foundation and National Natural Scientific Foundation of P. R. China

[2] Supported by Sanwa Systems Development Co. Ltd (Tokyo, Japan).

H.G.W. Begehr et al. (eds.), Proceedings of the Second ISAAC Congress, Vol. 1, 193–198.

where $\Omega \subset \mathcal{R}^n$ is a bounded domain with smooth boundary $\partial\Omega$. For given $\theta \in [0, T)$, we discuss a problem of determining $u(\theta, \cdot)$ from $u|_{(0,T)\times\omega}$, where $\omega \subset \Omega$ is an arbitrary subdomain. Our main result is the Lipschitz stability estimate in the case of $0 < \theta < T$ and conditional stability of logarithmic rate in the case of $\theta = 0$. The proof relies on a Carleman estimate for a parabolic equation.

1. Introduction

We consider a heat equation with the homogeneous Dirichlet boundary condition:

$$\begin{cases} u_t(t,x) = \Delta u(t,x), & 0 < t < T, \quad x \in \Omega \\ u(t,x) = 0, & 0 < t < T, \quad x \in \partial\Omega, \end{cases} \tag{1}$$

where $\Omega \subset \mathcal{R}^n$ is a bounded domain with smooth boundary $\partial\Omega$, Δ is the Laplacian operator. Let $\theta \in [0, T)$ be fixed. In this paper, we discuss a problem of determining $u(\theta, x)$, $x \in \Omega$ from the internal measurement data

$$u|_{(0,T)\times\omega}, \tag{2}$$

here $\omega \subset \Omega$ is an arbitrary subdomain.

The purpose of this paper is to establish a Lipschitz stability estimate in the case of $0 < \theta < T$ and conditional stability of logarithmic rate in the case of $\theta = 0$. This problem is related with state-estimation and we can refer to Cannon [1] for example, where various different types of state-estimation problems are discussed in the case of $n = 1$. Our argument is based on a Carleman estimate proved by Imanuvilov [2] and we can discuss similar problems for general parabolic equations with variable coefficients with various boundary conditions.

For stability in multidimensional inverse parabolic problems, we refer to Imanuvi- lov and Yamamoto [3], Yamamoto [5], for example.

2. Main Result

Let us set

$$C^{1,2}([0,T]\times\overline{\Omega}) = \{g(t,x) \mid g, \frac{\partial g}{\partial t}, \frac{\partial g}{\partial x_i},$$

$$\frac{\partial^2 g}{\partial x_i \partial x_j} \in C([0,T]\times\overline{\Omega}), 1 \le i,j \le n\}$$

and $H^2(\Omega), L^2(\Omega)$ denote a usual Sobolev space and the L^2-space, and we set

$$\| g \|_{L^2(0,T;L^2(\omega))} = \left(\int_0^T \int_\omega | g(t,x) |^2 \, dxdt\right)^{\frac{1}{2}}$$

and

$$\| g \|_{H^1(0,T;L^2(\omega))} = \left(\int_0^T \int_\omega (| g_t(t,x) |^2 + | g(t,x) |^2) dxdt\right)^{\frac{1}{2}}.$$

Now we can state our main result.

Theorem. (1) *Let $0 < \theta < T$ and $u \in C^{1,2}([0,T]\times\overline{\Omega})$ satisfy* (1). *Then there exists a constant $C = C(\Omega, T, \omega, \theta) > 0$ such that*

$$\| u(\theta, \cdot) \|_{L^2(\Omega)} \le C \| u \|_{L^2(0,T;L^2(\omega))}, \tag{3}$$

where the constant $C > 0$ is independent of u.

(2) *Let $\theta = 0$. We set*

$$U_M = \{a \in H^2(\Omega) \cap H_0^1(\Omega);\ \|a\|_{H^2(\Omega)} \le M\}$$

for an arbitrarily fixed $M > 0$. Then there exists a constant $C = C(\Omega, T, \omega, M) > 0$ such that

$$\| u(0, \cdot) \|_{L^2(\Omega)} \le C \left(\log \left(\frac{1}{\|u\|_{H^1(0,T;L^2(\omega))}} \right) \right)^{-1} \tag{4}$$

provided that $u(0,\cdot) \in U_M$.

The inequality (3) shows unconditional stability while (4) is conditional stability depending on the admissible set U_M.

3. Carleman Estimate and Proof of Theorem

For the Carleman estimate we need a suitable weight function. The existence of such a function is proved in [2].

Lemma 1. *Let ω_0 be an arbitrary subdomain of Ω such that $\overline{\omega}_0 \subset \omega$. Then there exists a function $\psi \in C^2(\overline{\Omega})$ such that*

$$\psi(x) > 0, \forall x \in \Omega, \psi|_{\partial\Omega} = 0, |\nabla\psi(x)| > 0, \forall x \in \overline{\Omega \setminus \omega_0}.$$

For $\lambda > 0$, we set

$$\varphi(t,x) = \frac{e^{\lambda\psi(x)}}{t(T-t)}, \quad \alpha(t,x) = \frac{e^{\lambda\psi(x)} - e^{2\lambda\|\psi\|_{C(\overline{\Omega})}}}{t(T-t)}.$$

We consider

$$\begin{cases} v_t(t,x) = \Delta v(t,x), & 0 < t < T, x \in \Omega \\ v(t,x) = 0, & 0 < t < T, x \in \partial\Omega. \end{cases} \tag{5}$$

The following Carleman estimate is valid (e.g. [2]).

Lemma 2. *Let $v(t,x) \in C^{1,2}([0,T] \times \overline{\Omega})$ be satisfied (5). Then for sufficiently large $\lambda > 0$, there exists $s_0 = s_0(\lambda)$ such that*

$$\int_{(0,T)\times\Omega} \{\frac{1}{s\varphi}(|v_t|^2 + \Sigma_{i,j=1}^n \left|\frac{\partial^2 v}{\partial x_i \partial x_j}\right|^2) + s\varphi|\nabla v|^2 + (s\varphi)^3|v|^2\} e^{2s\alpha} dxdt$$
$$\le C \int_{(0,T)\times\omega} (s\varphi)^3 |v|^2 e^{2s\alpha} dxdt \tag{6}$$

for all $s > s_0$. Here the constant $C > 0$ is independent of s and v.

Proof of (1) of Theorem. Direct application of Lemma 2 to the problem (1) gives the following inequality

$$\int_{(0,T)\times\Omega} \{\frac{1}{s\varphi}(|u_t|^2 + \Sigma_{i,j=1}^n \left|\frac{\partial^2 u}{\partial x_i \partial x_j}\right|^2) + s\varphi|\nabla u|^2 + (s\varphi)^3|u|^2\} e^{2s\alpha} dxdt$$
$$\le C_1 \int_{(0,T)\times\omega} (s\varphi)^3 |u|^2 e^{2s\alpha} dxdt. \tag{7}$$

Here the positive constant C_1 is independent of s and u. By the definition of α, for $\theta > 0$ and suffciently small $\varepsilon > 0$, from (7) we see that there exists a constant $C_2 = C_2(\Omega, T, \omega, \theta, \varepsilon) > 0$ such that

$$\|u\|_{H^1((\theta,T-\varepsilon)\times\Omega)} \le C_2 \parallel u \parallel_{L^2(0,T;L^2(\omega))} .$$

By the trace theorem in $H^1((\theta, T-\varepsilon) \times \Omega)$, we have

$$\|u(\theta,\cdot)\|_{L^2(\Omega)} \le C_3 \parallel u \parallel_{H^1((\theta,T-\varepsilon)\times\Omega)},$$

where $C_3 = C_3(\Omega, T, \theta, \varepsilon) > 0$. Thus the proof of (1) is complete.

Proof of (2) of Theorem. In place of u, we consider $u_1 = u_t$ and apply (1) of Theorem to obtain

$$\|u_1(\theta, \cdot)\|_{L^2(\Omega)} \le C\|u_1\|_{L^2(0,T;L^2(\omega))}, \tag{8}$$

where $C = C(\Omega, T, \omega, \theta) > 0$ is independent of u_1. Without loss of generality, we may assume that $\|u_1(\theta, \cdot)\|_{L^2(\Omega)} < 1$ because we can consider u_1/M_1 for sufficiently large $M_1 > 0$ if necessary. Applying the logarithmic convexity method (e.g. Payne [4]) to u_1, we have

$$\|u_1(t, \cdot)\|_{L^2(\Omega)} \le C_4^{(\theta-t)/\theta}\|u_1(\theta, \cdot)\|_{L^2(\Omega)}^{t/\theta}, \quad 0 \le t \le \theta. \tag{9}$$

Here $C_4 = C_4(\Omega) > 1$ is independent of u_1. Since

$$u(0, \cdot) = -\int_0^\theta u_1(\eta, \cdot)d\eta + u(\theta, \cdot),$$

by (8) and (9) we have

$$\begin{aligned}
&\|u(0, \cdot)\|_{L^2(\Omega)} \le \int_0^\theta \|u_1(\eta, \cdot)\|_{L^2(\Omega)}d\eta + \|u(\theta, \cdot)\|_{L^2(\Omega)} \\
&\le C_4 \int_0^\theta \|u_1(\theta, \cdot)\|_{L^2(\Omega)}^{\eta/\theta} d\eta + C\|u\|_{L^2(0,T;L^2(\omega))} \\
&\le C_4 \frac{1}{\frac{1}{\theta}\log \frac{1}{\|u_1\|_{L^2(0,T;L^2(\omega))}}} + C\|u\|_{L^2(0,T;L^2(\omega))}.
\end{aligned}$$

Hence, noting that $\|u_1\|_{L^2(0,T;L^2(\omega))} \le \|u\|_{H^1(0,T;L^2(\omega))}$, we can complete the proof of (2) of Theorem.

References

[1] Cannon J. R., *The One-dimensional Heat Equation*, Addison-Wesley, Reading, 1984.

[2] Imanuvilov O. Yu, *Controllability of parabolic equations,* Sbornik Math., 186 (1995), 879–900.

[3] Imanuvilov O. Yu. and Yamamoto M., *Lipschitz stability in inverse parabolic problems by the Carleman estimate,* Inverse Problems, 14 (1998), 1229–1245.

[4] Payne L. E., *Improperly Posed Problems in Partial Differential Equations*, SIAM, Philadelphia, 1975.

[5] Yamamoto M., *Conditional stability in determination of densities of heat sources in a bounded domain,* International Series of Numerical Mathematics, Birkhauser Verlag Basel, 118 (1994), 359-370.

26 PLASTIC ZONE AND OPENING DISPLACEMENT FOR AN ASYMMETRICAL FAST PROPAGATING SEMI-INFINITE CRACK IN A STRIP

Xiao-Chun YANG

Department of Mathematics, Ningxia University
Yinchuan 750021, China

li_x@nxu.edu.cn

Tian-You FAN
Beijing Institute of Technology, Beijing 100081, China

Abstract: In this paper, the problems of the generalized Dugdale model for a semi-infinite fast propagating crack asymmetrically in a strip have been discussed. The plastic zone and crack tip opening displacement are given in both cases.

1. Introduction

A semi-infinite asymmetrical crack in a strip is one of significant configurations for theoretical interesting and practical application in fracture mechanics. The symmetrical crack in the strip has been extensively studied by Refs.[1-6], in which the comformal mapping based on transcendental function is developed and many analytic solutions in closed form are found. The approach can be used in treating cracked strip

H.G.W. Begehr et al. (eds.), Proceedings of the Second ISAAC Congress, Vol. 1, 199-206.

problem of infiltrated porous medium, e. g. see Ref.[7]. Results obtained in Refs.[1-6] were applied for analysing the fault instability and explained the low stress drop observed in earthquake source[9]. In foundation engineering, hydraulic fracture, blasting engineering, petroleum geology and seismology, asymmetrical crack in a strip is more interesting model, which was reported briefly in Ref.[8], where the conformal mapping technique has been further developed. In the present paper, we extend the discussion in Ref.[8] into dynamic case, and an approximate conformal mapping will be used, and we also extend our discussion into material case, and put forward a generalized Dugdale model, the plastic zone sizes in front of the crack tip and the crack tip opening displacements for dynamic rapid propagating cases is determined. All results are given in explicit analytic form. The works mentioned above[8] and will be published in Ref.[18] show that the conformal mapping based on transcendental functions are effective in constructing analytic solutions of crack problems for some complicated configurations.

2. The formulation and solution of the dynamic propagating crack in a strip

As shown in Fig. 1, which can be obtained by a request to the e-mail address: li_x@nxu.edu.cn , an asymmetrical semi-infinite crack is propagating at a uniform speed V in the direction x_1. We set up a moving coordinate system (x, y), whose origin is at the propagating crack tip. Between the fixed coordinate system (x_1, y) and the moving one there is the relation

$$x = x_1 - Vt, \quad y = y \tag{2.1}$$

Assume that over the portion $y = 0, -a < x < 0$ of the semi-infinite crack subjected to an equilibrium pressure p, of vanishing magnitude over the reminder of the crack tip, there is a so-called cohesive force zone with length R, whose value is unknown temporarily, in this zone a traction $\sigma_{yy} = \sigma_Y$ is subjected, in which σ_Y denotes the yield limit of the materials. Evidently the problem is a linearized.

The problem may be considered as a superposition of the following two boundary value problems:

$$\begin{cases} y = -H_1 \text{or} H_2, -\infty < x < +\infty : & \sigma_{yy} = \sigma_{xy} = 0, \\ x = \pm\infty, -H_1 < y < H_2 : & \sigma_{xx} = \sigma_{xy} = 0, \\ y = 0, -\infty < x < a : & \sigma_{yy} = \sigma_{xy} = 0, \\ y = 0, -a < x < 0 : & \sigma_{yy} = -p, \sigma_{xy} = 0, \end{cases} \tag{2.2}$$

and

$$\begin{cases} y = -H_1 \text{or} H_2, -\infty < x < +\infty : & \sigma_{yy} = \sigma_{xy} = 0, \\ x = \pm\infty, -H_1 < y < H_2 : & \sigma_{xx} = \sigma_{xy} = 0, \\ y = 0, -\infty < x < R : & \sigma_{yy} = \sigma_{xy} = 0, \\ y = 0, -R < x < 0 : & \sigma_{yy} = -\sigma_Y, \sigma_{xy} = 0. \end{cases} \tag{2.3}$$

In the moving coordinate system, the governing equations (wave equations) may be reduced to the Laplace equations

$$\nabla_1^2 \phi(x, y_1) = 0, \quad \nabla_1^2 \phi(x, y_2) = 0. \tag{2.4}$$

where ϕ and ψ are Lame potentials

$$u_x = \frac{\partial \phi}{\partial x} + \frac{\partial \psi}{\partial y}, \quad u_y = \frac{\partial \phi}{\partial y} - \frac{\partial \psi}{\partial x}, \tag{2.5}$$

and

$$\nabla_1^2 = \frac{\partial^2}{\partial x^2} + \frac{\partial^2}{\partial y_1^2}, \quad \nabla_2^2 = \frac{\partial^2}{\partial x^2} + \frac{\partial^2}{\partial y_2^2},$$

$$y_1 = \alpha_1 y, \quad y_2 = \alpha_2 y,$$

$$\alpha_1 = \sqrt{1 - V^2/c_1^2}, \quad \alpha_2 = \sqrt{1 - V^2/c_2^2}, \tag{2.6}$$

$$c_1 = \sqrt{(\lambda + 2\mu)/\rho}, \quad c_2 = \sqrt{\mu/\rho}. \tag{2.7}$$

Introducing complex variables

$$z_1 = x + iy_1, \quad z_2 = x + iy_2, \tag{2.8}$$

the solutions of equations (2.4) may be written as[6]

$$\begin{cases} \phi(x, y_1) = F_1(z_1) + \overline{F_1(z_1)}, \\ \psi(x, y_2) = i[F_2(z_2) - \overline{F_2(z_2)}], \end{cases} \tag{2.9}$$

where $F_1(z_1)$ and $F_2(z_2)$ are analytic function of z_1, z_2 respectively, and $\overline{F_1(z_1)}$, $\overline{F_1(z_1)}$ are the complex conjugates of $F_1(z_1)$, $F_2(z_2)$.

We introduce the approximate conformal mapping function as follow

$$\frac{z_1}{\alpha_1},\ \frac{z_2}{\alpha_2} = \omega(\zeta) = \frac{H_1}{\pi}\ln(1-\zeta) + \frac{H_2}{\pi}\ln\left(1+\frac{H_1}{H_2}\zeta\right) \tag{2.10}$$

to transform the region in the physical plane onto the upper-half plane in the ζ plane, so that

$$F_1(z_1) = F_1[\alpha_1\omega(\zeta)] = f_1(\zeta),\ \ F_2(z_2) = F_2[\alpha_2\omega(\zeta)] = f_2(\zeta). \tag{2.11}$$

After some manipulation, these functions can be determined at the ζ plane, and the dynamic stress factor problem (2.3) are obtained as

$$K_I^P = \frac{\sqrt{\alpha_1}\sqrt{H_1}}{\pi}p\left(1+\frac{H_1}{H_2}\right)^{-\frac{1}{2}}\left[\ln\left(\frac{1+\frac{H_1\sigma_{-a}}{H_2}}{1-\sigma_{-a}}\right) - \ln\left(\frac{1+\frac{H_1\overline{\sigma}_{-a}}{H_2}}{1-\overline{\sigma}_{-a}}\right)\right], \tag{2.12}$$

and

$$K_I^{\sigma_Y} = \frac{\sqrt{\alpha_1}\sqrt{H_1}}{\pi}\sigma_Y\left(1+\frac{H_1}{H_2}\right)^{-\frac{1}{2}}\left[\ln\left(\frac{1+\frac{H_1\sigma_{-R}}{H_2}}{1-\sigma_{-R}}\right) - \ln\left(\frac{1+\frac{H_1\overline{\sigma}_{-R}}{H_2}}{1-\overline{\sigma}_{-R}}\right)\right] \tag{2.13}$$

respectively, where the approximate values of $\overline{\sigma}_{-a}$, σ_{-a},$\overline{\sigma}_{-R}$,σ_{-R} are

$$\begin{cases} \overline{\sigma}_{-a} \approx \dfrac{H_1-H_2}{2H_1} - \sqrt{\left(\dfrac{H_1+H_2}{2H_1}\right)^2 - \exp\dfrac{-2a\pi H_1}{(H_1+H_2)\alpha_1 H_2}}, \\ \sigma_{-a} \approx \dfrac{H_1-H_2}{2H_1} + \sqrt{\left(\dfrac{H_1+H_2}{2H_1}\right)^2 - \exp\dfrac{-2a\pi H_1}{(H_1+H_2)\alpha_1 H_2}}, \end{cases} \tag{2.14}$$

and

$$\begin{cases} \overline{\sigma}_{-R} \approx \dfrac{H_1-H_2}{2H_1} - \sqrt{\left(\dfrac{H_1+H_2}{2H_1}\right)^2 - \exp\dfrac{-2R\pi H_1}{(H_1+H_2)\alpha_1 H_2}}, \\ \sigma_{-R} \approx \dfrac{H_1-H_2}{2H_1} + \sqrt{\left(\dfrac{H_1+H_2}{2H_1}\right)^2 - \exp\dfrac{-2R\pi H_1}{(H_1+H_2)\alpha_1 H_2}}. \end{cases} \tag{2.15}$$

Substituting $(2.14)_1$,$(2.14)_2$ into (2.12),(2.13) respectively, then the expressions (2.12), (2.13) can be written to the explicit formulations as

$$K_I^P \approx \frac{2\sqrt{\alpha_1}\sqrt{H_1}}{\pi} p\left(1+\frac{H_1}{H_2}\right)^{-\frac{1}{2}} \ln\left[\frac{\frac{H_1+H_2}{2H_1}+\sqrt{\left(\frac{H_1+H_2}{2H_1}\right)^2-\exp\frac{-2a\pi H_1}{(H_1+H_2)\alpha_1 H_2}}}{\frac{H_1+H_2}{2H_1}-\sqrt{\left(\frac{H_1+H_2}{2H_1}\right)^2-\exp\frac{-2a\pi H_1}{(H_1+H_2)\alpha_1 H_2}}}\right], \tag{2.16}$$

$$K_I^{\sigma_Y} \approx \frac{2\sqrt{\alpha_1}\sqrt{H_1}}{\pi} \sigma_Y\left(1+\frac{H_1}{H_2}\right)^{-\frac{1}{2}} \ln\left[\frac{\frac{H_1+H_2}{2H_1}+\sqrt{\left(\frac{H_1+H_2}{2H_1}\right)^2-\exp\frac{-2aR\pi H_1}{(H_1+H_2)\alpha_1 H_2}}}{\frac{H_1+H_2}{2H_1}-\sqrt{\left(\frac{H_1+H_2}{2H_1}\right)^2-\exp\frac{-2aR\pi H_1}{(H_1+H_2)\alpha_1 H_2}}}\right]. \tag{2.17}$$

Because the stress at the crack tip must be finite, i.e., there is no singularity, then

$$K_I = K_I^p + K_I^{\sigma_Y} = 0. \tag{2.18}$$

This equation can be determined the plastic zone size R

$$\ln\left[\frac{1+\frac{H_1\sigma_{-R}}{H_2}}{1-\sigma_{-R}} \cdot \frac{1-\overline{\sigma}_{-R}}{1+\frac{H_1\sigma_{-R}}{H_2}}\right] = \frac{p}{\sigma_Y}\ln\left[\frac{1+\frac{H_1\sigma_{-a}}{H_2}}{1-\sigma_{-a}} \cdot \frac{1-\overline{\sigma}_{-a}}{1+\frac{H_1\sigma_{-a}}{H_2}}.\right]. \tag{2.19}$$

In the case of small scale yield, i.e., $R \ll a$, formula (2.18) gives

$$R \approx \frac{\pi}{16} \cdot \frac{H_1+H_2}{H_1}\left[1-\frac{2H_1}{H_1+H_2}\right]\frac{(K_I^p)^2}{\sigma_Y^2}, \tag{2.20}$$

where K_I^p is given by (2.15), and if $H_1 = H_2$, then

$$R \approx \frac{\pi}{8} \cdot \frac{(K_I^p)^2}{\sigma_Y^2}, \tag*{(2.20)$_1$}$$

it is just the result in Ref.[1]; and from the expression (2.19), we can also find that if $H_1 < H_2, H_1 = H$, then the value of R will be decrease; $H_1 < H_2, ?H_1 = H$, then the value of R will be increase, this proves that the results is consistent with those observed in practice.

Now, the crack opening displacement is

$$\delta_I \approx \begin{cases} \dfrac{\alpha_1(1-\alpha_2^2)}{(1-\nu)\left[4\alpha_1\alpha_2-(1+\alpha_2^2)^2\right]} \cdot \dfrac{(1-\nu^2)(K_I^p)^2}{E\sigma_Y} & \text{(plane strain)}, \\ \dfrac{(1+\nu)\alpha_1(1-\alpha_2^2)}{4\alpha_1\alpha_2-(1+\alpha_2^2)^2} \cdot \dfrac{(K_I^p)^2}{E\sigma_Y} & \text{(plane stress)}. \end{cases} \tag{2.21}$$

The discussion of Mode II problem is similar to that of Mode I. Like in Refs.[8,18] now, that requiring to do is replacing pressure stress p and tensile yield limit σ_Y to shear stress π and the shear yield limit τ_Y in corresponding conditions respectively, the stress intensity factors K_{II}^{τ} and $K_{II}^{\tau_Y}$ can be written out immediately as follows

$$K_{II}^{\tau} \approx \frac{2\alpha_2\sqrt{H_1}}{\pi\sqrt{\alpha_1}}\tau\left(1+\frac{H_1}{H_2}\right)^{-\frac{1}{2}}\ln\left[\frac{\frac{H_1+H_2}{2H_1}+\sqrt{\left(\frac{H_1+H_2}{2H_1}\right)^2-\exp\frac{-2a\pi H_1}{(H_1+H_2)\alpha_1 H_2}}}{\frac{H_1+H_2}{2H_1}-\sqrt{\left(\frac{H_1+H_2}{2H_1}\right)^2-\exp\frac{-2a\pi H_1}{(H_1+H_2)\alpha_1 H_2}}}\right], \tag{2.22}$$

$$K_{II}^{\tau_Y} \approx \frac{2\alpha_2\sqrt{H_1}}{\pi\sqrt{\alpha_1}}\tau_Y\left(1+\frac{H_1}{H_2}\right)^{-\frac{1}{2}}\ln\left[\frac{\frac{H_1+H_2}{2H_1}+\sqrt{\left(\frac{H_1+H_2}{2H_1}\right)^2-\exp\frac{-2R\pi H_1}{(H_1+H_2)H_2\alpha_1}}}{\frac{H_1+H_2}{2H_1}-\sqrt{\left(\frac{H_1+H_2}{2H_1}\right)^2-\exp\frac{-2R\pi H_1}{(H_1+H_2)H_2\alpha_1}}}\right], \tag{2.23}$$

$$R \approx \frac{\pi}{4}\cdot\frac{H_1+H_2}{H_1}\left(1-\frac{2H_1}{H_1+H_2}\right)\frac{(K_{II}^{p})^2}{\tau_Y^2}, \tag{2.24}$$

and

$$\delta_{II} \approx \begin{cases} \dfrac{\alpha_2(1-\alpha_2^2)}{(1-\nu)\left[4\alpha_1\alpha_2-(1+\alpha_2^2)^2\right]}\cdot\dfrac{(1-\nu^2)(K_{II}^{p})^2}{E\tau_Y} & \text{(plane strain)}, \\ \dfrac{(1+\nu)\alpha_2(1-\alpha_2^2)}{4\alpha_1\alpha_2-(1+\alpha_2^2)^2}\cdot\dfrac{(K_{II}^{p})^2}{E\tau_Y} & \text{(plane stress)}, \end{cases} \tag{2.25}$$

where R still represents the plastic zone size and δ_{II} represents the crack sliding displacement.

3. Conclusion and discussion

For the dynamic case an approximate conformal mapping is used and solutions in closed form are constructed. These solutions can reduce to those given by Refs.[1-2] as $H = H_1 = H_2$. The stress intensity factors K_I, K_{II} as well as crack tip opening and sliding displacements δ_I and δ_{II} are useful in fracture analysis in engineering and seismology.

Works have been done by us here and those given in Refs.[8,18] show that the conformal mapping based on the transcendental functions is a powerful tool for solving some complicated crack problems, and advance, in some extend, the Muskhelishvili complex potential method in constructing analytic solutions of crack problems.

Ones of significance of the work is that the results quantitatively reveal the effects of finite size H_1, H_2 of the cracked structure. The results for limit cases as H_1/a or $H_2/a \to \infty$ and H_1/a or $H_2/a \to 0$ are included by the proceeding results as special cases.

References

[1] Fan T. Y., Engine. Fact. Mech., 37(1990), 1085–1087.

[2] Fan T. Y., Engn. Fract. Mech., 44(1993), 243–246.

[3] Fan T. Y., Chin. Phys. Lett. 7(1990), 402–405.

[4] Fan T. Y., Chin. Phys. Lett. 7(1990), 445–448.

[5] Fan T. Y., Science in China, Ser.A, 21(1991), 262–269(in Chinese).

[6] Fan T. Y., Scientia Sinica, Ser.A, 31(1988), 950–960.

[7] Fan T. Y., Theoretical and Applied Fracture Mechanics, 16(1991), 237–243.

[8] Yang X. C., Wang. L., Fan. T. Y., Chin. Phys. Letts. 15(1998), 117–119.

[9] Fan T. Y., Science in China, Ser.E, 39(1996), 561–568.

[10] Muskhelishvili N. I, Mathematical Theory of Elasticity., P. Noordhoff, Amsterdam, 1956.

[11] Tada H, Paris C., Irwin R. G., The Stress Analysis of Crack Handbook, Pennsylvania, Hellertown, 1973.

[12] Sih G. C., Handbook of Stress Intensity Factors, Lehigh University, Bethehem, 1973.

[13] Rice J. R., Trans. AMSE, J. Appl. Mech, 35(1968), 379–386.

[14] Dugdele D. S., J. Mech. Phys. Solids, 8(1960), 100.?

[15] Barenblatt G. S., Appl. Mech. and Mech. Engn. Trans., 26(1962), 469.?

[16] Lu C. K., Boundary value Problem of Analytic Functions, Sci. Tech. Press, Shanghai (in Chinese), 1987.

[17] Yang X. C., Fan T. Y., Li H. X., Chin. Phys. Letts. 16(1999), 32–35.

[18] Yang X. C., Wang L., Fan T. Y., China Acta of Mechanics (to appear in Chinese).

27 CERTAIN CLASS OF HYPERANALYTIC HASEMAN BOUNDARY VALUE PROBLEMS

Yue Sheng ZENG[1]

Huaihua Teacher's College, Huaihua, Hunan 418008, China

Abstract: This paper deals with Haseman boundary value problems for hyperanalytic functions with power shift. Applying the method of conformal identification and regularization of pseudosolution, the explicit form of general solutions for a jump problem, homogeneous and nonhomogeneous Haseman problems with power shift are established.

1 The Statement of the Problem

The fundamental theory of hyperanalytic functions was studied in papers [1,2], several classes of boundary value problems (BVP) for hyperanalytic functions were studied in [3–9]. In this paper, we will discuss certain class of Haseman boundary value problems for hyperanalytic functions with power shift, i.e.

$$DW(z) = 0, \quad z \in \mathbb{C} \setminus R_+, \quad R_+ = (0, \infty), \tag{1}$$

$$W^+(x^d) = G(x)W^-(x) + g(x), \quad x \in R_+, \tag{2}$$

where

$$D = \frac{\partial}{\partial x} + i\frac{\partial}{\partial y} + e(a\frac{\partial}{\partial x} + b\frac{\partial}{\partial y})$$

[1] Supported by the Funds of the Hunan Province Education Commission of China (No. 98B129).

H.G.W. Begehr et al. (eds.), Proceedings of the Second ISAAC Congress, Vol. 1, 207-213.

is Douglis differential operator, and e is a nilpotent, such that $e^n = 0\,(n \in N)$, $G(x), g(x) \in H_\mu(R_+), G(x) \neq 0$ in R_+ and $d\,(\neq 1)$ is a given positive number.

If a solution $W(z)$ of the problem (1), (2) satisfies condition X in the neighborhood of $z = 0$ and condition Y in a neighborhood of $z = \infty$, then $W(z)$ is called XY-solution. As conditions X and Y are follows: I-integrability, i.e. $W(z) = O(|z|^{-r})(z \to 0)$ and $W(z) = O(|z|^{r})(z \to \infty)$, $r < 1$; B-$W(z)$ is bounded of $z = 0$ or $z = \infty$; C-$W(z)$ is continuous in a neighborhood of $z = 0$ or $z = \infty$; R-W(z) is bounded in a neighborhood of $z = 0$ or $z = \infty$ and Re$W(z)$ is continuous at $z = 0$ or $z = \infty$.

2 Reduction of the Problem to Riemann Problem

Applying the method of the conformal identification, the problem (1), (2) is reduced to hyperanalytic Riemann BPV. Moreover the gluing curve and the conformal identification are constructed in [10].

The gluing curve Λ lies in ring $\Delta = \{\zeta : e^{-q} < |\zeta| < e^{q}\}$, it is pirelli which coiled itself round the circles: $C_0 = \{\zeta : |\zeta| = e^{-q}\}$ and $C_\infty = \{\zeta : |\zeta| = e^{q}\}$. The conformal mapping is a $z = h(\zeta)$[10].

Let $\zeta = w(z)$ is inverse mapping of $z = h(\zeta)$. Then $\zeta = w(z)$ is the conformal identification mapping from $C \setminus R_+$ onto $\Delta \setminus \Lambda$, and $w^+(x^d) = w^-(x)$.

Lemma 1 [10] *Assume hypercomplex function $f(x) \in H_\mu(R_+)$, and define $f^*(\tau)$ on $\overline{\Lambda} = \Lambda \cup C_0 \cup C_\infty$ by the following formula:*

$$f^*(\tau) = \begin{cases} f(h^-(\tau)), & \tau \in \Lambda, \\ f(0), & \tau \in C_0, \\ f(\infty), & \tau \in C_\infty, \end{cases} \tag{3}$$

then $f^*(\tau) \in H_\mu(\overline{\Lambda})$.

Now we introduce a new unknown function

$$W(z) = W^*(w(z)),$$

then boundary value condition (2) turns into

$$W^{*+}(\tau) = G^*(\tau)W^{*-}(\tau) + g^*(\tau), \quad \tau \in \overline{\Lambda}, \tag{4}$$

where $G^*(\tau) = G(h^-(\tau))$, $g^*(\tau) = g(h^-(\tau))$. ¿From Lemma 1, we known that $G^*(\tau)$, $g^*(\tau) \in H_\mu(\overline{\Lambda})$. So hyperanalytic BVP (1), (2) with power shift is reduced to the hyperanalytic BVP:

$$\begin{cases} DW^*(\zeta) = 0, & \zeta \in \Delta \setminus \Lambda, \\ W^{*+}(\tau) = G^*(\tau)W^*(\tau) + g^*(\tau), & \tau \in \overline{\Lambda}. \end{cases} \tag{5}$$

The hyperanalytic Riemann BVP was studied in [4–8], its solutions $W^*(\zeta)$ are obtained.

Let

$$\varphi(z) = W^*(\omega(z)), \tag{6}$$

taking into account $h^+(\tau) = (h^-(\tau))^d$, we get the function (6), which is continuous in $C \setminus R_+$, and has a limit at $x \in R_+$, satisfying boundary value condition (2), i.e.

$$\varphi^+(x^d) = G(x)\varphi^-(x) + g(x), \quad x \in R_+.$$

Such hyperanalytic function $\varphi(z) = W^*(\omega(z))$ is called pseudo-solution of the BVP (1), (2), and $\zeta = w(z)$ approaches C_0 or C_∞ as z to 0 or ∞ respectively.

Theorem 1 *If $W^*(\zeta)$ is a solution of the Riemann Problem* (5), *then hypercomplex function $\varphi(z) = W^*(w(z))$ is a pseudo-solution of BVP(1),* (2). *At the same time, the condition* I *corresponds to* $W^*(\zeta) = O(|h(\zeta)|^{-r})$ *(in a neighborhood of C_0) or* $W^*(\zeta) = O(|h(\zeta)|^r)$ *(in a neighborhood of C_∞), $r < 1$;* B-$W^*(\zeta)$ *is bounded in a neighborhood of C_0 or C_∞; C-the limit of $W^*(\zeta)$ at C_0 or C_∞ is hypercomplex constants;* R-$W^*(\zeta)$ *is bounded in a neighborhood of C_0 or C_∞ and* Re$W^*(\zeta)$ *is a real hypercomplex constant.*

We construct the pseudo-solution regularization operator

$$K \equiv I - J(D), \tag{7}$$

with I being identify operator and J is a hypercomplex integral operator:

$$J\varphi(z) = \frac{1}{2\pi i}\iint_C \frac{t_\xi(\zeta)}{i + eb(\zeta)} \frac{\varphi(\zeta)}{t(\zeta) - t(z)} d\xi d\eta, \quad \zeta = \xi + i\eta, \tag{8}$$

then

$$W(z) = K\varphi = \varphi(z) - J(D\varphi)(z), \quad z \in C \setminus R_+ \tag{9}$$

is a solution of the BVP(1),(2)[5].

3 Jump Problem

The jump problem of hyperanalytic functions with power shift is

$$\begin{cases} DW(z) = 0, & z \in C \setminus R_+, \\ W^+(x^d) - W^-(x) = f(x), & x \in R_+. \end{cases} \tag{10}$$

where $f(x) \in H_\mu(R_+)$

According Theorem 1, we first find out a solution of following problem:

$$\begin{cases} DW^*(\zeta) = 0, & \zeta \in \Delta \setminus \Lambda, \\ W^{*+}(\tau) - W^{*-}(\tau) = f^*(\tau), & \tau \in \Lambda. \end{cases} \tag{11}$$

Let $k_d(\zeta) = -\frac{1}{2\pi i} \ln_d \zeta$. Then from (10), we get

$$k_d^+(\tau) - k_d^-(\tau) = 1,$$

it is easy to see that a pseudo-solution of problem (11) is

$$\psi(\zeta) = k_d(\zeta) f^{*w}(\zeta), \tag{12}$$

here f^{*w} is the Whiteny extension of $f^*(\tau) \in H_v(\Lambda)$ onto Δ [5], Furthermore $D\psi = k_d(\zeta) Df^{*w}(\zeta)$. By virtue of [5]

$$W_0^*(\zeta) = K\psi = k_d(\zeta) f^{*w}(\zeta) - J(k_d Df^{*w})(\zeta), \tag{13}$$

is a solution of jump problem (11). In the near of C_E, we have

$$W_0^*(\zeta) = f(E) k_d(\zeta) + O(1) = \frac{f(E)}{\ln d} \ln \mathrm{Sec}(\frac{\pi}{2q} \ln |\zeta| + O(1)), \tag{14}$$

here and below, E denotes 0 or ∞. An arbitrary solution of problem (11) can be expressed by

$$W^*(\zeta) = W_0^*(\zeta) + \Omega(\zeta),$$

here $\Omega(\zeta)$ is arbitrary hyperanalytic in Δ. This implies pseudo-solution of jump problem (10) is

$$\varphi(z) = W_0^*(\omega(z)) + \Omega(\omega(z)).$$

Thus, we obtain that general solution of jump problem (10) is

$$W(z) = W_0^*(\omega(z)) + \Omega(\omega(z)) - J([W_0^{*\prime}(\omega) + \Omega'(\omega)] \cdot D\omega)(z), \quad z \in \mathbb{C} \setminus R_+ \tag{15}$$

Theorem 2 *Suppose $f(x) \in H_\mu(R_+)$, $\mu > \frac{1}{2}$, then the general solution of* II*-class for jump problem* (10) *is* (15)*, where $\Omega(\zeta)$ is an arbitrary hyperanalytic function in Δ satisfying $\Omega(\zeta) = O(h^{-r}(\zeta))$ in the near of C_0 and $\Omega(\zeta) = O(h^r(\zeta))$, $r = r(\Omega) < 1$ in the near of C_∞. If $f(0) = f(\infty) = 0$, then the general solution of* BB*-class for jump problem* (10) *is* (15)*, where $\Omega(\zeta)$ is an arbitrary hyperanalytic function on $\overline{\Delta}$; and there exists* RR*-solution of jump problem* (10)*, and it is unique in the* *sense of exactness up to addition term of a real hypercomplex constant. If $f(0)$ or $f(\infty)$ is not equal to zero, then jump problem* (10) *has no solution bounded at 0 or ∞.*

4 Homogeneous Problem

We first solve the following homogeneous Riemann BVP for $W^*(\zeta)$, which is hyperanalytic in $\Delta \setminus \Lambda$ satisfying the boundary condition

$$W^{*+}(\tau) = G(\tau)W^{*-}(\tau), \quad \tau \in \Lambda. \tag{16}$$

Because of $G(x) \in H_\mu(R_+)$ and is non-zero everywhere, then $G(x) = \exp f(x)$, $f(x) \in H_\mu(R_+)$. Thus the problem (16) is reduced to the jump problem (10).

Consider the hyperanalytic function

$$X^*(\zeta) = \exp W_0^*(\zeta), \quad \zeta \in \Delta \setminus \Lambda, \tag{17}$$

where $W_0^*(\zeta)$ is determined by (13). Obviously, $X^{*+}(\tau) = G^*(\tau)X^{*-}(\tau)$. From (14) we have in the near of C_E

$$\ln|X^*(\zeta)| = \mathrm{Re} f(E)k_d(\zeta) + O(1) = \frac{\ln|G(E)|}{\ln d} \ln \mathrm{Sec}(\frac{\pi}{2q} \ln|\zeta|) + O(1). \tag{18}$$

According to Theorem 1, the homogeneous problem (1), (2) has the pseudo-solution

$$\varphi(z) = X^*(\omega(z))\Omega(\omega(z)). \tag{19}$$

hus general solution of the homogeneous problem (1), (2) ($g(x) \equiv 0$) is

$$W(z) = X^*(\omega(z))\Omega(\omega(z)) - J([X^*(\omega)\Omega(\omega) + X^*(\omega)\Omega(\omega)] \cdot D\omega)(z). \tag{20}$$

Theorem 3 *Suppose $\mu > \frac{1}{2}$, then the homogeneous problem* (1), (2) ($g(x) \equiv 0$) *always has linearly independent infinite* II*-solutions with the form* (20). *If the numbers*

$$n_0 = \frac{\ln|G(0)|}{\ln d}, \quad n_\infty = \frac{\ln|G(\infty)|}{\ln d}$$

are non-positive, then it has linearly independent infinite BB-solutions with the form of (20). *If n_0, n_∞ are non-negative, then it has infinite CC-solutions. If n_0 or n_∞ is positive, then it has no non-zero solution bounded in the near of $z = 0$ or $z = \infty$.*

Theorem 4 Suppose $\mu > \frac{1}{2}$, $G(0) = G(\infty) = 1$ and $\ln d_{R+} G(x) = \kappa$, then homogeneous problem (1), (2) ($g(x) \equiv 0$) has $2\kappa + 2$ linearly independent RR-solutions with the form (20) if $\kappa \geq 0$; and has no RR-solutions if $k < 0$.

5 Nonhomogeneous Problem

¿From second section, the BVP (1), (2) is reduced to hyperanalytic Riemann BVP (5). Similarly to [9], we can construct a pseudo-solution of problem (5): $\psi(\zeta) = \alpha(\zeta) + \beta(\zeta)k_d(\zeta)$ satisfying following conditions:

1^0 $D\psi \in L_p(\overline{\Delta})$, $2 < p < 1/(1-v)$, $(v > 1/2)$;

2^0 When $G^*(C_E) \neq 1, \psi(\zeta)$ is continuous on C_E and equal to $g^*(C_E)/(1-G^*(C_E))$;

3^0 When $G^*(C_E) = 1$, $\psi(\zeta)$ can be expressed as $\psi(\zeta) = g^*(C_E)k_d(\zeta) + \psi_E(\zeta)$ in the near of C_E, here ψ_E is continuous on C_E. From (18), we know $X^{*-1}(\zeta)$ is bounded on $\overline{\Delta}$, thus $X^{*-1}D\psi \in L_p(\overline{\Delta})$, $p > 2$, $\mu > 1/2$. Therefore $J(X^{*-1}D\psi)(\zeta)$ is continuous on $\overline{\Delta}$[2]. Thus

$$W^*(\zeta) = \psi(\zeta) - X^*(\zeta)J(X^{*-1}D\psi)(\zeta) \tag{21}$$

is a solution of problem (5)[5].

¿From Theorem 1, (21) and Lemma 1 imply that

$$\varphi(z) = W^*(\omega(z))$$

is II-pseudo-solution of problem (1), (2). Therefore, we get its II-solution, which can be expressed as

$$W(z) = W^*(\omega(z)) - J(W^{*'}(\omega) \cdot D\omega)(z), \tag{22}$$

here $W^*(\zeta)$ is given by (21).

Theorem 5 *Suppose $\mu > 1/2$, n_0, n_∞ are nonnegative, then nonhomogeneous problem* (1), (2) *has infinite dimensional solutions set of* II-*class with the form* (22). *If $|G(0)| = |G(\infty)| = 1$, $g(0) \neq 1$, $g(\infty) \neq 1$, then it has solutions set of* BB-*class with the form* (22).

Theorem 6 Suppose $mu > 1/2$, $G(0) = G(\infty) = 1$, $g(0) = g(\infty) = 0$ and $\ln d_{R_+} G = \kappa$, then general RR-solution of the problem (1), (2) contains $2\kappa + 2$ arbitrary hypercomplex constants, if $\kappa > 0$, and the RR-solution exists and is unique, if $\kappa < 0$ and $-2\kappa - 2$ real othogonality conditions are fulfilled.

References

[1] Douglis A., A functions theoretic approach to elliptic systems of equations in two variables. Comm. Pure Appl. Math., 6 (1953), 259–289.

[2] Gilbert R. P. and Hile G. N., Generalized hyperanalytic function theory. Tran. Amer. Math. Soc., 195 (1974), 1–9.

[3] Begehr H., Gilbert R. P., On Riemann boundary value problems for certain linear elliptic systems in the plane. J. Diff. Eq., 32 (1979), 1–14.

[4] Hou Zongyi, Haseman boundary value problems for certain elliptic systems of first order. Chinese Ann. Math., 5A(5): 634–664 (in Chinese).

[5] Zeng Yuesheng, Riemann problems for hyperanalytic functions on countable many nonsmooth closed curves. Acta Math. Sinica, 32 (1989), 358–368 (in Chinese).

[6] Zeng Yuesheng, On Riemann problems for hyperanalytic function in rectifiable open curve, J. Hunan Normal Univ. 12 (1989), 7–15.

[7] Zeng Yuesheng, Hyperanalytic Riemann boundary value problem in the irrectifiable closed curve. J. Math.(PRC), 10 (1990), 353–360 (in Chinese).

[8] Zeng Yuesheng, On Riemann problem of hyperanalytic functions in the Jordan open curve. Chinese Quart J. Math., 4 (1989), 64–72 (in Chinese).

[9] Zeng Yuesheng, On Riemann problem of hyperanalytic function in the Jordan closed curve. Hunan Ann. Math., 91, 11:1–2 (in Chinese).

[10] Carts B. A., On Riemann boundary value problem with power shift. Comm. Universities, Math., 8 (1987), 18–26 (Russian).

28 ON COMPOUND BOUNDARY VALUE PROBLEMS FOR NONLINEAR ELLIPTIC SYSTEMS OF FIRST ORDER

Chun ZHAO

Department of Mathematics, Ningxia University
Yinchuan 750021, China

li_x@nxu.edu.cn

Abstract: This paper deals with the compound boundary value problems for nonlinear elliptic systems in multiply connected domains. A priori estimates of solutions for modified boundary value problems is given, by using Leray-Schauder theorem, the existence of solutions for the modified boundary value problem is proved.

1. Introduction

Let D be an $(N+1)$-connected bounded domain in the plane C. Without loss of generality, we may assume that D is bounded by $N+1$ circles $L_k = |z - z_k| = \gamma_k$, $k = 1, \cdots, m$ and $L_0 = |z| = 1, z = 0 \in D^+$. Denote by $L_k(k = 1, \cdots, N)$ the boundary of D_k^-, and $D^- = \bigcup_{K=1}^N D_k^-$, $L = \bigcup_{k=1}^N L_k$, $D = D^+ \bigcup L \bigcup D^-$.

We consider the elliptic system of first order equations in the complex form

$$w_{\bar{z}} = F(z, w, w_z), F = Qw_z + A^{(1)}w + A^{(2)}\overline{w} + A^{(3)}, \qquad (1.1)$$

H.G.W. Begehr et al. (eds.), Proceedings of the Second ISAAC Congress, Vol. 1, 215–221.

where $w=(w_1,\cdots,w_m)'$, $w_k=u_k+iv_k$, $j=1,2$, $k=1,2,\cdots,m$,
$F=(F_1,\cdots,F_m)'$, $Q=(Q^{kl}(z,w,w_z))_{m\times m}$, $A^{(j)}=(A^{(j)}_{kl}(z,w))_{m\times m}$,
$A^{(3)}=(A^{(3)}_1(z,w),\cdots,A^{(3)}_m(z,w))'$.

We shall find the solution $w=(w_1(z),\cdots,w_m(z))$ of systems (1.1), which satisfies the boundary conditions

$$\begin{cases} \operatorname{Re}(\overline{\lambda}w)=g_1+h, & z\in L, \\ w^+=Gw^-+g_2, & z\in L_0. \end{cases} \tag{1.2}$$

where

$$G=\begin{pmatrix} G_1(z) & \cdots & 0 \\ \vdots & \ddots & \vdots \\ 0 & \cdots & G_m(z) \end{pmatrix},\ \lambda=\begin{pmatrix} \lambda_1(z) & \cdots & 0 \\ \vdots & \ddots & \vdots \\ 0 & \cdots & \lambda_m(z) \end{pmatrix},\ g_1=\begin{pmatrix} g_{11}(z) \\ \vdots \\ g_{1m} \end{pmatrix},$$

$$g_2=\begin{pmatrix} g_{21}(z) \\ \vdots \\ g_{2m}(z) \end{pmatrix},\quad h=\begin{pmatrix} h_1(z) \\ \vdots \\ h_m(z) \end{pmatrix},$$

$$h_k=\begin{cases} 0, & \kappa_k\geq 0, \\ h_{k0}+\operatorname{Re}\sum_{n=1}^{-\kappa_k+1}(h_{kn}+ih_{-kn})z^n, & \kappa_k<0. \end{cases}$$

$$\kappa_k=\frac{1}{2\pi}[\arg\lambda_k]_{L_0}+\frac{1}{2\pi}\sum_{n=1}^{N}[\arg G_n]_{L_n},\ k=1,\cdots,m,$$

in which $h_{\pm kn}(k=1,\cdots,m,\ n=1,\cdots,|\kappa_k|+1)$ are unknown real constants to be appropriately determined. We call $\kappa=(\kappa_1,\cdots,\kappa_m)$ as the index of the boundary problem. When $\kappa_k\geq 0$, we assume that the solution $w_k(z)$ satisfy the point conditions

$$\operatorname{Im}[\overline{\lambda_k(\xi_{kj})}w_k(\xi_{kj})]=a_{kj},\ j=1,2,\cdots,2\kappa_k+1.\ k=1,\cdots,m, \tag{1.3}$$

in which $\xi_{kj}(j=1,\cdots,2\kappa_k+1.\ k=1,\cdots,m)$ are the deffirent points in L_0, and a_{jk} are real constants.

Problem (1.1)–(1.3) is called Problem B. In particular if $m=1$, Problem B is called Problem B_0.

Suppose that Problem B satisfies the following condition.

1) Let $1<2\alpha<p$, real functions $g_{1k}\in C_\alpha(L_0)$, and complex functions $g_{2k},G_k\in C_\alpha(L)$, $\lambda_k\in C_\alpha(L_0)$, $\det\lambda\neq 0,\det G\neq 0$.

2) The functions $Q_{kl}(z,w,v)$, $A^{(j)}_{kl}(z,w)$, $A^{(3)}_k(z,w)$ for almost every point $z\in D$ and $v\in C^m$ are continuous with respect to $w\in C^m$, and

and are zero with $z \notin \overline{D}$. The functions $Q_{kl}^{(j)}(z, w(z), v(z))$, $A_{kl}^{(j)}(z, w(z), v(z))$, $A_k^{(3)}(z, w(z))$ are measurable in D, and satisfy the conditions

$$||A_{kl}^{(j)}||_{p_0} \le d_{kl} \le d, \ j = 1, 2. \tag{1.4}$$

$$||A_k^{(3)}||_{p_0} \le d_k \le d, \ d_{kl} \le \epsilon, \ 1 \le k < l \le m,$$

$$|F_k(z, w_1, v) - F_k(z, w_2, v)| \le M(z) \sum_{l=1}^{m} |w_{1l} - w_{2l}|, \tag{1.5}$$

$$|F_k(z, w, v_1) - F_k(z, w, v_2)| \le \sum_{l=1}^{m} q_{kl} |v_{1l} - v_{2l}|,$$

in which the nonnegative constants q_{kl} satisfy

$$\sum_{l=1}^{m} q_{kl} \le q_0 < \frac{1}{m}(1 \le k \le m), \ q_{kl} \le \epsilon < 1, \ 1 \le k < l \le m.$$

where $0 < M(z) \in L_p(D)$, and d_{kl}, d_k, q_{kl} are constants indepadent of w.

2. Basic lemma

When $m = 1$, we first consider the system of complex equations

$$w_{\overline{z}} = Q(z, w)w_z + A^{(1)}(z, w)w + A^{(2)}(z, w)\overline{w} + A^{(3)}(z, w) + C(z). \tag{2.1}$$

By using [3], we can prove the following lemma.

Lemma 2.1. *Suppose conditions* 1), 2) *are true, and* $C \in L_p(D)$. *Then Problem* B_0 *for system* (2.1) *has a unquie solution* $w \in \overline{w}_{p_0}^1(D)$, *and possesses the estimate*

$$|w|_{1,p_0} \le C(||C||_{p_0} + ||A^{(3)}(z, 0)||_{p_0} + ||g_1||_\alpha + ||g_2||_\alpha + \sum_{j=1}^{2\kappa+1} |a_{1j}|), \tag{2.2}$$

where constant $C = C(\alpha, D, ||G||_\alpha, ||\lambda||_\alpha, ||M||_p, q_0)$, $2 < p_0 < min(p, \frac{p}{1-\alpha})$. *The functin space are the same as* [3] *in* (2.2).

If $w = (w_1, \cdots, w_m) \in L_p(D)$, denote by

$$||w||_p = \sum_{i=1}^{m} ||w_i||_p$$

its norm. The definition of the norm of other vector of function space are similar as above. With repect to problem B, we have the following estimate.

Lemma 2.2. *Under conditions* 1), 2), *if* $||A^{(3)}||_{P_0} \leq d$ *is replaced by* $||A^{(3)}||_{P_0} \leq d_0$, *when* ϵ *in* (1.4), (1.5) *is small enough, then any solution of problem B for system* (1.1) *satisfies the estimate*

$$|w|_{1,p_0} \leq M'(d_0 + ||g_1||_\alpha + ||g_2||_\alpha + \sum_{k=1}^{m} \sum_{j=1}^{2\kappa_k+1} |a_{kj}|) = M_1, \tag{2.3}$$

in which $M' = M'(\alpha, D, ||G||_\alpha, ||\lambda||_\alpha, ||M||_p, q_0, \epsilon)$.

Proof. From (1.1) we can write equation of the first unknown function

$$w_{1\bar{z}} - Q_{11} w_{1z} = A_{11}^{(1)} w_1 + A_{11}^{(2)} \overline{w}_1 + A_1^*, \tag{2.4}$$

$$A_1^* = A_1^{(3)} + \sum_{l=2}^{m} (Q_{1l} w_{lz} + A_{1l}^{(1)} w_l + \overline{w}_l A_{1l}^{(2)}),$$

and $w_1(z)$ satisfies the boundary conditions

$$\text{Re}(\overline{\lambda}_1 w_1) = g_{11}(z) + h_1, \ z \in L_0,$$

$$w_1^+ = G_1 w_1^- + g_{21}(z), \ z \in L,$$

$$I_m(\overline{\lambda}_1 w_1)(\xi_{1j}) = a_{1j}, \ j = 1, \cdots, 2, \kappa_1 + 1, \ \kappa \geq 0.$$

By condition (2), it follows that

$$||A_1^*||_{p_0} \leq d_0 + \epsilon ||w_z||_{p_0} + 2\epsilon ||w||_0 \leq d_0 + 3\epsilon |w|_{1,p_0}.$$

By the same reason as (2.2), we have

$$|w_1|_{1,p_0} \leq c(d_0 + 3\epsilon |w|_{1,p_0} + ||g_{11}||_\alpha + ||g_{21}||_\alpha + \sum_{j=1}^{2\kappa_1+1} |a_{1j}|). \tag{2.5}$$

Let $N_1 = 1 + 3dc$. According to the same reason as above, we have

$$|w_l|_{1,p_0} \leq c(d_0 + 3\epsilon |w|_{1,p_0} + ||g_1||_\alpha + ||g_2||_\alpha + \sum_{k=1}^{m} \sum_{j=1}^{2\kappa_k+1} |a_{kj}|) N_{l-1}, \tag{2.6}$$

where $N_l = 1+3d(1+N_1+\cdots+N_{l-1})c,\ l = 2,\cdots,m-1$. By (2.5), (2.6), there is

$$|w|_{1,p_0} \le c(d_0 + 3\epsilon|w|_{1,p_0} + ||g_1||_\alpha + ||g_2||_\alpha + \sum_{k=1}^{m}\sum_{j=1}^{2\kappa_k+1}|a_{kj}|)N.$$

$$N = 1 + N_1 + \cdots + N_{m-1}.$$

When ϵ is small enough, such that $3\epsilon cN < 1$, then we have

$$|w|_{1,p_0} \le \frac{c}{1-3\epsilon cN}(d_0 + ||g_1||_\alpha + ||g_2||_\alpha + \sum_{k=1}^{m}\sum_{j=1}^{2\kappa_k+1}|a_{kj}|).$$

3. Main results

Now we discuss the solvability of Problem B for system (1.1).

Theorem 3.1. *Under condition* 1), 2), *if* ϵ *is small enough, then Problem B for system* (1.1) *is solvable.*

Proof. Let $D_n = \{z|z \in D, d(z, L\cup L_0) > 1/n\}, n$ is the positive integer. First of all, we prove that the following system is solvable

$$w_{\bar{z}} = F^n(z,w,w_z), F^n = \begin{cases} F(z,w,w_z), & z \in D_n, \\ 0, & z \in D - D_n. \end{cases} \tag{3.1}$$

Denote by B_{M_1} the bounded set in the Banach space $L_{p_0}(D)$:

$$B_{M_1} = \{w(z)|w(z) = (w_1(z),\cdots,w_m(z)) \in L_{p_0}(D), ||w||_{p_0} < M_1 + 1\},$$

where M_1 is a constant as in (2.3), the solution of problem B for equsten (3.1) can be written as

$$w(z) = \phi(z) + T\omega, T\omega = (T\omega_1,\cdots,T\omega_m),$$

where $\omega(z) = F^{(n)} = (\omega_1(z),\cdots,\omega_m(z))'$, $\phi(z) = (\phi_1(z),\cdots,\ \phi_z(z))'$ ia an analytic function vector in D. we arbitrarily choose $\omega \in B_{M_1}$, and find the analytic function vector, which satisfies the point condition (1.3) and the boundary condotion

$$\begin{cases} \phi^+ = G\phi^- + g_1 - T\omega + GT\omega, z \in L, \\ \text{Re}(\overline{\lambda}\phi + \overline{\lambda}T\omega) = g_2 + h, z \in L_0. \end{cases} \tag{3.2}$$

By Lemma (2.1), we see that above problem has a unique solution. Substitute $w = \Phi(z) + T\omega$ into (3.1), and consider the system with the parameter

$$\Omega_k(z) = F_k^n(z, w(z), \phi'(z) + \Pi\Omega),$$
$$\Pi\Omega = (T\Omega)_z, \ k = 1, \cdots, m. \ 0 \le t \le 1. \tag{3.3}$$

By the conditions 1) 2), there exists a unique solution $\Omega(z) \in B$. Denote by $\Omega(z) = g(\omega(z), t)(0 \le t \le 1)$ the mapping from $\omega(z)$ to $\Omega(z)$. On basis of Lemma 2.2 and the method in [1], we can prove that the mapping $\Omega = g(\omega, t)$ satisfies three conditions of Leray-Schauder theorem:

Hence for any $t \in [0,1]$, in particular $t = 1$, there exists $\Omega \in B_{M_1}$, such that $\Omega(z) = g[\Omega(z), t]$, i.e. $w(z) = \phi(z) + T\Omega$ is a solution of system (3.1).

Finally, we cancel the condition that the coefficients of system (3.1) equal zero. From (2.2), the solution $w^{(n)}(z)$ of Problem B for system (3.1) satisfies the estimate (2.3). Hence from $\{w^{(n)}(z)\}$ we can choose a subsequence, which uniformly converges to $w^{(0)}(z)$ on D^+, which satisfies the estimate (2.3), and from $\{w_{\bar{z}}^{(n)}(z)\}, \{w_z^{(n)}(z)\}$, we can choose the subsequences, which weakly converge to $u^{(0)}(z)$, $v^{(0)}(z)$ respectively. For convenience, we denote these subsequences by $\{w_z^{(n)}(z)\}$ and $\{w_{\bar{z}}^{(n)}(z)\}$.

By the defintion of generalized derivatives, for any $\phi \in D_0^1$, there is

$$\iint_D (w^{(n)}(z)\phi_z(z) + w_z^{(n)}(z)\phi(z))d\sigma_z = 0,$$

when $n \longrightarrow \infty$, we have

$$\iint_D (w^{(0)}(z)\phi_z(z) + v^{(0)}(z)\phi(z))d\sigma_z = 0,$$

hence $[w^{(0)}(z)]_z = v^{(0)}(z)$, $z \in D$. By the condition (1.4), we can obtain

$$||F^{(n)}(z, w^{(n)}, w_z^{(0)}) - F(z, w^{(0)}, w_z^{(0)})||p_0 \longrightarrow 0$$

and $(w^{(n)} - w^{(0)})_z$ in D weakly converges to zero. From (1.5) we see that $F^{(n)}(z, w^{(n)}, w_z^{(n)}) - F^{(n)}(z, w^{(n)}, w_z^{(0)})$ in D also weakly converges to zero, i.e. $F^{(n)}(z, w^{(n)}, w_z^{(n)})$ in D wealy converges to $F(z, w^{(0)}, w_z^{(0)})$. By the definition of generalized derivatives, when $n \longrightarrow \infty$ for any $\phi(z) \in D_0^1$, there is

$$\iint_D (w^0(z)\phi_{\bar{z}}(z) + F(z, w^{(0)}, w_z^{(0)})\phi(z))d\sigma_z = 0.$$

Thus we obtain $w_{\bar{z}}^{(0)} = F(z, w^{(0)}, w_z^{(0)})$, i.e. $w^{(0)}(z)$ is a solution of Problem B for system (1.1).

References

[1] Wen Guochun, Yang Guangwu, Huang Sha. *Generalized analytic functions and its generalizations.* Hebei: Hebei Education Press, 1987 (in Chinese).

[2] Wen Guochun. *Linear and nonlinear elliptic complex equations.* Shanghai: Shanghai Science & Technology Press, 1986 (in Chinese).

[3] Dai Daoqing. *On compound boundary value problem for nonlinear elliptic equation of first order.* Chinese Ann. of Math., 1994, 15A(3): 335–339.

[4] Vekua I. N., *Generalized analytic functions.* Oxford: Pergamon, 1962.

29 ON THE INTEGRAL OF CAUCHY TYPE AND THE GENERALIZED HARNACK THEOREM FOR BIANALYTIC FUNCTIONS[1]

Zhen ZHAO

Department of Mathemathematics
Beijing Normal University
Beijing 100875
The People's Republic of China

zhaozhen@bnu.edu.cn

Abstract: In this paper the integral of Cauchy type, its some properties and the generalized theorem Harnack for bianalytic functions are obtained. These results are very useful for applications, especially, for solvability of boundary value problems for bianalytic functions.

1. Introduction

In [1] we have suggested the concept of bianalytic functions and considered some their properties. Furthermore, some boundary value problems for them (i.e. Dirichlet's problem, Dirichlet's problem in the changed form, Riemann-Hilbert's Problem, Schwarz's problem, Problem A, etc.) are also considered [1–6]. In [7] we investigated the Cauchy formula and the integral of Cauchy type for bianalytic functions and

[1] Research supported by The National Natural Science Foundation of China.

H.G.W. Begehr et al. (eds.), Proceedings of the Second ISAAC Congress, Vol. 1, 223–230.

proved some useful theorems for them; finally, the Problem Hilbert for bianalytic functions is also considered. In this paper some properties of the Cauchy type integral and the generalized theorem Harnack are obtained.

The complex form of C-R equations is

$$\frac{\partial w}{\partial \overline{z}} = 0, \tag{1.1}$$

here $\frac{\partial}{\partial \overline{z}} = \frac{1}{2}(\frac{\partial}{\partial x} + i\frac{\partial}{\partial y})$.

In addition we also define the derivatives: $\frac{\partial}{\partial z} = \frac{1}{2}(\frac{\partial}{\partial x} - i\frac{\partial}{\partial y})$ and $\Delta \equiv \frac{\partial^2}{\partial \overline{z} \partial z} = \frac{\partial^2}{\partial z \partial \overline{z}}$.

Definition 1. Suppose G is a $p+1$-connected domain on the plane, the boundary of G is $L = L_0 + L_1 + \cdots + L_p$. We say that a complex function $w(z)$ in G having the second order derivative in $\overline{z}$, i.e. $\frac{\partial^2 w}{\partial \overline{z}^2}$, is called a bianalytic function, if $w(z)$ satisfies the following partial differential equation:

$$\frac{\partial^2 w}{\partial \overline{z}^2} = 0, \ z \in G. \tag{1.2}$$

We denote the set of all bianalytic functions by $\mathcal{D}_2(G)$.

We know that a bianalytic function can be represented by formula [1]

$$w(z) = \overline{z}\varphi_1(z) + \varphi_2(z), \ z \in G, \tag{1.3}$$

where $\varphi_1(z), \varphi_2(z)$ are all analytic functions in G. For simplicity we will call $\varphi_1(z)$ and $\varphi_2(z)$ the analytic factor and analytic addition of bianalytic function $w(z)$ respectively, in the later we denote all functions in the form $w^*(z) = \overline{z}\varphi(z)$, by $\mathcal{D}_2^*(G) \in \mathcal{D}_2(G)$, where $\varphi(z)$ is arbitrary analytic function and call it by a characteristic bianalytic function.

When we use the methods of function theory to solve problems of elasticity in the plane, the stress function $U(x, y)$ which is biharmonic, will play a very important role. We can successfully represent the stress function $U(x, y)$ as the real part of a certain bianalytic function, i.e. $U(x, y) = \text{Re}[w(z)]$, where $w(z)$ is a bianalytic function. This means that investigation of bianalytic functions is of important practical interest.

2. Integral of Cauchy type for bianalytic function

We know that in the formula Cauchy of bianalytic functions is contained two boundary values $w(t)$ and $\frac{\partial w}{\partial t}$, this is the difference from the formula Cauchy of analytic functions [7]. In the later for convenience we will reduce the integral of Cauchy type only for characteristic bianalytic functions, i.e. the functions belonging to class $\mathcal{D}_2^*(G)$, as the follows:

Definition 2. The integral of Cauchy type for the characteristic bianalytic function $w^*(z)$ belonging to the class $\mathcal{D}_2^*(G)$ is as follows:

$$\begin{aligned} w^*(z) &= \frac{\overline{z}}{2\pi i}\int_L \frac{\varphi(z)}{t-z}\,dt \\ &= \frac{1}{2\pi i}\int_L \frac{\overline{t}\varphi(t)}{t-z}\,dt - \frac{1}{2\pi i}\int_L \frac{\overline{(t-z)}\varphi(t)}{t-z}\,dt \qquad (2.1) \\ &= \frac{1}{2\pi i}\int_L \frac{w(t)}{t-z}\,dt - \frac{1}{2\pi i}\int_L \frac{\overline{(t-z)}w(t)}{\overline{t}(t-z)}\,dt,\ t\in L, \end{aligned}$$

where $\varphi(z)$ is arbitrary continuous function and $w(t)=\overline{t}\varphi(t)$. The first and the second terms of the right-hand side of (2.1) is an analytic and a bianalytic function respectively. Consequently, we have

Theorem 1. *If $w(z)$ is a bianalytic function in a bounded domain G, then $w(z)=w^*(z)+\Phi(z)$, $z\in G$, where $w^*(z)$ is a function as in (2.1) and $\Phi(z)$ is an arbitrary analytic function.*

Proof. We know that $w(z)=\overline{z}\varphi_1(z)+\varphi_2(z)$ [1], as soon as take $\varphi_1(z)=\frac{1}{2\pi i}\int_L \frac{\varphi(t)}{t-z}\,dt$, $\Phi(z)=\varphi_2(z)$, then will immediately get Theorem 1.

Definition 3. A piecewise bianalytic function with discontinuous curve L, we will call the function $\Psi(z)$, which is bianalytic in any bounded domain D not containing curve L, has a finite k-th order at the infinity and it is continuously extendible from left-hand side and right-hand side to L, where L is a closed smooth curve.

Definition 4. A piecewise bianalytic function $w^*(z)$ is called characteristic bianalytic, if it belongs to $\mathcal{D}_2^*(G)$ in any bounded domain D not containing curve L, has a finite k-th order at the infinity and it is continuously extendible from left-hand side and right-hand side to L, where L is a closed smooth curve.

Theorom 2. *Suppose L is a closed smooth curve, and $w(t)$ belongs to class H, (i.e. the class of Hölder continuous functions), then the integral of Cauchy type $w^*(z)$ ((2.1)) is a piecewise bianalytic function, which is bounded at infinity, and L is the discontinuous curve.*

By using the properties of the Cauchy type integral for analytic function [10], we can easily prove theorem 2.

Theorem 3. (Plemelj's formula) *Suppose L is a closed smooth curve, and $w(t)$ belongs to class H (i.e. the class of Hölder continuous functions). For the integral of Cauchy type for the characteristic bianalytic functions*

$$\begin{aligned} w^*(z) &= \frac{1}{2\pi i}\int_L \frac{\bar{t}\varphi(t)}{t-z}\,dt - \frac{1}{2\pi i}\int_L \frac{\overline{(t-z)}\varphi(t)}{t-z}\,dt \\ &= \frac{1}{2\pi i}\int_L \frac{w(t)}{t-z}\,dt - \frac{1}{2\pi i}\int_L \frac{\overline{(t-z)}w(t)}{\bar{t}(t-z)}\,dt \quad z \notin L, \end{aligned} \tag{2.2}$$

where $\varphi(t)$ is arbitrary continuous function and $w(t) = \bar{t}\varphi(t)$, we have the following formula:

$$\begin{aligned} w^{*+}(t) &= \frac{1}{2}w(t) + \frac{1}{2\pi i}\int_L \frac{w(\tau)}{\tau - t}\,d\tau - \frac{1}{2\pi i}\int_L \frac{\overline{(\tau-t)}w(\tau)}{\bar{\tau}(\tau-t)}\,d\tau, \\ w^{*-}(t) &= -\frac{1}{2}w(t) + \frac{1}{2\pi i}\int_L \frac{w(\tau)}{\tau - t} - \frac{1}{2\pi i}\int_L \frac{\overline{(\tau-t)}w(\tau)}{\bar{\tau}(\tau-t)}\,d\tau, \quad t \in L. \end{aligned} \tag{2.3}$$

By Plemelj's formula (2.3) we know its another form is as follows:

$$\begin{aligned} w^{*+}(t) + w^{*-}(t) &= \frac{1}{\pi i}\int_L \frac{w(\tau)}{\tau - t}\,d\tau - \frac{1}{\pi i}\int_L \frac{\overline{(\tau-t)}w(\tau)}{\bar{\tau}(\tau-t)}\,d\tau, \\ w^{*+}(t) - w^{*-}(t) &= w(\tau), \; t \in L. \end{aligned} \tag{2.4}$$

3. The generalized Harnack theorem

Suppose $w(t)$ is a continuous function given on the boundary $L = L_0 + L_1 + \cdots + L_p$, problem is that $w(t)$ can be as the boundary value of a bianalytic (characteristic) in G and continuous on $G + L$ function $w(z)$? Or what will be the conditions for that $w(t)$ can be as the boundary value of the function $w(z)$?

Firstly, we know the function $w(t)$ is the boundary value of a bianalytic (characteristic) in G and continuous on $G + L$ function $w(z)$, (when L_0 not exists, we require also $w(\infty)$ is bounded), then by the Cauchy theorem of bianalytic functions have: [7]

$$w^-(z) = \frac{1}{2\pi i}\int_L \frac{w(t)}{t-z}\,dt - \frac{1}{2\pi i}\int_L \frac{\overline{t-z}}{t-z}\frac{\partial w(t)}{\partial \bar{t}}\,dt = 0, \quad z \in S^-, \tag{3.1}$$

where S^- is the complementary set of $G + L$.

Inversely, if condition (3.1) is satisfied, then we consider the integral of Cauchy type:

$$w^*(z) = \frac{1}{2\pi i}\int_L \frac{w(t)}{t-z}\,dt - \frac{1}{2\pi i}\int_L \frac{\overline{t-z}}{\bar{t}(t-z)} w(t)\,dt, \; z \notin L. \tag{3.2}$$

By (3.1) and Plemelj's formula, obviously, we have $w^{*-}(t_0) = 0,\ t_0 \in L$, and

$$w(t_0) = w^{*+}(t_0) - w^{*-}(t_0) = w^{*+}(t_0), \quad t_0 \in L.$$

This just shows that $w(t_0)$ is the boundary value of a (characteristic) bianalytic in G and continuous on $G+L$ function $w^*(z)$. Thus we obtain the following theorem:

Theorem 4. *A given on L continuous function $w(t)$ can be as the boundary value of a (characteristic) bianalytic in G and continuous on $G+L$ function $w(z)$ (when L_0 not exists, we require also that $w(\infty)$ is bounded), if and only if the condition* (3.1) *is satisfied.*

If $w(t),\ t \in L$ satisfies the Hölder condition, then we can change the form of (3.1) as follows:

Let $z \longrightarrow t_0$, by the Plemelj's formula we can get

$$-\frac{1}{2}w(t_0) + \frac{1}{2\pi i}\int_L \frac{w(t)}{t-t_0}\,dt - \frac{1}{2\pi i}\int_L \frac{\overline{t-t_0}}{\bar{t}(t-t_0)} w(t)\,dt = 0, \; t_0 \in L. \tag{3.3}$$

Furthermore, from (3.3) we can also get (3.1). In fact, by the condition (3.3) we know that the function $w^*(z)$ given by (3.2) is bianalytic in S^-, where S^- is the complementary set of $G+L$ and continuously extendible to L. Obviously, we have

$$w^-(t_0) = -\frac{1}{2}w(t_0) + \frac{1}{2\pi i}\int_L \frac{w(t)}{t-t_0}\,dt - \frac{1}{2\pi i}\int_L \frac{\overline{t-t_0}}{\bar{t}(t-t_0)} w(t)\,dt = 0, \; t_0 \in L.$$

¿From the Cauchy theorem for bianalytic function we know $w(z) \equiv 0,\ z \in S^-$, i.e. the condition (3.1) is satisfied.

Theorem 5. *A function $w(t) \in H(L)$ with the boundary value of a (characteristic) bianalytic in G and continuous on $G+L$ function $w^*(z)$ (when L_0 not exists, we require that $w(\infty)$ is bounded), if and only if the condition* (3.3) *is satisfied.*

Analogously, we have that a given on L continuous function $w(t)$ can be as the boundary value of a (characteristic) bianalytic in S^- and continuous on $S^- + L$ function $w^*(z)$ (when L_0 exists, we require also

$w(\infty)$ is bounded), if and only if the following condition is satisfied:

$$w^*(z) = \frac{1}{2\pi i}\int_L \frac{w(t)}{t-z}\,dt - \frac{1}{2\pi i}\int_L \frac{\overline{t-z}}{\bar{t}(t-z)}w(t)\,dt \equiv 0, \ \ z \in G.$$

If the function $w(t)$ satisfies the Hölder condition, we can get also the analogous result of Theorem 5.

Theorem 6 (The generalized Harnack theorem). *Suppose* $\frac{w(t)}{\bar{t}}$ *is a real function, and*

$$w^*(z) = \frac{1}{2\pi i}\int_L \frac{w(t)}{t-z}\,dt - \frac{1}{2\pi i}\int_L \frac{\overline{t-z}}{\bar{t}(t-z)}w(t)\,dt \equiv 0, \ \ z \in G,$$

then we must have $w(t) = \bar{t}c_k$, $t \in L$, $k = 0,1,\cdots,p$, *where* c_k *is a (real) constant, and* $c_0 = 0$. *(If* L_0 *not exists, then* $k = 1,2,\cdots,p$*). But if* $w^*(z) \equiv 0, z \in S^-$, *then we must have* $w(t) = \bar{t}\,C$, *where* C *is a (real) constant.*

Proof. If $w^*(z) \equiv 0, z \in G$, then $w(t)$ is the boundary value of a bianalytic in S^- and continuous on $S^- + L$ function $w^{*-}(z) \in \mathcal{D}_2^*(S^-)$, where $w^{*-}(z)$ is the sum of functions $w_k^{*-}(z)$, which are bianalytic in $S_k^-, k = 0,1,\cdots,p$. When L_0 exists, we require also $w^{*-}(\infty)$ is bounded. Because $\frac{w(t)}{\bar{t}}$ is a real function and $\frac{w_k^{*-}(z)}{\bar{z}}$ are analytic functions (which is an analytic factor of bianalytic functions $w_k^{*-}(z), k = 0,1,\cdots,p$, then the imaginary part of $\frac{w_k^{*-}(z)}{\bar{z}}$ on L_k is equal to zero. Furthermore, $\mathrm{Im}[\frac{w_k^{*-}(z)}{\bar{z}}] \equiv 0, \ z \in S_k^-, \ k = 0,1,\cdots,p$. Finally we have $\frac{w_k^{*-}(z)}{\bar{z}} \equiv c_k, \ z \in S_k^-, \ k = 0,1,\cdots,p$. By using the condition $\frac{w_k^{*-}(z)}{\bar{z}} = 0$, if $z = \infty$, we have $c_0 = 0$. In the other word we have $w_k^{*-}(t) = \bar{t}c_k, \ k = 0,1,\cdots,p$. Analogously, we can prove the second part of Theorem 6.

Remark. In Theorem 6 instead of $w^*(z) \equiv 0$ we can use the condition $\mathrm{Re}[\frac{w^*(z)}{\bar{z}}] \equiv 0$, then the conclusion of Theorem 6 still is right.

In fact, we first suppose $\mathrm{Re}[\frac{w^*(z)}{\bar{z}}] \equiv 0, \ z \in G$, then $\frac{w^*(z)}{\bar{z}} = Ai$, where A is a (real) constant. Thus we have $\frac{w^{*+}(t)}{\bar{t}} = Ai$, by using the Plemelj's theorem we obtain

$$\frac{w^{*-}(t)}{\bar{t}} = \frac{w^{*+}(t)}{\bar{t}} - \frac{w(t)}{\bar{t}} = Ai - \frac{w(t)}{\bar{t}}.$$

Moreover $\mathrm{Im}[\frac{w^{*-}(t)}{\bar{t}}] = A, \ t \in L$, and $\mathrm{Im}[\frac{w^{*-}(z)}{\bar{z}}] \equiv A, \ z \in S_k^-, \ k = 0,1,\cdots,p$. In the other word $\frac{w^{*-}(z)}{\bar{z}} = Ai - c_k, \ z \in S_k^-$. Finally we have

:

$$\frac{w(t)}{\bar{t}} = Ai - \frac{w^{*-}(t)}{\bar{t}} = Ai - Ai + c_k = c_k, \ t \in L_k.$$

or $w(t) = \bar{t}c_k, \ t \in L_k, \ k = 0, 1, \cdots, p$. When L_0 exists we also have $c_0 = 0$. Analogously, If the condition $\mathrm{Re}[\frac{w^*(z)}{\bar{z}}] \equiv 0, \ z \in S_k^-$ is satisfied, then the second conclusion of Theorem 6 still is right ($w(t) = \bar{t}\,C$, where C is a real constant).

References

[1] Zhao Zhen, Bianalytic functions, complex harmonic functions and their basic boundary value problem, Journal of Beijing Normal University, Natural Sei., 31 (1995), No.2, 175–179.

[2] Zhao Zhen, Bianalytic functions and its applications, Proceedings of second Asian Math. Conference, (1995. 11, in Tailand), 1–5.

[3] Zhao Zhen, Riemman Hilbert's problem for bianalytic functions, Journal of Beijing Normal University, Natural Sei., 32 (1996), No. 3.

[4] Zhao Zhen, Schwarz's problems for complex partial differential equations of second order, Beijing Mathematics, part 1, 1996, No 2, 131–137.

[5] Zhao Zhen, Some Boundary problems for bianalytic and complex harmonic functions, Report on International conference of complex analyses, Ningxia University, China, 1996, 8.

[6] Zhao Zhen and Chenfang Quan, Dirichlet's problems for bianalytic functions, Journal of Beijing Normal University, Natural Sci., 34 (1998), No. 3, 174–178.

[7] Zhao Zhen, Cauchy formula, integral of Cauchy type and Hilbert problem for bianalytic functions, Proceedings of ISAAC, 1997, USA, Deleware University, Kluwer Academic Publishers, 1998, 211–218.

[8] Zhao Zhen, The Schwarz's problem for a class of complex partial differential equation of third order, Journal of Ningxia University, Natural Sci., 9 (1998), No 1, 6–8.

[9] Lu Jian-ke, Complex Methods of Plane Elasticity, Publ. house of Wuhan university, 1986 (in Chinese).

[10] Mushkelishvili N. I., Singular Integral Equation, Noordhoof, Groningen, 1953.

[11] Vekya I. N., Generalized Analytic Functions, Moskow, 1959 (in Russian).

30 THE GROWTH OF SPIRALLIKE MAPPINGS

Hidetaka HAMADA

Faculty of Engineering, Kyushu Kyoritsu University
1-8, Jiyugaoka, Yahatanishi-ku, Kitakyushu 807-8585, Japan

hamada@kyukyo-u.ac.jp

Gabriela KOHR

Faculty of Mathematics Babeş-Bolyai University
1 M.Kogălniceanu Str. 3400 Cluj-Napoca, Romania

gkohr@math.ubbcluj.ro

Abstract: In this paper, we give an upper estimate of a component of the growth for normalized spirallike mappings relative to a diagonal matrix on the Euclidean unit ball.

1 Introduction

Let f be a univalent mapping in the unit disc Δ with $f(0) = 0$ and $f'(0) = 1$. Then the classical growth theorem is as follows:

$$\frac{|z|}{(1+|z|)^2} \leq |f(z)| \leq \frac{|z|}{(1-|z|)^2}.$$

It is well known that the above growth theorem cannot be generalized to normalized biholomorphic mappings on the Euclidean unit ball $\mathbf{B}^n$ in $\mathbf{C}^n$ $(n \geq 2)$. Barnard, FitzGerald and Gong [1] and Chuaqui [2]

H.G.W. Begehr et al. (eds.), Proceedings of the Second ISAAC Congress, Vol. 1, 231–236.

extended the above growth theorem to normalized starlike mappings on $\mathbf{B}^n$. Dong and Zhang [3] generalized the above result to normalized starlike mappings on the unit ball in complex Banach spaces.

It is interesting to consider about the possibility to extend the above growth theorem to a larger family of biholomorphic mappings. Hamada and Kohr [5] generalized the above growth theorem to spirallike mappings of type α on the unit ball $\mathbf{B}$ in an arbitrary complex Banach space and gave an example of a normalized spirallike mapping in the sense of Suffridge [8] such that the same growth theorem does not hold. This example also shows that the growth of normalized spirallike mappings cannot be estimated from above.

In this paper, we will give an upper estimate of the growth of a component of spirallike mappings relative to a diagonal matrix.

2 Preliminaries

For complex Banach spaces X, Y, let $\mathcal{L}(X,Y)$ be the space of all continuous linear operators from X into Y with the standard operator norm. By I we denote the identity in $\mathcal{L}(X,X)$. Let G be a domain in X and let $f: G \to Y$. f is said to be holomorphic on G, if for any $z \in G$, there exists a $Df(z) \in \mathcal{L}(X,Y)$ such that

$$\lim_{h\to 0} \frac{\|f(z+h)-f(z)-Df(z)h\|}{\|h\|} = 0.$$

Let $\mathcal{H}(G)$ be the set of holomorphic mappings from a domain $G \subset X$ into X.

A mapping $f \in \mathcal{H}(G)$ is said to be locally biholomorphic on G if its Fréchet derivative $Df(z)$ as an element of $\mathcal{L}(X,X)$ is nonsingular at each $z \in G$. A mapping $f \in \mathcal{H}(G)$ is said to be biholomorphic on G if $f(G)$ is open in X, the inverse f^{-1} exists and is holomorphic on $f(G)$. Let $\mathbf{B}$ denote the unit ball with respect to the norm $\|\cdot\|$ on X. A mapping $f \in \mathcal{H}(\mathbf{B})$ is said to be normalized if $f(0)=0$ and $Df(0)=I$. For each $z \in X \setminus \{0\}$, we define

$$T(z) = \{z^* \in \mathcal{L}(X,\mathbf{C}) : \|z^*\| = 1, z^*(z) = \|z\|\}.$$

By the Hahn-Banach theorem, $T(z)$ is nonempty. Let

$$\mathcal{N} = \left\{g \in \mathcal{H}(\mathbf{B}) \, : \, g(0)=0, \ \mathrm{Re} z^*(g(z)) > 0 \text{ for all } z \in \mathbf{B}\setminus\{0\}, z^* \in T(z)\right\}.$$

For $h \in \mathcal{N}$, let

$$k = k(Dh(0)) = \inf\{\mathrm{Re} z^*(Dh(0)z) : \|z\| = 1, z^* \in T(z)\}$$

and

$$m = m(Dh(0)) = \sup\{\mathrm{Re} z^*(Dh(0)z) : \|z\| = 1, z^* \in T(z)\}.$$

By Lemma 4 of Gurganus [4],

$$\frac{1-\|z\|}{1+\|z\|}\mathrm{Re} z^*(Dh(0)z) \leq \mathrm{Re} z^*(h(z)) \leq \frac{1+\|z\|}{1-\|z\|}\mathrm{Re} z^*(Dh(0)z)$$

for all $z \in \mathbf{B} \setminus \{0\}$, $z^* \in T(z)$. Therefore, $0 \leq k \leq m < \infty$ and

$$k\frac{1-\|z\|}{1+\|z\|}\|z\| \leq \mathrm{Re} z^*(h(z)) \leq m\frac{1+\|z\|}{1-\|z\|}\|z\| \tag{2.1}$$

for all $z \in \mathbf{B} \setminus \{0\}$, $z^* \in T(z)$. ¿From (2.1), we obtain the following lemma by integration.

Lemma 2.1 *Let $h \in \mathcal{N}$ and let $z \in \mathbf{B} \setminus \{0\}$. If $v(t) = v(z,t)$ is a solution to the initial value problem*

$$\frac{\partial v}{\partial t} = -h(v), \quad v(0) = z$$

defined for all $t \geq 0$, then

$$\frac{\|v(z,t)\|}{(1-\|v(z,t)\|)^2} \leq e^{-kt}\frac{\|z\|}{(1-\|z\|)^2}$$

and

$$e^{-mt}\frac{\|z\|}{(1+\|z\|)^2} \leq \frac{\|v(z,t)\|}{(1+\|v(z,t)\|)^2}$$

holds for all $t \geq 0$.

The following definition is due to Suffridge [8].

Definition 2.1 Let $f : \mathbf{B} \to X$ be a normalized biholomorphic mapping. Let $A \in \mathcal{L}(X, X)$ such that

$$\inf\{\mathrm{Re} z^*(A(z)) : \|z\| = 1, z^* \in T(z)\} > 0. \tag{2.2}$$

We say that f is spirallike relative to A if $e^{-tA}f(\mathbf{B}) \subset f(\mathbf{B})$ for all $t \geq 0$, where

$$e^{-tA} = \sum_{k=0}^{\infty} \frac{(-1)^k}{k!}t^k A^k.$$

Remark When X is a finite dimensional complex Banach space, a normalized locally biholomorphic mapping f is spirallike relative to A if and only if $[Df(z)]^{-1}Af(z) \in \mathcal{N}$ (cf. [8]).

Lemma 2.2 *Let $A \in \mathcal{L}(X, X)$ which satisfies the condition* (2.2). *Let $f : \mathbf{B} \to X$ be a normalized spirallike mapping relative to A and let $v(z,t) = f^{-1}(e^{-tA}f(z))$. Then, for any $z \in \mathbf{B} \setminus \{0\}$ and $\varepsilon > 0$, there exists a t_0 such that*

$$\|f(v(z,t)) - v(z,t)\| \leq e^{-kt}\varepsilon$$

for all $t \geq t_0$.

3 The growth theorem

Let $\|\cdot\|$ be the Euclidean norm on $\mathbf{C}^n$ and let $\mathbf{B}^n$ be the Euclidean unit ball in $\mathbf{C}^n$. Hamada-Kohr [5] gave an example of a normalized spirallike mapping relative to a diagonal matrix such that its growth cannot be estimated from above. The example is as follows:

Example. Let

$$f(z_1, z_2) = (z_1 + az_2^2, z_2)$$

on the Euclidean unit ball in $\mathbf{C}^2$. Let A be a linear mapping such that

$$A(z_1, z_2) = (2z_1, z_2).$$

Then $[Df(z)]^{-1}Af(z_1, z_2) = (2z_1, z_2)$. Therefore, f is a normalized spirallike mapping relative to A for any $a \in \mathbf{C}$. Let $a \in \mathbf{R}$ and $a > 2\sqrt{15}$. Let $z^0 = (0, 1/2)$. Then $f(z^0) = (a/4, 1/2)$ and $\|f(z^0)\| > 2$. On the other hand,

$$\frac{\|z^0\|}{(1 - \|z^0\|)^2} = 2.$$

Therefore,

$$\|f(z^0)\| > \frac{\|z^0\|}{(1 - \|z^0\|)^2}.$$

Also, $\|f(z^0)\| \to \infty$ as $a \to \infty$. Therefore, the growth of normalized spirallike mappings cannot be estimated from above.

Using Lemmas 2.1 and 2.2, we obtain the following theorem.

Theorem 3.1 *Let* $A = (a_i^j)$ *be a diagonal matrix. Assume that*

$$0 < Rea_1^1 = \cdots = Rea_l^l < Rea_{l+1}^{l+1} \leq \ldots \leq Rea_n^n.$$

Let $f : \mathbf{B}^n \to \mathbf{C}^n$ *be a normalized spirallike mapping relative to* A. *Then*

$$\|(f_1, \ldots, f_l, 0, \ldots, 0)\| \leq \frac{\|z\|}{(1 - \|z\|)^2}.$$

By the same argument as in the proof of the above theorem, we obtain the following corollaries.

Corollary 3.1 *Let* A *be a normal matrix and let* $\lambda_1, \ldots, \lambda_n$ *be the eigenvalues of* A. *Assume that* $\mathrm{Re}\lambda_1 = \cdots = \mathrm{Re}\lambda_n > 0$. *Let* $f : \mathbf{B}^n \to \mathbf{C}^n$ *be a normalized spirallike mapping relative to* A. *Then*

$$\frac{\|z\|}{(1 + \|z\|)^2} \leq \|f(z)\| \leq \frac{\|z\|}{(1 - \|z\|)^2}.$$

Corollary 3.2 *Let* $\mathbf{B}$ *be the unit ball in an arbitrary complex Banach space* X. *Let* $A = aI$ *with* $\mathrm{Re}a > 0$. *Let* $f : \mathbf{B} \to X$ *be a normalized spirallike mapping relative to* A. *Then*

$$\frac{\|z\|}{(1 + \|z\|)^2} \leq \|f(z)\| \leq \frac{\|z\|}{(1 - \|z\|)^2}.$$

Corollary 3.2 was obtained by Hamada-Kohr [5]. Also, if we put $a = 1$ in Corollary 3.2, we obtain the growth theorem of normalized starlike mappings ([1], [2], [3]).

Acknowledgements. The first author is partially supported by Grant-in-Aid for Scientific Research (C) no.11640194 from Japan Society for the Promotion of Science, 1999.

References

[1] R.W. Barnard, C.H. FitzGerald and S. Gong, *The growth and* 1/4-*theorems for starlike mappings in* $\mathbf{C}^n$, Pacific J. Math. **150** (1991), 13–22.

[2] M. Chuaqui, *Application of subordination chains to starlike mappings in* $\mathbf{C}^n$, Pacific J. Math. **168** (1995), 33–48.

[3] D. Dong and W. Zhang, *Growth and* $1/4$*-theorem for starlike maps in the Banach space,* Chin. Ann. Math. Ser.A **13**, No.4 (1992), 417–423.

[4] K.R. Gurganus, *Φ-like holomorphic functions in* $\mathbf{C}^n$ *and Banach spaces*, Trans. Amer. Math. Soc. **205** (1975), 389–406.

[5] H. Hamada and G. Kohr, *Subordination chains and the growth theorem of spirallike mappings*, to appear in Mathematica(Cluj).

[6] T. Kato, *Nonlinear semigroups and evoluation equations,* J. Math. Soc. Japan **19** (1967), 508–520.

[7] J.A. Pfaltzgraff, *Subordination chains and univalence of holomorphic mappings in* $\mathbf{C}^n$, Math. Ann. **210** (1974), 55–68.

[8] T.J. Suffridge, *Starlikeness, convexity and other geometric properties of holomorphic maps in higher dimensions,* Lecture Notes in Mathematics, vol. 599, Springer, Berlin New York Heidelberg, 1976, pp.146–159.

31 SUBORDINATION PRINCIPLE TO FUNCTIONS OF SEVERAL COMPLEX VARIABLES[1]

Kwang Ho SHON
Gwang Min SON

Department of Mathematics, Pusan National University
Pusan 609-735, Korea

khshon@hyowon.pusan.ac.kr

MOS subject classification Categories: 32F15, 32H10

Keywords: Strongly starlike mapping, Bergman Kernel

1. Introduction.

Let D be a bounded schlicht domain such that $D \in \mathbf{C}^n$ is the union of an increasing sequance of strictly pseudoconvex domains

$$D_t = \{z \in D : K_D(z, \bar{z}) - t < 0\}$$

for some $t > 0$, where $z = (z_1, z_2, \cdots, z_n)' \in D \subset \mathbf{C}^n$. The Bergman kernel function $K_D(z, \bar{p})$ of D is a holomorphic function of $(z, \bar{p})$ in the product domain $D \times D$ and belong to the space $L^2(D)$ of square integrable holomorphic functions of D. We know that $K_D(z, \bar{z}) > 0$. We suppose that the kernel function becomes infinite everywhere on ∂D, $\frac{\partial K_D(z,\bar{z})}{\partial z} \neq 0$ for $z \neq 0$ and $K_D(0,0) = \inf_{z \in D} K_D(z, \bar{z})$ at only $z = 0$. Also, let denote by $\varphi(z) = K_D(z, \bar{z})$ and $\varphi_t(z) = \varphi(z) - t$, for all $z \in D$. For a holomorphic univalent function $w = f(z)$ of D, we let $\varphi_t(z) = \varphi_t(f^{-1}(w)) = \Phi_t(w)$, $t > K_D(0,0)$, and let $\Delta = f(D)$, $\Delta_t = f(D_t)$. Then we have $\Delta_t = \{w : \Phi_t(w) < 0, w \in \Delta\}$. We say that $f \in C^1(D)$ if for all j, k the functions $\mathrm{Re} f_j = u_j$, $\mathrm{Im} f_j = v_j$, have all

[1]The authors wish to acknowledge the financial support of the Korea Research Foundation made in the program year of 1998, Project No.1998-015-D00022.

H.G.W. Begehr et al. (eds.), Proceedings of the Second ISAAC Congress, Vol. 1, 237–242.

first oder partial derivatives with respect to x_k, y_k and they are continuous on D. If $f \in C^1(D)$ and

$$J_r f(z) = \left| \frac{\partial(u_1, v_1, \cdots, u_n, v_n)}{\partial(x_1, y_1, \cdots, x_n, y_n)} \right| \neq 0$$

for all $z \in D$, then f is a locally diffeomorphism of C^1class. For $f \in C^1(D)$ and $z \in D$, we denote by

$$D_w f^{-1}(w) = [\frac{\partial z_i}{\partial w_j}(w)]_{1 \leq i,j \leq n}$$

and $< z, w >= \sum_{j=1}^{n} z_j \bar{w}_j$. Henceforce, we suppose that $f(0) = 0$ on D. And we denote by z^* as the conjugate transposed of z. This work is based on the technique of K. Kikuchi [6].

2. Theorem for the starlike mappings.

For the starlike mapping on bounded schlicht domains, G. Kohr [8] proved the following theorem.

Theorem 2.1[8]. *Let $f \in C^1(D)$, $f(z) \neq 0$ for all $z(\neq 0) \in D$ and $J_r f(z) > 0$ for all $z \in D$. If*

$$Re \left\langle \frac{\partial z}{\partial w} f(z) + \frac{\partial z}{\partial \bar{w}} f(z), \ \frac{\partial K_D(z, \bar{z})}{\partial \bar{z}} \right\rangle > 0$$

for all $z(\neq 0) \in D$, then f is starlike on D.

Using the Theorem 2.1, we can prove the following theorem of K. Kikuchi [6], simply.

Theorem 2.2. *Let f be a pseudoconformal mapping on D,*

$$K_D(g(z), \overline{g(z)}) \leq K_D(z, \bar{z})$$

for all $z \in D$ and any holomorphic mapping g of D into D with $g(0) = 0$. Then f is stalike if and only if

$$Re \, \{ \frac{\partial K_D(z, \bar{z})}{\partial z} \left(\frac{\partial f(z)}{\partial z} \right)^{-1} f(z) \} > 0$$

for all $z(\neq 0) \in D$.

Proof. If f is a starlike mapping, then all images Δ_t are starlike. That is, for all $w^{(1)} \in \partial \Delta_t$ we have $w^{(0)} = (1 - \tau) w^{(1)} \in \Delta_t, \ \tau \in I = [0, 1]$. In fact, if we set

$$z^{(1)} = f^{-1}(w^{(1)}), \ K_D(z^{(1)}, \bar{z}^{(1)}) = t \text{ and } \psi(z) = f^{-1}((1 - \tau) f(z)),$$

then we have

$$K_D(z^{(1)}, \bar{z}^{(1)}) \geq K_D(\psi(z^{(1)}), \overline{\psi(z^{(1)})}) = K_D(g(w^{(0)}), \overline{g(w^{(0)})})$$

where $g(w) = f^{-1}(w) = z$, i.e., $w = f(z)$, because $\psi(z)$ is a mapping of D into D and $\psi(0) = 0$. Then it holds that $g(w_0) \in D_t$, which yields $w_0 \in \Delta_t$. Now, since $\Phi_t\left(w + \epsilon \frac{\partial \Phi_t}{\partial w^*}\right) = 2\epsilon \left|\frac{\partial \Phi_t}{\partial w^*}\right|^2 + O(\epsilon^2) > 0$ when $\epsilon > 0$ is sufficiently small and $w \in \partial \Delta_t$, then $\frac{\partial \Phi_t}{\partial w^*}$ is the outward normal vector at the point w. Hence $(1-\tau)w \in \Delta_t$ $(w \in \partial\Delta_t,\ 0 < \tau \leq 1)$ implies $\cos\left(-\frac{\partial \Phi_t}{\partial w^*},\ -w\right) = \frac{\mathrm{Re}\{\frac{\partial \Phi_t}{\partial w} w\}}{\left|\frac{\partial \Phi_t}{\partial w^*}\right||w|} > 0$ and

$$\frac{\partial \Phi_t}{\partial w} w = \frac{\partial \varphi_t}{\partial z}(z)\left(\frac{\partial z}{\partial w}\right) f(z) = \frac{\partial K_D(z,\bar{z})}{\partial z}\left(\frac{\partial f(z)}{\partial z}\right)^{-1} f(z),$$

thus $0 < \mathrm{Re}\{\frac{\partial \Phi_t}{\partial w} w\} = \mathrm{Re}\{\frac{\partial K_D(z,\bar{z})}{\partial z}\left(\frac{\partial f(z)}{\partial z}\right)^{-1} f(z)\}$. Next, we suppose that $\mathrm{Re}\,\{\frac{\partial K_D(z,\bar{z})}{\partial z}\left(\frac{\partial f(z)}{\partial z}\right)^{-1} f(z)\} > 0$. Then we have

$$\begin{aligned}
&\mathrm{Re}\left\langle \left(\frac{\partial z}{\partial w}\right) f(z) + \left(\frac{\partial z}{\partial \bar{w}}\right)\overline{f(z)}, \frac{\partial K_D(z,\bar{z})}{\partial \bar{z}} \right\rangle \\
&= 2\mathrm{Re}\,\{\left(\frac{\partial f(z)}{\partial z}\right)^{-1} f(z) \frac{\partial K_D(z,\bar{z})}{\partial \bar{z}}\}.
\end{aligned}$$

Since $\mathrm{Re}\,\{\frac{\partial K_D(z,\bar{z})}{\partial z}\left(\frac{\partial f(z)}{\partial z}\right)^{-1} f(z)\} > 0$, we have

$$\mathrm{Re}\left\langle \left(\frac{\partial z}{\partial w}\right) w + \left(\frac{\partial z}{\partial \bar{w}}\right)\bar{w}, \frac{\partial K_D(z,\bar{z})}{\partial \bar{z}} \right\rangle > 0.$$

By Theorem 2.1, the mapping f is starlike on D.

Remark 2.3. *Particularly if D is a unit hypersphere in $\mathbf{C}^n$, then we have*

$$K_D(z,\bar{z}) = \frac{n!}{\pi^n(1-|z|^2)^{n+1}}.$$

And we have

$$\frac{\partial K_D(z,\bar{z})}{\partial z} = \frac{(n+1)}{(1-|z|^2)} z^* K_D(z,\bar{z}).$$

Thus the equivalent condition of starlikeness in Theorem 2.2 becomes the following condition

$$Re\,\{z^*\left(\frac{\partial f(z)}{\partial z}\right)^{-1} f(z)\} > 0$$

for all $z(\neq 0) \in D$.

The following sets are defined by means of the kernel function.

$$m(D) = \{z \in D : K_D(z, \bar{z}) = \text{minimum}\}$$
$$c(D) = \{z \in D : K_D(z, \bar{z}) = \frac{1}{E(D)}\}$$

where $E(D)$ denotes the euclidean volume of D. The domain which the property $c(D) = m(D)$ is called a minimal domain with center at the origin. If D is a minimal domain, then there exists a point $z \in D$ such that the euclidean volume of the image domain Δ_f of D under any biholomorphic mapping $w = f(z)$, normalized at z by $f(0) = 0$ and $J_z f(z) = I$ is not smaller than the volume of D (see [5]). Consider the class $\mathcal{G}(D)$ of biholomorphic mappings $w = f(z)$, which are normalized at z. Every mapping $f \in \mathcal{G}(D)$ possesses a single valued Jacobian $J_z f(z)$ in D.

Theorem 2.4[5]. *Let z be a point in a bounded schlicht domain D. Then there exists a mapping $w = \varphi(z)$ in $\mathcal{G}(D)$ which maps D onto a domain Δ_φ with the minimum volume. Further,* $E(\Delta_\varphi) = \frac{1}{K_D(z,\bar{z})}$.

Theorem 2.5. *Let $f \in \mathcal{G}(D)$ and D be a minimal domain with $g(0) = 0$ for any holomorphic mappings g of D into D. If f is starlike if and only if*

$$Re\ \{\frac{\partial K_D(z, \bar{z})}{\partial z} \left(\frac{\partial f(z)}{\partial z}\right)^{-1} f(z)\} > 0$$

for all $z(\neq 0) \in D$.

Proof. Let $z \in c(D)$ and φ be a mapping in $\mathcal{G}(D)$ which maps D onto a domain Δ_φ with the minimum volume. Then we have $E(\Delta_\varphi) = \frac{1}{K_D(z,\bar{z})}$ and $E(\Delta_\varphi) \leq E(\Delta_f)$, from Theorem 2.4 . It implies that $E(\Delta_f) \geq E(D)$ for all f in $\mathcal{G}(D)$. Thus we have

$$K_D(z, \bar{z}) = \frac{1}{E(D)} \geq \frac{1}{E(\Delta_f)} = K_D(g(z), \overline{g(z)}).$$

Thus by Theorem 2.2, we have the result.

3. Strongly starlike mappings.

For the unit ball B^n, let $z \in \partial B^n$ and let $\zeta \in B^1$, $f(z,t)$ be a holomorphic function on $B^n \times [0, \infty)$ for each t, normalized so that $f(0,t) = 0$ and $D_z f(0,t) = e^t I$. The subordination chain $f(z,t) = e^t f(z)$ satisfying $\frac{\partial}{\partial t} f(z,t) = D_z f(z,t)(h(z,t))$ where $h(z,t)$ is holomorphic in z with

$h(0,t) = 0$, $D_z h(0,t) = I$ and Re $\langle z, h(z,t)\rangle \geq 0$ and $h(z,t) = w(z)$. Let $z \in \partial B^n$ and $\zeta \in B^1$. Then we have

$$\mathrm{Re}\langle\overline{\zeta z}, w(\zeta z)\rangle = \mathrm{Re}\sum_{i=1}^{n} \bar{\zeta}\bar{z}_i w_i(\zeta z) = |\zeta|^2 \mathrm{Re}\sum_{i=1}^{n} \bar{z}_i \frac{w_i(\zeta z)}{\zeta} \geq 0.$$

The function $g(\zeta) = \sum_{i=1}^{n} \bar{z}_i \frac{w_i(\zeta z)}{\zeta}$ is holomorphic. Since $D_z w(0) = I$, we have $g(0) = 1$.

Definition 3.1[2]. *We say that the mapping f is strongly starlike on B^n if the values of g lie on a fixed compact subset of the right half-plane, independent of z.*

Remark 3.2. *The definition of strongly starlike mappings is equivalent to saying that $g(\zeta) = \frac{1+\sigma(\zeta)}{1-\sigma(\zeta)}$, where $\sigma(0) = 0$ and $|\sigma(\zeta)| \leq c < 1, c$ an absolute constant. That is, f is a strongly starlike mapping on B^n, (see [2]).*

4. Example of strongly starlike mapping.

Example. *Let $f(z) = (z_1, z_2 - z_1 z_2, z_3 - z_1 z_3)$ then we obtain that*

$$D_z f = \begin{pmatrix} 1 & 0 & 0 \\ -z_2 & 1 - z_1 & 0 \\ -z_3 & 0 & 1 - z_1 \end{pmatrix}.$$

Therefore, since $(D_z f(z))^{-1} f(z) = w(z) = \left(z_1, \frac{z_2}{1-z_1}, \frac{z_3}{1-z_1}\right)$, we have $g(\zeta) = \frac{1-\zeta z_1 |z_1|^2}{1-\zeta z_1}$. It can be written in the form $\frac{1+\sigma}{1-\sigma}$ where $\sigma(\zeta) = \frac{\zeta z_1(1-|z_1|^2)}{2-\zeta z_1(1+|z_1|^2)}$. Hence

$$|\sigma(\zeta)| \leq \frac{|z_1|(1-|z_1|^2)}{2-|z_1|(1+|z_1|^2)} = \frac{|z_1|(1+|z_1|)}{2+|z_1|(1+|z_1|)} \leq \frac{1}{2} < 1.$$

It follows that f is strongly starlike.

Recall that a map $f : \Omega \subset R^m \to R^m$ is said to be quasiconformal if it is differentiable a.e., ACL(absolutely continuous on lines) for almost all lines, and $\|D_z f(z)\|^m \leq K|\det D_z f(z)|$ a.e. Ω. Here $\|D_z f(z)\|$ stands for the usual norm $\|D_z f(z)\| = \sup\{|D_z f(z)| : |z| = 1\}$ and the constant K is called quasiconformal distortion of the map f. The map f is Lipschitz continuous but not quasiconformal in B^3. In fact, $||D_z f(z)||^6 \leq K|\det D_z f(z)|$ if and only if $\frac{||D_z f(z)||^6}{|\det D_z f(z)|} \leq K$. Therefore, we have $|\frac{1}{(1-z_1)^2}| \not\leq K$.

References

1. R. Barnard, C. Fitzgerald and S. Gong, *The Growth and* 1/4*-theorems for stralike mappings in* $\mathbf{C}^n$, Pacific J. Math., **150** (1991), 13-22.
2. M. Chuaqui, *Applications of subordination chains to starlike mappings in* C^n, Pacific J. Math., **168** (1995), 33-48.
3. P. Duren, *Univalent Functions*, Springer-Verlag, New York, 1983.
4. R. C. Gunning and H. Rossi, *Analytic functions of several complex variables*, Prentice-Hall Inc., Englewood Cliffs, New Jersey, 1965.
5. K. T. Hahn, *Some properties of relative invariants on bounded domains*, Duke J. Math., **34** (1967), 325-332.
6. K. Kikuchi, *Starlike and convex mappings in several complex variables*, Pacific J. Math., **44(2)** (1973), 569-580.
7. G. Kohr, *On some conditions of spirallikess in* $\mathbf{C}^n$, Complex Variables, Theory Appl. **32** (1997), 79-88.
8. G. Kohr, *Some sufficient conditions of starlikeness for mappings of* C^1 *class*, Complex Variables, Theory Appl. **36** (1998), 1-9.
9. S. G. Krantz, *Function theory of several complex variables*, John Wiley and Sons, New York, 1982.
10. O. Lehto and K. Virtanen, *Quasiconformal Mappings in the Plane*, Springer-Verlag, Berlin, 1973.
11. P. Liczberski, *Jack lemma for holomorphic mappings in* $\mathbf{C}^n$, Ann. Univ. Mariae Curie-Sklodowska Sec. **A15, 13** (1986), 131-140.
12. T. Liu and G. Ren, *The Growth theorem for stralike mappings on bounded starlike circular domains*, Chin. Ann. of Math., **19B** (1998), 401-408.
13. T. Matsuno, *On star-like theorems and convex-like theorems in the complex vector space*, Sci., Rep. Tokyo, Kyoiku Daigaku, Sect **A5** (1955), 88-95.
14. J. Pfaltzgraff, *Subordination chains and univalence of holomorphic mappings in* $\mathbf{C}^n$, Math. Ann., **210** (1974), 55-68.
15. J. Pfaltzgraff and T. Suffridge, *Close-to-starlike holomorphic functions of several variables*, Pacific J. Math., **57** (1975), 271-279.
16. R. M. Range, *Holomorphic functions and integral representations in several complex variables*, GTM 108, Springer-Verlag, New York Inc., 1986.
17. M. S. Robertson, *Applications of the subordination principle to univalent function*, Pacific J. Math., **11** (1961), 315-324.
18. W. Rudin, *Function theory in the unit ball in* $\mathbf{C}^n$, Grundlehren der Mathematischen Wissenschaften in Einzeldarstellungen, Springer, Berlin, 1980.
19. T. Suffridge, *The principle of subordination applied to functions of several complex variables*, Pacific J. Math., **33** (1970), 241-248.
20. T. Suffridge, *Starlike and convex maps in Banach spaces*, Pacific J. Math., **48** (1973), 575-589.
21. Q. Yu, S. Wang and S. Gong, *The Growth and* 1/4*-theorems for stralike mappings in* B^p, Chin. Ann. of Math., **11B** (1990), 100-104.

32 P-ADIC NEVANLINNA THEORY AND FUNCTIONAL ANALYSIS

Abdelbaki BOUTABAA
Alain ESCASSUT

Mathématiques Pures
Université Blaise Pascal
Les Cézeaux
63177 Aubiere-Cedex, France

boutabaa@ucfma.univ-bpclermont.fr
escassut@ucfma.univ-bpclermont.fr

Abstract

Let K be an algebraically closed field of characteristic zero, complete for an ultrametric absolute value. We apply the p-adic Nevanlinna theory to functional equations of the form $g = R \circ f$, where $R \in K(x)$, f, g are meromorphic functions in K, or in an "open disk", g satisfying conditions on the order of its zeros and poles. In various cases we show that f and g must be constant when they are meromorphic in all K, or they must be quotients of bounded functions when they are meromorphic in an "open disk". In particular, we have an easy way to obtain again Picard-Berkovich's theorem for curves of genus 1 and 2. These results apply to equations $f^m + g^n = 1$, when f, g are meromorphic functions, or entire functions in K or analytic functions in an "open disk". We finally apply the method to Yoshida's equation $y'^m = F(y)$, when $F \in K(X)$, and we describe the only case where solutions exist: F must be a polynomial of the form $A(y-a)^d$ where $m - d$ divides m, and then the solutions are the functions of the form $f(x) = a + \lambda(x - \alpha)^{\frac{m}{m-d}}$, with $\lambda^{m-d}(\frac{m}{m-d})^m = A$.

H.G.W. Begehr et al. (eds.), Proceedings of the Second ISAAC Congress, Vol. 1, 243–250.

1 Introduction

Let K denote an algebraically closed field of characteristic zero, complete for an ultrametric absolute value, let a be a point of K, let R be a positive number and let $d(a, R^-)$ be the disk $\{x| \ |x-a| < R\}$. We denote by $\mathcal{A}(K)$ the K-algebra of entire functions in K and by $\mathcal{M}(K)$ the field of meromorphic functions in K, i.e. the field of fractions of $\mathcal{A}(K)$. In the same way, we denote by $\mathcal{A}(d(a, R^-))$ the K-algebra of analytic functions in $d(a, R^-)$, i.e., the set of power series converging inside $d(a, R^-)$, and by $\mathcal{M}(d(a, R^-))$ the field of meromorphic functions in $d(a, R^-)$, i.e., the field of fractions of $\mathcal{A}(d(a, R^-))$. Furthermore, here we consider the K-subalgebra $\mathcal{A}_b(d(a, R^-))$ of bounded analytic functions inside $d(a, R^-)$, and we denote by $\mathcal{M}_b(d(a, R^-))$ its field of fractions.

Given $(m, n) \in \mathbb{N} \times \mathbb{N} \setminus \{(0, 0)\}$ we denote by (m, n) the greatest common divisor of m and n.

For every $n \in \mathbb{Z}$, $|n|_\infty$ will denote the archimedean absolute value of n.

In [3] we stated an improvement to the p-adic Nevanlinna's second Main Theorem which is exactly what we need for applying this theory to the present study devoted to functional equations of the form $gQ(f) = P(f)$ in $\mathcal{M}(K)$ and in $\mathcal{M}(d(a, R^-))$. On the other hand, there does exist a Nevanlinna Theorem for p-adic unbounded analytic functions inside a disk such as $d(a, R^-)$. We will recall and use it in order to obtain similar results for such functions. However, most results are somewhat weaker because the p-adic improvement mentioned above when applying to meromorphic functions in all K, no longer works for functions just defined in $d(a, R^-)$.

Our study will be applied to certain curves and to differential equations. For more details, see [8].

2 Parametrization of curves

We denote by D an infinite bounded set included in $d(a, R^-)$. Theorems 1 and 2 may apply to curves of genus 0 as well as curves of genus ≥ 1.

Theorem 1: *Let P, $Q \in K[X]$ be two relatively prime polynomials of degrees s and t respectively, let n be the number of distinct zeros of Q, let $m \in \mathbb{N}^*$, and let $g \in \mathcal{M}(d(a, R^-))$ be a non constant function all poles of which have order $\geq m$. Suppose that there exists a function $f \in \mathcal{M}(d(a, R^-))$ satisfying $g(x)Q(f(x)) = P(f(x))$ for every $x \in D$ which is not a pole of f or g.*

i) Assume that $f \notin \mathcal{M}_b(d(a,R^-))$. *Then* $mn \le t + 2m$. *Moreover, if* $s > t$, *then* $mn \le \min(t+2m, s+m)$.

ii) Assume $f,\ g \in \mathcal{M}(K)$. *Then* $mn < t + 2m$. *Moreover, if* $s > t$, *then* $mn < \min(t+2m, s+m)$.

Examples: 1) Let Γ be the curve of equation $y^4(x-b')(x-b'') = (x-c)^3$ (with b', b'', c all distinct) and let $f,\ g \in \mathcal{M}(d(a,R^-))$ be such that $(f(u), g(u)) \in \Gamma$ for all $u \in D$. Then by Theorem 1 $f,\ g \in \mathcal{M}_b(d(a,R^-))$. 2) Let Γ be the curve of equation $y^3(x-b')(x-b'') = (x-c)^3$ (with b', b'', c all distinct) and let $f,\ g \in \mathcal{M}(K)$ be such that $(f(u), g(u)) \in \Gamma$ for all $u \in D$. Then by Theorem 1 $f,\ g$ are constant.

Theorem 2: *Let* $P(X) = A\prod_{i=1}^{k}(X-a_i)^{s_i}$, $Q(X) = B\prod_{j=1}^{n}(X-b_j)^{t_j}$ *be two relatively prime polynomials of* $K[X]$ *of respective degrees* s *and* t. *Let* $m \in \mathbb{N}^*$. *Let* $f,\ g \in \mathcal{M}(d(a,R^-))$ *satisfy* $(g(x))^m Q(f(x)) = P(f(x))$ *for all* $x \in D$.

α) *If* $k+n > 1 + \frac{1}{m}\Big((m, |s-t|_\infty) + \sum_{i=1}^{k}(m, s_i) + \sum_{j=1}^{n}(m, t_j)\Big)$ *then both* f *and* g *lie in* $\mathcal{M}_b(d(a,R^-))$.

Moreover, if f *lies in* $\mathcal{A}(d(a,R^-))$, *and if* $k+n > 1 + \frac{1}{m}\Big(\sum_{i=1}^{k}(m, s_i) + \sum_{j=1}^{n}(m, t_j)\Big)$ *then* $f \in \mathcal{A}_b(d(a,R^-))$ *and* $g \in \mathcal{M}_b(d(a,R^-))$.

β) *If both* $f,\ g$ *lie in* $\mathcal{M}(K)$, *and if* $k+n \ge 1 + \frac{1}{m}\Big((m, |s-t|_\infty) + \sum_{i=1}^{k}(m, s_i) + \sum_{j=1}^{n}(m, t_j)\Big)$, *then both* f *and* g *are constant.*

Finally, if $f,\ g \in \mathcal{M}(K) \setminus K$ *and if* $k+n \ge 1 + \frac{1}{m}\Big(1 + \sum_{i=1}^{k}(m, s_i) + \sum_{j=1}^{n}(m, t_j)\Big)$, *then* f *admits at least one pole of order* $< m$.

Examples: 3) Let $c',\ c'' \in K$ (with $c' \ne c''$) and let Γ be the curve of equation $y^3 = (x-c')^2(x-c'')$. Let $f,\ g \in \mathcal{M}(K)$ be such that $(f(u), g(u)) \in \Gamma$ for all $u \in D$. If f and g lie in $\mathcal{A}(K)$, they are constant. If f and g are not constant, then f admits at least one pole of order 1 or 2. Here the genus is clearly 0, therefore there exist $f,\ g \in K(u)$

satisfying $g^3 = (f-c')^2(f-c'')$. 4) In the same way, let Γ be the curve of equation $y^3(x-b)^2 = (x-c)$ (with $b \neq c$) and let $f,\ g \in \mathcal{M}(K)$ be such that $(f(u), g(u)) \in \Gamma$ for all $u \in D$. If f and g lie in $\mathcal{A}(K)$, they are constant. If f and g are not constant, then f admits at least one pole of order 1 or 2.

5) Let Γ be the curve of equation $y^3(x-b)^2 = (x-c')^2(x-c'')$ (with b, c', c'' all distinct)and let $f,\ g \in \mathcal{M}(d(a, R^-))$ be such that $(f(u), g(u)) \in \Gamma$ for all $u \in D$. Then by Theorem 2 $f,\ g \in \mathcal{M}_b(d(a, R^-))$.

6) Let Γ be the curve of equation $y^2(x-b')(x-b'') = (x-c)$ (with b', b'', c all distinct) and let $f \in \mathcal{A}(d(a, R^-))$ and let $g \in \mathcal{M}(d(a, R^-))$ be such that $(f(u), g(u)) \in \Gamma$ for all $u \in D$. Then by Theorem 2 $f \in \mathcal{A}_b(d(a, R^-))$ and $g \in \mathcal{M}_b(d(a, R^-))$.

Remark: The conclusion appears without determining the genus of curves of equation $y^m Q(x) = P(x)$. However, Theorem 2 lets us easily obtain Picard Berkovich's Theorem [1] as far as curves of genus 1 and 2 are concerned. Actually, Corollaries a) and b) are even more general because they also apply to analytic functions in $d(a, R^-)$.

Corollary a: *Let Γ be an algebraic curve on K of genus 1 or 2 and let $f,\ g \in \mathcal{M}(K)$ be such that $(f(u), g(u)) \in \Gamma\ \forall u \in D$. Then f and g are constant.*

Corollary b: *Let Γ be a non degenerate elliptic curve on K and let $f,\ g \in \mathcal{A}(d(a, R^-))$ be such that $(f(u), g(u)) \in \Gamma\ \forall u \in D$. Then f and g are bounded.*

Corollary c: *Let Γ be an algebraic curve on K of genus 2 and let $f,\ g \in \mathcal{M}(K)$ (resp. $f,\ g \in \mathcal{M}(d(a, R^-))$) be such that $(f(u), g(u)) \in \Gamma\ \forall u \in D$. Then both f and g are constant (resp. lie in $\mathcal{M}_b(d(a, R^-))$).*

Indeed, every algebraic curve of genus 1 (resp. 2) is birationally equivalent to a smooth elliptic (resp. hyperelliptic) curve [15]. So, we can apply Theorem 2 with $m = 2,\ t = 0,\ s = n = 3$ in Corollary a), and $s \geq 4,\ n \geq 4$ in Corollary c). Corollary b) is obvious.

Here we take this opportunity to recall that there exists no parametrization of conics with a center, by entire functions, on the field K. Such a result cannot be extended to bounded analytic functions as show the functions sin and cos defined in $d(0, (p^{-\frac{1}{p-1}})^-)$ when the residue characteristic of K is p (resp. $d(0,1^-)$ when the residue characteristic of K is 0).

Proposition A: *Let Γ be a non degenerate conic with a center in K, and let $f,\ g \in \mathcal{A}(d(a,R^-))$ be such that $\big(f(u),g(u)\big) \in \Gamma\ \forall u \in D$. Then f and g are bounded in $d(a,R^-)$. Moreover, if both f and g lie in $\mathcal{A}(K)$, then they are constant.*

Remark: There exist no generalization of the p-adic Nevanlinna Theorem to bounded analytic functions.

3 Equality $f^m + g^n = 1$

In [4] it was proven that the equation $f^m + g^n = 1$ in $\mathcal{M}(K)$ leads to $f,\ g \in K$ as soon as the least commun multiple q of m and n satisfies: $\frac{1}{m} + \frac{1}{n} + \frac{1}{q} \geq 1$ and that in $\mathcal{A}(K)$ it leads to $f,\ g \in K$ as soon as $\min(m,n) \geq 2$. Here we are now able to generalize these conclusions.

Theorem 3: *Let $f,\ g \in \mathcal{M}(d(a,r))$ satisfy $g^m + f^n = 1$, with* $\min(m,n) \geq 3$, $\max(m,n) \geq 4$. *Then both f and g lie in $\mathcal{M}_b(d(a,r))$. Moreover, if $f,\ g \in \mathcal{M}(K)$, and if* $\min(m,n) \geq 2$, $\max(m,n) \geq 3$ *then $f,\ g$ are constant.*

Theorem 4: *Let $f,\ g \in \mathcal{A}(d(a,R^-))$ satisfy $g^m + f^n = 1$, with* $\min(m,n) \geq 2$. *Then f and g are bounded in $d(a,R^-)$.*

On the other hand, in order to complete results in this domain, we have the following Proposition:

Proposition B: *Let $f,\ g \in \mathcal{A}(d(a,R^-))$ and let $h \in \mathcal{A}_b(d(a,R^-))$ satisfy $f^2 + g^2 = h$. Then f and g are bounded inside $d(a,R^-)$.*

Remark: The p-adic functions sin and cos are bounded inside $d(0,(p^{-\frac{1}{p-1}})^-)$ when the residue characteristic is p (resp. inside $d(0,1^-)$ when the residue characteristic is 0) and satisfy $\sin^2 x + \cos^2 x = 1$.

4 Applications to certain differential equations

As a corollary of Theorem 1, we have Theorem 5:

Theorem 5 : *Let $P,\ Q \in K[X]$ be relatively prime, let $q \in \mathbb{N}^*$, let $F \in K[X_0, X_1, ...X_q] \setminus K[X_0, X_1, ...X_{q-1}]$, let n be the number of distinct zeros of Q, $s = \deg(P)$ and $t = \deg(Q)$.*

i) Let $f \in \mathcal{M}(d(a, R^-)) \setminus \mathcal{M}_b(d(a, R^-))$. Suppose that the equation $F(y, y', y'', ...y^{(q)})Q(f) = P(f)$ admits a solution in $\mathcal{M}(d(a, R^-))$. Then $(q+1)n \leq t + 2q + 2$. Moreover, if $s > t$, then we have $(q+1)n \leq \min(t + 2q + 2, s + q + 1)$.

ii) Let $f \in \mathcal{M}(K) \setminus K$. Suppose that the equation $F(y, y', y'', ..., y^{(q)})Q(f) = P(f)$ admits a solution in $\mathcal{M}(K)$. Then $(q+1)n < t + 2q + 2$. Moreover, if $s > t$, then we have $(q+1)n < \min(t + 2q + 2, s + q + 1)$.

In [2], it was shown that if the p-adic Yoshida equation $(\mathcal{E})\quad (y')^m = F(x, y)$ (with $F(x, y) \in K(x, y)$) admits solutions in $\mathcal{M}(K) \setminus K(x)$, then $F \in K(x)[y]$, and $\deg_y(F) \leq 2m$. Moreover, in [4], it was shown that if $F \in K[y]$, then any solution of the equation lying in $\mathcal{A}(K)$ is actually a polynomial.

Here we extend this last result by studying all meromorphic solutions of the equation when $F \in K(y)$.

Theorem 6: *Let $F(y) \in K(y)$ and suppose that there exists a non constant solution $f \in \mathcal{M}(K)$ of the differential equation $(\mathcal{E})\quad (y')^m = F(y)$.*
Then F is a polynomial $A(X-b)^d$ $(0 \leq d \leq 2m)$ such that $m-d$ divides m. In that case the solutions $f \in \mathcal{M}(K) \setminus K$ of $(\mathcal{E})$ are the functions of the form $f(x) = b + \lambda(x-\alpha)^{\frac{m}{m-d}}$, where λ satisfies $\lambda^{m-d}(\dfrac{m}{m-d})^m = A$.

In particular, when $m = 1$, we obtain the solutions of Malmquist's equation when $F \in K(y)$:

Corollary d: *Let $F \in K(y)$ be such that the equation $y' = F(y)$ admits a non constant solution in $\mathcal{M}(K)$. Then either $F(y)$ is of the form $A(y-b)^2$ and then the solutions in $\mathcal{M}(K)$ are the functions of the form $-\dfrac{1}{A(x-\alpha)} + b$, or F is a constant A and then the solutions are the functions $y = Ax + c$, with $c \in K$.*

Now, what is true in the field K for meromorphic functions, (and in particular for rational functions), also holds in any algebraically closed field of characteristic zero, as far as rational functions are concerned, because we have an obvious isomorphism from a finite extension E of $\mathbb{Q}$ containing all coefficients of all rational functions involved in $\mathcal{E}$, into K.

Corollary e: *Let L be an algebraically closed field of caracteristic zero. Let $F \in L(y)$. If the equation $(y')^m = F(y)$ admits a non constant solution $f \in L(x)$, then F is a polynomial of the form $A(y-b)^d$ $(0 \leq d \leq 2m)$ such that $m-d$ divides m. In that case the solutions $f \in L(x) \setminus L$ of $(\mathcal{E})$ are the functions of the form $f(x) = b + \lambda(x-\alpha)^{\frac{m}{m-d}}$, where λ satisfies $\lambda^{m-d}(\frac{m}{m-d})^m = A$.*

Remarks: 1) In [13] (par. 31), two similar examples of Corollary e) are considered, when $m = 2$.

2) Generalizing Theorem 6 to analytic functions in $d(a, R^-)$ does not seem easy because there is no reason to think that F is a polynomial. Yet, should it be a polynomial, when applying the Nevanlinna inequality to such analytic functions, we no longer get a contradiction.

5 References

[1] **Berkovich, W.** *Spectral Theory and Analytic Geometry over Non-archimedean Fields.* AMS Surveys and Monographs 33, (1990).

[2] **Boutabaa, A.** *Théorie de Nevanlinna p-adique.* Manuscripta Mathematica 67, p.251-269, (1990).

[3] **Boutabaa, A. , Escassut, A.** *An Improvement of the p-adic Nevanlinna Theory and Application to Meromorphic Functions.* To appear in Lecture Notes in Pure and Applied Mathematics (Marcel Dekker).

[4] **Boutabaa, A.** *On some p-adic functional equations.* Lecture Notes in Pure and Applied Mathematics (Marcel Dekker), n. 192 p. 49-59 (1997).

[5] **Boutabaa, A. Escassut, A. and Haddad, L.** *On uniqueness of p-adic entire functions.* Indagationes Mathematicae 8 p.145-155, (1997).

[6] **Boutabaa, A., Escassut, A.** *Urs and ursim for p-adic unbounded analytic functions inside a disk* (Preprint).

[7] **Boutabaa, A. Escassut, A.** *Property $f^{-1}(S) = g^{-1}(S)$ for p-adic entire and meromorphic functions* To appear in Rendiconti del Circolo Matematico di Palermo.

[8] **Boutabaa, A. Escassut, A.** *Applications of the p-adic Nevanlinna theory to functional equations.* To appear in Annales de l'Institut Fourier.

[9] **Cherry, W.** *Non-archimedean analytic curves in Abelian varieties* Math. Ann. , 300, p. 393-404 (1994).

[10] **Escassut, A.** *Analytic Elements in p-adic Analysis.* World Scientific Publishing Co. Pte. Ltd. (Singapore, 1995).

[11] **Escassut, A., Haddad, and L. Vidal, R.** *Urs, ursim, and non-urs,* Journal of Number Theory 75, p. 133-144 (1999).

[12] **Gross, F.** *On the equation* $f^n + g^n = 1$, Bull. Amer. Math. Soc. 72 p. 86-88 (1966).

[13] **Kaplansky, I** *An Introduction to Differential Algebra,* Actualités Scientifiques et Industrielles 1251, Hermann, Paris, (1957)

[14] **Nevanlinna R.** *Le théorème de Picard-Borel et la théorie des fonctions méromorphes* Gauthiers-Villars, Paris (1929).

[15] **Picard, E.** *Traité d'analyse II.* Gauthier-Villars, Paris (1925).

33 UNIQUE RANGE SETS IN P-ADIC AND COMPLEX ANALYSIS

Alain ESCASSUT
Abdelbaki BOUTABAA

Mathématiques Pures
Université Blaise Pascal
Les Cézeaux
63177 Aubiere-Cedex, France

escassut@ucfma.univ-bpclermont.fr
boutabaa@ucfma.univ-bpclermont.fr

Abstract

We study unique range sets with and without multiplicity. Ursims of n points for polynomials and p-adic entire functions were found by the authors for every $n \geq 9$, and in the same way, ursims of n points for rational functions and p-adic meromorphic functions, for every $n \geq 16$. We recall how urs for polynomials were characterized, as the sets which are not preserved by any non trivial mapping of the form $ax+b$, and urs of n elements for p-adic entire functions were found by the authors, for all $n \geq 3$, and were finally characterized by W. Cherry and C.C. Yang: they are the same as urs for polynomials. In the same way, bi-urs based upon sets of n points, for meromorphic functions and rational functions were found for every $n \geq 5$. However, we show that certain sets which are not preserved by any non trivial linear fractional function, are

H.G.W. Begehr et al. (eds.), Proceedings of the Second ISAAC Congress, Vol. 1, 251–259.

not urs for rational functions, a contradiction to a previous conjecture. Using in Archimedean analysis a method somewhat similar to the p-adic methods, we show that complex Weierstrass products without exponential factors admits urs of n points for every $n \geq 4$.

1 Generalities and ursims

Definitions and notation: B is an algebraically closed field of characteristic zero, and $P_1(B)$ is the projective space of dimension 1 over B. We denote by K an algebraically closed field of characteristic zero, complete for an ultrametric absolute value. Let L be a field which is either $\mathbb{C}$ or K. In particular, all definitions involving the field B apply to L, hence to K and $\mathbb{C}$. We denote by $\mathcal{A}(L)$ the set of entire functions in L, and by $\mathcal{M}(L)$ the set of meromorphic functions in L, i.e. the field of fractions of $\mathcal{A}(L)$.

Given a family of functions $\mathcal{F}$ defined in B, with values in $P_1(B)$, a set S in B will be called an *ursim for* $\mathcal{F}$ if for any two non constant functions $f, g \in \mathcal{F}$ satisfying $f^{-1}(S) = g^{-1}(S)$, these functions are equal. In [12] C.C.Yang and H.X. Yi showed the existence of ursims for $\mathcal{M}(\mathbb{C})$ having only 17 points, and ursims for $\mathcal{A}(\mathbb{C})$ with only 10 points. In B, for every $n \in \mathbb{N}$, we put $u_n = \dfrac{(n-1)^{n-1}}{n^n}$. Let $k \in B \setminus \{0, u_n\}$. We put $Q_n(x) = x^n - x^{n-1} + k$ and denote by $T_n(k)$ the set of zeros of Q_n in B. In the same way, we choose $c \in B \setminus \{0,1\}$ and put $R_n(x) = \dfrac{(n-1)(n-2)}{2}x^n - n(n-2)x^{n-1} + \dfrac{n(n-1)}{2}x^{n-2} - c$ and we denote by $U_n(c)$ the set of zeros of R_n in B. Since B has characteristic zero, we can check that all zeros of R_n and Q_n are simple.

Theorem 1: [8] *Let $k \in K \setminus \{0, u_n\}$. Then $T_n(k)$ is an ursim for $\mathcal{A}(K)$ for every $n \geq 9$.*

Corollary 1: *Let $k \in B \setminus \{0, u_n\}$. Then $T_n(k)$ is an ursim for $B[x]$ for every $n \geq 9$.* **Theorem 2:** [4] *Let $c \in K \setminus \{0,1\}$. Then $U_n(c)$ is an ursim for $\mathcal{M}(K)$ for every $n \geq 16$ and is an ursim for $\mathcal{A}(K)$ for every $n \geq 9$.*

Corollary 2: *Let $c \in B \setminus \{0,1\}$. Then $U_n(c)$ is an ursim for $B(x)$ for every $n \geq 16$. Moreover, $U_n(c)$ is an ursim for $B[x]$ for every $n \geq 9$.*

Remark 1: Theorems 1 and 2 mainly use a generalization of a lemma given in [9] for complex meromorphic functions sharing the same poles, taking multiplicities into account. Here we consider functions sharing the same poles without multiplicity: :

Lemma J: *Let $F, G \in \mathcal{M}(L)$ be non constant and have the same poles, ignoring multiplicities. Let $c_1, \dots c_q$ be pairwise distinct elements of L with $q \geq 2$ and let $S = \{c_1, \dots c_q\}$. We assume that $F(0) \neq c_j$, $G(0) \neq c_j$, $\forall j = 1, \dots, q$, and $\omega_0(F') = \omega_0(G') = 0$. Then one of the following two cases occurs*

i) *F and G satisfy the inequality :* $(\mathcal{I})$ $(3q-5)\,(T(r,F) + T(r,G)) \leq$

$$\leq \sum_{j=1}^{q} 2\Big(N_{[2]}\Big(r, \frac{1}{F-c_j}\Big) + N_{[2]}\Big(r, \frac{1}{G-c_j}\Big)\Big) + \overline{N}\Big(r, \frac{1}{F-c_j}\Big) + \overline{N}\Big(r, \frac{1}{G-c_j}\Big) +$$

$$+2\Big(N_{[2]}(r,F) - \overline{N}(r,F) + N_{[2]}(r,G) - \overline{N}(r,G)\Big) - \Big(N_0^S\Big(r, \frac{1}{F'}\Big) + N_0^S\Big(r, \frac{1}{G'}\Big)\Big) + \psi(r),$$

where $\psi(r) = o(T(r, F + T(r,G))$ and if $L = K$, then $\psi(r) + \log r$ is a bounded function.

ii) *There exist $A, B \in L$ with $A \neq 0$ such that $F = AG + B$ and* $(\mathcal{J})$ $\#(\{c_1, \dots c_q\} \cap \{Ac_j + B \mid 1 \leq j \leq q\}) \geq 2$.

2 Urs' and bi-urs'

For a subset S of B (resp. L) and $f \in B(x)$, (resp. $f \in \mathcal{M}(L)$) we denote by $E(f,S)$ the set in $B\times \mathbb{N}^*$: $\bigcup_{a\in S}\{(z,q) \in W\times \mathbb{N}^* \mid z$ zero of order q of $f(z) - a\}$,
(resp. the set in $L\times \mathbb{N}^*$: $\bigcup_{a\in S}\{(z,q) \in L\times \mathbb{N}^* \mid z$ zero of order q of $f(z) - a\}$)

Definitions: A subset S of B is said to be *affinely rigid* if there exist no function $ax+b$, other than the identity, presserving it. And S is called a *unique range set* (*a urs* in brief) *for $\mathcal{F}$*, if for any two non constant elements f, g of $\mathcal{F}$ such that $E(f,S) = E(g,S)$, one has $f = g$.

Lemma A: *A set S of n elements in B is not affinely rigid if and only if it is of one the following two forms:*

i) $S = \{a + b\zeta^j \mid 0 \leq j \leq n-1\}$, where ζ is a primitive n-th root of 1,

ii) $S = \{a + b\xi^j \mid 0 \leq j \leq n-2\} \cup \{a\}$ where ξ is a primitive $(n-1)$-th root of 1.

bf Notation: Let (1), (2), (3), (4) be the following relations about any $a \in B$:

(1) $a^n \neq \dfrac{n^n}{m^m(n-m)^{n-m}}$

(2) $a^n \neq (1+\xi)^{n-m}\dfrac{n^n}{m^m(n-m)^{n-m}}$ for every $(n-m)$-th root ξ of 1 different from -1.

(3) $n,\ m$ are relatively prime,

(4) Either $(m < n-1)$, or $(m > 2)$, or $((n, m, a^3) \neq (3, 2, \frac{27}{2}))$

Notation: Given $n, m \in \mathbb{N}*$, we will denote by $Z_{n,m}(a)$ the set of zeros of the polynomial $P(X) = X^n - aX^m + 1$.

Here, by elementary considerations we considered the problems of urs for polynomials and for p-adic entire functions [2].

Theorem 3 (L. Haddad, A. Boutabaa, A. Escassut): *A finite set S in B is a urs for $B[x]$ if and only if it is affinely rigid.*

Theorem 4 (L. Haddad, A. Boutabaa, A. Escassut): *Let $n, m \in \mathbb{N}$ satisfy $n > m \geq 3$ and be such that the polynomial $P(x) = x^n - ax^m + 1$ satisfies Properties (1),(3),(4). Then $Z_{n,m}(a)$ is an urs for $\mathcal{A}(K)$.*

Example: Let $a \in K^*$ be such that $\dfrac{27a^4}{256}$ is different from 1. Then $Z_{4,3}(a)$ is an urs for $\mathcal{A}(K)$.

Theorem 5: *A set S of three different points in K is an urs for $\mathcal{A}(K)$ if and only if it is affinely rigid.*

Theorem 6 (W. Cherry- C.C. Yang): [6] *A set S in K is an urs for $\mathcal{A}(K)$ if and only if it is affinely rigid.*

Definition: Similarly, a pair of sets (S, T) in $P_1(B)$ (resp. $P_1(L)$) is called a *bi-urs* for a subset $\mathcal{F}$ of $B(x)$ (resp. $\mathcal{M}(L)$ if for any two non constant elements $f,\ g \in \mathcal{F}$ such that $E(f, S) = E(g, S)$, and $E(f, T) = E(g, T)$ we have $f = g$.

In particular it is interesting to look for bi-urs of the form $(S, \{\infty\})$, for $B(x)$ and $\mathcal{M}(L)$.

Theorem 7: [3] *Let $n, m \in \mathbb{N}$ be relatively prime and satisfy $n \geq m+2$, and $m \geq 3$. Let $a \in K^*$ satisfy (1) and (2). Let $P(u) =$*

$u^n - au^m + 1$. Then P admits n distinct zeros and the set $Z_{n,m}(a)$ is such that $(Z_{n,m}(a), \{\infty\})$ is a bi-urs for $\mathcal{M}(K)$.

Theorem 8:[3] *Let $a \in K^*$ satisfy $a^6 \neq \frac{729}{16}$ and $a^6 \neq \frac{729}{4}$. Then the polynomial $P(u) = u^6 - au^4 + 1$ admits 6 distinct zeros and $(Z_{n,m}(a), \{\infty\})$ is a bi-urs for $\mathcal{M}(K)$.*

Corollary 3: [3] *For every $n \geq 5$, there exist bi-urs' for $\mathcal{M}(K)$ of the form $(\{z_1, ..., z_n\}, \{\infty\})$.*

Corollary 4: *For every $n \geq 5$, there exist bi-urs' for $B(x)$ of the form $(\{z_1, ..., z_n\}, \{\infty\})$.*

Remark 2: According to Theorem 6, one could imagine that a set of n elements is a urs for $B(x)$ or $\mathcal{M}(K)$ if and only if it is unpermutable by any non constant linear fractional function other than the identity. In the same way, one could think that there exist bi-urs's for $\mathcal{M}(K)$ of the form $(\{z_1, z_2, z_3\}, \{\omega\})$, and that a finite set $(S, \{\omega\})$ is a bi-urs for $B(x)$ or $\mathcal{M}(K)$ if and only if S is unpermutable by any non constant linear fractional function (different from the identity), admitting ω as a fixed point. Actually, this is wrong.

Theorem 9: *Let $h \in B(x)$ be non constant, and let $n \in \mathbb{N}$ be such that $n \geq 3$. Let $g = \dfrac{h^{n-1} - 1}{h^n - 1}$, and let $f = gh$. Then $E(f, T_n(k)) = E(g, T_n(k))$. Moreover, f and g share the same poles, counting multiplicities.*

Theorem 10 (L. Haddad, R. Vidal, A. Escassut): *Let $k \in B \setminus \{0, u_n\}$. Then $T_n(k)$ is not an urs for $B(x)$ whenever $n \in \mathbb{N}^*$. Moreover, there exist no bi-urs for $W(x)$ of the form $(\{a, b, c\}, \{\omega\})$, with $a, b, c, \in W$, and $\omega \in P_1(B)$.*

Remark 3: Obviously, any set which is not an urs for $B(x)$, cannot be an urs for $\mathcal{M}(K)$, cannot be an ursim for $B(x)$, and cannot be an ursim for $\mathcal{M}(K)$. In fact we do not know any example of an urs for a family of functions $\mathcal{F}$ which is not also an ursim for $\mathcal{F}$. For example, sets of the form $\{a, a + h, a + 3h\}$, with $a \in B$, $h \in B \setminus \{0\}$, are urs for $B[x]$, but are they also ursims for $B[x]$?

Theorem 11 (L. Haddad, R. Vidal, A. Escassut):[8] *Let $k \in B \setminus \{0, u_n\}$. For every $n \geq 5$, there exist no linear fractional functions h permuting $T_n(k)$, except the identity.*

Remark 4: In Theorem 11, the hypothesis $n \geq 5$ is necessary because given any four distinct points $\alpha,\ \beta,\ \gamma,\ \delta \in B$, there does exist a linear fractional function h, different from the identity, satisfying $h(\alpha) = \beta,\ h(\beta) = \alpha,\ h(\gamma) = \delta,\ h(\delta) = \gamma$. Indeed, h is of the form: $h(x) = \dfrac{x(\alpha\beta - \gamma\delta) - \alpha\beta(\gamma+\delta) + \gamma\delta(\alpha+\beta)}{x(\alpha+\beta-\gamma-\delta) - \alpha\beta + \gamma\delta}$.

Conclusion: By Theorem 10, the sets $T_n(k)$ in B are not urs for $B(x)$ though, by Theorem 11, there exist no linear fractional functions h (different from the identity) permuting $T_n(k)$ as soon as $n \geq 5$. Thus, the set of finite urs for $B(x)$ is strictly smaller than the set of finite sets S such that there exist no linear fractional functions h preserving S. So, it is not easy to imagine a characterization of urs, or bi-urs for $B(x)$. And a characterization of urs, or bi-urs, for $\mathcal{M}(K)$ appears even more difficult.

3 Urs' for complex Weierstrass products

The behaviour of p-adic entire functions has suggested us considering complex Weierstrass products without exponential factor [5].

For any f in $\mathcal{A}(\mathbb{C})$, for any $r > 0$, we put $M(r, f) = \sup\limits_{|x| \leq r} |f(x)|$.

We will denote by $\mathcal{W}$ the set of polynomials and Weierstrass products of the form

$f(x) = Ax^q \lim\limits_{r\to\infty} \prod\limits_{|a_j|\leq r} (1 - \dfrac{x}{a_j})^{q_j}$ with $A \in \mathbb{C}$, $q \in \mathbb{N}$, and with $(a_j)_{j\in\mathbb{N}}$ an injective sequence in $\mathbb{C}$ such that $|a_j| \leq |a_{j+1}|$, $\lim\limits_{j\to\infty} |a_j| = +\infty$, and such that the limit of the infinite product is uniform in all bounded subsets of $\mathbb{C}$.

Let $f(x) = Ax^q \lim\limits_{r\to\infty} \prod\limits_{|a_j|\leq r} (1 - \dfrac{x}{a_j})^{q_j} \in \mathcal{W}$. Then each a_j is a zero of f of order q_j and f has no other zeros. Moreover, for every $k \in \mathbb{N}$, $\prod\limits_{|a_j|\leq r, k\leq j} (1 - \dfrac{x}{a_j})^{q_j}$ also lies in $\mathcal{W}$.

Lemma B is known and easily proven: **Lemma B:** *$\mathcal{W}$ is a multiplicative semi-group. Let $f(x) = Ax^q \lim\limits_{r\to\infty} \prod\limits_{|a_j|\leq r} (1 - \dfrac{x}{a_j})^{q_j} \in \mathcal{W}$. Then each a_j is a zero of f of order q_j and f has no other zeros. Moreover, for every $k \in \mathbb{N}$, $\prod\limits_{|a_j|\leq r, k\leq j} (1 - \dfrac{x}{a_j})^{q_j}$ also lies in $\mathcal{W}$.*

Similarly, we denote by $\mathcal{W}'$ the subset of $\mathcal{W}$ consisting of the f which are either polynomials or functions of the form $f(x) = Ax^t \lim_{n\to\infty} \prod_{j=0}^{n} (1 - \frac{x}{a_j})^{q_j}$ with $A \in \mathbb{C}$, and with $t \in \mathbb{N}$, $q_j \in \mathbb{N}^*$ $\forall j \in \mathbb{N}$, $(a_j)_{j\in \mathbb{N}}$ an injective sequence in $\mathbb{C}$ such that $|a_j| \leq |a_{j+1}|$, $\lim_{j\to\infty} |a_j| = +\infty$, such that the limit of the infinite product is uniform in all bounded subsets of $\mathbb{C}$.

Moreover, here we particularly consider the subset $\mathcal{V}$ (resp. $\mathcal{V}'$) of $\mathcal{W}$ consisting of the $f \in \mathcal{W}$ such that $\mathbb{C}[f] \subset \mathcal{W}$ (resp. $\mathbb{C}[f] \subset \mathcal{W}'$). So, by definition, we have $\mathcal{V}' \subset \mathcal{W}'$.

We denote by $\mathcal{E}$ the set of entire functions of order strictly inferior to 1, i.e. the entire functions satisfying $\overline{\lim_{r\to\infty}}\Big(\frac{\ln(\ln^+ M(r,f))}{\ln r}\Big) < 1$, and we denote by $\mathcal{H}$ the set of entire functions satisfying $\lim_{r\to\infty}\Big(\frac{\ln M(r,f)}{r}\Big) = 0$ (where $\ln^+ u$ denotes $\max(\ln u, 0)$). Inside $\mathcal{W}'$ one calls *Weierstrass products of genus zero* the functions which are polynomials or Weierstrass products of the form $Ax^t \lim_{n\to\infty} \prod_{j=0}^{n} (1 - \frac{x}{b_j})$, where the sequence b_j is not necessarily injective and where the series $\sum_{n=1}^{\infty} \frac{1}{|b_n|}$ converges. We will denote by $\mathcal{Z}$ the set of Weierstrass products of genus zero. In order to clarify relations between these different sets, we need to state Lemma C which is mainly based upon Borel-Hadamard's Theorems, and on counterexamples.

Lemma C: *Both $\mathcal{E}$ and $\mathcal{H}$ are $\mathbb{C}$- algebras, $\mathcal{E}$ is strictly included in $\mathcal{Z}$, $\mathcal{Z}$ is strictly included in $\mathcal{H}$, and $\mathcal{H}$ is strictly included in $\mathcal{V}'$. For every $\omega, \varphi \in \mathbb{C}$, both* $\sin(\omega x + \varphi)$, $\cos(\omega x + \varphi)$ *belong to $\mathcal{V}'$, but do not belong to $\mathcal{H}$. If f lies in $\mathcal{W}'$, then so does $f(\omega x + \varphi)$, for every ω, $\varphi \in \mathbb{C}$. Moreover, $\mathcal{V}$, $\mathcal{V}'$, $\mathcal{W}$, $\mathcal{W}'$ are not additive subgroups of $\mathcal{A}(\mathbb{C})$.*

Similarly to p-adic entire functions, if a function in $\mathcal{W}$ does not vanish in $\mathbb{C}$, it is a constant. We now apply this property to $\mathcal{V}$ and in particular to $\mathcal{H}$, which, in fact, is the biggest $\mathbb{C}$-subalgebra of $\mathcal{A}(\mathbb{C})$ included in $\mathcal{W}$ that we know. The reason why we are interested in $\mathcal{V}$ is that our method requires that, given any two entire functions f, $g \in \mathcal{W}$, both $\mathbb{C}[f]$ and $\mathbb{C}[g]$ be included in $\mathcal{W}$.

Properties (1) and (2) are those defined above.

Theorem 12: *Let $n, m \in \mathbb{N}^*$ satisfy $n > m \geq 3$. Assume that a satisfies* (1) *and (2). Then $Z_{n,m}(a)$ is an urs for $\mathcal{V}$.*

Corollary 4: *For every $n \geq 4$, there exist urs' for $\mathcal{V}$ of cardinal n.*

Examples: Let $a \in \mathbb{C}^*$ satisfy $a^5 \neq \frac{3125}{108}$ and $a^5 \neq \frac{3125}{27}$. Then $Z_{5,3}(a)$ is a urs for $\mathcal{V}$.
In the same way, if $a \in \mathbb{C}^*$ satisfies $a^4 \neq \frac{256}{27}$ and $a^4 \neq \frac{512}{27}$, then $Z_{4,3}(a)$, is a urs for $\mathcal{V}$.

Application: Let $a \in \mathbb{C}^*$ satisfy $a \neq 2$, $a \neq -2$, $a^4 \neq \frac{256}{27}$ and $a^4 \neq \frac{512}{27}$. Let $\Sigma = \sin^{-1}(Z_{4,3}(a))$, and let $f \in \mathcal{V}$ satisfy $f^{-1}(Z_{4,3}(a)) = \Sigma$, and $0 \notin f'(\Sigma)$. Then we can claim that $f(x) = \sin x \ \forall x \in \mathbb{C}$. Indeed, since $a \neq 2$ and $a \neq -2$, we check that both 1, -1 don't lie in $Z_{4,3}(a)$, and therefore no points of the form $\frac{\pi}{2} + n\pi$, $n \in \mathbb{Z}$ lie in Σ. As a consequence, we have $\cos x \neq 0 \ \forall x \in \Sigma$, and thereby $E(\sin, \ Z_{4,3}(a))$ is just $\{(\alpha, 1) | \alpha \in \Sigma\}$. But since $0 \notin f'(\Sigma)$, of course $E(f, \ Z_{4,3}(a))$ is also equal to $\{(\alpha, 1) | \alpha \in \Sigma\}$. Hence $E(f, \ Z_{4,3}(a)) = E(\sin, \ Z_{4,3}(a))$, and therefore $f = \sin$.

4 References

[1] **Boutabaa A.** *Théorie de Nevanlinna p-adique.* Manuscripta Mathematica 67, p. 251-269, (1990).

[2] **Boutabaa A., Escassut A. and Haddad L.** *On uniqueness of p-adic entire functions.* Indagationes Mathematicae 8 p. 145-155, (1997).

[3] **Boutabaa A., and Escassut A.** *On uniqueness of p-adic meromorphic functions.* Proceedings of the AMS 126, N.9, p. 2557-2568, (1998).

[4] **Boutabaa, A. Escassut, A.** *Property $f^{-1}(S) = g^{-1}(S)$ for p-adic entire and meromorphic functions* to appear in Rendiconti del Circolo Matematico di Palermo.

[5] **Boutabaa, A. Escassut, A.** *Urs' for Weierstrass products without exponential factors* to appear in Complex variable and applications.

[6] **Cherry, W. Yang, C.C.** *Uniqueness of non-archimedean entire functions sharing sets of values counting multiplicities*, Proceedings of the AMS, Vol 127, n.4, p. 967-971, (1998).

[7] **Escassut, A.** *Analytic Elements in p-adic Analysis.* World Scientific Publishing Co. Pte. Ltd. (Singapore, 1995).

[8] **Escassut A., Haddad L. and Vidal R.** *Urs, ursims and non urs for p-adic functions and polynomials*, Journal of Number Theory 75, p. 133-144 (1999).

[9] **Frank G. and Reinders M.** *A unique Range set for meromorphic functions with 11 elements*, Complex Variable Theory Appl. v 37 , p.185-193 (1998).

[10] **Gross, F. Yang C.C.** *On preimage and range sets of meromorphic functions.* Proc. Japan Acad. 58 (1):17 (1982).

[11] **Ostrovskii I.V., Pakovitch F.B., Zaidenberg M.G.** *A remark on complex polynomials of least deviation*, Internat. Math. Res. Notices, n.14, p. 699-703, (1996).

[12] **Yi H.X. and Yang C.C.** *Uniqueness theory of meromorphic functions*, Pure and applied Math. Monographs n. 32, Science Press (1995).

34 ON UNIQUENESS OF MEROMORPHIC FUNCTIONS SHARING FINITE SETS

Hirotaka FUJIMOTO

Department of Mathemathematics
Faculty of Science
Kanazawa University
Kanazawa 920-1192, Japan

fujimoto@kappa.s.kanazawa-u.ac.jp

1991 Mathematics Subject Classification: 30D35

Abstract: In this article, we first study polynomials $P(w)$ such that $P(f) = cP(g)$ implies $f = g$ for any nonzero constant c and nonconstant meromorphic functions f and g on $\mathbb{C}$. Next, we give some sufficient conditions for a finite set S to be a uniqueness range set, namely, to satisfy the condition that $f^{-1}(S) = g^{-1}(S)$ implies $f = g$ for any nonconstant meromorphic functions f and g on $\mathbb{C}$.

1 INTRODUCTION

We first explain some notations and terminology. In this article, we mean by a meromorphic function a meromorphic function on the complex plane $\mathbb{C}$ and by a divisor a map $\nu : \mathbb{C} \to \mathbb{Z}$ whose support $\overline{\{z; \nu(z) \neq 0\}}$ is discrete. For a divisor ν, we set $\bar{\nu}(z) := \min(\nu(z), 1)$.

H.G.W. Begehr et al. (eds.), Proceedings of the Second ISAAC Congress, Vol. 1, 261–269.

Let f be a nonconstant meromorphic function. For a point $a \in \mathbb{C}$, we define the divisor $\nu_f^a : \mathbf{C} \to \mathbf{Z}$ by

$$\nu_f^a(z) := \begin{cases} 0 & \text{if } f(z) \neq a \\ m & \text{if } f - a \text{ has a zero of order } m \text{ at } z, \end{cases}$$

and set $\nu_f^\infty := \nu_{1/f}^0$, $\nu_f := \nu_f^0 - \nu_f^\infty$. With a discrete set S in $\mathbb{C}$ we associate the divisors $\nu_f^S := \sum_{a \in S} \nu_f^a$ and $\bar{\nu}_f^S := \sum_{a \in S} \bar{\nu}_f^a$.

We call a set S a uniqueness range set for meromorphic functions if for any nonconstant meromorphic functions f and g the condition $\nu_f^S = \nu_g^S$ implies $f = g$, and a uniqueness range set for entire functions if $\nu_f^S = \nu_g^S$ implies $f = g$ for any nonconstant entire functions f and g . We call also a set S a uniqueness range set for meromorphic (or entire) functions ignoring multiplicities if $\bar{\nu}_f^S = \bar{\nu}_g^S$ implies $f = g$ for any nonconstant meromorphic (or entire)] functions f, g.

In 1982, F. Gross and C. C. Yang proved that the set $\{w \in \mathbb{C}; w + e^w = 0\}$ is a uniqueness range set for entire functions([4]). Afterwards, many efforts were made to seek uniqueness range sets which are as small as possible by several authors P. Li, H.X. Yi, B. Shiffman and so on. In fact, in their papers [7], [9] and [5], they showed that the set $\{w; w^q + aw^{q-r} + b = 0\}$ gives small uniqueness range sets for meromorphic functions or entire functions under suitable conditions on constants a, b and positive integers q, r. Recently, Frank and Reinders gave a uniqueness range set for meromorphic functions with 11 elements in [2].

The purpose of this article is to give some sufficient conditions for a finite subset S of $\mathbb{C}$ to be a uniqueness range set for meromorphic functions or entire functions by applying the arguments used in [2] to more general settings.

2 SOME CONDITIONS FOR UNIQUENESS POLYNOMIALS

For a given finite set $S = \{a_1, \ldots, a_q\}$, we study the polynomial

$$P_S(w) := (w - a_1)(w - a_2) \cdots (w - a_q)$$

instead of the set S itself. We say that a nonconstant polynomial $P(w)$ is a uniqueness polynomial if $P(f) = cP(g)$ implies $f = g$ for arbitrary nonconstant meromorphic functions f, g and $c \in \mathbb{C} - \{0\}$. If S is a uniqueness range set for meromorphic functions, then $P_S(w)$ is a uniqueness polynomial. Here, the converse is not necessarily valid.

In their papers [7] and [8], B. Shiffman, C.C. Yang and X. Hua studied polynomials $P(w)$ such that $P(f) = P(g)$ implies $f = g$ for arbitrary nonconstant meromorphic functions f and g. In this article, we call such a polynomial a uniqueness polynomial in the wider sense.

Let $P(w)$ be a nonconstant polynomial of degree q with simple zeros whose derivative is given by

$$P'(w) = q(w - d_1)^{q_1}(w - d_2)^{q_2} \cdots (w - d_k)^{q_k}, \tag{2.1}$$

where $d_1, \cdots, d_k$ are mutually distinct.

We first give conditions for uniqueness polynomials in the wider sense.

Theorem 2.2. *Assume that $k \geq 4$. If $P(w)$ satisfies the hypothesis*

(H) *$P(d_\ell) \neq P(d_m)$ for any ℓ, m with $1 \leq \ell < m \leq k$,*

then $P(w)$ is a uniqueness polynomial in the wider sense.

Theorem 2.3 *Assume that $k = 3$ and $\min(q_1, q_2, q_3) \geq 2$. If $P(w)$ satisfies the above hypothesis* (H) *and any one of three values d_1, d_2 and d_3 is not the arithmetic mean of two others, then $P(w)$ is a uniqueness polynomial in the wider sense.*

These theorems are proved by applying the second main theorem in Nevanlinna theory to meromorphic functions f and g with $P(f) = P(g)$ and the values $d_1, \cdots, d_k, d_{k+1} = \infty$. We omit the details.

We can prove also the following theorems on uniqueness polynomials.

Theorem 2.4 *Let $P(w)$ be a polynomial with $k \geq 4$ satisfying the hypothesis* (H). *If $P(w)$ is not a uniqueness polynomial, then there is some permutation $(t_1, t_2, \cdots, t_k)$ of $(1, 2, \cdots, k)$ such that*

$$\frac{P(d_{t_1})}{P(d_1)} = \frac{P(d_{t_2})}{P(d_2)} = \cdots = \frac{P(d_{t_k})}{P(d_k)} \neq 1.$$

Theorem 2.5 *Let $P(w)$ be a polynomial satisfying the hypothesis* (H) *and assume that $k = 3$ and $\min\{q_1, q_2, q_3\} \geq 2$. If $P(w)$ is not a uniqueness polynomial, then, after suitable changes of indices of d_j's, $d_3 = (d_1 + d_2)/2$ or*

$$\frac{P(d_2)}{P(d_1)} = \frac{P(d_3)}{P(d_2)} = \frac{P(d_1)}{P(d_3)}.$$

The outlines of the proofs of these theorems will be stated in §4.

As a consequece of Theorem 2.4, we have the following:

Corollary 2.6 *Let $P(w)$ be a polynomial with $k \geq 4$. If it satisfies the hypothesis* (H) *and*

$$P(d_1) + P(d_2) + \cdots + P(d_k) \neq 0, \tag{2.7}$$

then $P(w)$ is a uniqueness polynomial.

In fact, by Theorem 2.4, (2.7) deduces a contradiction

$$\frac{P(d_{t_1})}{P(d_1)} = \cdots = \frac{P(d_{t_k})}{P(d_k)} = \frac{P(d_{t_1}) + P(d_{t_2}) + \cdots + P(d_{t_k})}{P(d_1) + P(d_2) + \cdots + P(d_k)} = 1.$$

By $\mathcal{P}_q$ we denote the set of all monic polynomials of degree q. With each $P(w) = w^q + A_1 w^{q-1} + \cdots + A_q \in \mathcal{P}_q$ by associating the point $(A_1, \cdots, A_q) \in \mathbb{C}^q$, we can identify $\mathcal{P}_q$ with $\mathbb{C}^q$. We say that some fact holds for generic polynomials of degree q if it holds for all polynomials in $\mathcal{P}_q$ except the zero set of a nonzero polynomial in $A_1, \cdots, A_q$.

By using Corollary 2.6 we can prove the following improvement of the fact which was shown by B. Shiffman in [7].

Corollary 2.8 *Generic polynomials of degree at least five are uniqueness polynomials.*

PROOF. As is easily seen, for generic polynomials, their derivatives have no multiple zeros and so we may consider the only case $k = q-1 (\geq 4)$. Moreover, it is easily seen that the hypothesis (H) and (2.7) hold for generic polynomials of degree $q \geq 5$.

Here, the number five is best-possible. In fact, it is known that every polynomial of degree less than five is not a uniqueness polynomial([5]).

3 SOME CONDITIONS FOR UNIQUENESS RANGE SETS

Let $P(w)$ be a monic polynomial of degree q with simple zeros whose derivative is given by (2.1) and assume that $k \geq 2$. We can prove the following:

Theorem 3.1 *Let f, g be meromorphic functions satisfying the condition*

$$(\mathrm{C})_{m_0} \qquad \textstyle\sum_{j=1}^{q} \min(\nu_f^{a_j}(z), m_0) = \sum_{j=1}^{q} \min(\nu_g^{a_j}(z), m_0)$$

for a positive integer m_0 or $m_0 = \infty$. If $q > 2k + 12$ for the case $m_0 = 1$, if $q > 2k + 4/(m_0 - 1) + 6$ for the case $m_0 \geq 2$, or if $q > 2k + 6$ for the *case* $m_0 = \infty$*, then there are some constants* $c_0(\neq 0)$ *and* c_1 *such that*

$$\frac{1}{P(f)} = \frac{c_0}{P(g)} + c_1. \tag{3.2}$$

For the particular case where f *and* g *are holomorphic,* (3.2) *remains valid if* $q > 2k+5$ *in case* $m_0 = 1$*, if* $q > 2k + \dfrac{2}{m_0 - 1} + 2$ *in case* $m_0 \geq 2$*, or if* $q > 2k + 2$ *in case* $m_0 = \infty$*.*

The outline of the proof will be stated in §5.

By some improvements of the arguments in [2], we can show the following:

Proposition 3.3 *Let* $P(w)$ *be a polynomial of degree* $q \geq 5$ *without multiple zeros whose derivative is given by* (2.1)*. Assume that* $P(w)$ *satisfies the hypothesis* (H) *and* (3.2) *holds for two distinct nonconstant meromorphic functions* f *and* g*. If* $k \geq 3$*, or if* $k = 2$ *and* $\min\{q_1, q_2\} \geq 2$*, then* $c_1 = 0$*.*

Combining this with Theorem 3.1, we have the following theorem:

Theorem 3.4 *Suppose that* $P(w)$ *is a uniqueness polynomial satisfying the hypothesis* (H) *with* $k \geq 3$*, or with* $k = 2$ *and* $\min\{q_1, q_2\} \geq 2$*. For a positive integer* m_0 *or* $m_0 = +\infty$*, assume that*

(a) $q > 2k + 12$ *for the case* $m_0 = 1$*,*

(b) $q > 2k + \dfrac{4}{m_0 - 1} + 6$ *for the case* $m_0 \geq 2$ *or*

(c) $q > 2k + 6$ *for the case* $m_0 = \infty$*.*

Particularly, in case where f *and* g *are holomorphic, assume that*

(d) $q > 2k + 5$ *for the case* $m_0 = 1$*,*

(e) $q > 2k + \dfrac{2}{m_0 - 1} + 2$ *for the case* $m_0 \geq 2$ *or*

(f) $q > 2k + 2$ *for the case* $m_0 = \infty$*.*

Then, if two nonconstant meromorphic functions f *and* g *satisfy the condition* $(\mathrm{C})_{m_0}$*, then* $f = g$*.*

In [2], Frank and Reinders showed the following:

Theorem 3.5 *Consider the polynomial*

$$P^{FR}(w) = \frac{(q-1)(q-2)}{2} w^q - q(q-2)w^{q-1} + \frac{q(q-1)}{2} w^{q-2} - c,$$

where $c \neq 0, 1$*. If* $q > 7$ *then* $P^{FR}(w)$ *is a uniqueness polynomial and, if* $q \geq 7$*, then it is a uniqueness polynomial for entire functions.*

For the polynomial $P^{FR}(w)$ as in Theorem 3.5,

$$(P^{FR})'(w) = \frac{q(q-1)(q-2)}{2} w^{q-3}(w-1)^2.$$

This shows that we can apply Theorem 3.4 to the polynomial $P^{FR}(w)$ for the case $k = 2$.

By these observations, we can conclude that there exists a uniqueness range set S for meromorphic functions with $\#S \geq 11$([2]), a uniqueness range set S for meromorphic functions ignoring multiplicities with $\#S \geq 17$([1]), a uniqueness range set S for entire functions with $\#S \geq 7$, and a uniqueness range set S for entire functions ignoring multiplicities with $\#S \geq 10$, respectively.

4 THE PROOFS OF THEOREMS 2.4 AND 2.5

We state the outline of the proofs of Theorems 2.4 and 2.5. We use the standard terminology and notations $T(r, f)$, $m(r, f)$ and so on in Nevanlinna theory, and $S(r, f)$ means that $S(r, f) = o(T(r, f))$ holds for all r except a subset $E \subset [0, +\infty)$ with $\int_E dr < +\infty$.

The counting function of a divisor $\nu : \mathbb{C} \to \mathbb{Z}$ is defined by

$$N(r, \nu) := \int_0^r \left(\sum_{0<|z|\leq t} \nu(z) \right) \frac{dt}{t} + \nu(0) \log r.$$

Let ν be a divisor and E a discrete subset of $\mathbb{C}$ or some condition prescribing a set. Define

$$\nu|_E(z) := \begin{cases} \nu(z) & \text{if } z \in E \text{ or } z \text{ satisfies the condition } E \\ 0 & \text{otherwise,} \end{cases}$$

$$\nu^*_{f'}(z) := \begin{cases} \nu^0_{f'} & \text{if } f(z) \neq d_\ell \text{ for any } \ell \\ 0 & \text{otherwise,} \end{cases}$$

and $(\nu - 1)^+ := \nu - \bar{\nu}$. By assumption, there are a nonzero constant c and two distinct nonconstant meromorphic functions f and g such that $P(f) = cP(g)$, where $c \neq 1$ by Theorem 2.2. We set $\Lambda := \{(\ell, m); P(d_\ell) = cP(d_m)\}$ and $k_0 := \#\Lambda$. As is easily seen, the hypothesis (H) yields the following:

(4.1) *For (ℓ, m) and (ℓ', m') in Λ, if $\ell = \ell'$ or $m = m'$, then $(\ell, m) = (\ell', m')$.*

We choose labels $r(\ell), s(\ell) (\ell = 1, \cdots, k)$ of indices so that

$$\Lambda = \{(r(\ell), s(\ell)); 1 \leq \ell \leq k\}$$

and

$$\begin{aligned} \{1, 2, \cdots, k\} &= \{r(1), \cdots, r(k_0), r(k_0 + 1), \cdots, r(k)\} \\ &= \{s(1), \cdots, s(k_0), s(k_0 + 1), \cdots, s(k)\}. \end{aligned}$$

For our purpose, we need some lemmas.

Lemma 4.2 *If $g \not\equiv c_0 f + c_1$ for any $c_0 (\neq 0), c_1 \in \mathbb{C}$, then $k \leq 3 + k_0 - k^*$, where $k^* := \min(k_0, 2)$. If $k = 3 + k_0 - k^*$, then $N(r, \nu^*_{g'}|_{f=d_{r(\ell)}}) = S(r) (:= S(r, f) + S(r, g))$ for each ℓ with $q_{r(\ell)} \geq 2$ and, similarly, $N(r, \nu^*_{f'}|_{g=d_{s(\ell)}}) = S(r)$ for each ℓ with $q_{s(\ell)} \geq 2$.*

Lemma 4.3 *In the same situation as in Lemma 4.2, assume furthermore that $c \neq 1$ and $k = k_0 + 1 \geq 4$. Then,*

(1) $N(r, (\nu^\alpha_f - 1)^+) = S(r)$ for each α with $P(\alpha) = P(d_{r(\ell)})$ $(1 \leq \ell \leq k_0)$, and $N(r, (\nu^\beta_g - 1)^+) = S(r)$ for each β with $P(\beta) = P(d_{s(\ell)})$ $(1 \leq \ell \leq k_0)$,

(2) $q_{r(\ell)} = q_{s(\ell)}$ for $1 \leq \ell \leq k$.

These lemmas are proved by comparing the zeros of the both sides of the identity $P'(f)f' = cP'(g)g'$ and by applying the second main theorem to the values $d_1, \cdots, d_k, \infty$. We omit the details.

Using these lemmas, we can show that there are some $\varphi(\neq 0), \psi \in \mathcal{S}$ with $g = \varphi f + \psi$, where $\mathcal{S}$ denotes the field of all meromorphic functions φ with $T(r, \varphi) = S(r, f)$. ¿From the identity $P(w) = cP(\varphi w + \psi)$ in $\mathcal{S}[w]$, we obtain

$$(w - d_1)^{q_1} \cdots (w - d_k)^{q_k} = c\varphi^q \left(w - \frac{d_1 - \psi}{\varphi} \right)^{q_1} \cdots \left(w - \frac{d_k - \psi}{\varphi} \right)^{q_k}.$$

By unique factorization theorem, we have $d_{t_\ell} = (d_\ell - \psi)/\varphi$ $(1 \leq \ell \leq k)$ for a permutation $(t_1, t_2, \cdots, t_k)$ of $(1, 2, \cdots, k)$. This gives $P(d_{t_\ell}) = cP(d_\ell)$ $(1 \leq \ell \leq k)$, whence we get $k = k_0$. Thus, we have Theorem 2.4.

The proof of Theorem 2.5 is similar. To this end, we use the last part of Lemma 4.2. The details are omitted.

5 THE PROOF OF THEOREM 3.1

We state the outline of the proof of Theorem 3.1 for the only case where $m_0 = \infty$ and f, g are meromorphic, because the similar arugments are available for the other cases. For meromorphic functions f, g with $\nu_f^S = \nu_g^S$, we set $F := 1/P(f)$, $G := 1/P(g)$ and define $H(z) := F''(z)/F'(z) - G''(z)/G'(z)$. We can easily show that H has the following properties:

$$\begin{aligned}\nu_H^\infty(z) \leq \sum_{\ell=1}^{k}(\bar{\nu}_f^{d_\ell}(z) + \bar{\nu}_g^{d_\ell}(z)) + \bar{\nu}_f^\infty(z) + \bar{\nu}_g^\infty(z) \\ + \bar{\nu}|_{(f')^{-1}(0)\setminus f^{-1}(S)} + \bar{\nu}|_{(g')^{-1}(0)\setminus g^{-1}(S)},\end{aligned} \tag{5.1}$$

and

$$m(r, H) = S(r, f) + S(r, g). \tag{5.2}$$

If $H = 0$, then we easily have (3.2). Assume that $H \not\equiv 0$ in the following. For brevity, we set

$$N^S(r, \nu_{f'}^0) := N(r, \nu|_{(f')^{-1}(0)\cap f^{-1}(S)}),$$

$$N^{CS}(r, \nu_{f'}^0) := N(r, \nu|_{(f')^{-1}(0)\setminus f^{-1}(S)}).$$

Each $z_0 \in f^{-1}(S)$ is a zero of H for the case $\nu_f^{f(z_0)}(z_0) = \nu_g^{g(z_0)}(z_0) = 1$, and a zero of f' (and g') for the other cases. Therefore,

$$\begin{aligned}(q-1)T(r, f) &\leq N(r, \bar{\nu}_f^\infty) + \sum_{i=1}^{q} N(r, \bar{\nu}_f^{a_i}) - N^{CS}(r, \nu_{f'}^0) + S(r) \\ &\leq N(r, \bar{\nu}_f^\infty) + N(r, \bar{\nu}_H^0) + N^S(r, \nu_{f'}^0) - N^{CS}(r, \nu_{f'}^0) + S(r).\end{aligned}$$

On the other hand, by the help of the first main theorem, (5.1) and (5.2) yield

$$\begin{aligned}N(r, \bar{\nu}_H^0) &\leq T(r, H) + O(1) \leq N(r, \nu_H^\infty) + S(r) \\ &\leq \sum_{\ell=1}^{k}(N(r, \bar{\nu}_f^{d_\ell}) + N(r, \bar{\nu}_g^{d_\ell})) + N(r, \bar{\nu}_f^\infty) + N(r, \bar{\nu}_g^\infty) \\ &\quad + N^{CS}(r, \nu_{f'}^0) + N^{CS}(r, \nu_{g'}^0) + S(r).\end{aligned}$$

Therefore,

$$\begin{aligned}(q-1)T(r, f) \leq kT(r) + 2N(r, \bar{\nu}_f^\infty) + N(r, \bar{\nu}_g^\infty) + \\ + N^{CS}(r, \nu_{g'}^0) + N^S(r, \nu_{f'}^0) + S(r).\end{aligned}$$

To this inequality adding a similar inequality for g, we conclude

$$\begin{aligned}(q-1)T(r) &\leq 2kT(r) + 3(N(r,\nu_f^\infty) + N(r,\nu_g^\infty)) \\ &\quad + (N(r,\nu_{f'}^0) + N(r,\nu_{g'}^0)) + S(r), \\ &\leq (2k+5)\,T(r) + S(r).\end{aligned}$$

This gives Proposition 3.1.

References

[1] S. Bartels,*Meromorphic functions sharing a set with 17 elements ignoring multiplicities*, preprint.

[2] G. Frank and M. Reinders, *A unique range set for meromorphic functions with 11 elements*, Complex Var. Theory Appl.. **37**(1998), 185 – 193.

[3] H. Fujimoto, *Value distribution theory of the Gauss map of minimal surfaces*, Aspect of Math, Vol. E21, Vieweg, Wiesbaden, 1993.

[4] F. Gross and C.C. Yang, *On preimage and range sets of meromorphic functions*, Proc. Japan Acad. **58**(1982), 17 – 20.

[5] P. Li and C.C. Yang, *Some further results on the unique range sets of meromorphic functions*, Kodai Math. J., **18**(1995), 437 – 450.

[6] R. Nevanlinna, *Einige Eindeutigkeitssätze in der Theorie der meromorphen Funktionen*, Acta Math., **48**(1926), 367 – 391.

[7] B. Shiffman, *Uniqueness of entire and meromorphic functions sharing finite sets*, unpublished.

[8] C.C. Yang and X. Hua, *Unique polynomials of entire and meromorphic functions*, Matematicheskaya fizika, analiz, geometriya, **3**(1997), 391 – 398.

[9] H. Yi, The unique range sets of entire or meromorphic functions, Complex Var. Theory Appl., **28**(1995), 13 – 21.

35 DEFECT RELATIONS FOR MOVING TARGETS

Pei-Chu HU[1]
Department of Mathematics, Shandong University
Jinan 250100, Shandong, P. R. China

pchu@public.sdu.edu.cn

Chung-Chun YANG[2]

Department of Mathematics
The Hong Kong University of Science and Technology
Clear Water Bay, Kowloon, Hong Kong

mayang@uxmail.ust.hk

1991 Mathematics Subject Classification: Primary 32A22, 32H30

Abstract

In this paper, we extend the general defect relation of the associated curves of a non-degenerate holomorphic curve to moving targets. Our results also improve the corresponded defect relations of Stoll for the associated mappings.

---

[1] The work of first author was partially supported by N.N.S.F. of China.
[2] The work of second author was partially supported by a UGC grant of Hong Kong.

H.G.W. Begehr et al. (eds.), Proceedings of the Second ISAAC Congress, Vol. 1, 271–297.

1 Introduction

Let f be a non-constant meromorphic function on $\mathbb{C}$ and let $\mathcal{G}$ be a set of meromorphic functions on $\mathbb{C}$. Assume that each $g \in \mathcal{G}$ grows slower than f, i.e.,

$$\lim_{r\to\infty} \frac{T_g(r)}{T_f(r)} = 0. \tag{1}$$

If each $g \in \mathcal{G}$ is constant, Nevanlinna proved the famous defect relation as follows:

$$\sum_{g\in\mathcal{G}} \delta_f(g) \leq 2. \tag{2}$$

Further he [6] conjectured that (2) remains valid for the slow moving target $\mathcal{G}$. He proved the conjecture if $\#\mathcal{G} = 3$. If each $g \in \mathcal{G}$ is a polynomial of at most degree d, Dufresnoy [2] proved

$$\sum_{g\in\mathcal{G}} \delta_f(g) \leq 2 + d. \tag{3}$$

Chuang [1] proved

$$\sum_{g\in\mathcal{G}} \delta_f(g) \leq 1 + q(1 - \delta_f(\infty)), \tag{4}$$

where q is the dimension of the vector space generated by $\mathcal{G}$ over $\mathbb{C}$. In particular, this led to the conjecture if f is an entire function. If f has finite lower order, Yang [17] showed that there are at most countably many meromorphic target functions g growing slower than f with $\delta_f(g) > 0$. Shiffman [10],[11] proved the conjecture in several complex variables if $\mathrm{rank}(\{f\} \cup \mathcal{G}) = 1 + \mathrm{rank}(\mathcal{G})$. In one complex variable, the condition implies that all elements of $\mathcal{G}$ are constant. Frank and Weissenborn [3] proved the conjecture in the case where all $g \in \mathcal{G}$ are rational. Osgood [7], [8] claimed that he can prove the conjecture using number theoretical argument. Finally, Steinmetz [12] proved the conjecture using value distribution theory.

Next we consider a non-degenerate holomorphic curve $f : \mathbb{C} \longrightarrow \mathbb{P}(V)$, where V is a complex vector space of dimension $n+1$, and let $f_k : \mathbb{C} \longrightarrow G_k(V)$ $(0 \leq k \leq n-1)$ be its k-th associated curve. For a finite system $\mathcal{A} \subset G_p(V^*)$ $(0 \leq p \leq n-1)$ in general position, Hu [4] showed a defect relation:

$$\sum_{a\in\mathcal{A}} \delta_{f_k}(a) \leq \frac{n+1}{k+1} I_{pk}, \tag{5}$$

where

$$I_{pk} = \begin{cases} \binom{n}{p} & : \quad 0 \leq p \leq k \\ \frac{(n-p)(k+1)}{(n-k)(p+1)} \binom{n}{p} & : \quad k \leq p \leq n-1 \end{cases} \tag{6}$$

In particular, if $k = p = 0$, (5) is the Cartan-Ahlfors defect relation

$$\sum_{a \in \mathcal{A}} \delta_f(a) \leq n + 1. \tag{7}$$

Let $\mathcal{G}$ be a finite set of holomorphic curves $g : \mathbb{C} \longrightarrow \mathbb{P}(V^*)$. Take

$$0 \leq p \leq \min_{g \in \mathcal{G}} \ell_g,$$

where ℓ_g is the index of generality of g for $B = 1$. Assume that $\{g_p \mid g \in \mathcal{G}\}$ is in general position and assume that each g_p grows slower than f_k, i.e.,

$$\lim_{r \to \infty} \frac{T_{g_p}(r)}{T_{f_k}(r)} = 0. \tag{8}$$

Naturally the equivalent of the Nevanlinna conjecture for the associated curves is

$$\sum_{g \in \mathcal{G}} \delta_{f_k}(g_p) \leq \frac{n+1}{k+1} I_{pk}. \tag{9}$$

If $k = p = 0$, indeed (9) was proved by Mori[5] for the case $\#\mathcal{G} = n + 2$, and by Ru and Stoll [9]. If each pair (f_k, g_p) is obstruction term free for $g \in \mathcal{G}$ (see Definition 5.1), we can show that (9) is true. In fact, for $0 \leq l \leq \min\{p, k\}$ we will prove a general defect relation as follows (see Theorem 7.1):

$$\sum_{g \in \mathcal{G}} \delta[f_k \odot^l g_p] \leq \frac{n+1}{k+1} I^l_{pk}, \tag{10}$$

if, for each $g \in \mathcal{G}$, the pair (f_k, g_p) is free for $\odot^{l+1}$, and obstruction term free of order $l + 1$, where

$$I^l_{pk} = \sum_{i=k-l-1}^{k} \frac{k-i}{n-k} \binom{i}{l-k+i+1} \binom{n-i-1}{p+k-l-i}. \tag{11}$$

Obviously, (9) and (10) improve the defect relations of Stoll [14],[15].

2 Hermitian geometry

Let V be a complex vector space of dimension $n+1$ and let V^* be the dual vector space of V. The *inner product* between $\xi \in V$ and $\alpha \in V^*$ is defined by

$$\langle \xi, \alpha \rangle = \alpha(\xi).$$

Take integers $k, p \in \mathbb{Z}$ with

$$0 \leq k, p \leq n, \quad u = u_{kp} = \min\{k, p\}. \tag{12}$$

Indentify $V^{**} = V$ by $\langle \xi, \alpha \rangle = \langle \alpha, \xi \rangle$ and $(\bigwedge_{p+1} V)^* = \bigwedge_{p+1} V^*$ by

$$\langle \xi_0 \wedge ... \wedge \xi_p, \alpha_0 \wedge ... \wedge \alpha_p \rangle = \det(\langle \xi_i, \alpha_j \rangle).$$

Take $\xi \in \bigwedge_{k+1} V$ and $\alpha \in \bigwedge_{p+1} V^*$. If $p \leq k$, the *interior product* $\xi \angle \alpha \in \bigwedge_{k-p} V$ is uniquely defined by

$$\langle \xi \angle \alpha, \beta \rangle = \langle \xi, \alpha \wedge \beta \rangle \text{ for all } \beta \in \bigwedge_{k-p} V^*.$$

If $p = k$, then $\xi \angle \alpha = \langle \xi, \alpha \rangle \in \mathbb{C} = \bigwedge_0 V$ by definition. On the other hand, if $p > k$, we define $\xi \angle \alpha \in \bigwedge_{p-k} V^*$ such that if $\eta \in \bigwedge_{p-k} V$,

$$\langle \eta, \xi \angle \alpha \rangle = \langle \xi \wedge \eta, \alpha \rangle.$$

For $\xi \in V - \{0\}$, let $\mathbb{P}(\xi) = \mathbb{C}\xi$ be the complex line spanned by ξ. For $E \subseteq V$ define

$$\mathbb{P}(E) = \{\mathbb{P}(\xi) \mid 0 \neq \xi \in E\}.$$

Then $\mathbb{P}(V)$ is a compact connected complex manifold of dimension n called the complex *projective space* of V. Observe that

$$\mathbb{P} : V - \{0\} \longrightarrow \mathbb{P}(V),$$

is a surjective holomorphic mapping. The *Grassmann cone* of order k is the analytic subset

$$\tilde{G}_k(V) = \{\xi_0 \wedge ... \wedge \xi_k \mid \xi_j \in V \text{ for } j = 0, ..., k\}$$

of $\bigwedge_{k+1} V$. The *Grassmann manifold* $G_k(V) = \mathbb{P}(\tilde{G}_k(V))$ of order k is a connected compact smoothly imbedded complex submanifold of dimension

$$\dim G_k(V) = (n-k)(k+1)$$

of $\mathbb{P}(\bigwedge_{k+1} V)$. Observe $G_0(V) = \mathbb{P}(V)$ and that $G_n(V)$ consists of one and only one point denoted by ∞.

Take $x \in G_k(V)$. Then $x = \mathbb{P}(\xi)$ with $\xi = \xi_0 \wedge ... \wedge \xi_k \in \tilde{G}_k(V)$. Then

$$E(x) = \{\eta \in V | \eta \wedge \xi = 0\} = \mathbb{C}\xi_0 + ... + \mathbb{C}\xi_k,$$

is a complex linear subspace of dimension $k+1$ of V with base $\xi_0, ..., \xi_k$, and $\ddot{E}(x) = \mathbb{P}(E(x))$ is a k-dimensional projective plane in $\mathbb{P}(V)$. The mapping $x \longmapsto \ddot{E}(x)$ provides a bijective parametrization of the set of k-dimensional projective planes in $\mathbb{P}(V)$.

Take $\xi = \xi_0 \wedge ... \wedge \xi_k \in \bigwedge_{k+1} V$ and abbreviate

$$\hat{\xi}_i = (-1)^i \xi_0 \wedge ... \wedge \xi_{i-1} \wedge \xi_{i+1} \wedge ... \wedge \xi_k, \quad i = 0, ..., k.$$

If $k = 0$, this shall mean $\hat{\xi}_i = 1$. For $\alpha = \alpha_0 \wedge ... \wedge \alpha_p \in \bigwedge_{p+1} V^*$, define

$$\xi \odot \alpha = \frac{1}{u} \sum_{i=0}^{k} \sum_{j=0}^{p} \langle \xi_i, \alpha_j \rangle \hat{\xi}_i \otimes \hat{\alpha}_j, \tag{13}$$

which is well defined and extends to a linear mapping

$$\odot : \bigwedge_{k+1} V \otimes \bigwedge_{p+1} V^* \longrightarrow \bigwedge_{k} V \otimes \bigwedge_{p} V^*, \tag{14}$$

called contraction (see [14]). If $l \in \mathbb{Z}$ with $0 \leq l \leq u$, the operator can be iterated $\odot^{l+1} = \odot \circ \odot^l$. Here we let $\odot^0$ be the identity. Then

$$\odot^r = \odot^s \circ \odot^{r-s} \ (0 \leq s \leq r), \quad \odot^{u+1} = \angle.$$

The pair (ξ, α) is said to be free for $\odot^r$ if $\xi \odot^r \alpha \neq 0$.

Take a Hermitian metric (,) on V which induces on derived vector spaces such as tensor products, exterior products, and dual vector spaces, etc. If $\xi \in V$, the norm is $\|\xi\| = (\xi, \xi)^{1/2}$. Take $x = \mathbb{P}(\xi) \in \mathbb{P}(\bigwedge_{k+1} V)$ and $a = \mathbb{P}(\alpha) \in \mathbb{P}(\bigwedge_{p+1} V^*)$. For $0 \leq l \leq u$, define the distance

$$|x, a|_l = \frac{\|\xi \odot^{l+1} \alpha\|}{\|\xi\| \cdot \|\alpha\|}, \tag{15}$$

with $|x, a|_{-1} = 1$. Then

$$\|x, a\| = |x, a|_u = \frac{\|\xi \angle \alpha\|}{\|\xi\| \cdot \|\alpha\|}, \tag{16}$$

is the projective distance from x to a. If $x \in G_k(V), a \in G_p(V^*)$, then

$$|x, a|_l \leq \binom{u+1}{l+1}^{-\frac{1}{2}}. \tag{17}$$

Lemma 2.1 *Take $a \in G_p(V^*)$ and take $x \in G_k(V), y \in G_{k-1}(V)$ with $E(y) \subset E(x)$. Then*

$$|\infty, a|_l^2 = \binom{p+1}{l+1}^{-1}, \tag{18}$$

$$\frac{|x, a|_{l-1}}{|y, a|_{l-1}} \geq \frac{1}{k+1}. \tag{19}$$

Proof. Let $\xi_0, ..., \xi_n$ be an orthonormal base of V and let $\xi = \xi_0 \wedge \cdots \wedge \xi_n$. Then $\infty = \mathbb{P}(\xi) \in G_n(V)$. Take $\alpha_j \in V^*$ such that $\alpha = \alpha_0 \wedge ... \wedge \alpha_p$ with $a = \mathbb{P}(\alpha)$ and such that $\alpha_0, ..., \alpha_p$ is an orthonormal system. By Lemma 1.5 of Stoll [14], we have

$$\begin{aligned} |\infty, a|_l^2 &= \|\xi \odot^{l+1} \alpha\|^2 \\ &= \binom{p+1}{l+1}^{-2} \sum_{\mu \in J_l^p} \sum_{\nu \in J_l^n} |\langle \xi_\nu, \alpha_\mu \rangle|^2 \\ &= \binom{p+1}{l+1}^{-2} \sum_{\mu \in J_l^p} \|\alpha_\mu\|^2 = \binom{p+1}{l+1}^{-1}, \end{aligned}$$

where J_s^r is the set of all increasing injective mappings $[0, s] \cap \mathbb{Z} \longrightarrow [0, r] \cap \mathbb{Z}$, and $\xi_\nu = \xi_{\nu(0)} \wedge \cdots \wedge \xi_{\nu(l)}$ for $\nu \in J_l^n$.

Let $x = \mathbb{P}(\xi')$ and $y = \mathbb{P}(\eta)$ with $\xi' = \xi_0 \wedge \cdots \wedge \xi_k$ and $\eta = \xi_0 \wedge \cdots \wedge \xi_{k-1}$. By Lemma 1.5 of Stoll [14], we obtain

$$\begin{aligned} & \binom{u+1}{l} |x, a|_{l-1} = \binom{u+1}{l} \|\xi' \odot^l \alpha\| \\ =\ & \Big(\sum_{\nu \in J_{l-1}^k} \|\alpha \angle \xi_\nu\|^2\Big)^{\frac{1}{2}} \geq \Big(\sum_{\nu \in J_{l-1}^{k-1}} \|\alpha \angle \xi_\nu\|^2\Big)^{\frac{1}{2}} \\ =\ & \binom{u_{k-1,p}+1}{l} \|\eta \odot^l \alpha\| = \binom{u_{k-1,p}+1}{l} |y, a|_{l-1}, \end{aligned}$$

which implies (19). ♠

3 An algebraic identity

Take integers l, k, p, n with

$$0 \leq k \leq n-1, \quad 0 \leq p \leq n, \quad 0 \leq l \leq u = \min\{k, p\}.$$

According to Wu [16] and Stoll [14], define

$$D_p(k,l) = \binom{n+1}{p+1} - \sum_{j \geq 0} \binom{k+1}{l+j+1} \binom{n-k}{p-l-j}, \tag{20}$$

where the binomial coefficients are defined for all integers by the expansion

$$(1+x)^i = \sum_{j=-\infty}^{+\infty} \binom{i}{j} x^j.$$

Stoll [14] shown

$$D_p(k,l) = \sum_{j=0}^{l} \binom{k+1}{j} \binom{n-k}{p+1-j}. \tag{21}$$

Note that

$$\begin{aligned} D_p(n+1, l+1) &= \sum_{j=0}^{l+1} \binom{n+2}{j} \binom{-1}{p+1-j} \\ &= (-1)^{p+1} \sum_{j=0}^{l+1} (-1)^j \binom{n+2}{j} \\ &= (-1)^{p+1} \sum_{j=0}^{l+1} (-1)^j \{ \binom{n+1}{j} + \binom{n+1}{j-1} \} \\ &= (-1)^{p-l} \binom{n+1}{l+1} = \binom{n+1}{l+1} \binom{-1}{p-l}, \end{aligned} \tag{22}$$

and hence

$$D_p(n+1, l+1) - \binom{n}{l+1} \binom{-1}{p-l} = \binom{n}{l} \binom{-1}{p-l}. \tag{23}$$

We have proved the following:

Lemma 3.1 (Hu [4]) *If y_j are indeterminates over the ring $\mathbb{Z}$ with $y_j = 0$ for $j > n$, we have the algebraic identity*

$$\begin{aligned} D_p(k,l;y) &\equiv -\sum_{j=k}^{n-1}\sum_{i=0}^{l} D_p(j-i,l-i)(y_{j-i-1}-2y_{j-i}+y_{j-i+1}) \\ &= D_p(k+1,l+1)y_k - \binom{n}{l}\binom{-1}{p-l}y_n \\ &- \sum_{i=k-l-1}^{k} a_i y_i - \sum_{i=k+1}^{n} b_i y_i, \end{aligned} \quad (24)$$

where

$$a_i = \binom{i}{l-k+i+1}\binom{n-i-1}{p+k-l-i}, \quad b_i = \binom{i}{l}\binom{n-i-1}{p-l-1}. \quad (25)$$

Lemma 3.2 *Define*

$$I_{pk}^l = \sum_{i=k-l-1}^{k} \frac{k-i}{n-k}\binom{i}{l-k+i+1}\binom{n-i-1}{p+k-l-i}. \quad (26)$$

Then

$$\binom{n}{l}\binom{-1}{p-l} + \sum_{i=k+1}^{n} \frac{i-k}{n-k} b_i = I_{pk}^l, \quad (27)$$

$$D_p(k+1,l+1) - \sum_{i=k-l-1}^{k} \frac{i+1}{k+1} a_i - \sum_{i=k+1}^{n} \frac{n-i}{n-k} b_i = \frac{n+1}{k+1} I_{pk}^l, \quad (28)$$

$$\begin{aligned} &\sum_{j=k}^{n-1}\sum_{i=0}^{l} D_p(j-i,l-i) + \frac{1}{2}\sum_{i=k-l-1}^{k}(k-i)(i+1)a_i \\ + &\frac{1}{2}\sum_{i=k+1}^{n}(i-k)(n-i)b_i = \frac{1}{2}(n+1)(n-k)I_{pk}^l. \end{aligned} \quad (29)$$

Proof. In (24), we take

$$y_j = \begin{cases} j-k & : \quad j \le n \\ 0 & : \quad j > n \end{cases} \quad (30)$$

then $D_p(k,l;y) = 0$, so (27) follows.

In (24) taking

$$y_j = \begin{cases} j+1 & : \quad j \le n \\ 0 & : \quad j > n \end{cases} \tag{31}$$

then $D_p(k,l;y) = 0$, and

$$\begin{aligned} D_p(k+1,l+1) &- \sum_{i=k-l-1}^{k} \frac{i+1}{k+1} a_i \\ &= \frac{n+1}{k+1}\binom{n}{l}\binom{-1}{p-l} + \sum_{i=k+1}^{n} \frac{i+1}{k+1} b_i. \end{aligned} \tag{32}$$

Hence (28) follows from (27) and (32).

In (24) taking

$$y_j = \begin{cases} j(j+1) & : \quad j \le n \\ 0 & : \quad j > n \end{cases} \tag{33}$$

then

$$\begin{aligned} \sum_{j=k}^{n-1} \sum_{i=0}^{l} D_p(j-i, l-i) &= -\frac{1}{2} D_p(k,l;y) \\ = \ & \frac{1}{2}(-k(k+1) D_p(k+1,l+1) + n(n+1)\binom{n}{l}\binom{-1}{p-l} \\ + \ & \sum_{i=k-l-1}^{k} i(i+1)a_i + \sum_{i=k+1}^{n} i(i+1) b_i). \end{aligned} \tag{34}$$

Now (29) follows from (27), (32) and (34). ♠

Lemma 3.3 *If $l = u$, then*

$$I^u_{pk} \equiv I_{pk} = \begin{cases} \binom{n}{p} & : \quad 0 \le p \le k \\ \frac{(n-p)(k+1)}{(n-k)(p+1)}\binom{n}{p} & : \quad k \le p \le n-1 \end{cases} \tag{35}$$

and

$$\sum_{j=k}^{n-1} \sum_{i=0}^{k} D_p(j-i, k-i) = \frac{1}{2}(k+1)(k+2)\binom{n+1}{p+2}. \tag{36}$$

Proof. If $l = p$, (27) implies $I_{pk} = \binom{n}{p}$. If $l = k$, (26) implies

$$I_{pk} = \frac{k+1}{n-k}\binom{n}{p+1} = \frac{(n-p)(k+1)}{(n-k)(p+1)}\binom{n}{p}.$$

In (24), we let $l = k$ and take

$$y_j = \begin{cases} j & : \quad j \leq n \\ 0 & : \quad j > n \end{cases} \tag{37}$$

then we obtain

$$b_{kp} \equiv \sum_{i=k+1}^{n} ib_i = D_p(k+1, k+1)k - \binom{n}{k}\binom{-1}{p-k}n + \binom{n}{p+1}. \tag{38}$$

Since $a_i = 0$ for $i \geq 1$, and

$$\begin{aligned} \sum_{i=k+1}^{n} i(i+1)b_i &= (k+1)\sum_{i=k+1}^{n}\binom{i+1}{k+1}\binom{n-i-1}{p-k-1}i \\ &= (k+1)(b_{kp} + \binom{n-k-2}{p-k-1}(k+1) + b_{k+1,p+1}), \end{aligned} \tag{39}$$

then (36) follows from (34), (38), (39) and the following equality

$$D_p(k+1, k+1) = \binom{n+1}{p+1} - \binom{n-k-1}{p-k-1}. \tag{40}$$

See [14], p.228. ♠

Remark. If $l = p$, the double sum of (34) have been obtained in Hu [4].

4 Growth of associated mappings

Let M be a connected complex manifold of dimension m and let τ be a parabolic exhaustion function on M, i.e., τ is an exhaustion function on M with

$$\hat{\upsilon} \geq 0, \quad \hat{\upsilon}^m \equiv 0, \quad \upsilon^m \not\equiv 0 \tag{41}$$

on $M - M[0]$. Let C_τ be the set of critical points of τ and denote

$$\mathcal{E}_\tau = \{r \in \mathbb{R}_+ \mid C_\tau \cap \tau^{-1}(r^2) = \emptyset\}.$$

Then

$$\tau\upsilon^m = md\tau \wedge d^c\tau \wedge \upsilon^{m-1} = m\tau^m d\tau \wedge \sigma_{m-1}, \quad d\sigma_{m-1} = \hat{\upsilon}^m = 0. \tag{42}$$

Thus Stokes Theorem implies that

$$\int_{M\langle r\rangle} \sigma_{m-1} = \int_{M\langle s\rangle} \sigma_{m-1}$$

for $r > s > 0$ with $r, s \in \mathcal{E}_\tau$. Denote this common number by ς. Obviously,

$$\varsigma = r^{-2m} \int_{M[r]} \upsilon^m > 0 \quad (r > 0). \tag{43}$$

Let $\Theta > 0$ be any form of degree $2m$ and class C^∞ on M. Then Θ is the same as a metric in the canonical line bundle K_M. Let Ric $\Theta \in c_1(K_M)$ be the curvature form of the metric in K_M. Define a function $\theta \geq 0$ of class C^∞ on M by $\upsilon^m = \theta\, \Theta$. Then the set

$$\mathcal{E}_\tau^0 = \{r \in \mathcal{E}_\tau \mid M\langle r; \log\theta\rangle \text{ exists }\},$$

does not depend on the choice of Θ, and $\mathbb{R}_+ - \mathcal{E}_\tau^0$ has measure zero. Also the Ricci function of τ is defined by

$$\mathrm{Ric}_\tau(r) = M\langle r; \log\theta\rangle - M\langle r_0; \log\theta\rangle + M(r, r_0; \mathrm{Ric}\ \Theta),$$

for fixed $r_0 \in \mathcal{E}_\tau^0$ and for $r_0 < r \in \mathcal{E}_\tau^0$, and does not depend on the choice of Θ (see [13]).

A holomorphic $(m-1,0)$-form B on M is said to be majorized by τ if, for every $r > 0$ there exists a minimal constant $Y(r) \geq 1$ such that

$$m i_{m-1} B \wedge \overline{B} \leq Y(r) \upsilon^{m-1}, \tag{44}$$

on $M[r]$, where

$$i_p = \left(\frac{\sqrt{-1}}{2\pi}\right)^p (-1)^{\frac{p(p-1)}{2}} p!\ (p \in \mathbb{Z}_+).$$

The function Y is called a majorant.

Let V be a Hermitian vector space of dimension $n+1 > 1$. We will consider a meromorphic mapping

$$f : M \longrightarrow \mathbb{P}(V).$$

Denote the characteristic function of f relative to the Fubini-Study Kähler form Ω on $\mathbb{P}(V)$ by

$$T_f(r) = M(r, r_0; f^*\Omega).$$

Fix a holomorphic $(m-1,0)$-form B on M and take a neighborhood U in M with holomorphic coodinates $z=(z_1,...,z_m)$ such that there exists a representation $\tilde{f}:U\longrightarrow V$ with $\dim \tilde{f}^{-1}(0)\leq m-2$, called a reduced representation of f on U. The B-derivative $\tilde{f}':U\longrightarrow V$ is defined by

$$d\tilde{f}\wedge B=\tilde{f}'dz_1\wedge\cdots\wedge dz_m,$$

which can be iterated $\tilde{f}^{(k)}=(\tilde{f}^{(k-1)})'$. Thus we obtain

$$\tilde{f}_{k,z}=\tilde{f}\wedge\tilde{f}'\wedge\cdots\wedge\tilde{f}^{(k)}:U\longrightarrow\tilde{G}_k(V),\quad k=1,...,n.$$

According to Stoll [14], there exist a holomorphic line bundle L_f on M called the hyperplane section bundle of f over M, and a holomorphic section

$$F_k:M\longrightarrow\{\bigwedge_{k+1}(M\times V)\}\otimes L_f^{k+1}\otimes K_M^{\frac{k(k+1)}{2}},$$

called the k-th associated section of f such that the divisor $D_{F_k}\equiv(F_k)$ is given by $D_{F_k}\cap U=(\tilde{f}_{k,z})$. The mapping f is said to be general of order k for B if $\tilde{f}_{k,z}\not\equiv 0$ for some pair $(z,\tilde{f})$, which is equivalent to $F_k\not\equiv 0$. One and only one maximal integer $\ell_f(B)$, called the index of generality of f for B, exists such that f is general of order k for B if $0\leq k\leq\ell_f(B)$. If $\ell_f(B)=n$, f is said to be general for B. If $0\leq k<\ell_f(B)$, the k-th stationary divisor

$$D_{F_{k-1}}-2D_{F_k}+D_{F_{k+1}},$$

is effective, where $D_{F_{-1}}=0$ (see [14]). If f is general of order k for B, the k-th associated mapping

$$f_k:M\longrightarrow G_k(V),$$

is a meromorphic mapping defined by $f_k|_U=\mathbb{P}\circ\tilde{f}_{k,z}$. Observe that $f_0=f$ is the given mapping and f_k is constant if $k=n$.

Let Ω_k be the Fubini-Study form on $\mathbb{P}(\bigwedge_{k+1}V)$. Then

$$\mathbb{H}_k\equiv mi_{m-1}f_k^*(\Omega_k)\wedge B\wedge\overline{B}\geq 0, \tag{45}$$

for $0\leq k\leq\ell$ with $\mathbb{H}_\ell=0$, where $\ell=\ell_f(B)$. Define a non-negative function h_k on $M^+=\{x\in M\mid v>0\}$ by

$$\mathbb{H}_k=h_k^2v^m,$$

and put $h_k=0$ on $M-M^+$. Fix $r_0\in\mathcal{E}_\tau^0$. Assume that $0\leq k<\ell$ and define

$$N_k(r)=D_{F_k}(r,r_0;1)_{m-1},\quad S_k(r)=N_{k-1}(r)-2N_k(r)+N_{k+1}(r), \tag{46}$$

where $N_{-1}(r) = N_0(r) = 0$. The Plücker Difference Formula (see [14])

$$\begin{aligned} & S_k(r) + T_{f_{k-1}}(r) - 2T_{f_k}(r) + T_{f_{k+1}}(r) \\ = \; & M\langle r; \log h_k^2\rangle - M\langle r_0; \log h_k^2\rangle + \mathrm{Ric}_\tau(r), \end{aligned} \tag{47}$$

holds for $r \in \mathcal{E}_\tau^0$, where

$$T_{f_{-1}} = T_{f_\ell} = 0. \tag{48}$$

Abbreviate

$$T_k(r) = T_{f_k}(r) + N_k(r). \tag{49}$$

Lemma 4.1 *Assume that τ majorizes a holomorphic $(m-1,0)$-form B on M with the majorant Y and assume that f is general for B. Then*

$$N_n(r) - \frac{n+1}{k+1}N_k(r) = \hat{S}_k(r) \geq 0, \tag{50}$$

and

$$T_{k-1}(r) - 2T_k(r) + T_{k+1}(r) \leq Q_1(r) \quad \|, \tag{51}$$

for $\varepsilon > 0$, $k = 0, 1, ..., n-1$, where $\|$ means the inequality holds for $r \in [r_0, \infty) - E$, where E is a Borel subset of finite measure, and where

$$\hat{S}_k(r) = \frac{n-k}{k+1}\sum_{i=0}^{k-1}(i+1)S_i(r) + \sum_{i=k}^{n-1}(n-i)S_i(r), \tag{52}$$

$$\begin{aligned} Q_1(r) &= \mathrm{Ric}_\tau(r) + \frac{\varsigma}{2}(1+\varepsilon)^3\{\log T_f(r) + \log Y(r) \\ &+ \log^+ \mathrm{Ric}_\tau(r)\} + \varsigma\varepsilon \log r. \end{aligned} \tag{53}$$

Proof. Since $S_k(r) \geq 0$, upon (46) summing over k from 0 to $j-1$, we have

$$N_j(r) - N_{j-1}(r) = \sum_{i=0}^{j-1} S_i(r). \tag{54}$$

Upon summing over j from 1 to k, we obtain

$$N_k(r) = \sum_{j=0}^{k-1}(k-j)S_j(r), \tag{55}$$

which implies (50). The inequality (51) follows from Proposition 6.14 and 6.15 of Stoll [14]. ♠

Corollary 4.1 *If $i \leq k \leq n-1$, then*

$$\frac{i+1}{k+1}T_k(r) \leq T_i(r) + \frac{1}{2}(k-i)(i+1)Q_1(r) \quad \|. \tag{56}$$

If $n-1 \geq i \geq k$, then

$$\frac{n-i}{n-k}T_k(r) \leq T_i(r) - \frac{i-k}{n-k}T_n(r) + \frac{1}{2}(i-k)(n-i)Q_1(r) \quad \|. \tag{57}$$

Proof. Multiply (51) by $k+1$ and add these over k from 0 to k to obtain

$$(k+1)T_{k+1}(r) \leq (k+2)T_k(r) + \frac{1}{2}(k+1)(k+2)Q_1(r) \quad \|. \tag{58}$$

Multiply (51) by $n-k$ and add these over k from k to $n-1$ to obtain

$$(n-k)T_{k-1}(r) \leq (n-k+1)T_k(r) - T_n(r) + \frac{1}{2}(n-k)(n-k+1)Q_1(r) \quad \|, \tag{59}$$

by a straightforward induction from (58) and (59). ♠

Remark. If substituting $T_i(r)$ by $T_{f_i}(r)$, this corollary also holds. Also we can prove the following inequality

$$S_k(r) \leq \frac{n+1}{(k+1)(n-k)}T_{f_k}(r) + \frac{n+1}{2}Q_1(r) \quad \|, \tag{60}$$

for $k = 0, ..., n-1$. Hence if

$$R_{f_k} = \limsup_{r\to\infty} \frac{\mathrm{Ric}_\tau(r)}{T_{f_k}(r)} < \infty, \tag{61}$$

$$Y_{f_k} = \limsup_{r\to\infty} \frac{Y(r)}{T_{f_k}(r)} < \infty, \tag{62}$$

there exists a positive constant c such that

$$T_{f_k}(r) \leq T(r) \leq cT_{f_k}(r) \quad \|, \tag{63}$$

where

$$T(r) = \max\{T_{f_0}(r), T_{f_1}(r), ..., T_{f_{n-1}}(r)\}.$$

5 The Stoll-Ahlfors estimates

Assume that the meromorphic mapping $f : M \longrightarrow \mathbb{P}(V)$ is general for B and let $g : M \longrightarrow \mathbb{P}(V^*)$ be a meromorphic mapping. Assume that

$$0 \leq k \leq n-1, \quad 0 \leq p \leq \ell_g(B), \quad 0 \leq l \leq u = \min\{k, p\}.$$

Then the associated mappings $g_p : M \longrightarrow G_p(V^*)$ are defined. Define

$$\Phi_{pk}^l = \binom{u+1}{l+1}^2 |f_k, g_p|_l^2. \tag{64}$$

Here (f_k, g_p) is free for $\odot^{l+1}$ if and only if $\Phi_{pk}^l \not\equiv 0$.

Let Ω_p^* be the associated Fubini Study form on $G_p(V^*)$. Denote

$$\mathbb{K}_p = mi_{m-1} g_p^*(\Omega_p^*) \wedge B \wedge \overline{B} \geq 0, \tag{65}$$

where $\mathbb{K}_p \equiv 0$ if $p = \ell_g(B)$, and $\mathbb{K}_p > 0$ outside a thin analytic subset of M if $0 \leq p < \ell_g(B)$. Define a non-negative function $\hat{h}_p$ on M^+ by $\mathbb{K}_p = \hat{h}_p^2 \upsilon^m$, and put $\hat{h}_p = 0$ on $M - M^+$. Write

$$\sqrt{\mathbb{K}_p \mathbb{H}_k} = h_k \hat{h}_p \upsilon^m. \tag{66}$$

Take reduced representation $\tilde{f} : U \longrightarrow V$ of f and $\tilde{g} : U \longrightarrow V^*$ of g on an open subset U of M with coodinates $z = (z_1, ..., z_m)$. Assume that (f_k, g_p) is free for $\odot^{l+1}$. Then a mapping

$$f_k \odot^l g_p : M \longrightarrow \mathbb{P}((\bigwedge_{k-l} V) \otimes (\bigwedge_{p-l} V^*)),$$

is defined by

$$f_k \odot^l g_p|_U = \mathbb{P} \circ (\tilde{f}_{k,z} \odot^{l+1} \tilde{g}_{p,z}),$$

and a divisor $D_{f_k \odot^l g_p}$ is defined by

$$D_{f_k \odot^l g_p} \cap U = (\tilde{f}_{k,z} \odot^{l+1} \tilde{g}_{p,z}).$$

Stoll [14] obtained the First Main Theorem

$$T_{f_k}(r) + T_{g_p}(r) = T_{f_k \odot^l g_p}(r) + N_{f_k \odot^l g_p}(r) + m_{f_k \odot^l g_p}(r) - m_{f_k \odot^l g_p}(r_0), \tag{67}$$

where

$$\begin{aligned} N_{f_k \odot^l g_p}(r) &= D_{f_k \odot^l g_p}(r, r_0; 1)_{m-1}, & (68)\\ m_{f_k \odot^l g_p}(r) &= M\langle r; \log |f_k, g_p|_l^{-2}\rangle. & (69) \end{aligned}$$

Take reduced representation $\tilde{f} : U \longrightarrow V$ of f and $\tilde{g} : U \longrightarrow V^*$ of g on an open subset U of M with coodinates $z = (z_1, ..., z_m)$. On $U - D_{F_n}$, $\tilde{f}, \tilde{f}', ..., \tilde{f}^{(n)}$ is a base of V which we orthonormalize to $\xi_0, \xi_1, ..., \xi_n$, called the Frenet frame of f on U. Let $\xi_0^*, \xi_1^*, ..., \xi_n^*$ be the dual frame in V^*. At every point of U outside a thin analytic set, $\tilde{g}, \tilde{g}', ..., \tilde{g}^{(\ell)}$ ($\ell = \ell_g(B)$) are linearly independent in V^* and are orthonormalized to $\alpha_0, \alpha_1, ..., \alpha_\ell$. Define

$$\begin{aligned} \Lambda_{pk} &= (\alpha_p, \xi_k^*) \quad 0 \le p \le \ell;\ 0 \le k \le n, & (70)\\ S_{pk}^0 &= \frac{1}{2}(\Lambda_{p+1,k}\overline{\Lambda_{p,k+1}} + \Lambda_{p,k+1}\overline{\Lambda_{p+1,k}}). & (71) \end{aligned}$$

If $\mu \in J_l^p$ and $\nu \in J_l^n$, define the stress coefficient

$$\Lambda_{\mu\nu} = \det(\Lambda_{\mu(i)\nu(j)}) \quad (0 \le i, j \le l). \tag{72}$$

Then $|\Lambda_{\mu\nu}| \le 1$ and

$$\Phi_{pk}^l|_U = \sum_{\mu \in J_l^p} \sum_{\nu \in J_l^k} |\Lambda_{\mu\nu}|^2. \tag{73}$$

If $\mu \in J_s^r$ and if $t \in \mathbb{Z}$ with $t > r$, then $\nu = (\mu, t)$ is the unique mapping $\nu \in J_{s+1}^t$ where $\nu|_{[0,s]\cap\mathbb{Z}} = \mu$ and $\nu(s+1) = t$. For $l > 0$, define

$$S_{pk}^l = \sum_{\mu \in J_l^{p-1}} \sum_{\nu \in J_l^{k-1}} \frac{1}{2}(\Lambda_{(\mu,p+1)(\nu,k)}\overline{\Lambda_{(\mu,p)(\nu,k+1)}} + \Lambda_{(\mu,p)(\nu,k+1)}\overline{\Lambda_{(\mu,p+1)(\nu,k)}}). \tag{74}$$

For $0 \le l \le u$, define

$$\Xi_{pk}^l = \Phi_{p+1,k}^l \mathbb{K}_p + \Phi_{p,k+1}^l \mathbb{H}_k + 2S_{pk}^l \sqrt{\mathbb{K}_p \mathbb{H}_k}. \tag{75}$$

Stoll [15] shown the Stress Curvature Formula:

$$\begin{aligned} m i_{m-1} dd^c \Phi_{pk}^l \wedge B \wedge \overline{B} &= (\Phi_{p+1,k}^l - 2\Phi_{pk}^l + \Phi_{p-1,k}^l)\mathbb{K}_p \\ &+ (\Phi_{p,k+1}^l - 2\Phi_{pk}^l + \Phi_{p,k-1}^l)\mathbb{H}_k + 2S_{pk}^l \sqrt{\mathbb{K}_p \mathbb{H}_k}. \quad (76) \end{aligned}$$

Basing on this formula, if (f_k, g_p) is free for $\odot^{l+1}$, he ([14],[15]) given the Ahlfors Estimates:

$$\begin{aligned} &(1-\alpha)^2 \int_{r_0}^{r} \int_{M[t]} |f_k, g_p|_l^{-2\alpha} \Xi_{pk}^l \frac{dt}{t^{2m-1}} \\ \leq\ & 4(n+1)^{2l+3} Y(r)(T_{f_k}(r) + T_{g_p}(r) + \varsigma), \end{aligned} \tag{77}$$

with $0 < \alpha < 1$. Also define

$$\rho_{pk}^l = \frac{\Phi_{p+1,k}^l \hat{h}_p^2 + \Phi_{p,k+1}^l h_k^2 + 2S_{pk}^l h_k \hat{h}_p}{\Phi_{p,k+1}^l h_k^2}. \tag{78}$$

Then

$$\Xi_{pk}^l = \rho_{pk}^l \Phi_{p,k+1}^l \mathbb{H}_k.$$

Define the obstruction terms

$$P_{pk}^l(r, g) = M\langle r; \log^+ \frac{1}{\rho_{pk}^l} \rangle. \tag{79}$$

Definition 5.1 *The pair (f_k, g_p) is said to be obstruction term free of order $l+1$ if*

$$\lim_{r\to\infty} \frac{P_{p,j-i}^{l-i}(r, g)}{T_{f_k}(r)} = 0, \quad 0 \leq i \leq l; k \leq j \leq n-1. \tag{80}$$

The pair (f_k, g_p) is said to be obstruction term free if it is obstruction term free of order $u+1$.

Take a subset $A = \{x_0, x_1, ..., x_s\} \subset \mathbb{P}(V)$ $(0 \leq s < \infty)$ and take $\xi_i \in V - \{0\}$ with $x_i = \mathbb{P}(\xi_i)$. Define the gauge of A by

$$\Gamma(A) = \begin{cases} \frac{\|\xi_0 \wedge \cdots \wedge \xi_s\|}{\|\xi_0\| \cdots \|\xi_s\|} & : \quad s \leq n \\ \inf_{\mu \in J_n^s} \frac{\|\xi_{\mu(0)} \wedge \cdots \wedge \xi_{\mu(n)}\|}{\|\xi_{\mu(0)}\| \cdots \|\xi_{\mu(n)}\|} & : \quad s > n \end{cases} \tag{81}$$

with $0 \leq \Gamma(A) \leq 1$. Then A is in general position if and only if $\Gamma(A) > 0$.

Take $\emptyset \neq \mathcal{G} \subset G_p(V^*)$ $(0 \leq p < n)$. Then $\mathcal{G} \subset \mathbb{P}(\bigwedge_{p+1} V^*)$ and $\Gamma(\mathcal{G})$ is defined if V is replaced by $\bigwedge_{p+1} V^*$. For $t \in \mathbb{R}$ define $t^+ = \max\{0, t\}$.

Lemma 5.1 *Let $\mathcal{G} \neq \emptyset$ be a finite subset of $G_p(V^*)$ $(0 \leq p < n)$ in general position. Take $x \in G_k(V)$ and $y \in G_{k-1}(V)$ with $E(y) \subset E(x)$ and define*

$$\phi(a) = \frac{|y, a|_{l-1}^2}{|x, a|_{l-1}^2 |x, a|_l^{2\alpha(a)}}, \tag{82}$$

for $0 \leq l \leq u, a \in \mathcal{G}$ and for a function $\alpha : \mathcal{G} \longrightarrow (0, 1)$. Then

$$\begin{aligned} & D_p(k,l) c_{pk}(\mathcal{G}) \Big(\prod_{a \in \mathcal{G}} \rho(a)\phi(a)\Big)^{\frac{1}{D_p(k,l)}} \\ \leq \ & \sum_{a \in \mathcal{G}} \rho(a)\phi(a) + (D_p(k,l) - q)^+, \end{aligned} \tag{83}$$

where $q = \#\mathcal{G}$, $\rho : \mathcal{G} \longrightarrow [0,1]$ a function, and

$$c_{pk}(\mathcal{G}) = \Big(\frac{1}{k+1} 2^{-3n-3} \Gamma(\mathcal{G})\Big)^{\frac{2(q - D_p(k,l))^+}{D_p(k,l)}}. \tag{84}$$

By using Lemma 2.1, the proof of Lemma 5.1 can be completed in a similar fashion as Theorem 7.12 of Stoll [14].

A family $\mathcal{G} = \{w_1, ..., w_q\}$ of meromorphic mappings $w_j : M \longrightarrow \mathbb{P}(V^*)$ is said to be in general position if there is a point $z_0 \in M$ such that all w_j are holomorphic at z_0, such that $w_1(z_0), ..., w_q(z_0)$ are pairwise different and such that $\mathcal{G}(z_0) = \{w_1(z_0), ..., w_q(z_0)\}$ is in general position. Then $\Gamma(\mathcal{G}(z)) > 0$ for all $z \in M$ outside a thin analytic subset of M if and only if $\mathcal{G}$ is in general position. If so, the gauge function $\Gamma_{\mathcal{G}}$ is defined by

$$\Gamma_{\mathcal{G}}(r) = M\langle r; \log \Gamma(\mathcal{G})^{-2}\rangle \geq 0, \tag{85}$$

for all $r > 0$. Stoll ([14],[15]) showed

$$\Gamma_{\mathcal{G}}(r) \leq \binom{q-1}{n} \sum_{j=1}^{q} T_{w_j}(r) + O(1). \tag{86}$$

6 Second main theorems

Recall the well known result [6]: If $a(r) > 0$ is a monotone increasing function for $r \geq r_0$, then $a(r)$ is differentiable at almost all points, and for $\varepsilon > 0$

$$\frac{da(r)}{dr} \leq (a(r))^{1+\varepsilon} \quad \|. \tag{87}$$

Applying the result twice, we see that an inequality

$$\int_{r_0}^{r} t^{-p} dt \int_{0}^{t} s^p \exp(a(s)) ds \leq b(r),$$

implies

$$a(r) \leq (1+\varepsilon)^2 \log b(r) + p\varepsilon \log r \quad \|, \tag{88}$$

for $p \geq 0, \varepsilon > 0$, a real integrable function a and a monotone increasing function b.

Theorem 6.1 *Assume that τ majorizes a holomorphic $(m-1,0)$-form B on M with the majorant Y and assume that f is general for B. Let $\mathcal{G}$ be a finite family of meromorphic mappings $g : M \longrightarrow \mathbb{P}(V^*)$ such that for a fixed $p \in \mathbb{Z}$ with*

$$0 \leq p \leq \ell_{\mathcal{G}} = \min_{g\in\mathcal{G}} \ell_g(B),$$

the family $\mathcal{G}_p = \{g_p \mid g \in \mathcal{G}\}$ is in general position. Assume that

$$0 \leq k \leq n-1, \quad 0 \leq l \leq u = \min\{k, p\}.$$

Assume that for all $g \in \mathcal{G}$, (f_k, g_p) is free for $\odot^{l+1}$. Take $\varepsilon > 0$. Then

$$\begin{aligned} \sum_{g\in\mathcal{G}} m_{f_k \odot^l g_p}(r) &\leq I_{pk}^l\{\frac{n+1}{k+1}T_{f_k}(r) - \hat{S}_k(r) + \frac{1}{2}(n-k)(n+1)Q(r)\} \\ &+ \sum_{j=k}^{n-1}\sum_{i=0}^{l}\sum_{g\in\mathcal{G}} P_{p,j-i}^{l-i}(r,g) + q(l+1)(n-k)\Gamma_{\mathcal{G}_p}(r) \quad \|, \end{aligned} \tag{89}$$

where $q = \#\mathcal{G}$, and

$$\begin{aligned} Q(r) &= \mathrm{Ric}_\tau(r) + \frac{\varsigma}{2}(1+\varepsilon)^2\{3\log T_{f_k}(r) + 3\sum_g \log^+ T_{g_p}(r) \\ &+ (1+\varepsilon)\log Y(r) + (3+\varepsilon)\log^+ \mathrm{Ric}_\tau(r) + 2(m+1)\varepsilon \log r\}. \end{aligned} \tag{90}$$

Proof. We choose $\alpha = \alpha(r)$ satisfying

$$(1-\alpha)\sum_{g\in\mathcal{G}}(T_{f_k}(r) + T_{g_p}(r) + m_{f_k\odot^l g_p}(r_0)) = 1. \tag{91}$$

The First Main Theorem (67) implies

$$m_{f_k\odot^l g_p}(r) \leq T_{f_k}(r) + T_{g_p}(r) + m_{f_k\odot^l g_p}(r_0), \tag{92}$$

so that we have

$$-(1-\alpha)\sum_{g\in\mathcal{G}} m_{f_k\odot^l g_p}(r) \geq -1. \tag{93}$$

Setting

$$\Phi(g) = \frac{|f_{k-1}, g_p|_{l-1}^2}{|f_k, g_p|_{l-1}^2 |f_k, g_p|_l^{2\alpha}} \Phi_{p,k+1}^l,$$

then by Lemma 2.1,

$$\frac{1}{(k+1)^2}\Phi(g)\rho_{pk}^l \mathbb{H}_k \leq |f_k, g_p|_l^{-2\alpha} \Xi_{pk}^l. \tag{94}$$

Hence the Ahlfors Estimates imply

$$\begin{aligned} &(1-\alpha)^2 \int_{r_0}^r \int_{M[t]} \Phi(g)\rho_{pk}^l \mathbb{H}_k \frac{dt}{t^{2m-1}} \\ \leq\ & 4(n+1)^{2l+5} Y(r)(T_{f_k}(r) + T_{g_p}(r) + \varsigma). \end{aligned} \tag{95}$$

Set $\bar{\rho}_{pk}^l = \min\{1, \rho_{pk}^l\}$ and note that $\Phi_{p,k+1}^l \leq 1$. Then Lemma 5.1 gives

$$\begin{aligned} D_p(k,l)c_{pk}(\mathcal{G}_p)\Big(\prod_{g\in\mathcal{G}} \bar{\rho}_{pk}^l \Phi(g)\Big)^{\frac{1}{D_p(k,l)}} &\leq \sum_{g\in\mathcal{G}} \bar{\rho}_{pk}^l \Phi(g) + d_{pk}^l \\ &\leq \sum_{g\in\mathcal{G}} \rho_{pk}^l \Phi(g) + d_{pk}^l, \end{aligned} \tag{96}$$

where $d_{pk}^l = (D_p(k,l) - q)^+, q = \#\mathcal{G}$. Hence

$$\begin{aligned} \log\Big(\sum_{g\in\mathcal{G}} \rho_{pk}^l \Phi(g) + d_{pk}^l\Big) &\geq \frac{1}{D_p(k,l)}\Big\{\sum_{g\in\mathcal{G}} \log\Phi(g) \\ &+ \sum_{g\in\mathcal{G}} \log\bar{\rho}_{pk}^l + 2(q - D_p(k,l))^+ \log\Gamma(\mathcal{G}_p)\Big\} + e_{pk}^l, \end{aligned} \tag{97}$$

for a constant e_{pk}^l. Note that

$$P_{pk}^l(r,g) = -M\langle r; \log\bar{\rho}_{pk}^l\rangle.$$

By the concavity of the logarithm, we have

$$\begin{aligned} &\log\Big(\frac{1}{\varsigma}\int_{M\langle r\rangle} h_k^2\Big(\sum_{g\in\mathcal{G}} \rho_{pk}^l \Phi(g) + d_{pk}^l\Big)\sigma_{m-1}\Big) \\ \geq\ & \frac{1}{\varsigma}\int_{M\langle r\rangle} \log h_k^2 \sigma_{m-1} + \frac{1}{\varsigma}\int_{M\langle r\rangle} \log\Big(\sum_{g\in\mathcal{G}} \rho_{pk}^l \Phi(g) + d_{pk}^l\Big)\sigma_{m-1} \\ \geq\ & \frac{2}{\varsigma}(M\langle r; \log h_k^2\rangle - M\langle r_0; \log h_k^2\rangle) + \frac{2}{\varsigma D_p(k,l)}\Big\{\sum_{g\in\mathcal{G}} M\langle r; \log\Phi(g)\rangle \\ -\ & \sum_{g\in\mathcal{G}} P_{pk}^l(r,g) - (q - D_p(k,l))^+ \Gamma_{\mathcal{G}_p}(r)\Big\} + e_{pk}^l + c_1, \end{aligned} \tag{98}$$

for a constant c_1 and for $r \in \mathcal{E}_\tau^0$. Write

$$\psi(k,l) = \sum_{g\in\mathcal{G}} (m_{f_k \odot^l g_p}(r) - m_{f_{k+1}\odot^l g_p}(r)). \tag{99}$$

Then we have

$$\begin{aligned}\sum_{g\in\mathcal{G}} M\langle r; \log\Phi(g)\rangle &= \psi(k,l) - \psi(k-1,l-1)\\ &- (1-\alpha)\sum_{g\in\mathcal{G}} m_{f_k\odot^l g_p}(r) + \varsigma \log\binom{\hat{u}+1}{l+1}\\ &\geq \psi(k,l) - \psi(k-1,l-1) - 1 + \varsigma\log\binom{\hat{u}+1}{l+1}, \end{aligned}\tag{100}$$

where $\hat{u} = \min\{k+1, p\}$. Hence

$$\log(\int_{M\langle r\rangle} h_k^2(\sum_{g\in\mathcal{G}} \rho_{pk}^l \Phi(g) + d_{pk}^l)\sigma_{m-1}) \geq \frac{2}{\varsigma D_p(k,l)}\Theta(r) + e_{pk}^l + c_2, \tag{101}$$

where

$$\begin{aligned}\Theta(r) &= D_p(k,l)(M\langle r; \log h_k^2\rangle - M\langle r_0; \log h_k^2\rangle) - \sum_{g\in\mathcal{G}} P_{pk}^l(r,g)\\ &+ \psi(k,l) - \psi(k-1,l-1) - (q - D_p(k,l))^+\Gamma_{\mathcal{G}_p}(r). \end{aligned}\tag{102}$$

Noting that

$$\int_{r_0}^r t^{1-2m}(\int_{M[t]} \mathbb{H}_k)dt \leq Y(r)T_{f_k}(r),$$

then by (42), (95) and (91), there exists a positive constant c_3 such that

$$\begin{aligned}&\int_{r_0}^r t^{1-2m}(\int_0^t s^{2m-1}(\int_{M\langle s\rangle} h_k^2(\sum_{g\in\mathcal{G}} \rho_{pk}^l\Phi(g) + d_{pk}^l)\sigma_{m-1}ds)dt\\ = &\frac{1}{2m}\int_{r_0}^r t^{1-2m}(\int_{M[t]}(\sum_{g\in\mathcal{G}} \rho_{pk}^l \Phi(g) + d_{pk}^l)\mathbb{H}_k)dt\\ \leq &c_3 Y(r)(T_{f_k}(r) + \sum_{g\in\mathcal{G}} T_{g_p}(r) + 1)^3. \end{aligned}\tag{103}$$

Hence, by using (88), (47), (48), (101), and (103), we obtain

$$\begin{aligned}&\psi(k,l) - \psi(k-1,l-1) \leq -D_p(k,l)(T_{k-1}(r) - 2T_k(r) + T_{k+1}(r))\\ + &\sum_{g\in\mathcal{G}} P_{pk}^l(r,g) + D_p(k,l)Q_2(r)\\ + &(q - D_p(k,l))^+\Gamma_{\mathcal{G}_p}(r) + \frac{3\varsigma}{2}(1+\varepsilon)^2 D_p(k,l)\log T_{f_k}(r) \quad \|, \end{aligned}\tag{104}$$

where

$$Q_2(r) = \mathrm{Ric}_\tau(r) + \frac{\varsigma}{2}(1+\varepsilon)^2\{\log Y(r) + 3\sum_{g\in\mathcal{G}} \log^+ T_{g_p}(r) + 2m\varepsilon \log r\}. \tag{105}$$

In (104), replacing l, k by $l-i, k-i$, respectively, and adding it for $i = 0, 1, ..., l$, and noting that $\psi(k-l-1, -1) = 0$, we get

$$\begin{aligned}\psi(k,l) \leq & -\sum_{i=0}^{l} D_p(k-i, l-i)(T_{k-i-1}(r) - 2T_{k-i}(r) + T_{k-i+1}(r)) \\ & + \sum_{i=0}^{l}\sum_{g\in\mathcal{G}} P^{l-i}_{p,k-i}(r,g) + \sum_{i=0}^{l} D_p(k-i,l-i)Q_2(r) \\ & + \sum_{i=0}^{l}(q - D_p(k-i,l-i))^+\Gamma_{\mathcal{G}_p}(r) \\ & + \frac{3\varsigma}{2}(1+\varepsilon)^2\sum_{i=0}^{l} D_p(k-i,l-i)\log T_{f_{k-i}}(r) \quad \|. \end{aligned} \tag{106}$$

By Lemma 2.1, we have

$$m_{f_n \odot^l g_p}(r) = \varsigma \log \binom{p+1}{l+1}. \tag{107}$$

Thus adding (106) over k from k to $n-1$, we arrive at

$$\begin{aligned}\sum_{g\in\mathcal{G}} m_{f_k\odot^l g_p}(r) \leq & -\sum_{j=k}^{n-1}\sum_{i=0}^{l} D_p(j-i,l-i)(T_{j-i-1}(r) - 2T_{j-i}(r) + T_{j-i+1}(r)) \\ & + \sum_{j=k}^{n-1}\sum_{i=0}^{l}\sum_{g\in\mathcal{G}} P^{l-i}_{p,j-i}(r,g) + \sum_{j=k}^{n-1}\sum_{i=0}^{l} D_p(j-i,l-i)Q_2(r) \\ & + q(l+1)(n-k)\Gamma_{\mathcal{G}_p}(r) + q\varsigma\log\binom{p+1}{l+1} \\ & + \frac{3\varsigma}{2}(1+\varepsilon)^2\sum_{j=k}^{n-1}\sum_{i=0}^{l} D_p(j-i,l-i)\log T_{f_{j-i}}(r) \quad \|. \end{aligned} \tag{108}$$

By (57) and the Remark after Corollary 4.1, there exists a positive constant c_4 such that

$$T_f(r) \leq nk\mathrm{Ric}_\tau(r) + c_4\{T_{f_k}(r) + \log Y(r) + \log^+ \mathrm{Ric}_\tau(r) + \varepsilon \log r\} \quad \|, \tag{109}$$

because $T_f(r) \to \infty$ as $r \to \infty$ and

$$\frac{1}{4}nk\varsigma(1+\varepsilon)^2 \log T_f(r) < \frac{1}{2}T_f(r),$$

for sufficiently large r. Similarly, by (56) and the Remark after Corollary 4.1, we have

$$\begin{aligned} T_{f_j}(r) &\leq \frac{j(j+1)}{2}\mathrm{Ric}_\tau(r) + c_4\{T_f(r) + \log Y(r) \\ &+ \log^+ \mathrm{Ric}_\tau(r) + \varepsilon \log r\} \quad \|, \end{aligned} \tag{110}$$

for $j \geq 0$. Then (109) and (110) imply

$$\begin{aligned} \log T_{f_j}(r) &\leq \log T_{f_k}(r) + \log^+ \mathrm{Ric}_\tau(r) + \log^+ \log Y(r) \\ &+ \log^+ \log^+ \mathrm{Ric}_\tau(r) + \log^+ \log r + O(1) \quad \| \\ &\leq \log T_{f_k}(r) + \log^+ \mathrm{Ric}_\tau(r) + \frac{\varepsilon}{3}\{\log Y(r) \\ &+ \log^+ \mathrm{Ric}_\tau(r) + \log r\} \quad \|, \end{aligned} \tag{111}$$

for $j \geq 0$. By (108), (111) and (24) with $y_j = T_j(r)$, we obtain

$$\begin{aligned} \sum_{g\in\mathcal{G}} m_{f_k\odot^l g_p}(r) &\leq D_p(k+1,l+1)T_k(r) - \binom{n}{l}\binom{-1}{p-l}N_n(r) \\ &- \sum_{i=k-l-1}^{k} a_iT_i(r) - \sum_{i=k+1}^{n} b_iT_i(r) \\ &+ \sum_{j=k}^{n-1}\sum_{i=0}^{l} D_p(j-i,l-i)Q(r) \\ &+ \sum_{j=k}^{n-1}\sum_{i=0}^{l}\sum_{g\in\mathcal{G}} P^{l-i}_{p,j-i}(r,g) + q(l+1)(n-k)\Gamma_{\mathcal{G}_p}(r) \quad \|. \end{aligned} \tag{112}$$

Note that

$$Q_1(r) \leq Q(r) \quad \|,$$

by using (109). Then by (56) and (57), we have

$$\begin{aligned} -\sum_{i=k-l-1}^{k} a_iT_i(r) &\leq -\sum_{i=k-l-1}^{k} \frac{i+1}{k+1}a_iT_k(r) \\ &+ \frac{1}{2}\sum_{i=k-l-1}^{k}(k-i)(i+1)a_iQ(r) \quad \|, \end{aligned} \tag{113}$$

and

$$-\sum_{i=k+1}^{n} b_iT_i(r) \le -\sum_{i=k+1}^{n} \frac{n-i}{n-k}b_iT_k(r)$$
$$-\sum_{i=k+1}^{n} \frac{i-k}{n-k}b_iN_n(r) + \frac{1}{2}\sum_{i=k+1}^{n} (i-k)(n-i)b_iQ(r) \quad \|. \tag{114}$$

By (112), (113), (114) and Lemma 3.2, we obtain

$$\sum_{g\in\mathcal{G}} m_{f_k\odot^l g_p}(r) \le I_{pk}^l\{\frac{n+1}{k+1}T_k(r) - N_n(r) + \frac{1}{2}(n-k)(n+1)Q(r)\}$$
$$+ \sum_{j=k}^{n-1}\sum_{i=0}^{l}\sum_{g\in\mathcal{G}} P_{p,j-i}^{l-i}(r,g) + q(l+1)(n-k)\Gamma_{\mathcal{G}_p}(r) \quad \|. \tag{115}$$

Using (49) and (50), we finally arrive at (89). ♠

7 Defect relations

If one of f_k and g_p is not constant, then $T_{f_k}(r)+T_{g_p}(r) \to \infty$ for $r \to \infty$, and the defect is defined by

$$\delta(f_k \odot^l g_p) = \liminf_{r\to\infty} \frac{m_{f_k\odot^l g_p}(r)}{T_{f_k}(r) + T_{g_p}(r)}. \tag{116}$$

If f_k is not constant, define

$$\delta[f_k \odot^l g_p] = \liminf_{r\to\infty} \frac{m_{f_k\odot^l g_p}(r)}{T_{f_k}(r)}. \tag{117}$$

Then

$$\delta(f_k \odot^l g_p) \le \delta[f_k \odot^l g_p].$$

The First Main Theorem implies

$$0 \le \delta(f_k \odot^l g_p) = 1 - \limsup_{r\to\infty} \frac{T_{f_k\odot^l g_p}(r) + N_{f_k\odot^l g_p}(r)}{T_{f_k}(r) + T_{g_p}(r)} \le 1. \tag{118}$$

Define the types of growth of f_k for g_p by

$$\mathrm{typ}(f_k, g_p) = \liminf_{r\to\infty} \frac{T_{g_p}(r)}{T_{f_k}(r)}, \tag{119}$$

$$\mathrm{Typ}(f_k, g_p) = \limsup_{r\to\infty} \frac{T_{g_p}(r)}{T_{f_k}(r)}. \tag{120}$$

Then $0 \leq \operatorname{typ}(f_k, g_p) \leq \operatorname{Typ}(f_k, g_p)$. Also

$$\delta(f_k \odot^l g_p) = \delta[f_k \odot^l g_p],$$

if $\operatorname{Typ}(f_k, g_p) = 0$. Write

$$\delta_{f_k}(g_p) = \delta(f_k \odot^u g_p), \quad \delta_{f_k}[g_p] = \delta[f_k \odot^u g_p].$$

Theorem 7.1 *Assumptions as Theorem 6.1. Assume that for all $g \in \mathcal{G}$, the pair (f_k, g_p) is obstruction term free of order $l+1$, i.e.,*

$$\lim_{r\to\infty} \frac{P_{p,j-i}^{l-i}(r,g)}{T_{f_k}(r)} = 0, \quad 0 \leq i \leq l; k \leq j \leq n-1; g \in \mathcal{G}. \tag{121}$$

Assume that

$$R_{f_k} < \infty; \quad \operatorname{Typ}(f_k, g_p) < \infty \quad \text{for all } g \in \mathcal{G}, \tag{122}$$

$$\lim_{r\to\infty} \frac{\Gamma_{\mathcal{G}_p}(r)}{T_{f_k}(r)} = 0. \tag{123}$$

Then we have

$$\sum_{g\in\mathcal{G}} \delta[f_k \odot^l g_p] \leq I_{pk}^l \{\frac{n+1}{k+1} - \theta_k + \frac{1}{2}(n-k)(n+1)(R_{f_k} + \frac{\varsigma}{2} Y_{f_k})\}, \tag{124}$$

where

$$\theta_k = \lim_{r\to\infty} \inf \frac{\hat{S}_k(r)}{T_{f_k}(r)} \geq 0.$$

Proof. Note that (122) implies

$$\lim_{r\to\infty} \frac{\log^+ \operatorname{Ric}_\tau(r)}{T_{f_k}(r)} = \lim_{r\to\infty} \frac{\log^+ T_{\mathcal{G}_p}(r)}{T_{f_k}(r)} = 0.$$

Then Theorem 7.1 follows from Theorem 6.1. ♠

Remark. If $\operatorname{Typ}(f_k, g_p) = 0$, then (86) yields (123). If $l = \min\{k,p\}, R_{f_k} = Y_{f_k} = 0$, (124) reads

$$I_{pk}\theta_k + \sum_{g\in\mathcal{G}} \delta_{f_k}[g_p] \leq \frac{n+1}{k+1} I_{pk}. \tag{125}$$

Note that $R_{f_k} = 0$ if $M = \mathbb{C}^m$, and note that $Y_{f_k} = 0$ if $m = 1$. If each $g_p \in \mathcal{G}_p$ is constant, then (121) and (123) hold. Further if $m = 1$, Hu [4] has proved Theorem 6.1 and Theorem 7.1 (where the sup sign in the definitions of θ_k and Θ_k should be inf).

Theorem 7.2 *Assumptions as Theorem 6.1. Assume that (121) holds and assume that*

$$R_{f_k} = Y_{f_k} = 0; \quad \mathrm{Typ}(f_k, g) = 0 \quad \text{for all } g \in \mathcal{G}. \tag{126}$$

Then

$$I^l_{pk}\theta_k + \sum_{g\in\mathcal{G}} \delta[f_k \odot^l g_p] \leq \frac{n+1}{k+1} I^l_{pk}. \tag{127}$$

Proof. Proposition 6.15 of Stoll [14] implies

$$T_{g_p}(r) \leq 3^p T_g(r) + \frac{3^p - 1}{2}\{\varsigma \log Y(r) + \mathrm{Ric}_\tau(r) + \varsigma\varepsilon \log r\} \quad \|, \tag{128}$$

which gives

$$\log T_{g_p}(r) \leq \log^+ T_g(r) + \log Y(r) + \log^+ \mathrm{Ric}_\tau(r) + \varepsilon \log r) \quad \|. \tag{129}$$

Combining with (86) and Theorem 6.1, we can obtain (127). ♠

Theorem 7.1 and Theorem 7.2 refine the results of Stoll [15]. Let $l = k = 0$. Then (125) and (127) also improve a defect relation of Stoll in [14], p. 272-274.

References

[1] Chuang, C. T., Uné généralisation d'une inégalité de Nevanlinna, Scientia Sinica 13 (1964), 887-895.

[2] Dufresnoy, J., Sur les valeurs exceptionelles des fonctions méromorphes voisines d'une fonction méromorphe donnée, CR Acad. Sci. Paris 208 (1939), 255-257.

[3] Frank, G. and Weissenborn, G., Rational deficient functions of meromorphic functions, Bull. London Math. Soc. 18(1986), 29-33.

[4] Hu, P. C., A modified defect relation for holomorphic curves, Kodai Math. J. 13(1990), 349-362.

[5] Mori, S., Remarks on holomorphic mappings, Contempory Math. 25(1983), 101-114.

[6] Nevanlinna, R., Le théorème de Picard-Borel et la théorie des fonctions meromorphes, Grauthier Villars Paris, 1929.

[7] Osgood, C. F., Sometimes effective Thue-Siegel-Roth-Schmidt-Nevanlinna bounds or better, J. Number Theory 21(1985), 347-389.

[8] Osgood, C. F., A fully general Nevanlinna N-small function theorem and a Sometimes effective Thue-Siegel-Roth-Schmidt theorem for solution to linear differential equations, Contemp. Math. 25(1983), 129-130.

[9] Ru, M. and Stoll, W., The second main theorem for moving targets, J. of Geom. Analysis 1(1991), 99-138.

[10] Shiffman, B., New defect relations for meromorphic functions on $\mathbb{C}^m$, Bull. Amer. Math. Soc. 7(1982), 599-601.

[11] Shiffman, B., A general second main theorem for meromorphic functions on $\mathbb{C}^m$, Amer. J. Math. 106(1984), 509-531.

[12] Steinmetz, N., Eine Verallgemeinerung des zweiten Nevanlinnaschen Hauptsatzes, J. Reine Angew. Math. 368(1986), 134-141.

[13] Stoll, W., Value distribution on parabolic spaces, Lecture Notes in Math. 600, Springer, 1977.

[14] Stoll, W., Value distribution theory for meromorphic maps, Asp. Math. E7, 1985.

[15] Stoll, W., Value distribution theory for moving targets, Complex Analysis and Algebraic Geometry (Proceedings, Göttingen, 1985), Lecture Notes in Math. 1194, 214-235, Springer.

[16] Wu, H., The equidistribution theory of holomorphic curves, Annals of Math. Studies 64 (1970), Princeton Univ. Press, Princeton, NJ.

[17] Yang, L., Deficient functions of meromorphic functions, Scientia Sinica 24 (1981), 1179-1189.

36 THE GREEN-RESIDUE THEOREM AND THE NEVANLINNA'S SECOND THEOREM ON SMALL FUNCTIONS IN BANACH SPACES

Chuan-Gan HU
Xiu-Fang YE

Nankai University
Department of Mathematics
Tianjin 300071
P.R. China

chuganhu@public1.tpt.tj.cn

Abstract: In this paper, the Green-Residue Theorem with domain in a finite dimensional space and range in a Banach space is presented, and the Nevanlinna's Second Main Theorem on small functions in a Banach algebra is obtained via vector-valued differential polynomials as well.

1 INTRODUCTION

Project was supported by the National Natural Science Foundation of China

H.G.W. Begehr et al. (eds.), Proceedings of the Second ISAAC Congress, Vol. 1, 299–310.

The Residue Theorem in complex analysis was extended to the vector -valued case with domain in a complex plane C and range in a Banach space (see[5]).In 1973,Ph.Griffiths and J.King studied the Green-Residue Theorem with domain and range in finite dimensional spaces (see[4]). In this paper, the Green-Residue Theorem with domain in a finite dimensional space and range in a Banach space is presented in order to lay a foundation of value distribution theory in infinite dimensional spaces.

On the other hand,in 1925, Nevanlinna's first and second fundamental theorems were established by introducing characteristic function([10]). Later,the first main theorem for complex analytic mappings of several variables was researched by S.S.Chern (see [1,2]).Then, H.J.W.Ziegler extended the classicial Nevanlinna's theory of meromorphic functions to vector-valued meromorphic functions in finite dimensional spaces ([13]). In recent years,C.G.Hu and C.C.Yang made a contribution to value distribution theory of meromorphic function with range in an infinite dimensional Hilbert space (see[8]). In 1997, the results in [8] were extended to a Banach space by C.G. Hu ([6]).In this paper, we also research on the properties of vector-valued differential polynomials with range in a Banach algebra with a Schauder basis , where some useful symbols such as $m(r,f)$,$N(r,f)$, $T(r,f)$ and $\overline{N}(r,f)$ etc. are similar to corresponding cases in classical value distribution theory, and the E-valued Navanlinna's second main theorem on small functions is presented from the conclusions of the E-valued differential polynomials (see[3,11]).

2 THE GREEN-RESIDUE THEOREM

Assume that E and F are Banach spaces, and that E has a Schauder basis $\{e_n\}_{n=1}^{\infty}$. Let $\widehat{M}$ be a connected ,complex manifold over E ,and $\widehat{N}$ a connected Riemann domain over F. Suppose that M is a connected complex manifold of dimension m and that τ is an exhaustion of M.

Theorem 2.1 *On the each complex manifold the exterior derivative* d *operating on differential forms splits into* $d = \partial + \overline{\partial}$ *and twists to* $d^c = (i/4\pi)(\overline{\partial} - \partial)$.

Let $v = dd^c\tau > 0$ be the standard Kahler form on M and $\sigma = d^c \log \tau$.

Theorem 2.2 *A subset* A *of a complex manifold* $\widehat{M}$ *over* E *is said to be analytic if for every* $x \in \widehat{M}$ *has an open neighborhood* U *which carries* *finite many holomorphic mappings $f_1,\cdots,f_s$ from U into F such that*

$$A\bigcap U = f_1^{-1}(0)\bigcap\cdots\bigcap f_s^{-1}(0).$$

Theorem 2.3 *Let $\widehat{M}$ be a complex manifold over E, and $\widehat{N}$ a Riemann domain over F. Assume $f:\widehat{M}\to\widehat{N}$ is a holomorphic mapping. Let $A\neq\emptyset$ be an analytic subset of $\widehat{M}$.$f|A$ is called to be proper if $A\bigcap f^{-1}(K)$ is compact for every compact subset K of $\widehat{N}$.*

Let an injective holomorphic mapping

$$\omega = (z_1,\cdots,z_m,\cdots):U\to E$$

be a chart on $\widehat{M}$ and $z_1,\cdots,z_m,\cdots$ are holomorphic coordinates on U . Hence $U' = \omega(U)$ is open and $\omega : U \to U'$ is biholomorphic .If $a\in U$,then ω is a chart centered at a point a. If so , and if $f\not\equiv 0$ is a holomorphic mapping on $\widehat{M}$ to E, then there are a sequence $\{P_q\}_{q\in Z_+}(Z_+ = \{0,1,2,\cdots\})$ of homogeneous polynomials P_q of degree q on E and an open ,connected neighborhood W of a in U such that

$$f|W = \sum_{q=0}^{\infty}(P_q\circ\omega)|W$$

converges uniformly on W.

Since $f\not\equiv 0$, one and only one integer $\mu=\mu_f(a)\geq 0$ exists such that $P_q\not\equiv 0$ if $0\leq q<\mu$ and $P_\mu\not\equiv 0$. The number $\mu_f(a)$ depends on f and a but not on U, ω, W and is called the zero-multiplicity of f at a. The function $\mu_f:\widehat{M}\to Z_+$ is called the zero-divisor of f.

Theorem 2.4 *A function $\nu:\widehat{M}\to Z=\{0,\pm1,\pm2,\cdots\}$ is called a divisor on $\widehat{M}$ if for every $x\in\widehat{M}$ there is an open , connected neighborhood U of x with holomorphic mappings $g\not\equiv 0\not\equiv h$ on U to E such that $\nu|U=\mu_g-\mu_h$.*

Theorem 2.5 *A divisor ν is said to be non-negative if $\nu(x)\geq 0$ for all $x\in\widehat{M}$.*

Theorem 2.6 *Let A be a non-empty, open subset of M such that $S = M - A$ is analytic, where S is the support of ν. Let $f:A\to\widehat{N}$ be a holomorphic mapping with graph $\Gamma^\circ=\{(x,f(x))|x\in A\}$. Then $\overline{\Gamma^\circ}(=\Gamma)$ is the closure of Γ° in $M\times\widehat{N}$ and is called the closed graph of f.*

Define the projections $\pi : \Gamma \to M$ and $\tilde{f} : \Gamma \to \widehat{N}$ by $\pi(x, y) = x$ and $\tilde{f}(x, y) = y$ for all $(x, y) \in \Gamma$. For $x \in M$ define the scattering set

$$G = \{y \in \widehat{N} | (x, y) \in \Gamma\} = \tilde{f}(\pi^{-1}(x)).$$

Theorem 2.7 *The mapping f is said to be meromorphic if Γ is analytic in $M \times \widehat{N}$ and π is proper.*

For $t > 0$ the counting function n_ν of ν is defined by

$$n_\nu(t) = (1/t^{2m-2}) \int_{S[t]} \nu v^{m-1}, \quad \text{if } m > 1,$$

where $S[t] = \{x \in S | \tau(x) \leq t^2\}$.

For $0 < s < r$, the valence function N_ν of ν is defined by

$$N_\nu(r, s) = \int_s^r (n_\nu(t)/t) dt.$$

Let W^0 be a holomorphic vector bundle over $\widehat{N}$.The set of all holomorphic sections of W^0 over $\widehat{N}$ is a complex vector space $\Gamma(\widehat{N}, W^0)$. For $u \in \Gamma(\widehat{N}, W^0)$, let $Z(u) = \{x \in \widehat{N} | u(x) = O_x\}$ be the zero set of μ , where O_x is the zero element in W_x for each $x \in M$,and W_x is the fiber $\pi^{-1}(x)$.

Let $^{\#}G$ be the number of elements for G,and $I_G = \{x \in M | ^{\#}G > 1\}$.

Let $f : M \to \widehat{N}$ be a meromorphic mapping. Then $f_0 = f : M - I_G \to \widehat{N}$ is holomorphic. The pull back divisor $f_0^*(\nu)$ is defined on $M - I_G$.Let $f^*(W^0) = f_0^*(W^0)$ be the pull back bundle by $f_0 : M - I_G \to \widehat{N}$. It follows that

$$f^*(W^0) = \{(x, w) \in (M - I_G) \times W^0 | w \in W_{f(x)}\}.$$

Define the projections $\pi : f^*(W^0) \to M - I_G$ and $\tilde{f} : f^*(W^0) \to W^0$ by $\pi(x, w) = x$ and $\tilde{f}(x, w) = w$ respectively if $x \in M - I_G$ and $w \in W_{f(x)}$. A holomorphic section u_f of $f^*(W^0)$ is defined by $u_f(x) = (x, u(f(x)))$ for all $x \in M - I_G$.

Let W^0 be a holomorphic vector bundle with a Hermitian metric k along the fibers of W^0, where $k : W^0 \oplus W^0 \to C$ is a function of class C^∞ such that for each $x \in M$ the restriction $k_x : W_x \times W_x \to C$ is a positive definite Hermitian form. The norm of $\widehat{a} \in W_x$ is defined by $\|\widehat{a}\|_k = \sqrt{k(\widehat{a}, \widehat{a})} \geq 0$.

Let E_n be an n-dimensional section vector space with basis $\{e_j\}_{j=1}^{n}$. The projection operator $P_n : E \to E_n$ is a realization of E_n associated to the basis.A holo/mero-morphic mapping f on M is said to be of compact projection if for any given $\epsilon > 0$, $\|P_n f - f\| < \epsilon$ for n sufficiently large and any compact subset $D_0 \subset M$.

Theorem 2.1 *(Green-Residue Theorem) Let (M, τ) be a parabolic manifold of dimension m , $\widehat{N}$ a Riemann domain over E , and $f : M \to \widehat{N}$ meromorphic mapping with free of u(i.e.$f(M - I_G)$ is not contained in the zero set $Z(u)$ of u) and the property of compact projection. Then*

$$\int_s^r \int_{M[t]} dd^c \log \|u \circ f\|_k^2 \wedge v^{m-1} \frac{dt}{t^{2m-1}} + N_\nu(r, s)$$
$$= \int_{M\langle r\rangle} \log \|u \circ f\|_k \sigma - \int_{M\langle s\rangle} \log \|u \circ f\|_k \sigma$$

where $M\langle r\rangle = \{x \in M | \tau(x) = r^2\}$ and τ is an exhaustion.

Proof Let f_n be the projection of f on E_n.The property of compact projection yields $\|u \circ f_n - u \circ f\|_k < \epsilon$ for n sufficiently large, because $M[t]$,$M\langle r\rangle$ and $M\langle s\rangle$ are compact and u is a holomorphic section. It follows that the Green- Residue Theorem with domain and range in finite dimensional spaces implies required conclusion.

3 PRINCIPLE CONCEPTS AND THEOREMS

Assume that E_0 is a Banach algebra, that $f(x)$ is an E_0-valued transcendental meromorphic function,and that one of its differential polynomials is a polynomial of $f(z)$ and its first j-th derived function $f^{(j)}(z)(j = 1, 2, \cdots, q)$

$$P(f, f', \cdots, f^{(q)}) = \sum_{k=1}^{q} a_k(z)(f)^{s_{k_0}}(f')^{s_{k_1}} \cdots (f^{(q)})^{s_{k_q}},$$

where the coefficient $a_k(z)$ is a small function of $f(z)$ which means that the growth of $T(r, a_k)$ is slower than that of $T(r, f)$, and s_{k_j} is a non-negative integer.

Theorem 3.1 *A function has the condition (C) if $U(r)$ satisfies*

1^0 *$U(r)$ is continuous and positive for $r > 0$,and tends to infinity with r;*

2^0 *there exists a continuous function $\delta(r)$ for $r > 0$, such that*

$$\lim_{r\to+\infty} \frac{\log\{1 + 1/\delta(r)\}}{\log U(r)} = 0$$

and

$$\lim_{r\to+\infty} \frac{\log U(R)}{\log U(r)} = 1, \quad R = r\{1 + \delta(r)\}.$$

Theorem 3.2 *Assume that $f(z)$ and $g(z)$ are E_0-valued transcendental meromorphic functions, the growth of $f(z)$ and $g(z)$ are called almost the same and denoted by $f \sim g$,if there exist two positive numbers A,B , two positive integers p,q and a positive number r_0 such that*

$$T(r, f) < A(\frac{\lambda}{\lambda - 1})^p T(\lambda r, g) \quad \textit{for } r > r_0, \lambda > 1.$$

and

$$T(r, g) < B(\frac{\lambda}{\lambda - 1})^q T(\lambda r, f).$$

Let EMF be a set consisting of all E_0-valued meromorphic functions on C. Assume that $\Omega : EMF \to EMF$ is an operator with the growth of f and $\Omega(f)$ being almost same and with keeping transcendence.Using the method similar to classical one and methods in [6,8], we can make the following conclusions.

Theorem 3.2 *The set EMF is a Fréchet algebra.*

Proof Suppose that f is in EMF over C. Take a sequence $\{D_n\}_{n=1}^{\infty}$ of compact domains having no poles of f on the boundary ∂D_n of D_n for each n and satisfying $\lim_{n\to\infty} \partial D_n = C$ and $D_n \subset D_{n+1}$. Clearly,there are finite many poles of $\{a_j^{(n)}\}_{j=1}^k$ of f in D_n for any n.Make a punctured neighborhood $C(a_j^{(n)}; \delta^{(n)})$ at each $a_j^{(n)}$ with radius $\delta^{(n)}$ satisfying $\lim_{n\to\infty} \delta^{(n)} = 0$. Let

$$\eta_n(f) = \sup\{\|f(z)\| | z_n \in D_n \backslash \bigcup_{j=1}^{k} C(a_j^{(n)}; \delta^{(n)})\}.$$

By easy calculation we can show that η_n is a seminorm, and $H = \{\eta_n\}_{n=1}^{\infty}$ generates the topology of EMF.

On the other hand , for any f,$g \in EMF$ according to the multiplication of a Banach algebra we have $f \bullet g \in EMF$ and

$$\eta(f \bullet g) \leq \eta(f)\eta(g),$$

for any $\eta \in H$.

Moreover the set EMF forms a Fréchet algebra. In fact, let paranorm $\|\|f\|\|$ of f be

$$\sum_{n=1}^{\infty} \frac{1}{2^n} \frac{\eta_n(f)}{1+\eta_n(f)}.$$

It is easily checked that $\|\|f \bullet g\|\| \leq \|\|f\|\|\|\|g\|\|$. So we get the required result.

Theorem 3.3 *If f, g ,$h \in EMF$,and if $f(z)$,$g(z)$ and $h(z)$ are with transcendence, $f \sim g$ and $g \sim h$, then $f \sim h$.*

Theorem 3.4 *If ϕ and ψ :$EMF \to EMF$ are two operators with $\phi(f) \sim \psi(f)$, then the composite operator $\phi \circ \psi$ satisfies $\phi \circ \psi(f) \sim \phi(f)$.*

In order to extend the Theorem 4 to the general differential polynomial,the following lemmas are to be given.

Lemma 3.1 *Suppose that an E_0-valued transcendental holomorphic function $f(z)$ has the form in the neighborhood $N(z_0)$ of any point z_0*

$$f(z) = a_m(z-z_0)^m + a_{m+1}(z-z_0)^{m+1} + \cdots,$$

for $m \geq 0$,$a_k \in E$ then in $N(z_0)$,a_m^{-1} exists if and only if f^{-1} exists.

Proof Changing the form of $f(z)$,we get

$$\frac{f(z)}{(z-z_0)^m} = a_m + a_{m+1}(z-z_0) + \cdots,$$

then

$$\|\frac{f(z)}{(z-z_0)^m} - a_m\| = \|a_{m+1}(z-z_0) + \cdots\|.$$

Because the inverse element x^{-1} is a continuous funcion of x(see Lemma 3.1.3 in [7]), the conclusion is easily made.

Corollary 3.1 *Under the condition of Lemma 1, if f^{-1} exists in a punctured neighborhood of z_0, then so is $[f^{(p)}]^{-1}$ for any positive integer p.*

Proof We derive the required statement from the relation

$$\frac{f^{(p)}(z)}{p!}(z-z_0)^{m-p} = a_m + a_{m+1}\frac{(m+1-p)!}{p!}(z-z_0) + \cdots.$$

Corollary 3.2 *If the condition of Lemma 1 is satisfied in a punctured neighborhood for $m < 0$, then the results of Lemma 1 and Corollary 1 hold for any p.*

For non-linear E_0-valued meromorphic functions $\psi_k(z)(k = 1, 2. \cdots, p; p \geq 1)$ and an E_0-valued meromorphic function $f(z)$,$A_0 = W(\psi_1, \psi_2, \cdots, \psi_p)$ represents the Wronskian of $\psi_k(z)(k = 1, 2, \cdots, p)$ and $W(\psi_1, \psi_2, \cdots, \psi_p, f)$ represents Wronskian of $\psi_k(z)(k = 1, 2, \cdots, p)$ and f. Suppose ψ_k^{-1} exists. Define an linear operator

$$\begin{aligned} L(f) &= W(\psi_1, \psi_2, \cdots, \psi_p, f)W^{-1}(\psi_1, \psi_2, \cdots, \psi_p) \\ &= f^{(p)} + A_1 A_0^{-1} f^{(p-1)} + \cdots + A_p A_0^{-1} f \end{aligned}$$

where $\psi_k(z)(k = 1, 2. \cdots, p)$ are fixed. Suppose that the inverse function of

$$F(z) = L\{f(z)\} \tag{1}$$

is written as

$$f(z) = \sum_{k=1}^{p} c_k(z)\psi_k(z)$$

where $c_k(k = 1, , 2, \cdots, p)$ are E_0-valued meromorphic functions satisfying the following relations

$$c_k' = (-1)^{p+k} \Phi_k A_0^{-1} F,$$

for $k = 1, 2, \cdots, p$ and Φ_k is a determinant after we cancel the p-th row and the k-th column in

$$A_0 = \begin{vmatrix} \psi_1 & \psi_2 & \cdots & \psi_p \\ \psi_1' & \psi_2' & \cdots & \psi_p' \\ \cdots & \cdots & \cdots & \cdots \\ \psi_1^{(p-1)} & \psi_2^{(p-1)} & \cdots & \psi_p^{(p-1)} \end{vmatrix}.$$

Lemma 3.2 *Assume that $f(\not\equiv 0)$ is an E_0-valued meromorphic function, that $\psi_k(z)$ is a non-linear E_0-valued meromorphic function for $k = 1, 2, \cdots, p$, and that ψ_k^{-1}exists$(k = 1, 2, \cdots, p; p \geq 1)$, then*

$$T(r, F) \leq T(r, f) + p\overline{N}(r, f) + A\sum_{k=1}^{p}\{N(r, \psi_k) + N(r, \psi_k^{-1}\} + S(r)$$

for $r > 1$, where $S(r)$ satisfies

$$S(r) < a\{log^{+}T(R,f) + \sum_{k=1}^{p} log^{+}T(R,\psi_k)\}$$

$$+blog^{+}1/(R-r) + clogR + d, \ \ for\ 1 < r < R.$$

and A, a, b, c, d are positive constants.

The above lemmas yield the following conclusions.

Theorem 3.5 *Assume that $f(z)$ is an E_0-valued meromorphic function, that $\psi_k(z)(k = 1, 2, \cdots, p; p \geq 1)$ satisfies the condition of Lemma 2, and that $T(\lambda r, \psi_k) = \circ\{T(r, f)\}$ for some $\lambda > 1$ and $k = 1, 2, \cdots, p$, then $F(z)$ is transcendental and $F \sim f$.*

Consider an E_0-valued transcendental meromorphic function $f(z)$ and its derivatives $f^{(j)}(z)(j = 1, 2, \cdots, q)$. One of its differential polynomials can be represented as

$$P(z) = P(f, f', \cdots, f^{(q)}) = \sum_{k=1}^{p} a_k(z) \prod_{j=0}^{q} \{f^{(j)}(z)\}^{s_{k_j}} \tag{2}$$

where $a_k(z)$ $(k = 1, 2, \cdots, p)$ are E_0-valued meromorphic functions satisfying

$$T(r, a_k) = \circ\{T(r, f)\}, \text{for} \ \ k = 1, 2, \cdots, p,$$

and $s_{k_j}(k = 1, 2, \cdots, p; j = 0, 1, 2, \cdots, q)$ are non-negative integers.

From the two theorems, we derive the following result.

Theorem 3.6 *If f is an E_0–valued meromorphic function and $P(f, f', \cdots, f^{(q)})$ is a differential polynomial defined by (2) and if f^{-1} exists, then an E_0-valued function $Q(z) = fP(f, f', \cdots, f^{(q)})$ is transcendental and $f(z) \sim Q(z)$.*

Concepts for extending Navanlinna's second fundamental theorem to the case of E_0-valued small function are now in preparations as follows.

Theorem 3.3 *Let $f(z)$ be an E_0-valued meromorphic function and $\lambda \geq 1$, and let $\alpha(a, f)$ represents the order for every pole a of f. Let*

$$n_\lambda(r, f) = \sum_{|a|<r} \min\{\alpha(r, f), \lambda\}$$

where the right-hand side of the above formula is the sum of all poles of f in the disc $|z| \leq r$. Let

$$N_\lambda(r, f) = \int_0^r ((n_\lambda(t, f) - n_\lambda(0, f))/t)dt + n_\lambda(0, f)\log r.$$

Assume that $\psi_k(k = 1, \cdots, p; p \geq 1)$ are linearly independent E_0-valued meromorphic functions satisfying

$$T(r, \psi_k) = S(r, f), \ k = 1, 2, \cdots, p,$$

where $S(r, f)$ satisfies

$$\lim_{\substack{r\to+\infty \\ r\notin\sigma}} \frac{S(r, f)}{T(r, f)} = 0,$$

outside a set σ of the exceptional intervals of finite measure. Assume that $E_\nu(\psi_1, \cdots, \psi_p)$ is a subspace EMF_ν of EMF consisting of all E_0-valued meromorphic functions with the form

$$\prod_{k=1}^{p} \psi_k^{n_k}(n_k \geq 0), \ (k = 1, \cdots, p, \sum_{k=1}^{p} n_k = \nu).$$

Suppose that $I = J\{E_\nu(\psi_1, \cdots, \psi_p)\}$ is the dimension of EMF_ν and its basis is $\{\alpha_j(z)\}_{j=1}^{I}$. Suppose that $I' = J\{E_{\nu+1}(\psi_1, \cdots, \psi_p)\}$ is the dimension of $EMF_{\nu+1}$ and its basis is $\{\beta_j(z)\}_{j=1}^{I'}$. Let

$$P(f) = W(\beta_1, \cdots, \beta_{I'}, f\alpha_1, \cdots, f\alpha_I)(\not\equiv 0)$$

be the Wronskian of $\beta_j(j = 1, \cdots, I')$ and $f\alpha_j(j = 1, \cdots, I)$, and

$$\varphi_j(z) = \sum_{k=1}^{p} c_{jk}\psi_k(z), \quad j = 1, \cdots, q(q \geq 2),$$

where c_{jk} is constant coefficients.

Theorem 3.7 *If $(f - \varphi_j)^{-1}(j = 1, \cdots, q)$ and $P^{-1}(f)$ exist, then*

$$\sum_{j=1}^{q} m(r, (f - \varphi_j)^{-1}) \leq T(r, f) + \tfrac{I'}{I}N(r, f) - \tfrac{1}{I}N(r, (P(f))^{-1}) + S(r, f)$$

and

$$\sum_{j=1}^{q} m(r, (f - \varphi_j)^{-1}) \leq T(r, f) + (I' + \tfrac{(I-1)}{2})\overline{N}(r, f) - \tfrac{1}{I}N(r, (P(f))^{-1}) + S(r, f).$$

ACKNOWLEDGEMENT The authors would like to thank the Organizing Committee of Second Congress ISACC 1999 Fukuoka for its support ,Prof. S. S. CHERN for his guidance,encouragement and help, Prof. G. D. HU, Prof. J. KAJIWARA and Prof. C. C.YANG for his enduring help and support in this work.

References

[1] Chern, S.S., *The integrated form of the first main theorem for complex analytic mappings in several variables*, Ann. of Math., **77**(1960), 536–551.

[2] Chern, S.S., *Complex analytic mappings of Riemann surfaces I.*, Amer.J.Math., **82**(1960), 323–337.

[3] Chuang, C.T., Yang, C.C., He, Y.Z. & Wen, G.C., *Several Topics on Function Theory of One Complex Variable*, Science Press, Beijing, 1995.

[4] Griffiths, Ph. & King, J., *Nevanlinna theory and holomorphic mappings between algebraic varieties*, Acta Math., **130**(1973), 145–220.

[5] Guan, Z.Z., *Lecture Notes in Functional Analysis*, Science Press, Beijing, 1995.

[6] Hu, C.G., *Nevanlinna's theory in Banach spaces*, in "Proceedings of the Fifth International Colloquium on Finite or Infinite Dimensional Complex Analysis", 1997, 109–115.

[7] Hu, C.G., *Vector-Valued Functions and their Applications*, Kluwer Academic Publishers, 1992.

[8] Hu, C.G. and Yang, C.C., *Some remarks on Nevanlinna's theory in a Hilbert space*, Bulletin of the Hong Kong Mathematical Society, **1**(1997), 267–271.

[9] Mujica, J., *Complex Analysis in Banach Spaces*, in "North-Holland Math. Studies", Vol.**120**, North-Holland, Amsterdan, 1986.

[10] Nevanlinna, R., *Zur Treorie der meromorphen Funktionen*, Acta Math., **46**(1925), 1–99.

[11] Osgood, C.F., *Simplified Proofs of Generalizations of the n-Small Function Theorem*, Analysis of One Complex Variables, World Scientific Publishing Co., 1987.

[12] Stoll, W., *Value Distribution on Parabolic Spaces*, in "Lecture Notes in Mathematics" 600, Springer-Verlag, 1977.

[13] Ziegler, H.J.W., *Vector-Valued Nevanlinna Theory*, Pitman Advanced Publishing Program, Boston, London, Melbourne, 1982.

37 DYNAMICS OF TRANSCENDENTAL FUNCTIONS

Xinhou HUA[1]
Institute of Math., Nanjing University
Nanjing 210093, PRC

Xiaoling WANG
P.O.Box 10, Nanjing University of Economics
Nanjing 210003, PRC

Chung-Chun YANG[2]
Department of Mathematics
The Hong Kong University of Science and Technology
Clear Water Bay, Kowloon, Hong Kong

mayang@uxmail.ust.hk

[1] The first author's research was supported in part by NSFC and NSF of Jiangsu province, China
[2] The work of third author was partially supported by a UGC grant of Hong Kong.

H.G.W. Begehr et al. (eds.), Proceedings of the Second ISAAC Congress, Vol. 1, 311–316.

1 Introduction

Let f be a nonconstant meromorphic function. The sequence of the iterates of f is denoted by

$$f^0 = id, f^1 = f, \cdots, f^{n+1} = f^n(f), \cdots.$$

We define

$$F = F(f) = \{z \in \mathbb{C} : \text{the sequence } \{f^n\} \text{ is defined and normal at } z\}$$

and

$$J = J(f) = \mathbb{C} - F(f),$$

they were called the Fatou set and the Julia set of f respectively, here the concept "normal" is in the sense of Montel. According to the definition, F is open (possibly empty) and J is closed.

A point a is called a singular value if it is either a critical value or an asymptotic value. We denote by $sing(f^{-1})$ the set of all finite singular values of f. If the set $sing(f^{-1})$ is bounded, then we say f is of bounded type, and we denote it by $f \in \mathbb{B}$. In particular, if the set $sing(f^{-1})$ is finite, then f is called to be of finite type, and we write $f \in \mathbb{S}$ (cf. [2]).

2 Permutable functions

Let f and g be two nonconstant meromorphic functions. If

$$f(g) = g(f), \tag{1}$$

then we call f and g are permutable.

Fatou [7] proved the following result:

Theorem A *For two given rational functions R_1 and R_2, if they are permutable, then $F(R_1) = F(R_2)$.*

The following question is natural (see Baker [1]):

Question 1 *For two given permutable transcendental entire functions f and g, does it follows that $F(f) = F(g)$?*

This question was affirmatively resolved for some special cases and a complete answer to the question is yet to be found.

Theorem B (Baker [1]) *Suppose that f and g are permutable transcendental entire functions, and $f = g + c$ for some constant c. Then $F(f) = F(g)$.*

A point a is called a singular value of a function f if it is either a critical value or an asymptotic value of f. We denote by $sing(f^{-1})$ the set of all finite singular values of f. If the set $sing(f^{-1})$ is bounded, then we say f is of the bounded type, in particular, if the set $sing(f^{-1})$ is finite, then f is said to be of finite type. We denote them by $f \in B$ and $f \in S$ respectively (cf. [2]).

Poon and Yang [16] proved the following result.

Theorem C *Suppose that f and g are permutable transcendental entire functions. If both $sing(f^{-1})$ and $sing(g^{-1})$ are isolated in the finite complex plane, then $F(f) = F(g)$.*

Remark. Theorem C includes the case that f and g are of the finite type. This conclusion was also proved independently by Ren and Li [17].

For the functions of finite order, the following result is proved.

Theorem D ([15]) *Suppose that f and g are permutable transcendental entire functions of finite order, without wandering domains. Then $J(f) = J(g)$.*

By using a result due to Erememko and Lyubich [6], Hua and Wang [10] proved the following result:

Theorem 1 *Let f and g be two permutable transcendental entire functions. If f and g are of bounded type, then*

$$J(f) = J(g).$$

In general case, Hua and Wang [10] got the following conclusion:

Theorem 2 *If f and g are two permutable transcendental entire functions, then*

$$J(f \circ g) = J(f^n \circ g^m), \forall m, n \geq 1.$$

For the functions of finite order, Hua and Yang [13] proved the following result.

Theorem 3 *Let f and g be two permutable transcendental entire functions of finite order. Suppose that*

$$f(z) = p(z) + p_1(z)e^{q_1(z)} + p_2(z)e^{q_2(z)}, \tag{2}$$

where $p_1(z)$, $p_2(z)$, $q_1(z)$ and $q_2(z)$ are polynomials, $p(z)$ is a non-constant polynomial. Then $J(f) = J(g)$.

Remark. The function f in our theorem may have wandering domains. For example, let $p(z) = z + 2\pi i - 1$, $p_1(z) = 1$, $q(z) = -1$, $p_2(z) = 0$, then $f(z) = z + 2\pi i - 1 + e^{-z}$. This function f has wandering domain (see [9] or [12, p. 90]).

Remark. For two permutable transcendental functions, one cannot conclude that either g and f are linearly related or g is the iterates of f.

Example. Let

$$\begin{aligned} f(z) &= ic\left[\exp\left(\frac{(4k+3)\pi}{8c^2}iz^2\right) + \exp\left(-\frac{(4k+3)\pi}{8c^2}iz^2\right)\right], \\ g(z) &= c\left[\exp\left(\frac{(4k+3)\pi}{8c^2}iz^2\right) - \exp\left(-\frac{(4k+3)\pi}{8c^2}iz^2\right)\right]. \end{aligned}$$

Then $f(g) = g(f)$.

3 Semiconjugation of entire functions

Next we consider semiconjugation of entire functions which is a kind of generalization of permutable functions.

The following result was proved by Bergweiler [3] in 1995.

Theorem E *Let f be an entire function, g an analytic self-map of $\mathbb{C}^*$. Suppose that $E(f) = g(E)$, where $E(z) = \exp(z)$. If f is nonlinear, then $E^{-1}(J(g)) = J(f)$.*

Note that $\exp(z) \in \mathbb{S}$. A natural problem is whether we can replace $\exp(z)$ in Theorem A by a more general function.

We consider the following equation:

$$h \circ f = g \circ h. \tag{3}$$

Theorem 4 *([18]) Let f, h and g be transcendental entire functions such that $f \in \mathbb{B}$ and $h \in \mathbb{S}$. If (3) holds, then $h^{-1}(J(g)) = J(f)$.*

Note that the conditions $f \in \mathbb{B}$ and $h \in \mathbb{S}$ are somewhat strong. By using a different method, Hua and Wang [11] proved the following result.

Theorem 5 *Let f, h and g be transcendental entire functions. Assume that $h(z)$ is periodic with order 1 and mean type. If (3) holds, then $h^{-1}(J(g)) = J(f)$.*

Remark. We just found out that Bergweiler and Hinkkanen [4] obtained a related result:

If f, g and h satisfy (3), and $A(f) \subset J(f)$, where $A(f)$ is the set z such that there exits $L \in \mathbb{N}$ satisfies $|f^n(z)| > M(R, f^{n-L}$ for $n > L$. Then $h^{-1}(J(g)) = J(f)$.

References

[1] Baker, I. N., Wandering domains in the iteration of entire functions, Proc. London Math. Soc.(3)49(1984), 563-576.

[2] Bergweiler, W., Iteration of meromorphic functions, Bull. Amer. Math. Soc. 29(1993), 151-188.

[3] Bergweiler, W., On the Julia set of analytic self-maps of the punctured plane, *Analysis* 15(1995), 252-256.

[4] Bergweiler W. and Hinkkanen A., On semiconjugation of entire functions, Math. Proc. Camb. Phil. Soc. 126(1999), 565-574.

[5] Eremenko, A.E., M.Yu., On the iteration of entire functions, Dynamical Systems and Ergodic Theory, Banach Center Publ., Vol. 23, Polish Scientific Publishers, Warsaw 1989, 339-345.

[6] Eremenko, A.E. and Lyubich, M.Yu.: Dynamical properties of some classes of entire functions, Ann. Inst. Fourier 42(1992), 989-1020.

[7] Fatou, P., Sur les équations fonctionelles, *Bull. Soc. Math. France* 47(1919), 161-271; 48(1920), 33-94 and 208-314.

[8] Fatou, P., Sur l' itération des fonctions transcendentes entières, Acta Math. 47(1926), 337-370.

[9] Herman, M., Exemples de fractions rationelles ayant une orbite dense sur la sphere de Riemann, Bull. Soc. France 112(1984), 93-142.

[10] Hua X.,H. and Wang X.L., Dynamics of permutable transcendental entire functions, Proc. Int. Conf. at Vietnam 1998.

[11] Hua X.,H. and Wang X.L., Complex dynamics concerning functional equations, Preprint.

[12] Hua X.H. and Yang C.C., Dynamics of Transcendental Functions, Gordon and Breach Science Publishers 1998.

[13] Hua X.H. and Yang C.C., Dynamics of permutable functions, Preprint.

[14] Julia, G., Mémoire sur l' iteration des fractions rationnelles, J. Math. Pures Appl., (8)1(1918), 47-245.

[15] Langley J.K., Permutable entire functions and Baker domains, Math. Proc. Camb. Phil. Soc. 125(1999), 119-121.

[16] Poon, K.K. and Yang, C.C. Fatou sets of two permutable entire functions, Preprint.

[17] Ren, F.Y. and Li, W.S., An affirmative answer to a problem of Baker, J. Fudan Univ. 36(1997), 231-233.

[18] Wang X.L. and Hua X.H., Dynamics concerning functional equations, Preprint.

38 ON THE GROWTH OF SOLUTIONS OF ALGEBRAIC DIFFERENTIAL EQUATIONS

Liangwen LIAO

Department of Mathematics
The Hong Kong University of Science and Technology
Clear Water Bay, Kowloon, Hong Kong

maliao@ust.hk

1991 Mathematics Subject Classification: 34A20, 30D35
Key words: algebraic differential equation, order, Wiman-Valiron theory, normal family, differential polynomial, normal family

Abstract. In this paper, the growth of the meromorphic solutions of the equation

$$f'' = L(z,f)(f')^2 + M(z,f)f' + N(z,f),$$

where L, M, N are birational functions, is studied. We have proved that if $L(z,f)$ satisfies a quite general condition, then f must be of finite order. Furthermore, if $L(z,f) \equiv 0$, and $M(z,f)$, $N(z,f)$ are polynomials in f, then the order of any entire solution of the equation is a positive integral multiple of $\frac{1}{2}$. Furthermore, we give a criterion of the normality relating to algebraic differential equations and use it to derive the growth of some solutions of a certain kind of algebraic differential equations.

H.G.W. Begehr et al. (eds.), Proceedings of the Second ISAAC Congress, Vol. 1, 317–327.

1 Introduction

The growth of solutions of differential equations has always been an important topic in the study of differential equations. It is known that all single-valued solutions of linear differential equations with rational functions as the coefficients are meromorphic functions of finite and positive orders, (see e.g. Laine [12] and Gundersen etc. [9]). However, it is not easy to estimate the growth of an entire or meromorphic solution of a general nonlinear algebraic differential equation of the form:

$$P(z, f, f', \dots f^{(n)}) = 0, \tag{1}$$

where P is a polynomial in all its arguments. The first general result that relates to the first-order algebraic differential equations is the following one.

Theorem A. *(Gol'dberg [6]) Let $P(u_1, u_2, u_3)$ be a polynomial in all of its arguments u_1, u_2 and u_3. Consider the first-order algebraic differential equation*

$$P(z, f, f') = 0. \tag{2}$$

Then all meromorphic solutions of (2) are of finite order.

Later, different proofs of this result were given by Bank and Kaufman [1], Barsegian [2], etc. Recently, Barsegian [3] generalized Gol'dberg's result and obtained the following:

Theorem B. *Let f be a meromorphic solution of the following algebraic differential equation*

$$a_0(z, f)(f')^n = \sum_{i \in I} a_i(z, f)(f')^{i_1}(f'')^{i_2} \dots (f^{(m)})^{i_m},$$

where $I = \{i = (i_1, i_2, \dots, i_m), i_k \in N \bigcup \{0\}\}$ is a finite index set, $a_i(z, f)(\not\equiv 0)$ are polynomials in two variables. If $n > \max_{i \in I}\{i_1 + 2i_2 + \dots + m i_m\}$, then f is of a finite order.

Bergweiler [4] gave a new and simple proof of Theorem B by using the useful and well-known Zalcman's lemma [20], which is being used in this paper as Lemma 3. We refer the reader to [5] for further estimation of the orders. But little is known about the growth of solutions of algebraic differential equations of any order higher than one. We recall here the following challenging conjecture about the growth of solutions of algebraic differential equations.

Conjecture 1. *([7], [10]) If $f(z)$ is a meromorphic solution of (1), then*

$$T(r, f) < a\exp_{n-1}(br^c), \qquad 0 \leq r < \infty$$

where a, b, c are positive constants and $\exp_l(x)$ *is the l times iterated exponential.*

Recently, Bergweiler [4], Hayman [10], Liao-Yang [13] discussed the growth of entire and meromorphic solutions of some special classes of nonlinear algebraic differential equations. In this paper, we will discuss the growth of the solutions of certain types of Second-order, establish a normality criterion and discuss the growth of meromorphic solutions of a certain kind of algebraic differential equations.

2 The Growth of Meromorphic Solutions of Second Order Differential Equations

One of the traditional ways to study algebraic differential equations is to transform a given equation into an equation of standard form, for instance, Malmquist-Yosida type of equation. A classical and yet unsolved conjecture for the second-order differential equations

$$f'' = R(z, f, f'), \tag{3}$$

where R is rational in z, f and f', is the following:

Conjecture 2. *([12], p251) If the equation (3) has a transcendental meromorphic solution, then the equation can be transformed into the form:*

$$f'' = L(z, f)(f')^2 + M(z, f)f' + N(z, f), \tag{4}$$

where L, M, N are rational functions in their arguments.

In [4], Bergweiler proved that if a meromorphic solution f of the equation (4) is of infinite order and $l(w) = L(\infty, w)$ is well-define, then $l(w)$ assumes one of the six special forms. Clearly this result requires some very restricted conditions on the coefficients of $L(z, w)$. In this paper, we give, by using argument similar to that of [4], some estimations on the growth of meromorphic solutions of the equation (4) when $L(z, w)$ is much less restricted. But we do not give detail proof for the theorems in this section, we refer readers to [14] for detail proof.In order to prove our results, we need some lemmas.

Lemma 1. *(Zalcman [20]) Let $\mathcal{F}$ be a family of meromorphic functions on D: $|z| < 1$ and α be a real number satisfying $-1 < \alpha < 1$. Then $\mathcal{F}$ is not normal in D iff there exist*

(i) a number r, $0 < r < 1$;

(ii) a sequence points z_k, $|z_k| < r$;

(iii) a positive sequence ρ_k, $\rho_k \to 0$ and

(iv) a sequence $\{f\}_{n\in N} \subset \mathcal{F}$,

such that $\rho_k^\alpha f_k(z_k + \rho_k\zeta) \to g(\zeta)$ spherically uniformly on compact subsets of $\mathbb{C}$, where g is a non-constant meromorphic function.

Remark 1. *One can choose w_k and ρ_k (see, e. g.[15]) properly such that,*

$$\rho_k \leq \frac{2}{f_k^\#(w_k)^{\frac{1}{1+|\alpha|}}}, \quad f_k^\#(w_k) \geq f_k^\#(0).$$

Here $f^\#(z)$ denotes the spherical derivative $\frac{|f'|}{1+|f|^2}$. The following two lemmas in Wiman-Valiron theory are frequently used in the study of the growth of entire solutions of differential equations, see, e.g., [12].

Lemma 2. *If f is an entire function of order ρ, then*

$$\rho = \overline{\lim_{r\to\infty}} \frac{\log^+ \nu(r,f)}{\log r} = \overline{\lim_{r\to\infty}} \frac{\log\log^+ \mu(r,f)}{\log r}. \tag{5}$$

Lemma 3. *Let f be a transcendental entire function, and $0 < \delta < \frac{1}{4}$. Suppose that at the point z with $|z| = r$ the inequality*

$$|f(z)| > M(r,f)\nu(r,f)^{-\frac{1}{4}+\delta} \tag{6}$$

holds. Then there exists a set F in R^+ and of finite logarithmic measure, i.e.,$\int_F \frac{dt}{t} < +\infty$ such that

$$f^{(m)}(z) = \left(\frac{\nu(r,f)}{z}\right)^m (1+o(1))f(z) \tag{7}$$

holds whenever m is a fixed non negative integer and $r \notin F$.

Lemma 4. *(Wittich [19]) Given a Riccati differential equation:*

$$f' = R_0(z) + R_1(z)f + R_2(z)f^2, \tag{8}$$

where the $R_j(z)$ are rational functions. If f is a meromorphic solution of equation (8), then the order of f is a positive integral multiple of $\frac{1}{2}$.

Lemma 5. *(Steinmetz [17]) Let f be a transcendental meromorphic solution of the equation (10). Then either f satisfies a Riccati differential equation with rational functions as the coefficients, or (10) can be reduced to*

$$f'' = p_0(z) + p_1(z)f + p_2(z)f^2 + p_3(z)f^3 + q_0(z)f' + q_1(z)ff', \qquad (9)$$

where $p_j(z)$ and $q_j(z)$ are rational functions.

By using Lemma 1 and Remark 1, one can prove the following Theorem.

Theorem 1. *Let f be a solution of (4). Further assume that $L(z,f) \not\equiv 0$ in the equation (4) has the form:*

$$L(z,f) = \frac{P(z,f)}{Q(z,f)} = \frac{a_n(z)f^n + a_{n-1}(z)f^{n-1} + \ldots + a_s(z)f^s}{b_m(z)f^m + b_{m-1}(z)f^{m-1} + \ldots + b_r(z)f^r},$$

where $a_i(z), b_j(z) (s \le i \le n, r \le j \le m)$ are rational functions. If $m - n < 1$ or $r - s > 1$, then $\rho(f) < \infty$.

Remark 2. *Both of the conditions $m - n < 1$ and $r - s > 1$ can not be omitted simultaneously. For example, $f(z) = e^{e^z}$ satisfies the following algebraic differential equation*

$$ff'' - (f')^2 - ff' = 0.$$

Remark 3. *By similar argument as in the proof of Theorem 1, one can show that if f is a transcendental meromorphic solution of the equation:*

$$(f'')^n = L(z,f)(f')^{2n} + M_{2n-1}(z,f)(f')^{2n-1} + \ldots + M_1(z,f)f' + M_0(z,f),$$

where $L(z,f), M_j(z,f) (j = 0,1,\ldots,2n-1)$ are rational functions in z and f with $L(z,f)$ satisfying the condition in Theorem 1, then $\rho(f) < \infty$.

Next, we consider the case of $L(z,w) \equiv 0$ in equation (4). By using Lemma 2, 3 and 4, one can prove the following theorem.

Theorem 2. *Let f be a transcendental entire solution of the following equation*

$$f'' = M(z,f)f' + N(z,f), \qquad (10)$$

where $N(z,f)$, $M(z,f)$ are polynomials in f with rational functions as the coefficients. Then the order of f is a positive integral multiple of $\frac{1}{2}$.

Remark 4. *The conclusion of the theorem is not true for a transcendental meromorphic solution of (10). For example, the function $f(z) = e^z \tan(e^z)$ satisfies the equation*

$$f'' = (3 + 2f)f' - 2f - 2f^2.$$

3 Normal Families and Algebraic Differential Equations

Normal families is an important and useful concept in the complex analysis, several normality criteria have been given.

Theorem C. *(Marty's criterion, see [16]) A family $\mathcal{F}$ of meromorphic functions on a domain G is normal iff the spherical derivatives*

$$f^{\#}(z) = \frac{|f'(z)|}{1+|f(z)|^2}$$

are locally uniformly bounded for all $f \in \mathcal{F}$.

Theorem D. *(Montel's criterion, see [16]) Let $\mathcal{F}$ be a family of holomorphic functions in a domain $D \subset \hat{\mathbb{C}}$. Suppose that there exists $M > 0$ such that $|f(z)| \leq M$ for all $z \in D$ and all $f \in \mathcal{F}$. Then $\mathcal{F}$ is normal in D.*

Also some normality criteria relating to linear differential polynomial have been given,e.g. see [8], [16]. In order to prove our results, we need following lemmas,

Lemma 6. *(see,[8], P31) Let $f(z)$ be a meromorphic function in $|z| < R(0 < R \leq +\infty)$, and $f(0) \neq 0, \infty$. Suppose that k is an integral number, $0 < r < \rho < R$. Then*

$$m(r, \frac{f^{(k)}}{f}) <$$

$$K\{1 + \log^+ \log^+ \frac{1}{|f(0)|} + \log^+ \frac{1}{r} + \log^+ \frac{1}{\rho - r} + \log^+ \rho + \log^+ T(\rho, f)\},$$

where K is constant only dependent on k.

Lemma 7. *(see, [8], P42) Let $U(r)$ be a positive non decrease function in $0 < r < R$. And suppose that a, b, c are positive number such that for $0 < r < \rho < R$*

$$U(r) < a \log^+ U(\rho) + b(\log^+ \frac{1}{\rho - r} + \log^+ \rho + \log^+ \frac{1}{r}) + c.$$

Then for all r in $0 < r < R$,

$$U(r) < 4(a+b)(\log \frac{R}{R-r} + \log^+ R + \log^+ \frac{1}{R} + 20(a+b+1)^2 + 2c$$

Lemma 8. *Let f be an analytic function in $|z - z_0| < R \leq \infty$, and f satisfy the following non-homogeneous differential equation $P[z, f, f', ..., f^{(n)}] = 0$. But for some homogeneous part $P_q[z, f, f', ..., f^{(n)}]$ of $P[z, f, f', ..., f^{(n)})]$, f is not a solution of $P_q[z, f, f', ..., f^{(n)}] = 0$. Let $p = \max\{q, P_q \not\equiv 0\}$. Further suppose that $f(z) \neq 0$ in $|z - z_0| < R \leq \infty$, $P_p[z_0, f(z_0), f'(z_0), ..., f^{(n)}(z_0)] \neq 0$. Then*

$$m(r, f) \leq$$

$$K\{1 + \log^+ \log^+ \frac{1}{f(z_0)} + \log^+ \frac{1}{r} + \log^+ \frac{1}{\rho - r} + \log^+ \rho + \log^+ T(\rho, f)\}.$$

Proof. No loss generality, we assume $z_0 = 0$.By the condition of the Lemma, we have

$$P_p[z, f, f', ..., f^{(n)}] = - \sum_{t \leq p-1} P_t[z, f, f', ..., f^{(n)}]. \tag{11}$$

We denote the left side of (11) by $E(z)$ and right side by $F(z)$. By assumption $E(z) \not\equiv 0$ in $|z - z_0| \leq R$, so $F(z) \not\equiv 0$. By Lemma 6, for any monomial $M_t[z, f(z), f'(z), ...f^{(n)}(z)]$ of degree $t(t \leq p - 1)$, one can get

$$\frac{1}{2\pi} \int_{|f(z)| \geq 1} \log^+ \left| \frac{M_t[z, f(z), f'(z), ...f^{(n)}(z)]}{f(z)^{p-1}} \right| d\theta \leq$$

$$K\{\sum_{k=1}^{n} m(r, \frac{f^{(k)}}{f}\} + A \log^+ r + B$$

$$\leq K\{1 + \log^+ \log^+ \frac{1}{f(0)} + \log^+ \frac{1}{r} + \log^+ \frac{1}{\rho - r} + \log^+ \rho + \log^+ T(\rho, f)\}. \tag{12}$$

Now, we set $G = \frac{F}{f^{p-1}}$, then by (12) we can get

$$\frac{1}{2\pi} \int_{|f(z)| \geq 1} \log^+ |G(z)| d\theta \leq$$

$$K\{1 + \log^+ \log^+ \frac{1}{f(0)} + \log^+ \frac{1}{r} + \log^+ \frac{1}{\rho - r} + \log^\rho + \log^+ T(\rho, f)\}. \tag{13}$$

Now (11) can be written as

$$G(z) = \frac{E(z)}{f(z)^p} f(z). \tag{14}$$

Hence when $|f(z)| < 1$, $|G(z) < |\frac{E(z)}{f(z)^p}|$. We also have

$$\frac{1}{2\pi}\int_{|f(z)|\le 1}\log^+|G(z)|d\theta \le K\{1+\log^+\log^+\frac{1}{f(0)}+\log^+\frac{1}{r}+ \log^+\frac{1}{\rho-r}+\log^+\rho+\log^+T(\rho,f)\}. \tag{15}$$

Hence, it follows from (12) and (15) that

$$m(r,G)\le K\{1+\log^+\log^+\frac{1}{f(0)}+\log^+\frac{1}{r}+\log^+\frac{1}{\rho-r}+\log^+\rho+\log^+T(\rho,f)\}. \tag{16}$$

Let $H(z)=\frac{E(z)}{f(z)^p}$. Note $f(z)=\frac{G(z)}{H(z)}$, by the first fundamental theorem, Lemma 6 and (3), we have

$$\begin{aligned} m(r,f)&\le m(r,G)+m(r,\frac{1}{H})\le \\ &m(r,G)+m(r,H)+\log^+|H(0)|+N(r,H)\le \\ &K\{1+\log^+\log^+\frac{1}{f(0)}+\log^+\frac{1}{r}+\log^+\frac{1}{\rho-r}+ \\ &\log^+\rho+\log^+T(\rho,f)+\sum_{k=1}^{n}\log^+|\frac{f^k(0)}{f(0)}|\}. \end{aligned} \tag{17}$$

Noting that $f(z)$ has no poles, The estimations for $\log\left|\frac{f^k(0)}{f(0)}\right|$ follows from Poisson-Jensen formula.

The conclusion of the lemma follows from this and (17). □

Theorem 3. *Let $\mathcal{F}=\{f\}$ be a family of analytic functions in the domain D,*

$$P(z,f,f',...,f^{(n)})$$

be a nonhomogeneous algebraic differential polynomial and P_q be homogeneous part of P with degree q. If for any $f\in\mathcal{F}$, $f(z)$ has no zero in D and f satisfies the algebraic differential equation $P(z,f,f',...,f^{(n)})=0$, but for some $q>0$, $P_q(z,f,f',...,f^{(n)})\not\equiv 0$, then $\mathcal{F}$ is a normal family.

Proof. We only need to prove that for any point $z_0 \in D$, we can find a neighborhood of z_0:$N(z_0, \delta) = \{z : |z - z_0| < \delta\}$ and a constant M such that $f(z) \leq M$ in $N(z_0, \delta)$ for all $f \in \mathcal{F}$. We choose $1 > \delta > 0$ such that $N(z_0, 5\delta) \subset D$ and denote $p = \max\{q : M_q[z, f, f', ..., f^{(n)}] \not\equiv 0\}$, $E_f = \{z : M_q[z, f, f', ..., f^{(n)}] \not\equiv 0\}$. If $|f(z)| \leq 1$ in $|z - z_0| < \delta \backslash E_f$, then $|f(z)| \leq 1$ in $|z - z_0| < \delta$ by the continuity of $f(z)$. If there exists $z_1 \in N(z_0, \delta)\backslash E_f$, then by Lemma 3 we have for $|z - z_1| = r < \rho < 4\delta$,

$$T(r, z_1, f) = m(r, z_1, f) \leq$$

$$K\{1 + \log^+ \frac{1}{r} + \log^+ \frac{1}{\rho - r} + \log^\rho + \log^+ T(\rho, f)\}.$$

Then by Lemma 7, we have for $r < 4\delta$

$$T(r, z_1, f) < K\{\log \frac{4\delta}{4\delta - r} + \log^+ \frac{1}{4\delta} + 1\}.$$

Thus

$$\log^+ M(2\delta, z_1, f) \leq \frac{3\delta + 2\delta}{\delta} T(3\delta, z_1, f) \leq K.$$

Noting that $N(z_0, \delta) \subset N(z_1, 2\delta)$, we have

$$M(\delta, z_0, f) \leq M(2\delta, z_1, f) \leq e^K.$$

Thus, the proof is completed. □

Remark 5. *The condition $P_q(z, f, f', ..., f^{(n)}) \not\equiv 0$ for some q in the theorem is needed. For example, let $f(z) = e^{e^z}$. Since $\rho(f) = \infty$, we can find a sequence $\{a_n\}$ such that $f^{\#}(a_n) \to \infty$. Then $\{g_n(z) = f(a_n + z)\}$ are not normal at 0, and $g_n(z)$ satisfies the following algebraic differential equation in $|z| < 1$ for each n:*

$$ww'' - w'^2 - ww' = 0.$$

By using Theorem 3, we can show the following result.

Theorem 4. *Let f be a transcendental entire function with only finite zeros . If f satisfies a nonhomogeneous algebraic differential equation with constant coefficients*

$$P[f, f', ..., f^{(n)}] = 0,$$

but for some q, $P_q[f, f', ..., f^{(n)}] \not\equiv 0$, then we have $\rho(f) \leq 2$.

Remark 6. *:The condition f with only finite zeros can not be omitted. For example,$f(z) = e^{e^z} - a, a \neq 0$ satisfies the following non homogeneous algebraic differential equation with constant coefficients*

$$P[f, f', f''] = ff'' - (f')^2 - ff' + a(f'' - f') = 0$$

but $P_1[f, f', f''] \not\equiv 0, P_2[f, f', f''] \not\equiv 0$.

References

[1] S. Bank and R. Kaufman, On meromorphic solutions of first-order differential equations, Comment. Math. Helv., **51**(1976), 289-299.

[2] G. Barsegian, Estimates of derivatives of meromorphic functions on sets of a-points, J. London Math Soc. (2) **34**(1986), 534-540.

[3] G. Barsegian, On a method of study of algebraic differential equations, Bull. Hong Kong Math. Soc. **2**, 159-164.

[4] W. Bergweiler, On a theorem of Gol'dberg concerning meromorphic solutions of algebraic differential equations, Complex Variables, **37**(1998), 93-96.

[5] G. Frank and Y. Wang, On the meromorphic solutions of algebraic differential equations, Analysis, **18**(1998), 49-54.

[6] A. A. Gol'dberg On single-valued solutions of first-order differential equations, Ukrain. Mat. Zh., **8**(1956), 254-261.

[7] A. A. Gonchar, V.P. Havin and N. K. Nikolski, Encyclopaedia of Mathematical Sciences, volume 85: Complex Analysis I, Springer, Berlin-Heidelberg-New York 1997.

[8] Y. Gu, Normal families of meromorphic functions (Chinese), Sichuan Educational Press, Chengdu, China, 1991.

[9] G. Gundersen, M. Steinbart and S. Wang, The possible orders of linear differential equations with polynomial coefficients, Trans. Amer. Math. Soc., **350**(1998), 1225-1247.

[10] W. K. Hayman, The growth of solutions of algebraic differential equations, Rend. Mat. Acc. Lincei **7**(1996), 67-73.

[11] G. Jank and L. Volkmann, Einführung in die Theorie der ganzen und meromorphen Funktionen mit Anwendungen auf Differentialgleichungen, Birkhäuser, Basel-Boston-Stuttgart 1985.

[12] I. Laine, Nevanlinna Theory and Complex Differential Equations, Walter de Gruyter, Berlin-New York, 1993.

[13] L. W. Liao and C. C.Yang, On the growth and factorization of entire solutions of algebraic differential equations, to appear in Ann. Acad. Sci. Fenn. Ser. A I Math.

[14] L. W. Liao & C. C. Yang, On the growth of meromorphic and entire solutions of algebraic differential equations, to appear in Annalidi Matematica Pura ed Applicata.

[15] X. Pang, Bloch's principle and normal criterion, Sci. China Ser A, **32**(1989), 782-791.

[16] W. Schwick, Normality criteria for families of meromorphic functions, Journal d'Analyse Math. **52**(1989), 241-289.

[17] N. Steinmetz, Über eine Klasse von Painleveschen Differentialgleichungen, Arch. Math., **41**(1983), 261-266.

[18] G. Valiron, Lectures on the general theory of integral functions, translated by E. F. Collingwood, Chelsea, New York, 1949.

[19] H. Wittich, Eindeutige Lösungen der Differentialgleichungen $w' = R(z, w)$, Math. Z., **70**(1960), 278-288.

[20] L. Zalcman, Normal families: New perspectives, Bull. Amer. Math. Soc., **35**(1998), 215-230.

39 NUMERICAL CONFORMAL MAPPING BY THE CHARGE SIMULATION METHOD AND ITS APPLICATION TO POTENTIAL FLOW ANALYSIS [1]

Kaname AMANO
Dai OKANO
Hiromi SHIMOHIRA

Department of Computer Science
Faculty of Engineering
Ehime University
Matsuyama 790-8577, Japan

amano@cs.ehime-u.ac.jp
okano@cs.ehime-u.ac.jp

Masaaki SUGIHARA

Department of Computational Science and Engineering
Graduate School of Engineering
Nagoya University
Nagoya 464-8603, Japan

sugihara@cse.nagoya-u.ac.jp

[1]Supported by the Grant-in-Aid for Scientific Research of the Ministry of Education, Science, Sports and Culture in Japan (09440081, 10554003), and by the Japan Society for Promotion of Science.

H.G.W. Begehr et al. (eds.), Proceedings of the Second ISAAC Congress, Vol. 1, 329–336.

1 Introduction

Conformal mappings are familiar in science and engineering. But exact mapping functions are not known except for special domains. Particularly confomal mappings of a domain exterior to closed Jordan curves onto parallel, circular and radial slit domains are well known as a method of two-dimensional potential flow analysis. But no systematic method of computation has been established for these mappings. We here present a simple method of numerical conformal mappings of the multiply-connected domains and apply it to a typical flow problem. We use the charge simulation method for computation, which is known as a rapid solver of potential problems. A pair of conjugate harmonic functions are approximated by a linear combination of complex logarithmic potentials.

See Nehari [5], Henrici [3], Trefethen [6] and Kythe [4] for overviews of conformal mappings and their computational methods.

2 Problems and Mapping Theorems

The entire plane with slits or cuts is called a slit domain.

Consider the conformal mappings $w = f_{\mathrm{p}v}(z)$ and $w = f_{\mathrm{p}u}(z)$ of a domain D exterior to the closed Jordan curves $C_1, C_2, \ldots, C_n$ including the point at infinity in the $z(= x + \mathrm{i}y)$-plane onto parallel slit domains with rectilinear slits parallel to the v-axis and parallel to the u-axis in the $w(= u + \mathrm{i}v)$-plane, respectively. The suffices $()_{\mathrm{p}v}$ and $()_{\mathrm{p}u}$ are omitted hereafter if not indispensable. Then the mapping functions are uniquely determined by the normalization condition that the Laurent expansion at infinity is of the form

$$f(z) = z + \frac{a_1}{z} + \frac{a_2}{z^2} + \cdots, \tag{1}$$

i.e., $f(\infty) = \infty$, $f'(\infty) = 1$ and $a_0 = 0$. As a result $C_1, C_2, \ldots, C_n$ are mapped onto the parallel slits $S_{\mathrm{p}v1}, S_{\mathrm{p}v2}, \ldots, S_{\mathrm{p}vn}$ on $u = u_1, u_2, \ldots, u_n$ or $S_{\mathrm{p}u1}, S_{\mathrm{p}u2}, \ldots, S_{\mathrm{p}un}$ on $v = v_1, v_2, \ldots, v_n$.

Consider also the conformal mappings $w = f_{\mathrm{c}}(z)$ and $w = f_{\mathrm{r}}(z)$ of the domain D onto a circular slit domain with concentric circular slits having the center at the origin and onto a radial slit domain with radial rectilinear slits pointing at the origin, respectively. The suffices $()_{\mathrm{c}}$ and $()_{\mathrm{r}}$ are also omitted if not indispensable. Assume that the origin $z = 0$ lies in D, which does not lose generality. Then the mapping functions are uniquely determined by the normalization conditions $f(0) = 0$, and

$f(\infty) = \infty$ and $f'(\infty) = 1$. The conditions at infinity mean that the Laurent expansion is of the form

$$f(z) = z + b_0 + \frac{b_1}{z} + \cdots. \tag{2}$$

As a result $C_1, C_2, \ldots, C_n$ are mapped onto the circular slits $S_{c1}, S_{c2}, \ldots, S_{cn}$ with the radii $r_1, r_2, \ldots, r_n$ or onto the radial slits $S_{r1}, S_{r2}, \ldots, S_{rn}$ with the arguments $\theta_1, \theta_2, \ldots, \theta_n$.

3 Numerical Method

We express the mapping function $f_{pv}(z)$ as

$$f_{pv}(z) = z + g(z) + \mathrm{i}h(z), \tag{3}$$

where $g(z)$ and $h(z)$ are conjugate harmonic functions in D. They should satisfy the boundary condition $\mathrm{Re}\, f_{pv}(z) = u_m$ on C_m, i.e.,

$$x + g(z) = u_m \quad (z \in C_m, \quad m = 1, 2, \ldots, n) \tag{4}$$

and the normalization condition (1), i.e.,

$$g(\infty) + \mathrm{i}h(\infty) = 0. \tag{5}$$

From the existence and uniqueness of the mapping function, the problem is now to obtain $g(z)$ and $h(z)$ satisfying (4) and (5).

We apply the charge simulation method and approximate $g(z)+\mathrm{i}h(z)$ by a linear combination of complex logarithmic potentials as

$$G(z) + \mathrm{i}H(z) = \sum_{l=1}^{n} \sum_{i=1}^{N_l} Q_{li} \log(z - \zeta_{li}), \tag{6}$$

where N_l charge points $\zeta_{l1}, \zeta_{l2}, \ldots, \zeta_{lN_l}$ are placed inside the boundary curve C_l. The charges $Q_{11}, Q_{12}, \ldots, Q_{nN_n}$ are determined to satisfy the boundary condition (4) at the same number of collocation points placed on the boundary, i.e., N_m collocation points $z_{m1}, z_{m2}, \ldots, z_{mN_m}$ on C_m. That is to say they should satisfy the linear equations

$$\sum_{l=1}^{n} \sum_{i=1}^{N_l} Q_{li} \log|z_{mj} - \zeta_{li}| - U_m = -x_{mj} \tag{7}$$

$$(z_{mj} \in C_m, \quad j = 1, 2, \ldots, N_m, \quad m = 1, 2, \ldots, n)$$

called collocation condition. From the normalization condition (5),

$$\sum_{l=1}^{n}\sum_{i=1}^{N_l} Q_{li} = 0. \tag{8}$$

Moreover, since the mapping function should be single-valued in D,

$$\int_{\tilde{C}_l} dH(z) = 0, \quad \text{i.e.,} \quad \sum_{l=1}^{N_l} Q_{li} = 0 \quad (l = 1, 2, \ldots, n), \tag{9}$$

where $\tilde{C}_l$ is an arbitrary closed curve lying in D and surrounding only C_l. Equation (9) implies (8).

Equations (7) and (9) constitute a system of $N_1 + N_2 + \cdots + N_n + n$ simultaneous linear equations for $N_1 + N_2 + \cdots + N_n + n$ unknowns $Q_{11}, Q_{12}, \ldots, Q_{nN_n}$ and $U_1, U_2, \ldots, U_n$.

We use the principal value of logarithmic function in computation. If $C_1, C_2 \ldots, C_n$ are starlike with respect to their inside points $z_{10}, z_{20} \ldots$ $, z_{n0}$, we can change (6) to

$$\begin{aligned} G(z) + \mathrm{i}H(z) &= \sum_{l=1}^{n}\sum_{i=1}^{N_l} Q_{li} \log(z - \zeta_{li}) - \sum_{l=1}^{n}\sum_{i=1}^{N_l} Q_{li} \log(z - z_{l0}) \\ &= \sum_{l=1}^{n}\sum_{i=1}^{N_l} Q_{li} \log\left(\frac{z - \zeta_{li}}{z - z_{l0}}\right), \end{aligned} \tag{10}$$

which is continuous with the principal value on the closure $\bar{D}$.

In a similar way, we can construct the approximate mapping functions $F_{\mathrm{pu}}(z)$, $F_{\mathrm{c}}(z)$ and $F_{\mathrm{r}}(z)$ using the boundary conditions $\operatorname{Im} F_{\mathrm{pu}}(z) = v_m$, $|F_{\mathrm{c}}(z)| = r_m$ and $\arg F_{\mathrm{r}}(z) = \theta_m$ on C_m, respectively [1, 2]. They are integrated into the following algorithm.

Algorithm 1 *If $C_1, C_2 \ldots, C_n$ are starlike with respect to their inside points z_{10},*
$z_{20} \ldots, z_{n0}$, the approximate mapping functions are expressed as

$$F_{\mathrm{pv}}(z) = z + G(z) + \mathrm{i}H(z),$$

$$F_{\mathrm{pu}}(z) = z + \mathrm{i}(G(z) + \mathrm{i}H(z)),$$

$$F_{\mathrm{c}}(z) = z \exp(G(z) + \mathrm{i}H(z)),$$

$$F_{\mathrm{r}}(z) = z \exp(\mathrm{i}(G(z) + \mathrm{i}H(z))),$$

where

$$G(z) + \mathrm{i}H(z) = \sum_{l=1}^{n} \sum_{i=1}^{N_l} Q_{li} \operatorname{Log}\left(\frac{z - \zeta_{li}}{z - z_{l0}}\right).$$

The charges $Q_{11}, Q_{12}, \ldots, Q_{nN_n}$ *and the positions of the slits* $S_1, S_2, \ldots, S_n$ *are solutions of the* $N_1 + N_2 + \cdots + N_n + n$ *simultaneous linear equations*

$$\sum_{l=1}^{n} \sum_{i=1}^{N_l} Q_{li} \log \left| \frac{z_{mj} - \zeta_{li}}{z_{mj} - z_{l0}} \right| - S_m = -t_{mj}$$

$$(z_{mj} \in C_m, \quad j = 1, 2, \ldots, N_m, \quad m = 1, 2, \ldots, n),$$

$$\sum_{i=1}^{N_l} Q_{li} = 0$$

$$(l = 1, 2, \ldots, n),$$

where

$$S_m = U_m, \ t_{mj} = x_{mj} \quad \textit{for} \quad F_{\mathrm{pv}}(z),$$

$$S_m = V_m, \ t_{mj} = y_{mj} \quad \textit{for} \quad F_{\mathrm{pu}}(z),$$

$$S_m = \log R_m, \ t_{mj} = \log|z_{mj}| \quad \textit{for} \quad F_{\mathrm{c}}(z),$$

$$S_m = \Theta_m, \ t_{mj} = \operatorname{Arg} z_{mj} \quad \textit{for} \quad F_{\mathrm{r}}(z).$$

We can obtain another scheme if the boundary curves $C_1, C_2 \ldots, C_n$ are not starlike.

4 An Example

The problem domain D is the exterior of three disks with different radii:

$$|z - z_{l0}| > \rho_l, \quad z_{l0} = 2 \exp \frac{2(l-1)\pi \mathrm{i}}{3} \quad (l = 1, 2, 3),$$

$$\rho_1 = 1, \quad \rho_2 = 0.5, \quad \rho_3 = 1.5.$$

Collocation points and charge points are placed by

$$z_{lj} = z_{l0} + \rho_l \exp \frac{2(j-1)\pi \mathrm{i}}{N}, \quad \zeta_{lj} = z_{l0} + q\rho_l \exp \frac{2(j-1)\pi \mathrm{i}}{N}$$

$$(j = 1, 2, \ldots, N, \quad l = 1, 2, 3),$$

Table 1: Numerical results of $F_{\mathrm{p}v}(z)$, $F_{\mathrm{p}u}(z)$, $F_{\mathrm{c}}(z)$ and $F_{\mathrm{r}}(z)$ for the exterior of three disks ($N = 64$, $q = 0.8$).

	$E_{\mathrm{Re}l}$	E_{Ul}	U_l	cond
C_1	7.1×10^{-8}	2.0×10^{-8}	1.32203175	
C_2	2.2×10^{-8}	4.9×10^{-9}	−0.787052952	4.7×10^4
C_3	1.2×10^{-7}	9.7×10^{-9}	−0.697257057	
	$E_{\mathrm{Im}l}$	E_{Vl}	V_l	cond
C_1	5.0×10^{-8}	1.0×10^{-8}	−0.326576607	
C_2	3.0×10^{-8}	1.7×10^{-8}	0.99641470	4.7×10^4
C_3	8.1×10^{-8}	7.7×10^{-9}	−1.478338666	
	$E_{\mathrm{Mo}l}$	E_{Rl}	R_l	cond
C_1	2.1×10^{-7}	1.8×10^{-8}	2.69585239	
C_2	3.9×10^{-8}	2.3×10^{-8}	2.91217882	4.7×10^4
C_3	8.6×10^{-5}	6.2×10^{-9}	2.265373689	
	$E_{\mathrm{Ar}l}$	$E_{\Theta l}$	Θ_l	cond
C_1	6.7×10^{-8}	1.3×10^{-8}	−0.23582973	
C_2	2.1×10^{-8}	8.5×10^{-9}	2.246730504	4.7×10^4
C_3	2.5×10^{-5}	4.9×10^{-9}	−2.005025898	

where $0 < q < 1$ is a parameter for charge placement. From the maximum modulus theorem for analytic functions, it is easy to show that absolute errors of $F_{\mathrm{p}v}(z)$ and $F_{\mathrm{p}u}(z)$ and relative errors of $F_{\mathrm{c}}(z)$ and $F_{\mathrm{r}}(z)$ take their maximum values somewhere on the boundary. We here compute the following indications of errors on each C_l ($l = 1, 2, 3$):

$$E_{\mathrm{Re}l} = \max_{1 \le j \le N} |\operatorname{Re} F_{\mathrm{p}v}(z_{lj+1/2}) - U_l|, \quad E_{Ul} = |U_l - U_{\mathrm{d}l}|,$$

$$E_{\mathrm{Im}l} = \max_{1 \le j \le N} |\operatorname{Im} F_{\mathrm{p}u}(z_{lj+1/2}) - V_l|, \quad E_{Vl} = |V_l - V_{\mathrm{d}l}|,$$

$$E_{\mathrm{Mo}l} = \max_{1 \le j \le N} ||F_{\mathrm{c}}(z_{lj+1/2})| - R_l|, \quad E_{Rl} = |R_l - R_{\mathrm{d}l}|,$$

$$E_{\mathrm{Ar}l} = \max_{1 \le j \le N} |\operatorname{Arg} F_{\mathrm{r}}(z_{lj+1/2}) - \Theta_l|, \quad E_{\Theta l} = |\Theta_l - \Theta_{\mathrm{d}l}|,$$

where $z_{lj+1/2}$ is the middle point between z_{lj} and z_{lj+1}; $U_{\mathrm{d}l}, V_{\mathrm{d}l}, R_{\mathrm{d}l}$ and $\Theta_{\mathrm{d}l}$ are obtained by doubling the number of simulation charges.

Table 1 shows the numerical results, which suggest that high accuracy is obtained for $F_{\mathrm{p}v}(z)$, $F_{\mathrm{p}u}(z)$, $F_{\mathrm{c}}(z)$ and $F_{\mathrm{r}}(z)$ though their exact solutions are not known. U_l, V_l, R_l and Θ_l are shown till a non-zero digit appears in $U_l - U_{\mathrm{d}l}$, $V_l - V_{\mathrm{d}l}$, $R_l - R_{\mathrm{d}l}$ and $\Theta_l - \Theta_{\mathrm{d}l}$, respectively. The 'cond' is the L_1 condition number of the coefficient matrix of the linear equations to be solved.

Figure 1 illustrates the numerical mapping $F_\theta(z) = \mathrm{e}^{\mathrm{i}\theta}(\cos\theta F_{\mathrm{p}u}(z) - \mathrm{i}\sin\theta F_{\mathrm{p}v}(z))$ $(\theta = \pi/3)$ of D and contour lines of $\mathrm{Im}(\mathrm{e}^{-\mathrm{i}\theta}F_\theta(z))$, i.e., streamlines of a uniform flow past three cylindrical objects.

5 Concluding Remarks

We have presented a simple systematic method of numerical conformal mappings of the unbounded multiply-connected domains onto the parallel, the circular and the radial slit domains. The linear equations to be solved have the same coefficient matrix, and we need only one LU factorization to approximate these mappings. The method has the practical advantage that various types of flows can be obtained by superposition. It is easy to change shapes, numbers and positions of the obstacles.

References

[1] Amano, K. and Sugihara, M.: Numerical conformal mapping onto parallel slit domains with application to potential flow analysis, Theoretical and Applied Mechanics, Vol. 46 (eds. Yagawa, G. and Miki, C.), Hokusen-sha, Tokyo, pp. 295–305 (1997).

[2] Amano, K.: A charge simulation method for numerical conformal mapping onto circular and radial slit domains, SIAM J. Sci. Comput., Vol. 19, No. 4, pp. 1169–1187 (1998).

[3] Henrici, P.: Applied and Computational Complex Analysis, Vol. 3, John Wiley & Sons, New York (1986).

[4] Kythe, P.K.: Computational Conformal Mapping, Birkhaüser, Boston (1998).

[5] Nehari, Z.: Conformal Mapping, McGraw-Hill, New York (1952); Dover, New York (1975).

[6] Trefethen, L.N. (ed.): Numerical Conformal Mapping, North-Holland, Amsterdam (1986); reprinted from J. Comput. Appl. Math., Vol. 14, No. 1-2 (1986).

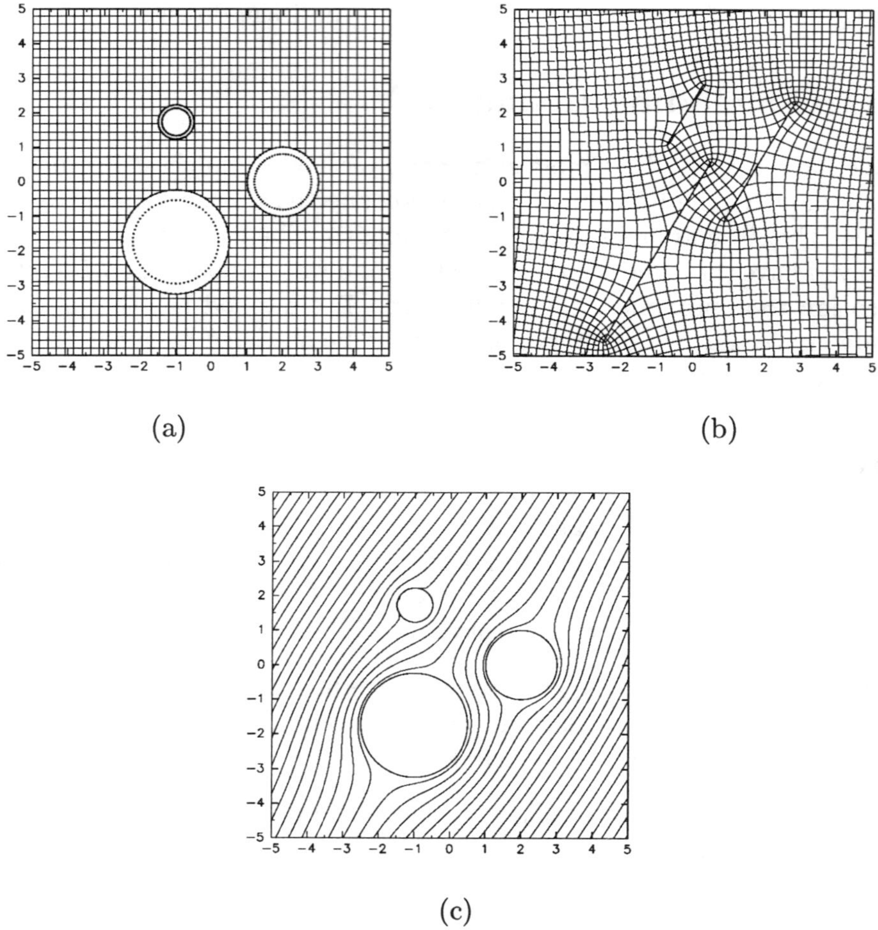

(a) (b) (c)

Figure 1: (a) A problem domain D, (b) the mapping $F_\theta(z)$ and (c) contour lines of $\mathrm{Im}(\mathrm{e}^{-\mathrm{i}\theta}F_\theta(z))$.

40 HÉNON MAPPING OF COMPLEX N-SPACE IN COMPLEX N-SPACE

Yukio FUKUSHIMA

Department of Applied Mathematics
Fukuoka University
FUKUOKA, 814-0180, Japan

yukio@bach.sm.fukuoka-u.ac.jp

1. Introduction

In $\mathbf{C}^2$, Steven G. Krantz[3] constructed the Fatou-Bieberbach domain. In this talk, we construct the Fatou-Bieberbach domain in $\mathbf{C}^n$. The method of Proof is similar to Krantz[3]. Also, we paint complex one dimensinal sections of the Fatou-Bieberbach domain in $\mathbf{C}^3$.

2. Definitions and lemmata

Let U be a proper domain of $\mathbf{C}^n$. U is said to be a *Fatou-Bieberbach domain*, if U is biholomorphic onto $\mathbf{C}^n$.

Let a be a complex number satisfying $0 < |a| < 1$. A mapping

$$\begin{array}{rcc} F = F_a : \mathbf{C}^n & \longrightarrow & \mathbf{C}^n \\ z = (z_1, z_2, \cdots, z_n) & \longmapsto & (z_{n-1}^2 + az_n, z_{n-2}^2 + az_{n-1}, \cdots, az_1) \end{array}$$

is said to be a *complex Hénon mapping* from $\mathbf{C}^n$ to $\mathbf{C}^n$.

We know that F is a biholomorpic self-mapping of $\mathbf{C}^n$ and that F(0)=0.

H.G.W. Begehr et al. (eds.), Proceedings of the Second ISAAC Congress, Vol. 1, 337–345.

Let A be the Jacobian matrix for F at the origin. We have

$$A = Jac_{\mathbf{C}} F(0) = \begin{pmatrix} 0 & 0 & \cdots & 0 & a \\ 0 & 0 & \cdots & a & 0 \\ \vdots & \vdots & \ddots & \vdots & \vdots \\ 0 & a & \cdots & 0 & 0 \\ a & 0 & \cdots & 0 & 0 \end{pmatrix}.$$

We shall use the notation F^k to denote the k-fold composition $\overbrace{F \circ \cdots \circ F}^{k-\text{times}}$.

Lemma 1 Let a be a complex constant satisfying $0 < |a| < 1$ and b be a positive constant satisfying $|a| < b < 1$. There is an $\varepsilon > 0$ such that if $|z| < \varepsilon$, then for any positive integer k we have $|F^k(z)| \leq \sqrt{n}\varepsilon b^k$.

Proof For any $z = (z_1, z_2, \cdots, z_n)$, we put

$$(z_1^{(k)}, z_2^{(k)}, \cdots, z_n^{(k)}) = F^k(z_1, z_2, \cdots, z_n).$$

There is an $\varepsilon > 0$ satisfying $|a| + \varepsilon < b$.
If $|z| = \sqrt{|z_1|^2 + |z_2|^2 + \cdots + |z_n|^2} < \varepsilon$, by mathematical induction we have that for any positive integer k

$$|z_j^{(k)}| < \varepsilon b^k \quad (j = 1, 2, \cdots, n).$$

Therefore we have

$$|F^k(z)| = \sqrt{|z_1^{(k)}|^2 + |z_2^{(k)}|^2 + \cdots + |z_n^{(k)}|^2} < \sqrt{n}\varepsilon b^k.$$

We have completed the proof.

Let $F = F_a$ be a complex Hénon mapping with complex constant a satisfying $0 < |a| < 1$. Let U be the set

$$\left\{ z = (z_1, z_2, \cdots, z_n) \in \mathbf{C}^n \;\middle|\; \lim_{n \to \infty} F^n(z) = 0 \right\}.$$

U is said to be the *basin of attraction* for the complex Hénon mapping F from $\mathbf{C}^n$ to $\mathbf{C}^n$. From lemma 1, we have that U is a non-empty subset of $\mathbf{C}^n$. Also, we know that $F(U) = U$ and $F^{-1}(U) = U$. Hence U is both forward and backward invariant under the complex Hénon mapping F.

Lemma 2 Let a be a complex constant satisfying $0 < |a| < 1$. Then we have that for all $z \in \mathbf{C}^n$,

$$|A^{-1}F(z) - z| \leq \frac{1}{|a|}|z|^2.$$

Proof For all $z \in \mathbf{C}^n$, we have

$$|A^{-1}F(z) - z| = \left|\frac{1}{a}(0, z_1^2, \cdots, z_{n-1}^2)\right| \leq \frac{1}{|a|}|z|^2.$$

We have completed the proof.

Lemma 3 Let a be a complex constant satisfying $0 < |a| < 1$. Then sequence $\{A^{-k}F^k\}_{k=1}^{\infty}$ converges uniformly on compact subsets of U.

Proof There is a positive number b satisfying $b^2 < |a| < b < 1$. By the lemma 1, there is an $\varepsilon > 0$ such that

$$B = \{z \in \mathbf{C}^n \| z| < \varepsilon\} \subset U.$$

We fix a compact subset K of U. By the continuity of F^k and the compactness of K, There is a positive integer N such that if $k \geq N$, then $F^k(K) \subset B$. For a positive integer k with $k \geq N$ and any $z \in K$, we have

$$\begin{aligned} |A^{-(k+1)}F^{k+1}(z) - A^{-k}F^k(z) &= |A^{-k}(A^{-1}F)F^k(z) - A^{-k}F^k(z)| \\ &= |A^{-k}(A^{-1}F - I)F^k(z)|, \end{aligned}$$

where I is the unit matrix of dimension n. By lemma 1 and lemma 2, we have

$$\begin{aligned} &|(A^{-1}F - I)F^k(z)| = |A^{-k}(A^{-1}F - I)F^{k-N}(F^N(z))| \\ \leq\ & \frac{1}{|a|}|F^{k-N}(F^N(z))|^2 \leq \frac{n\varepsilon^2}{|a|}b^{2(k-N)}. \end{aligned}$$

For the Jacobian matrix A, we have

$$A^{-k} = \begin{cases} \dfrac{1}{a^k}\begin{pmatrix} 1 & 0 & \cdots & 0 & 0 \\ 0 & 1 & \cdots & 0 & 0 \\ \vdots & \vdots & \ddots & \vdots & \vdots \\ 0 & 0 & \cdots & 1 & 0 \\ 0 & 0 & \cdots & 0 & 1 \end{pmatrix} & (k : \text{even number}) \\ \dfrac{1}{a^k}\begin{pmatrix} 0 & 0 & \cdots & 0 & 1 \\ 0 & 0 & \cdots & 1 & 0 \\ \vdots & \vdots & \ddots & \vdots & \vdots \\ 0 & 1 & \cdots & 0 & 0 \\ 1 & 0 & \cdots & 0 & 0 \end{pmatrix} & (k : \text{odd number}) \end{cases}$$

Therefore we have

$$\begin{aligned} & |A^{-(k+1)}F^{k+1}(z) - A^{-k}F^k(z)| \\ \leq \; & |A^{-k}(A^{-1}F - I)F^{k-N}(F^N(z))| \leq \frac{n\sqrt{n}\varepsilon^2}{|a|b^{2N}}\left(\frac{b^2}{|a|}\right)^k . \end{aligned}$$

For any positive integer k with $k \geq N$, any positive integer ℓ and any $z \in K$, we have

$$\begin{aligned} & |A^{-(k+\ell)}F^{k+\ell}(z) - A^{-k}F^k(z) \\ \leq \; & \sum_{j=0}^{\ell-1} \frac{n\sqrt{n}\varepsilon^2}{|a|b^{2N}}\left(\frac{b^2}{|a|}\right)^{k+\ell-1-j} \leq \frac{n\sqrt{n}\varepsilon^2}{b^{2N}(|a|-b^2)}\left(\frac{b^2}{|a|}\right)^k . \end{aligned}$$

Hence, we have completed the proof.

3. The main result

From lemma 3, we put

$$\Psi = \lim_{k\to\infty} A^{-k}F^k .$$

Proposition 4 Ψ is a biholomorphic mapping from U onto $\mathbf{C}^n$.

Proof By $Jac_{\mathbf{C}}A^{-1}F(0) = A^{-1}Jac_{\mathbf{C}}F(0) = A^{-1}A = I$, we have $Jac_{\mathbf{C}}A^{-k}F^k(0) = I$. Hence, we have

$$Jac_{\mathbf{C}}\Psi(0) = \lim_{k\to\infty} Jac_{\mathbf{C}}A^{-k}F^k(0) = I.$$

There is a neibourhood V of the origin with $V \subset U$ such that Ψ is the biholomorphic mapping from V onto $\Psi(V)$.

We prove that the mapping Ψ is injective. Let z and w be any elements of U with $\Psi(z) = \Psi(w)$. As U is the basin of attraction for the complex Hénon mapping F, there is a positive integer N such that $F^N(z) \in V$ and $F^N(w) \in V$. $\Psi(z) = \Psi(w)$ implies that

$$\lim_{k\to\infty} A^{-k}F^k(z) = \lim_{k\to\infty} A^{-k}F^k(w).$$

We have that for any positive integer k with $k \geq N$

$$A^{-k}F^k(z) = A^{-N}A^{-(k-N)}F^{k-N}(F^N(z)).$$

Hence we have

$$\lim_{k\to\infty} A^{-k}F^k(z) = A^{-N} \lim_{k\to\infty} A^{-(k-N)}F^{k-N}(F^N(z)) = A^{-N}\Psi(F^N(z)).$$

Similarly, we have

$$\lim_{n\to\infty} A^{-n}F^n(w) = A^{-N}\Psi(F^N(w)).$$

Since A is injective, We have

$$\Psi(F^N(z)) = \Psi(F^N(w)).$$

As Ψ is biholomorphic mapping, we have $F^N(z) = F^N(w)$.
Therefore $z = w$.

Next, we prove that Ψ is surjective. For any $z \in U$, we have that

$$\lim_{k\to\infty} A^{-k}F^k(F(z)) = \Psi(F(z))$$

implies

$$\Psi(z) = \lim_{k\to\infty} A^{-(k+1)}F^{k+1}(z) = \lim_{k\to\infty} A^{-1}(A^{-k}F^k)F(z) = A^{-1}\Psi(F(z)).$$

We already know that $F(U) = U$. We have that

$$\Psi(U) = A^{-1}\Psi F(U) = A^{-1}\Psi(U) = A^{-2}\Psi(U) = \cdots.$$

Therefore, we have that

$$\bigcup_{k=1}^{\infty} A^{-k}\Psi(U) = \mathbf{C}^n.$$

Since $\Psi(U)$ contains a neibourhood of the origin, we have

$$\Psi(U) = \mathbf{C}^n.$$

We have completed the proof.

We show that the basin of attraction U for the complex Hénon mapping F from $\mathbf{C}^n$ to $\mathbf{C}^n$ is the Fatou-Bieberbach domain.

Proposition 5 We set

$$W = \left\{ (z_1, z_2, \cdots, z_n) \in \mathbf{C}^n \;\middle|\; \begin{array}{c} |z_1| > 4(n-1),\ |z_1| + |z_2| > 4(n-1), \\ \cdots, \\ |z_1| + |z_2| + \cdots + |z_{n-1}| > 4(n-1), \\ |z_n| < 1 \end{array} \right\}.$$

Then

$$U \cap W = \phi.$$

Proof We put

$$\widetilde{W} = \left\{ (z_1, z_2, \cdots, z_n) \in \mathbf{C}^n \;\middle|\; \begin{array}{c} |z_1| + |z_2| + \cdots + |z_{n-1}| > 4(n-1) \\ |z_1| + |z_2| + \cdots + |z_{n-1}| > |z_n| \end{array} \right\}.$$

It is enough that we show $F(\widetilde{W}) \subset \widetilde{W}$. For $z = (z_1, z_2, \cdots, z_n) \in \widetilde{W}$, we have

$$\begin{aligned} & |z_{n-1}^2 + az_n| + |z_{n-2}^2 + az_{n-1}| + \cdots + |z_1^2 + az_2| - 4(n-1) \\ \geq\ & (|z_1| - 1)^2 + (|z_2| - 1)^2 + \cdots + (|z_{n-1}| - 1)^2 - 5(n-1) \\ \geq\ & \left(\frac{4(n-1)}{\sqrt{n-1}} - \sqrt{n-1} \right)^2 - 5(n-1) > 0. \end{aligned}$$

Similarly, we have

$$\begin{aligned} & |z_{n-1}^2 + az_n| + |z_{n-2}^2 + az_{n-1}| + \cdots + |z_1^2 + az_2| - |az_1| \\ \geq\ & (|z_1| - 1)^2 + (|z_2| - 1)^2 + \cdots + (|z_{n-1}| - 1)^2 - (n-1) \\ \geq\ & \left(\frac{4(n-1)}{\sqrt{n-1}} - \sqrt{n-1} \right)^2 - (n-1) > 0. \end{aligned}$$

We have completed the proof.

4. Computer Graphics

We paint the pictures of complex one dimensional section of the Fatou- Bieberbach domain in $\mathbf{C}^3$ by the computer. We use that Computer Language is Visual Basic Version 5.0 of Microsoft.

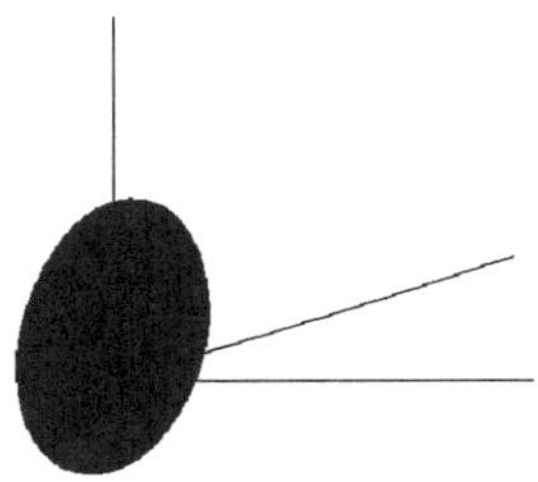

Figure 1

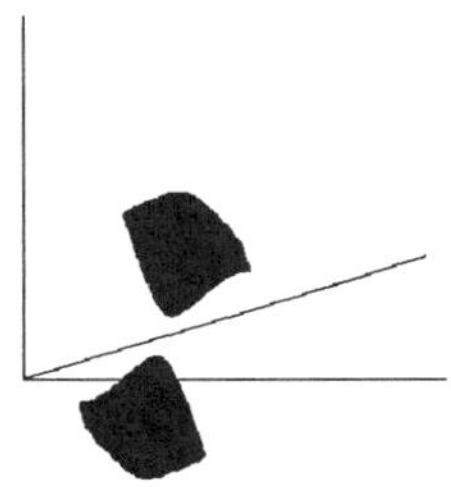

Figure 2

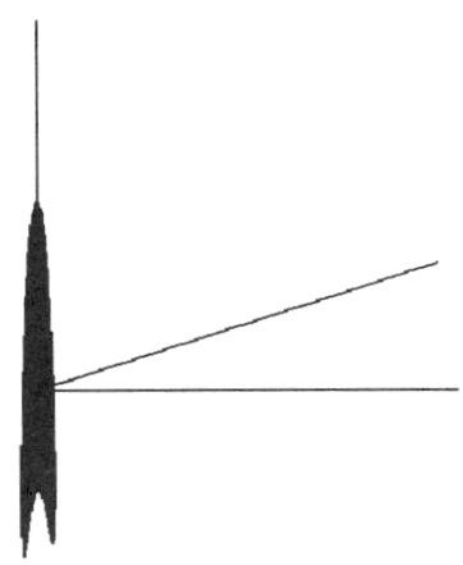

Figure 3

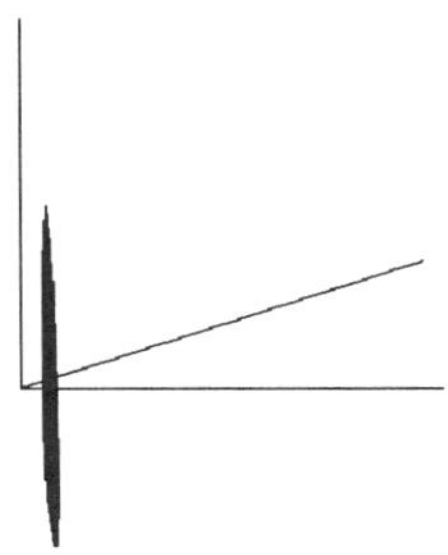

Figure 4

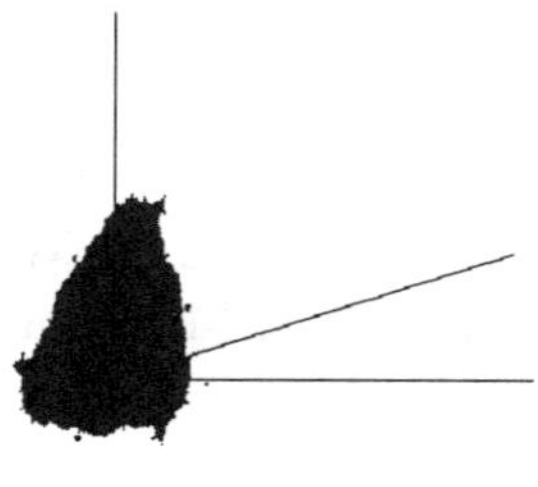

Figure 5

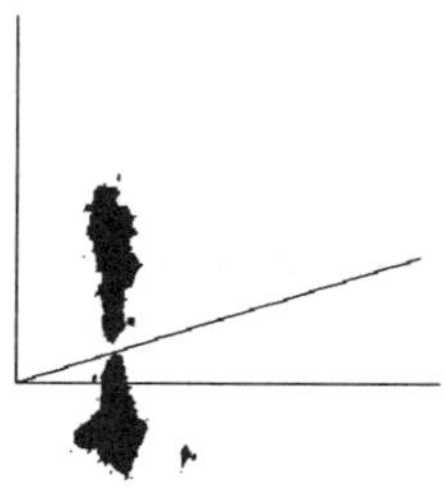

Figure 6

Data of Figure 1 is the following :

The black area is

$U \bigcap \{ (z_1, z_2, z_3) \in \mathbf{C}^3 \mid z_2 = z_3 = 0 \}.$

$$\left.\begin{array}{l} -1.5 \le Re\, z_1 \le 1.5 \\ -1.5 \le Im\, z_1 \le 1.5 \end{array}\right\} \text{Range}$$

$a = 0.1 + 0.1i$ ComplexConstant

Data of Figure 2 is the following :

The black area is

$$U \bigcap \left\{ (z_1, z_2, z_3) \in \mathbf{C}^3 \;\middle|\; \begin{array}{l} Re\, z_2 = 1,\ Im\, z_2 = 0 \\ z_3 = 0 \end{array} \right\}.$$

$$\left.\begin{array}{l} -1.5 \le Re\, z_1 \le 1.5 \\ -1.5 \le Im\, z_1 \le 1.5 \end{array}\right\} \text{Range}$$

$a = 0.1 + 0.1i$ ComplexConstant

Data of Figure 3 is the following :

The black area is

$$U \bigcap \left\{ (z_1, z_2, z_3) \in \mathbf{C}^3 \;\middle|\; \begin{array}{l} z_1 = 0 \\ Re\, z_2 = Re\, z_3 = 0 \end{array} \right\}.$$

$$\left.\begin{array}{l} -10 \le Im\, z_2 \le 10 \\ -10 \le Im\, z_3 \le 10 \end{array}\right\} \text{Range}$$

$a = 0.1 + 0.1i$ ComplexConstant

Data of Figure 4 is the following :

The black area is

$$U \bigcap \left\{ (z_1, z_2, z_3) \in \mathbf{C}^3 \;\middle|\; \begin{array}{l} Re\, z_1 = 1,\ Im\, z_1 = 0 \\ Re\, z_2 = Re\, z_3 = 0 \end{array} \right\}.$$

$$\left.\begin{array}{l} -10 \leq Im\, z_2 \leq 10 \\ -10 \leq Im\, z_3 \leq 10 \end{array}\right\} \text{Range}$$

$a = 0.1 + 0.1i$ ComplexConstant

Data of Figure 5 is the following :

The black area is

$$U \bigcap \{ (z_1, z_2, z_3) \in \mathbf{C}^3 \mid z_2 = z_3 = 0 \}.$$

$$\left.\begin{array}{l} -1.5 \leq Re\, z_1 \leq 1.5 \\ -1.5 \leq Im\, z_1 \leq 1.5 \end{array}\right\} \text{Range}$$

$a = 0.636 + 0.636i$ ComplexConstant

Data of Figure 6 is the following :

The black area is

$$U \bigcap \left\{ (z_1, z_2, z_3) \in \mathbf{C}^3 \;\middle|\; \begin{array}{l} Re\, z_2 = 0.75,\ Im\, z_2 = 0 \\ z_3 = 0 \end{array} \right\}.$$

$$\left.\begin{array}{l} -1.5 \leq Re\, z_1 \leq 1.5 \\ -1.5 \leq Im\, z_1 \leq 1.5 \end{array}\right\} \text{Range}$$

$a = 0.636 + 0.636i$ ComplexConstant

References

[1] Y. Fukushima, *On a complex Hénon mapping fron* $\mathbf{C}^3$ *to* $\mathbf{C}^3$, Proceeding of the fifth international colloquium on finite and infinite dimensinal complex analysis, 67-72(1997).

[2] Y. Fukushima, T. Sasaki and R. Tsukamoto., *Computer graphics by complex Hénon mapping from* $\mathbf{C}^2$ *to* $\mathbf{C}^2$., Proceeding of the fourth international colloquium of finite or infinite dimensional complex analysis, 115-120(1996).

[3] Krantz S. G., *Function theory of several complex variables, second edition.*, Wadsworth & Brooks/Cole Advanced Mathmatics Series,(1992).

[4] T. Ueda, M. Taniguchi and S. Morosawa, *Intoroduction of complex dynamical systems*, (in Japanese), Baifuukan,(1995).

41 THE CARATHÉODORY CONVERGENCE OF FATOU COMPONENTS OF POLYNOMIALS TO BAKER DOMAINS OR WANDERING DOMAINS

Shunsuke MOROSAWA[1]

Department of Mathematics and Information Science, Faculty of Science, Kochi University, Kochi, 780-8520, Japan

morosawa@math.kochi-u.ac.jp

1 Introduction

Let f be a rational function or a transcendental entire function. Consider X is a complex sphere $\widehat{\mathbb{C}}$ if f is a rational function and X is a complex plain $\mathbb{C}$ if f is a transcendental entire function. The maximal open subset of X on which the family $\{f^n\}$ is normal is called the Fatou set of f and we denote it by $F(f)$. The complement of $F(f)$ in $\widehat{\mathbb{C}}$ is called the Julia set of f and we denote it by $J(f)$. We remark that, in this note, ∞ is always contained in the Julia sets of transcendental entire functions. Fatou sets and Julia sets are completely invariant. Hence a component of $F(f)$ is mapped into some component of $F(f)$ by f. We say that a component of D of $F(f)$ is a cyclic component if $f^p(D)$ is contained in D

[1]Partially supported by Grant-in-Aid for Scientific Research 09640199

H.G.W. Begehr et al. (eds.), Proceedings of the Second ISAAC Congress, Vol. 1, 347–356.

for some natural number p. If $p = 1$, then it is called invariant. All the kinds of cyclic components are already known. They are an attracting component, a parabolic component, a Siege disk, a Herman ring and a Baker domain. The first three have relationship with cyclic points and the first four have relationship with singular values. On the other hand, Baker domains have relationship with neither cyclic point nor singular value. Baker domains are never appeared in the case of rational functions. Some facts on Baker domains are stated in § 3. We say that a component D is a wandering domain if $f^n(D) \neq f^m(D)$ for all natural numbers n and m with $n \neq m$. Due to Sullivan's theorem, we know that there exists no wandering domain in the case of rational functions. On the contrary, in the case of transcendental entire functions, there may exist wandering domains if it has infinitely many singular values.

Let f be a transcendental entire functions and $\{f_n\}_{n=1}^{\infty}$ be a sequence of polynomials converging to f locally uniformly on $\mathbb{C}$. Such sequences always exist. For example, the Taylor series of f is so. How about the convergence of the dynamics of f_n to that of f? Devaney, Goldberg and Hubbard ([3]) considered the exponential family $E_\lambda(z) = \lambda e^z$ and polynomials $P_{\lambda,n}(z) = \lambda[1 + (z/n)]^n$ as a convergent sequence and showed the convergence of hyperbolic components and external rays in the parameter space. After this result, Krauskopf([9]) showed that if E_λ has an attracting component, then $J(P_{\lambda,n})$ converges to $J(E_\lambda)$ in the Hausdorff metric. The definition of convergence in the Hausdorff metric is given in § 2. Further, Kisaka([8]) proved more general results.

Theorem 1 *Let f be a transcendental entire function and $\{f_n\}_{n=1}^{\infty}$ be a sequence of polynomials converging to f locally uniformly on $\mathbb{C}$. If $F(f)$ consists only of basins of attracting cycles, then $J(f_n)$ converges to $J(f)$ in the Hausdorff metric.*

The key fact for the results of Krauskopf and Kisaka is that an arbitrary cycle of f is approximated by some cycle of f_n, which immediately follows from the Hurwitz theorem.

In § 4, we consider examples of transcendental entire functions whose Fatou sets have Baker domains or wandering domains and whose Julia sets are approximated by Julia sets of polynomials.

2 Hausdorff convergence and Carathéodory convergence

First, we introduce an idea of convergence of Julia sets. Let ρ be the chordal metric on $\widehat{\mathbb{C}}$. The ϵ-neighborhood of a set A is denoted by $U_\epsilon(A)$. Let A and B be non-empty compact sets in $\widehat{\mathbb{C}}$. The Hausdorff distance $d(A,\ B)$ of A and B is defined by

$$d(A,\ B) = \inf\{\varepsilon > 0 \mid A \subset U_\varepsilon(B), \quad B \subset U_\varepsilon(A)\}.$$

This distance defines the Hausdorff metric on the set of all the non-empty compact sets in $\widehat{\mathbb{C}}$. Let K and K_n $(n \in \mathbb{N})$ be non-empty compact sets in $\widehat{\mathbb{C}}$. We say that K_n converges to K in the Hausdorff metric if $d(K_n,\ K) \to 0$ as $n \to \infty$. That is, for an arbitrary $\epsilon > 0$, there exists an N such that

$$K \subset U_\varepsilon(K_n) \quad \text{and} \quad K_n \subset U_\varepsilon(K)$$

for all $n \geq N$. We denote it by

$$K_n \to K \qquad\qquad \text{(Hausdorff metric)}.$$

Since the set of all the cyclic points are dense in the Julia set, it is easy to see the following.

Proposition 2 *If f_n converges to f locally uniformly, then, for an arbitrary $\epsilon > 0$, there exists an N such that*

$$J(f) \subset U_\varepsilon(J(f_n))$$

for all $n \geq N$.

This proposition implies the Hausdorff convergence of Julia sets is *lower semicontinuous* (see Douady [4]).

Next, we introduce an idea of convergence of open sets. Let O and O_n $(n \in \mathbb{N})$ be open sets in $\widehat{\mathbb{C}}$. We say that O_n converges to O in the Carathéodory sense if the following two conditions hold:
(1) for an arbitrary compact set $I \subset O$, there exits an N such that $I \subset O_n$ for all $n \geq N$ and
(2) if an open set U is contained in O_m for infinitely many m, then $U \subset O$.
We denote the convergence of O_n to O in the Carathéodory sense by

$$O_n \to O \qquad\qquad \text{(Carathéodory sense)}.$$

Two ideas of convergence defined above have the following relationship.

Lemma 3 *Non-empty closed set K_n converges to K in the Hausdorff metric if and only if the complement K_n^c of K_n converges to the complement K^c of K in the Carathéodory sense.*

Proposition 2 is rephrased as follows.

Lemma 4 *Assume f_n converges to f locally uniformly. If there exists an open set U such that $U \subset F(f_n)$ for infinitely many n, then $U \subset F(f)$.*

3 Baker domains

Let f be a transcendental entire function. An invariant component B of $F(f)$ is called a Baker domain if the limit function of $\{f^n\}$ is only ∞. Hence a Baker domains have no fixed point in it. Eremenko and Lyubich ([6]) showed that, if a transcendental entire function has a Baker domain, then the set of the singular values is unbounded. They ([5]) also gave an example of a Baker domain which has no singular value in it. Herman ([7]) also gave an example of a Baker domain having a property which the example given by Eremenko and Lyubich has and further its boundary is contained in the closure of the forward orbits of singular values. Moreover, Bergweiler ([2]) gave an example of a Baker domain such that the Euclidean distance between it and the set of all the singular values is positive.

A subdomain V of a Baker domain B is called a fundamental set of B, if V is simply connected, $f(V) \subset V$ and, for every compact set $K \subset B$, $f^n(K) \subset V$ for some n.

4 Examples

Example 1 *Let $f(z) = z + e^z - 1$. Then $F(f)$ has a Baker domain. Take*

$$f_n(z) = \left(1 - \frac{1}{n}\right) z + \frac{n-1}{n+1}\left(1 + \frac{z}{n}\right)^{n+1} - 1$$

as a locally uniformly convergent sequence to f. Then we have

$$J(f_n) \to J(f) \qquad \text{(Hausdorff metric)}.$$

Take

$$g_n(z) = z + \left(1 + \frac{z}{2n}\right)^{2n} - 1$$

as a locally uniformly convergent sequence to f. Then we have

$$J(g_n) \not\to J(f) \qquad \text{(Hausdorff metric)}.$$

First, we consider a transcendental entire function $h(z) = ze^{z-1}$. It has only two singular values 0 and $h(-1)$. Hence $F(h)$ has neither a wandering domain nor a Baker domain. It also has an attracting fixed point 0. The inverse image of sufficiently small open disk with center 0 consists of two components. Thus the attracting component of 0 contains -1 and $F(h)$ consists only of this component. The function $f(z)$ is a logarithmic lift of $h(z)$. Therefor $F(f)$ consists of only one Baker domain B, which is a lift of the attracting component of $F(f)$ (see [7], [1] and [10]). See Figure 1.

Figure 1: Julia set of $f(z) = z + e^z - 1$

Since

$$f_n'(z) = 1 - \frac{1}{n} + \frac{n-1}{n}\left(1 + \frac{z}{n}\right)^n,$$

we see that $-n$ is an attracting fixed point and finite critical points are

$$z = -n + n\exp(1+2k)\frac{\pi}{n}i \qquad (0 \le k \le n-1).$$

We denote the attracting component containing $-n$ by A_n. Since A_n contains at least one critical point and $f_n(z)$ is symmetric with respect to a rotation with center $-n$, all the finite critical points are contained in A_n. Hence $F(f_n)$ consists A_n and the superattracting component of the point at infinity. The boundary of A_n is a simple closed curve.

Set

$$U_n = \{z \mid |z+n| < n\}.$$

Then we have $f_n(U_n) \subset U_n$ and hence $U_n \subset A_n$. From Lemma 4,

$$\bigcap_{n\geq 1} \bigcup_{k\geq n} U_k = \{z \mid \mathrm{Re}\ z < 0\}$$

is contained in $F(f)$ and contains a fundamental set of B. Hence we have

$$A_n \to B \qquad \text{(Carathéodory sense)}.$$

By Lemma 3, we have

$$A_n^c \to B^c = J(f) \qquad \text{(Hausdorff metric)}.$$

Since $J(f)$ has no interior point, we obtain

$$J(f_n) \to J(f) \qquad \text{(Hausdorff metric)}.$$

See Figure 2.

Next, we consider the sequence $\{g_n\}$. The fixed points of g_n are

$$z_n^{(k)} = -2n + 2n \exp \frac{k}{n}\pi i \qquad (0 \leq k \leq 2n-1).$$

Since

$$g_n'(z) = 1 + \left(1 + \frac{z}{2n}\right)^{2n-1},$$

we have the critical points

$$w_n^{(k)} = -2n + 2n \exp \frac{2k+1}{2n-1}\pi i \qquad (0 \leq k \leq 2n-2)$$

and we see that, for $2n/3 < k < 4n/3$, $z_n^{(k)}$ are attracting fixed points of g_n. Since each attracting component has a fixed point on its boundary. The origin is a fixed point which is on the boundary of the attracting component of $z_n^{(n)} = -4n$. A closed interval $[-4n, 0]$ is a forward invariant curve $L_{n,1}$ under g_n and $L_{n,1} \setminus \{0\}$ is contained in the attracting

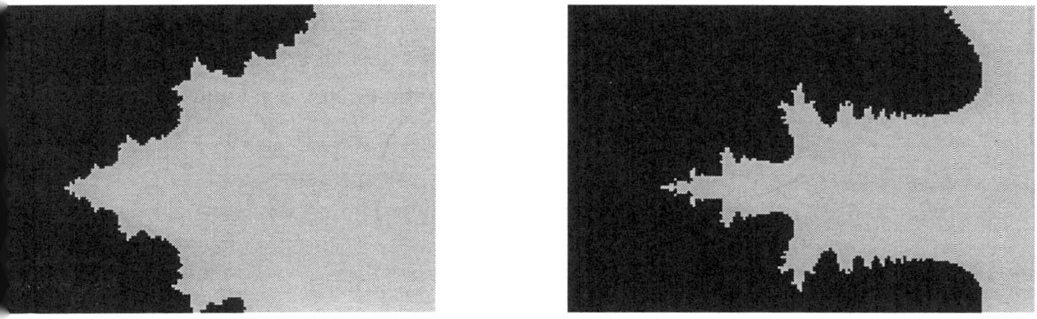

Figure 2: Fatou sets of $f_n(z) = [1 - (1/n)]z + [(n-1)/(n+1)][1 + (z/n)]^{n+1} - 1$

component of $-4n$. We can also see that $z_n^{(1)} = -2n + 2n\exp(\pi i/n)$ is a fixed point which is on the boundary of the attracting component of $z_n^{(n-1)} = -2n + 2n\exp[(n-1)\pi i/n]$. There exists a forward invariant curve $L_{n,2}$ joining $z_n^{(1)}$ and $z_n^{(n-1)}$ in the strip $\{z \mid \pi \leq \mathrm{Im}\, z \leq 3\pi\}$ and $L_{n,2} \setminus z_n^{(1)}$ is contained in the attracting component of $z_n^{(n-1)}$. There exist points of $J(g_n)$ between $L_{n,1}$ and $L_{n,2}$. Hence we have

$$J(g_n) \not\to J(f) \qquad \text{(Hausdorff metric)}.$$

See Figure 3.

Example 2 *Let $f(z) = z - e^z + 1 + 2\pi i$. Then $F(f)$ has wandering domains. Take*

$$f_n(z) = e^{i\frac{2\pi}{n}}\left(1 + \frac{1}{n}\right)z - e^{i\frac{2\pi}{n}}\left(1 + \frac{z}{n}\right)^{n+1} + n\left\{e^{i\frac{2\pi}{n}}\left(1 + \frac{1}{n}\right) - 1\right\}$$

as a locally uniformly convergent sequence to f. Then we have

$$J(f_n) \to J(f) \qquad \text{(Hausdorff metric)}.$$

First we consider a transcendental entire function $h(z) = z - e^z + 1$. Its singular values are $a_n = 2n\pi i$ $(n \in \mathbb{Z})$, which are also superattracting fixed points. Since $h(z)$ and $f(z)$ satisfy the following two formulae

$$h(z + 2\pi i) = h(z) + 2\pi i \qquad \text{and} \qquad f(z) = h(z) + 2\pi i,$$

we have $J(f) = J(h)$ (see [1] and [10]). Hence the component containing a_n is a wandering domain of f. We take

$$h_n(z) = \left(1 + \frac{1}{n}\right)z - \left(1 + \frac{z}{n}\right)^{n+1} + 1$$

as a locally uniformly convergent sequence to h. Since $F(h)$ consists only of attracting basins, $J(h_n)$ converges to $J(h)$ in the Hausdorff metric by Theorem 1. It is clear that $f_n^n(z) = h_n^n(z)$, which implies $J(f_n) = J(h_n)$. Therefor, we have

$$J(f_n) \to J(f) \qquad \text{(Hausdorff metric)}.$$

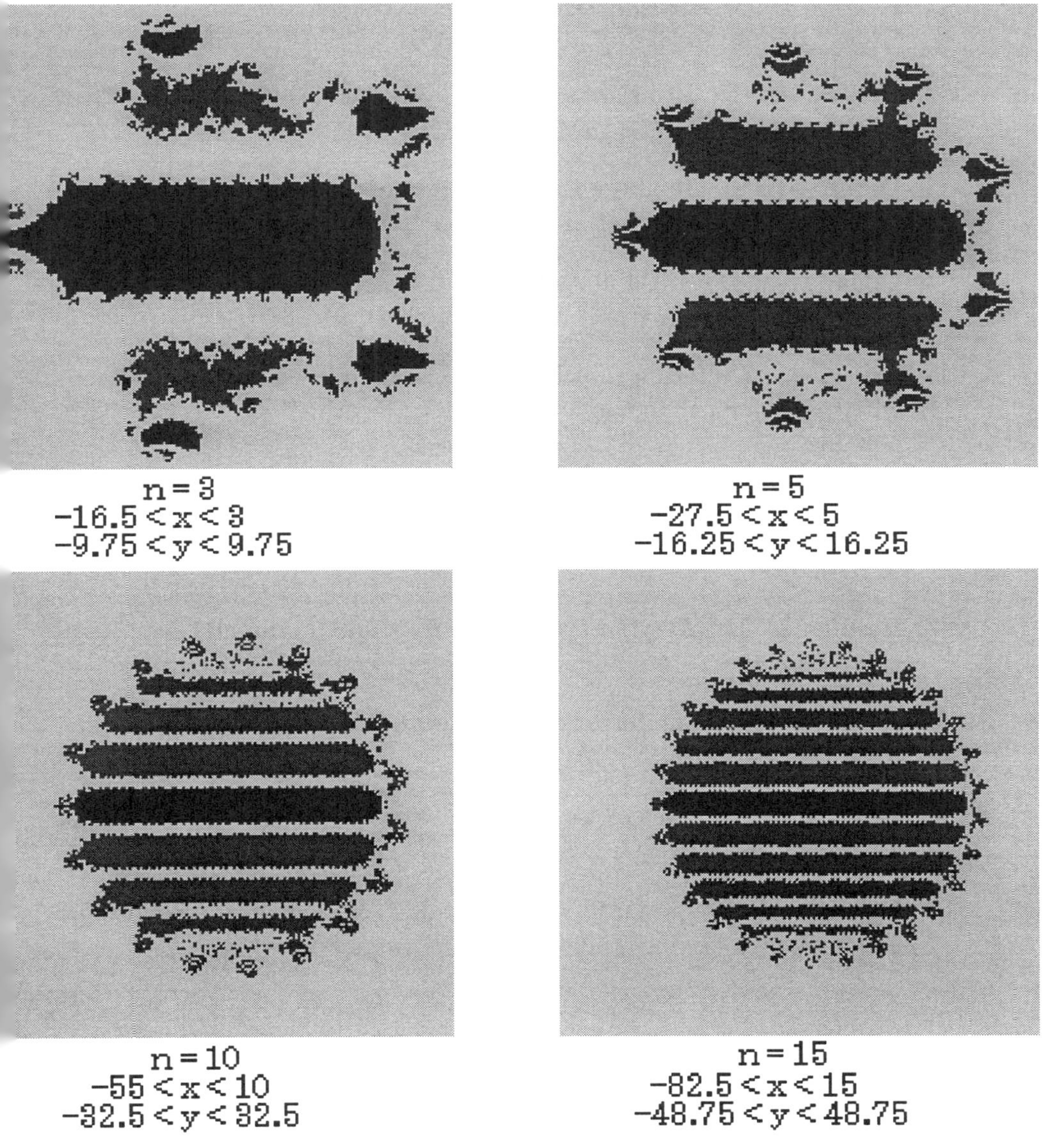

Figure 3: Fatou sets of $g_n = z + [1 + (z/2n)]^{2n} - 1$

References

[1] Baker I. N., Wandering domains in the iteration of entire functions, Proc. London Math. Soc. (3), 49(1984), 563-576.

[2] Bergweiler W., Invariant domains and singularities, Math. Proc. Camb. Phil. Soc., 117(1994), 525-532.

[3] Devaney R. L., Goldberg L. R. & Hubbard J., A dynamical approximation to the exponential map by polynomials, preprint.

[4] Douady A., Does a Julia set depend continuously on the polynomial?, Proc. Symp. Appl. Math., 49(1994), 91-137.

[5] Eremenko A. E. & Lyubich M. Yu., Examples of entire functions with pathological dynamics, J. London Math. Soc. (2), 36(1987), 458-468.

[6] Eremenko A. E. & Lyubich M. Yu., Dynamical properties of some classes of entire functions, Ann. Inst. Fourier, Grenoble, 42(1992), 989-1020.

[7] Herman M. R., Are there critical points on the boundaries of singular domains?, Commun. Math. Phys., 99(1985), 593-612.

[8] Kisaka M., Local uniform convergence and convergence of Julia sets, Nonlinearity, 8(1995), 273-281.

[9] Krauskopf B., Convergence of Julia sets in the approximation of λe^z by $\lambda[1+(z/d)]^d$, Int. J. Bif. Chaos, 3(1993), 257-270.

[10] Morosawa S., Nishimura Y., Taniguchi M. & Ueda T., *Holomorphic Dynamics*, to be published from Cambridge Univ. Press.

42 DE-TYPE QUADRATURE FORMULAE FOR CAUCHY PRINCIPAL-VALUE INTEGRALS AND FOR HADAMARD FINITE-PART INTEGRALS

Hidenori OGATA

Department of Computer Science,
Faculty of Engineering, Ehime University,
Matsuyama 790-8577, Japan

ogata@cs.ehime-u.ac.jp

Masaaki SUGIURA

Department of Computational Science and Engineering,
Graduate School of Engineering, Nagoya University,
Nagoya 464-8603, Japan

sugihara@na.cse.nagoya-u.ac.jp

Masatake MORI

Research Institute for Mathematical Sciences,
Kyoto University,
Kyoto 606-8502, Japan

mmori@kurims.kyoto-u.ac.jp

H.G.W. Begehr et al. (eds.), Proceedings of the Second ISAAC Congress, Vol. 1, 357–366.

1 INTRODUCTION

The double exponential formula (abbreviated to the DE formula) for numerical integration, which was first proposed by Takahasi and Mori[3] in 1974, is recognized as one of the most efficient quadrature formulae and has been widely used in a variety of area, physics, engineering, and so on. In this paper, DE-type quadrature formulae are proposed for evaluating the Cauchy principal-value integral p.v. $\int_{-1}^{1} f(x)/(x-\lambda)\mathrm{d}x$ and for evaluating the Hadamard finite-part integral f.p. $\int_{-1}^{1} f(x)/(x-\lambda)^2\mathrm{d}x$, where $f(x)$ is a function that is analytic on the interval $(-1,1)$, and where λ is a constant such that $-1<\lambda<1$.

For the evaluation of Cauchy principal-value integrals and for the evaluation of Hadamard finite-part integrals, there have already been proposed general variable transformation-type formulae in Bialecki's papers [1, 2], which are based on the so-called "Sinc Numerical Methods". It has been also shown there, that when employing the transformation $x=\tanh(t/2)$, which we call the single exponential transformation, the quadrature error is of the order $\mathrm{O}(\exp(-cN^{1/2}))$, where N is the number of nodes used, and where c is a positive constant independent of N. DE-type formulae are obtained simply by replacing the single exponential transformation with the double exponential transformation $x=\tanh((\pi/2)\sinh t)$. Yet they enjoy a good convergence property. In fact, it is demonstrated that the quadrature error is of the order $\mathrm{O}(\exp(-c_\lambda N/\log N))$, where c_λ is a constant dependent only on λ.

The paper is organized as follows: first, we derive DE-type formulae in 2; the quadrature error of the DE-type formulae is estimated theoretically in 3; a numerical example is shown in 4; we conclude the present paper in 5.

2 DE-TYPE FORMULAE

2.1 DE-TYPE FORMULA FOR CAUCHY PRINCIPAL-VALUE INTEGRALS

Let $f(z)$ be a function that is analytic on the interval $(-1,1)$ and that has poles of order one at the points $a_1,\ldots,a_M$, and let λ be a constant such that $-1<\lambda<1$. Bialecki [1] proposed a general variable transformation-type quadrature formula for the Cauchy principal-value integral p.v. $\int_{-1}^{1} f(x)/(x-\lambda)\mathrm{d}x$, which reads as follows:

$$
\begin{aligned}
&Q_{\mathrm{C}}(f,\lambda;h,N_1,N_2)\\
&= h\sum_{k=-N_2}^{N_1}\frac{f(\psi(kh))\psi'(kh)}{\psi(kh)-\lambda}+\pi\sum_{j=1}^{M}\frac{\mathrm{Res}\,(f,a_j)}{a_j-\lambda}\frac{\exp[\mathrm{i}(\pi/h)\phi(a_j)\mathrm{sgn}(\mathrm{Im}\,a_j)]}{\sin[(\pi/h)\phi(a_j)]}\\
&\quad+\pi\cot\left(\frac{\pi}{h}\phi(\lambda)\right)f(\lambda) \qquad (1)\\
&\qquad\qquad \text{if } \lambda\neq\psi(lh) \text{ for all } l\in\mathbf{Z},
\end{aligned}
$$

and

$$
\begin{aligned}
&Q_{\mathrm{C}}(f,\lambda;h,N_1,N_2)\\
&= h\sum_{\substack{k=-N_2\\k\neq l}}^{N_1}\frac{f(\psi(kh))\psi'(kh)}{\psi(kh)-\lambda}+\pi\sum_{j=1}^{M}\frac{\mathrm{Res}\,(f,a_j)}{a_j-\lambda}\frac{\exp[\mathrm{i}(\pi/h)\phi(a_j)\mathrm{sgn}(\mathrm{Im}\,a_j)]}{\sin[(\pi/h)\phi(a_j)]}\\
&\quad-\frac{h\phi''(\lambda)}{2[\phi'(\lambda)]^2}f(\lambda)+\frac{h}{\phi'(\lambda)}f'(\lambda) \qquad (2)\\
&\qquad\qquad \text{if } \lambda=\psi(lh) \text{ for some } l\in\mathbf{Z},
\end{aligned}
$$

where ψ is a variable transformation such that $\psi((-\infty,+\infty))=(-1,1)$, $\psi(-\infty)=-1$, $\psi(+\infty)=1$, ϕ is the inverse map of ψ, h is a positive constant, N_1,N_2 are positive integers and $\mathrm{Res}\,(f,a_j)$ $(j=1,\ldots,M)$ is the residue of $f(z)$ at the point a_j.

We here employ the double exponential transformation $\psi(t)=\psi_{\mathrm{DE}}(t)=\tanh((\pi/2)\sinh t)$, whereas Bialecki employed the single exponential transformation $\psi(t)=\psi_{\mathrm{SE}}(t)=\tanh(t/2)$ in [1]. In what follows we denote the quadrature formula thus obtained, i.e., the DE-type formula for the Cauchy principal-value integral, by
$Q_{\mathrm{C}}^{(\mathrm{DE})}(f,\lambda;h,N_1,N_2)$.

2.2 DE-TYPE FORMULA FOR THE HADAMARD FINITE-PART INTEGRALS

As is well known the Hadamard finite-part integral f.p. $\int_{-1}^{1} f(x)/(x-\lambda)^2\mathrm{d}x$ is given by the derivative of the Cauchy principal-value integral p.v. $\int_{-1}^{1} f(x)/(x-\lambda)\mathrm{d}x$, i.e.,

$$
\text{f.p.}\int_{-1}^{1}\frac{f(x)}{(x-\lambda)^2}\mathrm{d}x=\frac{\mathrm{d}}{\mathrm{d}\lambda}\text{p.v.}\int_{-1}^{1}\frac{f(x)}{x-\lambda}\mathrm{d}x. \qquad (3)
$$

Based on the fact, Bialecki [2] presented a general variable transformation-type quadrature formula for the Hadamard finite-part integral f.p. $\int_{-1}^{1} f(x)/(x-\lambda)^2 \mathrm{d}x$, which is given by

$$Q_{\mathrm{H}}(f,\lambda;h,N_1,N_2) = \frac{\partial}{\partial\lambda} Q_{\mathrm{C}}(f,\lambda;h,N_1,N_2). \tag{4}$$

We here take the double exponential transformation as the variable transformation $\psi(t)$, whereas Bialecki took the single exponential transformation in [2]. In what follows we denote the quadrature formula thus obtained, i.e., the DE-type formula for the Hadamard finite-part integral, by $Q_{\mathrm{H}}^{(\mathrm{DE})}(f,\lambda;h,N_1,N_2)$.

3 ERROR ANALYSIS

Before proceeding to theoretical error estimates of the DE-type formulae, we give some mathematical notations.

Let d be a constant such that $0 < d < \pi/2$. Let $D(d)$ be the strip domain $\{w \in \mathbf{C} \mid |\mathrm{Im}\, w| < d\}$. Let $\mathcal{D}_{\mathrm{DE}}(d)$ denote $\psi_{\mathrm{DE}}(D(d))$ (Note that $\mathcal{D}_{\mathrm{DE}}(d)$ is a Riemann surface, which boundary coils itself like a spiral around the points $z = \pm 1$ (See Figure 1)).

Let $B(\mathcal{D}_{\mathrm{DE}}(d))$ be the set of all functions that satisfy the following conditions.

1. $f(z)$ is analytic in $\mathcal{D}_{\mathrm{DE}}(d)$ except possibly for a finite number of points $a_1, a_2, \ldots, a_M$ in $\mathcal{D}_{\mathrm{DE}}(d)$ at which $f(z)$ has poles of order one.

2. $\mathcal{N}_{\mathrm{DE}}(f,d) = \liminf_{y\to d-0} \int_{\psi_{\mathrm{DE}}(L(y))} |f(z)\mathrm{d}z| < \infty$,

   where $L(y) = \{ z = u + \mathrm{i}y \mid u \in \mathbf{R} \}$.

3. $\int_{\psi_{\mathrm{DE}}(L(x))} |f(z)\mathrm{d}z| \to 0 \quad \text{as} \quad x \to \pm\infty$,

   where $L(x) = \{ z = x + \mathrm{i}v \mid -d < v < d \}$.

The following theorem gives error estimates for the DE-type formula, (1) and (2), for the Cauchy principal-value integral, and for the DE-type formula (4) for the Hadamard finite-part integral. For the simplicity, we consider the case where $N_1 = N_2$.

Theorem 1 *Let λ be a constant such that $-1 < \lambda < 1$ and d be a constant such that*

$$0 < d < \Delta(|\lambda|), \tag{5}$$

where $\Delta(\lambda)$ is the inverse function of

$$\Lambda(d) = \frac{\sinh[\pi \cos d\sqrt{(2/\sin d)^2 - 1}]}{\cosh[\pi \cos d\sqrt{(2/\sin d)^2 - 1}] + 1}. \tag{6}$$

Let $f(z)$ belong to $B(\mathcal{D}_{\mathrm{DE}}(d))$ and, furthermore, for all $x \in [-1, 1]$, let

$$|f(x)| \le C\left(\frac{1-|x|}{1+|x|}\right)^{2\alpha/\pi-1}\left[1+\left(\frac{1}{\pi}\log\left(\frac{1+|x|}{1-|x|}\right)\right)^2\right]^{-1/2} \tag{7}$$

with α a positive number. Then, taking $h = \log(2\pi d N_1/\alpha)/N_1$, we have

$$\left|\text{p.v.}\int_{-1}^{1}\frac{f(x)}{x-\lambda}\mathrm{d}x - Q_{\mathrm{C}}^{(\mathrm{DE})}(f,\lambda;h,N_1,N_1)\right| \le C_1\exp\left(-\frac{\pi d N}{\log(\pi d N/\alpha)}\right), \tag{8}$$

$$\left|\text{f.p.}\int_{-1}^{1}\frac{f(x)}{(x-\lambda)^2}\mathrm{d}x - Q_{\mathrm{H}}^{(\mathrm{DE})}(f,\lambda;h,N_1,N_1)\right| \le C_2\exp\left(-\frac{\pi d N}{\log(\pi d N/\alpha)}\right), \tag{9}$$

where $N = 2N_1 + 2$ is the number of nodes used, and C_1, C_2 are positive constants independent of N.

Remark 1 *The graph of the function $\Delta(\lambda)$ is shown in Figure 1. From (5), we know that d approaches 0 as $\lambda \to \pm 1$.*

Theorem 1 above tells us that the DE-type formulae are more efficient than the SE-type formulae presented in Bialecki's papers. In fact, although the quadrature error of Bialecki's SE-type formulae is of order $\mathrm{O}[\exp(-c\sqrt{N})]$, the quadrature error of our DE-type formulae is of order $\mathrm{O}[\exp(-c_\lambda N/\log N)]$.

To prove Theorem 1, we prepare a lemma.

Lemma 1 *Let λ be a constant such that $-1 < \lambda < 1$ and d be a constant such that $0 < d < \Delta(|\lambda|)$. Let $f(z)$ be a function that belongs to $B(\mathcal{D}_{\mathrm{DE}}(d))$. Then*

$$\left|\text{p.v.}\int_{-1}^{1}\frac{f(x)}{x-\lambda}\mathrm{d}x - Q_{\mathrm{C}}^{(\mathrm{DE})}(f,\lambda;h,+\infty,+\infty)\right|$$

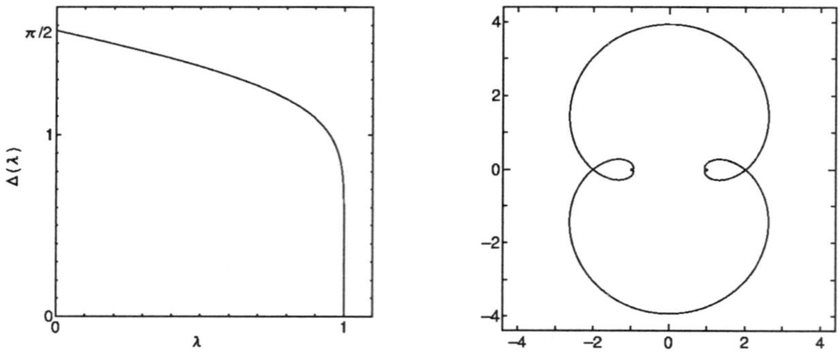

Figure 1: The graph of the function $\Delta(\lambda)$ and the domain $\mathcal{D}_{\mathrm{DE}}(d)$ ($d = 1$).

$$\leq \frac{\mathcal{N}_{\mathrm{DE}}(f,d)}{\mathrm{dist}(\lambda, \partial\mathcal{D}_{\mathrm{DE}}(d))} \frac{\exp(-2\pi d/h)}{1-\exp(-2\pi d/h)}, \tag{10}$$

$$\left| \text{f.p.} \int_{-1}^{1} \frac{f(x)}{(x-\lambda)^2} \mathrm{d}x - Q_{\mathrm{H}}^{(\mathrm{DE})}(f,\lambda;h) \right|$$

$$\leq \frac{\mathcal{N}_{\mathrm{DE}}(f,d)}{[\mathrm{dist}(\lambda, \partial\mathcal{D}_{\mathrm{DE}}(d))]^2} \frac{\exp(-2\pi d/h)}{1-\exp(-2\pi d/h)}, \tag{11}$$

where $\mathrm{dist}(\lambda, \partial\mathcal{D}_{\mathrm{DE}}(d))$ *is the distance between* λ *and the boundary* $\partial\mathcal{D}_{\mathrm{DE}}(d)$, *i.e.,*

$$\mathrm{dist}(\lambda, \partial\mathcal{D}_{\mathrm{DE}}(d)) = \inf_{w \in \partial D(d)} |\lambda - \psi_{\mathrm{DE}}(w)|. \tag{12}$$

(*Proof of Lemma 1*) The inequality (10) follows from Theorem 2.1 in [1] and the inequality (11) from Theorem 2.2 in [2].

It should here be noted that both Theorem 2.1 in [1] and Theorem 2.2 in [2] assumed implicitly that the equation $\psi(\mu) = \lambda$ has no other solutions than the trivial real one in $D(d)$. In our case, however, there is a possibility that the equation $\psi_{\mathrm{DE}}(\mu) = \lambda$ has many other solutions than the trivial real one in $D(d)$, because ψ_{DE} is not a one-to-one map. The seemingly strange inequality $0 < d < \Delta(|\lambda|)$ is for prohibitting this possibility. In fact, noting that the intersections of the real axis and the curve $\psi_{\mathrm{DE}}(\partial D_d)$ are given by

$$\pm x_n = \pm \frac{\sinh[\pi \cos d \sqrt{(n/\sin d)^2 - 1}]}{\cosh[\pi \cos d \sqrt{(n/\sin d)^2 - 1}] + (-1)^n} \quad (n = 1, 2, \ldots) \tag{13}$$

$$(0 < x_2 < x_4 < \cdots < x_{2n} < \cdots < 1 < \cdots < x_{2n-1} < \cdots < x_3 < x_1),$$

we can see that as far as the inequality $|\lambda| < x_2 = \Lambda(d)$, i.e., $0 < d < \Delta(|\lambda|)$ holds good, the equation $\psi_{\mathrm{DE}}(\mu) = \lambda$ has no other solutions than the trivial real one in $D(d)$. *q.e.d.*

(*Proof of Theorem 1*) We here prove only the inequality (8). The inequality (9) can be proved similarly.

The error of the formula is devided into two terms.

$$\begin{aligned}
&\left|\mathrm{p.v.}\int_{-1}^{1}\frac{f(x)}{x-\lambda}\mathrm{d}x - Q_{\mathrm{C}}^{(\mathrm{DE})}(f,\lambda;h,N_1,N_2)\right| \\
&\leq \left|\mathrm{p.v.}\int_{-1}^{1}\frac{f(x)}{x-\lambda}\mathrm{d}x - Q_{\mathrm{C}}^{(\mathrm{DE})}(f,\lambda;h,+\infty,+\infty)\right| \\
&\qquad + h\sum_{|k|\geq N_1+1}^{\infty}\frac{|f(\psi_{\mathrm{DE}}(kh))\psi_{\mathrm{DE}}'(kh)|}{|\psi_{\mathrm{DE}}(kh)-\lambda|}. \qquad (14)
\end{aligned}$$

The first term on the right hand side is the discretization error and is bounded as (10). The second term is the truncation error and is bounded in the following way. Note first that the inequality (7) implies

$$|f(\psi_{\mathrm{DE}}(t))\psi_{\mathrm{DE}}'(t)| \leq 2\pi C\mathrm{e}^{\alpha}\exp(-\alpha\exp|t|), \qquad (15)$$

which yields

$$h\sum_{|k|\geq N_1+1}^{\infty}\frac{|f(\psi_{\mathrm{DE}}(kh))\psi_{\mathrm{DE}}'(kh)|}{|\psi_{\mathrm{DE}}(kh)-\lambda|} \leq C'\exp(-\alpha\exp(N_1 h)) \qquad (16)$$

with C' a positive number, which is the demanded estimate for the second term. Then, substituting these estimates, (10) and (16), into (14), we have

$$\begin{aligned}
&\left|\mathrm{p.v.}\int_{-1}^{1}\frac{f(x)}{x-\lambda}\mathrm{d}x - Q_{\mathrm{C}}^{(\mathrm{DE})}(f,\lambda;h,N_1,N_1)\right| \\
&\leq \frac{\mathcal{N}_{\mathrm{DE}}(f,d)}{\mathrm{dist}(\lambda,\partial\mathcal{D}_{\mathrm{DE}}(d))}\,\frac{\exp(-2\pi d/h)}{1-\exp(-2\pi d/h)} + C'\exp(-\alpha\exp(N_1 h)) \qquad (17)
\end{aligned}$$

We should here choose h for a given N_1 to balance the magnitude of the first term with that of the second term on the right hand side of (17). This requirement leads us to the equation

$$\exp\left(-\frac{2\pi d}{h}\right) = \exp(-\alpha\exp(N_1 h)). \qquad (18)$$

Solving (18) for h, we have

$$h = \frac{1}{N_1}\log\left(\frac{2\pi d N_1}{\alpha}\right) + \mathrm{O}\left(\frac{1}{N_1}\log\log\left(\frac{2\pi d N_1}{\alpha}\right)\right). \qquad (19)$$

We here choose h as

$$h = \frac{1}{N_1}\log\left(\frac{2\pi d N_1}{\alpha}\right), \qquad (20)$$

which is the main term of the above asymptotic expansion for h, and substituting (20) into the first term and the second one on the right hand side of (17), we have (8). *q.e.d.*

4 A NUMERICAL EXAMPLE

In this section we give a numerical example. We considered

$$\text{p.v.}\int_{-1}^{1}\frac{(1-x)^{1/4}(1+x)^{-1/4}}{x-\lambda}\mathrm{d}x = \pi\left(\frac{1-\lambda}{1+\lambda}\right) - \sqrt{2}\pi, \qquad (21)$$

$$\text{f.p.}\int_{-1}^{1}\frac{(1-x)^{1/4}(1+x)^{-1/4}}{(x-\lambda)^2}\mathrm{d}x = -\frac{\pi}{2}(1+\lambda)^{-5/4}(1-\lambda)^{-3/4} \qquad (22)$$

which were taken as examples in [1, 2]. And we computed these integrals using the proposed DE-type formulae and Bialecki's SE-type formilae. Computations were carried out in double precision. Numerical results are presented in Figure 2 and 3. It can be seen from these figures that the proposed DE-type formulae are superior to Bialecki's SE-type ones.

5 CONCLUSION

We proposed DE-type quadrature formulae for Cauchy principal-value integrals and for Hadamard finite-part integrals. Through theoretical error analysis and numerical experiment, we showed that our DE-type formulae are more efficient than Bialecki's SE-type ones.

References

[1] Bialecki, B., *A Sinc-Hunter quadrature rule for Cauchy principal value integrals*, Math. Comput., **55**(1990), 665–681.

[2] Bialecki, B., *A Sinc quadrature rule for Hadamard finite-part integrals*, Numer. Math., **57**(1990), pp. 263–269.

[3] Takahasi, H., Mori, M., *Double exponential formulas for numerical integration*, Publ. RIMS Kyoto Univ., **9**(1974), pp. 721–724.

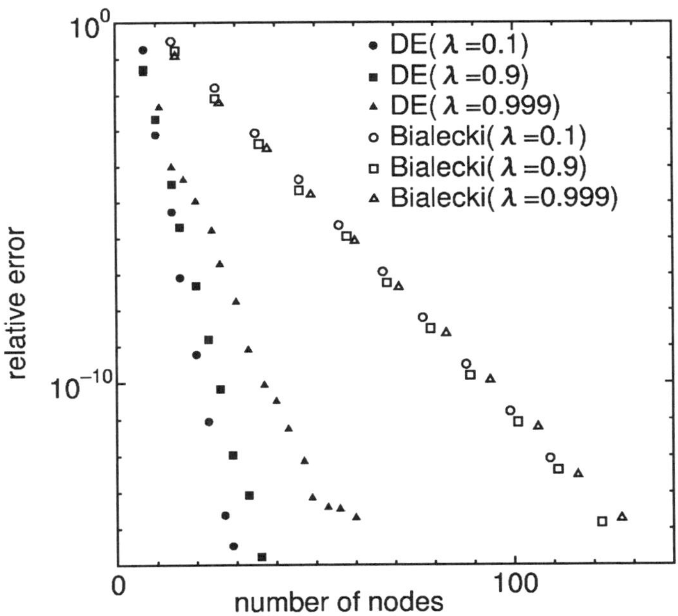

Figure 2: The relative errors of the numerical integration of the Cauchy principal-value integral p.v. $\int_{-1}^{1} f(x)/(x-\lambda)\mathrm{d}x$, where $f(x) = (1-x)^{1/4}(1+x)^{-1/4}$, by the present DE-type formula and Bialecki's SE-type one.

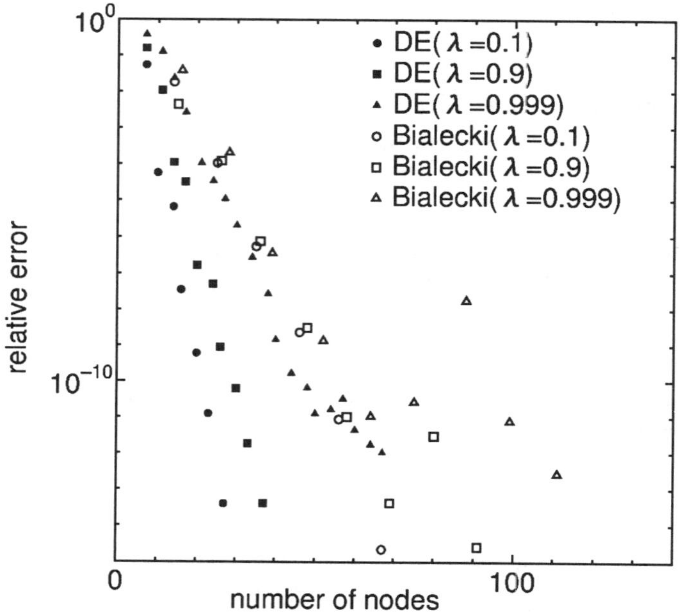

Figure 3: The relative errors of the numerical integration of the Hadamard finite-part integral f.p. $\int_{-1}^{1} f(x)/(x-\lambda)^2 \mathrm{d}x$, where $f(x) = (1-x)^{1/4}(1+x)^{-1/4}$, by the present DE-type formula and Bialecki's SE-type one.

43 MATHEMATICAL MODELS USED FOR SOUND SYNTHESIS

Ken'ichi OHYA

Department of Electronics and Information Science
Nagano National College of Technology
716 Tokuma, Nagano City, Nagano 381-8550, Japan

ohya@ei.nagano-nct.ac.jp

Kazumasa SHINJO

ATR Adaptive Communications Research Labs
2-2 Hikaridai, Seika-cho, Souraku-gun, Kyoto 619-0288, Japan

shinjo@acr.atr.co.jp

Abstract

Sound synthesis is one area of computer music research, and some mathematical models have been presented in its history. We present some mathematical models of sound synthesis, especially "Recurrent Neural Network Synthesis" and "Hamiltonian Algorithm Synthesis".

1 Introduction

Study of sound synthesis has a long history in computer music study since its early days of computer history, and many sound synthesis mod-

H.G.W. Begehr et al. (eds.), Proceedings of the Second ISAAC Congress, Vol. 1, 367–373.

els have been presented. In order to synthesize a new sound, some researchers have used mathematics to their models.

The author [2, 3] used recurrent neural network for sound synthesis as a mathematical framework. In this model, each neuron is continuous-time, continuous-variable model. Therefore total system consists of many nonlinear differential equations. The system produces a sound with natural fluctuations by this nonlinear effects.

Recently Hamiltonian Algorithm (HA) are presented. This algorithm generally finds solutions in optimization problems by [9, 10].

In this paper, some mathematical models for sound synthesis are shown.

2 History of Sound Synthesis

"Sound Synthesis" is one of the researches in computer music, and the history of sound synthesis dates back to the early beginning of computer history. The first experiments in synthesis of sound by computer began in 1957 by researchers (Roads [5], et al.) at Bell Telephone Laboratories by a giant IBM 704 computer fabricated with vacuum tube circuits.

Since then, various sound synthesis models have been presented and studied by many researhers in the world. One of the most famous model is "Frequency Modulation Synthesis" (FM Synthesis) by John Chowning at Stanford Univ. This model was adopted by YAMAHA and YAMAHA produced a synthesizer named DX7, which was the most popular synthesizer in the world in the early 1980's. Besides the FM Synthesis, "additive synthesis" was one of the major sound synthesis algorithm.

But after that, PCM technique became popular in 1990's. Although this technique needs much memory, it bacame a mainstream sound synthesis technique while memory price became cheaper and cheaper in 1990's.

However, this PCM technique also has some problems. One problem is its difficulties for users to edit a preset sound. And another is that the produced sound is somewhat monotonous. It is said that this monotonousness drives from loop playback of sampled sounds.

To avoid this monotonousness of simple playback, we present a new sound synthesis algorithm named "Hamiltonian Algorithm Sound Synthesis". This new technique uses "Hamiltonian Algorithm" (HA) by [9, 10]. HA uses differential equations, and it produces waveforms by numerically solving these equations. There is no limitation in time while producing sounds.

3 Physical Modeling

In this section, we will introduce "Physical Modeling Synthesis".

"Physical Modeling" is a sound synthesis model that uses mathematical models of the real physical acoustics. To produce a synthesized sound of a musical instrument, equations that describe its physical model are numerically solved.

YAMAHA was the first company that produced physical modeling synthesizers, VP-1 and VL-1 in 1993.

We will show an example of physical modeling synthesis of piano and guitar by Hikichi [1].

Fig.1 shows a diagram of this synthesis. It has three parts in synthesis, vibrator, exciter and resonator. the vibrator is a model of a string, the exciter is a model of a hammer in piano or a finger in guitar, and the resonator is a model of a body that resonates.

Each part is described as the following equations:

- vibrator (string):

$$\frac{\partial^2 y}{\partial t^2} = \frac{T}{\mu}\frac{\partial^2 y}{\partial x^2} - \frac{\kappa^2 ES}{\mu}\frac{\partial^4 y}{\partial x^4} - 2b_1\frac{\partial y}{\partial t} + 2b_3\frac{\partial^3 y}{\partial t^3} + \frac{F_H(t)g(x,x_0)}{\mu\int_{x_0-\delta x}^{x_0+\delta x} g(x,x_0)dx}$$

- exciter:

$$M_H\frac{d^2\eta}{dt^2} = -F_H(t)$$

$$F_H(t) = K|\eta(t) - y(x_0,t)|^p$$

- resonator:

$$\frac{\partial^2 z}{\partial t^2} = -\frac{\kappa^2 E}{\rho(1-\nu^2)}\nabla^4 z - 2b_1\frac{\partial z}{\partial t} + 2b_3\frac{\partial^3 z}{\partial t^3} + \frac{F_s(t)}{\rho h}\delta(x_1,y_1)$$

4 Recurrent Neural Network Synthesis

In this section, we will introduce "Recurrent Neural Network Synthesis" by the author [2, 3].

As a single neuron model, a continuous-time, continuous-variable neuron model is adopted. Therefore output value from any single neuron can be directly used for waveform.

Equation of dynamics of each neuron in a recurrent neural network is given as

$$\tau_i \frac{du_i}{dt} = -u_i + f(\sum_{j=1}^{n} W_{ij} u_j) + I_i$$

where $u_i(t)$ is the i-th unit output at a time t, τ_i a time delay constant, $f(x)$ a sigmoid function, I_i an external input of the i-th unit, W_{ij} a connection weight from the j-th unit to the i-th unit.

Recurrent Neural Network can generate very complex dynamics pattern because of its recurrent connections even if it is composed of relatively small number of neurons.

As the 1st variation of RNNS, sound synthesis by small recurrent neural network is shown. This model is useful for realtime software sound synthesis because its relatively less computation.

Fig. 3 is an example of "3 pair model". "3 pair model" is composed of 3 pairs of neurons and another output neuron. Each neuron of the pair neuron is connected each other and is also connected to itself. This pair neuron is the smallest recurrent neural network architecture. The output neuron receives 6 output values from the pair neuron and computes the final output value, which is a waveform of sound.

Dynamics of the pair neuron is completely described by eight parameters, 2 initial values, 2 time delay constants, 4 connection weight values. In an 8-dimension space of parameters, the region which gives this system lasting oscillation seems to be small. One of such region is given if weight values connecting each other are asymmetric. In this condition, 2 output values of a pair neuron spikes by turns, and this pair neuron functions as a kind of oscillator. It is also possible to use only one of the outputs from the pair neuron as waveform itself, but the third neuron, the output neuron, is used for producing more complex, rich waveform.

An example of a waveform of 3 pair model is shown in Fig. 2, and the parameters are shown in Fig. 3.

Frequency of the waveform mainly depends on the time delay constant τ, and the total waveform chiefly depends on values of connection weight.

In "Recurrent Neural Network Synthesis", resynthesis of any given sound is also possible with more complex architecture of recurrent neural network by the author [3], Pearlmutter [4] and Sato [6, 7, 8].

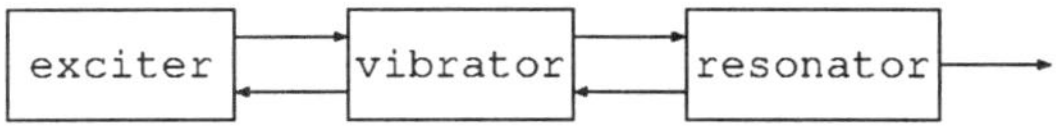

Figure 1: physical model of piano and guitar

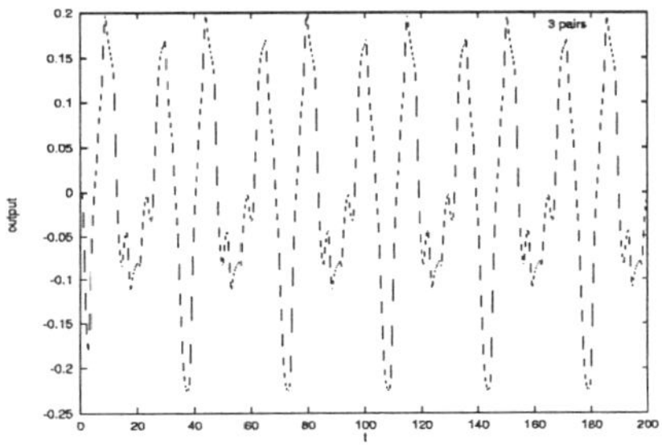

Figure 2: waveform example of 3 pair model

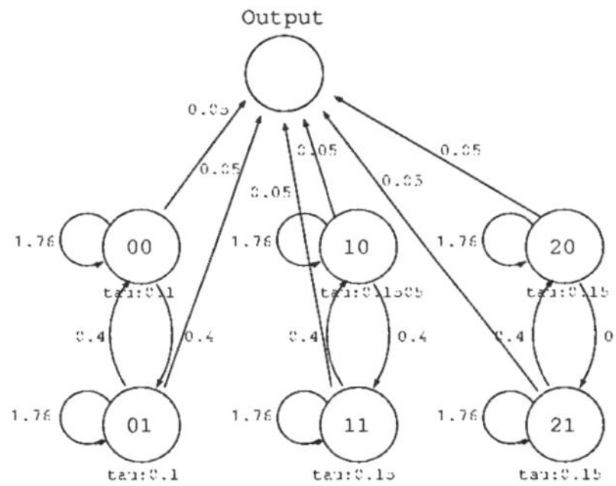

Figure 3: 3 pair model architecture and a parameter example

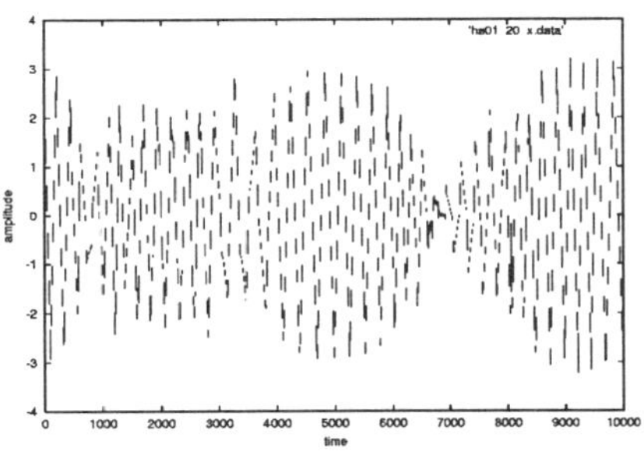

Figure 4: waveform: example

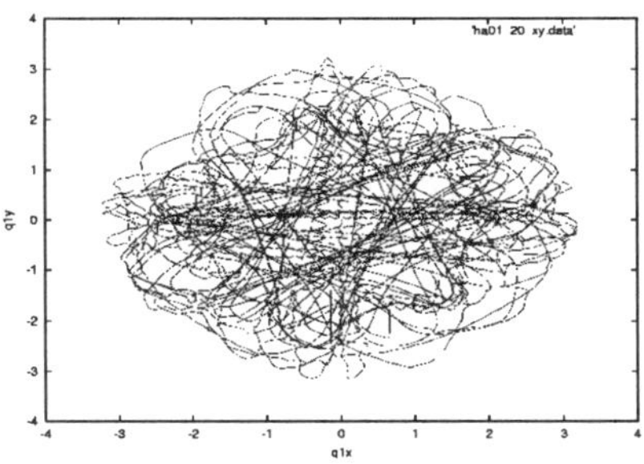

Figure 5: phase space trajectory: example

5 Hamiltonian Algorithm Sound Synthesis

"Hamiltonian Algorithm Sound Synthesis" is a new algorithm to create waveforms of sound using "Hamiltonian Algorithm" by [9, 10]. Following is one of examples of Hamiltonian Algorithm Sound Synthesis.

We set a potential curve described as the following formula:

$$V(q_{ix}, q_{iy}) = -3e^{-10(\sqrt{(q_{ix}^2+q_{iy}^2)}-2)^2} - 2e^{-10(\sqrt{(q_{ix}^2+q_{iy}^2)}-3)^2}$$

Hamiltonian is described as the following formula:

$$H = \sum_{i=1,2} \left\{ \frac{1}{2}(p_{ix}^2 + p_{iy}^2)^\gamma + V(q_{ix}, q_{iy}) \right\} + V_{12},$$

$$V_{12} = a\left\{ p_{1x}^\gamma p_{2x}^\gamma + p_{1y}^\gamma p_{2y}^\gamma \right\},$$

where V_{12} is a mixing term.

Given initial conditions of coordinates and velocities of the two phase points, the points move in this phase space.

Fig.5 is an example of the trajectory of one phase point. This figure shows that phase point not only moves in the region of one potential valley, but also transits one potential valley to the other, because of the mixing term effect.

Fig.4 is an example of the waveform using one projection of phase space trajectories. Nonmonotonous complex waveforms of sound can be seen in this figure.

6 Summary

Some mathematical models used for sound synthesis technique were shown, especially for physical modeling, recurrent neural network synthesis and hamiltonian algorithm synthesis.

7 Acknowledgment

We thank Joji Kajiwara and Yasuaki Nakano for giving us opportunities to study this research.

References

[1] T. Hikichi, N. Osaka, *Morphing between piano and guitar tones based on physical modeling.* IPSJ SIG Notes, Vol99, No.68, 19–26, 1999.

[2] K. Ohya, *A Sound Synthesis by Recurrent Neural Network.* Proceedings of the 1995 International Computer Music Conference, 420–423, 1995.

[3] K. Ohya, *Sound Variations by Recurrent Neural Network Synthesis.* Proceedings of the 1998 International Computer Music Conference, 280–283, 1998.

[4] B. A. Pearlmutter, *Learning state space trajectories in the recurrent neural network.* Neural Computation, **1** (2), 263–269, 1989.

[5] C. Roads, *The Computer Music Tutorial.* MIT Press, 1996.

[6] M. Sato, *A learning algorithm to teach spatiotemporal patterns to recurrent neural networks.* Biological Cybernetics, **62**, 259–263, 1990.

[7] M. Sato, Y. Murakami, K. Joe, *Chaotic dynamics by recurrent neural networks.* Proceedings of the International Conference on Fuzzy Logic and Neural Networks, 601–604, 1990.

[8] M. Sato, K. Joe, T. Hirahara, *APOLONN brings us to the real world: Learning nonlinear dynamics and fluctuations in nature.* Proceedings of the International Joint Conference on Neural Networks, San Diego, **I**, 581–587, 1990.

[9] K. Shinjo, T. Sasada, *Hamiltonian Systems with Many Degrees of Freedom: Asymmetric Motion and Intensity of Motion in Phase Space.* Phys. Rev., **E54**, 4686, 1996.

[10] K. Shinjo, S. Shimogawa, J. Yamada, K. Oida, *Strategy of Designing Routing Algorithms Based on Ideal Routings.* International Journal of Modern Physics C, **10**, 1–32, 1999.

44 AN INTEGRAL EQUATION METHOD OF NUMERICAL CONFORMAL MAPPING ONTO PARALLEL, CIRCULAR AND RADIAL SLIT DOMAINS [1]

Dai OKANO
Kaname AMANO

Department of Computer Science, Faculty of Engineering
Ehime University, Matsuyama 790-8577, Japan

okano@cs.ehime-u.ac.jp
amano@cs.ehime-u.ac.jp

1 Introduction

We recently proposed a method for numerical conformal mappings of multiply connected open domains onto parallel, circular, and radial slit domains [1, 2]. These three types of standard domains are important in two dimensional potential flow analysis. Our method uses the charge simulation method as the potential problem solver. The charge simulation method offers highly accurate approximation for the smooth, and

[1]Supported by the Grant-in-Aid for Scientific Research of the Ministry of Education, Science, Sports and Culture in Japan (09440081).

H.G.W. Begehr et al. (eds.), Proceedings of the Second ISAAC Congress, Vol. 1, 375–382.

round boundary conditions, but has some difficulties in application to exterior problems of narrow area and problems with concave corners.

In this paper, we propose an integral equation method for numerical conformal mapping onto the slit domains, which is expected to be effective to overcome the difficulties of the charge simulation method with some clever techniques [4, 5].

2 Integrated scheme for the three types of standard slit domains

We here call the entire plane with the slits a slit domain. To concern the problem of the computation of conformal mappings of the multiply connected domains onto the slit domains, we need to formulate standard slit domains properly. The standard slit domains of which we are going to have discussion have three types of slits, the parallel, circular, and radial slits [3].

Suppose, $w = f_\phi(z), f_c(z)$, and $f_r(z)$ are the mapping functions of the conformal mappings of the multiply connected problem domain D exterior to the closed Jordan curves $C_1, \dots, C_n$, which has the infinity point included, onto the parallel, circular, and radial slits domains, respectively. That is, $C_1, \dots, C_n$ are mapped, by f_ϕ, onto the slits parallel to the rectilinear line which intersects x-axis by angle ϕ, by f_c, onto the circular slits concentric to the origin with radii $r_1, \dots, r_n$, and by f_r, onto the radial slits radiated from the origin with the angles $\theta_1, \dots, \theta_n$, where the positions of the slits remain undetermined together with the mapping functions to be obtained later.

The mapping function onto the parallel slit domains is uniquely determined under the normalization condition which restricts the Laurent expansion at infinity to takes the form of

$$f_\phi(z) = z + \frac{a_1}{z} + \frac{a_2}{z^2} + \cdots, \tag{1}$$

i.e., $f_\phi(\infty) = \infty, f'_\phi(\infty) = 1,$ and $a_0 = 0$. Similarly, the mapping functions onto the circular slit domains, and the radial slit domains are uniquely determined under the normalization conditions, $f_c(0), f_r(0) = 0, f_c(\infty), f_r(\infty) = \infty, f'_c(\infty), f'_r(\infty) = 1$, which mean the Laurent expansion takes the form of

$$f_c(z), f_r(z) = z + b_0 + \frac{b_1}{z} + \cdots. \tag{2}$$

Then, we rewrite the functions with the conjugate harmonic functions $g(z), h(z)$ defined in D as

$$f_\phi(z) = z + \mathrm{i}e^{\mathrm{i}\phi}(g(z) + \mathrm{i}h(z)), \tag{3}$$
$$f_c(z) = z\exp(g(z) + \mathrm{i}h(z)), \tag{4}$$
$$f_r(z) = z\exp(\mathrm{i}(g(z) + \mathrm{i}h(z))). \tag{5}$$

If $-\phi$ rotation around the origin is given to the mapped image of D by f_ϕ, the imaginary part of the finally obtained image of $z \in C_l$ holds constant p_l over each $l = 1, \dots, n$,

$$\mathrm{Im}\,(e^{-\mathrm{i}\phi} f_\phi(z)) = \mathrm{Im}\,(e^{-\mathrm{i}\phi} z) + g(z) = p_l, \tag{6}$$

where $z \in C_l, l = 1, \dots, n$. Similarly, the boundary conditions for f_c and f_r are expressed as

$$|f_c(z)| = |z|\exp g(z) = r_l, \tag{7}$$
$$\arg f_r(z) = \arg z + g(z) = \theta_l, \tag{8}$$

by the constants, r_l, and θ_l, where $z \in C_l$, $l = 1, \dots, n$.

Thus, the boundary conditions for the three types of slit domains are rewritten into the very similar equations of harmonic function $g(z)$ and are different only in the right-hand side,

$$g(z) = -t(z) + t_l, \quad \text{for } z \in C_l, l = 1, \dots, n, \tag{9}$$

where $t(z) = \mathrm{Im}\,(e^{-\mathrm{i}\phi} z)$, $\log|z|$, $\arg z$, and $t_l = p_l$, $\log r_l$, θ_l for f_ϕ, f_c, f_r, respectively.

Furthermore, g, and h are uniquely determined for each problem of the mapping functions f_ϕ, f_c, f_r under the same normalization condition

$$g(\infty) + \mathrm{i}h(\infty) = 0. \tag{10}$$

Finally, the problem of computation of conformal mappings, which map the arbitrary given multiply connected open domain onto the three types of the standard slit domains, resolves itself into a boundary condition problem of a pair of the conjugate harmonic functions g, and h.

3 Formulation by the integral equation method for starlike domains

To compose the approximation of the functions g, and h, we here use the method of integral equations of Symm, i.e., we express the functions by the single layer potential,

$$g(z) + \mathrm{i}h(z) = \int_C \sigma(\zeta) \log(z - \zeta)|\mathrm{d}\zeta|, \tag{11}$$

where $z \in D$, $C = C_1 + \cdots + C_n$. Then the boundary conditions (9), the normalization condition (10) and the requirement for the mapping functions to be single valued in D end up to give the coupled integral equations

$$\int_C \sigma(\zeta) \log |z - \zeta||\mathrm{d}\zeta| = -t(z) + t_l \quad \text{for } z \in C_l, \tag{12}$$

$$\int_{C_l} \sigma(\zeta)|\mathrm{d}\zeta| = 0, \tag{13}$$

where $l = 1, \ldots, n$.

Practical calculation of the equations (12), (13), and the harmonic functions (11) needs to be given by the principal value of the complex logarithmic function. And a straight-forward implementation to solve the equations may have a difficulty on treating the discontinuity of the principal value of the complex logarithmic function.

If the Jordan curves $C_1, \ldots, C_n$ are starlike and have the center points $z_{10}, \ldots, z_{n0}$ for each, the equations (12) are transformed as

$$\int_C \sigma(\zeta) \log \left| \frac{z - \zeta}{z - z_{l0}} \right| |\mathrm{d}\zeta| = -t(z) + t_l \quad \text{for } z \in C_l, \tag{14}$$

where $l = 1, \ldots, n$. By this transformation discontinuous cut is excluded out of the problem domain D even if the logarithmic function is simply replaced by its principal value.

Finally we have the reformulation of the problem by the integral equations,

$$\int_C \sigma(\zeta) \log \left| \frac{z - \zeta}{z - z_{l0}} \right| |\mathrm{d}\zeta| = -t(z) + t_l \quad \text{for } z \in C_l, \tag{15}$$

$$\int_{C_l} \sigma(\zeta)|\mathrm{d}\zeta| = 0, \tag{16}$$

and the harmonic functions,

$$g(z) + \mathrm{i}h(z) = \int_C \sigma(\zeta) \mathrm{Log} \left(\frac{z - \zeta}{z - z_{l0}} \right) |\mathrm{d}\zeta|, \tag{17}$$

where $l = 1, \ldots, n$, and Log () is the principal value of log().

4 Numerical examples

We here provide examples of the computation for the problem domains with the smooth and starlike Jordan curves.

Suppose $C_l = \{\zeta_l(t) \mid 0 < t \leq 1\}$, i.e., the parameterization of $C_1, \dots, C_n$ is given by $\zeta_l(t)$, where $l = 1, \dots, n$. The density function σ is approximated by the linear combination of the B-spline basis of order q on the boundary,

$$\sigma(\zeta_l(t)) \sim \sum_{j=1}^{N_l} \sigma_{lj} b_{lj}^{(q)}(t), \tag{18}$$

where $l = 1, \dots, n$. The set of B-spline basis of order q is defined by the characteristic function $b^{(0)}$,

$$b^{(0)}(t) = 1 \quad \text{for } -\frac{1}{2} < t < \frac{1}{2}, \tag{19}$$

$$= 0 \quad \text{otherwise}, \tag{20}$$

$$b^{(q)}(t) = \int_{-q/2}^{+q/2} b^{(q-1)}(\tau) b^{(0)}(\tau - t) \mathrm{d}\tau, \tag{21}$$

$$b_{lj}^{(q)}(t) = \sum_{p=-\infty}^{\infty} N_l b^{(q)}(N_l(t+p) - (j-1)), \quad j = 1, \dots, N_l. \tag{22}$$

Then the approximation of the $g(z) + \mathrm{i}h(z)$,

$$G(z) + \mathrm{i}H(z) = \sum_{l=1}^{n} \sum_{j=1}^{N_l} \sigma_{lj} \int_0^1 b_{lj}^{(q)}(t) \mathrm{Log} \left(\frac{z - \zeta_l(t)}{z - z_{l0}} \right) \left| \frac{\mathrm{d}\zeta}{\mathrm{d}t} \right| |\mathrm{d}t|, \tag{23}$$

offers the linear equations, which give the discretization of the integral equations (15),(16),

$$\sum_{l=1}^{n} \sum_{j=1}^{N_l} \sigma_{lj} \int_0^1 b_{lj}^{(q)}(t) \log \left| \frac{z_{mk} - \zeta_l(t)}{z_{mk} - z_{l0}} \right| \left| \frac{\mathrm{d}\zeta}{\mathrm{d}t} \right| |\mathrm{d}t| = -t(z_{mk}) + T_m, \tag{24}$$

$$\sum_{j=1}^{N_m} \sigma_{mj} \int_0^1 b_{mj}^{(q)}(t) \left| \frac{\mathrm{d}\zeta}{\mathrm{d}t} \right| |\mathrm{d}t| = 0, \tag{25}$$

where $z_{mk} = \zeta((k-1)/N_m)$, $m = 1, \dots, n$, $k = 1, \dots, N_m$ are the sample points, and the undetermined variable T_m is the approximation of t_m.

Figure 1 shows the results of the approximation of the conformal mapping of D exterior to the unit disk centered at $z_0 = 2 + 2\mathrm{i}$ onto the parallel slit domain with a slit parallel to the rectilinear line which intersects x-axis by angle $\pi/3$. Maximum error estimation of the mapping function $f_{\pi/3}$ over the boundary circle is of the orders 10^{-2}, 10^{-3}, and 10^{-4} for discretization factors $N_1 = 17, 33$, and 129, respectively, where the order of the spline basis.

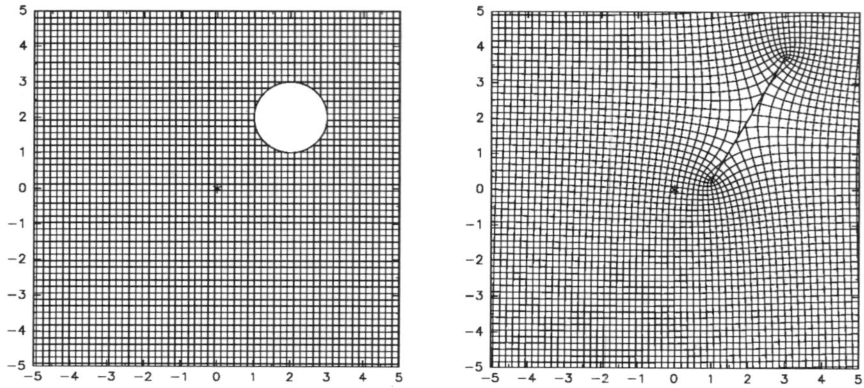

The unit disk centered at $2 + 2\mathrm{i}$ Parallel slit with $\theta = \pi/3$

Figure 1   Numerical example of D exterior to the single disk

Figure 2 shows the results of the approximation of the conformal mapping of D exterior to three disks with radii 0.5, 1, and 1.5 onto the parallel slit domain with slits parallel to the rectilinear line which intersects x-axis by angle $\pi/3$.

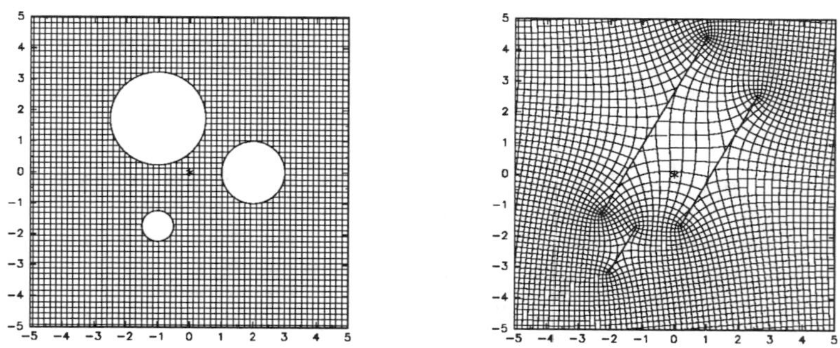

Three disks with radii $1, 0.5, 2$ Parallel slits with $\theta = \pi/3$

Figure 2   Numerical example of D exterior to the three disks

Since the exact solution of the multiply connected domain is unknown, we use the distance of the images of the boundary curves from the approximated positions of the slits, i.e., $E_l = \max|\mathrm{Im}(e^{-\pi/3}F(z)) - P_l|$ for $z \in C_l$, $l = 1, 2, 3$, where $F(z)$ and P_l are the approximation of the mapping function and p_l, respectively. Error estimation for the three disks is given in Table 1, where the order of the spline basis $q = 2$.

Table 1 Error estimation of the example of D exterior to the three disks

N	E_1	E_2	E_3	Cond.
17×3	3.6×10^{-3}	1.2×10^{-3}	7.3×10^{-3}	2.6×10^{2}
33×3	6.7×10^{-4}	6.9×10^{-4}	1.1×10^{-3}	3.2×10^{3}

5 Concluding remarks

We presented the integrated scheme for the problem of conformal mappings onto three types of different standard slit domains. The scheme is in part a modification and a generalization of Amano et al. [1, 2], but moreover it offers a natural formulation corresponding to the standard domains given by Nehari [3]. In fact, dual implementation by argument function which Amano et al. [1, 2] suggested may easily be reproduced.

We used the integral equation method of Symm to solve the potential problem which arises in the scheme. This may lead to the extension of Hough and Papamichael [4, 5] to the problem of multiply connected standard slit domains which is expected to overcome the difficulties associated with corner singularity.

References

[1] Amano, K.: A charge simulation method for numerical conformal mapping onto circular and radial slit domains, SIAM J. Sci. Comput., Vol. 19, No. 4, pp. 1169–1187 (1998).

[2] Amano, K., Okano, D., Shimohira, H. and Sugihara, M.: Numerical conformal mapping onto parallel, circular and radial slit domains, Advances in Numerical Mathematics (Proceedings of the Fourth Japan–China Joint Seminar on Numerical Mathematics, Edited by Kawarada, H., Nakamura, M. and Shi, Z.), Mathematical Sciences and Applications, Vol. 12, pp. 1–10 (1999).

[3] Nehari, Z.: Conformal Mapping, McGraw-Hill, New York (1952); Dover, New York (1975).

[4] Hough, D.M. and Papamichael, N.: The use of splines and singular functions in an integral equation method for conformal mapping, Numer. Math., Vol. 37, pp. 133–147 (1981).

[5] Hough, D.M., and Papamichael, N.: An integral equation method for the numerical comformal mapping of interior, exterior and doubly-connected domains, Numer. Math., Vol. 41, pp. 287–307 (1983).

45 UNIFORMLY PERFECT SETS, RATIONAL SEMIGROUPS, KLEINIAN GROUPS AND ITERATED FUNCTIONS SYSTEMS

Rich STANKEWITZ

Department of Mathematics, Texas A&M University
College Station, TX 77843, U. S. A.

rich.stankewitz@math.tamu.edu
http://www.math.tamu.edu/~richs

These are lecture notes based on the paper of the same title which is to appear in the Proceedings of the AMS. See this paper for more details and for a more extensive bibliography.

A **rational semigroup** G is a semigroup of nonconstant rational functions defined on the Riemann sphere $\overline{\mathbb{C}}$ with the semigroup operation being functional composition. When a semigroup G is generated by the functions $\{f_i : i \in I\}$ we write

$$G = \langle f_i : i \in I \rangle.$$

We allow G to consist of all Möbius maps (although we note that some of the literature considers only cases when the semigroup contains at least one element of degree at least two). In some sense we can consider semigroups as generalizations of

- Kleinian groups
- iteration of a rational function
- iterated function systems (IFS's).

H.G.W. Begehr et al. (eds.), Proceedings of the Second ISAAC Congress, Vol. 1, 383–388.

In particular, we will see that **limit sets of Kleinian groups, Julia sets of a rational function**, and self-similar **attractor sets generated by IFS's** are all special cases of **Julia sets of rational semigroups** as fixed points are always dense.

For a rational semigroup G we define (following Hinkkanen and Martin [8]) the **set of normality** of G by

$$N(G) = \{z \in \overline{\mathbb{C}} : \text{ there is a neighborhood of } z \text{ on which } G \text{ is a normal family}\}$$

and define the **Julia set** of G by

$$J(G) = \overline{\mathbb{C}} \setminus N(G).$$

When $G = \langle f \rangle$, we abuse the notation and write $N(f)$ and $J(f)$ for $N(\langle f \rangle)$ and $J(\langle f \rangle)$ respectively.

Properties:

• $f(N(G)) \subset N(G)$ for all $f \in G$ (forward invariance)

• $f^{-1}(J(G)) \subset J(G)$ for all $f \in G$ (backward invariance)

• The sets $N(G)$ and $J(G)$ are, however, not necessarily completely invariant under the elements of G. This is in contrast to the case of single function dynamics. For a discussion on completely invariant Julia sets the reader is referred to [14], [15] and [13].

• If $G = \langle g_1, \ldots, g_N \rangle$, then $J(G) = \cup_{i=1}^{N} g_i^{-1}(J(G))$ (backward self-similarity)

• $J(G)$ is perfect when G is non-elementary. We say that a rational semigroup G is **non-elementary** if $J(G)$ has three or more points.

Using an idea from Baker we have the following

Theorem(Hinkkanen and Martin [8]) If G is a non-elementary rational semigroup, then the **repelling fixed points of the elements of G are dense in** $J(G)$. Hence also

$$J(G) = \overline{\bigcup_{g \in G} J(g)}.$$

Möbius groups:

If a Möbius group G is such that there does not exist a finite G-orbit in $\mathbb{R}^3$ (for the group of Poincaré extension maps), i.e., G is non-elementary*, then G is non-elementary.

Since the **limit set** of a non-elementary* **Möbius group** (which need not be discrete) is the closure of the set of repelling fixed points we see that the notion of **limit set and Julia set coincide**, in particular, for non-elementary Kleinian groups.

IFS's:

An IFS generated by the (possibly infinite) set of contracting linear maps $\{g_i : i \in I\}$ on $\mathbb{C}$ corresponds to the semigroup $\langle g_i^{-1} : i \in I \rangle$. Defining the **attractor set** A for an IFS to be the closure of the attracting fixed points, we see that A is the **Julia set** for the **corresponding rational semigroup**.

Uniform perfectness:

A **conformal annulus** is an open subset $\mathcal{A}$ of $\overline{\mathbb{C}}$ that can be conformally mapped onto the genuine annulus $Ann(0; r_1, r_2) = \{z : 0 \le r_1 < |z| < r_2 \le \infty\}$ and the modulus of such a conformal annulus is given by $\text{mod}(\mathcal{A}) = \frac{1}{2\pi} \log \frac{r_2}{r_1}$. Note that $\text{mod}(\mathcal{A})$ is a conformal invariant.

A conformal annulus $\mathcal{A}$ is said to **separate** a set F if F intersects both components of $\overline{\mathbb{C}} \setminus \mathcal{A}$ and $F \cap \mathcal{A} = \emptyset$.

We say that a compact subset $F \subset \overline{\mathbb{C}}$ is **uniformly perfect** if F has at least two points and if there is a uniform upper bound on the moduli of all conformal annuli in $\overline{\mathbb{C}} \setminus F$ which separate F.

Uniformly perfect sets were introduced by A. F. Beardon and Ch. Pommerenke in 1978 in [1].

Theorem(Pommerenke [11]) An unbounded closed set $E \subset \mathbb{C}$ is uniformly perfect if, and only if, there exists a constant $\delta > 0$ such that

$$cap(E \cap \{w : |w - z| \le r\}) \ge \delta r$$

for all $z \in E$ whenever $0 < r < \infty$. We note that $cap(\{w : |w - z| \le r\}) = r$.

• In particular each component of the complement of a uniformly perfect set is regular for the Dirichlet problem.

• Uniformly perfect sets also must necessarily have positive Hausdorff dimension.

Results on uniform perfectness of Julia sets and limit sets:

• the **Julia set of a rational function** of degree at least two is uniformly perfect (Eremenko [6], Hinkkanen [7], and Mañé and da Rocha [10]).

• **Julia sets of finitely generated rational semigroups** are uniformly perfect when the semigroup is generated by maps all which have **degree greater than or equal to two** (Hinkkanen and Martin [9]).

• the limit set of a non-elementary **finitely generated Kleinian group** is uniformly perfect (Pommerenke [12])

We note that since each of the above results concerns finitely generated rational semigroups, these results follow as corollaries of the following Theorem 1.

For a rational function g we define Lip $g = \inf\{M : \sigma(g(z), g(w)) \leq M\sigma(z, w)$ for all $z, w \in \overline{\mathbb{C}}\}$ where σ denotes the spherical metric.

Theorem 1 Let $G = \langle g_i : i \in I \rangle$ be a non-elementary rational semigroup generated by the maps $\{g_i : i \in I\}$ such that $\sup_{i \in I}$ Lip $g_i \leq C < +\infty$. Then the Julia set $J(G)$ is uniformly perfect.

proof We follow the framework of Manẽ and da Rocha's (see [10]) proof for the result where $G = \langle f \rangle$ as presented in [5] by L. Carleson and T. Gamelin. Fix four points z_1, z_2, z_3 and z_4 in $J(G)$. Let $0 < \delta < \frac{4}{5} diam(J(G))$ be such that $\sigma(z_i, z_j) > C\delta$ for $i \neq j$.

Suppose $\{A_n\} \subset N(G)$ are conformal annuli that separate $J(G)$ such that we have $mod(A_n) \to \infty$. Let E_n be the component of $\overline{\mathbb{C}} \setminus A_n$ so that $diam(E_n) \to 0$. Without loss of generality we will assume $diam(E_n) < \delta$ for all n, where $diam$ denotes the diameter in the spherical metric. Since $J(G) = \overline{\{repellors\}}$, there exists $f_n \in G$ such that $diam(f_n(E_n)) > \delta$.

Let

$$f_n = g_{i_1} \circ g_{i_2} \circ \cdots \circ g_{i_m}$$

be of minimal word length such that $diam(f_n(E_n)) > \delta$.

Hence

$$\delta < diam(f_n(E_n)) \leq C\delta.$$

Let ϕ_n be a conformal map from the unit disk $\triangle$ onto $A_n \cup E_n$ such that $\phi_n(0) \in E_n$. Let $K_n = \phi_n^{-1}(E_n)$ and note that as the modulus of $\triangle \setminus K_n$ equals the modulus of A_n, we conclude that $diam(K_n) \to 0$. Let $h_n = f_n \circ \phi_n$.

Hence we have

$$\delta < diam(h_n(K_n)) \leq C\delta.$$

We also see by forward invariance that $h_n(\triangle \setminus K_n) \subset N(G)$. Hence each $h_n(\triangle)$ omits at least three of the points z_1, z_2, z_3 and z_4 in $J(G)$. Therefore $\{h_n\}$ is normal in $\triangle$ by Montel's theorem and so $diam(h_n(K_n)) \to 0$. **Contradiction**.

Let G' be a Möbius group generated (as a group) by $\{g_i : i \in I\}$, hence as a semigroup generated by $\{g_i : i \in I\} \cup \{g_i^{-1} : i \in I\}$. Since Lip $g_i =$ Lip g_i^{-1} when g_i is a Möbius map we have:

Corollary The limit set (Julia set) $J(G')$ of a non-elementary Móbius group G' is uniformly perfect when $\{g_i : i \in I\}$, the generators of G' (as a group), satisfy $\sup_{i \in I}$ Lip $g_i < +\infty$.

We note that Kleinian groups which possess a uniform bound on the Lipschitz constants of its generators must necessarily be finitely generated since otherwise one can show that the group is not discrete (see [3], p. 33 or [2], p. 42). Hence Theorem 1 can only be used to duplicate the following result due to Pommerenke (see [12]). Pommerenke's proof is based on more analytic methods as opposed to the more geometric view used in this paper.

Theorem(Pommerenke [12]) The limit set (Julia set) $J(G')$ of a non-elementary Kleinian group G' is uniformly perfect when G' is finitely generated.

Not all limit sets of Kleinian groups are uniformly perfect. One can easily construct a Schottky group whose limit set is not uniformly perfect.

Corollary The attractor set A of an IFS generated by the contracting linear maps $\{g_i : i \in I\}$ defined on $\mathbb{C}$ is uniformly perfect when $\sup_{i \in I}$ Lip $g_i < +\infty$.

Applications

Theorem(Hinkkanen and Martin [9]) Let G be a rational semigroup such that $J(G)$ is uniformly perfect. Suppose that z_0 is a superattracting fixed point of an element $h \in G$. Let A be the union of all the components of $N(h)$ in which the iterates of h tend to z_0. Then either $z_0 \in N(G)$ or $A \subset J(G)$. In particular, either $z_0 \in N(G)$ or z_0 lies in the interior of $J(G)$.

Corollary Let $G = \langle g_i : i \in I \rangle$ be a non-elementary rational semigroup generated by the maps $\{g_i : i \in I\}$ such that $\sup_{i \in I}$ Lip $g_i < +\infty$ and let z_0 be a superattracting fixed point of some element of G. Then either z_0 lies in (the interior of) the Fatou set of G or in the interior of the Julia set of G.

There exists an **infinitely generated** rational semigroup G (all of whose elements have degree at least two with the property that for any positive integer N, the semigroup G contains only finitely many elements of degree at most N) such that $J(G)$ is **not uniformly perfect**, and such that G contains an element g with a superattracting fixed point α with $\alpha \in \partial J(G) \subset J(G)$ (see [9]).

Questions

- What conditions imply that $J(G)$ has nonempty interior?
- What conditions imply that $J(G)$ has empty interior?
- When does $J(G)$ have nonempty interior yet is a proper subset of $\overline{\mathbb{C}}$?

In [17] and [16] Sumi uses an "open set condition" and the backward self-similarity of $J(G)$ to obtain some results in this direction and in particular shows that certain conditions imply $J(G)$ may be a generalized Cantor set, have no interior, have zero Lebesgue measure, or have Hausdorff dimension strictly less than 2.

To "see" what certain Julia sets of finitely generated rational semigroups may look like one may use computer algorithms to draw these Julia sets by obtaining a measure whose support is exactly the Julia set. Of course, the usual warnings must be heeded as certain situations exist that seem to provide considerable barriers to the method giving an accurate picture (see the work of David Boyd [4]).

REFERENCES

[1] A. F. Beardon and Ch. Pommerenke. The Poincarẽ metric of plane domains. *J. London Math. Soc.*, 18:475–483, 1978.

[2] Alan F. Beardon. *The geometry of discrete groups*. Springer-Verlag, New York, 1983.

[3] Alan F. Beardon. *Iterations of Rational Functions*. Springer-Verlag, New York, 1991.

[4] David Boyd. An invariant measure for finitely generated rational semigroups. *Complex Variables*, to appear.

[5] Lennart Carleson and Theodore W. Gamelin. *Complex Dynamics*. Springer-Verlag, New York, 1993.

[6] A. Eremenko. Julia sets are uniformly perfect. *Preprint, Purdue University*, 1992.

[7] A. Hinkkanen. Julia sets of rational functions are uniformly perfect. *Math. Proc. Camb. Phil.*, 113:543–559, 1993.

[8] A. Hinkkanen and G.J. Martin. The dynamics of semigroups of rational functions I. *Proc. London Math. Soc.(3)*, 73:358–384, 1996.

[9] A. Hinkkanen and G.J. Martin. Julia sets of rational semigroups. *Math. Z.*, 222(2):161–169, 1996.

[10] R. Mañé and L. F. da Rocha. Julia sets are uniformly perfect. *Proc. Amer. Math. Soc.*, 116:251–257, 1992.

[11] Ch. Pommerenke. Uniformly perfect sets and the poincare metric. *Arch. Math.*, 32:192–199, 1979.

[12] Ch. Pommerenke. On uniformly perfect sets and Fuchsian groups. *Analysis*, 4:299–321, 1984.

[13] Rich Stankewitz. Completely invariant sets of normality for rational semigroups. *Complex Variables*. to appear.

[14] Rich Stankewitz. *Completely invariant Julia sets of rational semigroups*. PhD thesis, University of Illinois, 1998.

[15] Rich Stankewitz. Completely invariant Julia sets of polynomial semigroups. *Proc. Amer. Math. Soc.*, to appear.

[16] Hiroki Sumi. On hausdorff dimension of Julia sets of hyperbolic rational semigroups. Preprint, 1998.

[17] Hiroki Sumi. On dynamics of hyperbolic rational semigroups. *Journal of Mathematics of Kyoto University*, to appear.

46 NON-REAL ZEROS OF DERIVATIVES OF REAL ENTIRE FUNCTIONS AND SOME PÓLYA-WIMAN CONJECTURES

Stephanie EDWARDS

Department of Mathematics and Computer Science
Bemidji State University, Bemidji, Minnesota 56601 USA

spedward1@vax1.bemidji.msus.edu
http://whitetail.bemidji.msus.edu/sedwards

Simon HELLERSTEIN

Department of Mathematics, University of Wisconsin– Madison
Madison, Wisconsin 53706, USA

hellerst@math.wisc.edu

spedward1@vax1.bemidji.msus.edu

1 Introduction

In this paper we will discuss relationships between the order of a real entire function with only real zeros and the number of non-real zeros of the derivatives of the function.

H.G.W. Begehr et al. (eds.), Proceedings of the Second ISAAC Congress, Vol. 1, 389–397.

Definition 1.1. $f \in V_{2p}$ iff

$$f(z) = cz^m e^{-az^{2p+2}+Q(z)} \prod_n \left(1 - \frac{z}{a_n}\right) e^{\left(\frac{z}{a_n} + \cdots + \frac{1}{q}\left(\frac{z}{a_n}\right)^q\right)}$$

where $c \in \mathbb{R}$, m and p are non-negative integers, $a \geq 0$, Q is a real polynomial of degree $\leq 2p+1$, $a_n \in \mathbb{R}$, $\sum |a_n|^{-q-1} < \infty$ and $q \leq 2p+1$.

Definition 1.2. The class U_{2p} is defined as follows:

$$U_0 = V_0,$$

$$U_{2p} = V_{2p} - V_{2p-2}, \text{ for } p \geq 1.$$

Definition 1.3. The entire function $f \in U^*_{2p}$ if and only if $f(z) = c(z)g(z)$ where $g \in U_{2p}$ and $c(z)$ is a real polynomial with no real zeros.

From definitions 1.1 and 1.2 we wee that $f \in U_0$ if and only if

$$f(z) = cz^m e^{-az^2+bz} \prod_n^{\infty} (1 - \frac{z}{a_n}) e^{\frac{z}{a_n}}$$

where $c \in \mathbb{R}, m$ is a non-negative integer, $a \geq 0$, $b \in \mathbb{R}, a_n \in \mathbb{R}$, and $\sum \frac{1}{a_n^2} < \infty$.

The functions in U_0 are frequently referred to as Laguerre - Pólya (LP) functions. E. Laguerre and G. Pólya proved the following:

Theorem 1.4. *A function f is in U_0 if and only if f is the uniform limit on compacts in $\mathbb{C}$ of a sequence of real polynomials having only real zeros.*

As an immediate corollary of this theorem, it follows that if $f \in U_0$ then $f^{(k)} \in U_0$ for all $k = 1, 2, \ldots$ and, in particular, that if f is not a polymonial, all the derivatives of f have only real zeros. In 1914, Pólya conjectured that the converse of this latter consequence was true. Namely, if f is a real entire function such that f, f', f'', $\ldots$ have all only real zeros, then $f \in U_0$.

For almost 50 years only partial progress was made in the attempt to prove this conjecture. The first significant break through was acheived by B. Ja. Levin and I.V. Ostrovskiĭ jointly in 1963. They showed that if f is real entire with only real zeros and $M(r, f)$, the maximum of $|f|$

on $|z| = r$, satisfies $\log\log M(r, f) \neq O(r)$, i.e. if f is of sufficiently high infinite order, then f'' has infinitely many non-reals zeros.

In 1977 S. Hellerstein and J. Williamson completely affirmed Pólya's conjecture in a series of two papers. The first resolved the problem for functions of finite order and the second extended the Levin-Ostrovskiĭ result to all real entire functions, f of infinite order, for which f and f' have only real zeros.

Theorem 1.5 (Hellerstein & Williamson [9]). *Let $f \in U_{2p}$. If f' has only real zeros, then f'' has exactly $2p$ non-real zeros.*

Theorem 1.5 also partially affirmed a long-standing conjecture of A. Wiman [1] which was finally proved in full by T. Sheil-Small in 1989 and may be stated as

Theorem 1.6 (Sheil-Small [20]). *If $f \in U_{2p}$, then f'' has at least $2p$ non-real zeros.*

The proof of Theorem 1.5 depends heavily on the assumption that both f and f' have only real zeros and is a purely analytcal study of the functions $\frac{f'}{f}$ and $\frac{f''}{f'}$ under that assumption for $f \in U_{2p}$. A critical lemma in [9] enumerates the number of zeros of f' for $f \in U_{2p}$ which are not consequences of Rolle's theorem.

In the proof of Theorem 1.6, Sheil-Small exploits this lemma of Hellerstein and Williamson in combination with a study of the auxiliary functions:

$$L(z) = \frac{f'(z)}{f(z)} \qquad \text{and} \qquad Q(z) = z - \frac{f(z)}{f'(z)}.$$

Noting that

$$Q'(z) = \frac{f(z)f''(z)}{(f'(z))^2}.$$

Sheil-Small investigates the components of $K = \{z \in H^+ : \text{Im}Q(z) > 0\}$ where $H^+ = \{z | \text{Im}z > 0\}$.

He shows, in particular, that each component U of K is mapped $\nu-$fold onto H^+ (ν depends on U). A clever study of the behavior of the components of positivity of these functions, coupled with a variant of a classical theorem of Riemann and Huritz, enables Sheil-Small to obtain a lower bound on the number of zeros of Q' in H^+. This lower bound yields Theorem 1.6 and thus Wiman's conjecture is proved.

Subsequently A. Hinkkanen [12], succeeded in characterizing all functions, f, meromorphic in $\mathbb{C}$ for which $f, f', \ldots, f^{(k)}, \ldots$ have only real zeros. His proof employs a closer study of Q, the associated components of K, and the components of $K^- = \{z \in H^+ : \mathrm{Im}Q(z) < 0\}$.

Following the approach of Sheil-Small and Hinkkanen we are able to prove the following extension of Theorem 1.6:

Theorem 1.7. *If $f \in U_{2p}^*$, then f'' has at least $2p$ non-real zeros.*

A classical result of E. Borel and Laguerre [4] together with an elementary growth argument implies that if $f \in U_{2p}^*$ so is $f' \in U_{2p}^*$. Thus we have the following (no doubt weak) assertion.

Corollary 1.8. *If $f \in U_{2p}^*$ and $k \geq 2$, then $f^{(k)}$ has at least $2p$ non-real zeros.*

The proof of Theorem 1.7 is ultimately obtain as a consequence of the following lemma which counts the number of critical points of Q' in U or U^- (components of K or K^-), hence the non- real zeros of f'', in terms of the degree of the map Q of each of these components onto H^+ and H^- respectively, and the number of "holes" (to be defined in the next section) in the component.

Lemma 1.9. *Suppose that U is a component of K (or U^- is a component of K^-) such that U (U^-) has exactly m holes. Suppose further that $Q' = 0$ exactly k times in U (U^-). Then Q maps U (U^-) onto H^+ (H^-) exactly ν times where $\nu = k + 1 - m$.*

Thus, in fact, a knowledge of the number of components of K and K^- and the geometry of each of the components is sufficient to determine the exact number of the non-real zeros of f''. In particular, a knowledge of the degrees of the maps of Q on all U and U^- yields a lower bound for the number of non-real zeros of f''.

In the next section we summarize some of the geometric aspects of this work and the relation between the number of zeros of L on the boundary of a component U or U^-, the geometry of the component, and the degree of Q on U (U^-).

In section 3, we have a summary of the results obtained when $f \in U_{2p}$ which extends Sheil-Small's observations to more precise connections between the geometry of the components and the number of non-real zeros of f''.

In section 4 we list a few open questions.

Detailed proofs for most of what is described in this summary are presented in the dissertation [6] of the first author under the direction of the second author. A full journal publication with proofs of all the results contained herein is forthcoming.

2 Geometry

2.1 Holes and Bumps - Bounded Components

In order to understand and apply the counting lemma 1.9, we must define some critical aspects of the geometry.

Definition 2.1. A **hole** in K is a bounded component of

$$H = \{z \in H^+ : \mathrm{Im}Q(z) \leq 0\}$$

where $C = \partial H$ satisfies $C \cap \mathbb{R} = \emptyset$, $C \cap H^+ \neq \emptyset$. (Recall $Q(z) = z - \dfrac{f(z)}{f'(z)}$.)

Note that a hole, which is a bounded component of $H^+ - U$ destroys the simple connectivity of the component U. Sometimes there are bounded components of $\overline{K^-}$ which do not destroy the simple connectivity of any component U. We call these components bumps.

Definition 2.2. A **bump** in K is a bounded component of

$$B = \{z \in H^+ : \mathrm{Im}Q(z) \leq 0\}$$

such that $(\partial B \cap \mathbb{R}) \neq \emptyset$ and $(\partial B \cap H^+) \neq \emptyset$.

2.2 Finding the "Degree" of Q

Recall that the counting lemma 1.9 connects the number of zeros of Q' in U (U^-) to the degree of the mapping Q on U (U^-). If we have information about the zeros of f and f' we can find information about the degree of Q in the component.

Proposition 2.3. *The degree of the map Q in a bounded component of either K or K^- is equal to the number of zeros of the function L on the boundary of the bounded component.*

Since we are now able to find the degree of Q in bounded components, we now move onto unbounded components.

Proposition 2.4. *Let U be an unbounded component of K. Suppose that there are n zeros of the function L on the boundary of U and there are m unbounded components of K^- which are inside U. Then the degree of the map Q in a unbounded component U of K is $n+m+1$.*

Similarily, let U^- be an unbounded component of K^-. Suppose that there are n zeros of the function L on the boundary of U^- and there are m unbounded components of K which are inside U^-. Then the degree of the map Q in an unbounded component U^- of K^- is $n+m+1$.

When there is only one unbounded component of K and no unbounded components of K^-, the count is simplified:

Corollary 2.5. *Let U be the sole unbounded component of K. Suppose that there are n zeros of the function L on the boundary of U and there are no unbounded components of K^-. Then the degree of the map Q in a unbounded component U of K is $n+1$.*

In certain situations there is only one unbounded component and we are able to simplify the count. When f has Borel exceptional zeros there is only one unbounded component of K.

Definition 2.6. Let $f \in U^*_{2p}$. Then, as in definition 1.3, $f(z) = c(z)g(z)$ with $g \in U_{2p}$ and c a real polynomial with no real zeros. Using the Hadamard factorization we have

$$f(z) = c(z)e^{Q_0(z)} \prod_n \left(1 - \frac{z}{a_n}\right) e^{\left(\frac{z}{a_n} + \cdots + \frac{1}{q}\left(\frac{z}{a_n}\right)^q\right)}$$

where $Q_0(z)$ is a real polynomial with $\deg Q_0(z) \leq 2p+2$, and the genus of the canonical product is $q \leq 2p+1$. We say that f has **Borel exceptional zeros** if $\deg Q_0(z) > q$.

A weaker version of the following lemma appears in [6]. In response to a query about the truth of Lemma 2.7, J. Miles provided a proof.

Lemma 2.7 (J. Miles). *Let $f \in U^*_{2p}$. Suppose that f has Borel exceptional zeros. Then the set*

$$K^- = \{z \in H^+ : ImQ(z) < 0\}$$

consists of only bounded components.

As a consequence of Lemma 2.7 we obtain :

Theorem 2.8 (A Result of Hellerstein - Williamson Type). *Let*

$$f \in U_{2p}^{*}.$$

Suppose that the function f has Borel exceptional zeros. Suppose further that the function f' has only real zeros. Then the function f'' has exactly $2p$ non-real zeros.

We are unable to prove Lemma 2.7 without the assumption that the zeros of f are Borel exceptional. We are unable, therefore, to obtain an alternate proof of Theorem 1.5 in its entirely, even for $f \in U_{2p}$.

3 RESULTS FOR $f \in U_{2p}$

Theorem 3.1. *Suppose $f \in U_{2p}$ and f has Borel exceptional zeros. Then f'' has exactly $2p$ non-real zeros iff each non-real zero of the function f' is the unique simple zero of the function L on the boundary of a bump in K.*

Theorem 3.2. *Suppose that $f \in U_{2p}$ and that the function f has Borel exceptional zeros. Let 2α represent the number of non-real zeros of the function f'. Let S represent the number of bumps in K which do not have a real zero of the function L on the boundary. Then the number of non-real zeros of the function f'' is $2p + 2\alpha - 2S$.*

Theorem 3.3. *Suppose that $f \in U_{2p}$.*

(i) Let 2α represent the number of non-real zeros of the function f'.

(ii) Let S represent the number of bumps in K without a real zero of the function L on the boundary.

(iii) Let m represent the number of unbounded components of K with boundaries that do not intersect the real axis.

(vi) Let r_0 represent the number of unbounded components of $\overline{K^-}$ with a real zero of the function L on the boundary.

(v) Let t repesent the number unbounded components of $\overline{K^-}$ with boundaries that do not intersect the real axis.

Then there are $2p+2\alpha-2S+2m+2r_0+2t$ non-real zeros of the function f''.

4 OPEN QUESTIONS

The authors had hoped that this work would enable them to answer the questions given below. Since $f \in U_{2p}$ implies that $f^{(k)} \in U_{2p}^*$ for all $k = 1, 2, \ldots$ there was the possibility that an understanding of the geometry of the components of K and K^- for $f \in U_{2p}^*$ might be sufficient to move inductively from f'' to higher derivatives to obtain sharp lower bounds on the number of non- real zeros of $f^{(k)}$ for $k > 2$. We have, so far, been unsuccessful in this attempt.

Conjecture 4.1 (Pólya). *Suppose $f \in U_{2p}$. Then the number of non-real zeros of $f^{(k)}$ tends to infinity as k increases for $p \geq 1$.*

Question 4.2. *If $f \in U_{2p}^*$, at least how many non-real zeros does $f^{(k)}$ have for $k \geq 2$? (Can we get a better lower bound?)*

Question 4.3. *If $f \in U_{2p}^*$, exactly how many non-real zeros does $f^{(k)}$ have for $k \geq 2$? (Can we get an exact number that does not depend on the geometry?)*

Question 4.4. *If $f \in U_{2p}$, does the number of non-real zeros of $f^{(k)}$ increase monotonically as k increases for $p \geq 1$.*

References

[1] M. Ålander, *Sur le déplacement des zéros fonctions entières par leur dérivation*, Thèse, Uppsala, 1914.

[2] ———, *Sur les zéros extraordinaires des dérivées des fonctions entières réelles*, Ark. Math. Astronom. Fys. **15** (1916), 11.

[3] A. F. Beardon, *Iteration of Rational Functions*, Springer, New York, 1991.

[4] Emile Borel, *Leçons sur les Fonctions Entières*, Gauthier - Villars et Cie, Paris, France, 1921.

[5] John B. Conway, *Functions of One Complex Variable*, Springer-Verlag, New York, 1978.

[6] S. Edwards, *On the non-real zeros of derivatives of real entire functions of finite order and the Pólya - Wiman conjectures*, Dissertation, University of Wisconsin – Madison, 1998.

[7] H. M. Farkas and I. Kra, *Riemann Surfaces*, Springer-Verlag, New York, 1980.

[8] W. K. Hayman, *Meromorphic Functions*, Oxford University Press, London, 1964.

[9] S. Hellerstein and J. Williamson, *Derivatives of entire functions and a question of Pólya*, Transactions of the A.M.S. **227** (1977), 227–249.

[10] ______, *Derivatives of entire functions and a question of Pólya II*, Transactions of the A.M.S. **234** (1977), 497–503.

[11] S. Hellerstein, J. Williamson, and L. Shen, *Reality of the zeros of an entire function and its derivatives*, Transactions of the A.M.S. **275** (1983), 319–331.

[12] A. Hinkkanen, *Iteration, level sets, and zeros of derivatives of meromorphic functions*, Preprint, 1997.

[13] E. Laguerre, *Oeuvres, Tome I.*

[14] B. Ja. Levin, *Distribution of Zeros of Entire Functions*, American Mathematical Society, Providence, Rhode Island, 1964.

[15] Zeev Nehari, *Conformal Mapping*, Dover Publications, New York, 1975.

[16] G. Pólya, *Über annäherung durch polynome mit lauter reellen wurzeln*, Rend. Circ. Mat. Palermo **36** (1913), 279–295.

[17] ______, *Sur une question concernant les fonctions entières*, C.R. Acad. Sci. Paris **158** (1914), 330–333.

[18] ______, *On the zeros of the derivatives of a function and its analytic character*, Bulletin of the A.M.S. **49** (1943), 179–191.

[19] G. Pólya and G. Szegö, *Aufgaben und Lehrsätze aus der Analysis*, Springer-Verlag, Berlin, 1925.

[20] T. B. Sheil-Small, *On the zeros of the derivatives of real entire functions and Wiman's conjecture*, Annals of Mathematics **129** (1989), 179–193.

[21] M. Tsuji, *Potential Theory in Modern Function Theory*, American Mathematical Society, Maruzen, Tokyo, 1959.

47 ON MEROMORPHIC SOLUTIONS OF LINEAR DIFFERENTIAL EQUATIONS WITH AT LEAST ONE TRANSCENDENTAL COEFFICIENTS

Kazuya TOHGE*

Faculty of Technology, Kanazawa University
2-40-24, Kodatsuno, Kanazawa, 920-8667, Japan

tohge@t.kanazawa-u.ac.jp

Abstract. Applying a result obtained in [TK], we investigate the zero-distribution of fundamental solutions of some linear differential equations with at least one transcendental coefficient, and compare the results with some of the well-known results on equations with the polynomial coefficients. Results corresponding to Frank's theorem [F] on exceptional fundamental sets and a theorem due to Frank and Hellerstein [F-H] on non-homogeneous equations are obtained.

1. Introduction

* The author was supported in part by the Grant-in-Aid for General Scientific Research from the Ministry of Education, Science and Culture 11640164 (N. Toda).

H.G.W. Begehr et al. (eds.), Proceedings of the Second ISAAC Congress, Vol. 1, 399–411.

In this note some special meromorphic solutions of differential equations of the form

$$w^{(n)} + \sum_{j=0}^{n-1} a_j(z) w^{(j)} = F(z)\,, \tag{1}$$

where the a_j and F are meromorphic functions in the whole plane, will be studied. We mainly use the language of the Nevanlinna theory, see, for example, [H], [LI], and [LS]. Let us say *meromorphic* concisely to express that a function is defined in the whole plane and globally meromorphic there.

The purpose of our consideration is to give some answers to the following problems:

Problem D. *To find a clear difference in the zero-distributions between solutions of equation* (1) *when all the coefficients a_j are polynomials and that when some of the coefficients are transcendental.*

and

Problem E. *To extend some results on the zero-distributions of solutions of equation* (1) *when all the coefficients a_j are polynomials to the case when some coefficients in* (1) *are transcendental.*

2. Polynomial coefficients

As we have learned from the book by I. Laine [LI], as well as its extensive bibliography, a lot of beautiful results on the oscillation of equation (1) have been obtained especially in the case of polynomial coefficients.

Let

$$L(w) = \Big(D^n + p_{n-1}(z) D^{n-1} + \cdots + p_0(z)\Big)\, w\,, \quad D := d/dz\,, \tag{2}$$

be a linear differential operator with the polynomial coefficients. The first result concerns the existence of an *exceptional* fundamental set of solutions of $L(w) = 0$. This was first proved by G. Frank [F] for the case of exceptional solutions in the sense of Picard, and extended independently by F. Brüggemann [BF] and N. Steinmetz [S] to exceptional solutions in the sense of Borel or Nevanlinna.

Theorem A.([BF], [S]) *Suppose that the coefficients of the equation $L(w) = 0$ are all polynomials, and that the equation possesses a fundamental set of solutions $\{f_1, f_2, \cdots, f_n\}$ with the property that $\delta(0, f_j) = 1$ for each $j = 1, 2, \ldots, n$. Then either the coefficients of L are* *constants, or there is a nonlinear polynomial* $Q(z)$ *(e.g.* $DQ = -p_{n-1}/n$*) such that all solutions of* $e^{-Q}L(e^Q y) = 0$ *have order less than* $\deg Q$.

We cite secondly the following result due to Frank and S. Hellerstein [F-H], treating of the zero-distribution of both an entire solution f of a non-homogeneous equation $L(w) = F$ and the differential polynomial $L(f)$.

Theorem B1.([F-H], also [LI]) *Let* f *be a solution of* $L(w) = F(z)$ *with polynomial coefficients* $p_0 \not\equiv 0$, $p_1, \ldots, p_{n-2}$, $p_{n-1} \equiv 0$, $n \geq 2$, *such that* $F \not\equiv 0$ *is entire. Then either*

$$m\left(r, \frac{f'}{f}\right) \leq 2\overline{N}\left(r, \frac{1}{f}\right) + 2\overline{N}\left(r, \frac{1}{F}\right) + O\left\{\log T\left(r, \frac{f'}{f}\right) + \log r\right\}$$

outside a set of r *of finite linear measure, or* f *and* F *both have no zeros and*

$$\frac{F}{f} = \left(\frac{n+1}{2n}\frac{f'}{f} + \frac{n-1}{2n}\frac{F'}{F}\right)^n .$$

As a consequence of this, Frank and Hellerstein ([F-H]) have been able to characterize an exceptional solution of a non-homogeneous equation in the sense of Picard.

Theorem B2.([F-H], also [LI]) *Let* $f = Pe^g$, *where* P *is a polynomial and* g *is entire, be a solution of a non-homogeneous linear differential equation* $L(w) = F$ *with* $F(z) = Q(z)\exp h(z)$, *where* $p_0 \not\equiv 0$, $p_1, \ldots, p_{n-2}$, $p_{n-1} \equiv 0$, $n \geq 2$, $Q \neq 0$ *are polynomials and* h *is an entire function. Then we have that either*

(1) h, g *are polynomials and* $h - g$ *is a constant, or*

(2) P, Q *are constants, and there exists a polynomial* S *such that* $h - g = nS$ *and that*

$$g'(z) = -\frac{n-1}{2}S'(z) + \omega e^{S(z)},$$

where ω *is an* n*th root of unity,* $\omega^n = 1$.

In the following sections, we explain briefly why we have considered the zero-distribution of the solutions and then investigate a possibility of extending the above results to equations with the transcendental p_0, by examples. We are to apply a theorem in [TK] in order to provide some answers to Problems D and E.

3. Zero-free solutions

When the coefficients of L are polynomials, any solution of $L(w) = 0$ is either polynomial or a transcendental entire function of positive finite order. The 0 is the unique candidate for finite exceptional values of the transcendental solutions in the sense of Nevanlinna. Theorem A determines the case when there will exist a fundamental set of $L(w) = 0$, such that each of the elements, f say, satisfies $\delta(0, f) = 1$. Especially, there is an exceptional fundamental set in the sense of Picard if and only if there exists a polynomial Q such that $L^*(y) := e^{-Q}L(e^Q y)$ is a linear differential operator with constant coefficients ([F]).

To see this more visibly, we now consider a holomorphic curve of the complex line $\mathbf{C}$ into the $(n-1)$-dimensional complex projective space $\mathbf{P}^{n-1}(\mathbf{C})$, which is formed by any fundamental system of $L(w) = 0$. If there exists a zero-free fundamental system, this curve misses exactly n hyperplanes of $\mathbf{P}^{n-1}(\mathbf{C})$ in general position. By Frank's theorem, the curve is of order exactly one. In the latter case, this also shows the sharpness of the second main theorem for hyperplanes, due to Cartan (see e.g. [LS]):

Cartan's Second Main Theorem. ([LS]) *Let $\boldsymbol{f} = (f_1 : \cdots : f_n) : \mathbf{C} \to \mathbf{P}^{n-1}(\mathbf{C})$ be a holomorphic map, with $f_1, \ldots, f_n$ entire without common zeros. Assume that the image of $\boldsymbol{f}$ is not contained in any hyperplane. Let $H_1, \ldots, H_q$ be hyperplanes in general position. Let $W := W(f_1, \ldots, f_n)$ be the Wronskian. Then*

$$(q-n)T_{\boldsymbol{f}} \leq \sum_{k=1}^{q} N_{\boldsymbol{f}}(r, H_k) - N(r, 1/W) + S(r),$$

where $S(r) = O(\log r + \log^+ T_{\boldsymbol{f}}(r))$, possibly outside a set of r of finite linear measure, is the usual Nevanlinna error term, and $N_{\boldsymbol{f}}(r, H) = N(r, 1/\langle \boldsymbol{f}, \boldsymbol{a} \rangle)$ with $\langle \boldsymbol{f}, \boldsymbol{a} \rangle := \sum_j a_j f_j$ for a hyperplane H : $\langle \boldsymbol{x}, \boldsymbol{a} \rangle = \sum_j a_j x_j = 0$.

Exponential curve [A]: An exponential curve is defined by $f_j(z) = e^{\lambda_j z}$, where the λ_j are different constants:

$$\boldsymbol{f} = (f_1 : \cdots : f_n) = (e^{\lambda_1 z} : \cdots : e^{\lambda_n z}),$$

with its Wronskian

$$W(f_1, \ldots, f_n) = V(\lambda_1, \cdots, \lambda_n)\, e^{(\lambda_1 + \cdots + \lambda_n)z},$$

where $V(\lambda_1, \cdots, \lambda_n)$ is the Vandermonde determinant. Then the characteristic function of this curve is

$$T_{\boldsymbol{f}}(r) := \int_0^{2\pi} \log \max_{1 \le j \le n} |f_j(re^{i\theta})| \frac{d\theta}{2\pi} - \log \max_{1 \le j \le n} |f_j(0)| = \frac{\ell}{2\pi} r + O(1) .$$

Here ℓ denotes the perimeter of the least convex polygon which contains the points λ_j: $\ell = \sum |\lambda_k - \lambda_{k+1}|$. Note taht this curve omits n hyperplanes $H_j : x_j = 0$.

It should be emphasized here that homogeneous linear differential equations with entire coefficients are good candidates of producing interesting examples of holomorphic curves. Especially if there exists a zero-free fundamental system of the equation, then we may construct a curve of infinite order with a property similar to an exponential curve, since $DW/W = -a_{n-1}$ is entire.

N. Toda [TN] has provided a method of finding a fundamental set of zero-free solutions.

4. The key result

We shall answer Problems D and E to a certain degree, more precisely, zero-free fundamental systems will be extracted and their equations will be determined in a certain family of homogeneous linear differential equations with one *dominant* coefficient.

Key Theorem. ([TK]) *Let $A(z)$ be a transcendental meromorphic function, k a positive integer, and the a_j $(0 \le j \le k-1)$ be meromorphic functions satisfying*

$$\sum_{j=0}^{k-1} T(r, a_j) = o\{T(r, A)\} , \qquad \text{as } r \to \infty , \ r \in I ,$$

where I is a set of r of infinite linear measure. Suppose that $f(z)$ is a non-trivial solution of the equation

$$D^k w + \sum_{j=0}^{k-1} a_j(z) D^j w = A(z) w . \tag{3}$$

Then either, as $r \to \infty$ in I possibly outside a set of finite linear measure,

$$m\left(r, \frac{Df}{f}\right) \le 2\left\{\overline{N}\left(r, \frac{1}{f}\right) + \overline{N}\left(r, \frac{1}{A}\right) + (r, A)\right\} + o\left\{T\left(r, \frac{Df}{f}\right)\right\} , \tag{4}$$

or, for a nonzero constant c,

$$\frac{Df}{f} = cA^{1/k} - \frac{k-1}{2k}\frac{DA}{A} - \frac{1}{k}a_{k-1}$$

in the whole plane, so that $A = \alpha^k$ *with* α *meromorphic.*

Particularly, if A *and the* a_j *are all entire functions, the second statement implies that both* A *and* f *are zero-free entire functions of the following forms :*

- $A(z) = \exp\{kH(z)\}$, $H(z)$ *a non-constant entire function,*
- $f \in \{f_1, \ldots, f_k\}$, *a fundamental system of equation* (3) *given by*

$$f_j(z) = c_j \exp\left\{\omega_j \int_0^z e^{H(t)}dt - \frac{k-1}{2}H(z) - \frac{1}{k}\int_0^z a_{k-1}(t)dt\right\}, \tag{5}$$

  where the c_j *are nonzero constants and the* ω_j *are mutually distinct* k*th roots of unity, i.e.,* $\omega_j^k = 1$.

Moreover, equation (3) *can be expressed as*

$$(D + \beta_1(z)) \cdots (D + \beta_k(z))w = e^{kH(z)}w,$$

where for $j = 1, \ldots, k$,

$$\beta_j(z) := \left(j - \frac{k+1}{2}\right) DH(z) - \frac{1}{k}a_{k-1}(z).$$

A result quite similar to ours, was proved by J. Rieppo in his thesis [R] by differential algebraic methods.

Remarks. 1) As mentioned in the proof of Proposition [TK], the terms

$$\overline{N}\left(r, \frac{1}{A}\right) + \overline{N}(r, A)$$

in the right-hand side of inequality (4) can be replaced by the terms

$$\overline{N}_0\left(r, \frac{1}{A}\right) + \overline{N}_0(r, A),$$

where $\overline{N}_0(r, g)$ counts only the distinct poles of g that are different from the zeros of f.

2) The number 2 in (4) cannot be replaced by any smaller number (see [TK, Example 3]).

One soon notices that this theorem could be applied in order to deduce a result similar to Thereoms B1-2, and particularly the second part would be closely connected to Theorem A. This has actually provided a starting-point of our consideration.

5. Theorem A' — for Problem D

Now we construct a holomorphic curve $\boldsymbol{f}$ by the fundamental system $\{f_1, \dots, f_n\}$ of an nth order homogeneous linear differential equation with entire coefficients, and call it a *fundamental curve* of the equation. Note that a fundamental curve of the equation $L^*(v) = e^{-Q}L(e^Q v) = 0$ is also a fundamental curve of $L(w) = 0$ itself. Theorem A states that if the equation $L(w) = 0$ with polynomial coefficients permits a fundamental system $\{f_1, \dots, f_n\}$ such that $\delta(0, f_j) = 1$ $(1 \le j \le n)$, then the fundamental curve $\boldsymbol{f} = (f_1 : \cdots : f_n)$ is a 'polynomial' curve or a holomorphic curve of positive finite order with n exceptional hyperplanes of $\mathbf{P}^{n-1}(\mathbf{C})$ in general position.

Applying our Key Theorem, we can obtain the following

Theorem A'. *Let $p(\not\equiv 0)$, q and $\alpha(\notin \mathbf{C})$ be polynomials. Suppose that the coefficients of the operator L in* (2) *are all polynomials, and that the equation*

$$L(w) = \left(p(z)e^{\alpha(z)} + q(z)\right) w \tag{6}$$

possesses a non-trivial solution f with the property that

$$\limsup_{r\to\infty, r\notin E} \frac{\overline{N}(r, 1/f)}{T(r, Df/f)} < \frac{1}{3}, \tag{7}$$

where E is a set of r of finite linear measure. Then there exists a non-constant polynomial Q $(DQ = D\alpha/n)$ such that equation (6) *has a form*

$$(D + \beta_1(z)) \cdots (D + \beta_n(z))w = e^{nQ(z)}w,$$

with $\beta_j(z) = (2j-n-1)DQ(z)/2 - p_{n-1}(z)/n$. Furthermore, there exists a zero-free fundamental system of this equation, which contains f as its entry and forms the fundamental curve

$$\left(e^{\omega_1 Z(z)} : \cdots : e^{\omega_n Z(z)}\right).$$

Here the ω_j are the nth roots of unity and $Z(z)$ is given by

$$Z(z) = \int_0^z e^{Q(t)}dt + c, \quad c, \text{ a constant.}$$

Proof. Put $A = pe^{\alpha}$ and apply our Key Theorem. Then we have the result.

This applies, of course, in a more general case when the coefficients p_j of the operator L, and also p, q and α are entire functions satisfying a dominance condition

$$\sum_{j=0}^{k-1} T(r, p_j) + T(r, p) + T(r, q) = S(r, e^{\alpha}) .$$

Answer to Problem D. While Theorem A' may be regarded as an extension that relates to Problem E, it also provides some answer to Problem D : Equations of the form (6) permit only zero-free fundamental systems as exceptional ones in the sense of (7). If a non-constant entire function f has a finite order, the condition (7) implies $\overline{N}(r, 1/f) = O(\log r)$. For the case of equations with polynomial coefficients, choices other than 'zero-free' are actually available for exceptional fundamental systems in the sense of Picard.

6. Value distribution of $L(f)$

Let us start this section with the following example given by S.B. Bank [BS] and also by Toda [TN]:

Example 1. *Let k be a positive integer greater than one, and let $c = -(k-1)/2$. Then the zero-free function*

$$\exp (cz + e^{z})$$

solves an equation

$$D^k w + \sum_{j=1}^{k-2} Q_j D^j w + \left(Q_0 + Re^{P} \right) w = 0$$

where $Q_0, \cdots, Q_{k-2}$ are constants, $R \equiv -1$ and $P(z) = kz$.

More generally, this equation reduces to

$$\prod_{j=1}^{k} \left(D + j - \frac{k+1}{2} \right) w = e^{kz} w$$

and permits a zero-free fundamental system $\{f_1, \ldots, f_k\}$, with

$$f_j(z) = \exp (cz + \omega_j e^{z}) ,$$

where the ω_j are the solutions of $\omega^k = 1$.

We mention again that one of the motivations to study zero-distribution for any non-trivial solution of linear differential equations lies in the fact (e.g. [LI]): *Let f be an admissible meromorphic solution of $\sum_{j=0}^{n} a_j(z)D^j w = 0$, where $a_0 \neq 0$, $a_1, \dots, a_n \neq 0$ are entire functions, in the sense that $T(r, a_j) = S(r, f)$ holds for $j = 0, \dots, n$. Then $\delta(a, f) = 0$ for all $a \neq 0, \infty$.*

Similar to this, a number of results on the value distribution of the linear differential operator $L(w)$ were obtained ([H]). J. Langley [LJ] proved *Hayman's alternative* as follows:

Theorem C. *Suppose that f is transcendental and meromorphic, and that*

$$F := L(f) = D^k f + \sum_{j=0}^{k-1} a_j(z) D^j f$$

is non-constant, where $a_0(z), \dots, a_{n-1}(z)$ are meromorphic and satisfy $T(r, a_j) = S(r, f)$ for $j = 0, \dots, k$. Then either

$$T(r, f) < 3N\left(r, \frac{1}{f}\right) + 4\overline{N}\left(r, \frac{1}{F-1}\right) + S(r, f) \tag{8}$$

or there exists a meromorphic function $H(z)$ such that $F = (D + \beta) \cdots (D + k\beta) f$ where

$$\beta = \frac{H''}{H'} = \frac{2a_{k-1}}{k(k+1)}; \qquad \text{and} \qquad f = \frac{H^{k+1} + HP_k(H) + (-1)^{k+1}}{k!H(H')^k},$$

where $P_k(x)$ is a polynomial of degree at most $(k-1)$.

Also, if $N(r, 1/f) + \overline{N}(r, \frac{1}{F-1}) = S(r, f)$, we must have $f = (H - \omega)^{k+1}/k!H(H')^k$ where $\omega^{k+1} = 1$.

Example 1 shows that this is not the case for the zeros of F, even if all the coefficients of $L(w)$ are constants : $F = e^{kz} f$, for $f = f_j$. Note that $a_{k-1} = 0$ there.

7. Zero-distribution of $L(f)$ — on Problem E

We show

Theorem C'. *Let f be a transcendental meromorphic function and F be a linear differential polynomial in f,*

$$F(z) := L(f)(z) = D^k f + \sum_{j=0}^{k-1} a_j(z) D^j f. \tag{9}$$

Suppose that $F \not\equiv 0$ *and the coefficients* a_j *of* L *are meromorphic functions satisfying*

$$T(r, a_j) = S\left(r, \frac{Df}{f}\right) + O(\log r) \quad \text{for } j = 0, \dots, k-1. \tag{10}$$

Then either there exists a set I *of infinite linear measure on which*

$$m\left(r, \frac{Df}{f}\right) \le 2\left\{\overline{N}\left(r, \frac{1}{f}\right) + \overline{N}_0\left(r, \frac{1}{F}\right) + \overline{N}(r, f)\right\} + o\left\{T\left(r, \frac{Df}{f}\right)\right\} + O(\log r) \tag{11}$$

holds, as $r \to \infty$, *or, for a non-zero constant* C,

$$\frac{F}{f} = C\left(\frac{k+1}{2k}\frac{Df}{f} + \frac{k-1}{2k}\frac{DF}{F} + \frac{1}{k}a_{k-1}\right)^k, \tag{12}$$

over the whole plane. Here $\overline{N}_0(r, 1/F)$ *counts the distinct zeros of* F *at which* $f \neq 0, \infty$.

Proof. First we note that if Df/f is a rational function, then inequality (11) is satisfied for $I = [0, +\infty)$. (The term $O(\log r)$ is essential!)

We assume that the function Df/f is transcendental. Put $A := F/f$. Define a non-increasing sequence of sets $I(n)$ of r by

$$I(n) := \{r \ge 0 \mid nT(r, A) \le T(r, Df/f)\}, \quad n = 1, 2, \dots.$$

Set $J(n) := [0, \infty) \setminus I(n)$. If there is a positive integer n_0 such that the measure of $J(n_0)$ is infinite, then take $I = J(n_0)$ and apply our Key Theorem together with its remark to obtain the statement in the second part of Theorem C'. Here f is regarded as a non-trivial solution of the equation $L(w) = Aw$ with $A = F/f$.

Otherwise, that is, if any $J(n)$ has possibly a finite measure, then (see page 282 of [TK])

$$\limsup_{r \to \infty, r \notin J} \frac{T(r, A)}{T(r, Df/f)} = 0,$$

where J is a set of r of finite linear measure, i.e., $T(r, A) = S(r, Df/f)$. But this also reduces to (11). In fact now $F/f = L(f)/f$ is a polynomial of Df/f over the ring

$$\mathcal{R} := \{a(z), \text{ meromorphic} \mid m(r, a) = S(r, Df/f)\}$$

(see, for example, [F-H], [H], [LI], or [R]). Therefore, if $F/f \in \mathcal{R}$, then $Df/f \in \mathcal{R}$, and thus $m(r, Df/f) = o(1)N(r, Df/f)$ possibly outside a set of r of finite linear measure. This proves our theorem.

Answer to Problem E. *We have extended* Theorem B1 *at the same time:* Theorem C'=Theorem B1'. *A so-called* Theorem B2' *would be automatically deduced from this result. Instead of going into it, we consider, however, a most specific case of* Theorem C' *when the* a_j *and* F/f *are all entire functions. Then* F *is also entire as well as* f, *and either*

$$m(r, Df/f) \leq 2 overline N(r, 1/F) + o\{T(r, Df/f)\} + O(\log r) \quad (r \in I),$$

or $C = 1$ *in* (12) *and both* f *and* F *are zero-free, which bears a remarkable resemblance to the conclusion of* Theorem B1.

Note that in any case that f is of finite order, inequality (11) will be satisfied on the positive real axis. Langley [LJ] gave two examples of f of finite order and F with polynomial coefficients, both of which have finitely many zeros, while $T(r, f) = (\alpha + o(1))$
$overline N(r, f)$ for a positive number α.

Example 2. ([LJ]) *Let* k *and* L *be positive integers.*

(i) If P *is a non-constant polynomial and if* $DH = e^P$, *then* $f = 1/[(DH)^k H^L]$ *satisfies*

$$(D + P') \cdots (D + kP')f = cH^{-L-k},$$

where c *is a constant. We can deduce identity* (12) *with* $C := (-1)^k(2L + k - 1)^k/2c$.

(ii) An equation with constant coefficient

$$(D + L + 1)(D + 1)f = -L(L + 1)(e^z + 1)^{-L-2}$$

has a solution $f(z) = e^{-z}(e^z + 1)^{-L}$. *In this case* (12) *does not hold:*

$F/f = -L(L + 1)e^z(e^z + 1)^{-2}$, *but*

$$\frac{3}{4} \cdot \frac{Df}{f} + \frac{1}{4} \cdot \frac{DF}{F} + \frac{L+2}{2} = -\frac{2L+1}{4} \cdot \frac{e^z - 1}{e^z + 1}.$$

Remark. The condition (10) is essential in Theorem C'.

It is very easy to give counter-examples of Theorem C', if $S(r, Df/f)$ is replaced by $S(r, f)$: Take $F = -Q_0 f$ in Example 1, for instance.

8. A concluding remark

Slightly more general homogeneous equations are considered in [TK]. The result applies to $DL(f) = (DF/F)L(f)$ with $L(f)$ given in (9) of Theorem C', that is, the equation

$$\begin{aligned} D^{k+1}f + a_{k-1}(z)D^k f \quad &+ \quad \sum_{j=1}^{k-1}\Big(Da_j(z) + a_{j-1}(z)\Big)D^j f + \Big(Da_0(z)\Big)f \\ &= \quad A(z)\left(D^k f + \sum_{j=0}^{k-1} a_j(z)D^j f\right), \end{aligned}$$

with $A := DF/F$. If $T(r, a_j) = S(r, DF/F)$ holds for $j = 1, 2, \ldots, k-1$, then Proposition in [TK] (with k and ℓ being replaced by $k+1$ and k, respectively) implies the following alternative: either

$$m(r, Df/f) \leq 2 + o(1)\left\{\overline{N}(r, 1/f) + \overline{N}(r, F) + \overline{N}(r, 1/F) + \overline{N}_0(r, 1/DF)\right\} \qquad (13)$$

at least outside a set of r of finite linear measure, or

$$\frac{Df}{f} \equiv cA - k\frac{DA}{A} = (c+k)\frac{DF}{F} - \frac{D^2F}{DF}$$

for some non-zero constant c.

The function $\overline{N}_0(r, 1/DF)$ in (13) counts the distinct zeros of DF not multiple zeros of F. This term seems to be added in surplus to our estimation. But its presence forces a_{k-1} to disappear from the second alternative.

In the special case when all the a_j and $A = DF/F$ are entire, we obtain as the latter alternative that $A = e^H$ with a non-constant entire function H and $c = 1$, so that

$$\frac{F(z)}{f(z)} = \exp\Big(kH(z) + \text{const.}\Big), \quad F(z) = \exp\left(\int^z e^{H(t)}dt\right),$$

which was discussed in the previous section as some answer to Problem E.

All the examples that the author has constructed indicate that $T(r, a_j) = S(r, DF/F)$ $(j = 1, 2, \ldots, k-1)$ as well as (10) are conditions indispensable to our studies of the zero-distribution of solutions of linear differential equations with transcendental coefficients.

Acknowledgements. The author wishes to thank Professor Chung-Chun Yang for reading the manuscript and making a number of helpful suggestions.

References

[A] Ahlfors, L., The theory of meromorphic curves, *Acta Soc. Sci. Fenn. N.S., A.*, **III** (1941), 1–31.

[BS] Bank, S. B., On the frequency of complex zeros of solutions of certain differential equations, *Kodai Math. J.*, **15**(1992), 165–184.

[BF] Brüggemann, F., On the zeros of fundamental systems of linear differential equations with Polynomial coefficients, *Complex Variables Theory Appl.*, **15** (1990), 159–166.

[F] Frank, G., Picardsche Ausnahmewerte bei Lösungen linearer Differentialgleichungen, *Manuscripta Math.*, **2** (1970), 181–190.

[F-H] Frank, G. and Hellerstein, S., On meromorphic solutions of nonhomogeneous linear differential equations with polynomial coefficients, *Proc. London Math. Soc.*(3), **53** (1986), 407–428.

[H] Hayman, W. K., *Meromorphic Functions*, Clarendon Press, Oxford, 1964.

[LI] Laine, I., *Nevanlinna Theory and Complex Differential Equations*, Walter de Gruyter, Berlin-New York, 1993.

[LJ] Langley, J., On Hayman's alternative, *Mathematika*, **32** (1985), 139–146.

[LS] Lang, S., *Introduction to Complex Hyperbolic Spaces*, Springer-Verlag, New York-Berlin-Heidelberg-London-Paris-Tokyo, 1987.

[R] Rieppo, J., Differential fields and complex differential equations, *Ann. Acad. Sci. Fenn. Ser. A I Math. Dissertations*, No. **118**, 1998.

[S] Steinmetz, N., Linear differential equations with exceptional fundamental sets, *Analysis*, **11** (1991), 119–128.

[TN] Toda, N., On the frequency of zeros of solutions of higher order linear differential equations with entire coefficients, *Complex Variables Theory Appl.*, **23** (1993), 249–255.

[TK] Tohge, K., Logarithmic derivatives of meromorphic or algebroid solutions of some homogeneous linear differential equations, *Analysis*, **19** (1999), 273–297.

48 HAUSDORFF DIMENSION OF THE CHAOTIC SET OF THE N-SHAPED MAPPING

Yeshun SUN
Chung-Chun YANG

Department of Mathematics, The Hong Kong University of Science and Technology
Clear Water Bay, Kowloon, Hong Kong

maysun@uxmail.ust.hk
mayang@uxmail.ust.hk

Lifeng XI

Institute of Mathematics, Zhejiang Wanli University
Ningbo, 315101, China

xilifengningbo@163.net

Abstract

For N-shaped mapping $g : [0,1] \to [0,1]$ given by $g(x) = 3x$, if $0 \le x \le 1/3$, $g(x) = 2 - 3x$, if $1/3 \le x \le 2/3$ and $g(x) = 3x - 2$ otherwise, there exists a chaotic set of g in the sense of Xiong, which is stronger than Li-Yorke's, with Hausdorff dimension one everywhere. On the other hand, it is shown that any chaotic set of g is of zero one-dimensional Hausdorff measure.

1 Introduction and Main Results

One of the objectives in investigating the chaotic phenomena in an iterative system is to describe a subset of the phase space with big magnitude and with the dynamical behavior of the orbit of each point of this subset wild. In fact, we do not have a unified formulation for the definitions of chaos.

T.Y. Li and J. Yorke [li-yorke75] gave a definition of chaos by comparing the entire process of the iterations of two different points while R. Devaney [devaney89] thought the SDI(*sensitive dependence on initial value*) condition and *topological transitivity* are the features of most chaotic systems.

In this paper, however, we use a much sharper definition due to J. Xiong [xiong95].

H.G.W. Begehr et al. (eds.), Proceedings of the Second ISAAC Congress, Vol. 1, 413–418.

Definition 1. Let (X, d) be a compact metric space, and $f : X \to X$ be a continuous map. A nonempty subset $C \subset X$ is said to be *chaotic with respect to* f, if for any subset A of C and arbitrary continuous map $F : A \to X$, there exists a strictly increasing sequence of positive integers $\{r_n\}$, such that $\lim_{n\to\infty} f^{r_n}(x) = F(x)$ for each $x \in A$.

The shift map σ on symbolic space Σ_n plays an important role in dynamical systems. Many people say a system chaotic when they detect a factor of it a shift map.

We make use of the Hausdorff dimension to measure the size of the chaotic set.

Definition 2. Let (X, d) be a compact metric space, a subset $D \subset X$ is said to be *Hausdorff dimension one everywhere*, if the intersection of D and any nonempty open subset U of X has Hausdorff dimension one.

J. Xiong[xiong95] constructed a chaotic subset with respect to the shift map σ on the symbolic space Σ_n. The Hausdorff dimension of this subset is one everywhere. This fact tells us that there exists extremely tremendous points with wild behavior.

We now consider a continuous map on unit interval. We denote $[0,1]$ by I.

By definition, a *N-shaped mapping* $g : I \to I$ is given by the following expression

$$g(x) = \begin{cases} 3x, & \text{if } 0 \le x \le 1/3; \\ 2 - 3x, & \text{if } 1/3 \le x \le 2/3; \\ 3x - 2, & \text{otherwise.} \end{cases}$$

Our main results in this paper are as follows.

Theorem 1. There exists a chaotic subset C of I w.r.t. the N-shaped mapping g with Hausdorff dimension one everywhere.

To illustrate our result is the best possible in some sense, we give

Theorem 2. The one-dimensional Hausdorff measure of any chaotic subset of I w.r.t. N-shaped mapping g is zero.

Our proofs will be applicable to a bigger class of mappings. We first give two definitions.

Definition 3. A continuous map f from I to itself is said to be *generalized N-shaped mapping*, if there exists an integer $n \ge 2$ such that f restricted on $[i/n, (i+1)/n]$ is a linear and onto map, where $i = 0, \cdots, n-1$.

Consequently, the well known tent map is also a generalized N-shaped mapping.

Definition 4. A continuous map $f : I \to I$ is said to be *Lipschitz conjugate* to $h : I \to I$ if there exists a homeomorphism s of I, with s and s^{-1} being Lipschitz maps, such that $f \circ s = s \circ h$.

Thus we can state the following result.

Theorem 3. For any continuous map f of I that is Lipschitz conjugate to some generalized N-shaped mapping, there exists a chaotic set C w.r.t. f with Hausdorff dimension one everywhere. Moreover, any chaotic set is of one-dimensional Hausdorff measure zero.

2 Notations and lemmas

We denote $\Sigma_3 = \prod_{i=1}^{\infty}\{0,1,2\}$ the one-side symbolic space consisting of three letters $0, 1$ and 2.

Write intervals $[0, 1/3]$, $[1/3, 2/3]$ and $[2/3, 1]$ as I_0, I_1 and I_2 respectively. The letters of the words involved in the following context all come from $\{0, 1, 2\}$.

Let $b = b_1 \cdots b_n$ be a word of length n. It follows from the graph of g that the set $\bigcap_{j=0}^{n-1} g^{-j}(I_{b_{j+1}})$ is exactly an closed interval of length $1/(3^n)$. Hence, for each $x = x_1x_2 \cdots \in \Sigma_3$, $\bigcap_{j=0}^{\infty} g^{-j}(I_{x_{j+1}})$ is a single point set. Denote this point $g^*(x)$. It is easy to verify that map g^* is continuous and satisfies the commutative digram

$$\begin{array}{ccc} \Sigma_3 & \xrightarrow{\sigma} & \Sigma_3 \\ {\scriptstyle g^*}\downarrow & & \downarrow{\scriptstyle g^*} \\ I & \xrightarrow{g} & I \end{array}$$

For convenience, define $g^*(b)$ to be $\bigcap_{j=0}^{n-1} g^{-j}(I_{b_{j+1}})$ for each word $b = b_1 \cdots b_n$ of finite length.

Lemma 1. Given positive integer n, $\bigcup g^*(b) = I$, where the union is taken for all $b \in \{0,1,2\}^n$. Each $g^*(b)$ is a closed interval of length $1/(3^n)$. Any two of them either have a common end point or have a gap of $1/(3^n)$.

Lemma 1.

The proof of Lemma 1 is straightforward.

Lemma 2. If $a = a_1 \cdots a_n$, $b = b_1 \cdots b_n$, $a \neq b$ and $g^*(a) \cap g^*(b) \neq \varnothing$, then there is one and only one positive integer k such that $a_k \neq b_k$. Meanwhile, I_{a_k} and I_{b_k} are adjacent to each other and either $a_j = b_j = 0$ for $j > k$ or $a_j = b_j = 2$ for $j > k$.

Proof. Suppose $a_k \neq b_k$ and $a_i = b_i$ for $i < k$. Take an $x \in g^*(a) \cap g^*(b)$. Then $g^{k-1}(x) \in I_{a_k} \cap I_{b_k} \neq \varnothing$ hence I_{a_k} and I_{b_k} are adjacent to each other. Meanwhile, we have either $g^{k-1}(x) = 1/3$ or $g^{k-1}(x) = 2/3$. In the case of $g^{k-1}(x) = 1/3$, we know $g^k(x) = g^{k+1}(x) = \cdots = g^{n-1}(x) = 1$. Thus, for $j > k$, $g^{j-1}(x) \in I_{a_j} \cap I_{b_j} = I_2$. Therefore, $a_j = b_j = 2$. In the latter case, the argument is similar.

For $I_0 \cap I_2 = \varnothing$, obviously we have

Corollary. Let $a = a_1 \cdots a_n$ and $b = b_1 \cdots b_n$. If there exists some $1 \leq k \leq n$ such that $a_k = 0$ and $b_k = 2$, then $g^*(a) \cap g^*(b) = \varnothing$.

Now we come to construct a continuous map from Σ_3 to itself. First, denote all the maps from $\{0,1,2\}^n$ to itself by $\varphi_1^{(n)}, \varphi_2^{(n)}, \cdots, \varphi_{l_n}^{(n)}$, where $l_n = (3^n)^{3^n}$ and specifically $\varphi_1^{(n)}(x_1x_2\cdots x_n) = \underbrace{11\cdots1}_{n}$, for each $(x_1x_2\cdots x_n) \in \{0,1,2\}^n$. Every word such as $\varphi_1^{(n)}(x_1x_2\cdots x_n)\varphi_2^{(n)}(x_1x_2\cdots x_n)\cdots\varphi_{l_n}^{(n)}(x_1x_2\cdots x_n)$ is of length $nl^n = n(3^n)^{3^n}$. Write this number as J_n.

Define $\Phi : \Sigma_3 \to \Sigma_3$ as follows. For any $x = x_1\cdots x_n \in \Sigma_3$, $\Phi(x) = x_1\cdots x_{J_1^3}\varphi_1^{(1)}(x_1)\cdots\varphi_{l_1}^{(1)}(x_1)x_{J_1^3+1}\cdots x_{J_2^3-J_1}\varphi_1^{(2)}(x_1x_2)\cdots\varphi_{l_2}^{(2)}(x_1x_2)\cdots$
$\varphi_1^{(n-1)}(x_1\cdots x_{n-1})\cdots\varphi_{l_{n-1}}^{(n-1)}(x_1\cdots x_{n-1})x_{J_{n-1}^3-J_{n-2}-\cdots-J_1+1}\cdots$
$x_{J_n^3-J_{n-1}-\cdots-J_1}\varphi_1^{(n)}(x_1\cdots x_n)\cdots\varphi_{l_n}^{(n)}(x_1\cdots x_n)\cdots$.

Let n be a positive integer and let $p_n : \Sigma_3 \to \{0,1,2\}^n$ be the nth projection map defined by $p_n(x) = x_1\cdots x_n$ whenever $x = x_1\cdots x_nx_{n+1}\cdots \in \Sigma_3$. It follows from the definition of Φ that the identity $p_n \circ \sigma^{J_n^3+(j-1)n} \circ \Phi = \varphi_j^{(n)} \circ p_n$ holds whenever $1 \le j \le l_n$. Moreover, $\Phi : \Sigma_3 \to \Sigma_3$ is continuous and one-to-one.

Lemma 3. Let x and y be elements in Σ_3 and $p_{J_n^3+J_n}(\Phi(x)) \ne p_{J_n^3+J_n}(\Phi(y))$. Then $g^*(p_{J_n^3+J_n}(\Phi(x))) \cap g^*(p_{J_n^3+J_n}(\Phi(y))) = \varnothing$.

Proof. In case $x_1\cdots x_n \ne y_1\cdots y_n$, we choose a $\varphi_s^{(n)} : \{0,1,2\}^n \to \{0,1,2\}^n$ such that $\varphi_s^{(n)}(x_1\cdots x_n) = \underbrace{00\cdots0}_{n}$ and $\varphi_s^{(n)}(y_1\cdots y_n) = \underbrace{22\cdots2}_{n}$. Then, by Lemma 1, $g^*(p_{J_n^3+J_n}(\Phi(x)))$ and $g^*(p_{J_n^3+J_n}(\Phi(y)))$ are disjoint and the distance between them is at least $3^{-(J_n^3+J_n)}$.

In case $x_1\cdots x_n = y_1\cdots y_n$, denoted b, we must have $p_{J_n^3}(\Phi(x)) \ne p_{J_n^3}(\Phi(y))$. Since $\varphi_1^{(n)}(b) = \underbrace{11\cdots1}_{n}$, by Lemma 2, $g^*(p_{J_n^3+n}(\Phi(x)))$ and $g^*(p_{J_n^3+n}(\Phi(y)))$ are disjoint and the distance between them at least $3^{-(J_n^3+n)}$. Obviously, for any $x \in \Sigma_3$, $g^*(p_{J_n^3+J_n}(\Phi(x))) \subset g^*(p_{J_n^3+n}(\Phi(x)))$. Thus the distance between $g^*(p_{J_n^3+J_n}(\Phi(x)))$ and $g^*(p_{J_n^3+J_n}(\Phi(y)))$ is at least $3^{-(J_n^3+n)}$.

3 Construction of Chaotic Set

Lemmas

Let $E_n = \bigcup g*(p_{J_n^3+J_n}(\Phi(x)))$, where the union is taken for all $x \in \Sigma_3$.

Claim 1. The Hausdorff dimension of $\bigcap_{n=1}^{\infty} E_n$ is one.

Proof. The sequence of positive integers $\{J_k\}$ satisfies the inequality $k + J_1 + \cdots + J_{k-2} < 2J_{k-1} < J_{k-1}^3$ when $k \ge 2$. This fact indicates that, for any integer $i \le k$, x_i will be some digit of $p_{J_{k-1}^3} \circ \Phi(x)$. Given $w \in \Sigma_3$, define $E_n(w) = g^*(p_{J_n^3+J_n} \circ \Phi(w))$. This is an interval of length $3^{-(J_n^3+J_n)}$ and contained in E_n.

Fix $z \in \Sigma_3$. If some $E_n(x)$ is contained in $E_{n-1}(z)$ then, by Lemma 3, $p_{J_n^3+J_n} \circ \Phi(x) = p_{J_n^3+J_n}\Phi(z)$ and hence $p_n(x) = p_n(z)$ from the argument in the preceding paragraph.Now suppose that $E_n(y)$ is an interval in $E_{n-1}(z)$ different from $E_n(x)$. By the proof of Lemma 3, we see that the distance between $E_n(x)$ and $E_n(y)$ is at least $3^{-(J_n^3+n)}$ and that the distance between any two intervals in E_{n-1} is at least $3^{-(J_{n-1}^3+J_{n-1})}$ which is bigger than $3^{-(J_n^3+n)}$. Therefore, the distance between any two different intervals in E_n is at least $3^{-(J_n^3+n)}$.

Two intervals $E_n(x)$ and $E_n(y)$ contained in $E_{n-1}(z)$ are different if and only if $x_{J_{n-1}^3+J_{n-1}+1}\cdots x_{J_n^3} \neq y_{J_{n-1}^3+J_{n-1}+1}\cdots y_{J_n^3}$. So $E_{n-1}(z)$ contains exactly $3^{J_n^3-J_{n-1}^3-J_{n-1}}$ intervals of E_n. Denote $m_n = 3^{J_n^3-J_{n-1}^3-J_{n-1}}$ and $\varepsilon_n = 3^{-(J_n^3+n)}$. By Example 4.6 in [falconer90],

$$\dim_H \bigcap_{i=1}^{\infty} E_i \geq \liminf_{n\to\infty} \log m_1 \cdots m_{n-1}/ - \log m_n\varepsilon_n =$$

$$\liminf_{n\to\infty}(J_{n-1}^3 - J_1^3 - \sum_{k=1}^{n-2} J_k)/(J_{n-1}^3 + J_{n-1} + n) \to 1$$

by the fact that $\sum_{k=1}^{n-2} J_k < 2J_{n-1}$.

In fact, $\bigcap_{n=1}^{\infty} E_n = g^*(\Phi(\Sigma_3))$.

Claim 2. $g^*(\Phi(\Sigma_3))$ is a chaotic set with respect to g.

Proof. Let b be a word of length n. Denote $V_b = Int(\bigcup g^*(a))$, where the union is taken for all such a that $g^*(a)\bigcap g^*(b) \neq \varnothing$. By Lemma 1, the diameter of V_b is at most $3^{-(n-1)}$. Particularly, if $b = 1$, then $V_b = I$.

For any $x \in \Sigma_3$, define $K_{p_n(x)} = \{y \in \Sigma_3 \mid y_i = x_i \text{ for } 1 \leq i \leq n\}$.

Suppsoe $A \subset g^*(\Phi(\Sigma_3))$ and $f : A \to I$ is an arbitrary continous map. Since Φ and $g^*|_{\Phi(\Sigma_3)}$ are one-to-one maps, we can take a subset $A_1 = (\Phi)^{-1} \circ (g^*|_{\Phi(\Sigma_3)})^{-1}(A)$ of Σ_3 and let $F = f \circ g^* \circ \Phi : A_1 \to I$.

For any $x \in A_1$, define $\psi_n(x) = \max\{j \leq n \mid$There exists a word b of lengthn such that$K_{p_n(x)} \bigcap A_1 \subset F^{-1}(V_b)\}$. Since $V_1 = I$, $\psi_n(x)$ is always well defined. We also have $\lim_{n\to\infty}\psi_n(x) = \infty$ for each $x \in A_1$ by the continuity of F.

Take a word b of the largest length $\psi_n(x)(\leq n)$, denoted $b_n(x)$, such that $K_{p_n(x)} \bigcap A_1 \subset F^{-1}(V_{b_n(x)})$.

Choose a $\varphi_{j_n}^{(n)}$ such that for any $x \in A_1, x = x_1 \cdots x_n \cdots \in \Sigma_3$,

$p_{\psi_n(x)}\varphi_{j_n}^{(n)}(x_1 \cdots x_n) = b_n(x)$. Let $r_n = J_n^3 + (j_n - 1)n$. Then

$p_{\psi_n(x)}(\sigma^{r_n}(\Phi(x))) = g^*(\sigma^{r_n}(\Phi(x))) \in g^*(b_n(x)) \subset V_{b_n(x)}$. On the other hand,

$F(x) \in F(K_{p_n(x)} \bigcap A_1) \subset V_{b_n(x)}$ and the diameter of $V_{b_n(x)}$ is no bigger than $3^{-(\psi_n(x)-1)}$. Therefore,

$d(g^*(\sigma^{r_n}(\Phi(x))),\ F(x)) \leq 3^{-(\psi_n(x)-1)}$. Since g^* semiconjugates σ to g and $\lim_{n\to\infty}\psi_n(x) = \infty$, this inequality implies $g^{r_n}(g^*(\Phi(x))) \to F(x) =$

$f(g^*(\Phi(x)))$, as $n \to \infty$. Thus $g^{r_n}(y) \to f(y)$ as $n \to \infty$ for any $y \in A$. This completes the proof.

Let $\bigcup_{k=1}^{\infty}\{0,1,2\}^k = \bigcup_{n=1}^{\infty} S(n)$. Here $S(n)$'s are distinct words of finite length. Further, define $K_n = \{y \in \Sigma_3 \mid y_{n+1} = 0 \text{ and} y_i = 2 \text{for} 1 \le i \le n\}$. Let $b = b_1 \cdots b_n$ be a word of length n. $\theta_b : \Sigma_3 \to \Sigma_3$ defined as $\theta_b(x) = b_1 \cdots b_n x_{n+1} x_{n+2} \cdots$ for $x = x_1 \cdots x_n x_{n+1} x_{n+2} \cdots \in \Sigma_3$.

Claim 3. $\bigcup_{n=1}^{\infty} g^*(\theta_{S(n)}(\Phi(K_n)))$ is a chaotic set with respect to g with Hausdorff dimension one everywhere.

4 Proofs of theorems and remarks

Proofs The chaoticity in Claim 3 can be obtained by combining Claim 2 and the corresponding proof in [xiong95], while the estimate of Hausdorff dimension is similar to Claim 1. These also prove Theorem 1.

Note that for N-shaped mapping we have the same result as the that of Lemma 5 in [xiong95]. This leads to Theorem 2.

As to the generalized N-shaped mapping, one just needs to imitate the construction for N-shaped mapping. Note that Hausdorff measure zero and Hausdorff dimension are invariants under bi-Lipschitz maps. This finishes the proof of Theorem 3.

Remark 1. Since N-shaped mapping g is *ergodic with respect to Lebesgue measure*, one can show that the set $\{x \in I \mid \text{The orbit of} x \text{ under} g \text{is dense in} I$ is of full Lebesgue measure, i.e., one. It follows from Theorem 2 that chaotic set is quite smaller than this set.

Remark 2. Actually, we have constructed a chaotic set in Σ_3 with its image of g^* having Hausdorff dimension one everywhere in I. Because $g^* : \Sigma_3 \to \Sigma_3$ is Lipschitz map, we know this chaotic set also has Hausdorff dimension one everywhere. This also produces another chaotic set for Σ_3 as described in [xiong95]. Furthermore, our arguments also apply to the case of Σ_n for $n \geq 2$.

References

1. Robert L. Devaney, *An introduction to chaotic dynamical systems*, 2nd ed., Addison Wesley, New York, 1989.
2. Kenneth Falconer, *Fractal geometry: Mathematical foundations and applications*, Wiley, New York, 1990.
3. T. Y. Li and James A. Yorke, *Period three implies chaos*, Amer. Math. Monthly. **82 (1975), no. 10, 985-992.**
4. **Ivan Mizera,** *Generic properties of one-dimensional dynamical systems*, **Proceedings of international conference on ergodic theory and related topics III (October 22-27, 1990, Gustrow, Germany)(K. Richter U. Krengel and V. Warstat, eds.), LNM, vol. 1514, Springer-Verlag, Berlin, Hong Kong, 1992, pp. 163-173.**
5. **incheng Xiong,** *Hausdorff dimension of a chaotic set of shift of a symbolic space*, **Science in China(Series A). 38 (1995), no. 6, 696-708.**

49 ALGEBRAIC DEPENDENCE OF MEROMORPHIC MAPPINGS WITH DEFICIENCIES INTO ALGEBRAIC VARIETIES

Yoshihiro AIHARA[1]

Numazu College of Technology, 3600 Ooka, Numazu
Shizuoka 410-8501, Japan

aihara@la.numazu-ct.ac.jp

1991 Mathematics Subject Classification: Primary 32H30

Introduction and Results.

In this paper we give theorems on the propagation of algebraic dependence for meromorphic mappings, especially under conditions on deficiencies. Let M be a projective algebraic manifold. Let $\pi : X \to \mathbf{C}^m$ be a finite analytic covering space, that is, X is a reduced irreducible normal complex space and $\pi : X \to \mathbf{C}^m$ is a proper surjective holomorphic mappings with discrete fibers. We denote by s_0 its sheet number and by B its ramification divisor. We first give the definition of algebraic dependence. Set $M^2 = M \times M$. For meromorphic mappings $f,\ g : X \to M$, we define a meromorphic mapping $f \times g : X \to M^2$ by

[1] Partly supported by the Grants-in-Aid for Scientific Research from the Ministry of Education, Science and Culture, Japan.

H.G.W. Begehr et al. (eds.), Proceedings of the Second ISAAC Congress, Vol. 1, 419–426.

$(f \times g)(z) = (f(z),\ g(z))$ for $z \in X - (I(f) \cup I(g))$, where $I(f)$ and $I(g)$ are the loci of the indeterminacy of f and g respectively.

Definition (0.1). Let S be an analytic subset of X. Nonconstant meromorphic mappings $f,\ g : X \to M$ are said to be *algebraically dependent* on S if there exists a proper algebraic subset Σ of M^2 such that Σ is not of the type $\Sigma_1 \times \Sigma_2$ with algebraic subsets $\Sigma_j \subseteq M$ for $j = 1,\ 2$ and $(f \times g)(S) \subseteq \Sigma$.

In this paper we study the case where S and Σ are hypersurfaces. Our main tool is the second main theorem for dominant meromorphic mappings $f : X \to M$. There have been several studies on the propagation of dependence. We notice here that a certain kind of unicity theorems such as results in [1] and [3] can be regarded as theorems on the propagation of dependence. Indeed, these theorems say that two meromorphic mappings $f,\ g : X \to M$ with the same inverse images of divisors as point sets on which $f = g$ are essentially identical under some conditions on given divisors. In this case we consider that the relation $f = g$ on the inverse image of divisors propagates to the whole space. We can prove some unicity theorems from this point of view. Details will be published elsewhere (cf. [2]).

Let $F \in \mathrm{Pic}(M) \otimes \mathbf{Q}$ and $\gamma \in \mathbf{Q}$. We simply write γF for $F^{\otimes \gamma}$. Then F is said to be *big* provided that a line bundle $\nu F \in \mathrm{Pic}(M)$ is big for some positive integer ν. We fix a big line bundle $L \to M$. Now we define $[F/L]$ to be the infimum of the set of rational numbers γ such that $\gamma L \otimes F^{-1}$ is big. Notice that $[F/L] < 0$ if and only if F^{-1} is big. A meromorphic mapping $f : X \to M$ is said to be *dominant* if $\operatorname{rank} f = \dim M$. Throughout this paper, we assume that there exists at least one dominant meromorphic mapping $f_0 : X \to M$. Notice that the canonical bundle K_M of M is not big in the case $X = \mathbf{C}^m$. Let $D_1, \ldots, D_q$ be divisors in $|L|$ such that $D_1 + \cdots + D_q$ has only simple normal crossings. For an effective divisor E on X and a positive integer k, we denote by $\mathrm{Supp}_k\, E$ the union of all irreducible components of E with the multiplicities at most k. Let $k_1, \ldots, k_q$ be fixed positive integers. Set $S_j = \mathrm{Supp}_{k_j} f_0^* D_j$ and assume that $\dim S_i \cap S_j \leq m - 2$ for any $i \neq j$. We define a hypersurface S in X by $S = S_1 \cup \cdots \cup S_q$.

Let F_1 and F_2 be big line bundles over M. Assume that, for one of $F_1,\ F_2$ (say F_0), $F_0 \otimes F_j^{-1}$ is either big or trivial. Set $\tilde{F} = \pi_1^* F_1 \otimes \pi_2^* F_2$, where $\pi_j : M^2 \to M$ are the natural projections on the j-th factor. Let

$\tilde{L}$ be a line bundle over M^2. Notice that, in general,

$$\tilde{L} \notin \pi_1^* \operatorname{Pic}(M) \oplus \pi_2^* \operatorname{Pic}(M).$$

For example, if M is a smooth algebraic curve of genus not less than one,then the diagonal in M^2 defines a line bundle not belonging to $\pi_1^* \operatorname{Pic}(M) \oplus \pi_2^* \operatorname{Pic}(M)$. Assume that $\tilde{\gamma}\tilde{F} \otimes \tilde{L}^{-1}$ is big for a fixed positive rational number $\tilde{\gamma}$. Let $\mathcal{R}$ be the set of all hypersurfaces Σ in X such that $\Sigma = \operatorname{Supp} \tilde{D}$ for some $\tilde{D} \in |\tilde{L}|$ and Σ is not of the type $\Sigma_1 \times \Sigma_2$. Let

$$\mathcal{F} = \mathcal{F}(f_0; (X, \{S_j\}); (M, \{D_j\}))$$

be the set of all *dominant* meromorphic mappings $f : X \to M$ such that $\operatorname{Supp}_{k_j} f^* D_j = S_j$ for all $1 \leq j \leq q$. We denote by

$$\mathcal{F}_0 = \mathcal{F}_0(f_0; \tilde{L}; (X, \{S_j\}); (M, \{D_j\}))$$

the family of all meromorphic mappings $f \in \mathcal{F}$ such that $(f_0 \times f)(S) \subseteq \Sigma$ for some Σ in $\mathcal{R}$.

Definition (0.2). Let Y be a compact complex manifold. We say that a meromorphic mapping $f : X \to Y$ *separates the fibers of* $\pi : X \to \mathbf{C}^m$ if there exists a point z in $\mathbf{C}^m - (\operatorname{Supp} \pi_* B \cup \pi(I(f)))$ such that $f(x) \neq f(y)$ for any distinct points $x,\ y \in \pi^{-1}(z)$.

Suppose that $f_0 : X \to M$ separates the fibers of $\pi : X \to M$. Then there exist a positive integer l and a pair of sections $\sigma_0, \sigma_1 \in H^0(M,\ lL)$ such that a meromorphic function $f^*(\sigma_0/\sigma_1)$ separates the fibers of $\pi : X \to \mathbf{C}^m$. We denote by l_0 the least integer that satisfies the above condition. Then we have the following ramification estimate due to J. Noguchi (cf. [4, p. 270]):

$$N(r,\ B) \leq 2l_0(s_0 - 1)\ T_{f_0}(r,\ L) + O(1).$$

This estimate plays an essential role in this paper. We now define $L_0 \in \operatorname{Pic}(M) \otimes \mathbf{Q}$ by

$$L_0 = \left(\sum_{j=1}^{q} \frac{k_j}{k_j + 1} - 4l_0(s_0 - 1) \right) L \otimes \left(-\frac{2k_0}{k_0 + 1} \right) \tilde{\gamma} F_0,$$

where $k_0 = \max\{k_1, \ldots, k_q\}$.

If L_0 is sufficiently big, we can conclude that the algebraic dependence on S propagates to the whole space X as follows (cf. [6, p. 336]):

Theorem (0.3). *Suppose that f_0 separates the fibers of $\pi : X \to \mathbf{C}^m$. If $L_0 \otimes K_M$ is big, then f_0 and f are algebraically dependent on X for each $f \in \mathcal{F}_0$.*

Set $\gamma_0 = [L_0^{-1} \otimes K_M^{-1}/L]$. If $L_0 \otimes K_M$ is big, we see $\gamma_0 < 0$. In the case where γ_0 is nonnegative, we cannot conclude the propagation of algebraic dependence in general. In the case of $\gamma_0 > 0$, the propagation of algebraic dependence does not occur even if we assume the existence of Picard's deficient divisors (cf. Remark 2.25 in [1]). However, in the case of $\gamma_0 = 0$, we have the following theorem on the propagation of algebraic dependence under a condition on the existence of Nevanlinna's deficient divisors:

Theorem (0.4). *Suppose that f_0 separates the fibers of $\pi : X \to \mathbf{C}^m$ and $\gamma_0 = 0$. If $\delta_{f_0}(D_j) > 0$ for at least one $1 \leq j \leq q$, then f_0 and f are algebraically dependent on X for each $f \in \mathcal{F}_0$.*

This theorem shows the existence of deficient divisors in the sense of Nevanlinna imposes a strong restriction on the behavior of meromorphic mappings.

As mentioned above, we can prove some unicity theorems for meromorphic mappings by taking line bundles F_j of special type in the theorems above. We deduce some unicity theorems from this point of view. Let $\Phi : M \to \mathbf{P}_n(\mathbf{C})$ be a meromorphic mapping with $\operatorname{rank} \Phi = \dim M$. Assume that a set $\{D_j\}_{j=1}^q$ of divisors is *generic with respect to* f_0 and Φ, that is,

$$R_j := f_0(X - I(f_0)) \cap \operatorname{Supp} D_j \cap \{w \in M;\ \operatorname{rank} d\Phi(w) = \dim M\} \neq \emptyset$$

for at least one $1 \leq j \leq q$. In what follows we assume that f_0 separates the fibers of $\pi : X \to \mathbf{C}^m$. We denote by H the hyperplane bundle over $\mathbf{P}_n(\mathbf{C})$. Now take $F_1 = F_2 = \Phi^* H$. We also take $\tilde{L} = \tilde{F}$. We denote by $\mathcal{F}_1$ the set of all mappings $f \in \mathcal{F}$ such that $f = f_0$ on S. Then we have the following unicity theorems, which are proved in [1] (cf. [1, Theorems 2.1 and 2.15]):

Theorem (0.5). *If $F_0 \otimes K_M$ is big, then the family $\mathcal{F}_1$ contains just one mapping f_0.*

Theorem (0.6). *Suppose that* $\gamma_0 = 0$. *If* $\delta_{f_0}(D_j) > 0$ *for at least one* $1 \leq j \leq q$, *then the family* $\mathcal{F}_1$ *contains just one mapping* f_0.

We show here how Theorems (0.5) and (0.6) are deduced from Theorems (0.3) and (0.4). Let $f \in \mathcal{F}_1$. Suppose that $\Phi \circ f \not\equiv \Phi \circ f_0$. Then we can take a hypersurface $\Sigma \in \mathcal{R}$ so that $(f_0 \times f)(S) \subseteq \Sigma$ and $(f_0 \times f)(X) \not\subseteq \Sigma$. Indeed, set $\tilde{H} = \pi_1^* H \otimes \pi_2^* H$. We denote by Δ the diagonal of $\mathbf{P}_n(\mathbf{C})^2$. Since $f = f_0$ on S, it is clear that $(f \circ \Psi \times f_0 \circ \Psi)(S) \subseteq \Delta$. It now follows from $\tilde{L} = (\Psi \times \Psi)^* \tilde{H}$ that the there exists a hypersurface Σ with the desired property. By the propagation of dependence on S to the whole space X, we now have a contradiction. Thus we see $\Phi \circ f \equiv \Phi \circ f_0$. Notice that some $R_j \neq \emptyset$. By uniqueness of analytic continuation, we thus conclude that $f \equiv f_0$ on X.

1. Preliminaries.

Let $\pi : X \to \mathbf{C}^m$ be as in the Introduction. Let $z = (z_1, \ldots, z_m)$ be the natural coordinate system in $\mathbf{C}^m$, and set

$$\|z\|^2 = \sum_{\nu=1}^{m} z_\nu \bar{z}_\nu, \qquad X(r) = \pi^{-1}\left(\{z \in \mathbf{C}^m;\ \|z\| < r\}\right),$$

$$d^c \frac{\sqrt{-1}}{4\pi}(\bar{\partial} - \partial), \qquad \alpha = \pi^* dd^c \|z\|^2.$$

For a (1,1)-current φ of order zero on X we set

$$N(r, \varphi) = \frac{1}{s_0} \int_1^r \langle \varphi \wedge \alpha^{m-1}, \chi_{X(t)} \rangle \frac{dt}{t^{2m-1}},$$

where $\chi_{X(r)}$ denotes the characteristic function of $X(r)$.

Let M be a compact complex manifold and $L \to M$ a big line bundle over M. Let $|\cdot|$ be a hermitian fiber metric in L, and ω its Chern form. Let $f : X \to M$ be a nonconstant meromorphic mapping. We set

$$T_f(r, L) = N(r, f^* \omega),$$

and call it the characteristic function of f with respect to L. We also define $T_f(r, F)$ for $F \in \mathrm{Pic}(M) \otimes \mathbf{Q}$ in the following way. If ν is a positive integer with $\nu F \in \mathrm{Pic}(M)$, then we set

$$T_f(r, F) = \frac{1}{\nu} T_f(r, \nu F).$$

It is easy to see that $T_f(r, F)$ is well-defined. We have Nevanlinna's inequality for meromorphic mappings as follows (cf. [4, p. 269]):

Theorem (1.1). *Let $f : X \to M$ be a nonconstant meromorphic mapping. Then*

$$N(r, f^*D) \leq T_f(r, L) + O(1)$$

for $D \in |L|$ with $f(X) \not\subseteq \operatorname{Supp} D$, where $O(1)$ stands for a bounded term as $r \to +\infty$.

Let $D \in |L|$. We define Nevanlinna's deficiency $\delta_f(D)$ by

$$\delta_f(D) = 1 - \limsup_{r \to +\infty} \frac{N(r, f^*D)}{T_f(r, L)}.$$

If $\delta_f(D) > 0$, then D is called a *deficient divisor in the sense of Nevanlinna.*

The following second main theorem plays a central role in the proofs of theorems (cf. [4, Theorem 1] and [5, Theorem 2]):

Theorem (1.2). *Let M be a projective algebraic manifold with $m \geq \dim M$ and L a big line bundle over M. Let $D_1, \ldots, D_q$ be divisors in $|L|$ such that $D_1 + \cdots + D_q$ has only simple normal crossings. Let $f : X \to M$ be a dominant meromorphic mapping. Then*

$$qT_f(r, L) + T_f(r, K_M) \leq \sum_{j=1}^{q} N(r, \operatorname{Supp} f^*D_j) + N(r, B) + S_f(r),$$

where $S_f(r) = O(\log T_f(r, L)) + o(\log r)$ except on a Borel subset $E \subseteq [1, +\infty)$ with finite measure.

§2. A sketch of proofs of Theorems (0.3) and (0.4).

In this section we give an outline of proofs of theorems given in the Introduction. By Theorem (1.2), we have the following lemma:

Lemma (2.1). *Let f be an arbitrary mapping in $\mathcal{F}$. Then*

$$qT_f(r, L) + T_f(r, K_M) \leq$$

$$\frac{k_0}{k_0+1} N(r, S) + \sum_{j=1}^{q} \frac{1}{k_j+1} N(r, f^*D_j) + N(r, B) + S_f(r).$$

Let $f_1 \in \mathcal{F}_0$. By the definition of $\mathcal{F}_0$, there exists $\Sigma \in \mathcal{R}$ such that $(f_0 \times f_1)(S) \subseteq \Sigma$. Suppose that $(f_0 \times f_1)(X) \not\subseteq \Sigma$. We first assume that $L_0 \otimes K_M$ is big. For brevity, we set $T(r, F) = T_{f_0}(r, F) + T_{f_1}(r, F)$ for $F \in \mathrm{Pic}(M) \otimes \mathbf{Q}$. We also set $S(r) = S_{f_0}(r) + S_{f_1}(r)$. By Theorem (1.1) and Lemma (2.1), we see

$$T(r, L_0) + T(r, K_M) \leq S(r).$$

Since $L_0 \otimes K_M$ is big, we have a contradiction.

Next we assume that $\gamma_0 = 0$. The following fact is a crucial step in the proof of this case: There exist positive constants C_1 and C_2 such that

$$C_1 \leq \frac{T_{f_1}(r, L)}{T_{f_0}(r, L)} \leq C_2$$

for all sufficiently large $r \notin E$, where E is an exceptional set for $S_f(r)$. Assume that $\delta_{f_0}(D_l) > 0$ for some l. By the definition of Nevanlinna's deficiency, for any $\varepsilon > 0$, there exists $r_1 > 0$ such that

$$N(r, f_k^* D_j) < (1 - \delta_{f_k}(D_j) + \varepsilon)\, T_{f_k}(r, L)$$

for all $r \geq r_1$ $(r \notin I)$, where $I \subseteq [1, +\infty)$ is a Borel subset with finite measure. By Lemma (2.1) and $\gamma_0 = 0$, we have

$$\delta_{f_0}(D_l)\, T_{f_0}(r, L) - (k_l + 1)(\gamma + q\varepsilon) T(r, L) \leq S(r)$$

for an arbitrary rational number $\gamma > 0$. Thus we see $\delta_{f_0}(D_l) = 0$, which contradicts the assumption. We have now the desired conclusion.

§3. The concluding remarks.

In this paper, we assume that at least one of the mappings $f_j : X \to M$ separates the fibers of $\pi : X \to \mathbf{C}^m$. In [6], W. Stoll proved some theorems on the propagation of dependence of meromorphic mappings $f : X \to M$ under condition on the growth of mappings in different settings. In his results at least one of the mappings f_j must grow quicker than the ramification divisor B. However the existence of such a mapping is a delicate matter even if both of X and M are Riemann surfaces. He already pointed out that in some cases his condition can be replaced by condition on the existence of mappings which separates the fibers of $\pi : X \to \mathbf{C}^m$ (cf. [6, p. 336]).

We finally give a remark on the case where some of $k_j = +\infty$. Set $1/(k_j + 1) = 0$ and $k_j/(k_j + 1) = 1$ if $k_j = +\infty$. Then Theorems (0.3) and (0.5) remain valid in this case. On the other hand, Theorems (0.4) and (0.6) do not hold without the additional condition on the existence of the deficient divisor D_j with $k_j \neq +\infty$. Indeed, there exists a counter example in the case where all $k_j = +\infty$ (cf. [1, Remarks 2.25]).

References

[1] Y. Aihara, The uniqueness problem of meromorphic mappings with deficiencies, Tohoku Math. J. **51** (1999), 315–328.

[2] Y. Aihara, Algebraic dependence of meromorphic mappings in value distribution theory,in preparation.

[3] S. Drouilhet, A unicity theorem for meromorphic mappings between algebraic varieties, Trans. Amer. Math. Soc. **265** (1981), 349–358.

[4] J. Noguchi, Meromorphic mappings of covering spaces over $\mathbf{C}^m$ into a projective variety and defect relations, Hiroshima Math. J. **6** (1976), 265–280.

[5] F. Sakai, Defect relations for equidimensional holomorphic maps, J. Faculty of Sci., Univ. Tokyo, Sec. IA, **23** (1976), 561–580.

[6] W. Stoll, Propagation of dependences, Pacific J. Math. **139** (1989), 311–336.

50 NEVANLINNA THEORY AND STOCHASTIC CALCULUS

–Some Applications of Stochastic First Main Theorem

Atsushi ATSUJI

Department of Mathematics, Graduate School of Science
Osaka University, Toyonaka, Osaka, 560-0043, Japan

atsuji@math.sci.osaka-u.ac.jp

Several relationships between classical function theory and analysis of Brownian motion have been studied by many authors (cf.[9], [5]). Davis gave an elegant proof of Picard's theorem in [8] which impressed us with their intimacy. It also gave some possibility to study value-distribution of meromorphic functions through Brownian motion or stochastic calculus. It is curious that there had been no article noting a simple relation between Nevanlinna theory and Brownian motion until Carne's paper [6]. He considered some probabilistic aspects of classical Nevanlinna theory. After the work the author improved it to several dimensional cases in [1]. In this note we exemplify this natural relation between stochastic calculus and Nevanlinna theory by considering a stochastic generalization of first main theorem of Nevanlinna and its applications.

H.G.W. Begehr et al. (eds.), Proceedings of the Second ISAAC Congress, Vol. 1, 427–432.

1 Probabilistic meaning of the characteristic function of holomorphic maps.

Let $f : \mathbf{C}^m \to \mathbf{P}^n(\mathbf{C})$ be a holomorphic map. Then the characteristic function of f is defined by

$$T(r) = \int_1^r \frac{dt}{t^{m-1}} \int_{|z|<t} f^*\omega \wedge (dd^c|z|^2)^{m-1},$$

where ω is Fubini-Study metric of $\mathbf{P}^n(\mathbf{C})$. It is well-known, what is called First Main Theorem, that

$$T(r) = m(r, D) + N(r, D) + O(1),$$

where D is a divisor on $\mathbf{P}^n(\mathbf{C})$, $m(r, D)$ is a proximity function of f and $N(r, D)$ is a counting function of f^*D. It seems only to be noted in usual function theory that this special form of the characteristic function is intended for validity of the above First Main Theorem. We emphasize here that this form has an intrinsic meaning as a probabilistic quantity.

We introduce some notions of probabilistic objects.

Definition 1 *We call a continuous stochastic process Y_t taking its value in a Kähler manifold N on a probability space $(\Omega, P, \mathcal{F}, \mathcal{F}_{\sqcup})$ a holomorphic martingale on N if Re $h(Y_t)$ is a local martingale for any holomorphic function h on each open set of N. Furthermore if it is a diffusion, it is called a holomorphic diffusion N.*

We note that Brownian motion defined from a Kähler metric is a holomorphic diffusion.

We give some elementary properties of holomorphic martingales.

Proposition 2 i) *If u is a plurisubharmonic function, then $u(Y)$ is a local submartingale.*
ii) *If $f : M \to N$ is a holomorphic map between Kähler manifolds and X is a holomorphic martingale on M, then $f(X)$ is a holomorphic martingale on N. In particular if X is a Brownian motion associated with the Kähler metric, $f(X)$ is a holomorphic martingale on N.*

We also define intrinsic time of a holomorphic martingale on N.

Definition 3 *Let Y be a holomorphic martingale on a Kähler manifold N and $\{g_{\alpha,\beta}\}$ the Kähler metric of N. We define intrinsic time of Y by*

$$[Y, \overline{Y}]_t = \int_0^t g_{\alpha,\beta} d < Y^\alpha, \overline{Y^\beta} >_t,$$

where $d < Y^\alpha, \overline{Y^\beta} >_t$ is the usual martingale quadratic variation process.

We note the following property which plays an important role in application to function theory.

Lemma 4 ([7]) *Let Y be a holomorphic martingale on a compact Kähler manifold N. If $[Y,\overline{Y}]_\infty < \infty$ a.s, then Y_t converges a.s. in N as t tends to infinity.*

We can interpret the characteristic function of holomorphic maps into probabilistic language.

Proposition 5 *Let $f : \mathbf{C}^m \to \mathbf{P}^n(\mathbf{C})$ be a holomorphic map and $T(r)$ its characteristic function. Then*

$$T(r) = E[[f(X), \overline{f(X)}]_{\tau_r}],$$

where X is a Brownian motion on $\mathbf{C}^m$ starting from the origin and $\tau_r = \inf\{t > 0 : |X_t| \geq r\}$.

We immediately obtain the following result on boundary behavior of holomorphic maps in Nevanlinna class by combining Lemma 4 with Proposition 5.

Proposition 6 *Let f be a holomorphic maps from a unit ball in $\mathbf{C}^m$ to $\mathbf{P}^n(\mathbf{C})$. If the characteristic function of f is bounded, then f has a limit along almost all Brownian paths. In particular, in case of $m = 1$, f has a non-tangential limit for almost all boundary points under this assumption.*

2 Stochastic First Main Theorem.

Let U be a non-negative local submartingale on a filtered probability space. It is well-known that Doob-Meyer decomposition implies

$$U_t - U_0 = \text{ a local martingale } + A_t,$$

where A_t is an increasing process. We define a probabilistic version of counting function by

$$N(T,U) = \lim_{\lambda \to \infty} \lambda P(\sup_{0 \leq t \leq T} U_t > \lambda),$$

for stopping time T. We have a stochastic first main theorem as follows.

Proposition 7 (cf.[10],[1]) *If a stopping time T satisfies $E[A_T] < \infty$, then*

$$E[U_T] - E[U_0] + N(T,U) = E[A_T].$$

The proof easily follows from integration of the above Doob-Meyer decomposition with stopping time argument.

3 Application I:Casorati-Weierstrass theorem

We apply the stochastic first main theorem to improve Casorati-Weierstrass theorem for holomorphic maps on general Kähler manifolds.

We introduce Molzon's capacity(cf.[11]) on $\mathbf{P}^n(\mathbf{C})$. Let p be a homogeneous polynomial of degree d on $\mathbf{P}^n(\mathbf{C})$. $\mathbf{S}_{n+1,d}$ denotes the set of all such polynomials. We can consider a potential function for $E \subset \mathbf{P}(\mathbf{S}_{n+1,d})$ in usual manner. It is defined by

$$U_{d,\nu}(z) = \int_E \log \frac{||z||^d ||p||}{|p(z)|} d\nu(p),$$

where ν is a probability measure supported by E. Molzon defined a capacity on $\mathbf{P}(\mathbf{S}_{n+1,d})$ using this function by

$$C_d(E)^{-1} = \inf_{\nu \in \mathcal{P}(E)} \sup_{z \in \mathbf{P}^n(\mathbf{C})} U_{d,\nu}(z),$$

where $\mathcal{P}(E)$ denotes a collection of all probability measures supported by E. We say that a set of algebraic varieties associated with $p \in E \subset \mathbf{P}(\mathbf{S}_{n+1,d})$ is of positive capacity if $C_d(E) > 0$.

Then we have the following generalization of Casorati-Weierstrass theorem.

Theorem 8 (cf. [2]) *Let f be a holomorphic map from a Kähler manifold M to $\mathbf{P}^n(\mathbf{C})$. If M has Liouville property(i.e. M does not admit any non-constant bounded harmonic functions.) and f omits a set of algebraic varieties of positive capacity in the above sense, then f is constant.*

The proof is similar to [2]. We only remark that a first main theorem in this case can be derived from our stochastic first main theorem in the previous section. Set $u(z) = \log \frac{||z||^d ||p||}{|p(z)|}$ and $U_t = u(f(Z_t))$ where Z is Brownian motion on M associated with its Kähler metric. If $f(Z_0) \notin Y^p$ (Y^p is the algebraic variety associated with p), then U_t is a non-negative local submartingale. Hence we can apply our stochastic first main theorem to this. We also note that if f omits Y^p, $N(T, U) = 0$ for suitable stopping times T.

We immediately have the following corollary with well-known Yau's result on Liouville property([13]).

Corollary 9 *Let f be a holomorphic map from a Kähler manifold M with $Ric_M \geq 0$ to $\mathbf{P}^n(\mathbf{C})$. If f omits a set of algebraic varieties of positive capacity in the above sense, then f is constant.*

4 Application II: minimal submanifold with Liouville property

Let f be an immersion from M to $\mathbf{R}^n$ and M a Riemannian manifold with the induced measure. We say that an immersion f is minimal if f is harmonic and isometric. We say that $f(M)$ is minimal submanifold for a minimal immersion f.

In order to obtain first main theorem, we define a chodal distance on $\mathbf{R}^n \cup \{\infty\}$.

$$[a, b] = \frac{|a - b|}{\sqrt{1 + |a|^2}\sqrt{1 + |b|^2}}$$

for $a, b \in \mathbf{R}^n$ and $[a, \infty] = \frac{1}{\sqrt{1+|a|^2}}$ (chodal distance on $\mathbf{R}^n \cup \{\infty\}$). Let $u(x) = \log[x, a]^{-1}$ and X a Brownian motion on M associated with the induced metric. Then $U_t = u(f(X_t))$ has a decomposition as

$$U_t - U_0 = \text{ a local martingale } + A_t^{(1)} - A_t^{(2)},$$

where $A^{(1)}, A^{(2)}$ are increasing processes. Since $U_t + A_t^{(2)}$ is a nonnegative local submartingale, we can apply stochastic first main theorem to this case. By the similar argument to the previous case, we have the following result for non-existence region of minimal submanifolds.

Theorem 10 ([3]) *Let $f : M \to \mathbf{R}^n$ be a minimal immersion. If M has Liouville property, then $f(M)$ cannot be contained in the domains having such a form as*

$$D = \{x \in \mathbf{R}^n : \log(1 + |x|) \leq k(x)\}$$

for a concave function.

Remark.

1. Omori showed a similar result in the case that M has bounded sectional curvature instead of Liouville property and D is a nondegenerate cone in [12]. His result can be generalized in [4] to the case of stochastically complete minimal submanifolds and more general D. We note that there is no relationship between stochastic completeness and Liouville property. Hence our result has a different meaning from these earlier works. We also remark that a nondegenerate cone is obviously included in the class of domains in the our theorem.

2. We remark that there is no assumption on (geodesically)completeness of M in our theorem.

References

[1] A. Atsuji, "Nevanlinna theory via stochastic calculus." Jour. Fanct. Anal. vol.132, No.2 1995, 437-510

[2] A. Atsuji "A Casorati-Weierstrass theorem for holomorphic maps and invariant $\sigma-$fields of holomorphic diffusions." Bull.Sci.math. 123(1999), 371-383.

[3] A. Atsuji, "Liouville property and minimal surfaces." in preparation.

[4] A. Atsuji, "Remarks on harmonic maps into a cone from stochastically complete manifolds." Proc. Japan Acad. Ser. A 75(1999), 105-108.

[5] R. F. Bass, *Probabilistic techniques in analysis.* Probability and its Applications. Springer-Verlag, New York, 1995.

[6] T. K. Carne, "Brownian motion and Nevanlinna theory." Jour Proc. London Math. Soc. vol 52, 1986, 349-368

[7] R. W. R. Darling, "Convergence of martingales on a Riemannian manifold." Publ.Res.Inst.Math.Kyoto Univ., 19(1983),753-763.

[8] B. Davis, "Picard's theorem and Brownian motion." Trans. Amer. Math. Soc. 213 (1975), 353-362.

[9] B. Davis, "Brownian motion and analytic functions." Ann. Probab. 7 (1979), no. 6, 913-932.

[10] K. D. Elworthy, X. M. Li and M. Yor, "On the tails of the supremum and the quadratic variation of strictly local martingales." Séminaire de Probabilités XXXI, Lect.Note Math.1655(1997), Springer.

[11] R. E. Molzon, "Potential theory on complex projective space: application to characterization of pluripolar sets and growth of analytic varieties." Ill.Jour.Math., 28(1984),103-119.

[12] H. Omori, "Isometric immersions of Riemannian manifolds." J. Math. Soc. Japan, 19 (1967), 205-214.

[13] S. T. Yau, "Harmonic functions on complete Riemannian manifolds." Comm.Pure.Appl.Math., 28(1975),201-228.

51 ELIMINATION OF DEFECTS OF NON-ARCHCHIMEDEAN HOLOMORPHIC CURVES

Pei-Chu HU[1]
Department of Mathematics, Shandong University
Jinan 250100, Shandong, P. R. China

pchu@public.sdu.edu.cn

Seiki MORI

Department of Mathematical Sciences, Yamagata University
Yamagata 990-8560,Japan

mori@sci.kj.yamagata-u.ac.jp

1991 Mathematics Subject Classification: Primary 11D88, 11E95, 11Q25. Secondary 30D35

In this paper, we will prove that for any transcendental holomorphic curve $f : \kappa \longrightarrow \mathbb{P}^n(\kappa)$ where κ is an algebraically closed field of characteristic zero, complete for a non-trivial non-Archimedean absolute value, we can eliminate all Nevanlinna defects by small deformations if there exists at least a hyperplane whose Valiron deficiency for f is zero.

[1] The work of first author was partially supported by a Grant of the Educational Office and NSFC, P. R. China.

H.G.W. Begehr et al. (eds.), Proceedings of the Second ISAAC Congress, Vol. 1, 433–449.

1 Introduction

Nevanlinna theory asserts that for each holomorphic (or meromorphic) mapping over complex spaces, Nevanlinna defects or Valiron defects of the mapping are very few. Recently S. Mori [3], [4] proved that for a transcendental horomorphic mapping g of $\mathbb{C}^m$ into $\mathbb{P}^n(\mathbb{C})$, we can eliminate all deficient hyperplanes (or deficient hypersurfaces of degree $\leq m$ for a given integer m) in $\mathbb{P}^n(\mathbb{C})$ by a small deformation of the mapping.

In this paper, we shall discuss about elimination theorems of defects of holomorphic curve over non-Archimedean fields by its small deformation. Here a small deformation f of g means that their order functions $T(r, g)$ and $T(r, f)$ satisfy $|T(r, f) - T(r, g)| \leq O(\log r)$, as r tends to infinity. We will prove that main theorems in [3] and [4] are true for non-Archimedean case. Main idea and methods also come from [3] and [4].

2 Preparatory facts

First of all, we need to introduce some notations and terminologies. Let κ be an algebraically closed field of characteristic zero, complete for a non-trivial non-Archimedean absolute value $|\cdot|$. Denote $\kappa - \{0\}$ by κ_*. Let $\mathcal{A}(\kappa)$ be the ring of entire functions on κ. Given a non-constant entire function

$$f(z) = \sum_{n=0}^{\infty} a_n z^n \in \mathcal{A}(\kappa), \quad (a_n \in \kappa)$$

on κ. Define the *maximum term:*

$$\mu(r, f) = \max_{n \geq 0} |a_n| r^n.$$

Let $n\left(r, \frac{1}{f}\right)$ denote the number of zeros (counting multiplicity) of f with absolute value $\leq r$. Fix a real $\rho_0 > 0$ and define the *valence function* of f for 0 by

$$N\left(r, \frac{1}{f}\right) = \int_{\rho_0}^{r} \frac{n\left(t, \frac{1}{f}\right)}{t} dt \quad (r > \rho_0).$$

For $f_1, f_2 \in \mathcal{A}(\kappa)$, the following properties are basic:

$$\mu(r, f_1 f_2) = \mu(r, f_1)\mu(r, f_2), \quad \mu(r, f_1 + f_2) = \max\{\mu(r, f_1), \mu(r, f_2)\}.$$

Let $\mathcal{M}(\kappa)$ be the field of meromorphic functions on κ. By a *meromorphic function* f on κ we will mean the quotient $\frac{g}{h}$ of two entire functions g and h such that g and h have not any common factors. We can uniquely extend μ to a meromorphic function $f = \frac{g}{h}$ by defining

$$\mu(r, f) = \frac{\mu(r, g)}{\mu(r, h)}.$$

Define the *counting function* $n(r, f)$ and the *valence function* $N(r, f)$ of f for poles respectively by

$$n(r, f) = n\left(r, \frac{1}{h}\right), \quad N(r, f) = N\left(r, \frac{1}{h}\right).$$

Then one has the *Jensen formula:*

$$N\left(r, \frac{1}{f}\right) - N(r, f) = \log \mu(r, f) - \log \mu(\rho_0, f). \tag{1}$$

Define the *proximity function*

$$m(r, f) = \log^+ \mu(r, f) = \max\{0, \log \mu(r, f)\},$$

and define the *characteristic function:*

$$T(r, f) = m(r, f) + N(r, f).$$

Then the *first main theorem*

$$m\left(r, \frac{1}{f-a}\right) + N\left(r, \frac{1}{f-a}\right) = T(r, f) + O(1) \quad (r \to \infty)$$

holds for every $a \in \kappa$. In particular, if f is entire, we have (cf. [2])

$$N\left(r, \frac{1}{f-a}\right) = T(r, f) + O(1) \ (a \in \kappa).$$

Now the *lemma of logarithmic derivative* has its stronger form:

$$\mu\left(r, \frac{f^{(k)}}{f}\right) \leq r^{-k},$$

see [1] or [2]. Thus if $\mathbf{W}(f_1, ..., f_k)$ is the Wronskian of meromorphic functions $f_1, ..., f_k$ on κ, then we have

$$\mu\left(r, \frac{\mathbf{W}(f_1, ..., f_k)}{f_1 \cdots f_k}\right) \leq r^{-\frac{k(k-1)}{2}}.$$

Take $\Phi \subset \mathcal{M}(\kappa)$. A meromorphic function $G \not\equiv 0$ is said to be a *universal denominator* of Φ if $G\varphi$ is an entire function for every $\varphi \in \Phi$, and such that for every $z \in \kappa$, there exists $\varphi \in \Phi$ with $(G\varphi)(z) \neq 0$. If Φ is a finite set of $\mathcal{M}(\kappa)$ with $\Phi - \{0\} \neq \emptyset$, then there exists a universal denominator of Φ. Further if $\Phi \cap \kappa_* \neq \emptyset$, universal denominators of Φ are entire functions (cf. [2]).

Let V be a vector space of finite dimension $n+1 > 0$ over κ. Write the projective space $\mathbb{P}(V) = V/\kappa_*$ and let $\mathbb{P} : V - \{0\} \longrightarrow \mathbb{P}(V)$ be the standard projection. If $S \subset V$, abbreviate

$$\mathbb{P}(S) = \mathbb{P}(S \cap V_*), \quad V_* = V - \{0\}.$$

The dual vector space V^* of V consists of all κ-linear functions $\alpha : V \longrightarrow \kappa$, and we shall call

$$\langle \xi, \alpha \rangle = \alpha(\xi)$$

the *inner product* of $\xi \in V$ and $\alpha \in V^*$. If $\alpha \neq 0$, the n-dimensional linear subspace

$$E[a] = E[\alpha] = \mathrm{Ker}(\alpha) = \alpha^{-1}(0)$$

depends only on $a = \mathbb{P}(\alpha) \in \mathbb{P}(V^*)$, and $\ddot{E}[a] = \mathbb{P}(E[a])$ is a *hyperplane* in $\mathbb{P}(V)$. Thus $\mathbb{P}(V^*)$ bijectively parameterizes the hyperplances in $\mathbb{P}(V)$. If $V = \kappa^{n+1}$, we also write

$$\mathbb{P}^n(\kappa) = \mathbb{P}(V), \quad \mathbb{P}^n(\kappa)^* = \mathbb{P}(V^*).$$

Take a base $e = (e_0, ..., e_n)$ for V. For $\xi = \xi_0 e_0 + \cdots + \xi_n e_n \in V$, define the norm

$$|\xi| = \max_{0 \le i \le n} \{|\xi_i|\}.$$

If $V = \kappa^{n+1}$, we will let e be the standard base of κ^{n+1}. Let $\epsilon = (\epsilon_0, ..., \epsilon_n)$ be the dual base of $e = (e_0, ..., e_n)$. Then the norm on V induces a norm on V^* defined by

$$|\alpha| = \max_{0 \le i \le n} \{|\alpha_i|\},$$

where $\alpha = \alpha_0 \epsilon_0 + \cdots + \alpha_n \epsilon_n$. Schwarz inequality

$$|\langle \xi, \alpha \rangle| \le |\xi| \cdot |\alpha|$$

holds for $\xi \in V,\ \alpha \in V^*$. For $x = \mathbb{P}(\xi),\ a = \mathbb{P}(\alpha) \in \mathbb{P}(V^*)$, define

$$0 \leq |x, a| = \frac{|\langle \xi, \alpha \rangle|}{|\xi| \cdot |\alpha|} \leq 1.$$

By a *(non-Archimedean) holomorphic curve*

$$f : \kappa \longrightarrow \mathbb{P}(V),$$

we mean an equivalence class of $(n+1)$-tuples of entire functions

$$(\tilde{f}_0, ..., \tilde{f}_n) : \kappa \longrightarrow \kappa^{n+1}$$

such that $\tilde{f}_0, ..., \tilde{f}_n$ have not common factors in the ring of entire functions on κ and such that not all of the $\tilde{f}_j$ are identically zero. Then

$$\tilde{f} = \tilde{f}_0 e_0 + \cdots + \tilde{f}_n e_n : \kappa \longrightarrow V$$

will be called a *reduced representation of* f. Define

$$\mu(r, \tilde{f}) = \max_{0 \leq k \leq n} \mu(r, \tilde{f}_k).$$

Then the *characteristic function*

$$T(r, f) = \log \mu(r, \tilde{f})$$

is well defined for all $r > 0$, up to $O(1)$.

Let $g : \kappa \longrightarrow \mathbb{P}(V^*)$ be a non-Archimedean holomorphic curve with a reduced representation

$$\tilde{g} = \tilde{g}_0 \epsilon_0 + \cdots + \tilde{g}_n \epsilon_n : \kappa \longrightarrow V^*,$$

where $\epsilon = (\epsilon_0, ..., \epsilon_n)$ is the dual base of e. The pair (f, g) is said to be *free* if

$$\langle \tilde{f}, \tilde{g} \rangle = \tilde{g}_0 \tilde{f}_0 + \cdots + \tilde{g}_n \tilde{f}_n \not\equiv 0.$$

Assume that the pair (f, g) is free. We can define the *proximity function*

$$m_f(r, g) = -\log \frac{\mu(r, \langle \tilde{f}, \tilde{g} \rangle)}{\mu(r, \tilde{f}) \mu(r, \tilde{g})},$$

and write the *valence function*

$$N_f(r,g) = N\left(r, \frac{1}{\langle \tilde{f}, \tilde{g}\rangle}\right).$$

Then the *first main theorem* reads

$$T(r,f) + T(r,g) = N_f(r,g) + m_f(r,g) + O(1). \tag{2}$$

The *defect of f for g* is defined by

$$\delta_f(g) = 1 - \limsup_{r\to\infty} \frac{N_f(r,g)}{T(r,f)+T(r,g)},$$

with $0 \le \delta_f(g) \le 1$. We say that g *grows slower than* f if

$$\lim_{r\to\infty} \frac{T(r,g)}{T(r,f)} = 0.$$

If so, we have

$$\delta_f(g) = 1 - \limsup_{r\to\infty} \frac{N_f(r,g)}{T(r,f)}.$$

In particular, if $g = a$ is a constant, then (2) becomes

$$T(r,f) = N_f(r,a) + m_f(r,a) + O(1), \tag{3}$$

and the defect of f for a is given by

$$\delta_f(a) = \liminf_{r\to\infty} \frac{m_f(r,a)}{T(r,f)} = 1 - \limsup_{r\to\infty} \frac{N_f(r,a)}{T(r,f)}.$$

For this case, the pair (f, a) is free if and only if $f(\kappa) \not\subset \ddot{E}[a]$.

Take non-negative integers a and b with $a \le b$. Let J_a^b be the set of all increasing injective mappings $\lambda : \mathbb{Z}[0,a] \longrightarrow \mathbb{Z}[0,b]$, where

$$\mathbb{Z}[m,n] = \{i \in \mathbb{Z} \mid m \le i \le n\}.$$

Let $f : \kappa \longrightarrow \mathbb{P}(V)$ be a non-constant holomorphic curve with a reduced representation

$$\tilde{f} = \tilde{f}_0 e_0 + \cdots + \tilde{f}_n e_n : \kappa \longrightarrow V_*.$$

Let $\{a_0, a_1, ..., a_q\}$ $(q \geq n)$ be a family of points $a_j \in \mathbb{P}(V^*)$ in general position. Take $\tilde{a}_j \in V^* - \{0\}$ with $\mathbb{P}(\tilde{a}_j) = a_j$. Write

$$\tilde{a}_j = \tilde{a}_{j0}\epsilon_0 + \cdots + \tilde{a}_{jn}\epsilon_n, \quad j = 0, ..., q,$$

where $\epsilon = (\epsilon_0, ..., \epsilon_n)$ is the dual of e. For $i = 0, 1, ..., q$, set

$$F_i = \left\langle \tilde{f}, \tilde{a}_i \right\rangle = \tilde{a}_{i0}\tilde{f}_0 + \tilde{a}_{i1}\tilde{f}_1 + \cdots + \tilde{a}_{in}\tilde{f}_n.$$

For any $\lambda \in J_n^q$, we obtain a non-constant holomorphic curve $\mathbf{f}_\lambda : \kappa \longrightarrow \mathbb{P}(V)$ with a reduced representation

$$\tilde{\mathbf{f}}_\lambda = F_{\lambda(0)}e_0 + \cdots + F_{\lambda(n)}e_n : \kappa \longrightarrow V_*.$$

Obviously, we have

$$T(r, \mathbf{f}_\lambda) = T(r, f) + O(1).$$

Note that

$$T(r, f) = T(r, \mathbf{f}_\lambda) + O(1) = \max_{0 \leq j \leq n} \log \mu(r, F_{\lambda(j)}) + O(1).$$

By using the Jensen formula, we obtain the formula (see [2]):

$$T(r, f) = \max_{0 \leq j \leq n} N_f(r, a_{\lambda(j)}) + O(1), \quad \lambda \in J_n^q. \tag{4}$$

One has the following result:

Theorem 2.1 ([1]) *Let $f : \kappa \longrightarrow \mathbb{P}(V)$ be a non-constant holomorphic curve and let $\{a_0, a_1, ..., a_q\}$ be a family of points $a_j \in \mathbb{P}(V^*)$ in general position such that*

$$\delta_f(a_j) > 0, \quad j = 0, 1, ..., q.$$

Then $q \leq n - 1$.

3 Main result and its proof

Lemma 3.1 *The following subset*

$$\mathcal{E} = \left\{ \left(1, a_1, ..., a_n, a_1a_2, ..., a_1a_n, a_2a_3, ..., a_{n-1}a_n, a_1a_2a_3, ..., \prod_{j=1}^{n} a_j \right) \mid a_j \in \kappa \right\} \tag{5}$$

of $\kappa^{2^n} - \{0\}$ contains vectors of the form $(1, a, a^2, ..., a^{2^n - 1})$ $(a \in \kappa)$ after some rearrangement.

The proof can be completed according to Proposition 2 in Mori [3].

Theorem 3.1 *Let $g : \kappa \longrightarrow \mathbb{P}^n(\kappa)$ be a transcendental holomorphic curve. Assume that there exists some $a \in \mathbb{P}^n(\kappa)^*$ such that*

$$\lim_{r\to\infty} \frac{N_g(r,a)}{T(r,g)} = 1. \tag{6}$$

Then there exists a matrix $L = (l_{ij})$ of the form $l_{ij}(z) = c_{ij}z^{k_i} + d_{ij}$ $(c_{ij}, d_{ij} \in \kappa,\ 0 \le i,j \le n)$ such that $\det(L) \neq 0$ and $f = L \circ g : \kappa \longrightarrow \mathbb{P}^n(\kappa)$ is a linearly non-degenerate holomorphic curve without Nevanlinna deficient hyperplanes. Further, we have

$$T(r,f) = T(r,g) + O(\log r). \tag{7}$$

Proof. There is a linear change $L_1 : \mathbb{P}^n(\kappa) \longrightarrow \mathbb{P}^n(\kappa)$ such that $h = L_1 \circ g : \kappa \longrightarrow \mathbb{P}^n(\kappa)$ has a reduced representation

$$\tilde{h} = (h_0, ..., h_n) : \kappa \longrightarrow \kappa^{n+1} - \{0\}$$

satisfying

$$N\left(r, \frac{1}{h_0}\right) = N_g(r,a).$$

Obviously, we have

$$T(r,h) = T(r,g) + O(1).$$

By the condition (6), it follows that

$$\lim_{r\to\infty} \frac{N\left(r, \frac{1}{h_0}\right)}{T(r,h)} = 1. \tag{8}$$

Now we choose integers $k_0, ..., k_n$ satisfying

$$0 = k_0 < k_1 < k_2 < \cdots < k_n,$$

and define a matrix $L_2 = (\zeta_{ij})_{0\le i,j\le n}$ by

$$\zeta_{ij} = \begin{cases} 1 & : \ i = j \\ 0 & : \ i \neq j,\ j \ge 1 \\ a_i z^{k_i} & : \ i \ge 1,\ j = 0\,. \end{cases}$$

Thus the matrix $L = (l_{ij}) = L_2 \cdot L_1$ is well defined such that the holomorphic curve $f = L \circ g = L_2 \circ h$ is given by the following reduced representation:

$$\tilde{f} = (f_0, ..., f_n) : \kappa \longrightarrow \kappa^{n+1} - \{0\},$$

where

$$f_0 = h_0; \quad f_i = h_i + a_i z^{k_i} h_0 \ (a_i \in \kappa - \{0\},\ i = 1, ..., n).$$

We will choose $a_1, ..., a_n$ in the sequel.

First of all, we prove that f is linearly non-degenerate. Suppose that there are $c_0, ..., c_n$ such that

$$c_0 f_0 + c_1 f_1 + \cdots + c_n f_n = 0,$$

that is

$$c_0 h_0 + c_1 h_1 + \cdots + c_n h_n = -h_0 \sum_{i=1}^{n} c_i a_i z^{k_i}. \tag{9}$$

Write

$$h_i(z) = \sum_{j=0}^{\infty} d_j^i z^j, \quad i = 0, ..., n.$$

Let l be the index of the lowest term of h_0. We compare with the coefficients of both sides of (9). Then we have

$$\begin{array}{lllll} d_l^0 c_0 \ + & d_l^1 c_1 & + \cdots + & d_l^n c_n & = 0 \\ d_{l+k_1}^0 c_0 + & (d_{l+k_1}^1 + a_1 d_l^0) c_1 & + \cdots + & d_{l+k_n}^n c_n & = 0 \\ \cdots & \cdots & \cdots & \cdots & \cdots \\ \multicolumn{5}{l}{d_{l+k_n}^0 c_0 + (d_{l+k_n}^1 + a_1 d_{l+k_n-k_1}^0) c_1 + \cdots + (d_{l+k_n}^n + a_n d_l^0) c_n = 0.} \end{array} \tag{10}$$

If the homogeneous equation (10) has nontrivial solutions $c_0, ..., c_n$, the determinant of the matrix consisting of these coefficients of (10) must be identically zero, namely,

$$P(a_1, .., a_n) := \begin{vmatrix} d_l^0 & d_l^1 & \cdots & d_l^n \\ d_{l+k_1}^0 & d_{l+k_1}^1 + a_1 d_l^0 & \cdots & d_{l+k_n}^n \\ \cdots & \cdots & \cdots & \cdots \\ d_{l+k_n}^0 & d_{l+k_n}^1 + a_1 d_{l+k_n-k_1}^0 & \cdots & d_{l+k_n}^n + a_n d_l^0 \end{vmatrix} = 0. \tag{11}$$

The term including $a_1 a_2 \cdots a_n$ in (11) appears only in the following term

$$d_l^0 (d_{l+k_1}^1 + a_1 d_l^0) \cdots (d_{l+k_n}^n + a_n d_l^0) = (d_l^0)^{n+1} a_1 a_2 \cdots a_n + \cdots + d_l^0 d_{l+k_1}^1 \cdots d_{l+k_n}^n.$$

Hence (11)$\not\equiv 0$ as a function of $a_1 a_2 \cdots a_n$ because $d_l^0 \neq 0$. If we take

$$a_1 = \alpha, \ a_i = \alpha^{m_i}, \ m_1 = 1, \ m_i = 1 + \sum_{j=1}^{i-1} m_j \ (i = 2, ..., n), \tag{12}$$

then (11) is a non-constant polynomial in α of degree $1 + \sum_{j=1}^{n} m_j = 2^n - 1$. Let $\mathcal{S} = \{\alpha_1, ..., \alpha_{2^n-1}\}$ be the solutions of (11). Therefore for $\alpha \in \kappa - \mathcal{S}$, (11) does not equal to zero, that is, (10) does not have a nontrivial solution. Thus $c_0 = c_1 = \cdots = c_n = 0$, i.e., f is linearly non-degenerate. In the sequel, we will assume that $a_1, ..., a_n$ satisfy $P(a_1, ..., a_n) \neq 0$.

Secondly, we prove (7). Since

$$\mu(r, \tilde{f}) = \max_{1 \le j \le n} \{\mu(r, f_0), \mu(r, f_j)\} \le \max_{1 \le j \le n} \{\mu(r, h_0), \mu(r, h_j), |a_j| r^{k_j} \mu(r, h_0)\},$$

we obtain

$$T(r, f) = \log \mu(r, \tilde{f}) \le k_n \log r + T(r, h) + O(1).$$

On the other hand, it follows that

$$\mu(r, \tilde{h}) = \max_{1 \le j \le n} \{\mu(r, h_0), \mu(r, h_j)\} \le \max_{1 \le j \le n} \{\mu(r, f_0), \mu(r, f_j), |a_j| r^{k_j} \mu(r, f_0)\},$$

we obtain

$$T(r, h) = \log \mu(r, \tilde{h}) \le k_n \log r + T(r, f) + O(1).$$

Therefore (7) is proved since $T(r, h) = T(r, g) + O(1)$.

Thirdly, we consider an auxiliary curve. Let $\mathbf{W}(f_0, ..., f_n)$ be the Wronskian of $f_0, ..., f_n$. Since f is linearly non-degenerate, then $\mathbf{W}(f_0, ..., f_n) \not\equiv 0$. Further, we have

$$\begin{aligned} \mathbf{W} \ &:= \ \mathbf{W}(f_0, ..., f_n) = \mathbf{W}(h_0, h_1 + a_1 z^{k_1} h_0, ..., h_n + a_n z^{k_n} h_0) \\ &= \ h_0^{n+1} \mathbf{W}(1, \mathbf{h}_1 + a_1 z^{k_1}, ..., \mathbf{h}_n + a_n z^{k_n}) \\ &= \ h_0^{n+1} \left(\mathbf{W}_0 + a_1 \mathbf{W}_1 + \cdots + \left(\prod_{i=1}^{n} a_i \right) \mathbf{W}_{2^n-1} \right), \end{aligned}$$

where $\mathbf{h}_i = h_i / h_0$ $(i = 1, ..., n)$ and $\mathbf{W}_k$ $(0 \le k \le 2^n - 1)$ is a Wronskian of $n+1$ elements in $\{1, \mathbf{h}_1, z^{k_1}, ..., \mathbf{h}_n, z^{k_n}\}$ by the multi-linearlity of a determinant. Let G be a universal denominator of $\{\mathbf{W}_0, ..., \mathbf{W}_{2^n-1}\}$. Then a holomorphic curve $F : \kappa \longrightarrow \mathbb{P}^{2^n-1}(\kappa)$ is defined with a reduced representation

$$\tilde{F} = (G\mathbf{W}_0, ..., G\mathbf{W}_{2^n-1}).$$

By (8), h_0 is a transcendental entire function, and so it has infinitely many zeros. Since $\tilde{h}$ is reduced, then $\mathbf{h}_{i_0}$ has infinitely many poles for some i_0. Let $\mathbf{W}_{k_0}$ be the Wronskian of $1, z^{k_1}, ..., z^{k_{i_0-1}}, \mathbf{h}_{i_0}, z^{k_{i_0+1}}, ..., z^{k_n}$. Then $\mathbf{W}_{k_0} \not\equiv 0$ and $\mathbf{W}_{k_0}$ does not proportionate to $\mathbf{W}_{2^n-1} = \mathbf{W}(1, z^{k_1}, ..., z^{k_n}) \not\equiv 0$. Thus F is not constant. Since F has at most 2^n Nevanlinna deficient hyperplanes, there exists an $\tilde{A} = (1, a_1, ..., a_n, a_1 a_2, ..., \prod_{i=1}^n a_i) \in \mathcal{E}$ such that

$$\liminf_{r\to\infty} \frac{m_F(r, A)}{T(r, F)} = 0, \tag{13}$$

where $A = \mathbb{P}(\tilde{A}) \in \mathbb{P}^{2^n-1}(\kappa)$. Then we have

$$\langle \tilde{F}, \tilde{A}\rangle = G\left(\mathbf{W}_0 + a_1\mathbf{W}_1 + \cdots + \left(\prod_{i=1}^n a_i\right)\mathbf{W}_{2^n-1}\right) = Gh_0^{-n-1}\mathbf{W}.$$

Since

$$\begin{aligned} \mu(r, \tilde{F}) &= \max_{0\le k\le 2^n-1} \mu(r, G\mathbf{W}_k) = \mu(r, G) \max_{0\le k\le 2^n-1} \mu(r, \mathbf{W}_k) \\ &\le \mu(r, G) r^{-\frac{n(n+1)}{2}} \max_{i_1,...,i_n; 1\le s\le n} \mu(r, \mathbf{h}_{i_1}\cdots\mathbf{h}_{i_s} z^{k_{i_{s+1}}}\cdots z^{k_{i_n}}), \end{aligned}$$

we obtain

$$\begin{aligned} T(r, F) &= \log\mu(r, \tilde{F}) \le \max_{1\le i_1,...,i_s\le n; 1\le s\le n} \log\mu(r, \mathbf{h}_{i_1}\cdots\mathbf{h}_{i_s}) \\ &\quad + \log\mu(r, G) + O(\log r) \\ &= \max_{1\le i_1,...,i_s\le n; 1\le s\le n} \{\log\mu(r, h_{i_1}) + \cdots + \log\mu(r, h_{i_s}) - s\log\mu(r, h_0)\} \\ &\quad + \log\mu(r, G) + O(\log r) \\ &\le nT(r, h) + \log\mu(r, G) + O(\log r) \\ &= nT(r, h) + N\left(r, \frac{1}{G}\right) - N(r, G) + O(\log r). \end{aligned}$$

Note that

$$N\left(r, \frac{1}{G}\right) \le N\left(r, \frac{1}{h_0^{n+1}}\right) < 2(n+1)T(r, h)$$

for r largely. Then

$$T(r, F) \le (3n+2)T(r, f) + O(\log r). \tag{14}$$

Finally, we prove that f has not Nevanlinna deficient hyperplanes. Take an arbitrary vector $\tilde{b} = (b_0, ..., b_n) \in \kappa^{n+1} - \{0\}$, which determines an element $b = \mathbb{P}(\tilde{b})$ in $\mathbb{P}^n(\kappa)^*$. W. l. o. g., we may assume $b_n \neq 0$. Set

$$w = \langle \tilde{f}, \tilde{b}\rangle = \sum_{i=0}^n b_i f_i.$$

Then $f_0, f_1, ..., f_{n-1}, w$ are linearly independent over κ. Thus we have

$$\begin{aligned}
m_f(r,b) &= \log\frac{\mu(r,\tilde{f})|\tilde{b}|}{\mu(r,w)} \\
&= \log\mu\left(r,\frac{\mathbf{W}}{f_0\cdots f_{n-1}w}\right)+\log\left\{\mu(r,\tilde{f})\mu\left(r,\frac{f_0\cdots f_{n-1}}{\mathbf{W}}\right)\right\}+O(1) \\
&\le \log\mu\left(r,\frac{\mathbf{W}(f_0,...,f_{n-1},w)}{f_0\cdots f_{n-1}w}\right)+(n+1)\log\frac{\mu(r,\tilde{f})}{\mu(r,h_0)} \\
&\quad +\log\mu\left(r,\frac{h_0^{n+1}}{\mathbf{W}}\right)+O(1) \\
&\le -\frac{n(n+1)}{2}\log r+(n+1)\left\{T(r,f)-N\left(r,\frac{1}{h_0}\right)\right\} \\
&\quad +\log\mu\left(r,\frac{G}{\langle\tilde{F},\tilde{A}\rangle}\right)+O(1) \\
&= \log\frac{\mu(r,\tilde{F})}{\mu(r,\langle\tilde{F},\tilde{A}\rangle)}-\max_{0\le k\le 2^n-1}\log\mu(r,\mathbf{W}_k)+o(T(r,f)) \\
&\le m_F(r,A)+\min_{0\le k\le 2^n-1}N(r,\mathbf{W}_k)+o(T(r,f)) \\
&= m_F(r,A)+o(T(r,f)).
\end{aligned}$$

By (13) and (14), we obtain

$$\delta_f(b)=\liminf_{r\to\infty}\frac{m_f(r,b)}{T(r,f)}=0.$$

Since b is arbitrary, this completes the proof of Theorem 3.1. □

4 First main theorem for hypersurfaces

Let $\mathcal{J}_m$ be the permutation group on $\mathbb{Z}[1,m]$ and let $\otimes_m V$ be the m-fold tensor product of V. For each $\lambda\in\mathcal{J}_m$, a linear isomorphism $\lambda:\otimes_m V\longrightarrow\otimes_m V$ is uniquely defined by

$$\lambda(\xi_1\otimes\cdots\otimes\xi_m)=\xi_{\lambda^{-1}(1)}\otimes\cdots\otimes\xi_{\lambda^{-1}(m)},\quad \xi_j\in V\ (j=1,...,m).$$

A vector $\xi\in\otimes_m V$ is said to be *symmetric* if $\lambda(\xi)=\xi$ for all $\lambda\in\mathcal{J}_m$. The set of all symmetric vectors in $\otimes_m V$ is a linear subspace of $\otimes_m V$, denoted by $\amalg_m V$, called the *m-fold symmetric tensor product of* V. Then

$$\dim\amalg_m V=\binom{n+m}{m}.$$

The linear mapping

$$S_m = \frac{1}{m!} \sum_{\lambda \in \mathcal{J}_m} \lambda : \otimes_m V \longrightarrow \otimes_m V$$

is called the *symmetrizer of* $\otimes_m V$ with $\mathrm{Im} S_m = \amalg_m V$. If $\xi \in \amalg_m V$ and $\eta \in \amalg_l V$, the *symmetric tensor product*

$$\xi \amalg \eta = \eta \amalg \xi = S_{m+l}(\xi \otimes \eta)$$

is defined. For $\xi_j \in V$ $(j = 1, ..., m)$, let $\xi_1 \amalg \cdots \amalg \xi_m$ be the symmetric tensor product $S_m(\xi_1 \otimes \cdots \otimes \xi_m)$. For $\xi \in V$, let $\xi^{\amalg m}$ be the m-th symmetric tensor power, and define

$$x^{\amalg m} = \mathbb{P}(\xi^{\amalg m})$$

for $x = \mathbb{P}(\xi)$. We can identify $\amalg_m V^* = (\amalg_m V)^*$ by

$$\langle \xi_1 \amalg \cdots \amalg \xi_m, \alpha_1 \amalg \cdots \amalg \alpha_m \rangle = \frac{1}{m!} \sum_{\lambda \in \mathcal{J}_m} \langle \xi_1, \alpha_{\lambda(1)} \rangle \cdots \langle \xi_m, \alpha_{\lambda(m)} \rangle$$

for all $x_j \in V, \alpha_j \in V^*, j = 1, ..., m$.

Let $J_{n,m}$ be the set of all mappings $\lambda : \mathbb{Z}[0, n] \longrightarrow \mathbb{Z}[0, m]$ such that

$$|\lambda| = \lambda(0) + \cdots + \lambda(n) = m.$$

For $\lambda \in J_{n,m}, e = (e_0, ..., e_n) \in V^{n+1}$, define

$$\lambda! = \lambda(0)! \cdots \lambda(n)!, \quad e^{\amalg\lambda} = e_0^{\amalg\lambda(0)} \amalg \cdots \amalg e_n^{\amalg\lambda(n)} \in \amalg_m V.$$

If $e = (e_0, ..., e_n)$ is a base of V, then $\{e^{\amalg\lambda}\}_{\lambda \in J_{n,m}}$ is a base of $\amalg_m V$, and $\{\frac{m!}{\lambda!}\epsilon^{\amalg\lambda}\}_{\lambda \in J_{n,m}}$ is the dual base of $\amalg_m V^*$, where $\epsilon = (\epsilon_0, ..., \epsilon_n)$ is the dual of e. The norm on V induces norms on $\amalg_m V$ and $\amalg_m V^*$ as follows: For $\eta \in \amalg_m V, \beta \in \amalg_m V^*$ with

$$\eta = \sum_{\lambda \in J_{n,m}} \eta_\lambda e^{\amalg\lambda}, \quad \beta = \sum_{\lambda \in J_{n,m}} \beta_\lambda \frac{m!}{\lambda!} \epsilon^{\amalg\lambda},$$

define

$$|\eta| = \max_{\lambda \in J_{n,m}} \{|\eta_\lambda|\}$$

and

$$|\beta| = \max_{\lambda \in J_{n,m}} \{|\beta_\lambda|\}.$$

Note that

$$\xi^{\amalg m} = \sum_{\lambda \in J_{n,m}} \xi_0^{\lambda(0)} \cdots \xi_n^{\lambda(n)} e^{\amalg \lambda}, \quad \alpha^{\amalg m} = \sum_{\lambda \in J_{n,m}} \alpha_0^{\lambda(0)} \cdots \alpha_n^{\lambda(n)} \frac{m!}{\lambda!} \epsilon^{\amalg \lambda},$$

for

$$\xi = \xi_0 e_0 + \cdots + \xi_n e_n \in V, \quad \alpha = \alpha_0 \epsilon_0 + \cdots + \alpha_n \epsilon_n.$$

Then

$$|\xi^{\amalg m}| = |\xi|^m, \quad |\alpha^{\amalg m}| = |\alpha|^m.$$

The holomorphic curve f induces a holomorphic curve

$$f^{\amalg m} : \kappa \longrightarrow \mathbb{P}\left(\amalg_m V\right)$$

with a reduced representation $\tilde{f}^{\amalg m} : \kappa \longrightarrow \amalg_m V$ defined by

$$\tilde{f}^{\amalg m}(z) = \tilde{f}(z)^{\amalg m}, \quad z \in \kappa.$$

Let $g : \kappa \longrightarrow \mathbb{P}(\amalg_m V^*)$ be a non-Archimedean holomorphic curve such that the pair $(f^{\amalg m}, g)$ is free. Then the characteristic function of $f^{\amalg m}$ satisfies

$$T\left(r, f^{\amalg m}\right) = mT(r, f).$$

Applying (2) to $f^{\amalg m}$ and g, we have

$$mT(r, f) + T(r, g) = N_{f^{\amalg m}}(r, g) + m_{f^{\amalg m}}(r, g) + O(1). \tag{15}$$

Define the *defect of f for g* by

$$\delta_f(g) = 1 - \limsup_{r \to \infty} \frac{N_{f^{\amalg m}}(r, g)}{mT(r, f) + T(r, g)}$$

with $0 \le \delta_f(g) \le 1$. If g is a constant a, then (15) becomes

$$mT(r, f) = N_{f^{\amalg m}}(r, a) + m_{f^{\amalg m}}(r, a) + O(1), \tag{16}$$

and the defect of f for a is given by

$$\delta_f(a) = 1 - \limsup_{r \to \infty} \frac{N_{f^{\amalg m}}(r, a)}{mT(r, f)}.$$

5 Elimination of deficient hypersurfaces

Theorem 5.1 *Let $g : \kappa \longrightarrow \mathbb{P}^n(\kappa)$ be a transcendental holomorphic curve. Assume that there exists some $a \in \mathbb{P}^n(\kappa)^*$ such that*

$$\lim_{r\to\infty} \frac{N_g(r,a)}{T(r,g)} = 1. \tag{17}$$

Take non-negative integers $d, k_0, k_1, ..., k_n$ satisfying

$$k_0 = 0,\ dk_{i-1} < k_i\ (i = 1, 2, ..., n).$$

Then there exists a matrix $L = (l_{ij})$ of the form $l_{ij}(z) = c_{ij}z^{k_i} + d_{ij}$ ($c_{ij}, d_{ij} \in \kappa$, $0 \le i, j \le n$) such that $\det(L) \ne 0$ and $f = L \circ g : \kappa \longrightarrow \mathbb{P}^n(\kappa)$ is a linearly non-degenerate holomorphic curve without Nevanlinna deficient hypersurfaces of degree $\le d$. Further, we have

$$T(r,f) = T(r,g) + O(\log r). \tag{18}$$

Proof. Define $L_1, h, \tilde{h}, L_2, L, f$ and $\tilde{f}$ as in the proof of Theorem 3.1. Then f is linearly non-degenerate, and satisfies (18). Take a positive integer m with $1 \le m \le d$. If $m > 1$, set

$$a_1 = \alpha,\ a_i = \alpha^{m(m^{i-1}-1)/(m-1)+1}\ (i = 2, ..., n).$$

According to the proof of Lemma 3 of Mori [4], $f^{\mathrm{II}m}$ is linearly non-degenerate except for finite numbers of α. Let $\mathbf{W}$ be the Wronskian of $\{f_0^{\lambda(0)} \cdots f_n^{\lambda(n)}\}_{\lambda \in J_{n,m}}$. Since $f^{\mathrm{II}m}$ is linearly non-degenerate, then $\mathbf{W} \not\equiv 0$. Further, we have

$$\begin{aligned}
\mathbf{W} &= \mathbf{W}(f_0^m, ..., f_0^{\lambda(0)} \cdots f_n^{\lambda(n)}, ..., f_n^m) \\
&= h_0^{ms}\mathbf{W}(1, ..., (\mathbf{h}_1 + a_1 z^{k_1})^{\lambda(1)} \cdots (\mathbf{h}_n + a_n z^{k_n})^{\lambda(n)}, ..., (\mathbf{h}_n + a_n z^{k_n})^m) \\
&= h_0^{ms}\left(\mathbf{W}_0 + a_1\mathbf{W}_1 + \cdots + \left(\prod_{i=1}^n a_i^m\right)\mathbf{W}_N\right),
\end{aligned}$$

where $\mathbf{h}_i = h_i/h_0$ $(i = 1, ..., n)$ and $\mathbf{W}_k$ $(0 \le k \le N)$ is a Wronskian of s elements in

$$\{\varphi_1 \cdots \varphi_m \mid \varphi_i \in \{1, \mathbf{h}_1, z^{k_1}, ..., \mathbf{h}_n, z^{k_n}\}, i = 1, ..., m\}$$

by the multi-linearlity of a determinant, and where

$$s = \binom{n+m}{m}.$$

Let G be a universal denominator of $\{\mathbf{W}_0, ..., \mathbf{W}_N\}$. Then a holomorphic curve $F : \kappa \longrightarrow \mathbb{P}^N(\kappa)$ is defined with a reduced representation

$$\tilde{F} = (G\mathbf{W}_0, ..., G\mathbf{W}_N).$$

Similar to the proof of Theorem 3.1, we can prove that F is not constant. Since F has at most N Nevanlinna deficient hyperplanes, there exists an $\tilde{A} = (1, a_1, ..., a_n, a_1a_2, ..., \prod_{i=1}^n a_i^m)$ such that

$$\liminf_{r\to\infty} \frac{m_F(r, A)}{T(r, F)} = 0, \tag{19}$$

where $A = \mathbb{P}(\tilde{A}) \in \mathbb{P}^N(\kappa)$. Then we have

$$\langle \tilde{F}, \tilde{A}\rangle = G\left(\mathbf{W}_0 + a_1\mathbf{W}_1 + \cdots + \left(\prod_{i=1}^n a_i^m\right)\mathbf{W}_N\right) = Gh_0^{-ms}\mathbf{W}.$$

According to the proof of Theorem 3.1, we also obtain

$$T(r, F) \leq O(T(r, f)). \tag{20}$$

Finally, we prove that f has not Nevanlinna deficient hypersurfaces of degree m. Take an arbitrary vector $\tilde{b} = (b_0, ..., b_{s-1}) \in \kappa^s - \{0\}$, which determines an element $b = \mathbb{P}(\tilde{b})$ in $\mathbb{P}^{s-1}(\kappa)^*$. W. l. o. g., we may assume $b_{s-1} \neq 0$. Set

$$w = \langle \tilde{f}^{\amalg m}, \tilde{b}\rangle = \sum_{i=0}^{s-1} b_i\tilde{f}_i.$$

Then $\tilde{f}_0, \tilde{f}_1, ..., \tilde{f}_{s-2}, w$ are linearly independent over κ. Thus we have

$$\begin{aligned}
m_{f^{\amalg m}}(r, b) &= \log\frac{\mu(r, \tilde{f}^{\amalg m})|\tilde{b}|}{\mu(r, w)}\\
&= \log\mu\left(r, \frac{\mathbf{W}}{\tilde{f}_0\cdots\tilde{f}_{s-2}w}\right) + \log\left\{\mu(r, \tilde{f})^m\mu\left(r, \frac{\tilde{f}_0\cdots\tilde{f}_{s-2}}{\mathbf{W}}\right)\right\} + O(1)\\
&\leq \log\mu\left(r, \frac{\mathbf{W}(\tilde{f}_0, ..., \tilde{f}_{s-2}, w)}{\tilde{f}_0\cdots\tilde{f}_{s-2}w}\right) + ms\log\frac{\mu(r, \tilde{f})}{\mu(r, h_0)}\\
&\quad + \log\mu\left(r, \frac{h_0^{ms}}{\mathbf{W}}\right) + O(1)\\
&\leq -\frac{s(s-1)}{2}\log r + ms\left\{T(r, f) - N\left(r, \frac{1}{h_0}\right)\right\}\\
&\quad + \log\mu\left(r, \frac{G}{\langle\tilde{F}, \tilde{A}\rangle}\right) + O(1)\\
&= \log\frac{\mu(r, \tilde{F})}{\mu(r, \langle\tilde{F}, \tilde{A}\rangle)} - \max_{0\leq k\leq N}\mu(r, \mathbf{W}_k) + o(T(r, f))\\
&\leq m_F(r, A) + o(T(r, f)).
\end{aligned}$$

By (13) and (14), we obtain

$$\delta_f(b) = \liminf_{r\to\infty} \frac{m_{f\amalg m}(r,b)}{mT(r,f)} = 0.$$

Since b is arbitrary, this completes the proof of Theorem 5.1. □

References

[1] Cherry, W. & Ye, Z., Non-Archimedean Nevanlinna theory in several variables and the Non-Archimedean Nevanlinna inverse problem, Trans. Amer. Math. Soc. 349(1997), 5043-5071.

[2] Hu, P. C. & Yang, C. C., Meromorphic Functions over non-Archimedean Fields, manuscript.

[3] Mori, S., Elimination of defects of meromorphic mappings of $\mathbb{C}^m$ into $\mathbb{P}^n(\mathbb{C})$, Ann. Acad. Sci. Fenn. 24(1999).

[4] Mori, S., Elimination of defects of meromorphic mappings by small deformation, Proc. of US-Japan Seminar of the Delaware Conference, Kluwer Acad. Publ. (1999).

52 DEFICIENCIES OF MEROMORPHIC FUNCTIONS

Xinhou HUA[1]

Institute of Math., Nanjing University
Nanjing 210093, China

mahua@netra.nju.edu.cn

Seiki MORI

Department of Mathematical Sciences, Yamagata University
Yamagata 990-8560,Japan

mori@sci.kj.yamagata-u.ac.jp

1991 Mathematics Subject Classification: 30D35

Abstract

Suppose that f is a meromorphic function and g is an entire function. In this paper, we study the relations between the deficiency of $f(g)$ and the one of f, g.

[1] Supported by NNSF of China and NSF of Jiangsu Province

H.G.W. Begehr et al. (eds.), Proceedings of the Second ISAAC Congress, Vol. 1, 451–457.

1 Introduction

Let f be a meromorphic function. We shall use the following standard symbols in Nevanlinna theory (see Hayman [3]):

$$m(r,f), n(r,f), N(r,f), T(r,f), S(r,f).$$

Let $a(z)$ be a function. The deficiency and Valiron deficiency are defined respectively by

$$\delta(a,f) = 1 - \overline{\lim_{r\to\infty}} \frac{N(r,f=a)}{T(r,f)}$$

and

$$\Delta(a,f) = 1 - \underline{\lim_{r\to\infty}} \frac{N(r,f=a)}{T(r,f)}.$$

Obviously,

$$0 \le \delta(a,f) \le \Delta(a,f) \le 1. \tag{1}$$

Valiron [7] proved the following result.

Theorem 1 *If $f(z)$ is a meromorphic function of finite order μ and lower order λ such that $\mu - \lambda < 1$, then all deficiencies of f are invariant under a change of origin.*

This theorem was extended by [5, 6] to the following cases:

Theorem 2 *Let $f(z)$ be a polynomial of degree n and $g(z)$ be a transcendental entire function. Then the following two conclusions hold:*

(i) If a is a deficient value of g, then $f(a)$ is a deficient value of $f(g)$.

(ii) If $g(z)$ is of finite order μ and lower order λ such that $\mu - \lambda < 1/n$, then for any w_0, it holds that $\delta(w_0, g(f)) = \delta(w_0, g)$.

Examples in Gol'dberg-Belinsky [2] and [6] show respectively that the condition $\mu - \lambda < 1$ in Theorem 1 and the condition $\mu - \lambda < 1/n$ in Theorem 2 are sharp.

Concerning Theorem 2 (i), a natural problem is whether the inverse holds. We shall prove the following result.

Theorem 3 *Let $f(z)$ be a polynomial of degree n and $g(z)$ be a transcendental entire function. If w is a deficient value of $f(g)$, then there exists a solution a of the equation $f(z) - w = 0$ such that a is a Valiron deficient value of g, and $\Delta(a,g) \ge \delta(w, f(g))$. Furthermore, if $\delta(w, f(g)) > 1 - \frac{1}{n}$, then there exists a solution a of the equation $f(z) - w = 0$ such that a is a deficient value of g.*

Remark. From the proof we see that the condition '$\delta(w, f(g)) > 1 - \frac{1}{n}$' can be weakened to '$\delta(w, f(g)) > \frac{n-1}{n^2} \sum_{j=1}^{n} \Delta(a_j, g)$'.

An example in [5] shows that Theorem 2 (i) does not hold if the polynomial f is replaced by a transcendental function f. We shall show that this is the case for a class of meromorphic function f.

Theorem 4 *Let $g(z)$ be a transcendental entire function and let $f(z)$ be a transcendental meromorphic solution of the Riccati equation*

$$f'(z) = a_1(z)f(z)^2 + a_2(z)f(z) + a_3(z), \tag{2}$$

where $a_1(z)$, $a_2(z)$ and $a_3(z)$ are rational functions such that $a_1c^2 + 2a_2c + a_3 \not\equiv 0$ for any constant c. Suppose that $m(r, f(g)) = S(r, f(g))$. Then $f(g)$ has no deficient values, and then any deficient value of f or g can not be a deficient value of $f(g)$.

Remark. Some results about the solutions of the Riccati equation can be found in Laine [4].

Remark. The following example describes some assumptions in our theorem.

Example. Take $f = z \tan z$ and $g(z) = e^z$. Then f satisfies the following Riccati equation

$$f'(z) = a_1(z)f^2(a) + a_2(z)f(z) + a_3(z),$$

where $a_1(z) = 1/z, a_2(z) = 1/z$ and $a_3(z) = z$. It is easy to see that

$$a_1c^2 + 2a_2c + a_3 \not\equiv 0$$

for any $c \in \mathbb{C}$.

Note that

$$f(g(z)) = e^z \tan e^z = \frac{e^z(e^{ie^z} - e^{-ie^z})}{i(e^{ie^z} + e^{-e^z})}.$$

Thus all the poles of $f(g(z))$ come from $e^{ie^z} + e^{-ie^z} = 0$, that is, roots of $e^{2ie^z} = -1$. On the other hand, roots of $\tan e^z = a$ are the roots of equation

$$e^{2ie^z} = \frac{1 + ai}{1 - ai}.$$

Hence for general $a \in \mathbf{C}$,

$$T(r, e^{2ie^z}) = (1 + o(1))N(r, \tan e^z = a) = (1 - o(1))T(r, \tan e^z),$$

and so

$$T(r, f(g)) = (1 + o(1))T(r, e^{2ie^z}) = (1 + o(1))T(r, \tan e^z).$$

Hence we get

$$T(r, f(g)) = (1 + o(1))T(r, e^{2ie^z}) = (1 + o(1))N(r, e^{2ie^z} = -1) =$$
$$(1 + o(1))N(r, f(g)),$$

because e^{2ie^z} is an entire function without zeros and $a(\neq 0)$ points are evenly distributed. Therefore we obtain $m(r, f(g)) = S(r, f(g))$.

Remark. It is possible to change rational coefficients to small functions of f. However, our method does work for this. Since we need to get $T(r, a(g)) = o\{T(r, f(g))\}$ from $T(r, a) = o\{T(r, f)\}$. Unfortunately, this does hold in general. For example, let $a(z) = \exp\exp(z^2)$, $f(z) = \exp\exp(z^2 - z)$, $g(z) = z^2 + z$. Then it is not difficult to deduce that $T(r, a) = o\{T(r, f)\}$, but $T(r, f(g)) = o\{T(r, a(g))\}$.

2 Preliminary results

A meromorphic function $a(z)$ is said to be a small proximity function or a small counting function of f if

$$m(r, a) = S(r, f) \tag{3}$$

or

$$N(r, a) = S(r, f) \tag{4}$$

respectively. Furthermore, $a(z)$ is said to be a small function of f if

$$T(r, a) = S(r, f). \tag{5}$$

Lemma 1 *([1]) Let f be a transcendental meromorphic function and g be a transcendental entire function. Then $T(r, g)/T(r, f(g)) \to 0$ as $r \to \infty$.*

3 Proof of Theorem 3

Let $a_j (j = 1, ..., n)$ be the n-roots of the equation $f(z) - w = 0$. Then we have

$$N\left(r, \frac{1}{f(g) - w}\right) = \sum_{j=1}^{n} N\left(r, \frac{1}{g - a_j}\right).$$

By Hayman [3, p. 54], we have

$$\frac{T(r,f(g))}{T(r,g)} \to n \tag{6}$$

as $r \to \infty$. It is easy to see that

$$N(r, f(g) = w) = \sum_{j=1}^{n} N(r, g = a_j).$$

Deviding both side by $T(r, f(g))$ we get

$$\frac{N(r, f(g) = w)}{T(r, f(g))} = \sum_{j=1}^{n} \frac{N(r, g = a_j)}{T(r, f(g))}. \tag{7}$$

It follows from (6) and (7) that

$$\begin{aligned} \overline{\lim_{r\to\infty}} \frac{N(r, f(g) = w)}{T(r, f(g))} &= \overline{\lim_{r\to\infty}} \left\{ \frac{1}{n} \sum_{j=1}^{n} \frac{N(r, g = a_j)}{T(r,g)} \right\} \\ &\geq \frac{1}{n} \overline{\lim_{r\to\infty}} \frac{N(r, g = a_i)}{T(r,g)} + \frac{1}{n} \sum_{j=1, j\neq i}^{n} \underline{\lim_{r\to\infty}} \frac{N(r, g = a_j)}{T(r,g)}, \end{aligned}$$

where $1 \leq i \leq n$. This implies that

$$1 - \delta(w, f(g)) \geq \frac{1}{n}(1 - \delta(a_i, g)) + \frac{1}{n} \sum_{j=1, j\neq i}^{n} (1 - \Delta(a_j, g)),$$

i.e.,

$$\delta(w, f(g)) \leq \frac{1}{n}\delta(a_i, g) + \frac{1}{n} \sum_{j=1, j\neq i}^{n} \Delta(a_j, g).$$

By summering the i from 1 to n we get

$$n\delta(w, f(g)) \leq \frac{1}{n} \sum_{i=1}^{n} \delta(a_i, g) + \frac{n-1}{n} \sum_{j=1}^{n} \Delta(a_j, g). \tag{8}$$

Combining this with (1) we get

$$n\delta(w, f(g)) \leq \sum_{j=1}^{n} \Delta(a_j, g).$$

Thus there exists some a_j such that $\Delta(a_j, g) \geq \delta(w, f(g))$.

Furthermore, if $\delta(w, f(g)) > 1 - 1/n$, then we rewrite (8) into the form

$$\sum_{i=1}^{n} \delta(a_i, g) \geq n^2 \left(\delta(w, f(g)) - \frac{n-1}{n^2} \sum_{j=1}^{n} \Delta(a_j, g) \right).$$

It is easy to see that the right hand side is positive. So there exists i such that $\delta(a_i, g) > 0$. This proves the theorem.

4 Proof of Theorem 4

For and constant c, set $F(z) = f(g(z)) - c$. Then we have

$$\begin{aligned} F' &= g'f'(g) = g'\left(a_1(g)f(g)^2 + a_2(g)f(g) + a_3(g)\right) \\ &= g'a_1(g)F^2 + g'(2ca_1(g) + a_2(g))F + b(g), \end{aligned}$$

where $b(g) = g'\left(a_1(g)c^2 + a_2(g)c + a_3(g)\right)$. We rewrite it in the form

$$\frac{1}{F} = \frac{1}{b(g)} \left(\frac{F'}{F} - g'a_1(g)F - g'(2ca_1(g) + a_2(g)) \right).$$

Thus, by our assumptions,

$$\begin{aligned} m(r, \frac{1}{F}) &= m\left(r, \frac{1}{b(g)}\right) + m\left(r, \frac{F'}{F}\right) + m(r, F) \\ &\quad + m(r, g'a_1(g)) + m(r, g'(2ca_1(g) + a_2(g))) + O(1) \\ &\leq kT(r, g) + S(r, g) + S(r, F), \end{aligned}$$

where k is a positive integer. Thus, by the Lemma,

$$\begin{aligned} T(r, F) &= N(r, 0, F) + m(r, 1/F) + O(1) \\ &= N(r, 0, F) + kT(r, g) + S(r, g) + S(r, F) \\ &\leq N(r, 0, F) + S(r, F). \end{aligned}$$

This implies that

$$\overline{\lim_{r\to\infty}} \frac{N(r, 0, F)}{T(r, F)} \geq 1.$$

Hence $\delta(0, F) = 0$, and so, $\delta(c, f(g)) = 0$. The conclusion follows.

References

[1] Clunie, J., The composition of entire and meromorphic functions, Mathematical Essays dedicated to A.J. McIntyre, Ohio Univ. Press, Athens, Ohio 1970.

[2] Gol'dberg, A.A. and Belinsky, P.P., Application d'un théorème sur les représentations conforms aux questions d'invariance des fonctions méromorphes, Ukrian.Mat.Zhurnal 3(1954), 263-269.

[3] Hayman, W.K., Meromorphic functions, Clarendon Press, Oxford,1964.

[4] Laine, I., Nevanlinna theory and complex differential equations, Walter de Gruyter, Berlin–New York, 1993.

[5] Mori, S., The deficiencies of a composite function of integral functions, Tohoku Math. J., 23(1971), 719-725.

[6] Mori, S., An example on defect of a composite function, Tohoku Math. J., 24(1972), 269-274.

[7] Valiron, G., Valeurs exceptionelles et valeurs deficientes des fonctions méromorphes, C.R.Acad.Sci.Paris 225(1947), 556-558.

53 AN EXAMPLE OF ANALOGY OF NEVANLINNA THEORY AND DIOPHANTINE APPROXIMATION

Takayuki HYUGA

Graduate School of Mathematics, Nagoya University
Chigusaku, Nagoya, 464-8602, Japan

1 Review of successive minima

1.1 "Usual" case

Let C be a convex, centrally symmetric body in $\mathbf{R}^n$ and let Λ be an n-dimensional lattice. Let $2^n V$ and $d(\Lambda)$ be their respective volumes. Then the successive minima of Λ relative to C are numbers $\lambda_1, \ldots, \lambda_n$ such that λ_j is the smallest real number for which $\lambda_j \cap \Lambda$ has at least j linearly independent elements.

Then Minkowski's first theorem, which states that a symmetric convex body of volume 2^n has at least one nontrivial lattice point, merely states that

$$\lambda^n V \leq d(\Lambda).$$

It is Minkowski's stronger second theorem, however, which is central to the study of successive minima. It states,

$$\frac{d(\Lambda)}{n!} \leq \lambda_1 \lambda_2 \cdots \lambda_n V \leq d(\Lambda).$$

H.G.W. Begehr et al. (eds.), Proceedings of the Second ISAAC Congress, Vol. 1, 459–465.

1.2 Number field case

Let k be a number field. Let M_k be a set of valuation of k, and $S \subset M_k$ be a finite set of valuations containing the set S_∞ of all archimedian valuations. In this case, we define the length function as follows. For each $v \in S$ let $L_{v,i}(1 \le i \le n)$ be n-linearly independent linear forms with coefficients in k , and let $A_{v,i}(1 \le i \le n)$ be positive real numbers with

$$\prod_{i=1}^{n} A_{v,i} = 1$$

for each $v \in S$. Let $d = [k : \mathbf{Q}]$. Then the length function be the following.

$$l(\mathbf{x}) := \left[\prod_{v \in S} \max_{1 \le i \le n} A_{v,i} \|L_{v,i}(\mathbf{x})\|_v\right]^{1/d}$$

Let $\mathcal{O}_{k,S}^n$ be the ring of S-integers of the number field k; i. e.

$$\mathcal{O}_{k,S}^n := \{\mathbf{x} \in k^n ; \|\mathbf{x}\|_v \le 1 \ (\forall v \notin S)\},$$

where $\|\mathbf{x}\|_v := \max_{1 \le i \le n} \|x_i\|_v$.

Relative to this length function, we can define the i-th successive minima of $\mathcal{O}_{k,S}^n$ as the smallest real number λ such that $l(\mathbf{x}) \le \lambda$ for at least i linearly independent points $x \in \mathcal{O}_{k,S}^n$. In the above, the word "linearly independent" mean linearly independent over k, so that there are n successive minima $\lambda_1, \ldots, \lambda_n$.

Then the next thorem holds by using idele argument.

Theorem 1.1 ([V,Thorem 6.1.11]) *In above situation,*

$$\left(\frac{1}{n!}\right)^{r_1} \left(\frac{2^n}{(2n)!}\right)^{r_2} \le \frac{(\lambda_1 \cdots \lambda_n)^d}{\prod_{v \in S} \|\det L_{v,i}\|_v} \le \frac{2^{n(r_1+r_2)}|D_k|^{n/2}}{C(r_1, r_2, n)},$$

where r_1 and r_2 are the number of real and complex places of k, respectively; $r_1 + 2r_2 = d$, D_k is discriminant, and $C(r_1, r_2, n)$ is the volume of the solid in $\mathbf{R}^{nr_1} \times \mathbf{C}^{nr_2}$ given by

$$\{\mathbf{x} \in (x_{v,i}), v \in S_\infty, 1 \le i \le n; \sum_v N_v \max |x_{v,i}| \le d\}$$

where $N_v = 1$ if real or 2 if complex.

In this note we will formula "$\log(\lambda_{i+1}/\lambda_i)$ is small" in terms of $m_D(P)$ and $m_{D^{(j)}}(P^{(j)})$'s, where the existence of such $P^{(j)}$'s is the consequence of the theory of successive minima. On the other hand, the secondary inequality of Theorem1.1 is used to bound $\log \lambda_n$.

2 The lemma on a logarithmic derivative

The argument of this section follows that observation due to R. Kobayashi([K]) and precise proof given by K. Yamanoi([Y]).

Consider a smooth projective variety X and a holomorphic curve $f : \mathbf{C} \to X$. And for simplicity let $D \subset X$ be a smooth divisor defined by the equation $s = 0$, where s is a holomorphic section of $\mathcal{O}_X(D)$. By a Weil function for D, we mean a continuous function $\lambda_D : X - \mathrm{supp} D \to \mathbf{R}$ such that $\lambda_D(P) - \log(1/\|\sigma(P)\|)$ is bounded, where $\|\cdot\|$ is a Hermitian metric of $\mathcal{O}_X(D)$. Suppose $f(\mathbf{C})$ is not contained in D. Take a Weil function λ_D for D and define the proximity function of f with respect to D by

$$m_{f,D}(r) = \int_0^{2\pi} \lambda_D(f(re^{i\theta}))\frac{d\theta}{2\pi} \quad \text{modulo}\, O(1).$$

Let (z, w) be a local coordinate system of TX, where z is a local coordinate system of X and w is chosen to be the components of $(1,0)$-vectors with respect to coordinate vector fields corresponding to z.

Then $D^{(1)}$ is a codimension two subvariety and (if D is smooth) $D^{(1)}$ is defined by two equations $s(z,w)(:= s(z)) = 0$ and $(ds)(z)(w) = 0$. For the holomorphic curve $f : \mathbf{C} \to X$, the proximity function of $f^{(1)}(z) = (f(z), f'(z))$ is defined in terms of $s(f(z), f'(z))(:= s(f(z)))$ (this measures how $f(z)$ is close to D) and $(ds)(f(z))(f'(z))$ (this measures how $f'(z)$ is tangent to D) and consider

$$m_{f^{(1)},D^{(1)}}(r) :=$$

$$-\frac{1}{2}\int_0^{2\pi} \log\{\|s(f(re^{i\theta}))\|^2 + \|(ds)(f(re^{i\theta}))(f'(re^{i\theta}))\|^2\}\frac{d\theta}{2\pi}.$$

(Of course these absolute values should be considered as a Weil function on $\overline{T}X = \mathbf{P}(TX \oplus \mathcal{O}_X)$; i. e. it takes ∞ along the $D^{(1)}$ and its exponential behaves like $-\mathrm{dist}(\cdot, D^{(1)})$).

We put $T^kX, \overline{T}^kX$ ($k \geq 0$ is integer) to be

$$T^kX = T(T^{k-1}X), \quad T^0X = X;$$

$$\overline{T}^kX = \overline{T}(\overline{T}^{k-1}X), \quad \overline{T}^0X = X.$$

For a holomorphic curve $f : \mathbf{C} \to X$, put $f^{(k)} : \mathbf{C} \to T^kX (\subset \overline{T}^kX)$ to be

$$f^{(k)} = (f^{(k-1)})^{(1)}, \quad f^{(0)} = f.$$

Moreover we put

$$D^{(k)} = (D^{(k-1)})^{(1)}, \quad D^{(0)} = D;$$

$$\overline{D^{(k)}} = \overline{(D^{(k-1)})^{(1)}}, \quad \overline{D^{(0)}} = D.$$

Theorem 2.1 ([Y,Theorem 3.4]) *Let X be a smooth projective variety. Let $D \subset X$ be a smooth divisor. Let $f : \mathbf{C} \to X$ be a holomorphic curve such that $f(\mathbf{C}) \not\subset D$. Then*

$$m_{f^{(i)},\overline{D^{(i)}}}(r) \leq m_{f^{(i+1)},\overline{D^{(i+1)}}}(r) + S_{f,E}(r),$$

where $S_{f,E}(r) = O(\log^+ T_{f,E}(r)) + O(\log^+ r) + O(1)$ and E is an ample line bundle.

3 Analogous phenomenon

In the case of multi-dimension, the key arguments are "the lemma on a logarithmic derivative for derived curve" in the proof of second main theorem in Nevanlinna theory and "successive minima" in the proof of Schmidt's subspace theorem in Diophantine approximation. In this section, we will show that both arguments are analogeous phenomenon.

We start from observing the following theorem lying in the arguments of successive minima.

Theorem 3.1 ([V,Thorem 6.4.3]) *Let $\mathbf{b}_0, \ldots, \mathbf{b}_N$ be a vector in $(k^{n+1})^*$ in general position and let $\epsilon > 0$. Then there exists a finite set $\mathcal{S} \subset k^{n+1}$ such that if $\mathbf{x} \in \mathcal{O}_{k,S}^{(n+1)}$ is not in a scalor multiple of some vectors in $\mathcal{S}$, then there exists an $\mathbf{x}' \in \mathcal{O}_{k,S}^{(n+1)}$ such that $\mathbf{x} \wedge \mathbf{x}' \neq 0$ and*

$$\sum_{v \in S} \log \frac{\|(\mathbf{x} \wedge \mathbf{x}') \cdot \mathbf{b}_i\|_v}{\|\mathbf{x}\|_v \|\mathbf{x} \cdot \mathbf{b}_i\|_v} < \epsilon \log H(\mathbf{x})$$

for $0 \leq i \leq n$ for which $\mathbf{x} \cdot \mathbf{b}_i \neq 0$, where $H(\mathbf{x}) = \prod_{v \in M_k} \|\mathbf{x}\|_v$, ($\|\mathbf{x}\|_v := \max_{0 \leq i \leq n} \|x_i\|_v$) and $(\mathbf{x} \wedge \mathbf{x}') \cdot \mathbf{b}_i$ means interior product, i. e. $(\mathbf{x} \wedge \mathbf{x}') \cdot \mathbf{b}_i \in \bigwedge^1 (k^{n+1}) \simeq k^{(n+1)}$ such that, for all dual vectors $\mathbf{z}$,

$$((\mathbf{x} \wedge \mathbf{x}') \cdot \mathbf{b}_i) \cdot \mathbf{z} = ((\mathbf{x} \wedge \mathbf{x}') \cdot (\mathbf{b}_i \wedge \mathbf{z})).$$

If $\mathbf{x} \cdot \mathbf{b}_i = 0$ then $\mathbf{x}' \cdot \mathbf{b}_i = 0$.

We consider a local affine open case. Let $U_0 = (Z_0 \neq 0)$ be a canonical affine open. On $U_0 \simeq k^n$, we have coordinates

$$(z_1, \ldots, z_n) = \left(\frac{Z_1}{Z_0}, \ldots, \frac{Z_n}{Z_0}\right).$$

Let D be a divisor in $\mathbf{P}_n(k)$ defined by

$$D = \{(z_1, \ldots, z_n) \in k^n; s(z) = 0\}$$

on U_0, now we define the "proximity function"

$$m_D(z) = \frac{1}{[k:\mathbf{Q}]} \sum_{v \in S} \log^+ \frac{1}{\|s(z)\|_v}.$$

Then we set

$$\overline{D^{(1)}} = \{(z, w) \in k^n \times k^n; s(z) = 0, \sum_{i=1}^{n} \frac{\partial s}{\partial z_i} w_i = 0\},$$

and define

$$m_{\overline{D^{(1)}}}(z, w) = \frac{1}{[k:\mathbf{Q}]} \sum_{v \in S} \log^+ \min\left\{\frac{1}{\|s(z)\|_v}, \frac{\|w_i\|_v}{\|\sum_{i=1}^{n} \frac{\partial s}{\partial z_i} w_i\|_v}\right\}.$$

Now then, we consider that $\overline{H_i^{(0)}}$ is linear divisor defined by the equation $s_i(z) = b_{0,i} + b_{1,i}z_1 + \cdots + b_{n,i}z_n$ for $z = (z_1, \ldots, z_n)$ on U_0. In this case since $ds_i(z) = b_{1,i}dz_1 + \cdots + b_{n,i}dz_n$, $\overline{H_i^{(1)}}$ is given by zeros of $s_i(z)$ and $ds_i(z)$.

By definition $m_D(z)$, for $P^{(0)} = (\mathbf{p}^{(0)}) = (P_0^{(0)} : \cdots : P_n^{(0)})$,

$$m_{\overline{H_i^{(0)}}}(P^{(0)}) = \frac{1}{[k:\mathbf{Q}]} \sum_{v \in S} \log \frac{\|\mathbf{p}^{(0)}\|_v}{\|\mathbf{p}^{(0)} \cdot \mathbf{b}_i\|_v}.$$

And by definition $m_{\overline{D^{(1)}}}(z, w)$,

$$m_{\overline{H_i^{(1)}}}(P^{(1)}) = \frac{1}{[k:\mathbf{Q}]} \sum_{v \in S} \log \min\left\{\frac{\|\mathbf{p}^{(0)}\|_v}{\|\mathbf{p}^{(0)} \cdot \mathbf{b}_i\|_v}, \frac{\|\mathbf{p}^{(1)}\|_v}{\|\mathbf{p}^{(1)} \cdot \mathbf{b}_i\|_v}\right\},$$

and also

$$m_{\overline{H_i^{(j)}}}(P^{(j)}) = \frac{1}{[k:\mathbf{Q}]} \sum_{v \in S} \log \min_{0 \le j \le n} \left\{\frac{\|\mathbf{p}^{(j)}\|_v}{\|\mathbf{p}^{(j)} \cdot \mathbf{b}_i\|_v}\right\}$$

Theorem 3.2 *Put* $m_{\overline{H_i^{(j)}}}(P^{(j)})$ *to be above. We can take* $P^{(j+1)}$ *such that* $P^{(j+1)}$ *satisfies the inequality*

$$m_{\overline{H_i^{(j)}}}(P^{(j)}) \leq m_{\overline{H_i^{(j+1)}}}(P^{(j+1)}) + \epsilon h(P) \quad (0 \leq j \leq n-1)$$

like as Thorem 2.1. Then either there exists a finite set S_j *of linear subspace such that for some* $P^{(j)}$ *lying in the subspace of* S_j, *or*

$$m_{\overline{H_i^{(j)}}}(P^{(j)}) < m_{\overline{H_i^{(j+1)}}}(P^{(j+1)}) + \epsilon h(P) \quad (0 \leq j \leq n-1),$$

where $h(P) = (1/[k:\mathbf{Q}]) \log H(P)$.

proof. Since $\|(\mathbf{x} \wedge \mathbf{x}') \cdot \mathbf{b}_i\|_v \gg\ll \max(\|\mathbf{x}\|_v \|\mathbf{x}' \cdot \mathbf{b}_i\|_v, \|\mathbf{x}'\|_v \|\mathbf{x} \cdot \mathbf{b}_i\|_v)$ holds (by[V,p.106]), where $f \gg\ll g$ means that both $f < cg$ and $g < c'f$ hold for some positive constants c, c',

$$\sum_{v \in S} \log \frac{\|(\mathbf{x} \wedge \mathbf{x}') \cdot \mathbf{b}_i\|_v}{\|\mathbf{x}\|_v \|\mathbf{x} \cdot \mathbf{b}_i\|_v} = \sum_{v \in S} \log \frac{\max(\|\mathbf{x}\|_v \|\mathbf{x}' \cdot \mathbf{b}_i\|_v, \|\mathbf{x}'\|_v \|\mathbf{x} \cdot \mathbf{b}_i\|_v)}{\|\mathbf{x}\|_v \|\mathbf{x} \cdot \mathbf{b}_i\|_v}.$$

If the first term in the above max is larger, then this right hand is

$$\sum_{v \in S} \log \frac{\|\mathbf{x}' \cdot \mathbf{b}_i\|_v}{\|\mathbf{x} \cdot \mathbf{b}_i\|_v}.$$

Hence by Theorem3.1,

$$\sum_{v \in S} \log \frac{\|\mathbf{x}' \cdot \mathbf{b}_i\|_v}{\|\mathbf{x} \cdot \mathbf{b}_i\|_v} < \epsilon \log H(\mathbf{x}).$$

Since $\|\mathbf{x}\|_v < c\|\mathbf{x}'\|_v$ holds for some constant c (by[V,p.106]),

$$\sum_{v \in S} \log \frac{\|\mathbf{x}\|_v}{\|\mathbf{x} \cdot \mathbf{b}_i\|_v} < \sum_{v \in S} \log \frac{\|\mathbf{x}'\|_v}{\|\mathbf{x}' \cdot \mathbf{b}_i\|_v} + \epsilon \log H(\mathbf{x}).$$

Here $\mathbf{x}'$ is only determined modulo $\mathbf{x}$, so it is reasonable that $\mathbf{x}$ and $\mathbf{x}'$ is taken as the vectors become first successive minima and second successive minima.

By the definition of $m_{\overline{H_i^{(j)}}}(P^{(j)})$ $(j = 0, 1)$,

$$m_{\overline{H_i^{(0)}}}(P^{(0)}) < m_{\overline{H_i^{(1)}}}(P^{(1)}) + \epsilon h(P)$$

Otherwise

$$\sum_{v \in S} \log \frac{\|\mathbf{x}'\|_v}{\|\mathbf{x}\|_v} < \epsilon \log H(\mathbf{x})$$

is trivial inequality.

By recursively applying the above argument on j, we get main theorem.

4 Conclusion

The property "Theorem2.1"

$$m_{f,D}(r) \leq m_{f^{(1)},D^{(1)}}(r) + S_f(r)$$

in Nevanlinna theory should be translated into algebraic setting (Diophantine approximation) as

$$m_D(P) \leq m_{D^{(1)}}(P^{(1)}) + \epsilon h_E(P) \tag{1}$$

where E is ample.

However, in algebraic setting, there is no a priori definition of $P^{(1)} \in \overline{T}X$. Then the conclusion of this note is that the theory of successive minima serves as an existence thorem for such $P^{(1)} \in \overline{T}X$ satisfying the property (1).

References

[K] R. Kobayashi, *Nevanlinna thory and number theory*, Sugaku **48**(1996), Iwanami Shoten, 113-127 (in Japanese).

[L] S. Lang, *Survey of Diophantine Geometry*, Springer, 1997.

[V] P. Vojta, *Diophantine Approximations and Value Distribution Theory*, Lecture Notes in Mathematics 1239, Springer-Verlag, 1987.

[Y] K. Yamanoi, *The second main theorem type inequality for holomorphic curves in abelian varieties*, preprint, 1999.

54 UNIQUENESS PROBLEMS ON MEROMORPHIC FUNCTIONS WHICH SHARE FOUR SMALL FUNCTIONS[1]

Katsuya ISHIZAKI

Department of Mathematics, Nippon Institute of Technology
4-1, gakuendai, Miyashiro
4-1 Minamisaitama, Saitama 345-8501, Japan

ishi@nit.ac.jp

1991 Mathematics Subject Classification: 30D35
Keywords: Meromorphic function, Sharing problem, Small functio

1. Introduction

Throughout this note the term "meromorphic" means meromorphic in the whole complex plane, and we shall use the standard notation of the Nevanlinna theory of meromorphic functions, see e.g. [3], [7]. In particular, $S(r, f)$ denote any quantity that satisfies $S(r, f) = o(1)T(r, f)$ as $r \to \infty$, possibly outside a set of r of finite linear measure for a transcendental meromorphic function f. A meromorphic function a is called "small" with respect to f if $T(r, a) = S(r, f)$ holds. We say that two transcendental meromorphic functions f and g share a small function a provided that $f(z) - a(z) = 0$ if and only if $g(z) - a(z) = 0$ in the case $a(z) \not\equiv \infty$. We also consider the case $a(z) \equiv \infty$. If every pole of f is

[1]Research partially supported by the Grant-in-Aid for General Scientific Research from the Ministry of Education, Science and Culture 11640164 (Toda).

H.G.W. Begehr et al. (eds.), Proceedings of the Second ISAAC Congress, Vol. 1, 467–472.

a pole of g and vice versa, we call f and g share ∞. For the sake of simplicity, we call z_0 is an a-point of f if $f(z_0)-a(z_0)=0$, and the multiplicity of a-point is derived from the multiplicity of zero of $f(z)-a(z)$. We state that f and g share a small function a CM (counting multiplicities), if a is shared by f and g and if a k-fold a-point z_0 of f is also a k-fold a-point of g, $k=k(z_0)$. In contract to sharing CM we also use the notation sharing IM (ignoring multiplicities).

We denote by $\overline{n}(r, f=a=g)$ the number of z_0 satisfying $f(z_0)-a(z_0)=0$ and $g(z_0)-a(z_0)=0$ in $|z|\le r$, which are counted only once ignoring multiplicities. Integrated counting function $\overline{N}(r, f=a=g)$ is defined in a usual manner.

Nevanlinna [10] proved that *if two meromorphic functions f and g share four values a_1, a_2, a_3, a_4 CM, then f is a Möbius transformation of g, two of the shared values, say a_1, a_2 are Picard values, and the cross ratio $(a_1, a_2, a_3, a_4)=-1$.* The condition f and g share four values a_1, a_2, a_3, a_4 IM does not implies that f is a Möbius transformation of g in general. The examples are given in e.g. [1], [11]. In this note we are concerned with the following question.

Question 1. Let f and g denote transcendental meromorphic functions sharing four distinct small functions a_1, a_2, a_3, a_4 with respect to f and g. Under what additive conditions we can say that f is expressed by a linear fractional transformation of g with small coefficients?

Suppose that transcendental meromorphic functions f and g share four distinct small functions a_1, a_2, a_3, a_4 with respect to f and g. By means of the Nevanlinna second fundamental theorem, we see that $T(r,f)/T(r,g)=O(1)$ and $T(r,g)/T(r,f)=O(1)$, as $r\to\infty$, $r\in\mathbb{R}^+\setminus E$, where E is an exceptional set with finite linear measure. This implies that $\varphi(r)=S(r,f)$ if and only if $\varphi(r)=S(r,g)$. Recently, Li and Qiao [9] proved that *if f and g share other small function a_5, then $f=g$.* This result asserts that Nevanlinna's five values theorem [10] is extended to small functions. The author [5] gave a slight generalization of Li and Qiao's result, i.e. *if there exists other small function a_5 such that $\overline{N}(r, f=a_5=g)\ne S(r,f)$, then $f=g$.* Toda and the author [6] showed that *if for other small function a_5 it holds that $\overline{N}\left(r,\frac{1}{f-a_5}\right)\le \frac{1}{19}T(r,f)+S(r,f)$ and $\overline{N}\left(r,\frac{1}{g-a_5}\right)\le\frac{1}{19}T(r,f)+S(r,f)$, then $f=g$.* We first mention the following theorem.

Theorem 1. *Let f and g denote transcendental meromorphic functions sharing four distinct small functions a_1, a_2, a_3, a_4 with respect to f and g. Then*

$$\overline{N}_{\mathrm{MDM}}(r, f=a_k=g)=S(r,f), \quad \text{for all } k\in\{1,2,3,4\}, \tag{1}$$

where $\overline{N}_{\mathrm{MDM}}(r, f = a_k = g)$ refers only to those a_k-points that are multiple for both of f and g but different multiplicities "count" them once each.

This theorem says that two transcendental meromorphic functions f and g cannot share four distinct small functions if f and g satisfy $\overline{N}_{\mathrm{MDM}}(r, f = b = g) \neq S(r, f)$ for a small function b. Proof of Theorem 1 is given in Section 2. We continue to mention other additive conditions which gives an answer for Question 1. *If two transcendental meromorphic functions f and g share four distinct small functions a_1, a_2, a_3, a_4 CM, then f is expressed by a linear fractional transformation of g whose coefficients are small functions with respect to f and g.* It is proved as follows. We may assume that $a_1(z) \equiv 0$, $a_2(z) \equiv \infty$, $a_3(z) \equiv 1$. If need, we define a linear fractional transformation L

$$
\begin{aligned}
L(x) &= \left(\frac{x - a_1(z)}{x - a_2(z)}\right) \Big/ \left(\frac{a_3(z) - a_1(z)}{a_3(z) - a_2(z)}\right), \quad a_2(z) \not\equiv \infty, \\
&= \frac{x - a_1(z)}{a_3(z) - a_1(z)}, \quad a_2(z) \equiv \infty.
\end{aligned}
$$

Put $F = L(f)$ and $G = L(g)$. Then we see that F and G share 0, ∞, 1 and $a(z) = L(a_4(z))$. We assume that a is not a constant and suppose that f can not be expressed by any linear fractional transformations, including the case $f = g$, having small coefficients. Let γ_0 and γ_1 be the functions defined by

$$\gamma_0 = \frac{f'}{f} - \frac{g'}{g} \quad \text{and} \quad \gamma_1 = \frac{f'}{f-1} - \frac{g'}{g-1}. \tag{2}$$

By means of the theorem on the logarithmic derivatives, we have $m(r, \gamma_j) = S(r, f)$, $j = 0, 1$. Since 0, ∞ are shared by CM, we get $N(r, \gamma_0) = S(r, f)$. This implies that γ_0 is a small function with respect to f and g. Similarly γ_1 is a small function with respect to f and g, since ∞, 1 are shared by CM. Since we assume that $f \neq g$, it follows from (2),

$$f' = \frac{f(f-1)((\gamma_0 - \gamma_1)g - \gamma_1)}{f - g}, \tag{3}$$

$$g' = \frac{g(g-1)((\gamma_0 - \gamma_1)f - \gamma_1)}{f - g}. \tag{4}$$

We also have that

$$\gamma_a = \frac{f' - a'}{f - a} - \frac{g' - a'}{g - a} \tag{5}$$

is a small function. In fact, by the theorem on the logarithmic derivatives, we see that $m(r, \gamma_a) = S(r, f)$. Since f and g share ∞, a CM, we have $N(r, \gamma_a) = S(r, f)$. Combining (3), (4) and (5), we get

$$(\gamma - 2a\gamma - \gamma_a)fg + (a^2\gamma + a\gamma_a + a')f + (a^2\gamma + a\gamma_a - a')g - a^2(\gamma + \gamma_a) = 0, \tag{6}$$

where $\gamma = \gamma_0 - \gamma_1$. We see that all coefficients of f, g and fg and constant term must vanish, otherwise f is expressed by a linear fractional transformation having small coefficients from (6). Observing the coefficients of f and g in (6), we obtain that $a' = 0$, a contradiction.

We mention some investigations when two transcendental meromorphic functions f and g share three distinct small functions a_1, a_2, a_3 CM and other small function a_4 IM. Li and Yang [8], and Hua and Fang [4] proved the following theorem.

Theorem A. *If two transcendental meromorphic functions f and g share three distinct small functions CM and other one small function IM, then f is expressed by a linear fractional transformation of g whose coefficients are small functions with respect to f and g.*

In Section 3, as an application of Theorem 1, we will give a proof of Theorem A which is slightly different from proofs in [8], [4].

2. Proof of Theorem 1

We may assume that $\overline{N}_{\mathrm{MDM}}(r, f = a_2 = g) \neq S(r, f)$. It follows from this $f \not\equiv g$. Using a linear fractional transformation L as stated in Section 1, we have that F and G share 0, ∞, 1 and $a(z) = L(a_4(z))$, where $F = L(f)$ and $G = L(g)$. By the definitions of F and G, we see that $\varphi(r) = S(r, f)$ if and only if $\varphi(r) = S(r, F)$, and $\varphi(r) = S(r, g)$ if and only if $\varphi(r) = S(r, G)$.

$$\overline{N}_{\mathrm{MDM}}(r, F = \infty = G) \neq S(r, F). \tag{7}$$

The condition F and G share 0, ∞, 1 and $a(z)$ gives that the function $\Psi(z)$ defined below is a small function, see [5],

$$\Psi = \frac{(F-G)\begin{vmatrix} F^2 & G^2 & a^2 & 1 \\ FF' & GG' & aa' & 0 \\ F & G & a & 1 \\ F' & G' & a' & 0 \end{vmatrix}}{F(F-1)(F-a)G(G-1)(G-a)}. \tag{8}$$

Let z_∞ be a common multiple pole of F and G that is neither a zero nor a pole of a and a'. Denote by p and q orders of poles at z_∞ of f and g, respectively. We may assume that $2 \leq p < q$ without loss of generality. The order of the numerator of the right-hand side of (8) at z_∞ is at most $q + (2p + 2q + 1) = 2p + 3q + 1$. On the other hand, the denominator has a pole of order $3p + 3q$ at z_∞. This concludes that z_∞ is a zero of Ψ of multiplicity at least $p - 1 \geq 1$. If it would hold that $\Psi(z) \not\equiv 0$, then by (7) we have

$$\overline{N}_{\mathrm{MDM}}(r, f = a_2 = g) = \overline{N}_{\mathrm{MDM}}(r, F = \infty = G) + S(r, F) \leq$$

$$N(r, \frac{1}{\Psi}) + S(r, F) = S(r, F),$$

a contradiction. Thus we have that $\Psi(z) \equiv 0$, and hence $F = G$ or the determinant in the right-hand side of (8) vanishes. It is obvious that $F = G$ implies $f = g$, a contradiction. Therefore we shall treat the latter case, assuming $F \not\equiv G$. We get from the condition that the the determinant vanishes

$$\frac{F'}{F(F-1)}(a'(F-a) + a(a-1)\frac{G'-a'}{G-a}) = a'\frac{F'-G'}{F-G}. \tag{9}$$

We assert that a is not a constant. If we assume the contrary then the determinant reduces to $a(1-a)(F-G)F'G'$, a contradiction. We see from (7) that there exists a common pole of z_∞ multiple for both of F and G which is neither zero, 1-point, nor a-point of F and G satisfying $a'(z_\infty) \neq 0$ and $a(z_\infty) \neq \infty$. Denote by p and q orders of poles at z_∞ of F and G, assuming that $2 \leq p < q$ without loss of generality. Both sides of (9) have poles at z_∞. The right-hand side of (9) has a simple pole at z_∞ with residue $-qa'(z_\infty)$. On the other hand the left-hand side of (9) also has a simple pole at z_∞, however the residue is $-pa'(z_\infty)$, a contradiction. We have thus proved Theorem 1.

3. Proof of Theorem A

We may assume that f and g share $0, \infty, 1$ CM and share a IM, where a is a nonconstant small function. It suffices to show that γ_a defined in (5) is a small function with respect to f and g, see Section 1. If $f = g$ then we have nothing to prove. Hence we suppose that $f \not\equiv g$, and hence from our assumption γ_0, γ_1 given in (2), (3) are small functions with respect to f and g, and (4) holds. In the case $\overline{N}(r, f = a = g) = S(r, f)$, we see that (5) holds. In fact, by the theorem on the logarithmic derivatives, we get that $m(r, \gamma_a) = S(r, f)$. Since f and g share ∞ CM and $\overline{N}(r, f = a = g) = S(r, f)$, we obtain $N(r, \gamma_a) = S(r, f)$. We consider the case $\overline{N}(r, f = a = g) \neq S(r, f)$. Since $f - g$ has zeros at common a-points of f and g, from (2) or (3) that $\gamma := \gamma_0 - \gamma_1 \not\equiv 0$ and $a = \gamma_1/\gamma$. Hence we can write (3) and (4) as

$$f' = \gamma\frac{f(f-1)(g-a)}{f-g} \quad \text{and} \quad g' = \gamma\frac{g(g-1)(f-a)}{f-g}. \tag{10}$$

From - Mon Nov 01 12:07:45 1999

$$(f-a)(g-a) = a(a-1) + \frac{f'-g'}{\gamma}. \tag{11}$$

Let z_a be a common a-points of f and g, and neither zero nor pole of a and γ. Denote by p and q orders of a-points of f and g at z_a, respectively. We may assume that $1 \leq p \leq q$ without loss of generality. By means of Theorem 1, almost all a-points have the same multiplicities when both of them are multiple a-points. Thus we may only treat the case $p = 1$. Assume that $q \geq 2$. We write $f - a$ and $g - a$ in a neighborhood of z_a,

$$f(z) - a(z) = c_1(z - z_a) + c_2(z - z_a)^2 + O(z - z_a)^3, \quad c_1 \neq 0, \tag{12}$$

$$g(z) - a(z) = d_q(z - z_a)^q + O(z - z_a)^{q+1}, \quad d_q \neq 0. \tag{13}$$

In view of (10), we see that $f'(z_a) = 0$. From (12), we have $a'(z_a) = c_1$. It follows from (12) and (13) that

$$\begin{aligned} f'(z) - g'(z) &= c_1 + 2(c_2 - d_q)(z - z_a) + O(z - z_a)^3, \quad q = 2 \\ &= c_1 + 2c_2(z - z_a) + O(z - z_a)^3, \quad q \geq 3. \end{aligned}$$

We observe the both sides of (11). The left-hand side has a zero at z_a. Hence we get $a(z_a)(a(z_a) - 1) - c_1/\gamma(z_a) = a(z_a)(a(z_a) - 1) - a'(z_a)/\gamma(z_a) = 0$. This implies that such z_a must be a zero of a small function $a(a-1) - a'/\gamma$. Therefore we conclude that almost all a-points are shared CM by f and g, which gives that $N(r, \gamma_a) = S(r, f)$. Obviously we have $m(r, \gamma_a) = S(r, f)$. Thus Theorem A is proved.

Acknowledgment The author would like to thank Professor Niro Yanagihara and Professor Tsuneo Sato for their careful reading of his manuscript.

References

[1] Gundersen, G G., Meromorphic functions that share four values, Trans. Amer. Math. Soc. 277 (1983), 545–567. Correction: 304 (1987), 847–850.

[2] ———, Meromorphic functions that share three values IM and a fourth value CM, Complex Variables 20 (1992), 99–106.

[3] Hayman, W. K., Meromorphic functions. Oxford at the Clarendon Press, 1964.

[4] Hua, X.-H. and M.-L. Fang, Meromorphic functions that share four small functions, Indian J. Pure Appl. Math. 28 (1997), 797–811.

[5] Ishizaki, K., Meromorphic functions sharing small functions, Preprint.

[6] Ishizaki, K. and N. Toda, Unicity theorems for meromorphic functions sharing four small functions, Kodai Math. J. 21 (1998), 350–371.

[7] Laine, I., Nevanlinna theory and complex differential equations, W.Gruyter, Berlin–New York, 1992.

[8] Li, P. and C.-C. Yang, On two meromorphic functions that share pairs of small functions, Complex Variables 32 (1997), 177–190.

[9] Li, Y.-H. and J.-Y. Qiao, On the uniqueness of meromorphic functions concerning small functions, (Chinese) Preprint 1998.
Complex Variables 12 (1989), 169–179.

[10] Nevanlinna, R., Einige Eindueutigkeitssätze in der Theorie der mermorphen Funktionen, Acta Math. 48 (1926), 367–391.

[11] Reinders, M., A new characterization of Gundersen's example of two meromorphic functions sharing four values, Results in Math. 24 (1/2) (1993), 174–179.

55 A SPACE OF MEROMORPHIC MAPPINGS AND DEFECTS OF A MEROMORPHIC MAPPING INTO $P^n(C)$

Seiki MORI

Department of Mathematical Sciences, Faculty of Science
Yamagata University, Yamagata, 990-8560 Japan

mori@ sci. kj. yamagata-u. ac. jp

1 Introduction

Nevanlinna defect relations ware established for various cases, for example, holomorphic (or meromorphic) mappings of $\mathbf{C}^m$ into a complex projective space $\mathbf{P}^n(\mathbf{C})$ for constant or moving targets of hyperplanes (arbitrary $m \geq 1$ and $n \geq 1$), or holomorphic mappings of an affine variety A of dimension m into a projective algebraic variety V of dimension n for divisors on V ($m \geq n \geq 1$), and so on. On the other hand, the size of a set of (Valiron) deficient hyperplanes or deficient divisors are investigated. (e.g. Mori[1], Sadullaev [4].) Nevanlinna theory asserts that for each holomorphic (or meromorphic) mapping, Nevanlinna defects or Valiron defects of a mapping are very few. Recently the author [1], [2]

This research was partially supported by Grant-in-Aid for Scientific Research (No.10640149),Ministry of Education, Science and Culture of Japan.

H.G.W. Begehr et al. (eds.), Proceedings of the Second ISAAC Congress, Vol. 1, 473–478.

and [3] proved that for a transcendental meromorphic mapping (or holomorphic curve) f of $\mathbf{C}^{\mathrm{m}}$ (or $m=1$) into $\mathbf{P}^n(\mathbf{C})$, we can eliminate all deficient hyperplanes, hypersurfaces of degree at most a given integer d or rational moving targets in $\mathbf{P}^n(\mathbf{C})$ by a small deformation of the mapping. Here a samll deformation $\tilde{f}$ of f means that their order functions $T_f(r)$ and $T_{\tilde{f}}(r)$ satisfy $|T_f(r)-T_{\tilde{f}}(r)| \leq o(T_f(r))$, as r tends to infinity. Therefore,it seems to me that mappings with positive deficiencies are very few in a space of meromorphic mappings into $\mathbf{P}^n(\mathbf{C})$.

We shall now discuss about an elimination theorems and also discuss that mappings without Nevanlinna defects are dense in a space of meromorphic mappings into $\mathbf{P}^n(\mathbf{C})$.

2 Preliminaries

2-1.Notation and Terminology

Let $z=(z_1,...,z_m)$ be the natural coordinate system in $\mathbf{C}^m$. Set

$$\langle z, \xi\rangle = \sum_{j=1}^{m} z_j\xi_j \quad \text{for} \quad \xi=(\xi_1,...,\xi_m), \qquad \|z\|^2 = \langle z, \overline{z}\rangle,$$

$$B(r)=\{z\in\mathbf{C}^m \mid \|z\|<r\}, \partial B(r)=\{z\in\mathbf{C}^m \mid \|z\|=r\},$$
$$\psi = dd^c\log\|z\|^2 \quad \text{and} \quad \sigma = d^c\log\|z\|^2\wedge\psi^{m-1},$$

where $d^c = \frac{\sqrt{-1}}{4\pi}(\overline{\partial}-\partial)$, and $\psi^k=\psi\wedge\cdots\wedge\psi$ (k-times).

Let f be a meromorphic mapping of $\mathbf{C}^m$ into $\mathbf{P}^n(\mathbf{C})$. Then f has a reduced representation $(f_0 : ... : f_n)$, where $f_0,...,f_n$ are holomorphic functions on $\mathbf{C}^m$ with $\mathrm{codim}_{\mathbf{C}}\{z\in\mathbf{C}^m \mid f_0(z)=\cdots=f_n(z)=0\}\geq 2$. We write $f=(f_0,...,f_n)$ as the same letter of the meromorphic mapping f.

The proximity function $m_f(r,H)$ and the counting function $N_f(r,H)$ of a hyperplane H in $\mathbf{P}^n(\mathbf{C})$ are given by:

$$m_f(r,H) := \int_{\partial B(r)} \log\frac{\|f\|\|\mathbf{a}\|}{|\langle f,\mathbf{a}\rangle|}\sigma, \quad \text{and}$$

$$N_f(r,H) := \int_{r_0}^{r}\frac{dt}{t}\int_{(f^*H)\cap B(t)}\psi^{m-1},$$

for some fixed $r_0>0$, where $H=\{w=(w_0,...,w_n)\in\mathbf{C}^{n+1}\setminus\{0\} \mid \sum_{j=0}^{n}a_jw_j=0\}$,

$\mathbf{a} = (a_0, ..., a_n) \in \mathbf{C}^{n+1} \setminus \{0\}$ and f^*H denotes the pullback of H under f. The Nevanlinna's order function $T_f(r)$ of f is given by:

$$T_f(r) := \int_{r_0}^{r} \frac{dt}{t} \int_{B(t)} f^*\omega \wedge \psi^{m-1},$$

where $\omega = \{\omega_\alpha\} = dd^c \log \sum_{j=0}^{n} (|w_j/w_\alpha|^2)$, in a neighborhood $U_\alpha := \{w_\alpha \neq 0\}$. We write

$$N(r, (\phi)) := \int_{r_0}^{r} \frac{dt}{t} \int_{(\phi)\cap B(t)} \psi^{m-1},$$

where (ϕ) denotes the divisor determined by a meromorphic function ϕ on $\mathbf{C}^m$. We observe that

$$T_f(r) = \int_{\partial B(r)} \log(\sum_{j=0}^{n} |f_j|^2)^{1/2}\sigma + O(1) = \int_{\partial B(r)} \log \sum_{j=0}^{n} |f_j|\sigma + O(1).$$

The Nevanlinna deficiency $\delta_f(H)$ and the Valiron deficiency $\Delta_f(H)$ of a hyperplane H for f are given by:

$$\delta_f(H) := \liminf_{r\to+\infty} \frac{m_f(r,H)}{T_f(r)} = 1 - \limsup_{r\to+\infty} \frac{N_f(r,H)}{T_f(r)}$$

and

$$\Delta_f(H) := \limsup_{r\to+\infty} \frac{m_f(r,H)}{T_f(r)} = 1 - \liminf_{r\to+\infty} \frac{N_f(r,H)}{T_f(r)}.$$

3 Elimination of defects of a meromorphic mapping of $\mathbf{C}^m$ into $\mathbf{P}^n(\mathbf{C})$.

For a meromorphic mapping g of $\mathbf{C}^m$ into $\mathbf{P}^n(\mathbf{C})$, we can eliminate all deficient hyperplanes by its samll deformation.

Theorem 1.[1] Let $g : \mathbf{C}^m \to \mathbf{P}^n(\mathbf{C})$ be a given transcendental meromorphic mapping. Then there exists a regular matrix $L = (l_{ij})_{0\leq i,j\leq n}$ of the form $l_{i,j} = c_{ij}\zeta_i + d_{ij}$, $(c_{ij}, d_{ij} \in \mathbf{C} :\ 0 \leq i, j \leq n)$, such that $\det L \neq 0$ and $f := L \cdot g : \mathbf{C}^m \to \mathbf{P}^n(\mathbf{C})$ is a meromorphic mapping without Nevanlinna deficient hyperplanes, where $\zeta_1, ..., \zeta_n$ are some monomials in $z_1, ..., z_m$ which are linearly independent over $\mathbf{C}$.

Here the mapping $f := L \cdot g : \mathbf{C}^m \to \mathbf{P}^n(C)$ means a product the matrix $L = (l_{ij})$ and a vector of a reduced representation $\tilde{g} =^t (g_0 : ... : g_n)$ of g which does not depend on a choice of $\tilde{g}$, and also a Nevanlinna deficient hyperplane H for f means a hyperplane with $\delta_f(H) > 0$.

Next we consider an elimination theorem for hypersurfaces in $\mathbf{P}^n(\mathbf{C})$. A hypersurface D of degree d in $\mathbf{P}^n(\mathbf{C})$ is given by a homogeneous polynomial $P(w)$ of degree d. The proximity function $m_g(r, D)$ of D and the Nevanlinna deficiency $\delta_g(D)$ of D for g are defined by

$$m_g(r, D) := \int_{\partial B} \log \frac{\|g\|^d}{|P(g)|}\sigma \quad \text{and} \quad \delta_g(D) := \liminf_{r\to\infty} \frac{m_g(r, D)}{dT_g(r)}.$$

Then we have the following theorem:

Theorem 2.[2] Let g be a given transcendental holomorphic mapping of $\mathbf{C}$ into $\mathbf{P}^n(\mathbf{C})$, and d be a given positive integer. Then there exists a regular matrix $L = (l_{ij})$ of the form $l_{ij} = c_{ij}z^{m_j} + d_{ij}, (c_{ij}, d_{ij} \in \mathbf{C}), |L| \neq 0$ and $f := L \cdot g : \mathbf{C} \rightarrow \mathbf{P}^n(\mathbf{C})$ is a holomorphic mapping without Nevanlinna deficient hypersurfaces of degree $\leq d$, where m_j (j=1,...,n) are some integers such that $m_1 < dm_1 < m_2 < \cdots < dm_{n-1} < m_n$.

Note that Theorem 2 can be extend to the case where meromorphic mappings of $\mathbf{C}^m$ into $\mathbf{P}^n(\mathbf{C})$ by using the similar method to Theorem 1.

4 Elimination of defects of holomorphic curves for rational moving targets.

We now discuss an elimination theorem of a holomorphic curve for rational moving targets. Let $f : \mathbf{C} \rightarrow \mathbf{P}^n(\mathbf{C})$ be a holomorphic curve with a reduced representation $(f_0, ..., f_n)$. Let $\phi : \mathbf{C} \rightarrow \mathbf{P}^n(\mathbf{C})^*$ be a holomorphic curve with a reduced representation $(\phi_0, ..., \phi_n)$ which is called a moving target for f. Then the proximity function $m_f(r, \phi)$ and the counting function $N_f(r, \phi)$ of a moving target ϕ into $\mathbf{P}^n(\mathbf{C})^*$ are given by:

$$m_f(r, \phi) := \frac{1}{2\pi}\int_0^{2\pi} \log \frac{\|f\|\|\phi\|}{|\langle f, \phi\rangle|}(re^{i\theta})d\theta \quad \text{and} \quad N_f(r, \phi) := \int_{r_0}^{r} n(t, A)\frac{dt}{t},$$

where $n(t, A)$ denotes the number of zeros of $A := \langle f, \phi\rangle = \sum_{j=0}^{n} \phi_j f_j$ in $\{z \in \mathbf{C} | |z| < t\}$ counting with multiplicities. Then the Nevanlinna deficiency $\delta_f(\phi)$ and the Valiron deficiency $\Delta_f(\phi)$ of a moving target ϕ for f are given by the same manner as in the case of constant targets. Then we have:

Theorem 3.[3] Let $f : \mathbf{C} \rightarrow \mathbf{P}^n(\mathbf{C})$ be a given transcendental holomorphic curve. Then there exists a regular matrix $L = (l_{ij})_{0\leq i,j\leq n}$, of the form $l_{i,j} = c_{ij}g_j + d_{ij}$, $(c_{ij}, d_{ij} \in \mathbf{C} : 0 \leq i, j \leq n)$, such that $\det L \neq 0$ and $\tilde{f} = L \cdot f : \mathbf{C} \rightarrow \mathbf{P}^n(\mathbf{C})$ is a holomorphic curve without Nevanlinna defects of rational moving targets and satisfies $|T_f(r) - T_{\tilde{f}}(r)| = o(T_f(r)), (r \rightarrow \infty)$, where g_j (j=1,...,n) are some transcendental entire functions satisfying $T_{g_j}(r) = o(T_{g_{j+1}}(r))$, (j=1,...,n-1) and $T_{g_n} = o(T_f(r)), (r \rightarrow \infty)$.

Note that we cannot replace transcendental eitire functions g_j by any rational functions.

5 Space of meromorphic mappings into $\mathbf{P}^n(\mathbf{C})$

We shall induce a distance on the space of meromorphic mappings into $\mathbf{P}^n(\mathbf{C})$.
For points $\mathbf{a} = (a_0 : ... : a_n)$ and $\mathbf{b} = (b_0 : ... : b_n)$ in $\mathbf{P}^n(\mathbf{C})$, we define the distance $d(\mathbf{a}, \mathbf{b})$ by

$$d(\mathbf{a}, \mathbf{b}) := \| \frac{\mathbf{a}}{\|\mathbf{a}\|} - \frac{\mathbf{b}}{\|\mathbf{a}\|} \|.$$

Then $d(\mathbf{a}, \mathbf{b})$ satisfies the condition of a distance.
Let $f = (f_0 : ... : f_n)$ and $g = (g_0 : ... : g_n)$ be reduced representations of meromorphic mappings of $\mathbf{C}^m$ into $\mathbf{P}^n(\mathbf{C})$. Then we define the distance $d(f(z), g(z))$ at $z \in \mathbf{C}^m$ as

$$d(f(z), g(z)) := \| \frac{f(z)}{\|f(z)\|} - \frac{g(z)}{\|g(z)\|} \| \leq 2.$$

and define the distance $d(f, g)$ as $d(f, g) := d_1(f, g) + d_2(f, g)$, where

$$d_1(f, g) := \sum_{n=1}^{\infty} \frac{1}{2^{n+1}} \int_n^{n+1} dt \int_{\partial B(t)} \| \frac{f(z)}{\|f(z)\|} - \frac{g(z)}{\|g(z)\|} \| \sigma \leq 1,$$

which is a distance and does not distinguish mappings of rational and transcendental, and

$$d_2(f, g) := \liminf_{\alpha \rightarrow +1} \limsup_{r \rightarrow \infty}$$

$$\{ | \frac{T_f(r)}{(\log r)^\alpha + T_f(r)} - \frac{T_g(r)}{(\log r)^\alpha + T_g(r)} | + \sum_{n=1}^{\infty} | \frac{T_f(r)}{r^n + T_f(r)} - \frac{T_g(r)}{r^n + T_g(r)} | \}.$$

which is a pseudodistance and distinguish mappings of rational and transcendental. Then $d(f, g)$ satisfies distance conditions on the space of meromorphic mappings into $\mathbf{P}^n(\mathbf{C})$.

Note that, if f is rational, $d_1(f,0) < 1$ and $d_2(f,0) = 0$. Hence $d(f,0) < 1$. If f is transcendental, $d_1(f,0) < 1$, while $d_2(f,0) \geq 1$. Hence $d(f,0) \geq 1$. Here 0 means a representation $(1,0,...,0)$. Therefore we can distinguish mappings of rational and transcendental by this distance.

Now,we consider a space of meromorphic mappings,

$$\mathcal{F} := \{f : \mathbf{C}^m \rightarrow \mathbf{P}^n(\mathbf{C}) | \text{ f is meromorphic}\}.$$

In Theorem 1, 2 and 3, a small deformation $\tilde{f} := L_2 \cdot h$ of f is represented as

$$\tilde{f} = (h_0 : h_1 + a_1\zeta_1 h_0 :, ..., : h_n + a_n\zeta_n h_0),$$

where $h = (h_0 : \cdots : h_n) := L_1 \cdot f$. Also, we can choose $(a_1, ..., a_n)$ such that $\|\mathbf{a}\| := |a_1| + \cdots + |a_n|$ is as small as possible. So,we can choose $\hat{f} := L_1^{-1} \cdot \tilde{f}$ which is a small deformation without Nevanlinna defects of f such that $d(\hat{f}, f)$ is as small as possible. Hence meromorphic mappings without Nevanlinna defects are dense in $\mathcal{F}$.

References

[1] S. Mori, *Elimination of defects of meromorphic mappings of* $\mathbf{C}^m$ *into* $\mathbf{P}^n(\mathbf{C})$, Ann. Acad. Sci. Fenn. vol. 24, (1999),89-104.

[2] S. Mori, *Elimination of defects of meromorphic mappings by small deformation,* Proc. of US-Japan Seminar of the Delaware Conference, Kluwer Acad. Publ. (1999).

[3] S. Mori, *Defects of holomorphic curves into* $\mathbf{P}^n(\mathbf{C})$ *for rational moving targets and a space of meromorphic mappings,* to appear.

[4] A. Sadullaev, *Deficient divisors in Valiron sense,* Math. USSR. Sb. 36(4), (1980), 535 - 547.

56 ON PICARD CONSTANTS OF ALGEBROID RIEMANN SURFACES

Kazunari SAWADA

Tokyo Metropolitan College of Technology
1-10-40 Higasi-Oi, Shinagawa, Tokyo, Japan

sawada@tokyo-tmct.ac.jp

1 Introduction

1.1 Picard constants

Let $\mathcal{M}(\mathcal{R})$ be the family of non–constant meromorphic functions on a Riemann surface $\mathcal{R}$. Let $p(f)$ be the cardinal number of values which are not taken by $f \in \mathcal{M}(\mathcal{R})$. Then we put

$$P(\mathcal{R}) = \sup_{f \in \mathcal{M}(\mathcal{R})} p(f),$$

which is called the Picard constant of $\mathcal{R}$. We can prove that $P(\mathcal{R}) \geq 2$ if $\mathcal{R}$ isopen and $P(\mathcal{R}) = 0$ if $\mathcal{R}$ is compact. Picard constant plays a very important role in the theory of analytic mappings of Riemann surfaces. Indeed, in 1965, Ozawa[8] proved the following result : If $P(\mathcal{X}) < P(\mathcal{R})$, then there exists no non–trivial analytic mapping of $\mathcal{X}$ into $\mathcal{R}$.

H.G.W. Begehr et al. (eds.), Proceedings of the Second ISAAC Congress, Vol. 1, 479–487.

1.2 algebroid Riemann surfaces

An n-sheeted algebroid Riemann surface is the proper existence domain of an n-valued algebroid function, which is defined by the following irreducible equation:

$$S_0(z)y^n - S_1(z)y^{n-1} + \cdots + (-1)^{n-1}S_{n-1}(z)y + (-1)^n S_n(z) = 0,$$

where $S_i(z)$ $(i = 0, 1, \cdots, n)$ are entire functions on $\mathbf{C}$ with no common zeros. An algebroid function y is called transcendental if at least one of $S_i(z)/S_0(z)$ $(i = 1, 2, \cdots, n)$ is transcendental and y is called entire if all the $S_i(z)/S_0(z)$ $(i = 1, 2, \cdots, n)$ are entire. If $\mathcal{R}$ is an n-sheeted algebroid surface, then $P(\mathcal{R}) \leq 2n$ by Selberg's theory of algebroid functions[18]. However it is very difficult in general to calculate $P(\mathcal{R})$ of a given open Riemann surface $\mathcal{R}$, even an algebroid surface. So we shall construct algebroid surfaces with a given Picard constant.

1.3 Aogai's result and some problems

An n-sheeted algebroid surface is called regularly branched if all its branch points are of order $n-1$. In 1973 Aogai proved the following

Theorem A (Aogai[1]) [1] *Let $\mathcal{R}$ be an n-sheeted regularly branched algebroid surface. If $P(\mathcal{R}) > 3n/2$, then $P(\mathcal{R}) = 2n$ and $\mathcal{R}$ can be defined by an algebroid function y such that*

$$y^n = \left(e^{H(z)} - \alpha\right)\left(e^{H(z)} - \beta\right)^{n-1}, \quad H(0) = 0, \ \alpha\beta(\alpha - \beta) \neq 0,$$

where $H(z)$ is a non-constant entire function and α and β are constants.

By Theorem A, it is easy to verify that there exists no 3-sheeted regularly branched surface with $P(\mathcal{R}) = 5$. Naturally we have the following question: Does there exist an 3-sheeted algebroid surface of Picard constant 5 ? And what can we say about the existence of non-trivial analytic mappings under the condition of $P(\mathcal{X}) \geq P(\mathcal{R})$?

In this paper we shall construct 3-sheeted algebroid surfaces of Picard constant 5. And we shall investigate the existence of non-trivial analytic mappings from $\mathcal{X}$ into $\mathcal{R}$ under the condition[2] of $P(\mathcal{X}) = 6$ and $P(\mathcal{R}) = 5$. And we shall introduce some open problems.

---

[1] Ozawa[9] and Hiromi–Niino[5] proved the above result in the case of $n = 2$ and $n = 3$ respectively.

[2] Ozawa[10] proved some interesting results in the case of $P(\mathcal{X}) = P(\mathcal{R}) = 6$.

2 Picard constants of 3-sheeted algebroid Riemann surfaces

In this section we shall construct 3-sheeted algebroid Riemann surfaces, defined by 3-valued entire algebroid functions, with $p(y) = 6, 5$ respectively.

2.1 algebroid surfaces with $p(y) = 6$

In 1994, Ozawa and the author proved the following

Theorem 1 (Ozawa–Sawada[11]) *Let $\mathcal{X}$ be a 3-sheeted algebroid Riemann surface defined by*

$$y^3 - U_1 y^2 + U_2 y - U_3 = 0.$$

If $p(y) = 6$, then we have

$$U_1(z) = x_0 e^{L(z)} + x_1, U_2(z) = b_1 x_0 e^{L(z)} + x_2, U_3(z) = x_3, \qquad \mathbf{x-1} \ (1)$$

where b_1 $(\neq 0)$, x_0 $(\neq 0)$, x_1, x_2, x_3 $(\neq 0)$ are constants and $L(z)$ is an entire function with $L(0) = 0$. And its discriminant $D_{\mathcal{X}}$ is

$$D_{\mathcal{X}} = -{b_1}^2 {x_0}^4 e^{4L} + \eta_3 {x_0}^3 e^{3L} + \eta_2 {x_0}^2 e^{2L} + \eta_1 x_0 e^L + \eta_0, \qquad (2)$$

where

$$\eta_3 = 4{b_1}^3 - 2{b_1}^2 x_1 - 2b_1 x_2 + 4x_3,$$

$$\eta_2 = 12x_1x_3 - 18b_1x_3 - {x_2}^2 - 4b_1x_1x_2 + 12{b_1}^2x_2 - {b_1}^2{x_1}^2,$$

$$\eta_1 = 12{x_1}^2x_3 - 18b_1x_1x_3 - 18x_2x_3 - 2x_1{x_2}^2 + 12b_1{x_2}^2 - 2b_1{x_1}^2x_2,$$

$$\eta_0 = 4{x_1}^3x_3 - {x_1}^2{x_2}^2 + 27{x_3}^2 - 18x_1x_2x_3 + 4{x_2}^3 \ (\neq 0). \qquad \mathbf{x-3} \ (3)$$

It is easy to verify that the surface $\mathcal{X}$ is of Picard constant 6.

2.2 algebroid surfaces with $p(y) = 5$

In 1994, Ozawa and the author proved the following

Theorem 2 (Ozawa–Sawada[11]) *Let $\mathcal{R}$ be a 3-sheeted algebroid surface defined by*

$$y^3 - S_1 y^2 + S_2 y - S_3 = 0.$$

We have $p(y) = 5$ if and only if $\mathcal{R}$ is defined by one of the following three cases:
Case (A)

$$S_1(z) = y_1, S_2(z) = y_0 e^{H(z)} + y_2, S_3(z) = y_3, \qquad \mathbf{a-1} \qquad (4)$$

where $y_0\ (\neq 0), y_1, y_2, y_3\ (\neq 0)$ are constants and $H(z)$ is a non–constant entire function with $H(0) = 0$. We denote this surface by $\mathcal{R}_A$. Furthermore its discriminant $D_{\mathcal{R}_A}$ is

$$D_{\mathcal{R}_A} = 4{y_0}^3 e^{3H} + \zeta_2 {y_0}^2 e^{2H} + \zeta_1 y_0 e^H + \zeta_0, \qquad (5)$$

where

$$\zeta_0 = 4{y_1}^3 y_3 - {y_1}^2 {y_2}^2 - 18 y_1 y_2 y_3 + 4{y_2}^3 + 27{y_3}^2\ (\neq 0),$$

$$\zeta_1 = 12{y_2}^2 - 18 y_1 y_3 - 2{y_1}^2 y_2, \zeta_2 = 12 y_2 - {y_1}^2. \qquad \mathbf{a-3} \qquad (6)$$

Case (B)

$$S_1(z) = y_0 e^{H(z)} + y_1, S_2(z) = y_2, S_3(z) = y_3, \qquad \mathbf{b-1} \qquad (7)$$

where $y_0\ (\neq 0), y_1, y_2, y_3\ (\neq 0)$ are constants and $H(z)$ is a non–constant entire function with $H(0) = 0$. We denote this surface by $\mathcal{R}_B$. Its discriminant $D_{\mathcal{R}_B}$ is

$$D_{\mathcal{R}_B} = 4 y_3 {y_0}^3 e^{3H} + \zeta_2 {y_0}^2 e^{2H} + \zeta_1 y_0 e^H + \zeta_0, \qquad (8)$$

where

$$\zeta_0 = 4{y_1}^3 y_3 - {y_1}^2 {y_2}^2 - 18 y_1 y_2 y_3 + 4{y_2}^3 + 27{y_3}^2\ (\neq 0),$$

$$\zeta_1 = 12{y_1}^2 y_3 - 2 y_1 {y_2}^2 - 18 y_2 y_3,$$

$$\zeta_2 = 12 y_1 y_3 - {y_2}^2. \qquad \mathbf{b-3} \qquad (9)$$

Case (G)

$$S_1(z) = y_0 e^{H(z)} + a_3, S_2(z) = y_1 \cdot y_0 e^{H(z)}, S_3(z) = y_2 \cdot y_0 e^{H(z)}, \qquad \mathbf{g-1} \qquad (10)$$

where $a_3\ (\neq 0), y_0\ (\neq 0), y_1, y_2\ (\neq 0)$ are constants $H(z)$ is a non-constant entire function with $H(0) = 0$. We denote this surface by $\mathcal{R}_G$. Its discriminant $D_{\mathcal{R}_G}$ is

$$D_{\mathcal{R}_G} = y_0 e^H \left(\zeta_3 {y_0}^3 e^{3H} + \zeta_2 {y_0}^2 e^{2H} + \zeta_1 y_0 e^H + \zeta_0\right), \qquad (11)$$

where

$$\zeta_3 = 4y_2 - {y_1}^2 \ (\neq 0), \quad \zeta_0 = 4y_2{a_3}^3 \ (\neq 0),$$

$$\zeta_1 = \left(12y_2 - {y_1}^2\right){a_3}^2 - 18y_1y_2a_3 + 27{y_2}^2,$$

$$\zeta_2 = \left(12y_2 - 2{y_1}^2\right)a_3 + 4{y_1}^3 - 18y_1y_2. \qquad \mathbf{g-3} \qquad (12)$$

In the next section we shall verify that $P(\mathcal{R}) = 5$.

2.3 Picard constants of 3-sheeted surfaces with $p(y) = 5$

Let us suppose that $P(\mathcal{R}) = 6$, then we have the following relation between $D_{\mathcal{R}}$ and $D_{\mathcal{X}}$ (see [11]):

$$D_{\mathcal{X}} = D_{\mathcal{R}}G^2, \qquad * \qquad (13)$$

where

$$G = {f_1}^3 + 2S_1{f_1}^2f_2 + \left({S_1}^2 + S_2\right)f_1{f_2}^2 + (S_1S_2 - S_3){f_2}^3,$$

S_i $(i = 1, 2, 3)$ are coefficients of defining equation of $\mathcal{R}$ and f_i $(i = 1, 2)$ are meromorphic functions on $\mathbf{C}$.

From (*) and Borel's unicity theorem (Niino-Tohge[7]), in 1995, Tohge and the author proved the following result;

Theorem 3 (Sawada–Tohge[17]) [3] *Let $\mathcal{R}$ be the surface defined by Theorem2 with $p(y) = 5$. If $(\zeta_1,\ \zeta_2) \neq (0,\ 0)$, then $P(\mathcal{R}) = 5$.*

To prove Theorem 3, Borel's unicity theorem played a very important role. However the condition $(\zeta_1,\ \zeta_2) \neq (0,\ 0)$ was unremovable in 1995. In the case of $\zeta_1 = \zeta_2 = 0$, we can also verify that $P(\mathcal{R}) = 5$. In fact we can prove the following

Theorem 4 (Sawada[16]) *Let $\mathcal{R}$ be the surface defined by Theorem 2 with $p(y) = 5$. Then we have $P(\mathcal{R}) = 5$ with no condition.*

To prove Theorem 4, the following result by Picard, plays a very important role:

Picard's theorem (Picard[14]) *If the curve $\varphi(x, w) = 0$ is of genus $g > 1$, then there exists no pair of meromorphic functions $x(z)$ and $w(z)$ such that $\varphi(x(z), w(z)) \equiv 0$.*

---

[3] Ozawa-Sawada[11] proved the above result under the condition that the surfaces $\mathcal{R}$ are of finite order.

In fact, let us assume that $P(\mathcal{R}) = 6$, then we can deduce that there exists a single valued meromorphic function f, satisfying the following equation(see [16]):

$$A(e^J)f^3 + B(e^J)f + C(e^J) = 0, \qquad ** \tag{14}$$

where $A(e^J)$, $B(e^J)$ and $C(e^J)$ are suitable polynomials with respect to only one exponential function e^J and J is a non-constant entire function. And let us put $x = e^J$ and $w = f(z)$, we have

$$\varphi(x, w) := A(x)w^3 + B(x)w + C(x) = 0.$$

Furthermore it is easy to prove that the curve $\varphi(x, w) = 0$ is of genus $g > 1$. By Picard's theorem, we can verify that there exists no single valued function f, satisfying [**]. This is absurd. Hence we have $P(\mathcal{R}) = 5$.

By the similar way of proving thorem[Theorem 4], we can investigate the existence of non-trivial analytic mappings between two algebroid Riemann surfaces.

3 Analytic mappings between two algebroid Riemann surfaces

Let $\mathcal{X}_n$ $(n \geq 2)$ and $\mathcal{R}_m$ $(m \geq 2)$ be algebroid surfaces defined by the following irreducible equations:

$$A_0 y^n + A_1(z)y^{n-1} + \cdots + A_n(z) = 0,$$

$$B_0 u^m + B_1(w)u^{m-1} + \cdots + B_m(w) = 0,$$

respectively, where $A_0, \cdots, A_n, B_0, \cdots, B_m$ are entire functions. And let $\mathcal{P}_{\mathcal{X}_n} : \mathcal{X}_n \to z$-plane and $\mathcal{P}_{\mathcal{R}_m} : \mathcal{R}_m \to w$-plane be the projection mappings. Now let $\varphi : \mathcal{X}_n \to \mathcal{R}_m$ be an analytic mapping, then its projection

$$h(z) := \mathcal{P}_{\mathcal{R}_m} \circ \varphi \circ \mathcal{P}_{\mathcal{X}_n}{}^{-1}(z)$$

is, in general, an n-valued algebroid function of z. An analytic mapping φ is called rigid if $h(z)$ is single valued.

In 1967, Hiromi and Mutō proved the following

Theorem B **(Hiromi-Mutō[4])** *If there is a non-trivial analytic mapping of $\mathcal{X}_n$ into $\mathcal{R}_m$, then its projection h reduces to a k-valued algebroid function of z, where k is a proper divisor of n. Furthermore if n is prime, then φ is rigid.*

3.1 Analytic mappings $\mathcal{X}$ into $\mathcal{R}$

In this section we shall cnsider the case of $\mathcal{X}_n = \mathcal{X}$ and $\mathcal{R}_m = \mathcal{R}$, defined by Theorem 1 and Theorem 2 respectively. In this case if there exists a non-trivial analytic mapping φ of $\mathcal{X}$ into $\mathcal{R}$, then φ is rigid. By the similar way of proving [*] we can prove the following relation:

$$D_{\mathcal{R}} \circ h = D_{\mathcal{X}} G^2,$$

where $D_{\mathcal{X}}$ and $D_{\mathcal{R}}$ are discriminants of $\mathcal{X}$ and $\mathcal{R}$, respectively, h is the projection of φ,

$$G = {f_1}^3 + 2U_1 {f_1}^2 f_2 + \left({U_1}^2 + U_2\right) f_1 {f_2}^2 + (U_1 U_2 - U_3) {f_2}^3,$$

U_i $(i = 1, 2, 3)$ are coefficients of defining equation of $\mathcal{X}$ and f_i $(i = 1, 2)$ are meromorphic functions on **C**. If the counting function of zeros of G is relatively small, by a similar way of proving theorem[Theorem 4], we can prove the following

Theorem 5 *Let $\mathcal{X}$ and $\mathcal{R}$ be the surfaces defined by theorem 1 and 2, respectively. If*

$$N_0\left(r, 0, D_{\mathcal{R}}\right) = S\left(r, e^H\right),$$

then there exists no non-trivial analytic mapping of $\mathcal{X}$ into $\mathcal{R}$, where $N_0\left(r, 0, D_{\mathcal{R}}\right)$ is the counting function of multiple zeros of $D_{\mathcal{R}}$.

4 Some open problems

At the end of this paper, we introduce some open problems:

1. In the case of $N_0\left(r, 0, D_{\mathcal{R}}\right) \neq S\left(r, e^H\right)$, what can we say about analytic mappings of $\mathcal{X}$ into $\mathcal{R}$?

2. If n is not prime and $P(\mathcal{X}_n) \geq P(\mathcal{R}_m)$, what can we say about non-trivial analytic mappings of $\mathcal{X}_n$ into $\mathcal{R}_m$ and its projection h ?

   For example how about $n = m = 4$?

References

[1] H. Aogai, *Picard constant of a finitely sheeted covering surface*, Kōdai Math. Sem. Rep. **25**(1973), 219–224.

[2] W. K. Hayman, *Meromorphic functions*, Clarendon Press, London, 1964

[3] G. Hiromi and H. Mutō, *On the existence of analytic mappings, I*, Kōdai Math. Sem. Rep. **19**(1967), 236–244.

[4] G. Hiromi and H. Mutō, *On the existence of analytic mappings, II*, Kōdai Math. Sem. Rep. **19**(1967), 439–450.

[5] G. Hiromi and K. Niino, *On a characterization of regularly branched three–sheeted covering Riemann surfaces*, Kōdai Math. Sem. Rep. **17**(1965), 250–260.

[6] G. Hiromi and M. Ozawa, *On the existence of analytic mappings between two ultrahyperelliptic surfaces*, Kōdai Math. Sem. Rep. **17**(1965), 281–306.

[7] K. Niino and K. Tohge, *Some functional equations and Picard constants of algebroid surfaces*, J. Math. Soc. Japan **48**(1996), no. 4, 649–665.

[8] M. Ozawa, *On complex analytic mappings*, Kōdai Math. Sem. Rep. **17**(1965), 93–102.

[9] M. Ozawa, *On ultrahyperelliptic surfaces*, Kōdai Math. Sem. Rep. **17**(1965), 103–108.

[10] M. Ozawa, *Analytic mappings between three-sheeted algebroid surfaces of maximal Picard constant*, in preparation

[11] M. Ozawa and K. Sawada, *Three–sheeted algebroid surfaces whose Picard constants are five*, Kōdai Math. J. **17**(1994), no.1, 101–124.

[12] M. Ozawa and K. Sawada, *Picard constants of four–sheeted algebroid surfaces, I*, Kōdai Math. J. **18**(1995), no. 1, 99-141.

[13] M. Ozawa and K. Sawada, *Picard constants of four–sheeted algebroid surfaces, II*, Kōdai Math. J. **18**(1995), no. 2, 199-233.

[14] E. Picard, *Démonstration d'un théorème général des fonctions uniformes liées par une relation algébrique*, Acta math. **11**(1887), 1-12.

[15] G. Rémoundos, *Extension aux fonctions algébroïdes multiformes du théorème de M. Picard et de ses généralisations*, Mém. Sci. Math. **23**, Gauthier-Villars, Paris, 1927.

[16] K. Sawada, *Picard constants of three-sheeted algebroid surfaces with* $p(y) = 5$, preprint

[17] K. Sawada and K. Tohge, *A remark on three–sheeted algebroid surfaces whose Picard constants are five*, Kōdai Math. J. **18**(1995), no. 1, 142-155.

[18] H. L. Selberg, *Algebroide Funktionen und Umkehrfunktionen Abelscher Integrale*, Avh. Norske Vid. Akad. Oslo **8**(1934), 1-72.

57 ON DEFICIENCIES AND RAMIFICATION INDICES FOR PAINLEVÉ TRANSCENDENTS OF THE SECOND KIND

Shun SHIMOMURA

Department of Mathematics, Keio University
3-14-1, Hiyoshi, Yokohama 223-8522, Japan

shimomurmath.keio.ac.jp

1. Introduction

Consider the second Painlevé equation

$$\text{(II)} \qquad w'' = 2w^3 + zw + \alpha$$

($' = d/dz$), where $\alpha \in \mathbf{C}$. Every solution of (II) is meromorphic in the whole complex plane ([1]). The rational solutions of (II) are completely known.

Theorem 1.1. ([5]). *If* $\alpha \in \mathbf{Z}$, *then* (II) has a unique rational solution. Conversely, if (II) has a rational solution, then $\alpha \in \mathbf{Z}$.

Let $w = \psi(z)$ be an arbitrary transcendental meromorphic solution of (II). For $a \in \mathbf{C} \cup \{\infty\}$, the deficiency and the ramification index

$$\delta(a, \psi) = \liminf_{r \to \infty} \frac{m(r, a, \psi)}{T(r, \psi)}, \qquad m(r, a, \psi) = \begin{cases} m(r, \psi) & (a = \infty), \\ m(r, 1/(\psi - a)) & (a \in \mathbf{C}), \end{cases}$$

$$\vartheta(a, \psi) = \liminf_{r \to \infty} \frac{N_1(r, a, \psi)}{T(r, \psi)}, \qquad N_1(r, a, \psi) = \begin{cases} N_1(r, \psi) & (a = \infty), \\ N_1(r, 1/(\psi - a)) & (a \in \mathbf{C}) \end{cases}$$

H.G.W. Begehr et al. (eds.), Proceedings of the Second ISAAC Congress, Vol. 1, 489–493.

are estimated as follows (for the notation of the value distribution theory, see [3]):

Theorem 1.2. ([6], [7], [8]). *If $\alpha \neq 0$, then, for every $a \in \mathbf{C} \cup \{\infty\}$, $\delta(a, \psi) = 0$. If $\alpha = 0$, then, for every $a \in \mathbf{C} \cup \{\infty\} - \{0\}$, $\delta(a, \psi) = 0$, and $\delta(0, \psi) \leq 1/2$.*

Theorem 1.3. ([2]). (1) *For every $a \in \mathbf{C} - \{0\}$, $\vartheta(a, \psi) \leq 1/4$, and $\vartheta(\infty, \psi) = 0$.*

(2) *If $\alpha \neq 0$, then $\vartheta(0, \psi) \leq 1/5$, and if $\alpha = 0$, then $\vartheta(0, \psi) = 0$.*

For a meromorphic function $g(z)$, we say that $g(z)$ is small with respect to $\psi(z)$, if $T(r, g) = S(r, \psi)$. The deficiency and the ramification index of $g(z)$ are defined by

$$\delta(g, \psi) = \liminf_{r \to \infty} \frac{m(r, 1/(\psi - g))}{T(r, \psi)}, \qquad \vartheta(g, \psi) = \liminf_{r \to \infty} \frac{N_1(r, 1/(\psi - g))}{T(r, \psi)},$$

respectively. The purpose of this paper is to examine $\delta(q, \psi)$ and $\vartheta(q, \psi)$, where $q(z)$ is a rational function.

2. Main results

Let $w = \psi(z)$ be an arbitrary transcendental meromorphic solution of (II). If $\alpha \in \mathbf{Z}$, then (II) possesses a unique rational solution, which we denote by $q_\alpha(z)$ (cf. Theorem 1.1).

Theorem 2.1. (1) *If $\alpha \notin \mathbf{Z}$, then, for every rational function $q(z)$, we have $\delta(q, \psi) = 0$.*

(2) *If $\alpha \in \mathbf{Z}$, then, for every rational function $q(z)$ other than $q_\alpha(z)$, we have $\delta(q, \psi) = 0$.*

Remark 2.1. If $\alpha = 0$, then $q_0(z) \equiv 0$ satisfies $\delta(q_0, \psi) = \delta(0, \psi) \leq 1/2$ (cf. Theorem 1.2).

For the rational solution $q_\nu(z)$ ($\nu \in \mathbf{Z}$) of

$$(\text{II},\nu) \qquad w'' = 2w^3 + zw + \nu,$$

a nontrivial estimate of $\vartheta(q_\nu, \psi)$ is obtained.

Theorem 2.2. *For every $\nu \in \mathbf{Z}$ satisfying $\nu \neq \alpha$, we have $\vartheta(q_\nu, \psi) \leq 2/5$. If $\alpha \in \mathbf{Z}$, then $\vartheta(q_\alpha, \psi) = 0$.*

Remark 2.2. If $\alpha \neq 0$, then $\vartheta(q_0, \psi) = \vartheta(0, \psi) \leq 1/5$ (cf. Theorem 1.3).

3. Lemmas

Substitution of the Laurent series expansion into (II) yields the expression of $\psi(z)$ around a pole.

Lemma 3.1. *Around a movable pole $z = z_\infty$ of $\psi(z)$,*

$$\psi(z) = \pm(z - z_\infty)^{-1} + O(z - z_\infty).$$

Furthermore recall the following lemmas.

Lemma 3.2. ([3; Lemma 2.4.2]). Let f be a transcendental meromorphic function such that $f^n P(z,f) = Q(z,f)$, $n \in \mathbf{N}$, where $P(z,u)$ and $Q(z,u)$ are polynomials in u and its derivatives with coefficients rational in z. If the total degree of $Q(z,u)$ as a polynomial in u and its derivatives is at most n, then $m(r, P(z,f)) = S(r,f)$.

Lemma 3.3. ([3; Proposition 9.2.3], [4]). Let $F(z,u)$ be a polynomial in u and its derivatives with coefficients rational in z. Assume that $u = f$ is a transcendental meromorphic solution of the differential equation $F(z,u) = 0$, and that $F(z,0) \not\equiv 0$. Then $m(r, 1/f) = S(r,f)$.

4. Proof of Theorem 2.1

Substituting $w = W + q(z)$ into (II), we have

$$W'' = 2W^3 + 6q(z)W^2 + (6q(z)^2 + z)W + Q(z), \tag{4.1}$$
$$Q(z) = 2q(z)^3 + zq(z) + \alpha - q''(z).$$

Note that $W = \psi(z) - q(z)$ is a solution of (4.1). By Theorem 1.1, if $q(z)$ is a rational function satisfying the supposition of the assertion (1) or (2), then $Q(z) \not\equiv 0$. Applying Lemma 3.3, we have

$$m(r, 1/(\psi - q)) = S(r, \psi - q) = S(r, \psi),$$

from which it immediately follows that $\delta(q, \psi) = 0$.

5. Proof of Theorem 2.2

Putting $w = q_\nu(z) + W$ $(\nu \in \mathbf{Z})$ in (II), we have

$$W'' = 2W^3 + 6q_\nu(z)W^2 + (6q_\nu(z)^2 + z)W + (\alpha - \nu), \tag{5.1}$$

which admits the solution $W = \Psi(z) = \psi(z) - q_\nu(z)$. Consider the set

$$\Delta = \{c \mid q_\nu(c) = \infty\} \subset \mathbf{C}, \qquad \sharp\Delta < +\infty. \tag{5.2}$$

Suppose that $\alpha = \nu \in \mathbf{Z}$. If there exists a point $z = z_0 \notin \Delta$ such that $\Psi(z_0) = \Psi'(z_0) = 0$, then $\Psi(z) = \psi(z) - q_\alpha(z) \equiv 0$, which contradicts the transcendency of $\psi(z)$. Hence z_0 is a simple zero of $\Psi(z)$, which implies $\vartheta(q_\alpha, \psi) = 0$.

In what follows, we suppose that $\nu \neq \alpha$. Since $\Psi(z)$ satisfies (5.1), we have

$$U'(z) = -4q_\nu'(z)\Psi(z)^3 - (12q_\nu(z)q_\nu'(z) + 1)\Psi(z)^2, \tag{5.3}$$

where

$$U(z) = \Psi'(z)^2 - \Psi(z)^4 - 4q_\nu(z)\Psi(z)^3 - (6q_\nu(z)^2 + z)\Psi(z)^2 - 2(\alpha - \nu)\Psi(z). \tag{5.4}$$

By (5.3),

$$U(z) = U(c_0) - \int_{c_0}^{z} \big(4q_\nu'(s)\Psi(s)^3 + (12q_\nu(s)q_\nu'(s) + 1)\Psi(s)^2\big)ds, \quad c_0 \notin \Delta. \tag{5.5}$$

Differentiating both sides of (5.5) successively, we obtain

$$U'(z) = R_{10}(\Psi)\Psi^2, \qquad U''(z) = R_{20}(\Psi)\Psi,$$
$$U^{(3)}(z) = R_{30}(\Psi)\Psi + R_{31}(\Psi)\Psi', \quad U^{(4)}(z) = R_{40}(\Psi)\Psi + R_{41}(\Psi)\Psi',$$

where $R_{ij}(\Psi)$ $(1 \leq i \leq 4,\ j = 0, 1)$ are polynomials in Ψ, q_ν and their derivatives. The relations above combined with (5.4) yield the following.

Lemma 5.1. *Suppose that* $\Psi(z_0) = \Psi'(z_0) = 0$, $z_0 \notin \Delta$. *Then*

(1) $\Psi''(z_0) = \alpha - \nu \neq 0$,

(2) $U(z_0) = U'(z_0) = U''(z_0) = U^{(3)}(z_0) = U^{(4)}(z_0) = 0$.

Substituting $W = \Psi(z)$ into (5.1), and applying Lemma 3.2, we obtain the following.

Lemma 5.2. $m(r, \Psi) = S(r, \Psi) = S(r, \psi)$, $m(r, \Psi') = S(r, \Psi) = S(r, \psi)$.

Lemma 5.1 implies that every double zero of $\Psi(z)$ is at least a quintuple zero of $U(z)$, and hence

$$N_1(r, 1/\Psi) \leq \frac{1}{5}N(r, 1/U) + O(\log r) \leq \frac{1}{5}T(r, U) + O(\log r). \tag{5.6}$$

By Lemma 5.2,

$$m(r, U) = S(r, \psi). \tag{5.7}$$

Around every movable pole $z = z_\infty \notin \Delta$, $\Psi(z) = \pm(z - z_\infty)^{-1} + O(1)$ (cf. Lemma 3.1). Substitution of this into (5.5) yields

$$N(r, U) = 2N(r, \Psi) + O(\log r) \leq 2T(r, \psi) + O(\log r). \tag{5.8}$$

$$N_1(r, 1/\Psi) \leq \frac{2}{5}T(r, \psi) + S(r, \psi).$$

Thus we obtain $\vartheta(q_\nu, \psi) \leq 2/5$, which is a desired estimate.

References

[1] A. Hinkkanen and I. Laine, *Solutions of the first and second Painlevé equations are meromorphic*, J. Anal. Math. (to appear).
[2] H. Kießling, *Zur Werteverteilung der Lösungen algebraischer Differentialgleichungen*, Dissertation, Fachbereich 3 Mathematik, Technischen Universität Berlin, 1996.
[3] I. Laine, *Nevanlinna Theory and Complex Differential Equations*, de Gruyter, Berlin, New York, 1993.
[4] A. Z. Mohon'ko and V. D. Mohon'ko, *Estimates for the Nevanlinna characteristics of some classes of meromorphic functions and their applications to differential equations*, Sibirsk Mat. Zh. **15** (1974), 1305–1322 [Russian]. Engl. Transl.: Siberian Math. J. **15** (1974), 921–934.
[5] Y. Murata *Rational solutions of the second and the fourth Painlevé equations*, Funkcial. Ekvac. **28** (1985), 1–32.
[6] H. Schubart *Zur Wertverteilung der Painlevéschen Transzendenten*, Arch. Math. **7** (1956), 284–290.
[7] H. Schubart and H. Wittich *Über die Lösungen der beiden ersten Painlevéschen Differentialgleichungen*, Math. Z. **66** (1957), 364–370.
[8] H. Wittich *Eindeutige Lösungen der Differentialgleichungen* $w'' = P(z, w)$, Math. Ann. **125** (1953), 355–365.

58 UNIQUENESS OF HOLOMORPHIC MAPPINGS WITH THE SAME INVERSE IMAGES OF HYPERSURFACES

Manabu SHIROSAKI

Department of Mathematical Sciences, College of Engineering
Osaka Prefecture University, Sakai, 599-8531 Japan

§1. Inroduction

Let $\mathcal{F}$ be a family of nonconstant holomorphic mappings of $\boldsymbol{C}$ into $\boldsymbol{P}^n(\boldsymbol{C})$ and S be a hypersurface of $\boldsymbol{P}^n(\boldsymbol{C})$. Here, we require neither non-singularity nor irreducibility of S. We call S have the *uniqueness property for* $\mathcal{F}$ if $f^*S = g^*S$ as divisors implies $f = g$ for any $f, g \in \mathcal{F}$. Also, if a discrete set in $\boldsymbol{C}$ (resp. $\boldsymbol{P}^1(\boldsymbol{C})$) without multiple points has the uniqueness property for nonconstant entire functions (resp. meromorphic functions on $\boldsymbol{C}$), then it is called a uniqueness range set for entire functions (resp. meromorphic functions). For example, the zero set of $e^z + z$ is a uniqueness range set for entire functions.

The following theorems by Nevanlinna([N]) are the starting point of us:

THEOREM 1.1. Let f and g be nonconstant meromorphic functions and a_j $(1 \leq j \leq 5)$ distinct points in $\overline{\boldsymbol{C}}$. If

$$f^{-1}(a_j) = g^{-1}(a_j) \quad (1 \leq j \leq 5),$$

then $f = g$.

H.G.W. Begehr et al. (eds.), Proceedings of the Second ISAAC Congress, Vol. 1, 495–500.

THEOREM 1.2. Let f and g be distinct nonconstant meromorohic functions and a_j $(1 \leq j \leq 4)$ be distinct points in $\overline{\boldsymbol{C}}$. If

$$f^{-1}(a_j) = g^{-1}(a_j) \text{ (counting multiplicity) } (1 \leq j \leq 4),$$

then g is a Möbius transformation of f, two of a_j (say a_3, a_4) are Picard exceptional values of f and g and the nonharmonic ratio $(a_1, a_2, a_3, a_4) = -1$.

For nonconstant meromorphic function f and a set $S \subset \overline{\boldsymbol{C}}$, put

$$E_f(S) := \cup_{a \in S}\{(z, m(f,z)); f(z) = a\},$$

where $m(f,z)$ is the multiplicity of zero of $f - f(z)$ at z, i.e., $E_f(S)$ is the pull-back of S by f as divisor.
We pick up four points $a_j \in \overline{\boldsymbol{C}}$ such that $(a_{\sigma(1)}, a_{\sigma(2)}, a_{\sigma(3)}, a_{\sigma(4)}) \neq -1$ for any $\sigma \in \mathfrak{S}_4$ and put $S_j = \{a_j\}$. Then by Theorem 1.2, if two nonconstant meromorphic functions f and g satisfy $E_f(S_j) = E_g(S_j)$ $(1 \leq j \leq 4)$, then $f = g$. For entire functions f and g, we take three points in $\boldsymbol{C}$ to get the same conclusion.

Gross([Gro]) gave the following problem:

Gross' Problem(1976). Can one find two (or possible even one) finite set $S_j \subset \boldsymbol{C}$ such that any two nonconstant entire functions f and g satisfing $E_f(S_j) = E_g(S_j) (j = 1, 2)$ must be indentical?

For this problem, H.-X. Yi([Y1]) showed the following (1994):

Take two positive integers n and m and put $w = \exp(2\pi i/n), u = \exp(2\pi i/m)$.

THEOREM 1.3. Let $S_1 = \{a+b, a+bw, \cdots, a+bw^{n-1}\}$ and $S_2 = \{c\}$, where $n > 4$, and a, b and c are constants such that $b \neq 0, c \neq a$ and $(c-a)^{2n} \neq b^{2n}$. If nonconstant entire functions f and g satisfy $E_f(S_j) = E_g(E_j)$ $(j = 1, 2)$, then $f = g$.

THEOREM 1.4. Let $S_1 = \{a_1 + b_1, a_1 + b_1 w, \cdots, a_1 + b_1 w^{n-1}\}$ and $S_2 = \{a_2 + b_2, a_2 + b_2 u, \cdots, a_2 + b_2 u^{m-1}\}$, where $n > 4, m > 4$, and a_1, b_1, a_2 and b_2 are constants such that $b_1 \neq 0, b_2 \neq 0$ and $a_1 \neq a_2$. If nonconstant entire functions f and g satisfy $E_f(S_j) = E_g(E_j)$ $(j = 1, 2)$, then $f = g$.

As vesions of Gross' problem for meromorphic functions, do there exist three sets $S_j \subset \overline{\boldsymbol{C}}$ which give unicity of meromorphic functions? Also, Yi gave

THEOREM 1.5. Let $S_1 = \{a_1 + b_1, a_1 + b_1 w, \cdots, a_1 + b_1 w^{n-1}\}, S_2 = \{a_2 + b_2, a_2 + b_2 u, \cdots, a_2 + b_2 u^{m-1}\}$ and $S_3 = \{\infty\}$, where $n > 6, m >$

6, and a_1, b_1, a_2 and b_2 constants such that $b_1 \neq 0, b_2 \neq 0$ and $a_1 \neq a_2$. If nonconstant meromorphic functions f and g satisfy $E_f(S_j) = E_g(E_j)$ $(j = 1, 2, 3)$ then, $f = g$.

In 1970's Fujimoto gave the following results:

Let f and g be holomorphic mappings of $\boldsymbol{C}$ into $\boldsymbol{P}^n(\boldsymbol{C})$ and H_j $(1 \leq j \leq q)$ hyperplanes in general position in $\boldsymbol{P}^n(\boldsymbol{C})$. We assume that

$$f^{-1}(H_j) = g^{-1}(H_j) \text{ (conunting multiplicity) } (1 \leq j \leq q).$$

THEOREM 1.6([F2]). Assume that f and g are linearly non-degenerate and $q \geq 3n + 2$. Then $f = g$.

THEOREM 1.7([F3]). Assume that f and g are algebraically non-degenerate and $q \geq 2n + 3$. Then $f = g$.

§2. Yi's uniqueness range set.

As mentioned in the previous section, $\{z : e^z + z = 0\}$ is a URS, where URS is an abbreviation of uniqueness range set for entire functions. Yi([Y2]) gave a URS with finite elements:

THEREM 2.1 (YI, 1995). Let n and m be relutively prime positive integers such that $n > 2m + 4$, and a and b nonzero constants such that $P(w) := w^n + aw^{n-m} + b = 0$ has no multiple root. Then, $S = \{w : P(w) = 0\}$ is a URS.

For example, the samllest n which satisfies the condition is 7 $(m = 1)$. Therefore, there is a URS with seven elements.

Examle. $S = \{w : w^7 + aw^6 + b = 0\}$, where $a, b \neq 0, -\frac{1}{6}\left(-\frac{6}{7}a\right)^7 + b = 0$.

§3. The case of holomorphic mappings into the complex projective spaces.

Let p and d be relutively prime intergers such that $d > 2p + 8$ and $p \geq 2$. Put $P(w_0, w_1) = P_1(w_0, w_1) = {w_0}^d + {w_0}^p {w_1}^{d-p} + {w_1}^d$.

THEOREM 3.1([S1]). Let f and g be algebraically non-degenerate holomorphic mappings of $\boldsymbol{C}$ into $\boldsymbol{P}^1(\boldsymbol{C})$ with reduced representaions $\tilde{f} = (f_0, f_1)$ and $\tilde{g} = (g_0, g_1)$, respectively. If

$$P(g_0, g_1) = \alpha P(f_0, f_1)$$

holds for some entire function α without zeros, then

$$g_0 = \beta f_0, \quad g_1 = \beta f_1,$$

and hence $f = g$. Here β is an entire function without zeros such that $\beta^d = \alpha$.

Next put $P_1(w_0, w_1) = P(w_0, w_1)$, and define inductively

$$P_n(w_0, \cdots, w_n) = P_{n-1}(P(w_0, w_1), P(w_1, w_2), \cdots, P(w_{n-1}, w_n)) \quad (n \geq 2).$$

Let S_n be the zero set of P_n. It is easy to see that S_n is reducible.

THEOREM 3.2([S1]). S_n has the uniqueness property ofr algebraically non-degenerate holomorphic mappings of $\boldsymbol{C}$ into $\boldsymbol{P}^n(\boldsymbol{C})$.

§4. Hyperbolicity.

In this section, we mention other the hyperbolicity of S_n in the previous section.

THEOREM 4.1([S2]). All holomorphic mappings $f : \boldsymbol{C} \to S_n$ are constant.

THEOREM 4.2([S2]). All holomorphic mappings $f : \boldsymbol{C} \to \boldsymbol{P}^n(\boldsymbol{C}) \setminus S_n$ are constant.

By Brody's theorem, we have

COROLLARY 4.3([S2]). S_n is Kobayashi hyperbolic.

Also, we get by Brody-Green's theorem

COROLLARY 4.4([S2]). $\boldsymbol{P}^n(\boldsymbol{C}) \setminus S_n$ is completely hyperbolic, and hyperbolically imbedded in $\boldsymbol{P}^n(\boldsymbol{C})$.

Hence, we can get hyperbolic hypersurfaces of degree 13^n with hyperbolic complements in $\boldsymbol{P}^n(\boldsymbol{C})$.

§5. Other hypersurfaces with the uniqueness property.

In this section we give a hypersurface with the uniqueness property for linearly non-degenerate holomorphic mappings.

THEOREM 5.1([Gre], [F1]). Let $f = (f_0 : \cdots : f_n)$ be a holomorphic mapping of $\boldsymbol{C}$ into $\boldsymbol{P}^n(\boldsymbol{C})$ such that any $f_j \not\equiv 0$. Define an equivalece relation $j \sim k$ in $\{0, \cdots, n\}$ by $f_j/f_k = \text{constant}$. Assume $d \geq n^2$. If

$$f_0{}^d + \cdots + f_n{}^d = 0,$$

then for each equivalence class I

$$\sum_{j \in I} f_j{}^d = 0.$$

Now define

$$Q_n(w_0, \cdots, w_n) = P(w_0, w_1)^q + \cdots + P(w_{n-1}, w_n)^q.$$

THEOREM 5.2([S3]). Assume $q \geq (2n-1)^2$. The zero set of $Q_n(w_0, \cdots, w_n)$ has the uniqueness property for linearly non-degenerate holomorphic mappings of $\boldsymbol{C}$ into $\boldsymbol{P}^n(\boldsymbol{C})$.

§6. Two hypersurfaces case.

As Gross' problem, we consider the problem of the uniqueness of holomorphic mappings f and g with $f^*(S_j) = g^*(S_j)$ $(j = 1, 2)$. Take $(a_{jk})_{0 \leq j,k \leq n} \in GL(n+1, \boldsymbol{C})$, and let S_1 and S_2 be the hypersurfaces in $\boldsymbol{P}^n(\boldsymbol{C})$ defined by

$$w_0{}^{p_1} + \cdots + w_n{}^{p_1} = 0$$

and

$$\sum_{j=0}^{n} \left(\sum_{k=0}^{n} a_{jk} w_k \right)^{p_2} = 0,$$

respectively, where p_1 and p_2 are positive integers.

THEOREM 6.1([SU]). *Let f and g be linearly non-degenerate holomorphic mappings of $\boldsymbol{C}$ into $\boldsymbol{P}^n(\boldsymbol{C})$ with reduced representations $\tilde{f} = (f_0, \cdots, f_n)$ and $\tilde{g} = (g_0, \cdots, g_n)$, respectively. Assume that $p_1, p_2 \geq (2n+1)^2$ and that*

$$(a_{jk})^{2p} \neq (a_{\mu\nu})^{2p} \text{ for any } (j,k) \text{ and } (\mu,\nu) \text{ such that } (j,k) \neq (\mu,\nu) \quad (*)$$

where p is the least common multiple of p_1 and p_2. If $f^*(S_j) = g^*(S_j)$ $(j = 1, 2)$, then $f = g$.

This theorem is an analogue of Theorem 1.4.

References

[F1] H. Fujimoto, *On meromorphic maps into the complex projective space*, J. Math. Soc. Japan**26** (1974), 272–288.

[F2] H. Fujimoto, *The uniqueness problem of meromorphic maps into the complex projective space*, Nagoya Math. J. **58** (1975), 1–23.

[F3] H. Fujimoto, *A uniqueness theorem of algebraically non-degenerate meromorphic maps into $\boldsymbol{P}^N(\boldsymbol{C})$*, Nagoya Math. J. **64** (1976), 117–147.

[Gre] M. L. Green, *Some Picard theorems for holomorphic maps to algebraic varieties,* Amer. J. Math. **97** (1975), 43–75.

[Gro] F. Gross, *Factoriztion of meromorphic functions and some open problems,* Complex Analysis (Proc. Conf. Univ. Kentucky, Lexington, Ky.1976), Lecture Notes in Math., vol.**599**, Springer–Verlag, Belrin, 1977, pp.51–69.

[N] R. Nevanlinna, *Einige Eindeutigkeitssäteze in der Theorie der meromrphen Funktionen,* Acta Math. **48** (1926), 367–391.

[S1] M. Shirosaki, *On polynomials which determine holomorphic mappings,* J. Math. Soc. Japan **49** (1997), 289–298.

[S2] M. Shirosaki, *On some hypersurfaces and holomorphic mappings,* Kodai Math. J. **21** (1998), 29–34.

[S3] M. Shirosaki, *On some hypersurfaces and holomorphic mappings,* Preprint.

[SU] M. Shirosaki and M. Ueda, *An analogue of Yi's theorem to holomorphic mappings,* Preprint.

[Y1] H.-X. Yi, *Unicity theorem for entire functions,* Kodai Math. J. **17** (1994), 133–141.

[Y2] H.-X. Yi, *A question of Gross on the uniqueness of entire functions,* Nagoya Math. J. **138** (1995), 169–177.

59 AN IMPROVEMENT OF THE SECOND FUNDAMENTAL THEOREM FOR HOLOMORPHIC CURVES

Nobushige TODA*

Department of Mathematics, Nagoya Institute of Technology
Gokiso, Showa-ku, Nagoya(466-8555), Japan

1. Introduction. Let $f = [f_1, \cdots, f_{n+1}]$ be a linearly non-degenerate and transcendental holomorphic curve from $\boldsymbol{C}$ in to the n-dimensional complex projective space $P^n(\boldsymbol{C})$ with a reduced representation $(f_1, \cdots, f_{n+1}) : \boldsymbol{C} \to \boldsymbol{C}^{n+1} - \{\boldsymbol{0}\}$, where n is a positive integer. In this paper, we use the same notation for f and for a vector $\boldsymbol{a} = (a_1, \cdots, a_{n+1}) \in \boldsymbol{C}^{n+1} - \{\boldsymbol{0}\}$ as in [6]:

$$\|f(z)\|,\ \|\boldsymbol{a}\|,\ (\boldsymbol{a}, f),\ (\boldsymbol{a}, f(z)),\ T(r,f),\ m(r,\boldsymbol{a},f),\ N(r,\boldsymbol{a},f),\ \delta(\boldsymbol{a},f),\ S(r,f)\ \textit{etc.}$$

It is well known that f is linearly non-degenerate over $\boldsymbol{C}$ if and only if the Wronskian $W = W(f_1, \cdots, f_{n+1})$ of $f_1, \cdots, f_{n+1}$ is not identically equal to zero.

For meromorphic functions in the complex plane we use the standard notation of the Nevanlinna theory of meromorphic functions([3]) and we use $\boldsymbol{e}_1, \cdots, \boldsymbol{e}_{n+1}$ as the standard basis of $\boldsymbol{C}^{n+1}$.

Let X be a subset of $\boldsymbol{C}^{n+1} - \{\boldsymbol{0}\}$ in N-subgeneral position; that is to say, $\#X \geq N+1$ and any $N+1$ elements of X generate $\boldsymbol{C}^{n+1}$, where N is an integer satisfying $N \geq n$, let $X(0) = \{\boldsymbol{a} = (a_1, \cdots, a_{n+1}) \in$

*Supported in part by Grant-in-Aid for Scientific Research (NO. 11640164), Ministry of Education, Science and Culture

H.G.W. Begehr et al. (eds.), Proceedings of the Second ISAAC Congress, Vol. 1, 501–510.

$X \mid a_{n+1} = 0\}$ and let s be the maximum number of linearly independent vectors in $X(0)$. Then, it is easy to see that $0 \le \#X(0) \le N$ and $0 \le s \le n$ since X is in N-subgeneral position. Further we put as follows (see Definition 1 in [5]):

$$u(z) = \max_{1\le j\le n} |f_j(z)|, \quad t(r,f) = \frac{1}{2\pi}\int_0^{2\pi} \{\log u(re^{i\theta}) - \log u(e^{i\theta})\}d\theta,$$

$$\text{and} \quad \Omega = \limsup_{r\to\infty} \frac{t(r,f)}{T(r,f)}.$$

It is easy to see the following inequalities (see [5]):
(a) $t(r,f) \le T(r,f) + O(1)$, (b) $0 \le \Omega \le 1$.

Recently we have given an amelioration of Nochka's positive answer([4]) to the conjecture of Cartan([1]) as follows (see Theorems 1 and 2 in [6]).

Theorem A. For any q elements $\boldsymbol{a}_j$ $(j = 1,\cdots,q)$ of X,

$$\sum_{j=1}^{q} m(r,\boldsymbol{a}_j,f) + \frac{N+1}{n+1}N(r,\frac{1}{W}) \le (N-n+n(s))T(r,f) + (N+1-n(s))t(r,f) + S(r,f),$$

where $2N - n + 1 < q < \infty$, W is the Wronskian of $f_1,\cdots,f_{n+1}$ and $n(s) = (s+1)(N+1)/(n+1)$.

Our main purpose of this paper is to give an improvement of Theorem A.

2. Preliminary and lemma. We shall give some lemmas for later use. Let $f = [f_1,\cdots,f_{n+1}]$, $X, X(0)$ etc. be as in Section 1 and let $\boldsymbol{a}_1,\cdots,\boldsymbol{a}_q$ be any elements of $X - X(0)$. Put

$$X(0) = \{\boldsymbol{b}_1,\cdots,\boldsymbol{b}_p \quad (0 \le p = \#X(0) \le N)\},$$
$$(\boldsymbol{b}_k, f) = G_k \ (k = 1,\cdots,p), \quad s = \dim H(X(0)),$$

where $H(X_1)$ is the vector space spanned by the elements of a subset X_1 of X, $N+1 \le q$ and $2N - n + 1 < q + p < \infty$.

Put $Q = \{1,2,\cdots,q+p\}$ and if $P \subset Q$, we denote

$H(P) =$ the vector space spanned by $\{\boldsymbol{a}_j \mid j \in P\}$ and $d(P) = \dim H(P)$.

Lemma 1(see [2], Theorem 0.3). For $\{\boldsymbol{a}_j \mid j \in Q\}$, there exist a Nochka weight function $\omega : Q \to (0,1]$ and a Nochka constant $\theta \ge 1$ such that

(a) $0 < \omega(j)\theta \le 1$ for all $j \in Q$;

(b) $q - 2N + n - 1 = \theta(\sum_{j=1}^{q+p} \omega(j) - n - 1)$;
(c) $(N+1)/(n+1) \le \theta \le (2N - n + 1)/(n+1)$;
(d) If $P \subset Q$ and $0 < \#P \le N+1$, then $\sum_{j \in P} \omega(j) \le d(P)$.

Lemma 2(see [2], Theorem 1.2). Let ω and θ be the same as in Lemma 1. Take $A \subset Q$ with $0 < \#A \le N+1$. Let

$$\{E_j \in \boldsymbol{R} \mid E_j \ge 1, j \in Q\}.$$

Then there exists a subset B of A such that

$$\{\boldsymbol{a}_j \mid j \in B\} \text{ is a basis of } H(A) \quad \text{and} \quad \Pi_{j \in A} E_j^{\omega(j)} \le \Pi_{j \in B} E_j.$$

Remark 1. If $\#A = N+1$, then $H(A) = \boldsymbol{C}^{n+1}$ and $\{\boldsymbol{a}_j \mid j \in B\}$ is a basis of $\boldsymbol{C}^{n+1}$.

Suppose that $1 \le p$ and put

$$\sum_{k=1}^{p} \omega(q+k) = d.$$

Then, by Lemma 1(d), $d \le s$ and if d is not an integer, $[d] + 1 \le s$. In this case we have the following

Lemma 3. Let z be any point in $\boldsymbol{C}$ satisfying $G_k(z) \equiv (\boldsymbol{b}_k, f(z)) \ne 0$ $(k = 1, \cdots, p)$.

(A) When d is an integer, there are linearly independent vectors $\boldsymbol{b}_{i_1}, \cdots, \boldsymbol{b}_{i_d}$ such that

$$|G_{i_1}(z) \cdots G_{i_d}(z)| \le \Pi_{k=1}^{p} |G_k(z)|^{\omega(q+k)}. \tag{1}$$

(B) When d is not an integer, there are linearly independent vectors $\boldsymbol{b}_{i_1}, \cdots, \boldsymbol{b}_{i_{[d]+1}}$ such that

$$|G_{i_1}(z) \cdots G_{i_{[d]+1}}(z)| \le \Pi_{k=1}^{p} |G_k(z)|^{\omega(q+k)} \cdot (cu(z))^{[d]+1-d}, \tag{2}$$

where $c = \max_{1 \le k \le p} \{n^{1/2} ||\boldsymbol{b}_k||\}$.

Proof. For a point z satisfying the condition, we suppose for brevity that

$$|G_1(z)| \le |G_2(z)| \le \cdots \le |G_p(z)|.$$

(A)(resp. (B)). We choose $i_1, \cdots, i_d$ (*resp.* $i_1, \cdots, i_{[d]+1}$) as follows:
(i) $i_1 = 1$.

(ii) Suppose that $i_1, \cdots, i_{\mu-1}$ are chosen for $\mu \geq 2$. Then we choose i_μ as follows ($\mu \leq d$ (resp. $\mu \leq [d]+1$)):

"i_μ is the least number in $\{i_{\mu-1}+1, \cdots, p\}$ such that $\boldsymbol{b}_{i_1}, \cdots, \boldsymbol{b}_{i_\mu}$ are linearly independent."

Then, $G_{i_1}(z), \cdots, G_{i_d}(z)$ (resp. $G_{i_1}(z), \cdots, G_{i_{[d]+1}}(z)$) satisfy the inequality (1) (resp. (2)).

In fact, put for $1 \leq m \leq d-1$ (resp. $1 \leq m \leq [d]$)

$$\sigma(m) = i_{m+1} - 1 \quad \text{and} \quad \varphi(m) = \sum_{k=1}^{\sigma(m)} \omega(q+k).$$

We first note that

$$\Pi_{k=1}^{\sigma(d-1)} |G_k(z)|^{\omega(q+k)} \cdot |G_{i_d}(z)|^{d-\varphi(d-1)} \leq \Pi_{k=1}^{p} |G_k(z)|^{\omega(q+k)} \tag{3}$$

$$(\text{resp.} \Pi_{k=1}^{\sigma([d])} |G_k(z)|^{\omega(q+k)} \cdot |G_{i_{[d]+1}}(z)|^{[d]+1-\varphi([d])} \leq \Pi_{k=1}^{p} |G_k(z)|^{\omega(q+k)} \cdot (cu(z))^{[d]+1-d}) \tag{4}$$

since

$$|G_{i_d}(z)| \leq |G_k(z)| \ (i_d < k)$$
$$(\textit{resp.} |G_{i_{[d]+1}}(z)| \leq |G_k(z)| \ (i_{[d]+1} < k) \ \text{ and } \ |G_p(z)| \leq cu(z) \ ).$$

Then, by using Lemma 1(d), we have

$$|G_{i_1}(z) \cdots G_{i_d}(z)| \leq \Pi_{k=1}^{\sigma(d-1)} |G_k(z)|^{\omega(q+k)} \cdot |G_{i_d}(z)|^{d-\varphi(d-1)} \tag{5}$$

which is equivalent to

$$\sum_{\nu=1}^{d} \log |G_{i_\nu}(z)| \leq \sum_{k=1}^{\sigma(d-1)} \omega(q+k) \log |G_k(z)| + (d - \varphi(d-1)) \log |G_{i_d}(z)|. \tag{6}$$

We prove (6) as follows. We first note that by Lemma 1(d)

$$\varphi(m) \leq m \quad (m = 1, \cdots, d-1 \ (\text{resp. } [d])). \tag{7}$$

By the choice of $\{i_1, \cdots, i_d\}$, we have the following inequalities.

$$\log |G_{i_1}(z)| \leq \sum_{k=1}^{\sigma(1)} \omega(q+k) \log |G_k(z)| + (1 - \varphi(1)) \log |G_{i_2}(z)|;$$

$$\log |G_{i_m}(z)| \leq \sum_{k=i_m}^{\sigma(m)} \omega(q+k) \log |G_k(z)| +$$

$$(m-\varphi(m))\log|G_{i_{m+1}}(z)|-(m-1-\varphi(m-1))\log|G_{i_m}(z)|$$
$$(m=2,\cdots,d-1);\qquad \log|G_{i_d}(z)|=\log|G_{i_d}(z)|.$$

Adding all these d inequalities side by side, we have

$$\sum_{\nu=1}^{d}\log|G_{i_\nu}(z)|\le\sum_{k=1}^{\sigma(d-1)}\omega(q+k)\log|G_k(z)|+(d-\varphi(d-1))\log|G_{i_d}(z)|,$$

which is the desired inequality. From (3) and (5) we have (1).

(resp. We can also prove

$$|G_{i_1}(z)\cdots G_{i_{[d]+1}}(z)|\le\sum_{k=1}^{\sigma([d])}|G_k(z)|^{\omega(q+k)}\cdot|G_{i_{[d]+1}}(z)|^{[d]+1-\varphi([d])}\tag{8}$$

as in (5). From (4) and (8), we have (2).)

3. Second fundamental theorem and defect relation. Let $\boldsymbol{a}_1,\cdots,\boldsymbol{a}_{q+p}$ and ω be as in Section 2. Then, we have the following

Theorem 1. Put $d=\sum_{k=1}^{p}\omega(q+k)$. Then we have the following inequality:

$$\sum_{j=1}^{q}\omega(j)m(r,\boldsymbol{a}_j,f)+N(r,\frac{1}{W})\le T(r,f)+(n-d)t(r,f)+$$

$$\sum_{k=1}^{p}\omega(q+k)N(r,\boldsymbol{b}_k,f)+S(r,f),$$

where $d=0$ when $p=0$.

Proof. We put $(\boldsymbol{a}_j,f)=F_j$ $(j=1,\cdots,q)$ and $(\boldsymbol{b}_k,f)=G_k$ $(k=1,\cdots,p)$. For any $z(\neq 0)$ arbitrarily fixed in $|z|<\infty$ for which $F_j(z)\neq 0$ $(j=1,\cdots,q)$ and $G_k(z)\neq 0$ $(k=1,\cdots,p)$, let $|F_{j_1}(z)|\le|F_{j_2}(z)|\le\cdots\le|F_{j_q}(z)|$, where $j_1,\cdots,j_q$ are distinct and $1\le j_1,\cdots,j_q\le q$. Then there is a positive constant K such that

$$\|f(z)\|\le K|F_{j_\nu}(z)|\ (\nu=N+1,\cdots,q),\tag{9}$$
$$|F_{j_\nu}(z)|\le K\|f(z)\|\ \ (\nu=1,\cdots,q)\tag{10}$$

and for any $j=1,\cdots,q$

$$\begin{aligned}\|f(z)\| &\le K(|f_1(z)|^2+\cdots+|f_n(z)|^2+|F_j(z)|^2)^{1/2}\\ &\le\begin{cases}K(n+1)^{1/2}u(z) & \text{if } |F_j(z)|\le u(z)\\ K(n+1)^{1/2}|F_j(z)| & \text{otherwise}\end{cases}\end{aligned}\tag{11}$$

since the $n+1$-th elements of $\boldsymbol{a}_j$ $(j=1,\cdots,q)$ are different from zero. (From now on we denote by K a positive constant, which may be different from each other when it appears.)

We have by (9)

$$\Pi_{j=1}^{q}\Big(\frac{\|\boldsymbol{a}_j\|\|f(z)\|}{|(\boldsymbol{a}_j,f(z))|}\Big)^{\omega(j)}\leq K\Pi_{\nu=1}^{N+1}\Big(\frac{\|\boldsymbol{a}_{j_\nu}\|\|f(z)\|}{|F_{j_\nu}(z)|}\Big)^{\omega(j_\nu)}\equiv M. \tag{12}$$

Here we estimate M by dividing into two cases: the case when $p=0$ and the case when $p>0$.

(I) In case $p=0$. This theorem is contained in Theorem 1(a)([6], p.126).

(II) In case $p>0$.

Case A. The case when $u(z)<|F_{j_1}(z)|$. We have $\|f(z)\|\leq K|F_j(z)|$ $(j=1,\cdots,q)$ from (11) in this case and so M in (12) satisfies the inequality:

$$M\leq K=K\Pi_{k=1}^{p}|G_k(z)|^{\omega(q+k)}\cdot\frac{1}{\Pi_{k=1}^{p}|G_k(z)|^{\omega(q+k)}}\equiv E_1. \tag{13}$$

When d is an integer, by Lemma 3(A)

$$E_1\leq K\Pi_{k=1}^{p}|G_k(z)|^{\omega(q+k)}\cdot\frac{1}{|G_{i_1}(z)\cdots G_{i_d}(z)|}=(*). \tag{14}$$

We can find $\boldsymbol{e}_{i_{d+1}},\cdots,\boldsymbol{e}_{i_n}$ $(i_{d+1},\cdots,i_n\leq n)$ such that the vectors

$$\boldsymbol{b}_{i_1},\cdots,\boldsymbol{b}_{i_d},\boldsymbol{e}_{i_{d+1}},\cdots,\boldsymbol{e}_{i_n},\boldsymbol{e}_{n+1}$$

are linearly independent. By using

$$|f_{n+1}(z)|\leq\|f(z)\|,\quad |f_{i_j}(z)|\leq u(z)\qquad (j=d+1,\cdots,n),$$

$$(*)\leq K\Pi_{k=1}^{p}|G_k(z)|^{\omega(q+k)}\cdot\frac{\|f(z)\|u(z)^{n-d}}{|W(z)|}\times$$
$$\frac{|W(G_{i_1},\cdots,G_{i_d},f_{i_{d+1}},\cdots,f_{i_n},f_{n+1})(z)|}{|G_{i_1}(z)\cdots G_{i_d}(z)\cdot f_{i_{d+1}}(z)\cdots f_{i_n}\cdot f_{n+1}(z)|}. \tag{15}$$

When d is not an integer, by Lemma 3(B)

$$E_1\leq Ku(z)^{[d]+1-d}\Pi_{k=1}^{p}|G_k(z)|^{\omega(q+k)}\cdot\frac{1}{|G_{i_1}(z)\cdots G_{i_{[d]+1}}(z)|}=(**) \tag{16}$$

We can find $e_{i_{[d]+2}}, \cdots, e_{i_n}$ $(i_{[d]+2}, \cdots, i_n \leq n)$ such that the vectors

$$b_{i_1}, \cdots, b_{i_{[d]+1}}, e_{i_{[d]+2}}, \cdots, e_{i_n}, e_{n+1}$$

are linearly independent. By using

$$|f_{n+1}(z)| \leq \|f(z)\|, \quad |f_{i_j}(z)| \leq u(z) \ (j = i_{[d]+2}, \cdots, i_n),$$

we have

$$(**) \leq K\Pi_{k=1}^{p}|G_k(z)|^{\omega(q+k)} \cdot \frac{u(z)^{[d]+1-d}u(z)^{n-[d]-1}\|f(z)\|}{|W(z)|} \times$$

$$\frac{|W(G_{i_1}, \cdots, G_{i_{[d]+1}}, f_{i_{[d]+2}}, \cdots, f_{i_n}, f_{n+1})(z)|}{|G_{i_1}(z) \cdots G_{i_{[d]+1}}(z) f_{i_{[d]+2}}(z) \cdots f_{i_n}(z) f_{n+1}(z)|}$$

$$\leq K\Pi_{k=1}^{p}|G_k(z)|^{\omega(q+k)} \cdot \frac{\|f(z)\|u(z)^{n-d}}{|W(z)|} \cdot \frac{|W(G_{i_1}, \cdots, f_{n+1})(z)|}{|G_{i_1}(z) \cdots f_{n+1}(z)|}. \quad (17)$$

Case B. The case when $|F_{j_1}(z)| \leq u(z)$. In this case the inequality

$$\|f(z)\| \leq Ku(z) \quad (18)$$

holds by (11). We have from (12)

$$M = K\Pi_{k=1}^{p}\Big(\frac{\|b_k\|\|f(z)\|}{|G_k(z)|}\Big)^{\omega(q+k)}\Pi_{k=1}^{p}\Big(\frac{|G_k(z)|}{\|b_k\|\|f(z)\|}\Big)^{\omega(q+k)} \times$$

$$\Pi_{\nu=1}^{N+1}\Big(\frac{\|a_{j_\nu}\|\|f(z)\|}{|F_{j_\nu}(z)|}\Big)^{\omega(j_\nu)} \quad (19)$$

Put

$$H_\nu = F_{j_\nu} \ (\nu = 1, \cdots, N+1), H_{N+1+k} = G_k \ (k = 1, \cdots, p)$$

Suppose that

$$|H_{i_1}(z)| \leq |H_{i_2}(z)| \leq \cdots \leq |H_{i_{N+1+p}}(z)|,$$

then

$$\|f(z)\| \leq K|H_{i_\nu}(z)| \ (\nu = N+2, \cdots, N+1+p)$$

and by Lemma 2 there are linearly independent elements $I_1, \cdots, I_{n+1}$ chosen from $H_{i_1}, \cdots, H_{i_{N+1}}$ satisfying

$$M \leq K\Pi_{k=1}^{p}|G_k(z)|^{\omega(q+k)} \cdot \frac{\|f(z)\|^{n+1-d}}{\Pi_{m=1}^{n+1}|I_m(z)|}. \quad (20)$$

We put

$$E_2 = \frac{\|f(z)\|^{n+1-d}}{\Pi_{m=1}^{n+1}|I_m(z)|},$$

where $I_m(z) = (\boldsymbol{h}_m, f(z))$ $(\boldsymbol{h}_m \in X; m = 1, \cdots, n+1)$.

By (18) we have for the Wroskian W_I of $I_1, \cdots, I_{n+1}$

$$E_2 \leq K\frac{u(z)^{n+1-d}}{|I_1(z)\cdots I_{n+1}(z)|} = K\frac{u(z)^{n+1-d}}{|W(z)|}\cdot\frac{|W_I(z)|}{|I_1(z)\cdots I_{n+1}(z)|}. \tag{21}$$

Since

$$u(z)^n\|f(z)\|^{1-d} \leq \|f(z)\|u(z)^{n-d} \quad \text{and} \quad u(z)^{n+1-d} \leq \|f(z)\|u(z)^{n-d},$$

we have from (12), (13), (14), (15), (16), (17), (19), (20) and (21)

$$\sum_{j=1}^{q}\omega(j)\log\Big(\frac{\|\boldsymbol{a}_j\|\|f(z)\|}{|(\boldsymbol{a}_j, f(z))|}\Big) \leq \log\|f(z)\| + (n-d)\log u(z)+$$

$$\sum_{k=1}^{p}\omega(q+k)\log|G_k(z)| - \log|W(z)| + \sum_J \log^+\frac{|W_J(z)|}{|J_1(z)\cdots J_{n+1}(z)|} + \log^+ K,$$

where the summation $\sum_J$ is taken over all $J = \{J_1, \cdots, J_{n+1}\}$ consisting of any $n+1$ linearly independent functions chosen from $\{F_j\}_{j=1}^q \cup \{G_k\}_{k=1}^p \cup \{f_j\}_1^{n+1}$ and W_J is the Wronskian of $J_1, \cdots, J_{n+1}$.

As this inequality holds for any $z \in \boldsymbol{C}$ but z contained in a set which is finite or countable without any accumulation point in $\boldsymbol{C}$, by integrating it from 0 to 2π with respect to φ $(z = re^{i\varphi})$, we obtain

$$\sum_{j=1}^{q}\omega(j)m(r, \boldsymbol{a}_j, f) + N(r, \frac{1}{W}) \leq$$

$$T(r,f) + (n-d)t(r,f) + \sum_{k=1}^{p}\omega(q+k)N(r, b_k, f) + S(r,f),$$

Theorem 2. For any $\boldsymbol{a}_1, \cdots, \boldsymbol{a}_q \in X$,

(A) $\sum_{j=1}^q \omega(j)m(r, \boldsymbol{a}_j, f) + N(r, \frac{1}{W}) \leq (1+d)T(r,f) + (n-d)t(r,f) + S(r,f)$;

(B) $\sum_{j=1}^q m(r, \boldsymbol{a}_j, f) + \frac{N+1}{n+1}N(r, \frac{1}{W}) \leq (N-n+n(d))T(r,f) + (N+1-n(d))t(r,f) + S(r,f)$

where $2N-n+1<q<\infty$, $\sum_{\boldsymbol{a}_j \in X(0)} \omega(j)=d$ and $n(d)=(1+d)(N+1)/(n+1)$.

Proof. When $d=0$, this theorem is contained in Theorem 1([6], p.126). We have only to prove this theorem when $d>0$.

(A) Without loss of generality let $\{\boldsymbol{a}_1,\cdots,\boldsymbol{a}_q\} \cap X(0) = \{\boldsymbol{a}_1,\cdots,\boldsymbol{a}_k\}$ $(1 \le k \le N)$.

We apply Theorem 1 to this case. Then we obtain

$$\sum_{j=k+1}^{q} \omega(j) m(r,\boldsymbol{a}_j,f) + N(r,\frac{1}{W}) \le$$

$$T(r,f)+(n-d)t(r,f)+\sum_{j=1}^{k}\omega(j)N(r,\boldsymbol{a}_j,f)+S(r,f).$$

Adding $\sum_{j=1}^{k}\omega(j)m(r,\boldsymbol{a}_j,f)$ to both sides of this inequality we obtain the desired inequality since

$$\sum_{j=1}^{k}\omega(j)m(r,\boldsymbol{a}_j,f)+\sum_{j=1}^{k}\omega(j)N(r,\boldsymbol{a}_j,f)=dT(r,f)+O(1).$$

(B) We first multiply (A) of this theorem by θ and then add $\sum_{j=1}^{q}(1-\theta\omega(j))T(r,f)$ to both sides of the result to obtain

$$\sum_{j=1}^{q} m(r,\boldsymbol{a}_j,f)+\theta N(r,\frac{1}{W})+\sum_{j=1}^{q}(1-\theta\omega(j))N(r,\boldsymbol{a}_j,f) \le$$

$$(q+\theta+\theta d-\sum_{j=1}^{q}\theta\omega(j))T(r,f)+\theta(n-d)t(r,f)+S(r,f). \tag{22}$$

By of Lemma 1(a), (b) and (22) we obtain

$$\sum_{j=1}^{q} m(r,\boldsymbol{a}_j,f)+\theta N(r,\frac{1}{W}) \le$$

$$(2N-n+1)T(r,f)-\theta(n-d)(T(r,f)-t(r,f))+S(r,f)$$

and by using (a) in Section 1 and Lemma 1(c) we obtain

$$\sum_{j=1}^{q} m(r,\boldsymbol{a}_j,f)+\frac{N+1}{n+1}N(r,\frac{1}{W}) \le (2N-n+1)T(r,f)-$$

$$\frac{N+1}{n+1}(n-d)(T(r,f)-t(r,f))+S(r,f)=$$
$$(N-n+n(d))T(r,f)+(N+1-n(d))t(r,f)+S(r,f),$$

where $n(d)=(d+1)(N+1)/(n+1)$.

Remark 2. This is an amelioration of Theorem 2 in [6] according to Lemma 1 (d).

Theorem 3(Defect relation). For any $\boldsymbol{a}_1,\cdots,\boldsymbol{a}_q\in X$, we have the following inequalities:

(A) $\sum_{j=1}^{q}\omega(j)\delta(\boldsymbol{a}_j,f)\leq d+1+(n-d)\Omega$;

(B) $\sum_{j=1}^{q}\delta(\boldsymbol{a}_j,f)\leq 2N-n+1-\dfrac{N+1}{n+1}(n-d)(1-\Omega)$,

where $d=\sum_{\boldsymbol{a}_j\in X(0)}\omega(j)$.

Proof. We easily obtain this theorem from Theorem 2.

Remark 3. (a) This is an amelioration of Theorem 4 in [6] due to Lemma 1 (d).

(b) $d+1+(n-d)\Omega\leq n+1$ and $2N-n+1-(N+1)(n-d)(1-\Omega)/(n+1)\leq 2N-n+1$.

References

[1] H. Cartan: Sur les combinaisons linéaires de p fonctions holomorphes données. Mathematica 7(1933), 5-31.

[2] W. Chen: Defect relations for degenerate meromorphic maps. Trans. Amer. Math. Soc., 319-2(1990), 499-515.

[3] R. Nevanlinna: Le théorème de Picard-Borel et la théorie des fonctions méromorphes. Gauthier-Villars, Paris 1929.

[4] E. I. Nochka: On the theory of meromorphic functions. Soviet Math. Dokl., 27-2(1983), 377-381.

[5] N. Toda: On the fundamental inequality for non-degenerate holomorphic curves. Kodai Math. J., 20-3(1997), 189-207.

[6] N. Toda: On the second fundamental inequality for holomorphic curves. Bull. of Nagoya Inst. of Tech., 50(1998), 123-135.

60. ON FACTORIZATION OF ENTIRE FUNCTIONS OF TWO COMPLEX VARIABLES

Hironobu URABE

Deptartment of Mathematics, Kyoto University of Education
Kyoto, 612-8522, Japan

1. Introduction Untill now, we have investigated the theory of factorization (under composition) for entire or meromorphic functions of one complex variable and obtained various results (cf. [9], [10]).

Concerning the factorization of entire or meromorphic functions of several complex variables, recently, B. Q. Li and C. C. Yang ([4]), and L. A. Rubel and C. C. Yang ([7]) have shown several interesting results.

In this note, we will study the factorization (under composition) of certain entire functions of two complex variables and prove some results.

For an entire function $F(z, w)$ of two complex variables, let us consider the factorization (in entire sense) of F of the following form:

$$F(z, w) = f(g(z, w)), \tag{1}$$

where $f(\zeta)$ (left-factor) is an entire function of one complex variable ζ and $g(z, w)$ (right-factor) is an entire function of two complex variables

H.G.W. Begehr et al. (eds.), Proceedings of the Second ISAAC Congress, Vol. 1, 511–521.

z and w.

First of all, we'd like to mention a Theorem due to L. A. Rubel and C. C. Yang (cf. [7], also [6]) as follows:

Theorem A. Let $A(z)$ and $B(w)$ be non-constant entire functions of one complex variable and suppose that

$$f(g(z, w)) = A(z) + B(w) \tag{2}$$

for all $z,\ w \in \mathbf{C}$, where f is an entire function of one complex variable and g is an entire function of two complex variables. Then f must be affine: $f(\zeta) = \alpha\zeta + \beta,\ \alpha,\ \beta \in \mathbf{C}$ and g must have the form

$$g(z, w) = a(z) + b(w)$$

for some entire functions $a(z)$ and $b(w)$ of one variable.

Relating to the above factorization, they posed a question on the equation such as

$$f(g(z, w)) = A(z)\,B(w), \tag{3}$$

where A and B are two non-constant entire functions ([7]).

Although it seems difficult to give the complete answer to this question, we want to show here some results relating to (3).

Here, let's give a definition concerning the factorization.

Definition An entire function $F(z, w)$ is said to be *prime* (*pseudo-prime, left-prime, right-prime*) if every factorization of the form (1) implies that either f is (bi-)linear in $\mathbf{C}$ or g is linear in $\mathbf{C}^2$ (either f is a rational function or g is a polynomial, f must be (bi-)linear whenever g is transcendental, g must be linear whenever f is transcendental).

In [4], as a generalization of the well-known fact that $z + e^z$ is prime in the case of one variable, B. Q. Li and C. C. Yang proved, for instance, that the function F (of finite order) defined by

$$F(z, w) = e^{z+w} + az + bw \qquad ((4)$$

is prime, if a, b are constants and not all zero.

Moreover, they proved the following result in [4].

Theorem B Let $F(z, w)$ be an entire function of two complex variables. If both of $F(z, c)$ and $F(c, w)$ are prime as the function of one complex variable z or w, where c is any fixed complex number and $F(sz, tz)$ is prime, where s, t are some fixed complex numbers with $st \neq 0$, then $F(z, w)$ is prime as the function of two variables (in $\mathbf{C}^2$).

We will also give some results relating to this Theorem B.

Now, let us note that the factorization of the form (3) should be natural and proper (as shown in [4], essentially).
Actually, one might consider another factorization, instead of (3), such as

$$F(z, w) = f(g(z, w), h(z, w)), \tag{5}$$

where $f(\xi_1, \xi_2)$, $g(z, w)$, $h(z, w)$ are all entire functions of two complex variables.
But, the factorization of this kind becomes incompatible with the theory in one variable.
Indeed, let

$$\xi_1 \equiv K_1(Z) = Z + H(Z), \quad \xi_2 \equiv K_2(Z) = e^Z e^{H(Z)} - H(Z), \tag{6}$$

with an entire function $H(Z)$. Then,

$$\xi_1 - e^{\xi_1} + \xi_2 = Z.$$

Hence, if we set

$$f(\xi_1, \xi_2) = \xi_1 - e^{\xi_1} + \xi_2 \tag{7}$$

and

$$\xi_1 = g(z, w) = K_1(z+w), \quad \xi_2 = h(z, w) = K_2(z+w),$$

we have the identical relation

$$(F(z, w) \equiv)\ z + w = f(g(z, w), h(z, w)). \tag{8}$$

We also have that $Z = f(K_1(Z), K_2(Z))$.

Thus, we do consider the factorization of the form (1) with only functions involved.

2. Main Results.

Theorem 1. Let $F(z, w)$ be an entire function of two complex varibles. If, for any fixed complex number c, both of $F(z, c)$ and $F(c, w)$ are pseudo-prime (left-prime) as the functions of one complex variable, then $F(z, w)$ is pseudo-prime (left-prime).

This assertion can be proved by using the fact:

Proposition 1. Let $g(z, w)$ be an entire function of two complex variables z and w. If, for any fixed complex number c, both of $g(z, c)$ and $g(c, w)$ are polynomials, then $g(z, w)$ is a polynomial of two complex variables.

This Proposition 1 can be verified by making use of the determinant of Van der Monde, along with the unicity theorem concerning holomorphic functions.

Remark. Concerning Proposition 1, if the assumption are satisfied for any c belonging to some set with uncountable cardinality, then the assertion (of Proposition 1 and hence Theorem 1) must hold true (in view of unicity theorem).

Example 1. Let's consider the function F, defined by

$$F(z, w) := P(z, w)\, e^{h(z,w)} + Q(z, w), \tag{9}$$

where P (not identically zero) and Q (non-constant with respect to both z and w) are polynomials. Then F is pseudo-prime.
Actually, this fact is seen to be valid by virtue of Theorem 1 and a Theorem due to Bergweiler which asserts that the function $G(z) = P_1(z) + P_2(z)\, e^{\phi(z)}$, where $P_j(z)$ are polynomials and $\phi(z)$ is entire

with P_1 : non-costant and P_2 : not identically zero, is pseudo-prime ([1]) (also cf. [2]).

In connection with Theorem A, we can prove the following fact.

Theorem 2. Set

$$F(z,w) := P(w)\,A(z) + Q(z)\,B(w), \tag{10}$$

where $P(w)$ and $Q(z)$ be two non-constant polynomials, $A(z)$ and $B(w)$ be two transcendental entire functions of one complex variable, then $F(z,w)$ is left-prime (in entire sense).

This assertion can be proved by using the argument due to Y. Noda ([5]) and above Proposition 1.

Example 2. Set

$$F(z,\, w) := P(w)\,e^{h(z)} + Q(z)\,e^{S(w)}, \tag{11}$$

where P, Q and S are non-constant polynomials and h is a transcendental entire function. Then F is prime (in entire sense).
Indeed, this F can be seen easily to be also right prime.

Next, we obtain several results in connection with Theorem B or the equation (3).

Theorem 3. Let $F(z,\, w)$ be an entire function of two complex variables. Assume that, for any fixed complex number c, $F(z,\, c)$ is prime as a function of z and that there exists a sequence $\{l(k)\}_{k=1}^{\infty}$ which constitutes an infnitely many points such that the function $G(w) := F(l(k)w,\, w)$ (putting $z = l(k)w$ in $F(z,\, w)$) is prime as a function of w for any k : a natural number. Then $F(z,\, w)$ is prime in entire sense.

Example 3. For an entire function $h(z,\, w)$ and any non-nozero constant a, the function, defined by

$$F(z,\, w) := e^{h(z,w)} + az + bw, \quad (|a| + |b| \neq 0) \tag{12}$$

is prime in entire sense (in view of Theorem 3, together with Bergweiler's Result mentioned above [1]).

Example 4. Assume that $e^H \equiv \exp[H]$, K and L are, periodic, entire functions of one complex variable with the same period $2\pi i$ (say) such that e^H is non-constant. Then the function F, defined by

$$F(z, w) \;:=\; (z + cw)\, e^{H(z+w)} \;+\; K(z)\, L(w), \tag{13}$$

is prime in entire sense, where c is a constant (cf. [10] in which we know that $z\, e^{H(z)} + K(z)$ is prime). (We may take $\{l(k)\}_{k=1}^{\infty}$ is a sequence of natural numbers.)

Theorem 4. Let

$$F(z, w) \;:=\; H(z)\, K(w), \tag{14}$$

where $H(z)$ and $K(w)$ are two non-constant entire functions of one complex variable. Assume that $H(z)$ is transcendental and prime. Then, F is prime in entire sense.
Moreover, under (1), $f(\zeta) = a\zeta + b$ $(a \neq 0)$ and $g(z, w) = \{F(z, w) - b\}/a$.

Example 5. Assume that $\exp[H_1]$ and H_2 are, periodic, entire functions with the same period $2\pi i$ (say) such that $\exp[H_1]$ is non-costant and K is a non-constant entire function. Then the function F, defined by

$$F(z, w) \;:=\; \left(z\, e^{H_1(z)} \;+\; H_2(z)\right) K(w), \tag{15}$$

is prime in entire sense (cf. [10], as mentioned above).
Theorem 5. Let

$$F(z,w) \;:=\; S(z)\, T(w)\, e^{\varphi(z)}\, e^{\psi(w)}, \tag{16}$$

where $S(z)$ and $T(w)$ be two entire functions of order less than one such that $S(z)$ has at least one simple zero or two zeros with co-prime multiplicities and let φ and ψ be also non-constant entire functions. Then, if either the hyper order of F is finite or the function $H(z) := S(z)\, e^{\varphi(z)}$ is pseudo-prime, $F(z, w)$ is prime in entire sense. Moreover, under (1), we have $f(\zeta) = a\zeta + b$ $(a \neq 0)$ and $g(z, w) = \{F(z, w) - b\}/a$.

Example 6. Let

$$S(z) := \prod_{n=1}^{\infty} \left(1 - \frac{z}{n!}\right)$$

and

$$T(w) := \prod_{n=1}^{\infty} \left(1 - \frac{z}{n^2}\right).$$

Assume that $\varphi(z)$ is a non-constant polynomial and $\psi(w)$ is a non-constant entire function. Then, since $H(z) = S(z)\,e^{\varphi(z)}$ is pseudo-prime by virtue of a theorem due to Edrei ([3]) (indeed, prime), the function $F(z, w)$ as in (16) is prime (in entire sense).

In the following sections, we will prove Theorems 3, 4 and 5. The proof of Theorems 1 and 2 will be given elsewhere.

3. Proof of Theorem 3

Assume that the equation (3) holds with non-constant entire functions f and g. Suppose that f is non-linear. Then, by the assumption that F is prime as a function of z with w fixed,

$$g(z, w) = \alpha(w) + z\,\beta(w) \tag{17}$$

for some entire functions α and β.
Further $F(l(k)w, w)$ is prime for any natural number k. Hence,

$$g(l(w), w) = \alpha(w) + l(k)w\,\beta(w) \equiv c(k)w + d(k) \tag{18}$$

with some constants $c(k)$ and $d(k)$ $(k \in \mathbf{N})$.
By making use of Taylor's expansion of $\alpha(w)$ and $\beta(w)$, we conclude that $\alpha(w) = a_0 + a_1 w$ is linear and $\beta(w) = b_0$ is constant. Therefore $g(z, w) = a_0 + a_1 w + b_0 z$ must be linear. This implies that F is prime (in entire sense).

4. Proof of Theorem 4

Under (3), asuume that f is non-linear. Then

$$g(z,\, w) \;=\; \alpha(w) \,+\, z\,\beta(w) \tag{19}$$

is linear with respect to z, where α and β are some entire functions. Note that $H(z)$ must possess at least one zero, since H is assumed to be prime.

The case (i) : $H(z)$ has at least two distinct zeros $z_1,\, z_2$.
In this case, $g(z_1,\, w)$ and $g(z_2,\, w)$ must be constant with respect to w. Hence, in view of (19) and $z_1 \neq z_2$, we see that both $\alpha(w)$ and $\beta(w)$ must be also constant, which shows that g is linear.

The case (ii) : $H(z)$ has only one (simple) zero z_1.
In this case, $g(z_1,\, w)$ is constant $= \gamma$ with respect to w. From (19), $\alpha(w) = \gamma - z_1\,\beta(w)$ so that

$$g(z,\, w) \;=\; \gamma + (z \,-\, z_1)\,\beta(w). \tag{20}$$

Hence,

$$F(z,\, w) \;=\; H(z)\,K(w) \;=\; f(\gamma \,+\, (z - z_1)\,\beta(w)). \tag{21}$$

Here,

$$H(z) \;=\; (z - z_1)\,e^{S(z)} \tag{22}$$

$$f(\zeta) \;=\; (\zeta - \gamma)\,e^{T(\zeta-\gamma)} \tag{23}$$

for some entire functions S and T.

$$K(w) \;=\; e^{-S(z)}\,\beta(w)\,e^{T((z-z_1)\,\beta(w))} \tag{24}$$

Now, f is non-linear, so that T is non-constant. In (24), take $z = z_1$. Then, $K(w) = d_1\,\beta(w)$ (d_1 : *a constant*, $\neq 0$). On the otherhand, take $z = z_3 \neq z_1$ in (24), we have

$$K(w) \;=\; d_2\,\beta(w)\,e^{T((z_3-z_1)\,\beta(w))} \quad (d_2 : \textit{a constant},\, \neq 0) \tag{25}$$

Thus $T((z_3 - z_1)\,\beta(w))$ must be constant, which is not in reason. Hence f must be affine.

5. Proof of Theorem 5

Let

$$H(z) \;:=\; S(z)\,e^{\varphi(z)}, \quad K(w) \;:=\; T(w)\,e^{\psi(w)}. \tag{26}$$

Suppose that (3) holds i.e.

$$F(z, w) = H(z)\,K(w) = f(g(z, w)) \tag{27}$$

with non-constant entire functions f and g.
Assuming that f is non-linear, we will deduce a contradiction.

Firstly, we wish to note that, under (26) and (27), the number of the zeros of f must be finite. In fact, differentiating (27) partially with respect z and w and eliminating H and K, we obtain the identity such as

$$U(g(z, w)) = -\frac{g_{zw}}{g_z\, g_w}, \tag{28}$$

where

$$U(\zeta) := \frac{f''(\zeta)}{f'(\zeta)} - \frac{f'(\zeta)}{f(\zeta)}. \tag{29}$$

Here, we may assume that $g_{zw} \not\equiv 0$, since otherwise f'/f is constant, in which case $f(\zeta) = a\,e^{b\zeta} \neq 0$.

By applying the Lemma of Logarithmic Derivative (cf. [7], [8]) and a Theorem of Clunie's type to (29) (cf. [7]), we see that U must be a rational function. Hence, f can possess only a finite number of zeros, as is to be shown.

Since f must have at least one zero, assume that $f(\zeta_0) = 0$. Then, noting the assumptions, we must have

$$f(\zeta) = (\zeta - \zeta_0)\, e^{q(\zeta - \zeta_0)} \tag{30}$$

and

$$g(z, w) = \zeta_0 + S(z)T(w)\, e^{h(z, w)}, \tag{31}$$

where q and h are some non-constant entire functions.
In fact : we can rule out the case where

$$f(\zeta) = (\zeta - \zeta_1)^{p_1} \cdots (\zeta - \zeta_m)^{p_m}\, e^{q(\zeta)}$$

and

$$g(z, w) = \zeta_1 + S_1(z)T_1(w),$$

where ζ_j $(j = 1, \cdots m)$ are mutually distinct zeros of f with multiplicities p_j and q, S_1 and T_1 are entire functions such that (order) $\rho(S_1) < 1$ and $\rho(T_1) < 1$. Then from $F = f(g)$, we have

$$\varphi(z) + \psi(w) = q(\zeta_1 + S_1(z)T_1(w)) + 2l\pi i$$

for some integer l. Then, taking z_0 with $S_1(z_0) = 0$ and z_1 with $S_1(z_1) \neq 0$ in the above equation, we easily obtain a contradiction, since $T_1(w)$ is non-constant.

Now, assume that

$$H(z) \equiv S(z)\, e^{\varphi(z)} \ : \ pseudo - prime. \tag{32}$$

Take any point w_0 with $T(w_0) \neq 0$ and consider the equation

$$f(g(z, w_0)) = H(z)\, K(w_0). \tag{33}$$

¿From (32), the function in (33) is pseudo-prime. Since q is non-constant, f is transcendental (cf. (30)), Hence $g(z, w_0)$ must be a polynomial so that $h(z, w_0)$ is constant and S is a polynomial. This implies that

$$h(z, w) \equiv A(w) \ \ (independant\ of\ z).$$

Therefore, from (27), we have

$$\varphi(z) + \psi(w) = A(w) + q(\left(S(z)T(w)\, e^{A(w)}\right). \tag{34}$$

By taking z_0 with $S(z_0) = 0$ and z_1 with $S(z_1) \neq 0$ in the equation (34), we conclude through subtraction that

$$q(S(z_1)T(w)\, e^{A(w)}) \ : \ constant$$

which is not valid. Hence, f must be linear.

When the hyper order of F is finite, we have the identity

$$\varphi(z) + \psi(w) = h(z, w) + q\left(S(z)T(w)\, e^{h(z,w)}\right). \tag{35}$$

In this case, we conclude that $h(z, w)$ is a polynomial.
By taking z_0 with $S(z_0) = 0$ and z_1 with $S(z_1) \neq 0$ in the equation (35), if $h(z_1, w)$ is non-costant, then we have a contradiction as above. On the other hand, if $h(z_1, w)$ is constant for any such z_1, then it must be that

$$h(z, w) \equiv B(z) \ \ (depending\ only\ of\ z),$$

in which case $\psi(w)$ must become constant. Since $T(w)$ is non-constant, this is clearly not in reason. Thus we also have a contradiction, if f is non-linear.

References

[1] W. Bergweiler, Proof of a Conjecture of Gross concerning Fix-Points, Math. Z. 204 (1990), 381-390.
[2] W. D. Brownawell, On the Factorization of Partial Differential Equations, Canad. J. Math. 39 (1987),825-834.
[3] A. Edrei, Meromorphic functions with three radially distributed values, Trans. Amer. Math. Soc. 78 (1955), 276-293.
[4] B. Q. Li and C. C. Yang, Factorization of Meromorphic Functions in Several Complex Variables, Contemporary Mathematics 142 (1993), 61-74.
[5] Y. Noda, On Factorization of Entire Functions, Kodai Math. J. 4 (1981), 480-491.
[6] L. A. Rubel, W. A. Squires and B. A. Taylor, Irreducibility of certain Entire Functions with the Applications to Harmonic Analysis, Ann. of Math. 108 (1978), 553-567.
[7] L. A. Rubel and C. C. Yang, The Factorization of $A(z) + B(w)$ under Composition, Illinois J. of Math. 39 (1995),258-270.
[8] W. Stoll, Value distribution and the lemma of logarithmic derivative on polydiscs, Internat. J. Math. Math. Sci. 6 (1983), 617-669.
[9] H. Urabe, Primeness in divisor sense for certain entire functions, J. Math. of Kyoto Univ. 24 (1984), 127-140.
[10] H. Urabe, On Factorization of certain Entire and Meromorphic Functions, *Ibid.* 26 (1986), 177-190.

61 BOREL REMOVABLE SETS IN VALUE DISTRIBUTION THEORY

Shengjian Wu[1]

School of Mathematical Sciences, Peking University
Beijing 100871, China

wusj@pku.edu.cn

1. INTRODUCTION

In complex analysis, Picard Theorem is one of the most significant results.

Picard Theorem *Suppose that $f(z)$ is a transcendental meromorphic function in the plane C. Then $f(z)$ assumes every complex number infinitely often with at most two exceptions.*

In 1958, Lehto generalized Picard Theorem by introducing the notation and proving the existence of Picard sets for entire and meromorphic functions (cf. [7]). Later on the Picard sets have been extensively studied for entire and meromorphic functions (cf. [1,2,6]). Most of the papers on Picard sets are dealing with entire functions. In general speaking, the best Picard set of entire functions known so far can only contain a sequence of disks with extremely small radii. In the following, we shall use the standard notations in value distribution theory of meromorphic functions (cf. [4]).

[1]Supported by SRF for ROCS, SEM and the NSF of China

H.G.W. Begehr et al. (eds.), Proceedings of the Second ISAAC Congress, Vol. 1, 523–530.

Now suppose that $f(z)$ is meromorphic in the plane C and that the order $\lambda(0 < \lambda \leq +\infty)$ is defined by

$$\lambda = \limsup_{r\to\infty} \frac{\log T(r.f)}{\log r}.$$

Then we have the following famous

Borel Theorem *Suppose that $f(z)$ is a transcendental meromorphic function with order λ in the plane. Then*

$$\limsup_{r\to\infty} \frac{\log n(r,a)}{\log r} = \lambda,$$

except for at most two values a for which the upper limit can be smaller.

Following the idea of Picard sets, L. Yang introduced the following notation (cf. [10]).

Definition *A set $E \subset C$ is called a Borel removable set for a family $\mathcal{F}$ of meromorphic functions, if for any meromorphic function $f(z) \in \mathcal{F}$ with nonzero order λ_f, Borel theorem always holds in $C\backslash E$. That is, we have*

$$\limsup_{r\to\infty} \frac{\log n\{(|z| \leq r)\backslash E, f = a\}}{\log r} = \lambda_f$$

for all $a \in \bar{C}$, except for at most two values , where $n\{D, f = a\}$ denotes the roots of $f(z) - a = 0$ in D including multiplicities.

It is obvious that if E is a Borel removable set for a family $\mathcal{F}$ of meromorphic functions, then E is also a Picard set for $\mathcal{F}$.

In this paper, we will discuss various Borel removable sets in value distribution theory and related subjects.

2. Entire Functions

Let $\mathcal{F}$ and $\mathcal{F}_\infty$ stand for the families of all the entire functions of finite and nonzero order, and infinite order, respectively.

In the sequel for real numbers $\varphi; \eta > 0; \rho > 0$ and $S > 1$, we shall define the angular domain $|\arg z - \varphi| < \eta$ by $G(\varphi, \eta)$, and the sectorial domain $\{z = re^{i\theta}, \rho \leq r \leq S\rho\}\backslash G(\varphi, \eta)$ by $D(\rho, S, \varphi, \eta)$.

Theorem 1 and 2 below are proved in [9].

Theorem 1 *Let (R_m) be a sequence of real numbers such that*

$$\lim_{m\to\infty} \frac{R_{m+1}}{R_m} = \infty$$

and let (φ_m) *be a sequence of real numbers. Suppose that* $\eta(0 < \eta < \pi)$ *and* $S(> 1)$ *are two constants. Then*

$$E = \bigcup_{m=1}^{\infty} D_m$$

is a Borel removable set for $\mathcal{F}$, *where* $D_m = D(R_m, S, \varphi_m, \eta)(m = 1, 2, \cdots)$.

Remark 1 The following example shows that Theorem 1 is essentially sharp.

Let $r_n = 16^n$ and set

$$f(z) = \prod_{n=1}^{\infty} (1 + \frac{z}{r_n})^{2^n}.$$

It follows from computation that

$$log|f(z)| \geq 2^n,$$

for all $2r_n \leq |z| \leq 8r_n = \frac{r_{n+1}}{2}$. Therefore

$$min\{|f(z)| : |z| = r\} \to \infty$$

as $r \to \infty$ through these ring domains. If we choose $D'_m = D(\frac{1}{2}16^m, 4, 0, \frac{1}{2}\pi)$, we deduce that Theorem 1 is not valid for entire functions of order $\frac{1}{4}$.

Let n be a positive integer and $g(z) = f(z^n)$. Choosing

$$D'_m = D((\frac{16^m}{2})^{\frac{1}{n}}, 4^{\frac{1}{n}}, 0, \frac{\pi}{2n}),$$

we see that Theorem 1 does not hold for functions of order $\frac{n}{4}$.

The Borel removable set for $\mathcal{F}_\infty$ is even bigger.

Theorem 2 *Let* (R_m) *be a sequence of real numbers such that*

$$R_{m+1} > \sigma R_m,$$

where $\sigma > 1$ *is a constant, and let* (φ_m) *be a sequence of real numbers such that* $0 \leq \varphi_m < 2\pi$. *Suppose that* $\eta(0 < \eta < \pi)$ *and* $S(1 < S < \sigma)$ *are two constants. Then*

$$E_\infty = \bigcup_{m=1}^{\infty} D'_m$$

is a Borel removable set for $\mathcal{F}_\infty$, where $D'_m = D(R_m, S, \varphi_m, \eta)(m = 1, 2, \cdots)$.

Remark 2 Combining Theorem 1 and 2, we see that the set E in Theorem 1 is a Borel removable set for $\mathcal{F} \cup \mathcal{F}_\infty$.

Remark 3 The following example is perhaps interesting too.

Let $r_n \to \infty$ and $\frac{r_{n+1}}{r_n} \to \infty$ as $n \to \infty$ and let $0 < \varepsilon_n < 1$ for any n. Set

$$f(z) = \sum_{n=1}^{\infty} (\frac{z}{r_n})^{t_n},$$

where t_n is a positive integer such that

$$t_n log \frac{r_n + \varepsilon_n}{r_n} > 4t_{n-1} log r_n.$$

for all large n. If $r_n + \varepsilon_n < |z| < r_{n+1}$, then

$$|(\frac{z}{r_n})^{t_n} - f(z)| < \frac{1}{4}|\frac{z}{r_n}|^{t_n}.$$

Therefore we deduce that $f(z) \to \infty$ as $n \to \infty$ through these ring domains.

This implies that in Theorem 2 the sets D'_m cannot be replaced by the ring domains

$$r_m < |z| < r_m + \varepsilon_m.$$

Thus Theorem 2 is essentially sharp, too.

Remark 4 Let f be a moromorphic function with non-zero order λ and let $D_n = \{z; |z - z_n| < \varepsilon_n |z_n|\}, (n = 1, 2, ...,)$ be a sequence of disks, where $z_n \to \infty$ is a sequence of complex numbers and $\varepsilon_n \to 0$ is a sequence of positive numbers. $\{D_n\}$ is called a sequence of filling disks of order λ for f if there exists a sequence of positive numbers $\varepsilon'_n, n = 1, 2, ...,$ with $\varepsilon'_n \to 0$ such that, in D_n, f takes every complex number at least $|z_n|^{\lambda - \varepsilon'_n}$ times, except some complex numbers which can be enclosed in two spherical circles with radius $e^{-|z_n|^{\lambda - \varepsilon'_n}}$. (if $\lambda = \infty$, then we replace $\lambda - \varepsilon'_n$ by a suquence of positive numbers $N_n \to \infty$).

It is not hard to see from the proof of the above results that we in fact have proved that for any entire function f with non-zero order λ_f, there exists a sequence of filling disks of order λ_f for f in the complement of the Borel removable sets (cf [9]).

3 GENERALIZATIONS

Let $\mathcal{F}'$ and $\mathcal{F}'_\infty$ stand for the families of all the meromorphic functions, which have a Nevanlinna deficient value, with finite and nonzero order, and infinite order, respectively. Then we have the following generalization of Theorem 1 and 2.

Theorem 3 *Let E and E_∞ be defined in Theorem 1 and 2, respectively. Then E and E_∞ are Borel removable sets for $\mathcal{F}'$ and $\mathcal{F}'_\infty$, respectively.*

The proof of Theorem 3 is the same as the proof of Theorem 2 in [8].

4. REMOVABLE SET FOR DIFFERENTIAL POLYNOMIALS

In this section, our starting point is the following result which was due to Hayman.

Theorem A[3] *Suppose that $f(z)$ is a transcendental entire function and $N(>1)$ is an integer. Then $g(z) = f^N(z)f'(z)$ assumes every finite nonzero complex number infinitely often and $\phi(z) = f'(z) - f^N(z)$ has infinitely many zeros.*

We can generalize Hayman's results as follows.

Theorem 4 *Let E be defined in Theorem 1. Then for any entire function $f(z)$ of order $\lambda(0 < \lambda < +\infty)$ and for any integer $N > 1$, we have*

$$\limsup_{r\to\infty} \frac{\log n\{(|z| < r)\backslash E, f'(z) - f^N(z) = 0\}}{\log r} = \lambda, \tag{4.1}$$

and for any finite nonzero complex number a, we have

$$\limsup_{r\to\infty} \frac{\log n\{(|z| < r)\backslash E, f'(z)f^N(z) = a\}}{\log r} = \lambda. \tag{4.2}$$

Remark 5 Similar results for functions of infinite order are also true.

Proof. If there exist an entire function f with order λ and an integer N such that (4.1) does not held, then we will deduce a contradiction by the following argument.

Consider $F = (N-1)f^{N-1}(z)$. Then f and F have the same order λ. Set $\psi = -\frac{1}{F}$. Then $\psi \neq 0$. Moreover $\psi'(z) = 1 \leftrightarrow f'(z) - f^N(z) = 0$.

Now we use the following lemma.

Lemma 1[11] *Let $\psi(z)$ be meromorphic in $|z| < 1$. If*

$$n(1, \psi = 0) + n(\psi' = 1) + 2 = M,$$

then

$$n(\frac{1}{8}, \psi = a) < C(M log M + log \frac{1}{|\psi(z_0), a|})$$

for any complex number a*, where* $|z_0| < 1$*,* $C > 0$ *is a absolute constant and* $|\psi(z_0), a|$ *is the spherical distance of* $\psi(z_0)$ *and* a*.*

On the other hand, it follows from Remark 4 that there is a sequence of filling disks $D_n, n = 1, 2, ...$, of order λ for F in the complement of E. By using Lemma 1, we see that D_n is not a sequence of filling disks of ψ. As ψ and $\frac{-1}{\psi}$ have the same sequence of filling disks, we get a contradiction which completes the proof of the first part of Theorem 4.

Now we prove the second part of Theorem 4. If our conclusion is not true, there must exist an entire function f of order $\lambda, 0 < \lambda < \infty$, an inter $N > 1$ and a complex number $a \neq 0$ such that (4.2) does not held.

Set

$$\psi = \frac{f^{N+1}}{(N+1)a}$$

and use a result in [12], we can get a contradiction in the same way as we did before.

5. MEROMORPHIC FUNCTIONS

In the case of meromorphic functions we cannot obtain an expectant result which is similar to Theorem 1 or 2. Actually we have the following result (cf [8]).

Theorem 5 *Let* (D_n) *be an abbitrary sequence of disks tending to the infinite and* $\lambda(0 < \lambda < 1)$ *be an arbitrary number. Then there exists a meromorphic function* $f(z)$ *of order* λ *such that* $f(z)$ *is uniformly bounded in* $C \backslash \{\bigcup D_n\}$*.*

Although Theorem 5 disproves that Borel removable set for meromorphic functions can contain a sequence of disks, we can easily get such a set from Definition 1. For example, we give the following result for the completeness of the paper.

Theorem 6 *Let* (r_n) *be a sequence of real numbers satisfying* $r_{n+1} > (1 + log^{-k} r_n) r_n > 1$*, where* $k(> 0)$ *is an arbitrary number, and let* $a_1^n.a_2^n, ..., a_{m_n}^n$ *be* m_n *points on* $|z| = r_n$*. If* $m_n < log^k r_n$*, then* $E = \{a_j^n; n = 1, 2, ..., j = 1, 2, ..., m_n.\}$ *is a Borel removable set for all meromorphic functions.*

6. APPLICATIONS

We consider the differential equation

$$f'' + Af = 0, \tag{6.1}$$

where $A(z)$ is an entire function, polynomial or transcendental. Recently there have been some papers on the distribution of zeros of solutions of (6.1). The main interests are that where the zeros of the solutions of (6.1) are distributed. In this direction we get the following result in [5].

Theorem 7 *Let E_∞ be defined in Theorem 2. Suppose $A(z)$ is an entire function and f_1 and f_2 are two linearly independent solutions of (6.1) with coefficient A. Set $F = f_1 f_2$. If A is transcendental and of finite order and*

$$\limsup_{r\to\infty} \frac{\log n\{(|z| < r), F = 0\}}{\log r} = \infty,$$

then we have

$$\limsup_{r\to\infty} \frac{\log n\{(|z| < r)\backslash E_\infty, F = 0\}}{\log r} = \infty.$$

If $A(z)$ is a polynomial of degree $n \geq 1$, then we have

$$\limsup_{r\to\infty} \frac{\log n\{(|z| < r)\backslash E_\infty, F = 0\}}{\log r} = \frac{n+2}{2}.$$

References

[1] Anderson J., Baker I. and Clunie J., *The distribution of values of certain entire and meromorphic functions,*Math. Z. 178(1981), 509-525.

[2] Baker I. and Liverpool L., *Picard sets of entire functions,* Math. Z., 126(1972), 230-238.

[3] Hayman W., *Picard values of meromorphic functions and their derivatives,* Ann. of Math., 70(1959), 9-42.

[4] Hayman W., *Meromorphic functions,* Oxford, 1964.

[5] Laine I. and Wu S., *Removable sets in the oscillation theory of complex differential equations,* J. Math. Anal. and Appl., 214(1997), 233-

244.

[6] Langley J., *The distribution of zeros of certain differential polynomials,* J. London Math. Soc., 29(1984), 485-498.

[7] Lehto O., *A generalization of Picard's theorem,* Ark. Mat. 3(1958), 495-500.

[8] Wu S., *On Borel removable sets of entire and meromorphic functions,* Sci. China, 35,12(1992),1444-1455.

[9] Wu S., *Further results on Borel removable sets of entire functions,* Ann. Acad. Sci. Fenn., Ser. A. I. Math., 19(1994), 67-81.

[10] Yang L., *Some results on the value distribution theory of entire and meromorphic functions,* Lecture at ZTH in 1978.

[11] Yang L. and Zhang. Q., *New singular directions of meromorphic functions,* Sci. China, 27(1984), 352-366.

[12] Yang L., *Angular distribution and multiple values between entire functions and their derivatives,* Sci. China, 23(1979), 731-751.

62 MEROMORPHIC SOLUTIONS OF ALGEBRAIC DIFFERENTIAL EQUATIONS AND α-YOSHIDA FUNCTIONS

Hasi WULAN[1]

Department of Mathematics, Shantou University
Guangdong, Shantou, 515063 P. R. China

wulan@mailserv.stu.edu.cn

MOS subject classification Categories: 30D50, 34A20

Keywords: α-Yosida functions, meromorphic solutions, algebraic differential equations

Abstract

A function f meromorphic on the complex plane $\mathbb{C}$ is called an α-Yosida function if $f^{\#}(z) = O(|z|^{\alpha})$ as $|z| \to \infty$, $\alpha \geq -1$, where $f^{\#}(z) = |f'(z)|/(1+|f(z)|^2)$. We give a sufficient and necessary condition for a meromorphic function f to be not an α-Yosida function by proving a relationship between α-Yosida functions and normal families. We also give some integral characterizations of α-Yosida functions. Applying our results we know that the meromorphic solutions of some higher order algebraic differential equations are α-Yosida functions. Moreover, some estimates for the order of growth of these meromorphic solutions by studying α-Yosida functions are given.

[1]The author was supported by a grant from the Väisälä Fund, Finland.

H.G.W. Begehr et al. (eds.), Proceedings of the Second ISAAC Congress, Vol. 1, 531–539.

1. Introduction

It is known that any meromorphic solution of a first order algebraic differential equation has finite order of growth, and this result is due to Gol'dberg [Go]. Also, Makhonjko [Mak] investigated the following first order differential equation

$$P_0(z,w)(w')^n = \sum_{j=1}^{n} P_j(z,w)(w')^{n-j}, \tag{1.1}$$

where

$$P_j(z,w) = \sum_{m=1}^{s_j} a_{m_j}(z)w^m, \ j = 1,2,...,n,$$

and $a_{m_j}(z), j = 1,2,...,n$, are rational functions with the asymptotic expansion $a_{m_j}(z) = c_{m_j}z^{m_j}(1+o(1)), z \to \infty$; $P_0(z,w)$ is a polynomial of z and w and $P_0(z,w) \not\equiv 0$.

It is natural to consider this topic for higher order case. Barsegian [Ba] considered the algebraic differential equations of the form

$$P_0(z,w)\left(\frac{dw}{dz}\right)^n = \sum_{j=1}^{m} P_j(z,w)D_j[w], \ n, \ m \in \mathbb{N}, \tag{1.2}$$

where $P_j(z,w)$ are polynomials in z and w and $P_0(z,w) \not\equiv 0$; $D_j[w]$ are differential monomials in w of the form

$$D_j[w] = \left(\frac{dw}{dz}\right)^{j_1}\left(\frac{d^2w}{dz^2}\right)^{j_2}\cdots\left(\frac{d^lw}{dz^l}\right)^{j_l}, \ 1 \le j \le m,$$

with $j_1, j_2, \cdots, j_l \in \mathbb{N}\cup\{0\}$. We call $\nu_j = j_1 + 2j_2 + \cdots + lj_l$ the weight of D_j, and the weight of

$$P[w](z) = \sum_{j=1}^{m} P_j(z,w)D_j[w]$$

is defined by $\nu(P) = \max_{1\le j\le m}\{\nu_j\}$. Let $d_j = \deg_z P_j(z,w)$, $j = 0, 1, \cdots, m$, and set

$$\alpha = \max_{1\le j\le m}\left\{\frac{d_j - d_0}{n-\nu_j}, -1\right\} \tag{1.3}$$

and

$$\beta = \max_{1 \leq j \leq m} \{d_j - d_0\}.$$

Barsegian [Ba] proved that Gol'dberg's result is also true for higher order algebraic differential equations of form (1.2) if $n > \nu(P)$. Recently, Frank and Wang [FW] gave some estimates for the growth of meromorphic solutions of (1.2) based on the assumption $\alpha \geq 0$ in (1.3). They showed that the order of meromorphic solution function of (1.2) has an upper bound $2\alpha + 2$ which is two at least since $\alpha \geq 0$. The ideas in their methods stem from the work of Bergweiler [Be], in which he gave a simple proof to Barsegian's result using Zalcman's Lemma [Za]. However, there are some differential equations with $\alpha < 0$. For example, let us look at the algebraic differential equation

$$(1.4) \quad 4z^2 \left(\frac{dw}{dz}\right)^6 = 4zw^3 \left(\frac{dw}{dz}\right)^2 \left(\frac{d^2w}{dz^2}\right) - w^5 \frac{d^2w}{dz^2} + zw^2 \left(\frac{dw}{dz}\right)^4.$$

It is easy to see that $n = 6$, $\nu(P) = 4$ and $\alpha = -\frac{1}{2}$. Thus we define α as (1.3), which satisfies $-1 \leq \alpha < \infty$. In [Wu2] we showed that meromorphic solutions of (1.2) are close connection with α-Yosida functions and that estimates given in [FW] can be extended to the case $\alpha > -1$. Now let us first introduce the definition of α-Yosida functions.

Let f be a meromorphic function on $\mathbb{C}$ and let $-1 \leq \alpha < \infty$. We say f an α-Yosida function, denoted by $f \in Y^\alpha$, if

$$f^\#(z) = O(|z|^\alpha), \ |z| \to \infty,$$

where $f^\#(z) = \frac{|f'(z)|}{1+|f(z)|^2}$ is the spherical derivative of f. In particular, a Yosida function [Yo] is the special case for $\alpha = 0$ of α-Yosida functions. The main result in [Wu2] is the following:

Theorem A. *Let f be a meromorphic solution of (1.2) with $n > \nu(P)$ and let α be defined by (1.3). Then $f \in Y^\alpha$.*

As a corollary of Theorem A, any meromorphic solution of (1.2) with $n > \nu(P)$ and $\alpha > -1$ has order $2\alpha + 2$ at most since any α-Yosida function has order $2\alpha + 2$ at most for $\alpha > -1$. It shows that this upper bound for the order of meromorphic solutions of (1.2) may near zero sufficiently.

Applying Theorem A we know that each meromorphic solution w of (1.4) satisfies that $w \in Y^{-\frac{1}{2}}$ and w has order 1 at most. Moreover, we know that the function $w(z) = \exp(\sqrt{z})$, $\sqrt{1} = 1$ and $z \neq 0$, satisfies the equation (1.4). Thus, $w \in Y^{-\frac{1}{2}}$ and w has order 1 at most.

For case $\alpha = -1$, α-Yosida function f is called a Julia's exceptional function. We know the fact from [Zi] that f is a Julia's exceptional function if and only if the family $\{f(az)\}$, $a \neq 0$, is normal in the sense of Montel. The results in [LV1] showed that a Julia's exceptional function has an upper estimate

$$T_0(r, f) = O(\log^2 r),$$

where $T_0(r, f)$ is the Ahlfors–Shimizu characteristic function of f. They also mentioned in [LV1] that this bound can be attained by the Julia's exceptional function

$$f(z) = \prod_{n=0}^{\infty} \frac{2^n - z}{2^n + z}.$$

As another corollary of Theorem A, it is immediately obtained that any meromorphic solution of (1.2) with $n > \nu(P)$ and $\alpha = -1$ is a Julia's exceptional function.

The following lemma, a principal tool, played an important role in the proof of Theorem A in [Wu2] as well as in the theory of meromorphic functions including the studying of α-Yosida functions.

Lemma. *Let f be a meromorphic function on $\mathbb{C}$ and let $-1 \leq \alpha < \infty$. Then $f \notin Y^\alpha$ if and only if there exist sequences $\{z_n\} \subset \mathbb{C}$ and $\{\rho_n\}$ of positive numbers with $\rho_n |z_n|^\alpha \to 0$ and $|z_n| \to \infty$ such that the sequence $\{g_n(\xi)\} = \{f(z_n + \rho_n \xi)\}$ converges spherically uniformly on each compact subset of $\mathbb{C}$ to a non-constant Yosida function.*

This lemma can be viewed as a version of the Lohwater's and Pommerenke's theorem [LP, Theorem1] for α-Yosida functions. A special case of this lemma for Yosida functions was given by Minda [Mi]. In section 3 of this paper, we write this lemma in a different way and give a new proof of it. Our new proof bases on an important result about α-Yosida functions, which will be showed in section 2 and assert a relationship between α-Yosida functions and normal families.

2. Criteria of α-Yosida functions

We begin with a result of independent interest.

Theorem 1. *Let f be a meromorphic function on $\mathbb{C}$ and let $-1 \leq \alpha < \infty$. Then $f \in Y^\alpha$ if and only if there exists R, $0 < R < \infty$, such that*

$\{f(a+|a|^{-\alpha}z)\}$ is a normal family in the unit desk $\mathcal{D} = \{z \in \mathbb{C} : |z| < 1\}$ for $|a| > R$.

Proof. Let $\{f(a+|a|^{-\alpha}z)\}$ be a normal family in $\mathcal{D}$ for $|a| > R$, where R is a fixed positive number. By Marty's result [Mar], for a compact subset K of $\mathcal{D}$ with $0 \in K$, there exists a constant C_K such that

$$(f(a+|a|^{-\alpha}z))^{\#} = f^{\#}(a+|a|^{-\alpha}z)|a|^{-\alpha} \leq C_K \tag{2.1}$$

for all a, $|a| > R$. Taking $z = 0$ in (2.1), we have

$$f^{\#}(a) \leq C_K|a|^{\alpha}, \ |a| > R.$$

It means that $f \in Y^{\alpha}$.

Conversely, let f be an α-Yosida function in $\mathbb{C}$. Then there exists δ, $1 \leq \delta < \infty$, such that

$$C = \sup_{\delta<|z|<\infty} \frac{f^{\#}(z)}{|z|^{\alpha}} < \infty.$$

Choose $R \in (0, \infty)$ such that $R - \frac{1}{R^{\alpha}} \geq \delta$. Then $R > 1$. We now show that the family $\{f(a+|a|^{-\alpha}z)\}$ is normal in $\mathcal{D}$ for $|a| > R$. For each $r \in (0, 1)$ and z, $|z| \leq r$, we have

$$|a+|a|^{-\alpha}z| \geq |a| - |a|^{-\alpha}|z| > R - \frac{1}{R^{\alpha}} \geq \delta.$$

If $\alpha \geq 0$, then

$$\begin{aligned}(f(a+|a|^{-\alpha}z))^{\#} &= \frac{f^{\#}(a+|a|^{-\alpha}z)}{|a+|a|^{-\alpha}z|^{\alpha}}\left(\frac{|a+|a|^{-\alpha}z|}{|a|}\right)^{\alpha} \\ &\leq C\left(1+\frac{|z|}{|a|^{\alpha+1}}\right)^{\alpha} \\ &\leq C\left(1+\frac{r}{R^{\alpha+1}}\right)^{\alpha},\end{aligned}$$

and if $-1 \leq \alpha < 0$,

$$\begin{aligned}(f(a+|a|^{-\alpha}z))^{\#} &= \frac{f^{\#}(a+|a|^{-\alpha}z)}{|a+|a|^{-\alpha}z|^{\alpha}}\left(\frac{|a+|a|^{-\alpha}z|}{|a|}\right)^{\alpha} \\ &\leq C\left(1-\frac{|z|}{|a|^{\alpha+1}}\right)^{\alpha} \\ &\leq C\left(1-\frac{r}{R^{\alpha+1}}\right)^{\alpha}.\end{aligned}$$

Therefore,

$$(f(a+|a|^{-\alpha}z))^{\#} \le C\Big\{\Big(1+\frac{r}{R^{\alpha+1}}\Big)^{\alpha}+\Big(1-\frac{r}{R^{\alpha+1}}\Big)^{\alpha}\Big\} \tag{2.2}$$

holds for all z, $|z| \le r$ and all a, $|a| > R$. By Marty's result $\{f(a+|a|^{-\alpha}z)\}$ is a normal family in $\mathcal{D}$ for $|a| > R$. This completes the proof of Theorem 1.

In view of some results on normal families (see [CL]), we first give some integral characterizations of α-Yosida functions.

Proposition 1. *Let f be a meromorphic function on $\mathbb{C}$ and let $-1 \le \alpha < \infty$. Then the following statements are equivalent:*

(a) $f \in Y^{\alpha}$.

(b) *There exists R, $0 < R < \infty$. For each $p > 2$, and each compact subset $K \subset \mathcal{D}$ there exists a positive constant $C_{K,p}$ such that*

$$\sup_{|a|>R} \iint_K ((f\circ g_{a,\alpha})^{\#}(z))^p dxdy \le C_{K,p},$$

where $g_{a,\alpha}(z) = a+|a|^{-\alpha}z$.

(c) *There exist $R \in (0,\infty)$ and $p > 2$ such that, for each compact subset $K \subset \mathcal{D}$ there exists a positive constant $C_{K,p}$ for which*

$$\sup_{|a|>R} \iint_K ((f\circ g_{a,\alpha})^{\#}(z))^p dxdy \le C_{K,p},$$

where $g_{a,\alpha}(z) = a+|a|^{-\alpha}z$.

Proposition 2. *Let f be a meromorphic function on $\mathbb{C}$, let $-1 \le \alpha < \infty$ and let $\mathcal{P}$ be a collection of paths in $\mathcal{D}$ such that each path $\gamma \in \mathcal{P}$ satisfies the conditions that γ ends at the unit circle $\partial\mathcal{D}$ and $\cup\{\gamma : \gamma \in \mathcal{P}\} = \mathcal{D}$. The following are equivalent:*

(a) $f \in Y^{\alpha}$.

(b) *There exists R, $0 < R < \infty$. For each $p > 1$ and each compact subset $K \subset \mathcal{D}$, there exists a positive constant $C_{K,p}$ such that*

$$\sup_{|a|>R}\sup_{\gamma\in\mathcal{P}} \int_{\gamma\cap K} ((f\circ g_{a,\alpha})^{\#}(z))^p |dz| \le C_{K,p},$$

where $g_{a,\alpha}(z) = a+|a|^{-\alpha}z$.

(c) *There exist $R \in (0,\infty)$ and $p > 1$ such that for each compact subset $K \subset \mathcal{D}$, there exists a positive constant $C_{K,p}$ for which*

$$\sup_{|a|>R}\sup_{\gamma\in\mathcal{P}} \int_{\gamma\cap K} ((f\circ g_{a,\alpha})^{\#}(z))^p |dz| \le C_{K,p}$$

where $g_{a,\alpha}(z) = a+|a|^{-\alpha}z$.

3. Lohwater–Pommerenke theorem for α-Yosida functions

The following theorem is due to Zalcman [Za]; it is a version of Lohwater–Pommerenke theorem [LP] for normal functions and can be used to get a well-known heuristic principle; see [Za]. We know that Lohwater–Pommerenke theorem is an essential reformulation of the concept of normal functions in the unit disk introduced by Lehto and Virtanen [LV2]. A non-trivial generalization of Lohwater–Pommerenke theorem has been given in [Wu1].

Theorem Z. *A family $\mathcal{F}$ of meromorphic functions is not a normal family in $\mathcal{D}$, then there exist a compact subset K of $\mathcal{D}$, a sequence $\{\rho_n\}$ of positive numbers with $\rho_n \to 0$, and a sequence $\{z_n\}$ in K and $\{f_n\}$ in $\mathcal{F}$, such that the sequence of functions $\{g_n(\xi)\} = \{f(z_n+\rho_n\xi)\}$ converges spherically uniformly on each compact subset of $\mathbb{C}$ to a non-constant meromorphic function.*

Using above theorem we prove the following:

Theorem 2. *Let f be a meromorphic function on $\mathbb{C}$ and let $-1 \leq \alpha < \infty$. Then the following statements are equivalent:*

(a) *f is not an α-Yosida function on $\mathbb{C}$.*
(b) *There exist a sequence $z_n \to \infty$ and a sequence $\{\sigma_n\}$ of positive numbers satisfying $\sigma_n \to 0$ and $\sigma_n|z_n|^\alpha \to 0$ such that the sequence of functions $\{g_n(\xi)\} = \{f(z_n + \sigma_n\xi)\}$ converges to a non-constant meromorphic function $g_0(\xi)$ spherically and locally uniformly on $\mathbb{C}$.*

Proof. (a) $\Rightarrow$ (b). Theorem 1 and the assumption that f is not an α-Yosida function assert that for any $R > 1$, the family $\{f(a+|a|^{-\alpha}z)\}$ is not normal in $\mathcal{D}$ for $|a| > R$. Thus $\{f(a+|a|^{-\alpha}z)\}$ is not normal at $z_0' \in \mathcal{D}$ for $|a| > R$. By Theorem Z there exist $z_n' \in \mathcal{D}$ with $z_n' \to z_0'$, positive numbers $\rho_n \to 0$ and a_n, $|a_n| > R$, such that $f(a_n + |a_n|^{-\alpha}(z_n' + \rho_n\xi))$ converges to a non-constant meromorphic function $g_0(\xi)$ spherically and locally uniformly on $\mathbb{C}$. Set

$$z_n = a_n + |a_n|^{-\alpha}z_n', \quad \sigma_n = |a_n|^{-\alpha}\rho_n, \quad n \in \mathbb{N}. \tag{3.1}$$

Then functions $g_n(\xi) := f(z_n + \sigma_n\xi)$ are defined on $\mathbb{C}$ for $n \in \mathbb{N}$ and we have

$$\sigma_n|z_n|^\alpha = \rho_n\left(\frac{|a_n + |a_n|^{-\alpha}z_n'|}{|a_n|}\right)^\alpha.$$

If $\alpha \geq 0$, then

$$\sigma_n|z_n|^\alpha \leq \rho_n\left(1+\frac{|z_n'|}{R^{1+\alpha}}\right)^\alpha,$$

and if $-1 \leq \alpha < 0$,

$$\sigma_n|z_n|^\alpha \leq \rho_n\left(1-\frac{|z_n'|}{R^{1+\alpha}}\right)^\alpha.$$

Thus $\sigma_n|z_n|^\alpha \to 0$ as $n \to \infty$ since $\rho_n \to 0$. Now we show that the sequence $\{z_n\}$ can be chosen to satisfy $z_n \to \infty$. In fact, since $R(>1)$ is arbitrary and $\{a_n\}$ satisfies $|a_n| > R$ for all $n \in \mathbb{N}$. Therefore, when R is chosen sufficiently large, we have $a_n \to \infty$. By (3.1), $z_n \to \infty$.

(b) $\Rightarrow$ (a). Suppose (b) is satisfied, that is, there exist a sequence $\{z_n\} \subset \mathbb{C}$ satisfying $z_n \to \infty$ and a sequence $\{\sigma_n\}$ of positive numbers with $\sigma_n \to 0$ and $\sigma_n|z_n|^\alpha \to 0$ such that the sequence of functions $g_n(\xi) = f(z_n + \sigma_n\xi)$ converges to a non-constant meromorphic function $g_0(\xi)$ spherically and locally uniformly on $\mathbb{C}$. We choose $\xi_0 \in \mathbb{C}$ such that $g_0^\#(\xi_0) \neq 0$. Assume that $f \in Y^\alpha$. Then there exists a constant K such that $\frac{f^\#(z)}{|z|^\alpha} \leq K$ for $|z|$ big enough. Thus for n sufficiently large, $|z_n + \sigma_n\xi_0| \to \infty$.

Therefore,

$$\begin{aligned} g_n^\#(\xi_0) &= \sigma_n f^\#(z_n + \sigma_n\xi_0) \\ &\leq K\sigma_n|z_n|^\alpha\left(\frac{|z_n + \sigma_n\xi_0|}{|z_n|}\right)^\alpha. \end{aligned}$$

Since $\sigma_n|z_n|^\alpha \to 0$ and $|z_n| \to \infty$, we have

$$g_n^\#(\xi_0) \leq K\sigma_n|z_n|^\alpha\left(1+\frac{\sigma_n|z_n|^\alpha|\xi_0|}{|z_n|^{1+\alpha}}\right)^\alpha \to 0, \ \alpha \geq 0,$$

and if $-1 \leq \alpha < 0$,

$$g_n^\#(\xi_0) \leq K\sigma_n|z_n|^\alpha\left(1-\frac{\sigma_n|z_n|^\alpha|\xi_0|}{|z_n|^{1+\alpha}}\right)^\alpha \to 0$$

as $n \to \infty$. These imply that $g_0^\#(\xi_0) = 0$, a contradiction. Therefore, $f \notin Y^\alpha$. We complete the proof of Theorem 2.

References

[Ba] G. Barsegian, *Geometrical theory of meromorphic functions*, manuscript.

[Be] W. Bergweiler, *On a theorem of Gol'dberg concerning meromorphic solutions of algebraic differential equations*, Complex Variables, to appear.

[CL] H. Chen and P. Lappan, *An integral bound for the spherical derivative*, Analysis, **12** (1992), 233–247.

[FW] G. Frank and Y. Wang, *On the meromorphic solutions of algebraic differential equations*, Analysis, **18** (1998), 49–54.

[Go] A. Gol'dberg, *On single-valued solutions of first-order differential equations (Russian)*, Ukrain Math. Zh., **8** (1956), 254–261.

[LP] A. J. Lohwater, and Ch. Pommerenke, *On normal meromorphic functions*, Ann. Acad. Sci. Fenn. Ser. A I Math., **550** (1973), 1–12.

[LV1] O. Lehto, and K. I. Virtanen, *On the behaviour of meromorphic functions in the neighbourhood of an isolated singularity*, Ann. Acad. Sci. Fenn. Ser. A I Math., **240** (1957), 1-9.

[LV2] O. Lehto, and K. I. Virtanen, *Boundary behaviour and normal meromorphic functions*, Acta Math., **97** (1957), 47–65.

[Mar] F. Marty, *Recherches sur la répartition des valeurs d'une fonction méromorphe*, Ann. Fac. Sci. Univ. Toulouse Math., **23** (1931), 183–261.

[Mak] Z. Makhonjko, *An application of a result of Clunie and Hayman to differential equations of the first order (Russian)*, Theory of Functions Funkt. Anal. and their Appl., Kharjkov **35** (1981), 74–75.

[Mi] D. Minda, *Yosida functions*, Lectures on complex analysis, Proceedings of the symposium on complex analysis, World Scientific, London, 1988.

[W1] H. Wulan, *On some classes of meromorphic functions*, Ann. Acad. Sci. Fenn. Math. Diss., **116** (1998), 1–57.

[W2] H. Wulan, *α-Yosida functions and meromorphic solutions of algebraic differential equations*, Aanalysis, to appear.

[Yo] K. Yosida, *On a class of meromorphic functions*, Proc. Phys-Math. Soc. Japan, **16** (1934), 227–235.

[Za] L. Zalcman, *A heuristic principle in complex function theory*, Amer. Math. Monthly, **82** (1975), 813–817.

[Zi] T. Zinno, *Some properties of Julia's exceptional functions and an example of Julia's exceptional function with Julia's direction*, Ann. Acad. Sci. Fenn. Ser. A I Math., **464** (1970), 1–12.

63 MEROMORPHIC SOLUTIONS OF DIFFERENCE EQUATIONS

Niro YANAGIHARA

Minami-Iwasaki 617-18, Ichihara, 290-0244, Japan

1 Difference equations and differential equations

Of course, difference equations and differential equations are in intimate relations. But there are several dissimilarities between them.

The differential equation

$$\frac{dy}{dx} = F(x, y), \quad x, y \in \mathbb{C}, \tag{1.1}$$

admits the unique analytic solution $y = f(x)$ with the initial values (x_0, y_0), when $F(x, y)$ is analytic there. In general, the solution exists only locally, can not be continued analtically to the whole plane, except for very restricted cases.

Consider the difference equation

$$y(x + 1) = F(x, y(x)), \quad x, y \in \mathbb{C}. \tag{1.2}$$

To obtain a solution $y = f(x)$ such that $y_0 = f(x_0)$, we have to prescribe data $y(x)$ for $x \in S$, where $S = \{x = \xi + i\eta \mid a \leq \xi < a + 1, \ \eta \in (-\infty, \infty)\}$ is a vertical strip, which contains $x_0 : \ a \leq \Re[x_0] < a + 1$. Then, we are requested that the prescribed values should be continued analytically on the vertical lines $\xi = a, \ \xi = a + 1$, a condition difficult to be met. So, we usually proceed as follows: Let V be a domain such

H.G.W. Begehr et al. (eds.), Proceedings of the Second ISAAC Congress, Vol. 1, 541–550.

that $x \in V$ implies $x - t \in V$ $(0 < t \le 1)$. For example, angular domain $\{x \mid |\arg[x] - \pi| < \pi/2\}$ will do. We will seek solutions with prescribed behaviours in the domain V, e.g. an asymptptic expansion. If this local solution is obtained, then we get a meromorphic solution, by analytic continuation using the equation.

Thus: for differential equations, initial value problems are easily formulated and analytic continuations are difficult, while for difference ones the situations stand in reverse.

Further, if $F(x, y)$ is rational, then any meromorphic solution of (1.1) is of finite order with normal type, while in (1.2) we have

THEOREM 1.1 *Suppose $F(x, y)$ is rational. If (1.2) admits a meromorphic solution of finite order, then either the solution is rational, or $F(x, y)$ is of the form $F(x, y) = \dfrac{a(x)y + b(x)}{c(x)y + d(x)}$, with polynomials $a(x), b(x), c(x), d(x)$.*

Further consider $(y(x+1) - y(x) - 1)^2 = 4y(x)$, algebraically irreducible. Putting $y(x) = u(x)^2$, we get solutions $y(x) = (x + C)^2$ and $y(x) = (Ce^{\pi i x} + \frac{1}{2})^2$. Compare with the correponding differential one $(y' - 1)^2 = 4y$. [2], p.74, p.179.

2 The case where $F(x, y) = F(y)$

Contrary to differential equations, there is no general theorem which guarantees existence of solutions for difference equations. In a sense, we should testify it case by case.

2.1 Existence of solutions

Here we consider the equation

$$y(x+1) = F(y(x)), \tag{2.1.1}$$

where $F(y)$ is either a rational (not linear) function or a transcendental entire one.

THEOREM 2.1.1 *Suppose there is a fixed point y^* with $|F'(y^*)| > 1$. Write $F'(y^*) = \lambda$. Then there is a meromorphic solution $\Psi_\lambda(x)$ which tends to y^* as $\Re[x] \to -\infty$ and is expanded as*

$$\Psi_\lambda(x) = y^* + \sum_{k=1}^{\infty} c_k \lambda^{kx} \quad (\lambda = F'(y^*)) \quad \textit{for } |\lambda^x| < \rho \tag{2.1.2}$$

with a sufficiently small $\rho > 0$, in which $c_1 \neq 0$ may be prescribed arbitrarily and $c_k (k \geq 2)$ are determined uniquely if c_1 is fixed.

On the other hand, when $F'(y^*) = 1$ for the fixed point y^*, we have

$$\begin{aligned} F(y) - y^* &= (y - y^*) + a_m (y - y^*)^{m+1} + \cdots \\ &= (y - y^*)\{1 + a_m (y - y^*)^m + \cdots\}, \quad a_m \neq 0,\ m \geq 1, \end{aligned}$$

$$\frac{-1}{m a_m} \frac{1}{(F(y) - y^*)^m} = \frac{-1}{m a_m} \frac{1}{(y - y^*)^m} \{1 + a_m (y - y^*)^m + \cdots\}^{-m}.$$

Put

$$u = \frac{-1}{m a_m} \frac{1}{(y - y^*)^m},$$

then we have the equation

$$u(x+1) = G(u(x)). \quad G(u) = u + 1 + \sum_{k=1}^{\infty} b_k u^{-k/m}. \tag{2.1.3}$$

The equation (2.1.3) was studied by Kimura[4]. Let $d \in \mathbb{C}$ be arbitrarily given. Kimura obtained a meromorphic solution $\phi(x, d)$ which is asymptotically expanded as

$$u = \phi(x, d) \sim x + d + \sum_{j+k \geq 1} p_{jk} x^{(-j+m)/m} \left(\frac{\log x}{x} \right)^k, \quad p_{m0} = d, \tag{2.1.4}$$

in a domain $D(\epsilon, R)$, where

$$\begin{aligned} &D(\epsilon, R) \\ =&\{|x| > R, |\arg[x] - \pi| < \frac{\pi}{2} - \epsilon\} \cup \{\Im[x e^{-i\epsilon}] > R\} \cup \{\Im[x e^{i\epsilon}] < -R\} \end{aligned}$$

in which $\epsilon \in (0, \pi/2)$ is arbitrarily given and $R > 0$ is determined by d and ϵ. Coefficients p_{jk} are determined uniquely if d is fixed. Thus we get

THEOREM 2.1.2 (Kimura) *Suppose there is a fixed point y^* with $F'(y^*) = 1$. Then there is a meromorphic solution $\Phi(x)$ which is expressed in $D(\epsilon, R)$ as*

$$\Phi(x) = y^* + \frac{1}{(-m a_m \phi(x, d))^{1/m}}, \tag{2.1.5}$$

where $\phi(x, d)$ is asymptotically expanded as (2.1.4) in a domain $D(\epsilon, R)$.

Thus we know that, if there is a fixed point $y^*, F(y^*) = y^*$, for $F(y)$ such that

$$\text{either} \quad |F'(y^*)| > 1 \quad \text{or} \quad F'(y^*) = 1, \tag{2.1.6}$$

we obtain a meromorphic solution. When $F(y)$ is rational, y^* may be ∞, in which case we define $F'(y^*) = \lim_{y\to\infty} 1/F'(y)$. The following lemma is due to Julia [3].

LEMMA 2.1.3 *When $F(y)$ is a rational function of degree not less than 2, then there is a fixed point $y^* \in \mathbb{C} \cup \{\infty\}$ for which (2.1.6) holds.*

By similar method, we can prove

LEMMA 2.1.4 *When $F(y)$ is transcendental entire with lower order $\mu < 1/2$, then there is a fixed point $y^* \in \mathbb{C}$ for which (2.1.6) holds.*

Proof. First note that $F(y)$ admits infinitely many fixed points $\{a_n\}$. If there is some a_{n_0} of multiplicity ≥ 2, we have $F'(a_{n_0}) = 1$. If each a_n is of multiplicity 1, we have $F'(a_n) \neq 1$. Then, denoting by C a circle $C = \{|x| = R\}$,

$$I(R) = \frac{1}{2\pi} \int_C \frac{d\zeta}{F(\zeta) - \zeta} = \sum_{|a_n|<R} A_n, \quad A_n = \frac{1}{F'(a_n) - 1},$$

supposed $a_n \notin C$ for any n. Write $F'(a_n) = \xi_n + i\eta_n$.

Assume that $|F'(a_n)| \leq 1$ for any n. Since $F'(a_n) \neq 1$, we have $\xi_n < 1, \Re[A_n] < 0$. By the well known theorem of Wimann[6], there is R such that $|I(R)| < |\Re[A_1]|$. Then we have

$$0 > \sum_{|a_n|<R, n\neq 1} \Re[A_n] = |\Re[A_1]| + \sum_{|a_n|<R} \Re[A_n] = |\Re[A_1]| + \Re[I(R)] > 0,$$

which is absurd. q. e. d.

The condition $\mu < 1/2$ is indispensable. In fact,

$$F(y) = y + A\left\{\frac{\sin(y^{1/2})}{y^{1/2}} - 1\right\}$$

is of order $1/2$ and, if $A > 0$ is small, $|F'(a_k)| < 1$ for any fixed point a_k.

In this respect, Bergweiler [1] obtained a remarkable result:

THEOREM 2.1.5 (Bergweiler) *Let $F(y)$ be transcendental entire. For any $p \geq 2$, the p-th iterate $F^p(y)$ admits infinitely many repelling fixed points.*

There may exist meromorphic solutions even if (2.1.6) is not satisfied. For example, consider the equation

$$y(x+1) = y(x) + 1 - e^{y(x)}. \tag{2.1.7}$$

Fixed points of $F(y) = y + 1 - e^y$ are $a_k = 2k\pi i$ and $F'(a_k) = 0$. But it admits a meromorphic solution $y(x)$ which is expanded as

$$y(x) = x + \sum_{n=1}^{\infty} c_n e^{nx} \tag{2.1.7$'$}$$

in a domain $D = \{|e^x| < \delta\}$ for sufficiently small $\delta > 0$. Further, for the equation

$$y(x+1) = y(x) + e^{y(x)}, \tag{2.1.8}$$

there are no fixed points. But there is a meromorphic solution which is asymptotically expanded as

$$y(x) \sim \log \frac{1}{x} + \pi i + \sum_{j+k \geq 1} c_{jk} x^{-j} \left(\frac{\log x}{x} \right)^k, \tag{2.1.8$'$}$$

in an angular domain $S(\epsilon, R) = \{x \mid \Re[x] \leq -R, |\arg[x] - \pi| < \frac{\pi}{2} - \epsilon\}$.

By Lemmas 2.1.3 and 2.1.4, we know that meromorphic solutions exist always if $F(y)$ is a rational function of degree ≥ 2, or a transecendental entire function of lower order $\mu < 1/2$. For other cases, we should consider separately.

2.2 Representation of solutions

Let $a, c_1, c_2 \in \mathbb{C} \cup \{\infty\}$. We define as follows:

(mf) : a is *(mf)-value* for $F(y)$, if $F(y) = a \Rightarrow y = a$.

(mp) : $(c_1, c_2), c_1 \neq c_2$, is *(mp)-pair* for $F(y)$, if $F(y) = c_j \Rightarrow y = c_{j+1}$.

(ex) : For transcendental $F(y)$ only. $a \in \mathbb{C}$ is *(ex)-value*, if $F(y) - a \neq 0$.

When $F(y)$ is rational: There may be at most two *(mf)-values*. There may exist at most one *(mp)-pair*.

When $F(y)$ **is transcendental entire**: There is no (mp)-*pair.* If b is (mf)- or (ex)-*value,* then F(y) is of the form

$$F(y) = b + (y - b)^p e^{f(y)}, \quad p \geq 0. \tag{2.2.2}$$

Thus, (mf)- or (ex)-*value* does not appear for $F(y)$ of lower order less than 1.

Obviously, if $F(y)$ admits a (mf)-*value* b, then $F(y)$ has a fixed point $b_1 \neq b$.

The following Theorems 2.2.1–2.2.7 are easily proved:

THEOREM 2.2.1 *Let* $F(y)$ *be entire (either polynomial or transcendental). Then any meromorphic solution* $y(x)$ *is entire.*

THEOREM 2.2.2 *Suppose* $F(y)$ *is rational and has* (mf)-*value* b, $|b| \leq \infty$. *Then any solution* $y(x)$ *does not take* b.

THEOREM 2.2.3 *Suppose* $F(y)$ *is transcendental entire and has* (mf)-*value* $b \in \mathbb{C}$. *If* $p \geq 2$ *in* (2.2.2), *then solution* $y(x)$ *does not take* b.

THEOREM 2.2.4 *Suppose* $F(y)$ *is transcendental entire and has* (ex)-*value* b, *which means that* $p = 0$ *in* (2.2.2). *Then any* $y(x)$ *does not take* b.

THEOREM 2.2.5 *Suppose* $F(y)$ *is transcendental entire and has* (mf)-*value* b. *If* $p = 1$ *in* (2.2.2), *then* b *may be taken by some (not any) solutions.*

THEOREM 2.2.6 *Any meromorphic solution* $y(x)$ *of* (2.1.1) *is transcendental, supposed* $F(y)$ *is not a linear (fractional) function.*

THEOREM 2.2.7 *Let* a *be neither* (mf)-*value, nor one of* (mp)-*pair, nor* (ex)-*value for* $F(y)$. *Then any solution* $y(x)$ *takes* a.

THEOREM 2.2.8 *Let* $F(y)$ *be a rational or a transcendental entire function. For a solution (if any)* $y(x)$ *of* (2.1.1), *there is* $y^* \in \mathbb{C} \cup \{\infty\}$ *such that*

(2.2.3) $y(x - t) \to y^*$ *uniformly on any compact set, as* $t (\in \mathbb{R}) \to \infty$.

1. *If* $y^* \in \mathbb{C}$, *then* y^* *is a fixed point of* $F(y)$ *with* $|F'(y^*)| \geq 1$.

2. *If* $y^* = \infty$, *then* $F(y)/y$ *has asymptotic value* $A, |A| \leq 1$. *Hence, when* $F(y)$ *is rational, then* ∞ *is a fixed point, and* $A = \lim_{y\to\infty} F'(y) \neq 0$.

Thus, if $F(y)$ *is either a polynomial or transcendental entire of lower order* $\mu < 1/2$, *then we have* $y^* \in \mathbb{C}$.

THEOREM 2.2.9 ([10]) *Let* $F(y)$ *be rational and* $y(x)$ *be a solution of* (2.1.1). *Let* y^* $(|y^*| \leq \infty)$ *be a fixed point for* $F(y)$ *for which* (2.2.3) *holds. Then we have*

$$\text{either} \quad |F'(y^*)| > 1 \quad \text{or} \quad F'(y^*) = 1. \tag{2.2.4}$$

Conversely, if a fixed point y^* meets (2.2.4), there is a solution $y(x)$ which satisfies (2.2.3), as seen by $\Psi_\lambda(x)$ in Theorem 2.1.1 and $\Phi(x)$ in Theorem 2.1.2.

THEOREM 2.2.10 *Suppose* $y(x)$ *satisfies* (2.2.3) *for some* y^*, $|F'(y^*)| > 1$. *Write* $F'(y^*) = \lambda$. *Then there is a periodic entire function* $\pi(x)$, $\pi(x+1) = \pi(x)$, *such that*

$$y(x) = \Psi_\lambda\left(x + \frac{\log \pi(x)}{\log \lambda}\right). \tag{2.2.5}$$

THEOREM 2.2.11 *Suppose* $y(x)$ *satisfies* (2.2.3) *for some* y^*, $F'(y^*) = 1$. *Then there is a periodic entire function* $\pi(x)$, $\pi(x+1) = \pi(x)$, *such that*

$$y(x) = \Phi(x + \pi(x)). \tag{2.2.6}$$

Thus, **for the case where** $F(y)$ **is rational**, any solution of (2.1.1) is represented in the form either (2.2.5) or (2.2.6). That is:
we obtain all solutions of (2.1.1) in the form either (2.2.5) or (2.2.6).

2.3 Value distribution of solutions

When $F(y)$ is rational, any solution $y(x)$ of (2.1.1) is of order ∞ by Theorems 1.1 and 2.2.6. For $\eta \in \mathbb{R}$ and $\delta > 0$, we put

$$L_\eta = \{x \mid \Im[x] = \eta\}, \quad \text{(a horizontal line)} \tag{2.3.1}$$

$$H(\eta, \delta) = \{x \mid \Re[x] > 0,\ |\Im[x] - \eta| \leq \delta\} \quad \text{(a half strip)}. \tag{2.3.2}$$

THEOREM 2.3.1 ([7]) *Let $F(y)$ be rational and $y(x)$ be a solution of* (2.1.1). *There is a horizontal line L_{η_0} such that $y(x)$ takes every value a (neither (mf)-value nor (mp)-pair) infinitely often in the half strip $H(\eta_0, \delta)$ for any $\delta > 0$. More precisely, if $x_n(a, \eta_0, \delta)$ are a-points of $y(x)$ in $H(\eta_0, \delta)$, then we have*

$$\sum_n \frac{1}{|x_n(a, \eta_0, \delta)|^\rho} = \infty, \quad \text{for any } \rho > 0. \tag{2.3.3}$$

For the proof, we follow the method of Tsuji [6] who proved an analogue of the second fundamental Theorem in an angular domain.

L_{η_0} is called *"Borel" line* for the solution $y(x)$, if $y(x)$ takes any value a (not (*mf*)-, (*mp*)-) infinitely often in the half strip $H(\eta_0, \delta)$ for any $\delta > 0$. It is natural to ask how many "Borel" lines may exist. In this respect, we have

THEOREM 2.3.2 ([8]) *Let $y = \phi(x)$ be a solution of the equation of* Kimura:

$$y(x+1) = y(x) + 1 + \frac{b_1}{y(x)}, \quad (\textit{see } (2.1.3)) \tag{2.3.4}$$

which is expanded asymptotically as in (2.1.4). *Suppose $b_1 \in \mathbb{R} \setminus \{0\}$.*

1. *If $b_1 \le 1/4$, there is only one "Borel" line for $\phi(x)$,*
2. *If $b_1 > 1/4$, there are at least two "Borel" lines for $\phi(x)$.*

THEOREM 2.3.3 ([9]) *Let $F(y)$ be rational. Then the solution $y(x)$ of* (2.1.1) *admits no Nevanlinna deficiencies other than (mf)-value and (mp)-pair.*

2.4 Factorization of solution

In (2.2.5) or (2.2.6), solutions of (2.1.1) are represented as compositions of $\Psi_\lambda(x)$ or $\Phi(x)$ with some periodic function $\pi(x) : \pi(x+1) = \pi(x)$.

$\Psi_\lambda(x)$ in (2.2.5) is factored as $\Psi_\lambda(x) = \psi(\lambda^x)$, where

$$\psi(x) = y^* + \sum_{n=1}^{\infty} c_n x^n, \tag{2.4.1}$$

which satisfies the Schröder equation

$$\psi(\lambda x) = F(\psi(x)), \quad \lambda = F'(y^*). \tag{2.4.2}$$

It would be natural to ask whether $\psi(x)$ and $\Phi(x)$ may be further factored into two transcendental functions.

THEOREM 2.4.1 ([11]) *Let $F(y)$ be a polynomial. Suppose there is a fixed point y^* with $F'(y^*) = 1$, and $\Phi(x)$ is the solution in Theorem 2.1.2. Then $\Phi(x)$ cannot be factored as $\Phi = f \circ g$ with transcendental f and g, where $g(x)$ is a solution of a difference equation*

$$g(x+1) = G(g(x)), \quad \text{with an entire } G(y). \tag{2.4.3}$$

THEOREM 2.4.2 *Let $F(y)$ be a polynomial. Suppose there is a fixed point y^* with $|F'(y^*)| > 1$, and $\Psi_\lambda(x)$ is a solution in Theorem 2.1.1, and $\psi(x)$ is the function in* (2.4.1). *Then $\psi(x)$ cannot be factored as $\psi = f \circ h$ with transcendental f and h, where $h(x)$ is a solution of*

$$h(\lambda x) = G(h(x)) \quad \text{with an entire } G(x). \tag{2.4.4}$$

References

[1] W. Bergweiler: *Periodic points of entire functions. Proof of a conjecture of Baker.* Complex Variables, Theory Appl. 18 (1991), 57–72.

[2] R.M. Cohn: *Difference Algebra.* Robert E. Krieger Publ., Huntington, New York 1979.

[3] G. Julia: *Sur l'iteration des fonctions rationelles.* Jour. Math. Pure et Appl. (7) 4 (1918), 47–245.

[4] T. Kimura: *Iteration of analytic functions.* Funkcial. Ekvac. 14 (1971), 197–238.

[5] I. Laine: *Nevanlinna Theory and Complex Differential Equations.* Walter de Gruyter, Berlin–New York 1993.

[6] M. Tsuji: *Potential Theory in Modern Function Theory.* MARUZEN, Tokyo 1959.

[7] N. Yanagihara: *Meromorphic solutions of some difference equations. II* Funkcial. Ekvac. 24 (1981), 113–124.

[8] N. Yanagihara: *"Borel" lines for meromorphic solutions of difference equation $y(x+1) = y(x) + 1 + \lambda/y(x)$.* Proc. Japan Acad. 57 (1981), 352–355.

[9] N. Yanagihara: *Exceptional values for meromorphc solutions of some difference equations.* J. Math. Soc. Japan. 34 (1982), 491–499.

[10] N. Yanagihara: *Meromorphic solutions of some difference equations and conformal mapping of infinite strips.* Funkcial. Ekvac. 26 (1983), 17–35.

[11] N. Yanagihara: *Factorization of entire solutions of some difference equations.* Proc. Japan Acad. 60 (1984), 185–188.

64 SOME RECENT PROGRESS IN THE UNIQUESS THEORY OF MEROMORPHIC FUNCTIONS[1]

Lian-Zhong YANG

Department of Mathematics, Shandong University
Jinan, Shandong, 250100, P.R. China

lzyang@sdu.edu.cn

Abstract

In this paper, we mention some recent results and problems in the uniqueness theory of meromorphic functions with shared values or sets.

0. Introduction

In this paper a meromorphic function will mean meromorphic in the whole complex plane. We say that two meromorphic functions f and g share a finite value a IM (ignoring multiplicities) when $f - a$ and $g - a$ have the same zeros. If $f-a$ and $g-a$ have the same zeros with the same multiplicities, then we say that f and g share the value a CM (counting multiplicities). We say that f and g share ∞ CM (IM) provided that $1/f$ and $1/g$ share 0 CM (IM). Now Let h be a nonconstant meromorphic

[1] This work was supported by the NSF of China.

H.G.W. Begehr et al. (eds.), Proceedings of the Second ISAAC Congress, Vol. 1, 551–564.

function, and let S be a subset of distinct elements in the whole complex plane. We define (see [1])

$$E_h(S) = \bigcup_{a \in S} \{z \mid h(z) = a, \quad \text{counting multiplicities}\},$$

$$\overline{E}_h(S) = \bigcup_{a \in S} \{z \mid h(z) = a, \quad \text{ignoring multiplicities}\}.$$

Let f and g be two nonconstant meromorphic functions, and let S be a subset of distinct elements in the whole complex plane. If $E_f(S) = E_g(S)$, we say f and g share the set S CM (counting multiplicity). If $\overline{E}_f(S) = \overline{E}_g(S)$, we say f and g share the set S IM (ignoring multiplicity). As a special case, let $S = \{a\}$. If $E_f(\{a\}) = E_g(\{a\})$, then f and g share the value a CM (If $\overline{E}_f(\{a\}) = \overline{E}_g(\{a\})$, then f and g share the value a IM) (see [2]). It is also assumed that the reader is familiar with the standard symbols and fundamental results of Nevanlinna theory, as found in [3].

The uniqueness theory of meromorphic functions is basically a study of the ways in which two given meromorphic functions must be identical under some conditions. In recent years, the theory of meromorphic functions with shared values have been studied by many authors in various aspects and many interesting and remarkable results have been obtained. In this lecture, I would like to mention some recent results and problems which are based on my own interests.

1. Nevanlinna's Five Value Theorem

In 1929, R. Nevanlinna proved the following theorem [4].

Theorem 1.1. *Let f and g be two nonconstant meromorphic functions, and let $a_j (j = 1, 2, \cdots, 5)$ be five distinct values in $C \cup \{\infty\}$. If f and g share $a_j (j = 1, 2, \cdots, 5)$ IM, then $f \equiv g$.*

let $f(z)$ and $\phi(z)$ be meromorphic functions. If

$$T(r, \phi) = S(r, f),$$

we say that $\phi(z)$ is a "small" function of f. Corresponding to results in the theory of value-distribution, we can ask some questions with respect to small functions. Some of them are very difficult and significant. For instance, R. Nevanlinna himself asked if the second fundamental theorem can be extended to small functions. About ten years ago, Osgood and Steinmetz independently settled this problem with a positive answer (For the case of reduced counting functions, this problem is still open). The Theorem 1.1 can be proved by Nevanlinna's second fundamental theorem with the reduced counting function directly.

Regarding Theorem 1.1, a natural question is:

Question 1. If the values $a_j (j = 1, 2, \cdots, 5)$ are replaced by five small functions $\phi_j(z)(j = 1, 2, \cdots, 5)$, is the Theorem 1.1 still true ?

Concerning this Question, many results have been obtained in recent years(See [5] [6] [34]), but no one can give an affirmative answer. Since the second fundamental theorem for small functions have not been completely solved, this Question seems difficult to answer. A few years ago, Q.-D. Zhang [5] answered this question under some additional conditions by proving the following theorem.

Theorem 1.2. *Let f and g be two nonconstant meromorphic functions, and let $\phi_j(j = 1, 2, \cdots, 5)$ be five distinct small functions of f and g. If f and g share $\phi_j(j = 1, 2, \cdots, 5)$ IM, and every mutiple zeros of $f - \phi_j(j = 1, 2, \cdots, 5)$ is also a multiple zero of $g - \phi_j$, then $f \equiv g$.*

In recent two years, Y.-H. Li has made a great progress in his research on Question 1. In 1998, He proved the following Theorem [6] which provided a positive answer to Question 1 for the case of entire functions.

Theorem 1.3. *Let f and g be two nonconstant entire functions, and let $\phi_j(j = 1, 2, 3, 4)$ be four distinct small functions of f and g. If f and g share $\phi_j(j = 1, 2, 3, 4)$ IM, then $f \equiv g$.*

Very recently, by using the same methods and more careful computations, Y.-H. Li and J.-Y. Qiao [7] settled the Question 1 completely and obtained the next result.

Theorem 1.4. *Let f and g be two nonconstant meromorphic functions, and let $\phi_j(j = 1, 2, \cdots, 5)$ be five distinct small functions of f and g. If f and g share $\phi_j(j = 1, 2, \cdots, 5)$ IM, then $f \equiv g$.*

At last, we want to mention a result of Ishizaki and Toda[36] to end this section.

Theorem 1.5. *Let f and g be two transcendental meromorphic functions satisfying*

$$\overline{N}(r, f) \leq uT(r, f) + S(r, f) \quad \text{and} \quad \overline{N}(r, g) \leq uT(r, g) + S(r, g)$$

for a constant u in [0,1/19), and let $\phi_j (j = 1, 2, 3, 4)$ be four distinct small functions of f and g. If f and g share $\phi_j (j = 1, 2, 3, 4)$ IM. then $f \equiv g$.

2. Meromorphic Functions That Share Sets

In 1976, F. Gross asked the following question (see [8, Question 6]):

Question 2. Can one find two finite sets $S_j \quad (j = 1, 2)$ such that any two entire functions f and g satisfying $E_f(S_j) = E_g(S_j)$ for $j = 1, 2$ must be identical? (If the answer is affirmative, it would be interesting to know how large both sets would have to be.)

In 1994, H.-X. Yi [9] proved the following theorem, which affirmatively answered the above Question 2.

Theorem 2.1. *Let $S_1 = \{\omega | \ \omega^n - 1 = 0\}$ and $S_2 = \{a\}$, where $n \geq 5$, $a \neq 0$ and $a^{2n} \neq 1$. If f and g are entire functions such that $E_f(S_j) = E_g(S_j)$ for $j = 1, 2$, then $f \equiv g$.*

Now it is natural to ask the following question:

Question 3. What are the smallest cardinalities of S_1 and S_2 respectively, where S_1 and S_2 are two finite sets such that any two entire functions f and g satisfying $E_f(S_j) = E_g(S_j)$ for $j = 1, 2$ must be identical?

Recently, H.-X. Yi proved the following two theorems [10].

Theorem 2.2. *Let $S_1 = \{0\}$ and $S_2 = \{\omega | \ \omega^2(\omega + a) - b = 0\}$, where a and b are two nonzero constants such that the algebraic equation*

$\omega^2(\omega+a)-b=0$ has no multiple roots. If f and g are two entire functions satisfying $E_f(S_j)=E_g(S_j)$ for $j=1,2$, then $f\equiv g$.

Theorem 2.3. *If S_1 and S_2 are two finite sets such that any two entire functions f and g satisfying $E_f(S_j)=E_g(S_j)$ for $j=1,2$ must be identical, then $\max\{\#(S_1),\#(S_2)\}\geq 3$, where $\#(S)$ denotes the cardinality of the set S.*

Obviously, the above two Theorems imply a sharp answer to Question 3. For example, let $S_1=\{0\}$ and

$$S_2=\{2,-3,-6\}=\{\omega|\ \omega^2(\omega+7)-36=0\}.$$

Closely related to Question 3, we have the following question.

Question 4. Can one find a finite set S such that any two meromorphic functions f and g satisfying $E_f(S)=E_g(S)$ must be identical? (See F.Gross [8]).

Concerning this Question, many authors obtained many results which implies the answer is positive(See Yi[21], Li-Yang[35] etc.). We mention two results of Professor Yi[21][1] here.

Theorem 2.4. *Let n and m be two positive integers such that n and m have no common factors and $n>2m+4$, and let $S=\{\omega\mid\ =\omega^n+a\omega^{n-m}+b=0\}$, where a and b are two non-zero constants such that the algebraic equation $\omega^n+a\omega^{n-m}+b=0$ has no multiple roots. If f and g are entire functions satisfying $E_f(S)=E_g(S)$, then $f\equiv g$.*

Theorem 2.5. *Let n and m be two positive integers such that n and m have no common factors and $n>2m+8$, $m\geq 2$, and let $S=\{\omega\mid\ \omega^n+a\omega^{n-m}+b=0\}$, where a and b are two non-zero constants such that the algebraic equation $\omega^n+a\omega^{n-m}+b=0$ has no multiple roots. If f and g are meromorphic functions satisfying $E_f(S)=E_g(S)$, then $f\equiv g$.*

We recall some definitions (cf [22]). A set S such that for any two nonconstant entire functions f and g the condition $E_f(S)=E_g(S)$ implies $f\equiv g$ is called a unique range set of entire functions (URSE in brief), and a set S is called a unique range set for meromorphic functions (URSM), if for any two nonconstant meromorphic functions f and

g the condition $E_f(S) = E_g(S)$ implies $f \equiv g$. Let

$$\lambda_E = inf\{\#(S) \mid \text{ S is an URSE}\}$$

$$\lambda_M = inf\{\#(S) \mid \text{ S is an URSM}\}$$

, where $\#S$ is the cardinality of S. The Theorem 2.4 and Theorem 2.5 imply that $\lambda_E \leq 7$ and $\lambda_M \leq 13$. Recently G. Frank and M.Reinders [23] showed that $\lambda_M \leq 11$. Examples show that $\lambda_E \geq 5$ and $\lambda_M \geq 6$ [1]. Therefore, we have the following Theorem.

Theorem 2.6. $5 \leq \lambda_E \leq 7, \quad 6 \leq \lambda_M \leq 11.$

The above results concern with functions with shared sets CM. For the functions with shared sets IM, we have the following Theorems.

Theorem 2.7 [10]. *Let* $S_1 = \{0\}$ *and* $S_2 = \{\omega | \omega^n(\omega + a) - b = 0\}$, *where* $n(\geq 2)$ *is an integer,* a *and* b *are two nonzero constants such that the algebraic equation* $\omega^n(\omega + a) - b = 0$ *has no multiple roots. If* f *and* g *are two entire functions satisfying* $\overline{E}_f(S_1) = \overline{E}_g(S_1)$ *and* $E_f(S_2) = E_g(S_2)$, *then* $f \equiv g$.

Theorem 2.8 [24]. *Let* $S_1 = \{\omega \mid \omega^n - 1 = 0, \ n \in N\ \}$ *and* $S_2 = \{\infty\}$, *If* f *and* g *are two meromorphic functions satisfying* $\overline{E}_f(S_1) = \overline{E}_g(S_1)$ *and* $E_f(S_2) = E_g(S_2)$, *and if* $n \geq 11$, *then* $f \equiv tg$ *or* $fg \equiv s$ *for some constant* t *and* s *satisfying* $s^n = 1$.

Recently, H.-X. Yi obtained some results on the uniqueness of meromorphic functions that share one set IM (See [1] and [31]).

Theorem 2.9 [1][31]. *Let* n *and* m *be two positive integers such that* n *and* m *have no common factors and* $n > 2m + 7$, *and let* $S = \{\omega \mid = \omega^n + a\omega^m + b = 0\ \}$, *where* a *and* b *are two non-zero constants such that the algebraic equation* $\omega^n + a\omega^m + b = 0$ *has no multiple roots. If* f *and* g *are entire functions satisfying* $\overline{E}_f(S) = \overline{E}_g(S)$, *then* $f \equiv g$.

Theorem 2.10 [1][31]. *Let* n *and* m *be two positive integers such that* n *and* m *have no common factors and* $n > 2m + 14$, $m \geq 2$, *and let* $S = \{\omega \mid \omega^n + a\omega^{n-m} + b = 0\ \}$, *where* a *and* b *are two non-zero constants such that the algebraic equation* $\omega^n + a\omega^{n-m} + b = 0$ *has no multiple roots. If* f *and* g *are meromorphic functions satisfying* $\overline{E}_f(S) = \overline{E}_g(S)$, *then*

$f \equiv g$.

According to the definitions in [31], the set S such that for any two nonconstant entire functions f and g the condition $\overline{E}_f(S) = \overline{E}_g(S)$ implies $f \equiv g$ is called a reduced unique range set of entire functions (RURSE in brief), and a set S is called a reduced unique range set for meromorphic functions (RURSM), if for any two nonconstant meromorphic functions f and g the condition $E_f(S) = E_g(S)$ implies $f \equiv g$. Let

$$\overline{\lambda}_E = inf\{\#(S) \mid \text{ S is an RURSE}\}$$

$$\overline{\lambda}_M = inf\{\#(S) \mid \text{ S is an RURSM}\}$$

, where $\#S$ is the cardinality of S. The Theorem 2.9 and Theorem 2.10 imply that $\overline{\lambda}_E \leq 10$ and $\overline{\lambda}_M \leq 19$. Recently H.-X. Yi [32] showed that $\overline{\lambda}_M \leq 17$. Combining the above results, we have the following open question.

Question 5. What exactly the number λ_E, λ_M, $\overline{\lambda}_E$ and $\overline{\lambda}_M$ are ?

3. Meromorphic Functions And Their Derivatives

In 1976, L. Rubel and C. C. Yang proved the following result.

Theorem 3.1 [11]. *Let f be a nonconstant entire function. If f and f' share two finite, distinct values CM, then $f \equiv f'$.*

In 1979, Mues and Steinmetz [12] generalized Theorem 3.1 to the case that f and f' share two values IM. In 1986, Frank and Weissenborn [13] obtained the following result.

Theorem 3.2. *Let f be a nonconstant meromorphic function. If f and $f^{(k)}$ share two finite, distinct values a and b CM, where k is an positive integer. Then $f \equiv f^{(k)}$.*

We know that the conditions in Theorem 3.2 that f and $f^{(k)}$ share two finite, distinct values a and b CM cannot be replaced by the condition that f and $f^{(k)}$ share a CM, and share b IM. (see [1]). But we have the following question.

Question 6. What can be said when a nonconstant entire function f shares one finite value CM with one of its derivative $f^{(k)}$ $(k \geq 1)$?

In 1986, Jank, Mues and Volkmann proved the next two results.

Theorem 3.3 [14]. *Let f be a nonconstant meromorphic function, and let $a \neq 0$ be a finite constant. If f, f', and f'' share the value a CM, then $f \equiv f'$.*

Theorem 3.4 [14]. *Let f be a nonconstant entire function, and let $a \neq 0$ be a finite constant. If f and f' share the value a IM, and if $f''(z) = a$ whenever $f(z) = a$, then $f \equiv f'$.*

In 1995, Zhong generalized Theorem 3.4 by proving the next result.

Theorem 3.5 [15]. *Let f be a nonconstant entire function, let $a \neq 0$ be a finite constant, and let n be a positive integer. If f and f' share the value a CM, and if $f^{(n)}(z) = f^{(n+1)}(z) = a$ whenever $f(z) = a$, then $f \equiv f^{(n)}$.*

The Theorem 3.3 suggests the following question asked by Yi-Yang.

Question 7 [1, p. 458]. Let f be a nonconstant meromorphic function, let $a \neq 0$ be a finite constant, and let n and m be positive integers satisfying $n < m$. If f, $f^{(n)}$, and $f^{(m)}$ share the value a CM, where n and m are not both even or both odd, must $f \equiv f^{(n)}$?

The following example [17] shows that the answer to Question 7 is, in general, negative. Let n and m be positive integers satisfying $m > n+1$, and let b be a constant which satisfies $b^n = b^m \neq 1$. Set $a = b^n$ and $f(z) = e^{bz} + a - 1$. Then f, $f^{(n)}$, and $f^{(m)}$ share the value a CM, and $f \not\equiv f^{(n)}$. On the other hand, we proved the next result [18], which gives a positive answer to Question 7 in the case when f is entire functions of finite order and $m = n + 1$.

Theorem 3.6. *Let f be a nonconstant entire function of finite order, let $a \neq 0$ be a finite constant, and let n be a positive integer. If the value a is shared by f, $f^{(n)}$, and $f^{(n+1)}$ IM, and shared by $f^{(n)}$ and $f^{(n+1)}$*

CM, then $f \equiv f'$.

Based on the Theorem 3.6, we have the following open question.

Question 8. Let f be a nonconstant meromorphic function, $a \neq 0$ be a finite constant, and let n be a positive integer. If the value a is shared by f, $f^{(n)}$, and $f^{(n+1)}$ CM, can we have $f \equiv f^{(n)}$?

In 1996, Brück [16] made the following conjecture.

Conjecture. Let f be a nonconstant entire function satisfying $\rho_1(f) < \infty$, where

$$\rho_1(f) = \limsup_{r \to \infty} \frac{log\, log\, T(r, f)}{log\, r}$$

is not a positive integer. If f and f' share one finite value a CM, then

$$\frac{f' - a}{f - a} \equiv c$$

for some constant $c \neq 0$.

Brück [16] showed that the conjecture does not hold when $\rho_1(f)$ is a positive integer n or ∞, by considering solutions of the following differential equations:

$$\frac{f' - 1}{f - 1} = e^{z^n} \qquad \text{and} \qquad \frac{f' - 1}{f - 1} = e^{e^z}.$$

He also proved that the Conjecture is true under the condition that $N(r, 0, f') = S(r, f)$. In 1998, Gundersen and Yang [18] proved this Conjecture is true for entire functions of finite order. Recently, the author proved [19]

Theorem 3.7. *Let f be a nonconstant entire function of finite order, k be a positive integer. If f and $f^{(k)}$ share the finite value $a \neq 0$ CM, then*

$$\frac{f^{(k)} - a}{f - a} = c,$$

where c is a non-zero constant.

Corollary. *Let f be a nonconstant entire function of finite order, k be a positive integer, and let $a \neq 0$, b be two distinct finite values. If f and $f^{(k)}$ share the value a CM, and there exists at least one point z_0 such that $f(z_0) = f^{(k)}(z_0) \neq a$, then $f \equiv f^{(k)}$.*

4. Meromorphic Functions That Share Four Values

In section one, we showed that two nonconstant meromorphic functions must be identical, if they share five distinct values. Now what can be said if two meromorphic functions share four distinct values ? In 1929, R. Nevanlinna [4] proved that if two meromorphic functions f and g share four distinct values CM, then f is a fractional linear transformation of g. In 1979, Gundersen [25] improved this result and proved the following theorem.

Theorem 4.1. *Let f and g be two nonconstant meromorphic functions, and let a_j ($j = 1, 2, 3, 4$) be four distinct values in $C \cup \{\infty\}$. If f and g share a_j ($j = 1, 2, 3$) CM, and share a_4 IM, then f is a fractional linear transformation of g.*

In 1983, Gundersen [26][27] improved Theorem 4.1 and proved that the same conclusion holds if f and g share a_j (j=1, 2) CM, and share the other two values a_j (j=3, 4) IM. A few years later, E. Mues [28], S. Wang [29],H.-X. Yi [30] improved Gundersen's result respectively. To state their results, we need the following notation. Let f and g be nonconstant meromorphic functions, a be a complex number. If f and g share a IM, we set

$$\tau(a) = \begin{cases} \liminf\limits_{r \to \infty} \frac{N_E(r,a)}{\overline{N}(r,1/(f-a))}, & \text{if } \overline{N}(r, \frac{1}{f-a}) \not\equiv 0, \\ 1, & \text{if } \overline{N}(r, \frac{1}{f-a}) \equiv 0. \end{cases}$$

where $N_E(r, a)$ is the counting function of common zeros of $f - a$ and $g - a$ with the same multiplicities. According to the definition of $\tau(a)$, we know that $\tau(a) = 1$, if f and g share a CM.

Theorem 4.2 [29]. *Let f and g be two nonconstant meromorphic functions such that $f \not\equiv g$, and let $a_j (j = 1, 2, 3, 4)$ be four distinct values* *in* $C \cup \{\infty\}$. *If* f *and* g *share* $a_j (j = 1, 2, 3, 4)$ *IM, and* $\tau(a_i) > 4/5 (j = 1, 2)$, *then* f *and* g *share* $a_j (j = 1, 2, 3, 4)$ *CM.*

In 1989, E. Mues [28] proved that the conclusion of Theorem 4.2 is also true if the condition $\tau(a_i) > 4/5 (j = 1, 2)$ is replaced by $\tau(a_1) = 1$, $\tau(a_2) > 2/3$. In 1996, H.-X. Yi generalized Theorem 4.2 and Mues's result by proving the next result.

Theorem 4.3 [30]. *Let* f *and* g *be two nonconstant meromorphic functions such that* $f \not\equiv g$, *and let* $a_j (j = 1, 2, 3, 4)$ *be four distinct values in* $C \cup \{\infty\}$. *If* f *and* g *share* $a_j (j = 1, 2, 3, 4)$ *IM, and* $\tau(a_2) > 2/3$, $\tau(a_1) > \frac{2\tau(a_2)}{5\tau(a_2) - 2}$, *then* f *and* g *share* $a_j (j = 1, 2, 3, 4)$ *CM.*

In 1979, Gundersen [25] made two meromorphic functions f and g which share four distinct values IM, but f is not fractional linear transformation of g. Combining this and the above results, we have the following open question.

Question 9. Let f and g be two nonconstant meromorphic functions,and let $a_j (j = 1, 2, 3, 4)$ be four distinct values. If f and g share $a_j (j = 1, 2, 3)$ IM, and share the fourth value a_4 CM, can we conclude that f is a fractional linear transformation of g ?.

In 1992, Gundersen [33] proved the following partial result on the Question 9.

Theorem 4.4. *Let* f *and* g *be two nonconstant meromorphic functions that share* $a_1, a_2, a_3 IM$ *and* $a_4 CM$. *Suppose that there exists some real constant* $\lambda > 4/5$ *and some set* $I \subset (0, \infty)$ *that has infinite linear measure such that*

$$\frac{N(r, a_4, f)}{T(r, f)} \geq \lambda$$

for all $r \in I$. *Then* f *and* g *share all four values CM.*

References

[1] H.X. Yi and C.C. Yang, "Uniqueness Theory of Meromorphic Functions", Science Press, Beijing, 1995 [Chinese].

[2] G.G. Gundersen, Meromorphic functions that share three or four values, *J. London Math. Soc.* **20** (1979), 457-466.

[3] W.K. Hayman, "Meromorphic Functions", Clarendon Press, Oxford, 1964.

[4] R. Nevanlinna, "Le Theoreme de Picard-Borel et la theorie des fonctions meromorphes", Gauthier-Villars, Paris, 1929.

[5] Q.-D. Zhang, Uniqueness theorems of meromorphic functions with respect to small functions, *Acta Math. Sinica,* **36** (1993), 826-833.

[6] Y.-H. Li, Entire functions with four shared small functions IM, *Acta Math. Sinica,* **41** (1998), 249-260.

[7] Y.-H. Li and J.-Y. Qiao, Uniqueness of meromorphic functions dealing with small functions, *Sci. Sinica.* (to appear).

[8] F. Gross, Factorization of meromorphic functions and some open problems, *in* "Complex Analysis, Kentucky 1976" (Proc.Conf.), Lecture Notes in Mathematics, Vol. 599, pp. 51-69, Springer-Verlag, Berlin, 1977.

[9] H.-X. Yi, Uniqueness of meromorphic functions and a question of Gross, *Sci. Sinica,* (Series A), **37** (1994), 802-813.

[10] H.-X. Yi, ON a question of Gross concerning uniqueness of entire functions, *Bull. Austral. Math. Soc.*, **57** (1998), 343-349.

[11] L. Rubel and C.C. Yang, Values shared by an entire function and its derivative, *in* "Complex Analysis, Kentucky 1976" (Proc.Conf.), Lecture Notes in Mathematics, Vol. 599, pp. 101-103, Springer-Verlag, Berlin, 1977.

[12] E. Mues and N. Steinmetz, Meromorphe Funktionen, die mit ihrer Ableitung Werte teilen, *Manuscripta Math.* **29** (1979), 195-206.

[13] G. Frank and G. Weissenborn, Meromorphe Funktionen, die mit ihrer Ableitung Werte teilen, *Complex Variables,* **7** (1986),33-43.

[14] G. Jank, E. Mues, and L. Volkmann, Meromorphe Funktionen die mit ihrer ersten und zweiten Ableitung einen endlichen Wert teilen, *Complex Variables* **6** (1986), 51-71.

[15] H. Zhong, Entire functions that share one value with their derivatives, *Kodai Math. J.* **18** (1995), 250-259.

[16] R. Brück, On Entire functions which share one value CM with their first derivative, *Results in Mathematics* **30** (1996), 21-24.

[17] L.-Z. Yang, Further results on entire functions that share one value with their derivatives, *J. Math. Anal. Appl.* **212** (1997), 529-536.

[18] G. Gundersen and L.-Z. Yang, Entire functions that share one value with one or two of their derivatives, *J. Math. Anal. Appl.* **223** (1998), 88-95.

[19] L.-Z. Yang, Solutions of a differential equation and its applications, *Kodai Math. J.* (to appear).

[20] L.-Z. Yang, Meromorphic functions that share two values, *J. Math. Anal. Appl.* **209** (1997), 542-550.

[21] H.-X. Yi, A question of Gross and the uniqueness of entire functions, *Nagoya Math. J.* **138** (1995), 169-177.

[22] F. Gross and C.C. Yang, On preimage range sets of meromorphic functions, Proc. Japan Acad. **58** (1982), 17-20.

[23] G. Frank and M. Reinders, A unique range set for meromorphic functions with 11 elements, *Complex Variables,*

[24] H.-X. Yi and L.-Z. Yang, Meromorphic functions that share two sets,*Kodai Math. J.* **20** (1997), 127-134.

[25] G. Gundersen, Meromorphic functions that share three or four values, *J. London Math. Soc.* **20** (1979), 457-466.

[26] G. Gundersen, Meromorphic functions that share three four values, *Trans. Amer. Math. Soc.* **277** (1983), 545-567.

[27] G. Gundersen, Correction to "Meromorphic functions that share four values", *Trans. Amer. Math. Soc.* **304** (1987), 847-850.

[28] E. Mues, Meromorphic functions sharing four values, *Complex Variables,* **12** (1989), 169-179.

[29] S. Wang, On meromorphic functions that share four values, *J. Math. Anal. Appl.* **173** (1993), 359-369.

[30] H.-X. Yi and C.-T. Zhou, Meromorphic functions that share four values, *Journal of Shandong University,* **31** (1996), 121-128.

[31] H.-X. Yi, The reduced unique range sets for entire or meromorphic functions, *Complex Variables,* **32** (1997), 191-198.

[32] H.-X. Yi, On the reduced unique range sets of meromorphic functions, *Journal of Shandong University,* **33** (1998), 361-368.

[33] G. Gundersen, Meromorphic functions that share three values IM and a fourth value CM,*Complex Variables,* **20** (1992), 99-106.

[34] N. Toda, Some generalizations of the unicity theorem of Nevanlinna, *Proc. Japan Acad. Ser.A,* **69** (1993), 61-65.

[35] P. Li and C.C. Yang, On the unique range set of meromorphic functions, *Proc. Amer. Math. Soc.,* **124** (1996), 177-185.

[36] K. Ishizaki and N. Toda, Unicity theorems for meromorphic functions sharing four small functions, *Kodai Math. J.* **21** (1998), 350-371.

65 ON THE UNIQUESS OF MEROMORPHIC FUNCTIONS CONCERNING FIX-POINTS [1]

YUAN Wenjun

Department of Mathematics, Guangzhou Normal University
Guangzhou 510405, P. R. China

FANG Mingliang

State Key Lab.Millimeter Waves
Southeast University, Nanjing 210096, P. R. China

1991 Mathematics Subject Classification: Primary 30D35
Key words:meromorphic function, unicity theorem, fix-point

ABSTRACT: In this paper, we prove two unicity theorems for meromorphic functions concerning fix-points.

§1. Introduction and Main Results

In this paper, meromorphic function always means a function meromorphic in the whole complex plain. If not otherwise stated, functions $f(z)$ and $g(z)$ appearing in this paper are supposed to be nonconstant.

[1]This work was supported by the National Natural Science Foundation of China

H.G.W. Begehr et al. (eds.), Proceedings of the Second ISAAC Congress, Vol. 1, 565–573.

We will use standard notation in Nevanlinna theorey (See e.g [3,5]). Let $f(z)$ be a meromorphic function, as usual, $m(r,f), N(r,f)$ and $T(r,f)$ denote the proximity function, the counting function, and the characteristic function of $f(z)$, respectively. In particular, $S(r,f)$ means any quantity of growth $o\{T(r,f)\}$ as $r \to \infty$ outside of a possible exceptional set of finite linear measure.

On the uniqueness for meromorphic function, a lot of results seem to answer the sharing value problems, Yang's problem and Gross' problem (See [1], [2], [4],[6] and [7] etc.). Now we pose new problems and answer them partly.

For convenience we give the following notations and definitions:

$$E(a,f) := \{(z,k)|f(z)-a=0 \quad \text{with multiplicity} \quad k \geq 1\}.$$

$N_{(2}(r, \frac{1}{f-a})$ is the counting function which only includes multiple zero of $f(z)-a$, $N_2(r,\frac{1}{f-a}) := \overline{N}(r,\frac{1}{f-a}) + \overline{N}_{(2}(r,\frac{1}{f-a})$ and $N_{1)}(r,\frac{1}{f-a}) := N(r,\frac{1}{f-a}) - N_{(2}(r,\frac{1}{f-a})$.

A polynomial $P(z)$ is called a unique polynomial concerning fix-points for entire (meromorphic) functions [UPFE (UPFM)] if for any two transcendental entire (meromorphic) functions $f(z)$ and $g(z)$ the condition $E(0,P(f)-z) = E(0,P(g)-z)$ implies $f(z)=g(z)$.

Our new problem is whether there exists a polynomial $P(z)$ such that it is a UPFE or UPFM ?

As far as we know that nobody posed above problem or proved unicity theorem for entire (meromorphic) functions concerning fix-points. In section 2 we will prove the following two theorems.

Theorem 1. *Let* $P(z) = z^6(z-1)$, *then* $P(z)$ *is a UPFE.*

Theorem 2. *Let* $P(z) = z^{11}(z-1)^2$, *then* $P(z)$ *is a UPFM.*

The following example shows that any polynomial of degree one is not a UPFE.

Example 1. *Suppose that* $P(z) = az+b$ *where* $a(\neq 0)$ *and* b *are constants. Let* $f(z) = \frac{1}{a}[z+H(z)-b]$ *and* $g(z) = z+cH(z)-b$ *with a constant* $c(\neq 1)$ *and a transcendental entire function* $H(z)$. *Then* $E(0,P(f)-z) = E(0,P(g)-z) = E(0,H)$. *However,* $f(z)-g(z) = \frac{1}{a}(1-c)H(z) \not\equiv 0$. *i.e.,* $f(z) \neq g(z)$.

An open problem is whether there exists a polynomial $P(z)$ with degree $\leq 6(12)$ such that $P(z)$ is a UPFE (UPFM).

§2. The Proofs of Theorem 1 and Theorem 2

In order to prove our results, we need the following Lemmas.

lemma 1. *Let $f(z)$ be a meromorphic function. Then*

$$\begin{aligned} T(r, \tfrac{1}{f-a}) &= T(r, f) + O(1), \\ T(r, \tfrac{f^{(k)}}{f^{(l)}}) &= S(r, f), \\ (q-1)T(r, f) &\leq N(r, f) + \textstyle\sum_{i=1}^{q} N(r, \tfrac{1}{f-a_i}) - N_1(r, f) + S(r, f) \end{aligned}$$

where $k > l \geq 0, a_i (i = 1, \cdots, q)$ are q distinct constants and

$$N_1(r, f) = N(r, \frac{1}{f'}) + 2N(r, f) - N(r, f').$$

In Lemma 1 three conclusions are called the first main theorem, the fundamental lemma of logrithmic derivative and the second main theorem respectively (see [3], [5]).

The following lemma is a corollary of the Valiron-Monkho'ko theorem proved by K. Valiron and A. Z. Monkho'ko independently (see [5]).

Lemma 2. *Let $R(z, \omega)$ be a rational function in z and ω. If $f(z)$ is a meromorphic function, then*

$$T(r, R) = nT(r, f) + O(\log r)$$

where $R = R(z, f), n = \deg_{\omega} R(z, \omega)$.

Proof of Theorem 1. Set

$$F(z) := \frac{P(f(z))}{z}, \quad G(z) := \frac{P(g(z))}{z}.$$

It follows from conditions of the theorem that

$$E(1, F) = E(1, G), \tag{2.1}$$
$$N_2(r, F) = N_2(r, G) \leq \log r. \tag{2.2}$$

If $z_0 (\neq 0)$ is a zero of $F(z)$, then z_0 is a zero of $f(z)$ or $f(z) - 1$. Thus

$$\overline{N}(r, \frac{1}{F}) \leq \overline{N}(r, \frac{1}{f}) + \overline{N}(r, \frac{1}{f-1}). \tag{2.3}$$

If $z_1 (\neq 0)$ is a multiple zero of $F(z)$, then z_1 is a zero of $f(z)$ or a multiple zero of $f(z) - 1$. Hence

$$\overline{N}_{(2}(r, \frac{1}{F}) \leq \overline{N}(r, \frac{1}{f}) + \overline{N}_{(2}(r, \frac{1}{f-1}). \tag{2.4}$$

$$\begin{aligned} N_2(r,\frac{1}{F}) & +N_2(r,F)=\overline{N}(r,\tfrac{1}{F})+\overline{N}_{(2}(r,\tfrac{1}{F})+N_2(r,F)\\ & \le 2\overline{N}(r,\tfrac{1}{f})+\overline{N}(r,\tfrac{1}{f-1})+\overline{N}_{(2}(r,\tfrac{1}{f-1})+S(r,F)\\ & \le 2\overline{N}(r,\tfrac{1}{f})+N(r,\tfrac{1}{f-1})+S(r,F)\\ & \le 3T(r,f)+S(r,F)\\ & \le (\tfrac{3}{7}+o(1))T(r,F). \end{aligned} \tag{2.5}$$

X-Mozilla-Status: 0001 Content-Length: 9706

We can also get in the same manner that

$$N_2(r,\frac{1}{G})+N_2(r,G)\le(\frac{3}{7}+o(1))T(r,G). \tag{2.6}$$

Since (2.5) and (2.6), therefore, we have

$$N_2(r,\frac{1}{F})+N_2(r,F)+N_2(r,\frac{1}{G})+N_2(r,G)\le(\frac{6}{7}+o(1))T(r) \tag{2.7}$$

where $T(r):=\max\{T(r,F),T(r,G)\}$.

We claim that $F(z)=G(z)$ or $F(z)G(z)=1$.

Set

$$\Phi(z)=\frac{F''(z)}{F'(z)}-2\frac{F'(z)}{F(z)-1}-\frac{G''(z)}{G'(z)}+2\frac{G'(z)}{G(z)-1}. \tag{2.8}$$

Suppose that $\Phi(z)\not\equiv 0$. Obviously, $m(r,\Phi)=S(r,F)+S(r,G)$.

If $z_@$ is a simple 1-point of $F(z)$ or $G(z)$, substituting their Taylor series at z_2 into (2.8), we see that z_2 is a zero of $\Phi(z)$. Thus by Lemma 1 we have

$$\begin{aligned} N_{1)}(r,\frac{1}{F-1}) & =N_{1)}(r,\tfrac{1}{G-1})\le\overline{N}(r,\tfrac{1}{\Phi})\\ & \le T(r,\Phi)+O(1)\\ & \le N(r,\Phi)+S(r,F)+S(r,G). \end{aligned} \tag{2.9}$$

It is easy to know that $\Phi(z)$ is analytic at the simple pole or the multiple 1-point of $F(z)$ or $G(z)$. Hence if z_3 is a pole of $\Phi(z)$ and not a multiple pole of $F(z)$ or $G(z)$, then z_3 is a zero of $F'(z)$ or $G'(z)$. Noting that z_3 is not a simple 1-point of $F(z)$ or $G(z)$, we know that If z_3 is not a multiple zero of $F(z)$ or $G(z)$ then $F'(z_3)=0, F(z_3)\ne 0,1$ or $G'(z_3)=0, G(z_3)\ne 0,1$. That is

$$\begin{aligned} \overline{N}(r,\Phi) & \le\overline{N}_{(2}(r,F)+\overline{N}_{(2}(r,G)+\overline{N}_{(2}(r,\tfrac{1}{F})+\overline{N}_{(2}(r,\tfrac{1}{G})+\\ & +\overline{N}_0(r,\tfrac{1}{F'})+\overline{N}_0(r,\tfrac{1}{G'}) \end{aligned}$$

where

$$\begin{aligned}\overline{N}_0(r,\tfrac{1}{F'}):&=\{z|F'(z)=0,F(z)\neq 0,1\},\\ \overline{N}_0(r,\tfrac{1}{G'}):&=\{z|G'(z)=0,G(z)\neq 0,1\}.\end{aligned}$$

Substituting above inequality into (2.9) we have

$$\begin{aligned}\overline{N}(r,\tfrac{1}{F-1})\ &\leq \overline{N}_{(2}(r,F)+\overline{N}_{(2}(r,G)+\overline{N}_{(2}(r,\tfrac{1}{F})+\overline{N}_{(2}(r,\tfrac{1}{G})+\\ &+\overline{N}_0(r,\tfrac{1}{F'})+\overline{N}_0(r,\tfrac{1}{G'})+S(r,F)+S(r,G).\end{aligned} \tag{2.10}$$

By the second main theorem, we have

$$T(r,F)\leq \overline{N}(r,F)+\overline{N}(r,\frac{1}{F})+\overline{N}(r,\frac{1}{F-1})-N_0(r,\frac{1}{F'})+S(r,F), \tag{2.11}$$

$$T(r,G)\leq \overline{N}(r,G)+\overline{N}(r,\frac{1}{G})+\overline{N}(r,\frac{1}{G-1})-N_0(r,\frac{1}{G'})+S(r,G). \tag{2.12}$$

Therefore, we deduce from (2.10) to (2.12) that

$$\begin{aligned}T(r,F)\ &+T(r,G)\leq \overline{N}(r,F)+\overline{N}(r,\tfrac{1}{F})+\overline{N}_{1)}(r,\tfrac{1}{F-1})-N_0(r,\tfrac{1}{F'})+S(r,F)\\ &+\overline{N}(r,G)+\overline{N}(r,\tfrac{1}{G})+\overline{N}(r,\tfrac{1}{G-1})+\overline{N}_{(2}(r,\tfrac{1}{G-1})-N_0(r,\tfrac{1}{G'})+S(r,G)\\ &\leq N_2(r,F)+N_2(r,\tfrac{1}{F})+N_2(r,G)+N_2(r,\tfrac{1}{G})+N(r,\tfrac{1}{G-1})\\ &\quad +S(r,F)+S(r,G).\end{aligned} \tag{2.13}$$

Without loss of generality, we assume that $T(r,G)\leq T(r,F)$ for $r\in I$ which is of infinite measure. Thus, (2.13) implies

$$T(r)\leq N_2(r,\frac{1}{F})+N_2(r,F)+N_2(r,\frac{1}{G})+N_2(r,G)+S(r,F)$$

for $r\in I$. It contradicts (2.7). Hence $\Phi(z)\equiv 0$. i.e.,

$$\frac{F''(z)}{F'(z)}-2\frac{F'(z)}{F(z)-1}=\frac{G''(z)}{G'(z)}-2\frac{G'(z)}{G(z)-1}. \tag{2.14}$$

Solving (2.14), we have

$$F(z)=\frac{(b+1)+G(z)+(a-b-1)}{bG(z)+(a-b)} \tag{2.15}$$

where $a(\neq 0)$ and b are two constants.

If $b+1\neq 0, a-b-1\neq 0$, then

$$\overline{N}(r,\frac{1}{F})=\overline{N}(r,\frac{1}{G+\frac{a-b-1}{b+1}}).$$

By the second main theorem and $T(r,F)=T(r,G)+O(1)$, we have

$$\begin{aligned} T(r) = T(r,G) &+ O(1) \le \overline{N}(r,G) \le \overline{N}(r,\frac{1}{G}) + \overline{N}(r,\frac{1}{G+\frac{a-b-1}{b+1}}) + S(r,G) \\ &\le \overline{N}(r,G) + \overline{N}(r,\tfrac{1}{G}) + \overline{N}(r,\tfrac{1}{F}) + S(r,G) \end{aligned}$$

which contradicts (2.7). Hence $b+1=0$ or $a-b-1=0$.
If $b+1=0$, then (2.15) becomes

$$F(z) = \frac{a}{-G(z)+a+1}.$$

Clearly

$$\overline{N}(r,F) = \overline{N}(r,\frac{1}{G-a-1}).$$

Using the same argument to this case, we can deduce that $a=-1$ which implies $G(z)F(z)=1$.

If $a-b-1=0$, then (2.15) becomes

$$F(z) = \frac{aG(z)}{bG(z)+1}.$$

Suppose that $b \neq 0$. It follows that

$$\overline{N}(r,F) = \overline{N}(r,\frac{1}{G-1/b}).$$

Using the former method once more, we can obtain a contradiction. Hence $b=0$ and then $a=1$ which implies $F(z)=G(z)$. Our claim holds.

Now we can prove $f(z)=g(z)$.

We see that $G(z)F(z)=1$ implies

$$f^6(z)(f(z)-1)g^6(z)(g(z)-1) = z^2.$$

It follows from above equation and conditions of this theorem that $f(z)$ has no zero and 1-point. By the second main theorem and $f(z)$ being transcendental we have

$$\begin{aligned} T(r,f) &\le \overline{N}(r,f) + \overline{N}(r,\tfrac{1}{f}) + \overline{N}(r,\tfrac{1}{f-1}) + S(r,f) \\ &= S(r,f) \end{aligned}$$

which is a contradiction. It shows that $F(z)G(z) \neq 1$. Hence $F(z) = G(z)$. i.e.,

$$f^6(z)(f(z)-1) = g^6(z)(g(z)-1).$$

If $f(z) \neq g(z)$. Set $h(z) := \frac{f(z)}{g(z)} \neq 1$. Substituting $h(z)$ into above equation, we have

$$g(z) = \frac{1 + h + \cdots + h^5}{1 + h + \cdots + h^6}$$

which implies that $g(z)$ must have pole or be a constant. This is impossible. Hence $f(z) = g(z)$.

The proof of Theorem 1 is complete.

Proof of Theorem 2 The program of the proof is analogous to the proof of Theorem 1. We still use the former notations. Now, (2.2), (2.5) and (2.6) become

$$\begin{aligned} N_2(r, F) &\leq 2\overline{N}(r, f) + \log r, \\ N_2(r, G) &\leq 2\overline{N}(r, g) + \log r. \end{aligned} \tag{2.2'}$$

$$N_2(r, \frac{1}{F}) + N_2(r, F) \leq (\frac{6}{13} + o(1))T(r, F), \tag{2.5'}$$

and

$$N_2(r, \frac{1}{G}) + N_2(r, G) \leq (\frac{6}{13} + o(1))T(r, G). \tag{2.6'}$$

And then (2.7) becomes

$$N_2(r, \frac{1}{F}) + N_2(r, F) + N_2(r, \frac{1}{G}) + N_2(r, G) \leq (\frac{12}{13} + o(1))T(r). \tag{2.7'}$$

(2.8)-(2.15) still hold. Therefore, it holds that

$$F(z)G(z) = 1 \qquad \text{or} \qquad F(z) = G(z).$$

Now we prove $f(z) = g(z)$.

Suppose that $F(z)G(z) = 1$. i.e.,

$$f^{11}(z)(f(z) - 1)^2 g^{11}(z)(g(z) - 1)^2 = z^2. \tag{2.16}$$

If $z_0(\neq 0)$ is a zero or 1-point of $f(z)$ with multiplicity τ, then z_0 must be a pole of $g(z)$ with multiplicity μ. Comparing multiplicity at z_0 in (2.16), we have $11\tau = 13\mu$ or $2\tau = 13\mu$. Thus $\tau \geq 13$, and then

$$\begin{aligned} \overline{N}(r, \tfrac{1}{f}) + \overline{N}(r, \quad \tfrac{1}{f-1}) &\leq \tfrac{1}{13}N(r, \tfrac{1}{f}) + \tfrac{1}{13}N(r, \tfrac{1}{f_1}) \\ &\leq \tfrac{2}{13}T(r, f) + O(1). \end{aligned} \tag{2.17}$$

If z_1 is a pole of $f(z)$ with multiplicity μ_1, then z_1 must be a zero or 1-point of $g(z)$ with multiplicity τ_1. Comparing multiplicity at z_1 in (2.16), we have $11\tau_1 = 13\mu_1$ or $2\tau_1 = 13\mu_1$. Thus $\mu_1 \geq 2$, and then

$$\overline{N}(r, f) \leq \frac{1}{2}N(r, f) \leq \frac{1}{2}T(r, f). \tag{2.18}$$

By the second main theorem, (2.17) and (2.18), we have

$$\begin{aligned} T(r,f) &\le \overline{N}(r,f)+\overline{N}(r,\tfrac{1}{f})+\overline{N}(r,\tfrac{1}{f-1})+S(r,f) \\ &\le \tfrac{15}{26}T(r,f)+S(r,f) \end{aligned}$$

which is impossible. Hence $F(z)G(z) \neq 1$ and then $F(z) = G(z)$. It shows that

$$f^{11}(z)(f(z)-1)^2 = g^{11}(z)(g(z)-1)^2. \tag{2.19}$$

By Lemma 2, we have $T(r,f) = T(r,g) + S(r,g)$.

If $f(z) \neq g(z)$, set $h(z) : \frac{f(z)}{g(z)}$, then (2.19) becomes

$$(h^{12}+h^{11}+\cdots+1)g^2 - 2(h^{11}+h^{10}+\cdots+1)g + (h^{10}+h^9+\cdots+1) = 0.$$

Solving it, we hava

$$g(z) = 1 + \frac{-h^6(h^6 + \quad \text{or} \quad -1)}{h^{12}+h^{11}+\cdots+1}. \tag{2.20}$$

And then

$$f(z) = h + \frac{-h^7(h^6 + \quad \text{or} \quad -1)}{h^{12}+h^{11}+\cdots+1}. \tag{2.21}$$

By (2.20) and Lemma 2, we deduce

$$T(r,g) = 12T(r,h) + O(1). \tag{2.22}$$

Hence $h(z)$ is transcendental.

Let $e_i (i = 1, \cdots, 6)$ be roots of $x^6 + 1 = 0$ or $x^6 - 1 = 0$. By (2.20)-(2.22) and the second main theorem we infer that there exists a $e_i \neq 1$ (say e_1) such that $h(z)$ has infinite e_1-points (say α). Thus $g(\alpha) = 1$ and $f(\alpha) = e_1 (\neq 0, 1)$ which contradicts (2.19). Therefore $f(z) = g(z)$.

The proof of Theorem 2 is complete.

REFERENCES

[1] Fang Mingliang and Guo Hui, On the unique range set of meromorphic functions, Acta Math. Sinica, to appear.

[2] G. Gundersen, Meromorphic functions that share three four values, Trans, Amer. Math. Soc. 304 (1983), 545-567.

[3] W. K. Hayman, Meromorphic Functions, *Clarendon Press, Oxford,*1964.

[4] Hua Xinhou, Sharing values and a conjecture due to C. C. Yang, Pacific J. Math. 275(1)(1996), 75-84,

[5] I. Laine, Nevanlinna Theory and Complex Differential Equations, *W.de Gruyter Berlin,* 1993.

[6] E. Mues, Shared value problems for meromorphic functions, Value distribution theory and comple differential equations, Joensuu 1994, 17-43.

[7] Yi Hongxun and C. C. Yang, Unicity theory of meromorphic functions, Science Press, Beijing, 1995.

66 NAVIER-STOKES EQUATIONS AND TIME-GLOBAL STRONG SOLUTIONS

Hisako KATO

Graduate School of Mathematics, Kyushu University
Ropponmatsu 4-2-1, Chuo-ku, Fukuoka 810-8560, Japan

kato@rcmath.rc.kyushu-u.ac.jp

1991 Mathematics Subject Classification: 35Q30, 76D05

1 Introduction

The problem we will consider is the Navier-Stokes initial-boundary value problem. This problem is based on hydrodynamics.

Assume that a domain Ω is filled with some viscous incompressible fluid like water. Further, assume that an initial velocity $a(x)$ is given to the fluid. Then, what will the velocity $u(x,t)$ be at time t ?

The motion of viscous incompressible fluids is described by the Navier-Stokes equations. The Navier-Stokes equations are as follows:

$$u_t - \nu_0 \Delta u + u \cdot \nabla u = -\nabla p \quad (x \in \Omega, \quad t > 0), \tag{1}$$

$$\text{div } u \equiv \sum_{i=1}^{3} \frac{\partial u^i}{\partial x_i} = 0 \quad (x \in \Omega, \quad t > 0) \tag{2}$$

H.G.W. Begehr et al. (eds.), Proceedings of the Second ISAAC Congress, Vol. 1, 575–580.

with initial-boundary conditions

$$u\mid_{\partial\Omega}=0 \quad (t>0), \tag{3}$$

$$u(x,0)=a(x) \quad (x\in\Omega). \tag{4}$$

Here, Ω is a bounded domain in R^3, $u=(u^1(x,t),u^2(x,t),u^3(x,t))$ is the unknown velocity, $p=p(x,t)$ is the unknown pressure, $\nu_0>0$ is the viscosity and $a=(a^1(x),a^2(x),a^3(x))$ is the initial velocity.

The known results about the existence of solutions of the Navier-Stokes problem (1)–(4) are as follows. The global (in time) existence of weak solutions and the local (in time) existence of regular solutions have been proved. Furthermore, the global existence of regular solutions in the case where the initial velocity $a(x)$ is small, namely

$$\|\nabla a\|\|a\| \quad \text{is small} \qquad (\|\cdot\|=\|\cdot\|_{L_2(\Omega)})$$

has been shown.

Our talk is on

1. the global existence of strong solutions $u=u^{(K)}(x,t)$ of Eq.(K*) (mentioned below), that is a new reguralization of the Navier-Stokes equations,

2. for any $K>0$, the existence of a number $T_K\geq 0$ such that

$$\|\nabla u^{(K)}(T_K)\|^2_{L_2(\Omega)}=K$$

and that $u=u^{(K)}(x,t)$ satisfies the Navier-Stokes equations in $[T_K,\infty)$.

Thus, we will be able to observe "large solutions" of the Navier-Stokes equations.

Now, we have found a regularization (K*) of the Navier-Stokes equations:

$$(\mathrm{K}^*) \qquad u_t-\frac{\nu_0}{2}(1+\nu(\|\nabla u\|^2)\,)\Delta u+u\cdot\nabla u=-\nabla p$$

$$(x\in\Omega, \quad t>0)$$

with a function

$$\nu(s)=\frac{1}{K}\max\{s,K\}$$

for any $K > 0$.

The term of viscosity in Eq.(K*) is different from the term of viscosity in the Navier-Stokes equations. Really, by the definition of $\nu(s)$ it follows

$$\nu(\|\nabla u(t)\|^2) = \begin{cases} 1, & t : \|\nabla u(t)\|^2 \le K \\ \frac{1}{K}\|\nabla u(t)\|^2, & t : \|\nabla u(t)\|^2 \ge K. \end{cases}$$

Therefore, the term of viscosity in Eq. (K*) is as follows:

$$\frac{\nu_0}{2}(1 + \nu(\|\nabla u\|^2)\ )\Delta u$$
$$= \begin{cases} \nu_0 \Delta u, & \|\nabla u(t)\|^2 \le K \\ \frac{\nu_0}{2}(1 + \frac{1}{K}\|\nabla u(t)\|^2)\Delta u & \|\nabla u(t)\|^2 \ge K. \end{cases}$$

Hence, a solution $u^{(K)}(x,t)$ of Eq.(K*) satisfies Eq.(1) for all t when the velocity gradient is below a given constant K, and satisfies 'a non-Newtonian equation' instead of Eq.(1) for all t when the velocity gradient is above the K, and further satisfying Eq.(1) for all large t.

2 Theorem

We consider the following function spaces and notion.

$C^\infty_{0,\sigma} = \{\varphi \in C^\infty_0(\Omega)\ ;\ \text{div}\ \varphi = 0\}$,

$H_\sigma =$ the closure of $C^\infty_{0,\sigma}$ in $L_2(\Omega)$,

$H^1_{0,\sigma} =$ the closure of $C^\infty_{0,\sigma}$ in $H^1(\Omega)$,

$P =$ the orthogonal projection from $L_2(\Omega)$ onto H_σ,

Stokes operator $A = -P\,\Delta$ in H_σ(:Friedrichs extension),

with $D(A) = H^2(\Omega) \cap H^1_0(\Omega) \cap H_\sigma$.

Further,

$H^m(\Omega) =$ the Sobolev space of vector-functions in $L_2(\Omega)$ together with their derivatives up to order m,

$H^m_0(\Omega) =$ the completion of $C^\infty_0(\Omega)$ in $H^m(\Omega)$,

$L_p((0,T);H)$ $(1 \le p \le \infty)$ = the set of H-valued functions such that

$$\| f \|_{L_p((0,T);H)} = (\int_0^T \| f(t) \|_H^p\, dt)^{1/p} < \infty,$$

$$\| f \|_{L_\infty((0,T);H)} = \text{ess} \sup_{0<t<T} \| f(t) \|_H < \infty,$$

and

$W^{m,p}((0,T);H) =$ the set of functions in $L_p((0,T);H)$ together with their t-derivatives up to order m.

Next, we consider the following:

Eq.(K*) with conditions (2)–(4) is formulated as an equation with the initial condition in the Hilbert space H_σ:

$$\frac{du}{dt} + \frac{\nu_0}{2}(1 + \nu(\| \nabla u \|^2)\ )Au + Pu \cdot \nabla u = 0 \qquad (t > 0), \qquad (5)$$

$$u(0) = a, \qquad (6)$$

by the orthogonal decomposition $L_2(\Omega) = H_\sigma \bigoplus \{\nabla p\ ;\ p \in H^1(\Omega)\}$.

Similarly, the Navier-Stokes equations (1)–(2) with (3)–(4) are formulated as

$$\frac{du}{dt} + \nu_0 Au + Pu \cdot \nabla u = 0 \qquad (t > 0), \qquad (7)$$

$$u(0) = a. \qquad (8)$$

We recall the well known estimate of the nonlinear term of (1):

$$\| Pv \cdot \nabla v \| \leq c_1 \| A^{1/4} v \| \| Av \| \quad \text{for any } v \in D(A) \qquad (9)$$

with an absolute constant c_1 Here, A^α represents the fractional power of the Stokes operator A.

Finally, we state our result more precisely.

Theorem 1. *Let $a \in H^1_{0,\sigma} \cap H^2(\Omega)$. Then there exists a unique strong solution $u = u^{(K)}(x,t)$ of Eq.(5) in $[0,\infty)$ with Eq.(6) satisfying*

$$u \in L_\infty((0,T);D(A)),$$

$$u \in W^{1,\infty}((0,T);H_\sigma)$$

$$u \in W^{1,2}((0,T);H^1_{0,\sigma})$$

for any $T > 0$. In addition, $u(t)$ is strongly continuous in $[0,\infty)$ as an element of $H^1_{0,\sigma}$, and $u_t(t)$ is weakly continuous in $[0,\infty)$ as an element of H_σ. Further, u satisfies the energy inequality:

$$\|u(t)\|^2 + \nu_0 \int_0^t \|\nabla u\|^2 dt + \frac{\nu_0}{K}\int_0^t \|\nabla u\|^4 dt \leq \|a\|^2 \quad (\ t > 0\ ).$$

Theorem 2. *Let $u = u^{(K)}(x,t)$ be the solution obtained in Theorem 1. Then, for any K there exists a number $T_K \geq 0$ such that*

$$\|\nabla u^{(K)}(T_K)\|^2 = K$$

and that $u = u^{(K)}(x,t)$ satisfies Eq.(7) (: Navier-Stokes equations) in $[T_K, \infty)$.

Furthermore, we have proved that there exists a subsequence $\{u^{(K')}\}$ converging to a weak solution w of Eqs.(1)–(4) ($K' \to \infty$), and a weak solution w satisfies

either

$$\text{(I): } \|w(t)\| > 0 \text{ for any } t \in [0, \infty),$$

or

$$\text{(II): } \|w(t)\| \equiv 0 \text{ for any } t \in [\alpha, \infty) \text{ with some } \alpha \geq 0.$$

In Theorem 2, if we choose an initial velocity $a(x)$ for which the weak solution has property (I), the set

$$\{\|\nabla u^{(K)}(T_K)\| \ \|u^{(K)}(T_K)\| \mid K = 1, 2, \cdots\}$$

is unbounded. This implies the existence of "large solutions" of the Navier-Stokes equations.

References

[1] R. Courant and D. Hilbert, "Methoden der Mathematishen Physik II", Springer, Berlin Heidelberg New York, 1968.

[2] H. Fujita and T. Kato, On the Navier-Stokes initial value problem 1, *Arch. Rational Mech. Anal.* **16**(1964), 269-315.

[3] D. Fujiwara and H. Morimoto, An L_r-theorem of the Helmholtz decomposition of vector fields, *J. Fac. Sci. Univ. Tokyo* **24**(1977), 685-700.

[4] Y. Giga and T. Miyakawa, Solutions in L_r of the Navier-Stokes initial value problem, *Arch. Rational Mech. Anal.* **89**(1985), 267-281.

[5] E. Hopf, Über die Anfangswertaufgabe für die hydrodynamishen Glundgleichungen, *Math. Nachrichten* **4**(1951), 213-231.

[6] D. Kinderlehrer and G. Stampacchia, "An Introduction to Variational Inequalities and Their Applications", Academic Press, New York London Toronto Sydney San Francisco, 1980.

[7] A. A. Kiselev and O. A. Ladyzhenskaia, On the existence and uniqueness of the solution of the nonstationary problem for a viscous incompressible fluid, *Izv. Akad. Nauk SSSR, Ser. Mat.* **21**(1957), 655-680.

[8] O. A. Ladyzhenskaia, "The Mathematical Theory of Viscous Incompressible Flow," Gorden and Breach, Revised English edn. New York London, 1969.

[9] J. Leray, Sur le mouvement d'un liquide visqueux emplissant l'espace, *Acta Math.* **63**(1934), 193-248.

[10] J. L. Lions, "Quelque Methodes de Résolution des Problèmes aux Limites Non Linéaires", Dunod, Paris, 1969.

[11] K. Masuda, Weak solutions of the Navier-Stokes equations, *Tohoku Math. J.* **36**(1984), 623-646.

[12] J. Serrin, The initial value problem for the Navier-Stokes equations, *in* "Nonlinear Problem"(R. E. Langer ed), The University of Wisconsin Press, Madison, 1963.

[13] M. Shinbrot, "Lectures on Fluid Mechanics", Gordon-Breach, New York London Paris, 1973.

[14] H. Sohr and W. von Wahl, On the singular set and the uniqueness of weak solutions of the Navier-Stokes equations, *Manuscripta Math.* **49**(1984), 27-59.

[15] R. Temam, "Navier-Stokes Equations", North-Holland, Amsterdam New York Oxford, 1984.

[16] W. von Wahl, "The Equations of Navier-Stokes and Abstract Parabolic Equations", Vieweg, Braunschweig-Wiesbaden, 1985.

67 SET-VALUED MAPPINGS ON LCA GROUP AND APPROXIMATE IDENTITY

H. C. LAI

Department of Applied Mathematics, I-Shou University
Ta-Hsu, Kaohsiung, Taiwan

J. C. LEE

Department of Applied Mathematics, Chung Yuang Christian University
Chung-Li, Taiwan

Abstract Let X be a separable Banach space, G a LCA group, and let $\mathcal{L}^1(G, \mathcal{P}_{KC}(X))$ be the space of all compact convex set-valued functions defined on a locally compact abelian group G, where $\mathcal{P}_{KC}(X)$ denotes the family of compact convex sets of X. $\mathcal{L}^1$ is a metric space under the metric

$$D_1(F_1, F_2) := \int_G h(F_1(t), F_2(t))dt$$

for any F_1, $F_2 \in \mathcal{L}^1(G, \mathcal{P}_{KC}(X))$ where h denotes the Hausdorff metric on $\mathcal{P}_{KC}(X)$. We formulate the convolution of $\mathcal{L}^1$ and $L^1(G)$, and prove that any set-valued mapping $F \in \mathcal{L}^1(G, \mathcal{P}_{KC}(X))$ is approximated by the convolution of an approximate identity $\{e_\alpha\}$ for $L^1(G)$ and F self under a specific metric topology D_1 in $\mathcal{L}^1$. This property will be extended to study Fourier analysis of set-valued maps.

H.G.W. Begehr et al. (eds.), Proceedings of the Second ISAAC Congress, Vol. 1, 581–584.

1. Introduction

Let $\mathcal{L}^1(G, \mathcal{P}_{KC}(X))$ be the space of all integrably bounded set-valued maps from a LCA group G into $\mathcal{P}_{KC}(X)$ where $\mathcal{P}_{KC}(X)$ is the family of compact convex subsets of a separable Banach space X. It is known that $\mathcal{L}^1(G, \mathcal{P}_{KC}(X))$ is not a linear space but a complete metric space with a suitable metric. We can define the convolution of a scalar function and a set-valued mapping, and establish that the approximate identity $\{e_\alpha\} \subset C_c(G)$ for $L^1(G)$ will play an approximate identity for $\mathcal{L}^1(G, \mathcal{P}_{KC}(X))$ by convolution under a specific metric D_1-topology. Several properties for $\mathcal{L}^1(G, \mathcal{P}_{KC}(X))$ as well as the Hausdorff metric on $\mathcal{P}_{KC}(X)$ and the D_1-metric for $\mathcal{L}^1(G, \mathcal{P}_{KC}(X))$ are discussed. The Fourier analysis on $\mathcal{L}^1(G, \mathcal{P}_{KC}(X))$ will be based on the property of convolution and approximate identity.

2. Preliminaries

We need the following notations:

- $\mathcal{P}_f(X)$: the collection of all closed subsets in X.
- $\mathcal{P}_{KC}(X)$: the collection of all compact convex subsets in X.
- h : the Hausdorff distance on $\mathcal{P}_f(X)$, that is, for $A, B \in \mathcal{P}_f(X)$

$$h(A,\, B) := \max\{\sup_{a\in A} d(a,\, B),\, \sup_{b\in B} d(b,\, A)\},$$

  and let

$$|A| := \sup\{\|a\|_X : a \in A \subset X\}, h(A,\, \{0\}).$$

- $\mathcal{M}(G, X)$: the set of all measurable set-valued maps
  $F : G \to \mathcal{P}_f(X)$.
- $\mathcal{S}_F^p$:=$\{f \in L^p(G, X) | f(t) \in F(t) \text{ a.e.}\}$, $1 \le p \le +\infty$:
  the family of all L^p-selection for $F \in \mathcal{M}(G, X)$.

- $\mathcal{L}^p(G,X)$: $thespaceofall$p$-integrablyboundedset$ $-$ $valuedmaps$F$\in$ $\mathcal{M}(G,X)$. That is, there exists a function $\varphi \in L^p(G)$ such that

$$|F(t)| \leq \varphi(t) \quad \text{for a. a. } t \in G.$$

- D_p : the metric on $\mathcal{L}^p$, that is for $F_1, F_2 \in \mathcal{L}^p$, D_p is defined by

$$D_p(F_1, F_2) := \{\int_G |h(F_1(t),\, F_2(t))|^p\, dt\}^{1/p}, \qquad 1 \leq p < +\infty.$$
$$D_\infty(F_1, F_2) := \operatorname*{ess\,sup}_{t\in G} h(F_1(t),\, F_2(t)).$$

- $\mathcal{S}(G, \mathcal{P}_{KC}(X))$: the set of all $\mathcal{P}_{KC}(X)$-valued measurable set value mappings.

Some preliminary results are known (cf. Lai and Lee [4, 5] and its references) which we summary as follows.

1. $(\mathcal{P}_f(X), h)$ is a complete metric space.

2. $(\mathcal{P}_{KC}(X), h)$ is a separable complete metric space.

3. For any F_1 and F_2 in $\mathcal{L}^p$, the function

$$t \in G \longrightarrow h(F_1(t),\, F_2(t))$$

is measurable and

$$h(F_1(t), F_2(t)) \leq |F_1(t)| + |F_2(t)| \quad \text{for any } t \in G\ ,$$

so that $h(F_1(\cdot),\, F_2(\cdot)) \in L^p(G)$, $1 \leq p < \infty$.

4. $(\mathcal{L}^p, D_p)$ is a complete metric space.

5. The family of simple set mappings in $\mathcal{S}(G,\, \mathcal{P}_{KC}(X))$ is D_p-dense in $\mathcal{L}^p(G,\, \mathcal{P}_{KC}(X))$.

6. $\mathcal{L}^p(G, \mathcal{P}_{KC}(X))$ is a D_p-closed subspace of $\mathcal{L}^p(G, \mathcal{P}_f(X))$.

By the facts of the above results, without loss of generality we can confine our discussion on the $\mathcal{P}_{KC}(X)$-valued set mappings. For convenience, we often write

$$\mathcal{L}^p(G, X) \text{ instead of } \mathcal{L}^p(G, \mathcal{P}_{KC}(X))$$

if there has not any confusion.

Full version of the paper will appear elsewhere.

68 MINKOWSKI'S INEQUALITY FOR EXTENDED MEAN VALUES

Zheng LIU

Department of Mathematics and Physics
Anshan 114002, Liaoning
People's Republic of China

asgykyc@mail.asptt.ln.cn

Abstract

In this paper, Minkowski's inequality and its reverse are considered for the extended mean values, including most of the classical two variable means as special cases. We obtain new proof for the case of pth power mean, find the whole results for the case of generalized logarithmic mean, and give some sufficient conditions for the general case.

1 Introduction

First defined by Stolarsky[1], then investigated by Leach and Scholander[2, 3], as well as various other authors [4, 5, 6], the so-called extended mean values $E_{p,q}(x, y)$ is a two-parameter mean of two positive numbers x and y defined as follows:

$$E_{p,q}(x,y) = \left(\frac{p}{q}\frac{x^q - y^q}{x^p - y^p}\right)^{\frac{1}{q-p}}, \qquad pq(p-q)(x-y) \neq 0; \tag{1}$$

$$E_{p,0}(x,y) = \left(\frac{1}{p}\frac{x^p - y^p}{\ln x - \ln y}\right)^{\frac{1}{p}}, \qquad p(x-y) \neq 0; \tag{2}$$

H.G.W. Begehr et al. (eds.), Proceedings of the Second ISAAC Congress, Vol. 1, 585–592.

$$E_{p,p}(x,y) = e^{-\frac{1}{p}}(\frac{x^{x^p}}{y^{y^p}})^{\frac{1}{x^p - y^p}}, \qquad p(x-y) \neq 0; \tag{3}$$

$$E_{0,0}(x,y) = \sqrt{xy}, \qquad x \neq y; \tag{4}$$

$$E_{p,q}(x,x) = x, \qquad x = y \tag{5}$$

Most of the classical two variable means are special cases of $E_{p,q}(x,y)$; for instance,

$E_{p,2p}(x,y) = M_p(x,y)$ is the pth power mean,
$E_{1,p}(x,y) = S_p(x,y)$ is the generalized logarithmic mean,
$E_{1,2}(x,y) = A(x,y)$ is the arithmetic mean,
$E_{0,0}(x,y) = G(x,y)$ is the geometric mean,
$E_{-2,-1}(x,y) = H(x,y)$ is the harmonic mean,
$E_{0,1}(x,y) = L(x,y)$ is the logarithmic mean,
$E_{1,1}(x,y) = I(x,y)$ is the identric mean.

The purpose of this paper is to study Minkowski's inequality

$$E_{p,q}(x_1+x_2, y_1+y_2) \leq E_{p,q}(x_1,y_1) + E_{p,q}(x_2,y_2) \tag{6}$$

and its reverse

$$E_{p,q}(x_1+x_2, y_1+y_2) \geq E_{p,q}(x_1,y_1) + E_{p,q}(x_2,y_2) \tag{7}$$

where $x_1, x_2, y_1, y_2 \in \mathbf{R}_+$.

It was reported in [7] that (6) and (7) for the special case $q - p = 1$ was treated by H.Alzer and one has

$E_{p,p+1}(x_1+x_2, y_1+y_2) \leq E_{p,p+1}(x_1,y_1) + E_{p,p+1}(x_2,y_2)$ if $p \geq 1$,

and

$E_{p,p+1}(x_1+x_2, y_1+y_2) \geq E_{p,p+1}(x_1,y_1) + E_{p,p+1}(x_2,y_2)$ if $p \leq 1$.

In order to find more sufficient conditions for the inequality (6) or (7) to hold, we would like to do some analyses and calculations at the beginning.

Since $E_{p,q}(x,y) = yE_{p,q}(\frac{x}{y},1) = yf_{p,q}(\frac{x}{y})$, where

$$f_{p,q}(t) = (\frac{p}{q})^{\frac{1}{q-p}}\sqrt{t}(\frac{\sinh(q\ln\sqrt{t})}{\sinh(p\ln\sqrt{t})})^{\frac{1}{q-p}}, \qquad pq(p-q) \neq 0; \tag{8}$$

$$f_{p,0}(t) = \sqrt{t}(\frac{\sinh(p\ln\sqrt{t})}{p\ln\sqrt{t}})^{\frac{1}{p}} \qquad p \neq 0 \tag{9}$$

$$f_{p,p}(t) = e^{-\frac{1}{p}}\sqrt{t}\exp[(\ln\sqrt{t})(\coth(p\ln\sqrt{t}))], \qquad p \neq 0 \tag{10}$$

$t \in \mathbf{R}_+$, we can write (6) as

$$f_{p,q}\left(\frac{y_1}{y_1+y_2}\frac{x_1}{y_1}+\frac{y_2}{y_1+y_2}\frac{x_2}{y_2}\right) \leq \frac{y_1}{y_1+y_2} f_{p,q}\left(\frac{x_1}{y_1}\right)+\frac{y_2}{y_1+y_2} f_{p,q}\left(\frac{x_2}{y_2}\right)$$

or, with the notations $\frac{x_1}{y_1} = u \in \mathbf{R}_+$, $\frac{x_2}{y_2} = v \in \mathbf{R}_+$, $\frac{y_1}{y_1+y_2} = \lambda \in (0,1)$,

$$f_{p,q}(\lambda u + (1-\lambda)v) \leq \lambda f_{p,q}(u) + (1-\lambda) f_{p,q}(v) \tag{11}$$

which is exactly the convexity of $f_{p,q}$. If $f_{p,q}$ is strictly convex, then equality stands in (11)(or in (6)) if and only if $u = v$ (see e.g.[8]).

A similar statement holds for the reverse inequality (7).

Let I be an open subinterval of $\mathbf{R}$ and $g\colon I \to \mathbf{R}$ be a function of class $C^2(I)$. It is well known(see e.g.,[8]) that g is convex if and only if $g''(x) \geq 0$ for $x \in I$ or g is concave if only if $g''(x) \leq 0$ for $x \in I$.

If $g''(x) > 0$ for $x \in I/A$, where A is a finite subset of I, then g is strictly convex, that is in the convex inequality

$$g(\lambda u + (1-\lambda)v) \leq \lambda g(u) + (1-\lambda) g(v)$$

equality holds if and only if $\lambda \in \{0,1\}, u, v \in I$ or $\lambda(0,1), u = v \in I$.

If $g''(x) < 0$ for $x \in I/A$, then g is strictly concave.

Hence (6) or (7) holds if and only if $f''_{p,q}(t) \geq 0$ or $f''_{p,q}(t) \leq 0$ for $t \in \mathbf{R}_+$.

Using logarithmic differentiation, we obtain, after some calculations, that

$$\frac{4t^2 f''_{p,q}(t)}{f_{p,q}(t)} = \frac{q^2(1+p-q)}{(q-p)^2}(\coth^2 qs - 1) + \frac{p^2(1-p+q)}{(q-p)^2}(\coth^2 ps - 1) \tag{12}$$

$$-\frac{2pq}{(q-p)^2}(\coth ps \coth qs - 1) \tag{13}$$

if $pq(p-q) \neq 0$ and

$$\frac{4t^2 f''_{p,0}(t)}{f_{p,0}(t)} = (1-p)(\coth^2 ps - 1) + \frac{1+p}{p^2 s^2} - \frac{2\coth ps}{ps} \tag{14}$$

if $p \neq 0$ and

$$\frac{4t^2 f''_{p,p}(t)}{f_{p,p}(t)} = (1-\coth^2 ps)(2p-1+2p(1-p)s\coth ps + p^2 s^2(1-\coth^2 ps)) \tag{15}$$

if $p \neq 0$, where $s = \ln\sqrt{t} \in \mathbf{R}$.

Multiplying (13),(14),(15) by $2\sinh^2 ps \sinh^2 qs$, after some calculations, by using the identities $q^2(1+p-q)-pq = q(1-q)(q-p), p^2(1-p+q) = -p(1-p)(q-p)$ and some product-to-sum identities for hyperbolic functions, the resulting product is an analytic function of s which can be written as

$$\begin{aligned} g_{p,q}(s) &:= \frac{8t^2 f''_{p,q}(t)}{f_{p,q}(t)} \sinh^2 ps \sinh^2 qs \\ &= \frac{q(1-q)}{q-p}(\cosh 2ps - 1) - \frac{p(1-p)}{q-p}(\cosh 2qs - 1) + \\ &\quad \frac{pq}{(q-p)^2}(\cosh 2(p-q)s - 1) \end{aligned} \tag{16}$$

if $pq(p-q) \neq 0$ and

$$g_{p,0}(s) = 2\sinh^2 ps[(1-p) + \frac{1+p}{2p^2s^2}(\cosh 2ps - 1) - \frac{\sinh 2ps}{ps}] \tag{17}$$

if $p \neq 0$ and

$$g_{p,p}(s) = 2p^2s^2 + (1-2p)(\cosh 2ps - 1) - 2p(1-p)s \sinh 2ps \tag{18}$$

if $p \neq 0$.

From (16),(17),(18), by using the series expansions of hyperbolic functions, we can easily obtain

$$g_{p,q}(s) = \sum_{n=1}^{\infty} \frac{pq}{q-p}[p^{2n-1}(1-q) - q^{2n-1}(1-p) + (q-p)^{2n-1}] \frac{(2s)^{2n}}{(2n)!} \tag{19}$$

if $pq(p-q) \neq 0$ and

$$g_{p,0}(s) = 2\sinh^2 ps \sum_{n=1}^{\infty} (\frac{1+p}{n+1} - 2) \frac{(2ps)^{2n}}{(2n+1)!} \tag{20}$$

if $p \neq 0$ and

$$g_{p,p}(s) = \sum_{n=2}^{\infty} [(2n-2)p - (2n-1)] \frac{(2ps)^{2n}}{(2n)!} \tag{21}$$

if $p \neq 0$.

Clearly, inequality (6) or (7) holds if and only if $g_{p,q}(s) \geq 0$ or $g_{p,q}(s) \leq 0$ for $s \in \mathbf{R}$.

2 Minkowski's Inequality and its Reverse

We start from considering some special cases of $E_{p,q}(x,y)$. In what follows, we always assume that $x_1, x_2, y_1, y_2 \in \mathbf{R}_+$.

For the pth power mean we havet the inequality

$$M_p(x_1 + x_2, y_1 + y_2) \leq M_p(x_1, y_1) + M_p(x_2, y_2) \tag{22}$$

holds if and only if $p \geq 1$ and the inequality

$$M_p(x_1 + x_2, y_1 + y_2) \geq M_p(x_1, y_1) + M_p(x_2, y_2) \tag{23}$$

holds if and only if $p \leq 1$.

If $p = 1$, then equality holds in (22) or (23) for all x_1, x_2, y_1, y_2.

If $p \neq 1$, then equality holds in (22) or (23) if and only if $\frac{x_1}{x_2} = \frac{y_1}{y_2}$.

Proof. It is evident that for $p \neq 0$ from (19)

$$g_{p,2p}(s) = (p-1)\sum_{n=1}^{\infty}(2^{2n} - 4)\frac{(2ps)^{2n}}{(2n)!}$$

and for $p = 0$ is the well known case of $G(x,y)$.

$H(x_1+x_2, y_1+y_2) \geq H(x_1, y_1)+H(x_2, y_2)$ and the sign of equality holds if and only if $\frac{x_1}{x_2} = \frac{y_1}{y_2}$.

For the generalized logarithmic mean $S_p(x,y)$ we have the inequality

$$S_p(x_1 + x_2, y_1 + y_2) \leq S_p(x_1, y_1) + S_p(x_2, y_2) \tag{24}$$

holds if and only if $p \geq 2$ and the inequality

$$S_p(x_1 + x_2, y_1 + y_2) \geq S_p(x_1, y_1) + S_p(x_2, y_2) \tag{25}$$

holds if and only if $p \leq 2$.

If $p = 2$, then equality holds in (24) or (25) for all x_1, x_2, y_1, y_2.

If $p \neq 2$, then equality holds in (24) or (25) if and only if $\frac{x_1}{x_2} = \frac{y_1}{y_2}$.

Proof. It is evident that for $p \neq 0$ from (19)

$$g_{1,p}(s) = p\sum_{n=1}^{\infty}[(p-1)^{2n-2} - 1]\frac{(2s)^{2n}}{(2n)!}$$

and for $p = 0$ from (20)

$$g_{1,0} = 2\sinh^2 s\sum_{n=1}^{\infty}(\frac{2}{n+1} - 2)\frac{(2ps)^{2n}}{(2n+1)!} < 0$$

$L(x_1+x_2, y_1+y_2) \geq L(x_1, y_1) + L(x_2, y_2)$ and the sign of equality holds if and only if $\frac{x_1}{x_2} = \frac{y_1}{y_2}$.
$I(x_1+x_2, y_1+y_2) \geq I(x_1, y_1) + I(x_2, y_2)$ and the sign of equality holds if and only if $\frac{x_1}{x_2} = \frac{y_1}{y_2}$.

Now we turn to the general case. For brevity, write (19) as

$$g_{p,q}(s) = \sum_{n=2}^{\infty} h_n(p,q) \frac{(2s)^{2n}}{(2n)!} \tag{26}$$

where

$$h_n(p,q) = \frac{pq}{q-p}[p^{2n-1}(1-q) - q^{2n-1}(1-p) + (q-p)^{2n-1}] \tag{27}$$

for $n \geq 2$.

The condition

$$\max\{p,q\} - 1 \geq \min\{p,q\} \geq 1 \tag{28}$$

is sufficient for (6) to hold. If $(p,q) = (1,2)$ or $(p,q) = (2,1)$, then equality holds in (6) for all x_1, x_2, y_1, y_2. For all other (p,q), equality holds in (6) if and only if $\frac{x_1}{x_2} = \frac{y_1}{y_2}$.

Proof. By the symmetry $E_{p,q} = E_{q,p}$, we may assume that $q \geq p$. Denote $r = q - p$. Then $r \geq 0$. It is not difficult to find that

$$h_n(p,p+r) = p^2(p+r)[(r-p)\sum_{k=0}^{2n-3} p^k r^{2n-3-k} + (p-1)\sum_{k=0}^{2n-3} C_{2n-1}^{k+1} p^k r^{2n-3-k}] \tag{29}$$

for $n \geq 2$.

If $r \geq p \geq 1$, then $h_n(p,p+r) \geq 0$ for $n \geq 2$.

If $1 \leq r < p$, then $p-1 \geq p-r$ and so from (29)

$$h_n(p,p+r) \geq p^2(p^2-r^2)\sum_{k=0}^{2n-3}(C_{2n-1}^{k+1} - 1)p^k r^{2n-3-k} > 0$$

for $n \geq 2$.

Thus from (26) we can conclude that $g_{p,q}(s) \geq 0$ for $p \geq 1$ and $q \geq p+1$. Hence (28) is sufficient for (6) to hold.

The condition

$$0 \le \min\{p, q\} \le 1, \qquad \min\{p, q\} \le \max\{p, q\} \le \min\{p, q\} + 1 \quad (30)$$

is sufficient for (7) to hold. If $(p, q) = (1, 2)$ or $(p, q) = (2, 1)$, then equality holds in (7) for all x_1, x_2, y_1, y_2. For all other (p, q), equality holds in (7) if and only if $\frac{x_1}{x_2} = \frac{y_1}{y_2}$.

Proof. We may assume that $q \ge p$ and denote $r = q - p$.

For $0 < p < 1$, using (29), we see that $h_n(p, p + r) < 0$ for $n \ge 2$ if $r \le p$. If $p < r \le 1$ then $p - 1 \le p - r < 0$ and so

$$h_n(p, p + r) \le p^2(p^2 - r^2) \sum_{k=0}^{2n-3} (C_{2n-1}^{k+1} - 1)p^k r^{2n-3-k} < 0$$

For $p = 0$, from (20) we have

$$g_{0,q}(s) = 2\sinh^2 qs \sum_{n=1}^{\infty} (\frac{1+q}{n+1} - 2)\frac{(2qs)^{2n}}{(2n+1)!} < 0$$

for $q \le 3$.

For $p = 1$, we see $g_{1,q}(s) < 0$ for $q < 2$ from Theorem 2.

Thus from (26) we can conclude that $g_{p,q}(s) \le 0$ for $0 \le p \le 1$ and $p \le q \le p + 1$. Hence (30) is sufficient for (7) to hold.

The condition

$$p + q \le 0 \qquad (31)$$

is sufficient for (7) to hold and equality holds in (7) if and only if $\frac{x_1}{x_2} = \frac{y_1}{y_2}$.

Proof. For $p + q = 0$ and $p \ne 0$, we have from (19)

$$g_{p,-p}(s) = \sum_{n=2}^{\infty} (1 - 2^{2n-2}) \frac{(2ps)^{2n}}{(2n)!} < 0$$

For $q = p \ne 0$, we have from (21)

$$g_{p,p}(s) = \sum_{n=2}^{\infty} [(2n - 2)p - (2n - 1)] \frac{(2ps)^n}{(2n)!} < 0$$

if $p < 0$

For $q = 0$, we have from (20) $g_{p,0}(s) < 0$ for $p < 0$.

For $q = ap, a \ne 0, \pm 1$, from (19) we have

$$g_{p,ap}(s) = \sum_{n=2}^{\infty} \frac{a}{a-1} [1 - a^{2n-2} + (a - 1)^{2n-1} + (a^{2n-1} - a)p] \frac{(2ps)^{2n}}{(2n)!}$$

By elementary calculus, we see that $\varphi(a) = 1 - a^{2n-1} + (a-1)^{2n-1}$ is positive if $a \in (0,1)$ and is negative if $a \in (-\infty, 0) \cup (1, +\infty)$, $\Psi(a) = a^{2n-1} - a$ is positive if $a \in (-1,0) \cup (1,+\infty)$ and negative if $a \in (-\infty,-1) \cup (0,1)$.

Write

$$h_n(p, ap) = \frac{a}{a-1}[\varphi(a) + \Psi(a)p]$$

for $n \geq 2$.

It is not difficult to find that in case $a > -1$ we have $h_n(p, ap) < 0$ for $n \geq 2$ if $p < 0$ and in case $a < -1$ we have $h_n(p, ap) < 0$ for $n \geq 2$ if $p > 0$.

Thus from (26) we can conclude that $g_{p,q}(s) < 0$ for $p+q \leq 0$. Hence (31) is sufficient for (7) to hold.

References

[1] K B Stolarsky. Generalization of the logarithmic mean. Math. Mag. 48(1975), 87-92.

[2] E B Leach and M C Scholander. Extended mean values. Amer. Math. Monthly 85(1978), 84-90.

[3] E B Leach and M C Scholander. Extended mean values II. J. Math. Anal. Appl. 92(1983),207-223

[4] Z Páles. Inequalities for differences of powers. J.Math.Anal,Appl, 131(1988), 271-281.

[5] M B Sun. Inequalities for two-parameter mean of convex function. J.Math. Res. Expo. 17(1997), 631-632.

[6] J Pecaric, F Qi, V Simic and S L Xu. Refinements and extensions of an inequality III. J.Math. Anal. Appl. 227(1998), 439-448.

[7] D S Mitrinovic, J E Pecaric and A M Fink. Classical and new inequalities in analysis. Kluwer Academic Publishers, 1993, Dordrecht.

[8] L Losonczi and Z Páles. Minkowski's inequality for two variable Gini means. Acta Sci. Math, (Szeged) 62(1996), 413-425.

69 ON SOME ESTIMATES OF SOLUTIONS OF SOME DEGENERATE NONLINEAR PARABOLIC EQUATIONS

Tokumori NANBU

Department of Mathematics, Toyama Medical and Pharmaceutical University
Sugitani 2630, Toyama, 930-0194,Japan

toku@ms.toyama-mpu.ac.jp

1991 Mathematics Subject Classification: 35K65,35B40
Key words: Degenerate nonlinear parabolic equation, Decay estimate

Abstract

We study the existence and some estimates of solution of the initial- boundary value problem for some degenerate nonlinear parabolic equations

$$u_t = \Delta(|u|^{m-1}u) + \sum_{k=1}^{N} b_k \frac{\partial(|u|^l u)}{\partial x_k} - c(x,t)A(u) + f(x,t)$$

where $u(x,t)$ is a scalar function of the spatial variable $x(\in \Omega)$ and time $t(> 0)$, and Ω is a regular unbounded domain in R^N.

H.G.W. Begehr et al. (eds.), Proceedings of the Second ISAAC Congress, Vol. 1, 593–604.

1 Introduction

In this work we consider the existence and some estimates of solutions of the following initial-boundary value problem $(P)_0$:
(1.1)

$$u_t = \Delta(|u|^{m-1}u) + \sum_{k=1}^{N} b_k \frac{\partial(|u|^l u)}{\partial x_k} - c(x,t)A(u) + f(x,t) \text{ in } Q = \Omega \times R^+,$$

$$u(x,t) = 0 \qquad \text{on} \quad \Gamma = \partial\Omega \times R^+, \tag{1.2}$$

$$u(x,0) = u_0(x) \qquad \text{in} \quad \Omega. \tag{1.3}$$

Here Ω is a regular unbounded domain in $R^N (N > 2)$ with the smooth boundary $\partial\Omega$, $R^+ = (0, +\infty)$, Δ denotes the N-dimensional Laplace operator, $b = (b_1, b_2, \cdots .b_N) \in R^N$ ($b \neq O$) is a real constant N-dimensional vector with $\|b\| = 1$, $m > 1$, $l > 0$, and $c(x,t)$, $A(u)$, $f(x,t)$ and $u_0(x)$ satisfy some hypotheses which will be given later.

At the First Internatinal ISAAC Congress we ([9]) gave some results on the existence and decay-estimates of solution to the equation

$$u_t = \Delta(|u|^{m-1}u) + \sum_{k=1}^{N} b_k \frac{\partial(|u|^l u)}{\partial x_k} - q(t)A(u) \quad \text{in} \quad Q = \Omega \times R^+.$$

In this work we consider the equation (1.1) with the force term $f(x,t)$.

2 Assumptions and theorems

Throughout this paper, the notations of function spaces and their norms are as usual. Denote R^+=$(0, +\infty)$, $Q_T = \Omega \times (0,T]$, $Q = \Omega \times (0, +\infty)$, $\Gamma_T = \partial\Omega \times (0,T]$, $\Gamma = \partial\Omega \times (0, +\infty)$ and $\nabla u = (u_{x_1}, u_{x_2}, ..., u_{x_N})$. We denote $L^p(D)$ norm by $\| \cdot \|_{p,D}$.

Definition 1. A *solution* u of the problem (1.1) - (1.3) on [0,T] is a function with the following properties :

$$u \in C([0,T] : L^2_{loc}(\Omega)), \tag{i}$$

$$\text{(ii)} \quad \int_\Omega u(t)\varphi(t)dx - \int_0^t \int_\Omega [u\varphi_t + |u|^{m-1}u\Delta\varphi - \sum_{k=1}^{N} b_k |u|^l u \varphi_{x_k}]dxd\tau$$

$$= \int_\Omega u_0(x)\varphi(x,0)dx - \int_0^t \int_\Omega c(x,t)A(u)\varphi dx d\tau + \int_0^t \int_\Omega f(x,t)\varphi(x,t)dx d\tau$$

for any $t \in (0, T]$ and $\varphi \in C^2(\overline{Q_T})$ such that $\varphi = 0$ on $\partial\Omega \times [0, T]$.

Definition 2. A *solution* u of the problem (1.1)-(1.3) on $[0, +\infty)$ means a solution on *each* $[0, T]$.

Now we shall give precise hypotheses on $c(x,t), A(u), f(x,t)$ and $u_0(x)$.

Assumption on the function $c(x,t)$:

(H.c) $$c(x,t) \in C^1(\overline{\Omega \times R^+}) \cap L^\infty(\Omega \times R^+),$$

$$c(x,t) \geq 0, \frac{\partial c(x,t)}{\partial t} \leq 0 \quad ((x,t) \in \overline{\Omega \times R^+}).$$

Assumption on the function $A(u)$:
(H.A)
$$A(u) \in C^1(R),\ a(u) \equiv A'(u) > 0 \quad (u \neq 0), \text{and} \quad A(0) = A'(0) = 0$$

Assumption on the function $f(x,t)$:
(H.f) $f(x,t) \in L^\infty(\Omega \times R^+) \cap L^1(\Omega \times R^+)$ and f satisfies

$$\int_0^\infty \|f(t)\|_{r_1,\Omega}^{\frac{m+2q+1}{m}} dt \leq C_0 < \infty,$$

$$\int_0^\infty \|f(t)\|_{2r_2,\Omega}^{\frac{4m}{m+1}} dt \leq C_0 < \infty,$$

$$\int_0^\infty \|f(t)\|_{\frac{2N}{N+2},\Omega}^2 dt \leq C_0 < \infty.$$

Here $r_1 = \frac{N(m+2l+1)}{mN+2(2l+1)} (> 1)$ and $r_2 = \frac{2Nm}{N(m+1)+2(m-1)} (> 1)$.

Assumption on the function $u_0(x)$:
(H.u_0) $|u_0(x)|^{m-1}u_0(x) \in H_0^1(\Omega)$ and $u_0(x) \in L^\infty(\Omega)$ has a bounded support.

We state our main results as follows:

Theorem 1. (Existence of solution)
Suppose that (H.c), (H.A), (H.f) and (H.u_0) are satisfied. Then the problem (1.1) - (1.3) has a solution on [0,+∞) such that

(i) $$\|\nabla(|u|^{m-1}u)(t)\|_{2,\Omega} \leq c_0(f, u_0) \quad \text{on} \quad [0, +\infty),$$

(ii) $$\|(\frac{\partial(|u|^{m-1}u)}{\partial t})\|^2_{L^2(0,T:L^2(\Omega))} \le c_0(f,u_0).$$

Here $c_0(f,u_0)$ is a constant depending only on the f and u_0.

Theorem 2. (Decay of solution)
Suppose that (H.c), (H.A), (H.f), (H.u_0), $f \in L^1(R^+;L^1(\Omega))$ and

$$\lim_{t\to\infty} \|f(t)\|_{\frac{2N}{N+2},\Omega} = 0.$$

Then on any compact domain $D \subset \Omega$ the solution $u(x,t)$ of the problem (1.1)-(1.3) satisfies

$$\lim_{t\to\infty} \int_D |u(x,t)|^{m+1}dx = 0.$$

3 Proof of Theorem 1

Proof of Theorem 1. Suppose that the support of $u_0(x)$ is contained in a ball $B_\rho \equiv \left\{x \in R^N : |x| < \rho\right\}$. For brevity, we assume that $u_0(x)$ is of class $C_0^\infty(B_\rho \cap \Omega)$. Set $\Omega_n = \Omega \cap B_n$, $Q_n = \Omega_n \times R^+$, $\Gamma_n = \partial\Omega_n \times R^+$ and $Q_{n,T} = \Omega_n \times [0,T]$.

Assume that $\rho < n < \infty$. Set $\Phi_n(s)=\frac{1}{n}s+|s|^{m-1}s$ and $\Phi(s)=|s|^{m-1}s$. Let us consider the following regularized problem:

$$(3.1)_n \quad u_t = \Delta(\Phi_n(u)) + \sum_{k=1}^N b_k\frac{\partial(|u|^l u)}{\partial x_k} - c(x,t)A(u) + f(x,t) \text{ in } Q_n,$$

$$(3.2)_n \qquad u(x,t) = 0 \quad \text{on} \quad \Gamma_n,$$

$$(3.3)_n \qquad u(x,0) = u_0(x) \quad \text{in} \quad \Omega_n.$$

By the standard theory, for each $T \in R^+$ this regularized problem $(3.1)_n - (3.3)_n$ has a unique classical solution $u_n(x,t) \in C_0^{2,1}(\overline{Q_{n,T}})$ ([**6**],Ch.V).

Meanwhile we write u for u_n.
Multiplying $(3.1)_n$ by $|u|^{2q}u$ $(q \ge 0)$ and integrating it over Ω_n, we have

$$(3.4) \qquad \int_{\Omega_n} u_t\{|u|^{2q}u\}dx =$$

$$\int_{\Omega_n} \left\{ \left[\Delta(\Phi_n(u)) + \sum_{k=1}^{N} b_k \frac{\partial(|u|^l u)}{\partial x_k} - c(x,t)A(u) \right] |u|^{2q}u \right\} dx$$

$$+ \int_{\Omega_n} f(x,t)|u|^{2q}u \, dx$$

$$= - \int_{\Omega_n} \{\nabla \Phi_n(u) \cdot \nabla \left(|u|^{2q}u\right)\} dx - \int_{\Omega_n} \{c(x,t)A(u)|u|^{2q}u\} dx$$

$$+ \int_{\Omega_n} f(x,t)|u|^{2q}u \, dx.$$

We recall the well-known Sobolev's lemma (Cf.[4]). If $v \in W_0^{1,2}(\Omega_n)$, then

$$\|v\|_{\frac{2N}{N-2},\Omega_n} \le \gamma \|\nabla v\|_{2,\Omega_n} \tag{3.5}$$

where the constant γ is independent of Ω_n.

Putting $p_q = \frac{N(m+2q+1)}{mN+2(2q+1)} (> 1)$, then we have

$$\int_{\Omega_n} f(x,t)|u|^{2q}u dx \le \||u|^{\frac{m-1}{2}+q}u\|^{\frac{2(2q+1)}{m+2q+1}}_{\frac{2N}{N-2},\Omega_n} \|f\|_{p_q,\Omega_n}$$

$$\le \left\{\gamma \|\nabla(|u|^{\frac{m-1}{2}+q}u)\|_{2,\Omega_n}\right\}^{\frac{2(2q+1)}{m+2q+1}} \|f(t)\|_{p_q,\Omega_n}.$$

Thus we can conclude that

$$\frac{1}{2(q+1)} \frac{d}{dt} \|u(t)\|^{2q+2}_{2q+2,\Omega_n} + \int_{\Omega_n} \left\{\nabla \left(|u|^{m-1}u\right) \nabla \left(|u|^{2q}u\right)\right\} dx$$

$$\le \gamma^{\frac{2(2q+1)}{m+2q+1}} \|\nabla(|u|^{\frac{m-1}{2}+q}u\|^{\frac{4q+2}{m+2q+1}}_{2,\Omega_n} \|f(t)\|_{p_q,\Omega_n}$$

$$\equiv C_1 \left\{\int_{\Omega_n} \nabla \left(|u|^{m-1}u\right) \cdot \nabla \left(|u|^{2q}u\right) dx\right\}^{\frac{2q+1}{m+2q+1}} \|f(t)\|_{p_q,\Omega_n}.$$

Here $C_1 \equiv \gamma^{\frac{2(2q+1)}{m+2q+1}} (\frac{(m+2q+1)^2}{4m(2q+1)})^{\frac{2q+1}{m+2q+1}}$ and $p_q = \frac{N(m+2q+1)}{mN+2(2q+1)} (> 1)$.

Using Young's inequality, we have

$$\frac{1}{2(q+1)} \frac{d}{dt} \|u(t)\|^{2q+2}_{2q+2,\Omega_n} + \frac{1}{2} \int_{\Omega_n} \left\{\nabla \left(|u|^{m-1}u\right) \cdot \nabla \left(|u|^{2q}u\right)\right\} dx \tag{3.6}$$

$$\le \frac{m}{m+2q+1} (\frac{2(2q+1)}{m+2q+1})^{\frac{2q+1}{m}} \{C_1 \|f(t)\|_{p_q,\Omega_n}\}^{\frac{m+2q+1}{m}}.$$

Multiplying $(3.1)_n$ by $(|u|^{m-1}u)_t$ and integrating it over Ω_n, we have

$$m\int_{\Omega_n}|u|^{m-1}(u_t)^2dx+\frac{1}{2}\frac{d}{dt}\left(\int_{\Omega}|\nabla(|u|^{m-1}u)|^2dx\right)$$

$$\leq \int_{\Omega_n}\left\{\left[\sum_{k=1}^{N}b_k\frac{\partial(|u|^lu)}{\partial x_k}-c(x,t)A(u)+f(x,t)\right](|u|^{m-1}u)_t\right\}dx.$$

Since it holds that

$$\int_{\Omega_n}f^2(x,t)|u|^{m-1}dx\leq\left[\int_{\Omega_n}||u|^{m-1}u|^{\frac{2N}{N-2}}dx\right]^{\frac{N-2}{2N}\frac{m-1}{m}}\times\|f(t)\|^2_{2r_2,\Omega_n}$$

where $r_2=\frac{2Nm}{N(m+1)+2(m-1)}(>1)$, we have

$$m\int_{\Omega_n}|u|^{m-1}(u_t)^2dx+\frac{1}{2}\frac{d}{dt}\left(\int_{\Omega_n}|\nabla(|u|^{m-1}u)|^2dx\right)$$

$$\leq m(l+1)^2\int_{\Omega_n}|u|^{m-1}|\nabla u|u^{2l}|\nabla u|dx+\frac{m}{2}\int_{\Omega_n}|u|^{m-1}u_t^2dx$$

$$+m\int_{\Omega_n}\left[f^2(x,t)|u|^{m-1}\right]dx$$

$$\leq\frac{(l+1)^2}{(2l+1)}\int_{\Omega_n}\nabla(|u|^{m-1}u)\cdot\nabla(u^{2l}u)dx+\frac{m}{2}\int_{\Omega_n}|u|^{m-1}u_t^2dx$$

$$+m\ \||u|^{m-1}u\|^{\frac{m-1}{m}}_{\frac{2N}{N-2},\Omega_n}\|f(t)\|^2_{2r_2,\Omega_n}$$

$$\leq\frac{(l+1)^2}{(2l+1)}\int_{\Omega_n}\nabla(|u|^{m-1}u)\nabla(u^{2l}u)dx+\frac{m}{2}\int_{\Omega_n}|u|^{m-1}u_t^2dx$$

$$+m(\gamma)^{\frac{m-1}{m}}\|\nabla(|u|^{m-1}u)\|^{\frac{m-1}{m}}_{2,\Omega_n}\|f(t)\|^2_{2r_2,\Omega_n}.$$

Therefore we have

$$\frac{m}{2}\frac{2l+1}{(l+1)^2}\int_{\Omega_n}|u|^{m-1}(u_t)^2dx+\frac{1}{2}\frac{2l+1}{(l+1)^2}\frac{d}{dt}\left(\int_{\Omega_n}|\nabla(|u|^{m-1}u)|^2dx\right)$$

$$\leq\int_{\Omega_n}\nabla(|u|^{m-1}u)\cdot\nabla(u^{2l}u)dx+$$

$$\frac{2l+1}{(l+1)^2}m(\gamma)^{\frac{m-1}{m}}\|\nabla(|u|^{m-1}u)\|^{\frac{m-1}{m}}_{2,\Omega_n}\|f(t)\|^2_{2r_2,\Omega_n}.$$

Using Young's inequality , we have

(3.7)

$$\frac{m}{2}\frac{2l+1}{(l+1)^2}\int_{\Omega_n}|u|^{m-1}(u_t)^2dx+\frac{1}{2}\frac{2l+1}{(l+1)^2}\frac{d}{dt}\left(\int_{\Omega_n}|\nabla(|u|^{m-1}u)|^2dx\right)$$

$$\leq\int_{\Omega_n}\nabla(|u|^{m-1}u)\nabla(u^{2l}u)dx+\frac{1}{4}\|\nabla(|u|^{m-1}u)\|^2_{2,\Omega_n}+C_2(\|f\|^2_{2r_2,\Omega_n})^{\frac{2m}{m+1}}.$$

Here $r_2=\frac{2Nm}{N(m+1)+2(m-1)}(>1)$ and $C_2\equiv(\frac{2l+1}{(l+1)^2})^{\frac{2m}{m+1}}\frac{m+1}{2}[2(m-1)\gamma^2]^{\frac{(m-1)}{m+1}}$.

Finally, by multiplying $(3.1)_n$ by $|u|^{m-1}u$ and integrating it over Ω_n, we have

$$\frac{d}{dt}\left\{\frac{1}{m+1}\int_{\Omega_n}|u|^{m+1}dx\right\}+\frac{1}{2}\int_{\Omega_n}|\nabla(|u|^{m-1}u)|^2dx \tag{3.8}$$

$$\leq\frac{1}{2}(\gamma)^2\|f\|^2_{\frac{2N}{N+2},\Omega_n}.$$

From (3.6) with $q=l$, (3.7) and (3.8) it follows that

$$\frac{d}{dt}\left\{\frac{1}{m+1}\|u(t)\|^{m+1}_{m+1,\Omega_n}+\right. \tag{3.9}$$

$$\left.\frac{2l+1}{2(l+1)^2}\|\nabla(|u|^{m-1}u)\|^2_{2,\Omega_n}+\frac{1}{(l+1)}\|u(t)\|^{2l+2}_{2l+2,\Omega_n}\right\}+$$

$$\frac{m}{2}\frac{2l+1}{(l+1)^2}\int_{\Omega_n}|u|^{m-1}(u)^2_tdx+\frac{1}{4}\int_{\Omega_n}|\nabla(|u|^{m-1}u)|^2dx$$

$$\leq C_2(\|f(t)\|^2_{2r_2,\Omega_n})^{\frac{2m}{m+1}}+C_3\|f(t)\|^{\frac{m+2l+1}{m}}_{r_1,\Omega_n}+C_4\|f(t)\|^2_{\frac{2N}{N+2},\Omega_n}.$$

Here $C_2\equiv(\frac{2l+1}{(l+1)^2})^{\frac{2m}{m+1}}\frac{m+1}{2}[2(m-1)\gamma^2]^{\frac{(m-1)}{m+1}}$, $C_3\equiv\gamma^{\frac{2(2l+1)}{m+2l+1}}(\frac{(m+2l+1)^2}{4m})^{\frac{2l+1}{m+2l+1}}$, $C_4=\frac{1}{2}(\gamma)^2$, $r_1=\frac{N(m+1+2l)}{mN+2(2l+1)}(>1)$ and $r_2=\frac{2Nm}{N(m+1)+2(m-1)}(>1)$.

By integrating (3.9) in t on $[0,T]$, for the function sequence of $\{u_n\}$ we have

$$\|u_n(t)\|_{m+1,\Omega_n}+\|u_n(t)\|_{2l+2,\Omega_n}+\|\nabla(|u_n|^{m-1}u_n(t))\|_{2,\Omega_n}\leq C(u_0,f),$$

on $[0,+\infty)$ and

$$(|u_n|^{m-1}u_n)_t\in L^2(0,T:L^2(\Omega_n)).$$

Here the constant $C(u_0, f)$ depends only on u_0 and f, but not on n and T.

We set $v_n(x,t)=|u_n|^{m-1}u_n(x,t)$. By Ascoli-Arzela's theorem, we can show that there exist a function $v(x,t)$ and a subsequence $\{v_{\epsilon_{k_n}}(x,t)\}$ of $\{v_n(x,t)\}$ such that

$$v(x,t) \in C([0,T] : L^2_{loc}(\Omega)),$$

and

$$v_{k_n}(x,t) \to v(x,t) \quad \text{as} \quad k_n \to \infty \quad \text{in} \quad C([0,T] : L^2_{loc}(\Omega)).$$

Put $u_n(x,t) = u_{k_n}(x,t) = \Phi_n^{-1}(v_{k_n}(x,t))$ and $u(x,t) = \Phi^{-1}(v(x,t))$. And $u_n(x,t)$ satisfies

$$\int_\Omega u_n(t)\phi(t)dx-$$

$$\int_0^t \int_\Omega [u_n\phi_t + \left(\epsilon_n u_n + |u_n|^{m-1}u_n\right)\Delta\phi - \sum_{k=1}^N b_k B(u_n)\phi_{x_k}]dxd\tau$$

$$= \int_\Omega u_0(x)\phi(x,0)dx - \int_0^t \int_\Omega (-c(x,t)A(u_n) + f(x,t))\phi dxd\tau$$

for any $t \in (0,T]$ and $\phi \in C^2(\overline{Q_T})$ such that $\phi = 0$ on $\partial\Omega \times [0,T]$.

Then we can easily verify that $u_{k_n}(x,t)$ converges to $u(x,t)$ which is of class $C([0,\infty); L^2_{loc}(\Omega))$ and a solution of the problem (1.1)-(1.3) on $[0,+\infty)$. For any domain D ($\subset \Omega$) the solution $u(x,t)$ satisfies

$$\|u(t)\|_{m+1,D} + \|u(t)\|_{2l+2,D} + \|\nabla(|u|^{m-1}u(t))\|_{2,D} \leq C(u_0, f)$$

on $[0,+\infty)$ and

$$(|u|^{m-1}u)_t \in L^2(0,\infty : L^2_{\rm loc}(\Omega)).$$

Here the constant $C(u_0, f)$ depends only on u_0 and f, but not on T.

4 Proof of Theorem 2

Proof of Theorem 2. Let $u_n(x,t)$ be a solution of the regularized problem $(3.1)_n - (3.3)_n$. We denote $u_n(x,t)$ by $u(x,t)$. Since $u(x,t) \in C_0^{2,1}(Q_n)$, we have

$$\int_{\Omega_n} \Delta(\Phi_n(u))\zeta(u(t,x))dx = -\int_{\Omega_n} \zeta'(u)\Phi_n'(u)|\nabla u|^2 dx \leq 0$$

for every nonincreasing $\zeta(u) \in C^1(R)$.
Multiplying $(3.1)_n$ by sgn(u(t,x)) and integrating in Ω_n it follows that

$$\frac{d}{dt}\int_{\Omega_n} |u(x,t)|dx \leq \int_{\Omega_n} |f(x,t)|dx \ (\ t \in \overline{R^+} \equiv [0,+\infty))$$

and then for any t $(\in (0,\infty))$

$$\int_{\Omega_n} |u(x,t)|dx = \|u(t)\|_{1,\Omega_n} \leq \|u_0\|_{1,\Omega} + \int_0^{+\infty} |f(s)|_{1,\Omega}ds \equiv M_0. \tag{4.1}$$

Multiply $(3.1)_n$ by $|u|^{p-1}u$ $(m \leq p)$ and integrate it in Ω_n. Then we have

$$\int_{\Omega_n} |u|^{p-1}uu_t dx =$$

$$\int_{\Omega_n} \{\Delta(\Phi_n(u)) + \sum_{k=1}^N b_k \frac{\partial(B(u))}{\partial x_k} - c(x,t)A(u) + f(x,t)\}|u|^{p-1}udx$$

$$= -\int_{\Omega_n} \{\nabla\Phi_n(u)\cdot\nabla(|u|^{p-1}u)\}dx - \int_{\Omega_n} \{c(x,t)A(u)|u|^{p-1}u\}dx$$

$$+\int_{\Omega_n} f(x,t)|u|^{p-1}udx$$

$$\leq -\int_{\Omega_n} \{\nabla\Phi_n(u)\cdot\nabla\left(|u|^{p-1}u\right)\}dx + \int_{\Omega_n} f(x,t)|u|^{p-1}udx.$$

On the other hand we have

$$\int_{\Omega_n} \{\nabla\Phi_n(u)\cdot\nabla\left(|u|^{p-1}u\right)\}dx = mp\int_{\Omega_n} \left(|u|^{m-1}|u|^{p-1}(\nabla u)\cdot(\nabla u)\right)\}dx$$

$$= \frac{4mp}{(p+m)^2}\int_{\Omega_n} |\nabla(|u|^{\frac{(p+m-2)}{2}}u)|^2 dx.$$

By putting $l_p = \frac{N(m+p)}{mN+2p}(\geq 1)$ and Sobolev's inequality, we have

$$\int_{\Omega_n} f(x,t)|u|^{p-1}udx \leq \||u|^{\frac{p+m-2}{2}}u\|^{\frac{2p}{m+p}}_{\frac{2N}{N-2},\Omega_n}\|f\|_{l_p,\Omega_n}$$

$$\leq \left\{\gamma\|\nabla(|u|^{\frac{p+m-2}{2}}u)\|_{2,\Omega_n}\right\}^{\frac{2p}{m+p}}\|f(t)\|_{l_p,\Omega_n}.$$

Then we obtain

$$\frac{d}{dt}\{\frac{1}{p+1}\int_{\Omega_n} |u|^{p-1}u^2 dx\}$$

$$\leq -\frac{4mp}{(p+m)^2}\int_{\Omega_n} |\nabla(|u|^{(p+m-2)/2}u)|^2 dx$$

$$+\left\{\gamma\|\nabla(|u|^{\frac{p+m-2}{2}}u)\|_{2,\Omega_n}\right\}^{\frac{2p}{m+p}}\|f(t)\|_{l_p,\Omega_n}.$$

Using Young's inequality , we have

$$\frac{d}{dt}\{\frac{1}{p+1}\int_{\Omega_n}|u|^{p-1}u^2dx\} \tag{4.2}$$

$$\leq -\frac{2mp}{(p+m)^2}\int_{\Omega_n}|\nabla(|u|^{(p+m-2)/2}u)|^2dx$$

$$+\frac{m}{m+p}\gamma^{\frac{2p}{m}}(\frac{m+p}{2m})^{\frac{p}{m}}\|f(t)\|_{l_p,\Omega_n}^{\frac{m+p}{m}}.$$

Since $v \equiv |u|^{(p+m-2)/2}u \in W_0^{1,2}(\Omega_n)$, by Sobolev's inequality we have

$$\|v\|_{\frac{2N}{N-2},\Omega_n} \leq \gamma\|\nabla v\|_{2,\Omega_n}$$

where the constants γ is independent of Ω_n.
Then we get

$$(\|u\|_{\frac{N(p+m)}{N-2},\Omega_n})^{\frac{p+m}{2}} \leq \gamma\|\nabla(|u|^{(p+m-2)/2}u)\|_{2,\Omega_n}. \tag{4.3}$$

Furthermore we remark that

$$|v|^{\alpha} = |u|^{\frac{p+m}{2}\alpha} \quad \textit{and} \quad \|v\|_{\alpha,\Omega_n} = (\|u\|_{\frac{p+m}{2}\alpha,\Omega_n})^{\frac{p+m}{2}}.$$

We set

$$\alpha = \frac{N(m-1)+2p+2}{N(m-1)+Np+2} \text{ and } \beta = \frac{pN(p+m)}{N(m-1)+Np+2}.$$

By $p+1 = \alpha+\beta$ and Schwarz's inequality, we have

$$\int_{\Omega_n}|u|^{p+1}dx = \int_{\Omega_n}|u|^{\alpha+\beta}dx \leq (\int_{\Omega_n}|u|dx)^{\alpha} \times (\int_{\Omega_n}|u|^{\frac{\beta}{1-\alpha}}dx)^{1-\alpha}$$

$$= (\int_{\Omega_n}|u|dx)^{\frac{N(m-1)+2p+2}{N(m-1)+Np+2}} \times (\int_{\Omega_n}|u|^{\frac{N(p+m)}{N-2}}dx)^{\frac{(N-2)p}{N(m-1)+Np+2}}.$$

Then from (4.1), we have

$$\int_{\Omega_n}|u(t)|^{p+1}dx \leq \|u(t)\|_1^{\frac{N(m-1)+2p+2}{Np+N(m-1)+2}} \times (\int_{\Omega_n}|u(t)|^{\frac{N(p+m)}{N-2}}dx)^{\frac{(N-2)p}{Np+N(m-1)+2}}$$

$$\leq (M_0)^{\frac{N(m-1)+2p+2}{Np+N(m-1)+2}} \times \|u(t)\|_{\frac{N(p+m)}{N-2},\Omega_n}^{\frac{pN(p+m)}{Np+N(m-1)+2}}.$$

Hence by (4.3) we have

$$(4.4) \qquad (M_0)^{-\frac{N(m-1)+2p+2}{Np}} \times \|u\|_{p+1}^{\frac{(p+1)(pN+N(m-1)+2)}{pN}} \leq \|u\|_{\frac{N(p+m)}{N-2},\Omega_n}^{p+m}$$

$$\leq \gamma^2 \|\nabla(|u|^{(p+m-2)/2}u)\|_{2,\Omega_n}^2.$$

Then from (4.2) and (4.4) we have

$$(4.5) \qquad \frac{d}{dt}\{\int_{\Omega_n} |u|^{p+1} dx\}$$

$$\leq -\frac{2mp(p+1)}{(p+m)^2}\frac{1}{\gamma^2}(M_0)^{-\frac{N(m-1)+2p+2}{Np}} \times (\|u\|_{p+1,\Omega_n}^{p+1})^{\frac{(pN+N(m-1)+2)}{pN}}$$

$$+\frac{m}{m+p}\gamma^{\frac{2p}{m}}(\frac{m+p}{2m})^{\frac{p}{m}}\|f(t)\|_{l_p,\Omega_n}^{\frac{m+p}{m}}.$$

Here $l_p = \frac{N(m+p)}{mN+2p} > 1$ $(p \geq m)$.
Putting $p = m$ in (4.5), we obtain $l_m = \frac{2N}{N+2}$ and

$$(4.6) \qquad \frac{d}{dt}(\int_{\Omega_n} |u_n(x,t)|^{m+1} dx)$$

$$\leq -(m+1)\frac{1}{\gamma^2}(M_0)^{-\frac{N(m-1)+2m+2}{Nm}} \times (\|u_n(t)\|_{m+1,\Omega_n}^{m+1})^{\frac{(mN+N(m-1)+2)}{mN}}$$

$$+\frac{1}{2}\gamma^2\|f(t)\|_{\frac{2N}{N+2},\Omega_n}^2.$$

Therefore from (4.6) and the assumption on $f(x,t)$ we can easily show that for all u_n $(n > \rho)$

$$\lim_{t\to\infty}(\int_{\Omega_n} |u_n(x,t)|^{m+1} dx) = 0.$$

Then it is easy to prove that on any compact domain $D \subset \Omega$ the solution $u(x,t)$ of the problem (1.1)-(1.3) satisfies

$$\lim_{t\to\infty}\int_D |u(x,t)|^{m+1} dx = 0.$$

References

[**1**]. N.D.Alikakos, L^p bounds of solutions of reaction-diffusion equations, *Comm. Partial Diff. Eq.*, **4**, 827-868, (1979).

[**2**]. D.G.Aronson M.Crandall and L.A.Peletier, Stabilization of solutions of a degenerate nonlinear diffusion problem, *Nonlinear Analysis T. M. A.*, **6** , 1001-1022, (1982).

[**3**]. M.Bertsch, T.Nanbu and L.A.Peletier, Decay of solutions of a degenerate nonlinear diffusion equation , *Nonlinear Analysis T. M. A.*, **6**, 539-554, (1982).

[**4**]. E.DiBenedetto, *Degenerate Parabolic Equations*, Springer-Verlag, (1991).

[**5**]. Y.Ebihara and T.Nanbu, Global classical solutions to $u_t - \Delta(u^{2m+1}) + \lambda u = f$, *Jour. Diff. Eq.*, **38**, 260-277, (1980).

[**6**]. O.A.Ladyzhenskaya, V.A.Solonikov and N.N.Ura'lceva, *Linear and quasilinear equations of parabolic type*, (Translation M. M. 23, Amer. Math. Soc. (1968)).

[**7**]. M.Nakao, On solutions to the initial-boundary value problem for an equation in the theory of infiltration, *Nonlinear Analysis T. M. A.* 8, 39 - 48,(1984).

[**8**]. T.Nanbu, Some degenerate nonlinear parabolic equations, *Math. Rep. Coll. Gen. Ed. Kyushu Univ.* 14, 91-110, (1984).

[**9**]. T.Nanbu, Existence and Decay of Solutions of Some Nonlinear Degenerate Parabolic Equations, *R.P.Gilbert, J. Kajiwara and Y. S. Xu, Direct and Inverse Problems of Mathematical Physics,* Vol.6, Chap. 17, ISAAC Series, Kluwer Academic Pub.,(1999).

70 HOLOMORPHIC FUNCTIONS OF A COMPLEX VARIABLE IN CLIFFORD ANALYSIS

Kiyoharu NÔNO

Department of Mathematics, Fukuoka University of Education
Akama 729-1, Munakata, 811-4192, Japan

nouno@fukuoka-u.ac.jp

1. Introduction

R.Delanghe([2]) gave a regularity (monogenicity) of functions with valued in the Clifford algebra as smooth solutions of the generalized Cauchy-Riemann equations

$$\frac{\partial u_0}{\partial x_0} - \sum_{i=1}^{n} \frac{\partial u_i}{\partial x_i} = 0, \quad \frac{\partial u_i}{\partial x_k} = \frac{\partial u_k}{\partial x_i}, \quad \frac{\partial u_0}{\partial x_k} = -\frac{\partial u_k}{\partial x_0} \quad (i, k = 1, 2, ..., n)$$

based on the Laplace equation, and developed the Clifford monogenic function theory([1]). Also,this function theory have been developed by many authors.

In the paper [4], H.Leutwiler considered the following system :

$$x_n\left(\frac{\partial u_0}{\partial x_0} - \sum_{i=1}^{n} \frac{\partial u_i}{\partial x_i}\right) + (n-1)u_n = 0,$$

H.G.W. Begehr et al. (eds.), Proceedings of the Second ISAAC Congress, Vol. 1, 605–611.

$$\frac{\partial u_i}{\partial x_k} = \frac{\partial u_k}{\partial x_i}, \quad \frac{\partial u_0}{\partial x_k} = -\frac{\partial u_k}{\partial x_0}, \quad (i, k = 1, 2, ..., n)$$

and studied the properties of solutions of this system.

In the paper [6,7], we considered the following system of generalized axial ly symmetric potential theory equations:

$$x_k \sum_{i=0}^{n} \frac{\partial^2 u}{\partial x_i^2} - (n-1)\frac{\partial u}{\partial x_k} = 0 \qquad (k = 1, 2, ..., n).$$

Using the following system:

$$x_k(\frac{\partial u_0}{\partial x_0} - \sum_{i=1}^{n} \frac{\partial u_i}{\partial x_i}) + (n-1)u_k = 0, \qquad (k = 1, 2, ...n),$$

$$\frac{\partial u_i}{\partial x_k} = \frac{\partial u_k}{\partial x_i}, \quad \frac{\partial u_0}{\partial x_k} = -\frac{\partial u_k}{\partial x_0} \quad (i, k = 1, 2, ..., n),$$

we defined a regularity of functions and developed a function theory based on the system of generalized axially symmetric potential equations.

The aim of this paper to give some connections between monogenic functions, regular functions and c-holomorphic functions.

2. Clifford Algebra

Let $A_n(R)$ be an universal Clifford algebra over an n-dimensional vector space with basis $\{e_1, e_2, ..., e_n\}$. Then, it is well known that $A_n(R)$ is a real 2^n-dimensional associative, but non-commutative algebra and its basis $\{e_0, e_1, ..., e_n, e_1e_2, ..., e_1e_2 \dots e_n\}$ satisfy the rule $e_ie_j + e_je_i = -2\delta_ij$ $(i, j = 1, 2, ..., n)$, $e_0 = 1$ is the identity of $A_n(R)$. We denote the subspace of $A_n(R)$ spanned by the base elements $\{e_0, e_1, ..., e_n\}$ by R_n. Also, we formally identify the subspace R_n with R^{n+1}.

Any element x in R_n is the form $x = \sum_{i=0}^{n} e_ix_i$. For elements $x = \sum_{i=0}^{n} e_ix_i + \sum_{i,j=1,i<j}^{n} e_ie_jx_{ij}$ in $A_n(R)$, the absolute value is defined by the form:

$$|x| = \left(\sum_{i=0}^{n} x_i^2 + \sum_{i,j=1,i<j}^{n} (x_{ij})^2\right)^{\frac{1}{2}}.$$

For any elements $x = \sum_{i=0}^{n} e_ix_i$ in R_n,the conjugate number x^* and x^{-1} are given by the forms:

$$x^* = x_0 - \sum_{i=1}^{n} e_ix_i, \quad x^{-1} = \frac{x^*}{|x|^2}, \quad (|x| \neq 0).$$

3. Holomorphic functions in Clifford Analysis

Let G be a domain in R^{n+1} and $f = \sum_{i=0}^{n} e_i f_i$ be a function defined in G with values in R^{n+1},where $x = \sum_{i=0}^{n} e_i x_i$.

We consider the following system of first order partial differential equations:

$$x_k Df = -(n-1) f_k \quad (k = 1, 2, ..., n),$$

where $D = \sum_{i=0}^{n} e_i \frac{\partial}{\partial x_i}$.

DEFINTION 1([2]). f is monogenic in G iff
(1) f_i is smooth in G $(i = 0, 1, 2, ..., n)$,
(2) $Df = 0$ in G .

DEFINITION 2([6,7]). f is regular in G iff
(1) f_i is smooth in G , $(i = 0, 1, 2, ..., n)$,
(2) $x_k Df = -(n-1) f_k$ in G , $(k = 1, 2, ..., n)$.

For a domain G with the origin in R^{n+1}, put

$$G_k = \{(x_0, 0, ..., 0, x_k, 0, ..., 0) \in R^{n+1}; (x_0, ..., x_n) \in G\}, (k = 1, 2, ..., n).$$

Also, for every element $x = x_0 + e_1 x_1 + ... + e_n x_n$ in R^{n+1},we put

$$z_k = x_0 + e_k x_k, \qquad (k = 1, 2, ..., n).$$

Since $e_k^2 = -1$, it is possible that each z_k is considered a complex number $(k = 1, 2, ..., n)$.

DEFINITION 3. Let G be a domain with the origin in R^{n+1} and $f = \sum_{i=0}^{n} e_i f_i$ be a function defined in G. We say that f is c-holomorphic in G iff

(1) f_i is smooth in G , $(i = 0, 1, 2, ..., n)$,

(2) $\frac{\partial f_0}{\partial x_0} = \frac{\partial f_k}{\partial x_k}, \frac{\partial f_k}{\partial x_0} = -\frac{\partial f_k}{\partial x_0}$ in G_k , $(k = 1, 2, ..., n)$.

REMARK 1. The condition (2) in Definition 2 is equivalent to the system of the following differential equations:

(1) $x_k(\frac{\partial f_0}{\partial x_0} - \sum_{i=1}^{n} \frac{\partial f_i}{\partial x_i}) = -(n-1) f_k$, $(k = 1, 2, ..., n)$,

(2) $\frac{\partial f_i}{\partial x_0} = -\frac{\partial f_0}{\partial x_i}$, $\qquad (i = 1, 2, ..., n)$,

(3) $\frac{\partial f_j}{\partial x_i} = \frac{\partial f_i}{\partial x_j}$, $\qquad (i, j = 1, 2, ..., n)$

From (1) of Remark 1,we obtain the following Lemma.

LEMMA 1. *If* $f=\sum_{i=0}^{n} e_i f_i(x)$ *is regular, then*

$$\frac{f_j}{x_j}+\frac{x_i}{x_j}\frac{\partial f_j}{\partial x_i}=\frac{\partial f_i}{\partial x_i}, \quad (i\neq j, i, j=1,2,...,n)$$

THEOREM 1. *Let* G *be a domain with the origin in* R^{n+1}. *If* $f=\sum_{i=0}^{n} e_i f_i(x)$ *be regular in* G, *then* $f(x_0,x_1,...,x_k,0,...,0)$ *is monogenic in*

$$\Omega_k=\{(x_0,x_1,...,x_k,0,...,0)\in R^{n+1};(x_0,...,x_n)\in G\},(k=1,2,...,n).$$

PROOF. From Lemma 1, we have that

$$\frac{\partial f_i}{\partial x_i}-\frac{f_k}{x_k}=\frac{x_i}{x_k}\frac{\partial f_k}{\partial x_i}, \quad (i\neq k, i, k=1,2,...,n)$$

Theorefore, we have that

$$\sum_{i=k}^{n}\frac{\partial f_i}{\partial x_i}-(n-1)\frac{f_1}{x_1}=\sum_{i=k}^{n}\frac{x_i}{x_1}\frac{\partial f_1}{\partial x_i},$$

From (1) of Remark 1, we have that

$$\sum_{i=k}^{n}\frac{\partial f_i}{\partial x_i}+\frac{\partial f_0}{\partial x_0}-\sum_{i=1}^{n}\frac{\partial f_i}{\partial x_i}=\sum_{i=k}^{n}\frac{x_i}{x_1}\frac{\partial f_1}{\partial x_i}$$

$$(k=1,2,...,n).$$

Putting $x_k=x_{k+1}=...=x_n=0$,

$$\frac{\partial f_0}{\partial x_0}(x_0,x_1,...,x_k,0,...,0)-\sum_{i=1}^{k}\frac{\partial f_k}{\partial x_k}(x_0,x_1,...,x_k,0,...,0)=0,$$

$$(k=1,2,...,n).$$

Also,from (2) of Remark 1, we have that

$$\frac{\partial f_i}{\partial x_0}(x_0,x_1,...,x_k,0,...,0)=-\frac{\partial f_0}{\partial x_i}(x_0,x_1,...,x_k,0,...,0),$$

$$(i=1,2,...,n),$$

$$\frac{\partial f_i}{\partial x_j}(x_0, x_1, ..., x_k, 0, ..., 0) = -\frac{\partial f_j}{\partial x_i}(x_0, x_1, ..., x_k, 0, ..., 0),$$

$$(i, j = 1, 2, ..., n).$$

Therefore, $f(x_0, x_1, ..., x_k, 0, ..., 0)$ is monogenic in Ω_k.

THEOREM 2. *Let G be a domain with the origin in R^{n+1}. If $f = \sum_{i=0}^{n} e_i f_i(x)$ be regular in G, then f is c-holomorphic in G .*

PROOF. By the similar way with Theorem 1, we have that

$$\frac{\partial f_i}{\partial x_i} - \frac{f_k}{x_k} = \frac{x_i}{x_k}\frac{\partial f_k}{\partial x_i}, \quad (i \neq k, i, k = 1, 2, ..., n)$$

Theorefore, we have that

$$\sum_{i=1, i\neq k}^{n} \frac{\partial f_i}{\partial x_i} - (n-1)\frac{f_k}{x_k} = \sum_{i=1, i\neq k}^{n} \frac{x_i}{x_k}\frac{\partial f_k}{\partial x_i} \quad (k = 1, 2, ..., n),$$

From (1) of Remark 1, we have that

$$\sum_{i=1, i\neq k}^{n} \frac{\partial f_i}{\partial x_i} + \frac{\partial f_0}{\partial x_0} - \sum_{i=1}^{n} \frac{\partial f_i}{\partial x_i} = \sum_{i=1, i\neq k}^{n} \frac{x_i}{x_k}\frac{\partial f_k}{\partial x_i} \quad (k = 1, 2, ..., n).$$

Putting $x_2 = x_3 = ... = x_n = 0$,

$$\frac{\partial f_0}{\partial x_0}(x_0, x_1, 0, ..., 0) = \frac{\partial f_k}{\partial x_k}(x_0, x_1, 0, ..., 0)$$

$$(k = 1, 2, ..., n).$$

Also,from (2) of Remark 1, we have that

$$\frac{\partial f_k}{\partial x_0}(x_0, x_1, 0, ..., 0) = -\frac{\partial f_0}{\partial x_k}(x_0, x_1, 0, ..., 0) \quad (k = 1, 2, ..., n).$$

Therefore, $f(x)$ is c-holomorphic in G.

Let G be a domain including the origin in R^{n+1} and $f = \sum_{i=0}^{n} e_i f_i(x)$ be a function defined in G. Put

$$F_k(x_0, x_k) \equiv f_0(x_0, 0, ..., 0, x_k, 0, ..., 0) + e_k f_k(x_0, 0, ..., 0, x_k, 0, ..., 0)$$

in G_k $(k=1,2,...,n)$. Also, put

$$F(x)=\sum_{k=1}^{n}F_k(x_0,x_k)$$

in G.

THEOREM 3. *If* $f=\sum_{i=0}^{n}e_if_i(x)$ *be regular in* G*, then* $F(x)$ *is monogenic in* G *.*

PROOF. Put

$$g_0(x_0,x_1,...,x_n)=\sum_{k=1}^{n}f_0(x_0,0,...,0,x_k,0,...,0)$$

$$g_k(x_0,x_1,...,x_n)=f_k(x_0,0,...,0,x_k,0,...,0),\ (k=1,2,...,n).$$

From Theorem 2, we have that

$$\begin{aligned}\frac{\partial g_0}{\partial x_0}-\sum_{i=k}^{n}\frac{\partial g_k}{\partial x_k}&\\ =\ &\sum_{k=1}^{n}\frac{\partial f_0}{\partial x_0}(x_0,0,...,0,x_k,0,...,0)-\sum_{k=1}^{n}\frac{\partial f_k}{\partial x_k}(x_0,0,...,0,x_k,0,...,0)\\ =\ &\sum_{k=1}^{n}\{\frac{\partial f_0}{\partial x_0}(x_0,0,...,0,x_k,0,...,0)-\frac{\partial f_k}{\partial x_k}(x_0,0,...,0,x_k,0,...,0)\}\\ =\ &0,\end{aligned}$$

$$\begin{aligned}\frac{\partial g_0}{\partial x_i}&=\sum_{k=1}^{n}\frac{\partial f_0(x_0,0,...,0,x_k,0,...,0)}{\partial x_i}\\ &=\frac{\partial f_0(x_0,0,...,0,x_i,0,...,0)}{\partial x_i}=-\frac{\partial f_i(x_0,0,...,0,x_i,0,...,0)}{\partial x_0}\\ &=-\frac{\partial g_i}{\partial x_0}\quad(i=1,2,...,n),\end{aligned}$$

$$\begin{aligned}\frac{\partial g_j}{\partial x_i}&=\frac{\partial f_j(x_0,0,...,0,x_j,0,...,0)}{\partial x_i}=0=\frac{\partial f_i(x_0,0,...,0,x_i,0,...,0)}{\partial x_j}\\ &=\frac{\partial g_i}{\partial x_j}\quad(i\neq j,\ i,j=1,2,...,n).\end{aligned}$$

Hence, $F(x)$ is monogenic in G.

References

[1] F.Brackx,R.Delanghe and F.Sommen , *Clifford Analysis* ,Research Notes in Mathematics, 76 Pitman Books Ltd. 1982.

[2] R.Delanghe, *On regular analytic functions with values in a Clifford algebra*, Math. Ann.,**185** (1970), 91–111

[3] R.P.Gilbert , *Function theoretic methods differential equations*, Academic Press Press , New York and London, 1969.

[4] H.Leutwiler, *Modified Clifford Analysis*, Complex Variables, **17**(1992), 153–171.

[5] K.Nôno , *On the linearization of a partial differential operator*, Bull. Fukuoka Univ. of Ed. part III **40**(1991),1–11

[6] K.Nôno , *Regularity of functions with values in Clifford algebra based on a generalized axially symmetric potential theory operator*, Proc. Conf. on Clifford algebra and their applications in Mathematical Physics (1993),159–166 , Kluwer Academic , Netherland

[7] K.Nôno , *On two approaches to regular function theory*, Bull. Fukuoka Univ. of Ed. part III **44**(1995), 1–13

[8] K.Nôno , *Generalized Euler-Poisson-Darboux's equations in Clifford analysis*, Bull. Fukuoka Univ. of Ed. part III **48**(1999), 9–31

[9] F.Sommen , *Some connections between Clifford analysis and complex analysis* , Complex Variables, **1**(1982), 97–118

71 ALGEBRAS OF HYPERCOMPLEX MONOGENIC FUNCTIONS AND AXIAL-SYMMETRICAL POTENTIAL FIELDS

Sergiy PLAKSA

Institute of Mathematics of the
National Academy of Sciences of Ukraine
3, Tereshchenkivska str.
GSP, 252601 Kiev-4
Ukraine

plaksa@imath.kiev.ua

System of equations

$$y\frac{\partial\varphi(x,y)}{\partial x}=\frac{\partial\psi(x,y)}{\partial y},\qquad y\frac{\partial\varphi(x,y)}{\partial y}=-\frac{\partial\psi(x,y)}{\partial x}\tag{1}$$

describes the spatial potential solenoid field, which is symmetrycal with respect to the axis Ox, in its meridianal plane xOy. Here φ is the axial-symmetrical potential and ψ is the Stokes flow function. The function ψ also should satisfy the condition

$$\psi(x,0)\equiv 0.\tag{2}$$

For exemple, for the model of steady flow of the ideal incompressible inviscid fluid withou sourcecs and vortexes the condition (2) means that the fluid does not flow through the axis Ox.

H.G.W. Begehr et al. (eds.), Proceedings of the Second ISAAC Congress, Vol. 1, 613–622.

From the system (1) it follows that the potential φ satisfies the equation

$$y\Delta\varphi(x,y) + \frac{\partial\varphi(x,y)}{\partial y} = 0 \tag{3}$$

and the Stokes flow function ψ satisfies the equation

$$y\Delta\psi(x,y) - \frac{\partial\psi(x,y)}{\partial y} = 0, \tag{4}$$

where $\Delta := \frac{\partial^2}{\partial x^2} + \frac{\partial^2}{\partial y^2}$ is the Laplace operator.

The qualitative theory is well developed for solutions of system (1). The base of this theory is in the fact that the system (1) has eliptic type outside arbitrary neighbourhood of the axis Ox. For this reason the functions φ and ψ are components of quasi-conformal mapping outside the neighbourhood mentioned above.

However, the quantitative theory is developed considerably worse for solutions of system (1) than for the plane harmonic functions. This is connected with degeneration of system type on the axis Ox. It was also connected with the fact that mathematical methods, which were analogous to the analytic functions methods in the complex plane, were absent for research of system (1).

Construction of methods for description of spatial potential solenoid fields, which are analogous to the analytic functions methods in the complex plane for the description of plane potential fields, is the problem, which was formulated by M.A. Lavrent'ev [1, pp. 205, 18].

The theory of monogenic functions in commutative assosiative Banach algebras presents a new approch and new methods for solving this problem.

In the work [2] I. Mel'nichenko has constructed a family of bases in a three-dimensional algebra with the indicated properties such that components of monogenic functions taking values in this algebra satisfy the three-dimensional Laplace equation. At the same time it is impossible to select a special class of axial-symmetrical potentials in the collection of spatial harmonic functions constructed in [2]. In the work [3] author has discovered an infinite-dimensional algebra H such that components of polynomials taking values in algebra H generate the solutions of equations (1).

We study algebraic-analytical properties of monogenic functions of a vector variable taking values in the commutative assosiative Banach algebra $H_{\mathbf{C}}$, which is the complexification of the algebra H. An isomorphism between this algebra $H_{\mathbf{C}}$ and the algebra of even absolutely convergent trigonometric Fourier series is a remarkable and important fact for applications. This isomorphism enabled to construct explicitly components of analytic functions of hypercomplex variable and to prove that these components generate axial-symmetrical potentials and Stockes flow functions. In such a way new integral expressions for solutions of system (1) were obtained and a new method for investigating spatial potential solenoid fields with axial symmetry was found. These new integral expressions for axial-symmetrical potentials and Stockes flow functions are analogous to expressions of harmonic functions on the plane by the Cauchy integral, and the mentioned method for investigating axial-symmetrical potential fields is analogous to analytic function method in the complex plane for plane potential fields.

Thus, now a chance appears to work out for solutions of system (1) a quantitative theory, which were as complete as the classic theory of harmonic functions on the plane. In particular, the mentioned new integral expressions for axial-symmetric potential functions and Stockes flow functions make it possible to work out methods for effective solving boundary problems for these functions having various applications in the mathematical physics.

We formulate some obtained results.

Let $\mathbf{R}$ be algebra of real numbers and $\mathbf{C}$ be algebra of complex numbers. A domain $D \subset \mathbf{R}^2$ is called a *proper* domain if

$(x, y) \in D \Longrightarrow \{(x, y') : |y'| \leq |y|\} \subset D.$

Then the domains $D_z := \{z = x + iy : (x, y) \in D\} \subset \mathbf{C}$ is also called a *proper* domain.

Theorem 1. *If F is an analytic function in D_z then the functions*

$$\begin{aligned} \varphi(x,y) &= \begin{cases} \frac{1}{2\pi i}\int\limits_{\gamma} \frac{F(t)}{\sqrt{(t-z)(t-\bar{z})}}dt, & \text{when } y \neq 0, \\ F(t), & \text{when } y = 0, \end{cases} \\ \psi(x,y) &= \begin{cases} -\frac{1}{2\pi i}\int\limits_{\gamma} \frac{F(t)(t-x)}{\sqrt{(t-z)(t-\bar{z})}}dt, & \text{when } y \neq 0, \\ 0, & \text{when } y = 0, \end{cases} \end{aligned} \tag{5}$$

are the solutions of system (1) in D.

Here $z = x + iy$, γ is arbitrary closed rectifiable Jordan curve in $\mathbf{C}$, *which bounds a proper domain D'_z such that $\overline{D'_z} \subset D_z$ and $z \in D'_z$, $\sqrt{(t-z)(t-\bar{z})}$ is a continuous branch of this function with the cut* $\{x + i\eta : |\eta| \leq |y|\}$.

We also have proved that every solution of equation (3) and every solution of equation (4) in the domain D can be expressed by the formulas (5) if certain natural conditions are satisfied.

The functions (5) are connected with the principal extension of analytic function F of the complex variable to a domain of special two-dimensional submanifold of certain infinitely dimensional commutative associative Banach algebra.

Let $H := \{a = \sum_{k=1}^{\infty} a_k e_k : a_k \in \mathbf{R}, \sum_{k=1}^{\infty} |a_k| < \infty\}$ be a commutative associative Banach algebra over $\mathbf{R}$ with the norm $\|a\|_H := \sum_{k=1}^{\infty} |a_k|$ and with the basis $\{e_k\}_{k=1}^{\infty}$. The multiplication table

$$e_n e_1 = e_n, \quad e_m e_n = \frac{1}{2}\left(e_{m+n-1} + (-1)^{n-1} e_{m-n+1}\right) \quad \forall m \geq n \geq 1$$

for elements of basis was offered by I. Mel'nichenko in the work [3].

Denote by $H_{\mathbf{C}} := H \oplus iH \equiv \{c = a + ib : a, b \in H\}$ the comlexification of algebra H such that the norm in $H_{\mathbf{C}}$ is given by the relation $\|c\|_{H_{\mathbf{C}}} := \sum_{k=1}^{\infty} |c_k|$, where $c = \sum_{k=1}^{\infty} c_k e_k, c_k \in \mathbf{C}$.

Consider the Cartesian plane $\mu := \{\zeta = xe_1 + ye_2 : x, y \in \mathbf{R}\}$ in H and the domain $D_\zeta := \{\zeta = xe_1 + ye_2 : (x, y) \in D\}$ in μ, which is congruent to the domain $D \subset \mathbf{R}^2$.

A function $\Phi : D_\zeta \longrightarrow H_{\mathbf{C}}$ is called a *monogenic* function, if Φ is differentiable in accordance with Hato.

The set $I_0 := \{c \in H_{\mathbf{C}} : \sum_{k=1}^{\infty} (-1)^k (\operatorname{Re} c_{2k-1} - \operatorname{Im} c_{2k}) = 0,$
$\sum_{k=1}^{\infty} (-1)^k (\operatorname{Re} c_{2k} + \operatorname{Im} c_{2k-1}) = 0\}$ is a maximum ideal of algebra $H_{\mathbf{C}}$.

Let $f_{I_0} : H_{\mathbf{C}} \longrightarrow \mathbf{C}$ be the linear functional such that I_0 is its kernel. Let A be the linear operator, which assigns to every function $\Phi : D_\zeta \longrightarrow H_{\mathbf{C}}$ the function $F : D_z \longrightarrow \mathbf{C}$ by the formula $F(z) := f_{I_0}(\Phi(\zeta))$, where $z = x + iy$ and $\zeta = xe_1 + ye_2$. If Φ is a monogenic function in D_ζ then F is an analytic function in D_z.

Theorem 2. *Every monogenic function $\Phi : D_\zeta \longrightarrow H_{\mathbf{C}}$ can be expressed in the form*

$$\Phi(\zeta) = \frac{1}{2\pi i} \int_\gamma (te_1 - \zeta)^{-1} (A\Phi)(t) dt + \Phi_0(\zeta), \qquad \forall \zeta \in D_\zeta,$$

where the curve γ has the same properties as in theorem 1, the function Φ_0 is a monogenic function taking values in the ideal I_0.

Note that in this case for analytic function $F = A\Phi$ the integral $\frac{1}{2\pi i} \int_\gamma (te_1 - \zeta)^{-1} F(t) dt, \zeta \in D_\zeta$, is the principal extension of function F to the domain D_ζ. Thus in the domain D_ζ the algebra of monogenic functions can be decomposed into the direct sum of the algebra of principal extension for analytic function of the complex variable and the algebra of monogenic functions taking values in the ideal I_0.

Theorem 3. *If F is an analytic function in D_z then the first and the second components of principal extension*

$$\frac{1}{2\pi i}\int_\gamma (te_1-\zeta)^{-1}F(t)dt=\sum_{k=1}^{\infty}U_k(x,y)e_k, \qquad \forall\zeta=xe_1+ye_2\in D_\zeta,$$

of function F to the domain D_ζ generate the solutions φ and ψ of system (1) in D by the formulas

$$\varphi(x,y)=U_1(x,y), \quad \psi(x,y)=\frac{y}{2}U_2(x,y).$$

The isomorphism between $H_{\mathbf{C}}$ and the algebra of even absolutely convergent trigonometric Fourier series is important for applications. In this case the basic elements e_k are isomorphic to functions $i^{k-1}\cos(k-1)\tau$. This isomorphism enabled to construct explicitly the principal extensions for analytic functions of the complex variable that is to obtain explicitly the components U_k and the solutions φ and ψ of system (1).

For example, if $\zeta=xe_1+ye_2$ then the decomposition of power function on the basic elements is of the form

$$\zeta^n=(x^2+y^2)^{n/2}\left(P_n(\cos\theta)e_1+2\sum_{k=1}^{n}\frac{n!}{(n+k)!}(\operatorname{sgn}y)^kP_n^k(\cos\theta)e_{k+1}\right),$$

where n is a positive integer, $\cos\theta := x(x^2+y^2)^{-1/2}$, $P_n(t) := \frac{1}{2^n n!}\frac{d^n}{dt^n}(t^2-1)^n$ is the Legendre polinomial, $P_n^k(t):=(1-t^2)^{k/2}\frac{d^k}{dt^k}P_n(t)$ is the assosiated Legendre polinomial. In this case the decomposition of exponential function on the basic elements is of the form

$$e^\zeta=e^x\left(J_0(y)e_1+2\sum_{k=1}^{\infty}J_k(y)e_{k+1}\right),$$

where $J_n(t) := \frac{(-1)^n}{\pi}\int_0^\pi e^{it\cos\tau}\cos k\tau\, d\tau$ is the Bessel function. For the function ζ^{-1} we obtain

$$\zeta^{-1}=\begin{cases} \frac{1}{\sqrt{x^2+y^2}}\left(e_1+2\sum_{k=1}^{\infty}(-1)^k\left(\frac{\sqrt{x^2+y^2}-x}{y}\right)^k e_{k+1}\right), & x>0,\\ -\frac{1}{\sqrt{x^2+y^2}}\left(e_1+2\sum_{k=1}^{\infty}(-1)^k\left(\frac{\sqrt{x^2+y^2}+x}{y}\right)^k e_{k+1}\right), & x<0.\end{cases}$$

Notice that for $x=0$ the function ζ^{-1} is not definite because e_2 have not the inverse element in algebra $H_{\mathbf{C}}$.

Introduce an element $e_0 \notin H_{\mathbf{C}}$ which satisfies the rules for the multiplication: $e_0e_1 = e_0, e_0e_2 = -e_1, e_0e_{2k+1} = e_0 - 2\sum\limits_{m=1}^{k} e_{2m}, e_0e_{2k+2} = -e_1 - 2\sum\limits_{m=1}^{k} e_{2m+1}$ for $k = 1, 2, \ldots$.

Include the algebra $H_{\mathbf{C}}$ into a Banach space $\tilde{H}_{\mathbf{C}} := \{d = \sum\limits_{k=0}^{\infty} d_k e_k : d_k \in \mathbf{C}, \sum\limits_{k=0}^{\infty} |d_k| < \infty\}$ with the norm $\|d\|_{\tilde{H}_{\mathbf{C}}} := \sum\limits_{k=0}^{\infty} |d_k|$. Notice that $\tilde{H}_{\mathbf{C}}$ is only an extension of linear space of algebra $H_{\mathbf{C}}$ and is not an algebra because the product e_0e_0 is not definite.

Consider the Cartesian plane $\tilde{\mu} := \{\tilde{\zeta} = xe_1 + ye_0 : x, y \in \mathbf{R}\}$ in $\tilde{H}_{\mathbf{C}}$ and the domain $D_{\tilde{\zeta}} := \{\tilde{\zeta} = xe_1 + ye_0 : (x, y) \in D\}$ in $\tilde{\mu}$, which is congruent to the domain $D \subset \mathbf{R}^2$.

A function $\Psi : \tilde{\mu} \setminus \overline{D_{\tilde{\zeta}}} \longrightarrow H_{\mathbf{C}}$ is called a *monogenic* function, if Ψ is differentiable in accordance with Hato.

Let $\tilde{f}_{I_0} : \tilde{H}_{\mathbf{C}} \longrightarrow \mathbf{C}$ be the linear functional such that $\tilde{f}_{I_0}(e_k) = f_{I_0}(e_k)$
$\forall k \geq 1$ and $\tilde{f}_{I_0}(e_0) = i$. Let B be the linear operator, which assigns to every function $\Psi : \tilde{\mu} \setminus \overline{D_{\tilde{\zeta}}} \longrightarrow H_{\mathbf{C}}$ the function $G : \mathbf{C} \setminus \overline{D_z} \longrightarrow \mathbf{C}$ by the formula $G(z) := \tilde{f}_{I_0}(\Psi(\tilde{\zeta}))$, where $z = x + iy$ and $\tilde{\zeta} = xe_1 + ye_0$. If Ψ is a monogenic function in $\tilde{\mu} \setminus \overline{D_{\tilde{\zeta}}}$ then G is an analytic function in $\mathbf{C} \setminus \overline{D_z}$.

Theorem 4. *If the domain $D_{\tilde{\zeta}}$ is bounded then every monogenic function $\Psi : \tilde{\mu} \setminus \overline{D_{\tilde{\zeta}}} \longrightarrow H_{\mathbf{C}}$, which satisfies the condition*

$$\lim_{\|\tilde{\mu}\|_{\tilde{H}_{\mathbf{C}}} \to \infty, \tilde{\zeta} \in \tilde{\mu} \setminus \overline{D_{\tilde{\zeta}}}} \Psi(\tilde{\zeta}) = 0,$$

can be expressed in the form

$$\Psi(\tilde{\zeta}) = -\frac{1}{2\pi i} \int\limits_{\Gamma} (te_1 - \tilde{\zeta})^{-1} (B\Psi)(t)dt + \Psi_0(\tilde{\zeta}) \qquad \forall \tilde{\zeta} \in \tilde{\mu} \setminus \overline{D_{\tilde{\zeta}}}.$$

Here Γ is arbitrary closed bounded rectifiable Jordan curve in $\mathbf{C}$ which bounds a proper domain D''_z such that $\overline{D_z} \subset D''_z$ and $z = \tilde{f}_{I_0}(\tilde{\zeta}) \in \mathbf{C} \setminus \overline{D''_z}$, the function Ψ_0 is a monogenic function taking values in the ideal I_0.

In this case components of integral

$$-\frac{e_0}{2\pi i} \int\limits_{\Gamma} (te_1 - \tilde{\zeta})^{-1} (B\Psi)(t)dt = \sum_{k=1}^{\infty} V_k(x, y)e_k, \qquad \tilde{\zeta} = xe_1 + ye_0,$$

generate the solutions φ and ψ of system (1) in $\mathbf{R}^2 \setminus \overline{D}$ by the formulas

$$\varphi(x,y) = \begin{cases} \pm V_1(x,y) \equiv \frac{1}{2\pi i} \int\limits_{\Gamma} \frac{(B\Psi)(t)}{\sqrt{(t-z)(t-\bar{z})}} dt, & \text{when } \pm y > 0, \\ (B\Phi)(t), & \text{when } y = 0, x > \sup\limits_{(x,0)\in D} x, \\ -(B\Phi)(t), & \text{when } y = 0, x < \inf\limits_{(x,0)\in D} x, \end{cases}$$

$$\psi(x,y) = \pm y \sum_{k=1}^{\infty} V_{2k}(x,y) \equiv \begin{cases} -\frac{1}{2\pi i} \int\limits_{\Gamma} \frac{(B\Psi)(t)(t-x)}{\sqrt{(t-z)(t-\bar{z})}} dt, \text{when } \pm y > 0, \\ 0, \quad \text{when } y = 0, \end{cases} \tag{6}$$

where $z = x + iy$, $\sqrt{(t-z)(t-\bar{z})}$ is a continuous branch of this function with the cut $\{x + i\eta : |\eta| \geq |y|\}$.

We have proved that every solution of equation (3) and every solution of equation (4) in $\mathbf{R}^2 \setminus \overline{D}$ can be expressed by the formulas (6) if certain natural conditions are satisfied.

The theorems 1 – 4 was proved by author jointly with I. Mel'nichenko in papers [4 – 6].

It is well known that new methods for solving of boundary problems for eliptic type equations first test when the Dirichlet boundary problem is solved by these methods.

Let's consider the *Dirichlet boundary problem for axial-symmetrical potential* on finding a solution $\varphi(x,y)$ of equation (3) in D (or in $\mathbf{R}^2 \setminus \overline{D}$), if its boundary values $\varphi_{\partial D}(x,y)$ are defined on the boundary ∂D, i.e. $\varphi(x,y) = \varphi_{\partial D}(x,y)$ for all $(x,y) \in \partial D$.

Consider also the *Dirichlet boundary problem for Stokes flow function* on finding a solution $\psi(x,y)$ of equation (4) in D (or in $\mathbf{R}^2 \setminus \overline{D}$), if its boundary values $\psi_{\partial D}(x,y)$ are defined on the boundary ∂D, i.e. $\psi(x,y) = \psi_{\partial D}(x,y)$ for all $(x,y) \in \partial D$.

Using the expressions (5) and (6) for the functions φ and ψ, we have established that the solutions of these Dirichlet problems can be obtained by solving the integral equations

$$\frac{1}{2\pi i} \int\limits_{\partial D_z} \frac{F(t)}{\sqrt{(t-z)(t-\bar{z})}} dt = \varphi_{\partial D}(x,y),$$
$$-\frac{1}{2\pi i} \int\limits_{\partial D_z} \frac{F(t)(t-x)}{\sqrt{(t-z)(t-\bar{z})}} dt = \psi_{\partial D}(x,y), \qquad (x,y) \in \partial D, z = x + iy, \tag{7}$$

where the sought-for function F would be analytic function in D_z (or in $\mathbf{C} \setminus \overline{D_z}$ respectively).

Author worked out transition methods for equations (7) to singular integral Cauchy equations. If the boundary ∂D is a smooth curve, which satisfies certain supplementary conditions, then these equations are reduced to Fredholm integral equations. If the domain D is the disk, then solutions F of equations (7) were obtained explicitly.

Let H_α be a class of functions $g : \partial D \longrightarrow \mathbf{C}$, which satisfy the Hölder condition

$$|g(x_1, y_1) - g(x_2, y_2)| \le c((x_1 - x_2)^2 + (y_1 - y_2)^2)^{\alpha/2}$$
$$\forall (x_1, y_1), (x_2, y_2) \in \partial D,$$

where $\alpha \in (0; 1]$, a constant c does not depend on x_1, y_1, x_2, y_2.

Theorem 5. *Let $D = \{(x, y) : x^2 + y^2 < 1\}$, the function $\varphi_{\partial D}$ belong to $H_\alpha, 1/2 < \alpha \le 1$, and satisfies the condition $\varphi_{\partial D}(x, -y) = \varphi_{\partial D}(x, y)$ for all $(x, y) \in \partial D$.*

Then the solution of Dirichlet boundary problem for axial-symmetrycal potential is expressed by the formula

$$\varphi(x,y) = \begin{cases} \varphi_{\partial D}(1;0) + \frac{1}{2\pi i} \int\limits_{\partial D} \frac{F_*\left(i\frac{1+t}{1-t}\right)}{\sqrt{(t-z)(t-\bar{z})}} dt, & \text{when } y \ne 0, \\ \varphi_{\partial D}(1;0) + \frac{1}{2\pi i} \int\limits_{\partial D} \frac{F_*\left(i\frac{1+t}{1-t}\right)}{t-x} dt, & \text{when } y = 0, \end{cases}$$

where $z = x + iy$,

$$F_*(\xi) = -i(\xi + i) f_*(|\xi|) - \frac{2\xi(\xi + i)}{\pi} \int\limits_0^\infty \frac{f_*(\tau)}{\tau^2 - \xi^2} d\tau \quad \forall \xi \in \mathbf{R} \setminus \{0\},$$

$$f_*(\tau) = \varphi_*(\tau) - \tau \int\limits_0^\tau \frac{s(\varphi_*(s) - \varphi_*(\tau))}{(\tau^2 - s^2)^{3/2}} ds,$$

$$\varphi_*(\tau) = \frac{1}{\sqrt{\tau^2 + 1}} \left(\varphi_{\partial D} \left(\frac{\tau^2 - 1}{\tau^2 + 1}, \frac{-2\tau}{\tau^2 + 1} \right) - \varphi_{\partial D}(1;0) \right) \quad \forall \tau > 0.$$

Theorem 6. *Let $D = \{(x, y) : x^2 + y^2 < 1\}$, the function $\psi_{\partial D}$ belong to $H_\alpha, 1/2 < \alpha \le 1$, and satisfies the conditions $\psi_{\partial D}(x, 0) = 0$ for all $(x, 0) \in \partial D$ and $\psi_{\partial D}(x, -y) = \psi_{\partial D}(x, y)$ for all $(x, y) \in \partial D$.*

Then the solution of Dirichlet boundary problem for Stokes flow function is expressed by the formula

$$\psi(x,y)=\begin{cases} -\frac{1}{2\pi i}\int\limits_{\partial D}\frac{F_*\left(i\frac{1+t}{1-t}\right)(t-x)}{\sqrt{(t-z)(t-\bar z)}}dt, & \text{when } y\neq 0,\\ 0, & \text{when } y=0,\end{cases}$$

where $z=x+iy$,

$$F_*(\xi)=(\xi+i)^2U_*(|\xi|)+\frac{2\xi(\xi+i)^2}{\pi i}\int\limits_0^\infty\frac{U_*(\tau)}{\tau^2-\xi^2}d\tau\quad\forall\xi\in\mathbf{R}\setminus\{0\},$$

$$U_*(\tau)=\frac{2\tau}{(\tau^2+1)^2}g_*(\tau)+\frac{\tau^2}{\tau^2+1}g_*'(\tau)+$$

$$+\frac{2(\tau^2-1)}{\pi(\tau^2+1)^2}\int\limits_0^\infty\frac{sg_*(s)}{s^2-\tau^2}ds-\frac{2}{\pi(\tau^2+1)}\int\limits_0^\infty\frac{s^2g_*'(s)}{s^2-\tau^2}ds,$$

$$g_*(\tau)=\frac{1}{2}\int\limits_0^\tau\frac{s\psi_*(s)}{\sqrt{\tau^2-s^2}}ds,$$

$$\psi_*(\tau)=\frac{\sqrt{\tau^2+1}}{\tau^2}\psi_{\partial D}\left(\frac{\tau^2-1}{\tau^2+1},\frac{-2\tau}{\tau^2+1}\right)\quad\forall\tau>0.$$

References

[1] Lavrent'ev M.A. and Shabat B.V., *Problems of hydrodynamics and theirs mathematical models* (in Russian), Nauka, Moskow, 1977, 408 p.

[2] Mel'nichenko I.P., "On expression of harmonic mappings by monogenic functions", *Ucr. Math. J.*, **27**, No. 5 (1975), 606-613.

[3] Mel'nichenko I.P., "On a method for description of potential fields with axial symmetry" (in Russian) in: *Contemporary Questions of Real and Complex Analysis*, Institut of Mathematics, Ukrainian Academy of Sciences, Kiev, 1984, 98-102.

[4] Mel'nichenko I.P. and Plaksa S.A., "Potential fields with axial symmetry and algebras of monogenic functions of vector variable, I", *Ucr. Math. J.*, **48**, No. 11 (1996), 1717 – 1730.

[5] Mel'nichenko I.P. and Plaksa S.A., "Potential fields with axial symmetry and algebras of monogenic functions of vector variable, II", *Ucr. Math. J.*, **48**, No. 12 (1996), 1916 – 1926.

[6] Mel'nichenko I.P. and Plaksa S.A., "Potential fields with axial symmetry and algebras of monogenic functions of vector variable, III", *Ucr. Math. J.*, **49**, No. 2 (1997), 253 – 268.

72 LOCAL CONTOUR-SOLID THEOREMS FOR FINELY HYPOHARMONIC FUNCTIONS

Promarz M. TAMRAZOV

Institute of Mathematics of
National Ukrainian Academy of Sciences
Tereshchenkivska str., 3
Kiev, 252601 MSP, Ukraine

ptamraz@capt.kiev.ua

We establish the purely fine local contour-solid results for finely hypoharmonic functions containing refined, strengthened and extended theorems for the mentioned functions in finely open sets of the complex plane with preservable majorants from the maximal class of such majorants.

1. Introduction.

Initial particular results in the contour-solid problems for usual holomorphic functions were given by G.H. Hardy, J.E. Littlewood, S. Warschawski, J.L. Walsh, W.E. Sewell and inspired the formulation of several open problems on the topic in the Sewell's monograph [Sew, p. 31 - 32] published in 1942.

In 1971 the author [T: 1] has completely solved the mentioned open problems (see also [T: 2, 3] where strengthened and more general results were established as well). The obtained results found various applications in geometric and constructive function theory, in theory of singular integral operators, in boundary problems and in other topics.

H.G.W. Begehr et al. (eds.), Proceedings of the Second ISAAC Congress, Vol. 1, 623–634.

Methods developed for holomorphic functions allowed to pose and solve analogous problems for subharmonic functions as well, despite essential differences between the results valid for these classes of functions.

In 1983 the author had established [T: 4] (see also [T: 5 - 8]) extended and refined local contour-solid theorems for subharmonic functions in open sets of the complex plane with arbitrary real majorants concave with respect to the logarithm of the independent variable, and these majorants form the maximal class of preservable majorants in the theorems for subharmonic functions. In the mentioned works, one can find references to earlier and other publications on the topics.

In [TS: 1, 2] certain analogues of results from [T: 4 - 8] were proved for finely hypoharmonic functions in finely open sets on **C**.

In the present work we establish refined, strengthened and extended contour-solid theorems for finely hypoharmonic functions in finely open sets of the complex plane with preservable majorants from the above mentioned maximal class of them.

While in contour-solid results of [TS: 1, 2] some of requirements relate to the standard, Euclidean topology, in results of this paper all requirements relate to the fine topology. So, we get purely fine contour-solid theory for finely hypoharmonic functions. One more difference is that in the present paper restrictions on functions concerning boundedness requirements and majorization are either avoided or essentially weakened. We have completely avoided the upper boundedness and finely inner majorization restrictions on functions (usual for related problems and being assumed on certain parts of sets in earlier publications) and essentially relaxed the majorization requirements to their boundary limit values.

The results of this work contain new assertions even for usual subharmonic functions in (standard) open sets.

The work is based on various new arguments and on a new approach. One of the tools is a new extended maximum principle for finely hypoharmonic functions free of any (upper) boundedness restrictions on the functions and of any global majorization requirements to them.

An essential part of this investigation was carried out at the Linkoping University (see preprint [T: 9]) where I was on a kind invitation and under warm and encouraging hospitality of Professor Lars-Inge Hedberg, to whom I am greatly thankful.

2. Some notions and notations.

Let $\check{C}$ be the compact Riemann sphere.

We refer to [B], [F: 1 - 4] concerning the fine topology and related notions such as thinnes, the fine boundary and the fine closure of a set, fine limits of functions, fine superior and fine inferior limits of functions, finely hypoharmonic, finely hyperharmonic, finely subharmonic, finely harmonic functions, the Green's function for a fine domain and so on.

Let $E \subset \check{\mathcal{C}}$. Denote by $\check{\mathcal{E}}$ the standard closure of E in $\check{\mathcal{C}}$, and by $\overline{E}$ the standard closure of a set $E \subset \mathbf{C}$ in $\mathbf{C}$. The set of all points $x \in \check{\mathcal{C}}$ in which E is not thin is called *the base of the set* E *in* $\check{\mathcal{C}}$ and is denoted by $b(E)$. The set $\tilde{E} := E \cup b(E)$ is called *the fine closure of the set* E *in* $\check{\mathcal{C}}$. Clearly, $\tilde{E} \subset \check{\mathcal{E}}$. Denote by $\check{\partial}_f E$ the fine boundary of E in $\check{\mathcal{C}}$. Let $\partial_f E := \mathbf{C} \cap \check{\partial}_f E$, $(E)_i := E \setminus b(E)$, $(E)_r := E \setminus (E)_i$. Points $x \in (E)_r$ and $x \in (E)_i$ are called *regular* and *irregular* points, respectively, of the set E.

For a set $E \subset \check{\mathcal{C}}$ let us denote $\check{\mathcal{C}} \setminus E =: FE$, and for a set $E \subset \mathbf{C}$, denote also $\mathbf{C} \setminus E =: CE$.

Let $G \subset \check{\mathcal{C}}$ be a finely open set.

Let $z \in G$. If the set FG is non-polar, then we denote by ω_z^G the generalized harmonic measure relative to G and z $(\in G)$, see [F: 1, p. 1]. If FG is polar, then we introduce the harmonic measure relative to G and $z \in G$ by the equality

$$\omega_z^G = 0.$$

Notice that if FG is non-polar, then $\omega_z^G \neq 0$. In any case for ω_z^G we use also the term *the harmonic measure relative to* G *at the point* $z \in G$.

If z belongs to a finely connected component T of G, Y is a subset of the set FT and $\omega_z^T(Y) = 0$, then $\omega_\zeta^T(Y) = 0$ at every point $\zeta \in T$ (see [F: 1, pp. 150 - 151]).

A set $Q \subset FG$ will be called *nearly negligible relative to* G if for every finely connected component T of G the set $Q \cap \partial_f T$ contains no compact subset K of the harmonic measure $\omega_z^T(K) > 0$ at some (and therefore at any) point $z \in T$. This requirement is equivalent to the following alternative: either FG is polar (and then Q is also polar), or for every finely connected component T of G both $\partial_f T$ is non-polar and Q is a set of inner harmonic measure zero relative to T and any point $z \in T$.

If a set $E \subset FG$ is such that for every finely connected component T of G it contains no compact subset $K \subset \partial_f T$ of logarithmic capacity $\operatorname{Cap} K > 0$, then E is nearly negligible relative to G.

In particular, any set $E \subset FG$ of inner logarithmic capacity zero is nearly negligible relative to G.

If T is a finely connected component of G and $z \in T$, then $\omega_z^G = \omega_z^T$.

Let D be a fine domain in $\check{C}$, i. e. a finely open, finely connected set. Then $\tilde{D} = \check{\tilde{D}}$. In particular, then D is finely separable from a point $z \in \check{C}$ if and only if it is separable from z in the standard topology. So in such a situation we may speak of separability not specifying in what sense.

Let G be a finely open set in $\check{C}$, and $z \in \check{\partial}_f G$. Given any functions $u : G \to [-\infty, +\infty]$ and $h : G \to \check{C}$, we introduce the following notations for fine superior limits of functions:

$$\text{fine lim sup}_{\zeta \to z,\ \zeta \in G}\ u(\zeta) =: (u)\check{}_{G,f}(z) =: \check{u}_{G,f}(z). \qquad (1)$$

Remark 1. Let $X \subset \check{C}$ and $Y \subset \check{C}$ be non-empty sets. For a function $v : X \to [-\infty, +\infty)$ let us consider two requirements:

(a) v is upper bounded on every part of X finely separable from Y;

(b) v is upper bounded on every part of X separable (in standard topology) from Y.

The requirement (a) is more restrictive than (b). It is the reason why in our results we use (when possible) restrictions of the type (b) rather than of the type (a). Such results are valid with any of these types of restrictions, but with (b) they are more general.

3. Maximum principle for finely hypoharmonic functions.

In this work we prove and use the following maximum principle for finely hypoharmonic functions that is free of any (upper) boundedness restrictions on them, as well as of any global majorization requirements.

Lemma 1. *Let $D \subset \check{C}$ be a finely open set, $E \subset FD$ be a set nearly negligible relative to D, $(\check{\partial}_f D) \setminus E \neq \emptyset$; u be a finely hypoharmonic in D function with*

$$\check{u}_{D,f}(z) < +\infty \quad \forall\ z \in \check{\partial}_f D,$$

$$\check{u}_{D,f}(z) \leq 0 \quad \forall\ z \in (\check{\partial}_f D) \setminus E.$$

Then $u \leq 0$ in D.

In R^n for $n \geq 3$ such a result is not true.

Remark 2. Under additional requirement that E is a (polar) set of irregular fine boundary points of D), a particular case of Lemma 1 was established in [F: 2, p.82] (c.f. results of [F: 3] where u in D is assumed to be upper bounded or majorized by certain potential).

4. Contour-solid theorems for finely hypoharmonic functions.

In [T: 5] (see also [T: 6 - 8]) we had introduced the following notions. Let L be the class of all functions $\lambda : (0, +\infty) \to [-\infty, +\infty)$ for each of which the set $I_\lambda := \{x : \lambda(x) > -\infty\}$ is connected and the restriction of λ to I_λ is concave with respect to $\log x$. Let L^* be the class of all $\lambda \in L$ for which I_λ is non-empty. For $\lambda \in L^*$ let us denote by x_λ^- and x_λ^+, respectively, the left and the right ends of the interval I_λ. Obviously, $0 \le x_\lambda^- \le x_\lambda^+ \le +\infty$. When $x_\lambda^- < x_\lambda^+$, the mentioned concavity condition is equivalent to the combination of the following conditions: the function $\lambda(x)$ is concave with respect to $\log x$ (and therefore continuous) in the interval $(x_\lambda^-, x_\lambda^+)$ and lower semicontinuous on I_λ. For $\lambda \in L$ the limits

$$\lambda^0 := \lim_{x \to 0} \frac{\lambda(x)}{\log x}, \quad \lambda^\infty := \lim_{x \to +\infty} \frac{\lambda(x)}{\log x}$$

exist, and we have

$$\lambda^0 \ge \lambda^\infty, \quad \lambda^0 > -\infty, \quad \lambda^\infty < +\infty.$$

For every fixed $\alpha \in \mathbf{R}$, $\theta \in \mathbf{R}$ the function $\lambda(x) := \alpha \log x + \theta$ belongs to L^*, and for it we have $\lambda_0 = \lambda_\infty = \alpha$.

Let G be a finely open set in $\mathbf{C}$, and u be a finely hypoharmonic in G function. Then for $\zeta \in G$ we denote

$$\gamma_G(u, \zeta) := \inf_v \{v(\zeta) : \ v \text{ finely hyperharmonic in } G, \ v \ge u \text{ in } G.\}.$$

Clearly, $u(\cdot) \le \gamma_G(u, \cdot)$. If $\gamma_G(u, \cdot) < +\infty$, then on the basis of Lemma 10.3 and Theorems 9.14 and 12.9 from [F: 1, pp. 103 - 104, 96, 158] one can show that $\gamma_G(u, \zeta)$ is finely hypoharmonic in G, and for every finely connected component T of G the following alternative holds: either $\gamma_G(u, \zeta) = u(\zeta) = -\infty$ in T, or the functions $\gamma_G(u, \zeta)$ and $u(\zeta)$ are finely subharmonic in T and $\gamma_G(u, \zeta)$ is finely harmonic on the set $\{\zeta \in T : \gamma_G(u, \zeta) \ne -\infty\}$.

Let $a \in CG$ be a fixed point.

Now we are going to introduce the following fine analogues of some notions from [T: 4 - 8].

Let us denote:

$$u^a_{G,f} := \begin{cases} \text{fine } \limsup_{\zeta\to a,\ \zeta\in G} \frac{u(\zeta)}{|\log\ |\zeta-a||} & \text{when } a\in\partial_f G, \\ 0 \quad \text{when } a\notin \partial_f G, \end{cases}$$

$$u^\infty_{G,f} := \begin{cases} \text{fine } \limsup_{\zeta\to \infty,\ \zeta\in G} \frac{u(\zeta)}{\log\ |\zeta|} & \text{when } \infty\in\check{\partial}_f G, \\ 0 \quad \text{when } \infty\notin \check{\partial}_f G, \end{cases}$$

$$M_{G,a}(u,r) := \inf_{p,q}\{p\ \log r + q :\ p\in(-\infty,+\infty),\ q\in(-\infty,+\infty],$$
$$u(\zeta)\le p\ \log|\zeta-a| +\ q\ \ \forall\,\zeta\in G\} \qquad (0<r<+\infty).$$

Obviously, the following alternative holds: either

$$M_{G,a}(u,r) = +\infty \quad \forall\, r>0,$$

or

$$M_{G,a}(u,r) < +\infty \quad \forall\, r>0.$$

If the latter case holds true, then the function $M_{G,a}(u,r)$ of $r>0$ belongs to the class L. Suppose that $M_{G,a}(u,\cdot)\not\equiv -\infty$, and denote by r^- and r^+ the left an the right ends of the maximal interval where this function is $>-\infty$. Then the function $\zeta\mapsto M_{G,a}(u,|\zeta-a|)$ is superharmonic under $r^-<|\zeta-a|<r^+$, and $M_{G,a}(u,|\zeta-a|)=-\infty$ under $|\zeta-a|<r^-$ and under $|\zeta-a|>r^+$. Furthermore, in this case u has in G a finely harmonic majorant and

$$u(\zeta)\le\gamma_G(u,\zeta)\ \le M_{G,a}(u,|\zeta-a|)\ \ \forall\,\zeta\ \in\ G.$$

These inequalities are valid also in the case when $M_{G,a}(u,|\zeta-a|)\equiv-\infty$.

Denote

$$\rho^- := \inf_{\zeta\in G}|\zeta|, \quad \rho^+ := \sup_{\zeta\in G}|\zeta|,$$

Then

$$\rho^-\le r^-\le r^+\le\rho^+.$$

Let $\lambda\in L$.

Consider the following conditions:

(A$'$, ∞) $\infty \in b(CG)$ and for every finely connected component T of G with $\infty \in b(T)$ there holds $u^{\infty}_{T,f} < +\infty$;

(B$'$, ∞) $\infty \notin b(CG)$ and there exist a constant $t \in \mathbf{R}$ and a fine neighbourhood V of ∞ for which

$$u(\zeta) \leq \lambda(|\zeta - a|) + t \quad \forall \zeta \in G \cap V;$$

(A$'$, a) $a \in b(CG)$ and for every finely connected component T of G with $a \in b(T)$ there holds $u^{a}_{T,f} < +\infty$;

(B$'$, a) $a \notin b(CG)$ and there exist a constant $t \in \mathbf{R}$ and a fine neighbourhood V of a for which

$$u(\zeta) \leq \lambda(|\zeta - a|) + t \quad \forall \zeta \in G \cap V.$$

Using these notations and (1), we get the following result.

Theorem 1. *Let $a \in \mathbf{C}$ be a fixed point; $G \subset \mathbf{C} \setminus \{a\}$ be a finely open set; $\lambda \in L$; $u : G \to [-\infty, +\infty)$ be a finely hypoharmonic function for which*

$$\check{u}_{G,f}(z) \leq \lambda(|z - a|) \quad \forall\, z \in (\partial_f G) \setminus \{a\}.$$

Denote $z_1 := a$, $z_2 := \infty$ and suppose that for each $s = 1, 2$ (independently from each other) one of the conditions (A$'$, z_s) or (B$'$, z_s) is satisfied. Then u has in G a finely harmonic majorant, and one and only one of the following two possibilities holds true: either the estimates

$$u(\zeta) \leq \gamma_G(u, \zeta) \leq \lambda(|\zeta - a|) \quad \forall\, \zeta \in G, \tag{2}$$

or the following exceptional case: $G = \mathbf{C} \setminus \{a\}$ and

$$u(\zeta) = \gamma_G(u, \zeta) = M_{G,a}(u, |\zeta - a|) = \nu \log|\zeta - a| + t \quad \forall\, \zeta \in G, \tag{3}$$

$$\lambda(x) = \nu \log x + l \quad \forall\, x > 0 \tag{4}$$

with constants $\nu, t \in \mathbf{R}$, $l \in [-\infty, t)$.

Moreover, if the exceptional case is not valid, then one of the following two possibilities holds true: either

$$M_{G,a}(u, r) = -\infty \quad \forall r > 0, \tag{5}$$

or $\lambda \in L^*$, $x^-_\lambda < x^+_\lambda$ and

$$u(\zeta) = \gamma_G(u, \zeta) = -\infty \quad \forall\, \zeta \in G : |\zeta - a| \notin (x^-_\lambda, x^+_\lambda), \tag{6}$$

$$M_{G,a}(u,r) \leq \lambda(r) \quad \forall\, r:\; x_\lambda^- \neq r \neq x_\lambda^+, \tag{7}$$

$$M_{G,a}(u,r) = -\,\infty \quad \forall\, r \notin [x_\lambda^-, x_\lambda^+]. \tag{8}$$

Remark 3. For every fixed $s = 1,\ 2$ the following statements are true.

(a) If there exists a finely connected component T of G with $z_s \in b(T)$, then the conditions $z_s \in b(CT)$ and $z_s \in b(CG)$ are equivalent.

(b) $z_s \in b(CG)$ if and only if for every finely connected component T of G we have $z_s \in b(CT)$.

(c) If $z_s \notin b(CG)$, then there exists one and only one finely connected component T of G with $z_s \in b(T)$, and for this T we have $z_s \notin b(CT)$.

(d) The condition $(\mathrm{B}', \mathrm{z_s})$ implies the inequality $u_{G,f}^{z_s} < +\infty$.

(e) If $z_s \notin b(CG)$, then the condition $(\mathrm{B}', \mathrm{z_s})$ in Theorem 1 may not be omitted.

(f) Now suppose that there exist more than one finely connected components T of G with $z_s \in b(T)$. Then for every such T we have $z_s \in b(CT)$, and in Theorem 1 in this case we may assume the condition $u_{T,f}^{z_s} < +\infty$ for all such T instead of the alternative containing the conditions $(\mathrm{A}', \mathrm{z_s})$ and $(\mathrm{B}', \mathrm{z_s})$.

Theorem 1 is contained in the following more general and sharper theorems.

Theorem 2.1. *Let* $a \in \mathbf{C}$ *be a fixed point;* $G \subset \mathbf{C} \setminus \{a\}$ *be a finely open set;* Q *be a set contained in* $\check{\partial}_f G$ *and containing the points* a *and* ∞; $\lambda \in L$; $u : G \to [-\,\infty, +\,\infty)$ *be a finely hypoharmonic function. Suppose that for every finely connected component* T *of* G *the following conditions are satisfied:*

(a2) $$\check{u}_{T,f}(z) < +\infty \quad \forall\, z \in (\partial_f T) \setminus \{a\};$$

(b2) $$\check{u}_{T,f}(z) \leq \lambda(|z-a|) \quad \forall\, z \in (\partial_f T) \setminus Q;$$

(c2) Q *is nearly negligible relative to* T;

(d2) $$u_{T,f}^{a} < +\infty, \quad u_{T,f}^{\infty} < +\infty.$$

Then $u(\zeta)$ *is upper bounded both on every part of* G *separable from* a *and* ∞, *and on every finely connected part of* G *finely separable from* a *and* ∞.

For G, a, u, λ under consideration and $s = 1, 2$ we introduce the quantities $\sigma_f^s = \sigma_f^s(G, a, u, \lambda)$ defined by the following conditions. If $\lambda \in L^*$, then we denote

$$\sigma_f^1 := \begin{cases} (u(\cdot) - \lambda(|\cdot - a|))_{G,f}^a & \text{when } x_\lambda^- = 0, \\ 0 \quad \text{when } x_\lambda^- > 0, \end{cases}$$

$$\sigma_f^2 := \begin{cases} (u(\cdot) - \lambda(|\cdot - a\,|))_{G,f}^\infty & \text{when } x_\lambda^+ = +\,\infty, \\ 0 \quad \text{when } x_\lambda^+ < +\,\infty. \end{cases}$$

If $\lambda \equiv -\,\infty$, then we assume

$$\sigma_f^1 = \sigma_f^2 = 0.$$

For $\lambda \in L^*$ the following equalities are valid: if $\lambda^0 \neq +\,\infty$, then $\sigma_f^1 = u_{G,f}^a + \lambda^0$, and if $\lambda^\infty \neq -\,\infty$, then $\sigma_f^2 = u_{G,f}^\infty - \lambda^\infty$.

Theorem 2.2. *Let $a \in \mathbf{C}$ be a fixed point; $G \subset \mathbf{C} \setminus \{a\}$ be a finely open set; Q be a set contained in $\check{\partial}_f G$ and containing the points a and ∞; $\lambda \in L$; $u : G \to [-\,\infty, +\,\infty)$ be a finely hypoharmonic function. Suppose that for every finely connected component T of G the conditions (a2) - (c2) of Theorem 2.1 are satisfied. Denote $z_1 := a$, $z_2 := \infty$ and suppose that for each $s = 1, 2$ (independently from each other) one of the conditions (A, z_s) or (B, z_s) is valid. Then u has in G a finely harmonic majorant, and one and only one of the following two possibilities holds true: either the estimates (2) and*

$$-\infty \leq \sigma_f^1 \leq 0, \quad -\infty \leq \sigma_f^2 \leq 0,$$

or the following exceptional case: $Q = \check{C} \setminus G$, Q is polar and (3), (4) hold with constants $\nu, t \in \mathbf{R}$, $l \in [-\,\infty, t)$.

Theorem 2.3. *Let all assumptions of Theorem 2.2 be satisfied, but its exceptional case and (5) be not valid. Then $\lambda \in L^*$, $x_\lambda^- < x_\lambda^+$ and (6) - (8) are true. Moreover, for every finely connected component T of G we have either*

$$u(\zeta) = \gamma_G(u, \zeta) = -\infty \quad \forall \zeta \in T, \tag{9}$$

or

$$\lambda(|\zeta - a|) > -\infty \quad \forall \zeta \in T.$$

For any finely open set $G \subset \mathbf{C}$ with non-polar compliment CG, and $w \in \check{\mathcal{C}}$, $\zeta \in G$, $w \neq \zeta$, there exists the Green's function $g_G(w, \zeta)$, and it satisfies the equality

$$g_G(w,\zeta) - g_G(\infty,\zeta) = \int \log \left| \frac{w-z}{w-\zeta} \right| d\omega_\zeta^G(z).$$

Let us admit the following agreements concerning possible indefinite commutative expressions:

$$\infty \cdot 0 = 0, \ -\infty + \infty = -\infty.$$

Theorem 2.4. *Let the assumptions and notations of Theorem 2.2 be valid. Suppose that* CG *is non-polar. Then for every finely connected component* T *of* G *we have*

$$\gamma_G(u,\zeta) \le \lambda(|\zeta - a|) + \sum_{s=1}^{2} \sigma_f^s g_G(z_s, \zeta) \quad \forall\, \zeta \in T,$$

$$M_{G,a}(u(\cdot) - \sum_{s=1}^{2} \sigma_f^s g_G(z_s, \cdot), |\zeta - a|) \ \le \lambda(|\zeta - a|) \ \forall\, \zeta \in T.$$

If some of the points z_s *is not in* $b(CT)$ *and* $\sigma_f^s = -\infty$ *for it, then (9) is true.*

Theorems 1 - 2.4 are purely fine refinements and extensions of results from [T: 4 - 8], [TS: 1, 2]. In particular, in our present notations, results given in [TS: 1, 2] were established under additional requirements among which there were the following: the restrictions onto a finely hypoharmonic function u to be upper bounded on each part of G separable from the points ∞ and a, see [TS: 1, 2].

We use the term *solid inequalities* for naming the estimates for functions in G and on $\tilde{G}$ given by theorems of the type under consideration. The question about the equality sign in solid inequalities at a finely inner point of G is treated in [TS: 2] (see also [T: 5 - 8] concerning the same questions for hypoharmonic functions in standard open sets).

Let D be a finely open set in $\check{\mathcal{C}}$ and $z \in \check{\partial}_f D$. For any function $v : D \to [-\infty, +\infty]$ denote

$$\text{fine lim inf}_{\zeta \to z, \zeta \in D}\ v(\zeta) =: (v\check{)}_{D,f}(z) =: \check{v}_{D,f}(z).$$

We have established the following statement used for proving Lemma 1.

Lemma 2. *Let* T *be a fine domain in* $\check{C}$ *with non-polar complement and* Z *be a non-empty subset of the fine boundary* $\check{\partial}_f T$ *such that the harmonic measure* $\omega_x^T(Z)$ *is zero at some fixed point* $x \in T$. *Then there exists a finely superharmonic function* $v : T \to (0, +\infty]$ *for which* $v(x) = 1$ *and*

$$\hat{v}_{T,f}(z) = +\infty \quad \forall\, z \in Z.$$

It has the following extension for higher dimensions.

Lemma 2^n. *Let* T *be a fine domain in* $\mathbf{R}^n$ *with non-polar complement,* $Z \subset \partial_f T$ *be a non-empty set whose (generalized) harmonic measure* $\omega_x^T(Z)$ *is zero at some fixed point* $x \in T$. *Then there exists a finely superharmonic function* $v : T \to (0, +\infty]$ *for which* $v(x) = 1$ *and*

$$\hat{v}_{T,f}(z) := \text{ fine lim inf}_{\zeta \to z, \zeta \in T}\ v(\zeta) = +\infty \quad \forall\, z \in Z.$$

References

M. Brelot:

[B] On topologies and boundaries in potential theory, Lecture Notes in Mathematics, vol. 175, Springer, Berlin etc., 1971. 176 p.

B. Fuglede:

[F: 1] Finely harmonic functions, Lecture Notes in Mathematics, vol. 289, Springer, Berlin etc., 1972. 188 p.

[F: 2] Fonctions harmoniques et fonctions finement harmoniques, Ann. Inst. Fourier, Grenoble, vol. 24, N 4, 1974, p. 77-91.

[F: 3] Boundary minimum principles in potential theory, Math. Ann., vol. 210, 1974, 213-226.

[F: 4] Sur la fonction de Green pour un domaine fin, Ann. Inst. Fourier, Grenoble, vol. 25, N 3-4, 1975, p. 201-206.

S.W. Sewell:

[Sew] Degree of approximation by polynomials in the complex domain, Ann. of Math. Studies, No. 9, Princeton Univ. Press, Princeton, N.J., 1942.

P.M. Tamrazov:

[T: 1] Constructive and structural properties of functions on compact sets of the complex plane, Abstracts of Reports Presented at the

All-Union Conference on Theory of Functions of a Complex Variable, Kharkov, 1971, p. 206 - 208. (Russian)

[T: 2] Boundary and solid properties of holomorphic functions in a complex domaun, Dokl. Akad. Nauk SSSR, vol. 204 (1972), No. 3, p. 565 - 568. (Russian)

[T: 3] Contour and solid structural properties of holomorphic functions of a complex variable, Uspekhi Mat. Nauk, vol. 28 (1973), No. 1 (169), p. 131 - 161. (Russian)

[T: 4] Contour-solid properties of holomorphic functions and mappings, Akad. Nauk Ukrain. SSR Inst Mat., Preprint: 1983, N 33. 17 p. (Russian)

[T: 5] Local contour-solid problem for subharmonic functions, Akad. Nauk Ukrain. SSR Inst. Mat., Preprint: 1984, N 52. 17 p. (Russian)

[T: 6] Subharmonic functions in a contour-solid problem, Akad. Nauk Ukrain. SSR Inst Mat., Preprint: 1985, N 47. 24 p. (Russian)

[T: 7] Local contour-and-solid problem for subharmonic functions, Complex Variables Theory Appl., vol. 7, N 1-3, 1986, p. 231 - 242.

[T: 8] Sharpened contour-solid theorems for subharmonic functions, Ukrain. Mat. Zh., vol. 40, N 2, 1988, p. 210 - 219. (Russian)

[T: 9] Contour-solid theorems for finely holomorphic and finely hypoharmonic functions, Linkoping University Report LITH-MAT-R-99-07, 1999,
29 p.

P.M. Tamrazov, A.A. Sarana:

[TS: 1] Contour-solid properties of finely subharmonic functions, Preprint 94.33, National Academy of Sciences of Ukraine, Institute of Mathematics, 1994, Kiev. 14 p. (Ukrainian)

[TS: 2] Contour-solid properties of finely hypoharmonic functions, Ukrain. Mat. Zh., vol. 49, N 8, 1997, p. 1114 - 1125. (Russian)

73 ANALYTIC SMOOTHING EFFECTS FOR NONLOCAL NONLINEAR SHRÖDOINGER EQUATIONS IN TWO SPACE DIMENSIONS

Nakao HAYASHI and Hidetake UCHIDA

Department of Applied Mathematics
Science University of Tokyo
Tokyo 162-8601, Japan

nhayashi@rs.kagu.sut.ac.jp

Pavel I. NAUMKIN

Instituto de Física y Matemáticas
Universidad Michoacana
AP 2-82, CP 58040 Morelia, Michoacan, Mexico

pavelni@zeus.ccu.umich.mx

1 Introduction

The Davey-Stewartson (D-S) systems are written

$$\begin{cases} i\partial_t u + c_0\partial_{x_1}^2 u + \partial_{x_2}^2 u = c_1|u|^2 u + c_2 u\partial_{x_1}\varphi, & (t,x)\in \mathbf{R}\times\mathbf{R}^2, \\ \partial_{x_1}^2\varphi + c_3\partial_{x_2}^2\varphi = \partial_{x_1}|u|^2, \\ u(0,x) = u_0(x), \quad x\in\mathbf{R}^2, \end{cases}$$

H.G.W. Begehr et al. (eds.), Proceedings of the Second ISAAC Congress, Vol. 1, 635–641.

where $c_0, c_3 \in \mathbf{R}$, $c_1, c_2 \in \mathbf{C}$. The (D-S) systems were derived by Davey-Stewartson [8], Benney-Roskes [5] and Djordjevic-Redekopp [9] and model the evolution of weakly nonlinear water waves that travel predominantly in one direction, but in which the wave amplitude is modulated slowly in horizontal directions. Ablowitz and Haberman [2] and Cornille [7] obtained a particular form of (D-S) which is considered an example of a completely integrable model which generalizes the one-dimensional Schrödinger equation. By Djordjevic-Redekopp [9] it was shown that the parameter c_3 can become negative when capillary effects are important. When $(c_0, c_1, c_2, c_3) = (1, -1, 2, -1)$, $(-1, -2, 1, 1)$ or $(-1, 2, -1, 1)$ the system (D-S) is referred in the inverse scattering literature as the DSI, DSII defocusing and DSII focusing, respectively. In these cases several results concerning the existence of solitons or lump solutions, the Cauchy problem have been established ([1], [2], [3], [11], [12]) by the inverse scattering techniques. Ghidaglia and Saut [13] classified (1) as elliptic-elliptic, elliptic-hyperbolic, hyperbolic-elliptic and hyperbolic-hyperbolic according to the respective sign of $(c_0, c_3) : (+,+), (+,-)$, $(-,+)$ and $(-,-)$. In the elliptic-hyperbolic and hyperbolic-hyperbolic cases one has to assume that $\varphi(\cdot)$ satisfies the radiation condition, i.e.,

$$\varphi(x_1, x_2, t) \to 0 \quad \text{as} \quad x_1 + x_2 \quad \text{and} \quad x_1 - x_2 \to +\infty.$$

Under the radiation condition and a rotation in the x_1x_2-plane and rescaling the elliptic-hyperbolic system (D-S) can be written as

$$\begin{cases} i\partial_t u + (\partial_{x_1}^2 + \partial_{x_2}^2)u \\ = c_1|u|^2u + c_2 u \int_{x_2}^{\infty} \partial_{x_1}|u|^2(x_1, x_2')dx_2' + c_3 u \int_{x_1}^{\infty} \partial_{x_2}|u|^2(x_1', x_2)dx_1', \\ u(0,x) = u_0(x), \end{cases} \tag{1}$$

where c_1, c_2, c_3 are arbitrary constants.

Our purpose in this note is to study analytic smoothing effect and global existence of small solutions to (1). In order to treat the derivatives of unknown functions smoothing properties of Schrödinger equations were used in Linares-Ponce [21]. They proved local existence in time of small solutions to (1) in higher order Sobolev spaces. Their result was improved by [6], [17], [15] for (1), and [14] for (1). Global existence of small solutions to (1) was shown by [16], [18].

In order to state our results precisely, we introduce

Notations and function spaces. We define the weighted Sobolev space as follows

$$\mathbf{H}^{m,l} = \{f \in \mathbf{L}^2; \|(1-\partial_{x_1}^2 - \partial_{x_2}^2)^{m/2}(1+{x_1}^2+{x_2}^2)^{l/2} f\| < \infty\},$$

$$\mathbf{H}^{m,l}(\mathbf{R}_{x_j}) = \{f \in \mathbf{L}^2(\mathbf{R}_{x_j}); \|(1-\partial_{x_j}^2)^{m/2}(1+{x_j}^2)^{l/2} f\|_{L^2(\mathbf{R}_j)} < \infty\},$$

where $\|\cdot\|$ denotes the usual $\mathbf{L}^2$ norm. We let $\partial = (\partial_{x_1}, \partial_{x_2})$, $J = (J_{x_1}, J_{x_2})$, $J_{x_j} = x_j + 2it\partial_{x_j}$. We write $\mathbf{L}^p_{x_j} = \mathbf{L}^p(\mathbf{R}_{x_j})$, $\mathbf{L}^p_{x_1}\mathbf{L}^q_{x_2} = \mathbf{L}^p(\mathbf{R}_{x_1}; \mathbf{L}^q(\mathbf{R}_{x_2}))$, $\mathbf{H}^{m,l}_{x_j} = \mathbf{H}^{m,l}(\mathbf{R}_{x_j})$. For each $r > 0$ we denote $S(r)$ the strip $\{z; -r < \Im z_j < r; 1 \le j \le 2\}$ in the complex plane $\mathbf{C}^2$. For $x \in \mathbf{R}^2$, if a complex-valued function $f(x)$ has an analytic continuation to $S(r)$, then we denote this by the same letter $f(z)$ and if $g(z)$ is an analytic function on $S(r)$, then we denote the restriction of $g(z)$ to the real axis by the same letter $g(x)$.

We let

$$\mathbf{A}^{m,l}(|\beta|) = \{f(z); f(z) \quad \text{is analytic on} \quad S(r), \|f\|_{\mathbf{A}^{m,l}(|\beta|)} < \infty\},$$

where

$$\begin{aligned}\|f\|_{\mathbf{A}^{m,l}(|\beta|)} &= \sup_{y \in (-|\beta|, |\beta|)^2} \|f(\cdot + iy)\|^2_{L^2} \\ &= \sum_{l,m=1}^{2} \left(\int |f(x_1 + i(-1)^l \beta, x_2 + i(-1)^m \beta)|^2 dx_1 dx_2 \right)^{1/2}.\end{aligned}$$

We denote $\partial_{x_j}^{-1} = \int_{-\infty}^{x_j} dx'_j$, $x_1 \equiv x$, $x_2 \equiv y$, $\partial_x^{-1} = \partial_1^{-1}$, $\partial_y^{-1} = \partial_2^{-1}$. Let $\mathcal{F}_j\phi$ or $\hat{\phi}$ be the Fourier transform of $\phi(x_j)$. We denote by $\mathcal{F}_j^{-1}\phi$ or $\check{\phi}$ the inverse Fourier transform of the function $\phi(\xi_j)$. We have the following relation $|\partial_j| = \mathcal{F}_j^{-1}|\xi_j|\mathcal{F}_j = -\mathcal{H}_j\partial_j$, where the Hilbert transformation with respect to the variable x is defined as follows

$$\mathcal{H}_1\phi(x,y) \equiv \mathcal{H}_x\phi(x,y) = \frac{1}{\pi}\mathrm{Pv}\int_{\mathbf{R}} \frac{\phi(z,y)}{x-z} dz = -i\mathcal{F}_x^{-1}\frac{\xi_x}{|\xi_x|}\mathcal{F}_x\phi,$$

where Pv means the principal value of the singular integral. The operator $\mathcal{H}_2$ is defined in the same way. We use the fact that the Hilbert transformations $\mathcal{H}_j$ are bounded operators from $\mathbf{L}^2_{\mathbf{j}}$ to $\mathbf{L}^2_{\mathbf{j}}$, $j = x, y$. The fractional derivative $|\partial_x|^\alpha$, $\alpha \in (0,1)$, is equal to

$$|\partial_x|^\alpha \phi = \mathcal{F}_x^{-1}|\xi_x|^\alpha \mathcal{F}_x\phi = C\int_{\mathbf{R}} (\phi(x+z,y) - \phi(x,y))\frac{dz}{|z|^{1+\alpha}}$$

with some constant C, see [22]. The derivative $|\partial_y|^\alpha$ is defined analogously.

We state our results.

Theorem 1.1 *We assume that*

$$\left\| \left(\prod_{j=1}^{2} \cosh \beta x_j \right) \phi \right\|_{3,0} + \left\| \left(\prod_{j=1}^{2} \cosh \beta x_j \right) \phi \right\|_{0,3}$$

is sufficiently small. Then there exists a unique global solution $u(t,x)$ of (1) such that $u(t,x)$ has an analytic continuation to $S(|\beta t|)$ and

$$\exp(-iz^2/2t)u(t) \in \mathbf{A}^{2,0}(|\beta t|), \quad \textit{for any} \quad t.$$

Remark 1. From the fact that $\exp(-iz^2/2t)u(t) \in \mathbf{A}^{2,0}(|\beta t|)$ it shows that the analytical domain of solutions increase with time. This implies the smoothing effect of solutions to (1).

2 Outline of the proof of Theorem 1.1

We need the analytic version of Lemmas 2.1 and 2.2 in [18]. An analytic version of Lemma 2.1 in [18] which is written as

$$\begin{aligned} &\frac{d}{dt}\|\mathcal{S}h\|^2 + \left\|\omega_1 \mathcal{S}\sqrt{|\partial_x|}h\right\|^2 + \left\|\omega_2 \mathcal{S}\sqrt{|\partial_y|}h\right\|^2 \\ &\le 2\left|\operatorname{Im}(\mathcal{S}h, \mathcal{S}\mathcal{U}(t)(1+x^2+y^2)^{m/2}e^{\beta x+\gamma y}\mathcal{U}(-t)(1-\partial_x^2-\partial_y^2)^{l/2}f)\right| \\ &\quad + C\|h\|^2 e^{2\|\varphi\|_\infty}(\|\omega\|_\infty^4 + \|\omega\|_{1,0,\infty}\|\omega\|_\infty + \|\varphi_t\|_\infty), \end{aligned} \tag{2}$$

where $h = \mathcal{U}(t)(1+x^2+y^2)^{m/2}e^{\beta x+\gamma y}\mathcal{U}(-t)(1-\partial_x^2-\partial_y^2)^{l/2}u$, u is a solution of

$$\begin{aligned} &i\partial_t u + (\partial_x^2 + \partial_y^2)u \\ &= c_1|\tilde{u}|^2\tilde{u} + c_2\tilde{u}\int_y^\infty \partial_x|\tilde{u}|^2(x,y')dy' + c_3\tilde{u}\int_x^\infty \partial_y|\tilde{u}|^2(x',y)dx', \end{aligned}$$

$\mathcal{S}(\varphi) = \mathcal{X}(\varphi)\mathcal{Y}(\varphi)$ with

$$\mathcal{X}(\varphi) = \cosh(\varphi_x) + i\sinh(\varphi_x)\mathcal{H}_x, \mathcal{Y}(\varphi) = \cosh(\varphi_y) + i\sinh(\varphi_y)\mathcal{H}_y,$$

and $\tilde{u}$ is in some analytic function space. We choose φ as follows:

$$\varphi(t,x,y) = \varphi_{\beta,\beta}(t,x,y) + \varphi_{\beta,-\beta}(t,x,y) + \varphi_{-\beta,\beta}(t,x,y) + \varphi_{-\beta,-\beta}(t,x,y),$$

where

$$\varphi_{\beta,\beta}(t,x,y) = \partial_x^{-1}\Big(\left|\mathcal{U}(t)e^{\beta(x+y)}\mathcal{U}(-t)\tilde{u}(t,x,y)\right|^2 + \left|\mathcal{U}(t)e^{\beta(x+y)}\mathcal{U}(-t)\tilde{u}_x(t,x,y)\right|^2 \Big),$$

$\omega_j(x_j) = \sqrt{(\partial_j \varphi_j)}$, and $\omega = (\omega_1, \omega_2)$.

An analytic version of the first estimate of ([18], Lemma 2.2) is given as

$$\begin{aligned}
&\left|(\mathcal{S}\mathcal{F}^{-1}e^{\beta\xi_x+\gamma\xi_y}\mathcal{F}u, \mathcal{S}\mathcal{F}^{-1}e^{\beta\xi_x+\gamma\xi_y}\mathcal{F}\phi\psi\partial_x u)\right| \\
&\le 2e^{2\|\varphi\|_\infty}\Big(\left\| \|\phi(\cdot+i\beta,\cdot+i\gamma)\|_{\mathbf{L}_y^\infty}\mathcal{S}\sqrt{|\partial_x|}u(\cdot+i\beta,\cdot+i\gamma)\right\|^2 \\
&\quad + \left\| \|\psi(\cdot+i\beta,\cdot+i\gamma)\|_{\mathbf{L}_y^\infty}\mathcal{S}\sqrt{|\partial_x|}u(\cdot+i\beta,\cdot+i\gamma)\right\|^2 \Big) \\
&\quad + Ce^{6\|\varphi\|_\infty}\|u(\cdot+i\beta,\cdot+i\gamma)\|^2\Big(\|\phi(\cdot+i\beta,\cdot+i\gamma)\|_{1,0,\infty}^2 \\
&\quad + \|\psi(\cdot+i\beta,\cdot+i\gamma)\|_{1,0,\infty}^2\Big)\left(1+\|\varphi\|_{1,0,\infty}^2\right). \qquad (3)
\end{aligned}$$

We will obtain the result by applying the contraction mapping principle and (2), (3) to a linearized equation of (1) in a suitable analytic function space involving the norm

$$\|u(t)\|_{X^k(t)} = \sum_{l+m\le k} \|e^{i|\cdot|^2/4t}(1-\partial_x^2-\partial_y^2)^{l/2}u\|_{\mathbf{A}^{m,0}(|\beta t|)}.$$

We choose an analytic function space

$$X^k = \left\{ f \in C(\mathbf{R}; L^2); \sup_{t\in\mathbf{R}} \|u(t)\|_{X^k(t)} < \infty \right\}.$$

By the analytic version of Lemma 2.1 we get

$$\|u(t)\|_{X^3(t)}^2 \le C\epsilon + C\epsilon^{1/2}\int_0^t (1+s)^{-1}\|\tilde{u}(s)\|_{X^3(s)}^2 ds.$$

Usual energy estimates and the structure of nonlinear term such that

$$u\int_{-\infty}^y \partial_x|u|^2 dy' = u\frac{1}{2it}\int_{-\infty}^y \bar{u}J_x u - u\overline{J_x u} dy'$$

yield

$$\|u(t)\|_{X^2(t)}^2 \le C\epsilon + C\epsilon^{1/2}\int_0^t (1+s)^{-2+C\epsilon}\|\tilde{u}(s)\|_{X^3(s)}^2 ds.$$

These integral inequalities give us the desired results. For details of the proof, see [19].

References

[1] Ablowitz, J.M. & Fokas, A.S., 1984, On the inverse scattering transform of multidimensional nonlinear equations, *J. Math. Phys.,* **Vol. 25**, pp. 2494-2505.

[2] Ablowitz, J.M. & Haberman, R.,1975, Nonlinear evolution equations in two and three dimensions, *Phys. Rev. Lett.,* **Vol. 35**, pp. 1185-1188.

[3] Ablowitz, J.M. & Segur, A., 1981, Solitons and Inverse Scattering Transform, *PA: SIAM,* .

[4] Anker, D. & Freeman, N.C., 1978, On the soliton solutions of the Davey-Stewartson equation for long waves, *Proc. R. Soc. A,* **Vol. 360**, pp. 529-540.

[5] Benney, D.J. & Roskes, G.L., 1969, Wave instabilities, *Stud. Appl. Math.,* **Vol. 48**, pp. 377-385.

[6] Chihara, H., 1995, The initial value problem for the elliptic-hyperbolic Davey-Stewartson equation, *preprint.*

[7] Cornille, H., 1979, Solutions of the generalized nonlinear Schrödinger equation in two spatial dimensions, *J. Math. Phys.,* **Vol. 20**, pp. 199-209.

[8] Davey, A. & Stewartson, K., 1974, On three-dimensional packets of surface waves, *Proc. R. Soc. A.,* **Vol. 338**, pp. 101-110.

[9] Djordjevic, V.D. & Redekopp, L.G. 1977, On two-dimendional packets of capillary-gravity waves, *J. Fluid Mech.,* **Vol. 79**, pp. 703-714.

[10] Doi, S., 1994, On the Cauchy problem for Schrödinger type equations and regularity of solutions, *J. Math. Kyoto Univ.,* **Vol. 34**, pp. 319-328.

[11] Fokas, A.S. & Santini, P.M., 1988, Recursion operators and bi-Hamiltonian structures in multidimensions I, *Comm. Math. Phys.,* **Vol. 115**, pp. 375-419.

[12] Fokas, A.S. and Santini, P.M., 1988, Recursion operators and bi-Hamiltonian structures in multidimensions II, *Comm. Math. Phys.,* **Vol. 115**, pp. 449-474.

[13] Ghidaglia, J.M. & Saut, J.C., 1990, On the initial value problem for the Davey-Stewartson systems, *Nonlinearity,* **Vol. 3**, pp. 475-506.

[14] Hayashi, N., 1996, Local existence in time of small solutions to the Davey-Stewartson system, *Annales de l'I.H.P. Physique Theorique,* **Vol. 65**, pp. 313-366.

[15] Hayashi, N., 1997, Local existence in time of solutions to the elliptic-hyperbolic Davey-Stewartson system without smallness condition on the data, *J. Analysé Mathématique,* **Vol. 73**, pp.133-164.

[16] Hayashi, N. & Hirata, H., 1996, Global existence and asymptotic behavior in time of small solutions to the elliptic-hyperbolic Davey-Stewartson system, *Nonlinearity,* **Vol. 9**, pp. 1387-1409.

[17] Hayashi, N. & Hirata, H., 1997, Local existence in time of small solutions to the elliptic-hyperbolic Davey-Stewartson system in the usual Sobolev space, *Proceedings of the Edinburgh Mathematical Society,* **Vol. 40**, pp. 563-581.

[18] Hayashi, N. & Naumkin, P.I., 1999, On the Davey-Stewartson and Ishimori systems, 1999 *Math. Phy., Anal. and Geometry,* **Vol. 2**, pp. 53-81

[19] Hayashi, N., Naumkin, P.I. & Uchida, H., in preparation.

[20] Kenig, C.E., Ponce, G. & Vega, L., 1993, Well-posedness and scattering results for the generalized Korteweg-de Vries equation via contraction principle, *Comm. Pure Appl. Math.,* **Vol. 46**, pp. 527-620.

[21] Linares, F. & Ponce, G., 1993, On the Davey-Stewartson systems, *Ann. Inst. Henri Poincaré, Anal. non linéaire,* **Vol. 10**, pp. 523-548.

[22] Stein, E.M., 1970, Singular Integral and Differentiability Properties of Functions, *Princeton Univ. Press, Princeton Math. Series 30.*

74 REPRESENTATIONS OF INITIAL HEAT DISTRIBUTIONS BY MEANS OF THEIR HEAT DISTRIBUTIONS[1]

Gen NAKAMURA
Sabrou SAITOH
Admi SYARIF[2]

Department of Mathematics
Faculty of Engineering
Gunma University
Kiryu 376-8515, Japan

gnaka@cc.gunma-u.ac.jp
ssaitoh@cc.gunma-u.ac.jp
sadmi@cc.gunma-u.ac.jp

This full paper will be published in Inverse Problems and so, we shall present the results and outlines of the proofs.

1. Introduction and result

[1]This research was partially supported by the Japanese Ministry of Education, Science, Sports and Culture, Grant-in-Aid Scientific Research, Kiban Kenkyuu (A) (1) 10304009.

[2]The work of the third author was also supported by LPIU DUE-Project, The University of Lampung, Indonesia.

H.G.W. Begehr et al. (eds.), Proceedings of the Second ISAAC Congress, Vol. 1, 643–649.

We shall consider the solutions $u(t, x)$ of the heat equation

$$\partial_t u = \Delta u \text{ on } \mathbf{R}^+ \times \mathbf{R}^n$$

satisfying the initial condition

$$u(0, x) = F(x) \text{ on } \mathbf{R}^n$$

for $L_2(\mathbf{R}^n, dx)$ functions F. Using the Fourier transform, we have the formal integral representation of the solution $u(t, x)$

$$u(t,x) = \frac{1}{(4\pi t)^{\frac{n}{2}}} \int_{R^n} F(\xi) \exp\{-\frac{1}{4t}|x-\xi|^2\} d\xi. \tag{1}$$

Then, we have

$$\lim_{t \to 0} \|u(t,x) - F(x)\|_{L_2(\mathbf{R}^n, dx)} = 0.$$

The second author ([1], [2], [3]) examined this integral transform using the general theory of integral transforms and using the theory of reproducing kernels. In particular, he established isometrical identities between F and $u(t, x)$ as functions in x for any fixed $t > 0$ and inversion formulas in this transform. However, as functions $u(t, x)$ of t and for fixed point x, he established only norm inequalities in the transform. In particular, he was not able to give an inversion formula such that

$$u(t, x) \to F(x) \text{ for fixed } x.$$

Here, we shall show an exact and analytical representation of the initial heat distribution F by means of the observations

$$u(t, x_1, x') \text{ and } \frac{\partial u(t, x_1, x')}{\partial x_1} \tag{2}$$

for $x' = (x_2, \cdots, x_n) \in \mathbf{R}^{n-1}$ and $t > 0$, at any fixed point x_1.

2. One-dimensional case

We shall consider the one-dimensional case, that is, the Weierstrass transform

$$u(t,x) = \int_R F(\xi)\ k(x - \xi, t) d\xi \tag{3}$$

for

$$k(x,t) = \frac{1}{\sqrt{4\pi t}} \exp[-\frac{x^2}{4t}]$$

and for functions $F \in L_2(\mathbf{R},\ dx)$. For fixed x, we shall decompose F as follows :

$$F(\xi) = F_{e(x)}(\xi) + F_{o(x)}(\xi);$$

where

$$F_{e(x)}(\xi) = \frac{F(\xi) + F(2x-\xi)}{2}$$

and

$$F_{o(x)}(\xi) = \frac{F(\xi) - F(2x-\xi)}{2}.$$

Note that for $q > 0$, in general

$$K_q(\tau, \overline{\tau'}) = \frac{\Gamma(2q)}{(\tau + \overline{\tau'})^{2q}} \text{ on } R^+ \times \overline{R^+} \tag{4}$$

is the Bergman-Selberg reproducing kernel on R^+. The reproducing kernel Hilbert space H_{K_q} is composed of all analytic functions $f(\tau)$ on R^+ satisfying

$$\sum_{n=0}^{\infty} \frac{1}{n!\Gamma(n+2q+1)} \int_0^{\infty} |\partial_t^n[tf'(t)]|^2\, t^{2n+2q-1} dt < \infty$$

and its norm $\|f\|_{H_{K_q}}$ is given by

$$\|f\|^2_{H_{K_q}} = \sum_{n=0}^{\infty} \frac{1}{n!\Gamma(n+2q+1)} \int_0^{\infty} |\partial_t^n[tf'(t)]|^2\, t^{2n+2q-1} dt$$

$$= \frac{1}{2\pi} \int_{-\infty}^{\infty} |(\mathcal{M}f)(q-it)|^2 \frac{dt}{|\Gamma(q+it)|^2}, \tag{5}$$

where $\mathcal{M}f$ denotes the Mellin transform

$$(\mathcal{M}f)(q-it) = \int_0^{\infty} f(\xi)\xi^{q-it-1} d\xi.$$

Then, for $K^{(1)}(\tau, \bar{\tau}') = \frac{K_{\frac{1}{4}}(\tau,\bar{\tau}')}{2\pi}$, and $K^{(2)}(\tau, \bar{\tau}') = \frac{K_{\frac{3}{4}}(\tau,\bar{\tau}')}{2\pi}$, we have the representations

$$\|u(t,x)\|^2_{H_{K^{(1)}}} = 2\pi \sum_{n=0}^{\infty} \frac{1}{n!\Gamma(n+\frac{3}{2})} \int_0^{\infty} |\partial_t^n[t\partial_t u(t,x)]|^2 \, t^{2n-\frac{1}{2}} dt, \quad (6)$$

and

$$\left\|\frac{\partial u(t,x)}{\partial x}\right\|^2_{H_{K^{(2)}}} = 2\pi \sum_{n=0}^{\infty} \frac{1}{n!\Gamma(n+\frac{5}{2})} \int_0^{\infty} \left|\partial_t^n[t\partial_t \frac{\partial u(t,x)}{\partial x}]\right|^2 t^{2n+\frac{1}{2}} dt. \quad (7)$$

Now, since $q > \frac{1}{2}$, the norm $H_{K^{(2)}}$ is also given by the complex form

$$\left\|\frac{\partial u(t,x)}{\partial x}\right\|^2_{H_{K^{(2)}}} = \sqrt{\frac{2}{\pi}} \int \int_{R^+} \left|\frac{\partial u(\tau,x)}{\partial x}\right|^2 \frac{1}{\sqrt{t}} dt \, dt',$$

where $\tau = t + t'$ and $\partial_x u(t,x)$ is extended analytically onto R^+ in the form $\partial_x u(\tau, x)$. Then, we have the isometrical identity in (3)

$$\begin{aligned} \int_{-\infty}^{\infty} |F(\xi)|^2 \, d\xi &= \int_{-\infty}^{\infty} \left|F_{e(x)}(\xi)\right|^2 d\xi + \int_{-\infty}^{\infty} \left|F_{o(x)}(\xi)\right|^2 d\xi \\ &= \|u(t,x)\|^2_{H_{K^{(1)}}} + \left\|\frac{\partial u(t,x)}{\partial x}\right\|^2_{H_{K^{(2)}}}. \end{aligned} \quad (8)$$

The isometrical identity (8) implies that :

[A] The image identifications in (3) for L_2 even functions $F_{e(x)}(\xi)$ and odd functions $F_{o(x)}(\xi)$, respectively, that is, $u(t,x)$ and $\partial_x u(t,x)$ are characterized as the members of the Hilbert spaces $H_{K^{(1)}}$ and $H_{K^{(2)}}$, respectively.

[B] Inversion formulas are given in the forms :

$$\begin{aligned} F_{e(x)}(\xi) &= 2\pi \; s - \lim_{N\to\infty} \sum_{n=0}^{N} \frac{1}{n!\Gamma(n+\frac{3}{2})} \\ &\quad \cdot \int_0^{\infty} \partial_t^n[t\partial_t u(t,x)] \partial_t^n[t\partial_t k(x-\xi,t)] t^{2n-\frac{1}{2}} dt \\ &= 2\pi \; s - \lim_{N\to\infty} \sum_{n=0}^{N} \frac{(-1)^{n+1}}{n!\Gamma(n+\frac{3}{2})} \\ &\quad \cdot \int_0^{\infty} u(t,x) \partial_t [t\partial_t^n (\partial_t^n (t\partial_t k(x-\xi,t)) t^{2n-\frac{1}{2}})] dt, \end{aligned} \quad (9)$$

and

$$\begin{aligned} F_{o(x)}(\xi) &= 2\pi \ s - \lim_{N\to\infty} \sum_{n=0}^{N} \frac{1}{n!\Gamma(n+\frac{5}{2})} \\ &\quad \cdot \int_0^\infty \partial_t^n [t\partial_t\partial_x u(t,x)] \partial_t^n [t\partial_t\partial_x k(x-\xi,t)] t^{2n+\frac{1}{2}} dt \\ &= 2\pi \ s - \lim_{N\to\infty} \sum_{n=0}^{N} \frac{(-1)^{n+1}}{n!\Gamma(n+\frac{5}{2})} \\ &\quad \cdot \int_0^\infty \partial_x u(t,x) \partial_t [t\partial_t^n (\partial_t^n (t\partial_t\partial_x k(x-\xi,t)) t^{2n+\frac{1}{2}})] dt, \quad (10) \end{aligned}$$

respectively. Here, $s - \lim_{N\to\infty}$ means the strong convergence in $L_2(\mathbf{R}, d\xi)$.

3. Higher dimensional case

By taking Fourier's transform in x' in (1)

$$(\mathcal{F}_{n-1}u)(t,x_1,X') = \frac{1}{(\sqrt{2\pi})^{n-1}} \int_{R^{n-1}} e^{-ix'\cdot X'} u(t,x_1,x')dx',$$

we have

$$\begin{aligned} (\mathcal{F}_{n-1}u)(t,x_1,X') &= \frac{1}{\sqrt{4\pi t}} \int_{-\infty}^{\infty} \exp\Big[-\frac{(x_1-\xi_1)^2}{4t}\Big] d\xi_1 \\ &\quad \cdot (\mathcal{F}_{n-1}F)(\xi_1,X') \cdot \exp\{-t|X'|^2\} \end{aligned}$$

and so,

$$\begin{aligned} &(\mathcal{F}_{n-1}u)(t,x_1,X') \exp\{t|X'|^2\} \\ &\quad = \frac{1}{\sqrt{4\pi t}} \int_{-\infty}^{\infty} (\mathcal{F}_{n-1}F)(\xi_1,X') \exp[-\frac{(x_1-\xi_1)^2}{4t}] d\xi_1. \quad (11) \end{aligned}$$

Hence from the result in one dimensional case, we see that

$$(\mathcal{F}_{n-1}F)(\xi_1,X') \quad (12)$$

is expressible in terms of the functions

$$(\mathcal{F}_{n-1}u)(t,x_1,X') \text{ and } \frac{\partial(\mathcal{F}_{n-1}u)(t,x_1,X')}{\partial x_1} \quad (13)$$

for $t>0$ and for any fixed point x_1. Hence, by taking Fourier's inversion formula in X', we have the desired representation:

$$F(\xi)=\Big[\mathcal{F}_{n-1}^{-1}\{(\mathcal{F}_{n-1}F)(\xi_1,X')\}\Big](\xi)$$

and

$$\begin{aligned}
&(\mathcal{F}_{n-1}F)(\xi_1,X')\\
&=2\pi\ s-\lim_{N\to\infty}\sum_{n=0}^{N}\frac{(-1)^{n+1}}{n!\Gamma(n+\frac{3}{2})}\cdot\int_0^\infty(\mathcal{F}_{n-1}u)(t,x_1,X')\exp\{t|X'|^2\}\\
&\quad\cdot\partial_t[t\partial_t^n(\partial_t^n(t\partial_t k(x_1-\xi_1,t))t^{2n-\frac{1}{2}})]dt\\
&\quad+2\pi\ s-\lim_{N\to\infty}\sum_{n=0}^{N}\frac{(-1)^{n+1}}{n!\Gamma(n+\frac{5}{2})}\int_0^\infty\partial_{x_1}(\mathcal{F}_{n-1}u)(t,x_1,X')\\
&\quad\cdot\exp\{t|X'|^2\}\partial_t[t\partial_t^n(\partial_t^n(t\partial_t\partial_{x_1}k(x_1-\xi_1,t))t^{2n+\frac{1}{2}})]dt, \qquad (14)
\end{aligned}$$

where

$$\begin{aligned}
&P_n(p,t)\\
&=\sum_{\nu=0}^{n}\binom{n}{\nu}(-p^2)^{n+\nu}\exp\{-p^2t\}\Big(\frac{1}{2}\Big)^{n-\nu}t^{\frac{2n+2\nu-1}{2}}\Big[\frac{(4n+1)!!}{(2n+2\nu+1)!!}p^4t^2\\
&\quad-\Big(\frac{n(4n-1)!!}{(2n+2\nu-1)!!}+\frac{(4n+1)!!}{2(2n+2\nu-1)!!}+\frac{(4n+1)!!}{((2n+2\nu+1)!!}\Big)p^2t\\
&\quad+\Big(\frac{n(4n-1)!!)}{2(2n+2\nu-3)!!}+\frac{n(4n-1)!!)}{(2n+2\nu-1)!!}\Big)\Big]. \qquad (15)
\end{aligned}$$

and

$$\begin{aligned}
&Q_n(p,t)\\
&=\sum_{\nu=0}^{n}\binom{n}{\nu}(-p^2)^{n+\nu}\exp\{-p^2t\}\Big(\frac{1}{2}\Big)^{n-\nu}t^{\frac{2n+2\nu+1}{2}}\cdot\\
&\Big[\frac{(4n+3)!!}{(2n+2\nu+3)!!}p^4t^2-\Big(\frac{n(4n+1)!!}{(2n+2\nu+1)!!}+\frac{(4n+3)!!}{2(2n+2\nu+1)!!}+\\
&\frac{(4n+3)!!}{(2n+2\nu+3)!!}\Big)p^2t+\Big(\frac{n(4n+1)!!)}{2(2n+2\nu-1)!!}+\frac{n(4n+1)!!)}{(2n+2\nu+1)!!}\Big)\Big] \qquad (16)
\end{aligned}$$

Therefore (14) can be represented as

$$(\mathcal{F}_{n-1}F)(\xi_1,X')=\ s-\lim_{N\to\infty}$$

$$\sum_{n=0}^{N} \frac{(-1)^{n+1}}{n!\Gamma(n+\frac{3}{2})} \int_0^\infty \int_{-\infty}^{\infty} (\mathcal{F}_{n-1}u)(t,x_1,X') \exp\{t|X'|^2\}$$
$$\cdot \exp\{-ip(x_1-\xi_1)\} P_n(p,t)dpdt$$
$$+ s - \lim_{N\to\infty} \sum_{n=0}^{N} \frac{(-1)^{n+1}}{n!\Gamma(n+\frac{5}{2})} \int_0^\infty \int_{-\infty}^{\infty} \partial_{x_1}(\mathcal{F}_{n-1}u)(t,x_1,X') \cdot$$
$$\exp\{t|X'|^2\}(ip)\exp\{-ip(x_1-\xi_1)\} Q_n(p,t)dpdt. \tag{17}$$

References

[1] Saitoh, S. (1983) The Weierstrass transform and an isometry in the heat equation, Applicable Analysis **16,** pp 1-6.

[2] Saitoh, S. (1992) Inequalities for the solutions of the heat equation, *General Inequalities* **6,** (Basel : Birkhäuser).

[3] Saitoh, S. (1997) Integral Transforms, Reproducing Kernels and their Applications, Pitman Research Notes in Mathematics Series, **369**, Addison Wesley Longman, UK.

[4] Widder D. V., (1972) The Laplace transform, (Princeton University Press).

75 LOCALLY CONVEX *-ALGEBRAS AND THE THERMODYNAMICAL LIMIT OF QUANTUM MODELS

Fabio BAGARELLO

Dipartimento di Matematica ed Applicazioni
Fac. Ingegneria, Università di Palermo
I-90128 Palermo, Italy

bagarell@unipa.it

Abstract

We show that the thermodynamical limit of several physical models is naturally obtained within the framework of topological quasi *-algebras. In particular, the relevance of the algebra $\mathcal{L}^+(\mathcal{D})$ is shown explicitly by concrete examples.

MSC Numbers: 46H35, 47N50

1 Introduction

The problem of the mathematical description of physical systems has an ancient origin. Already for classical mechanics many different possibilities have been developed during the years. One of these is the phase space description, where the dynamics of the system is governed, for instance, by the Hamilton equations. In ordinary quantum mechanics the

H.G.W. Begehr et al. (eds.), Proceedings of the Second ISAAC Congress, Vol. 1, 651–659.

particles are described by vectors of an Hilbert space $\mathcal{H}$, while the operations performed over the system are described by self-adjoint operators acting on $\mathcal{H}$. As for the dynamics, again we can use several equivalent strategies: in the Heisenberg picture, for instance, the vectors (often called wave functions) are constant in time, while the operators evolve in the following fashion: $A \to A_t := e^{iHt}Ae^{-iHt}$. Here H is the hamiltonian operator, which describes the energy of the system. Opposite is the situation for the Schröedinger picture: here the operators are independent of time, while the wave function at time t is given by $\Psi_t := e^{iHt}\Psi$, Ψ being the initial condition.

The situation is by far more intricate for quantum mechanics for systems with infinite degrees of freedom, QM_∞. A possible approach, known as the constructive quantum field theory, was proposed several years ago by Wightmann, see [1] for an overview. Here we will be more interested in the algebraic approach to QM_∞, as proposed by R. Haag and D. Kastler in 1964, [2]. As it is well known, this approach is based on the quasi local C*-algebra of the observables of the system, $\mathcal{A}$.

Even if this approach has proven to be so powerful during all these years, it is also well known that not all the physically relevant models fit into this framework. In particular, for instance, Coulomb systems or, more generally, systems with long range interactions cannot be described in a (quasi-)local C*-algebraic setting, [3, 4, 5, 6, 7]. In fact, for suchmodels, the time evolution of a local observable A, $\alpha^t(A)$, in general depends on variables completely delocalized in space. Two are the major consequences of this fact: $\alpha^t(A)$ may not belong to $\mathcal{A}$, and the natural topology where the thermodynamical limit must be discussed is not the uniform one.

Nevertheless, the models in [3, 4, 6, 7] can be analyzed in a framework quite close to that of Haag and Kastler: it is enough to use von Neumann algebras instead of C*-algebras and to replace the uniform topology with the strong one, [6]. However, the time evolution for other physical systems can be described in none of these algebras. In the following we will describe some models whose thermodynamical limit is more naturally discussed in an algebra $\mathcal{L}^+(\mathcal{D})$ or, more in general, in a topological quasi *-algebra.

In Section II we give some mathematical definitions. In Section III we describe several spin models and their thermodynamical limit. An approach to the existence of the algebraic dynamics for free and interacting bosons is contained in Section IV. Section V contains our the conclusions and plans for the future.

2 Mathematical Preliminaries

We begin this Section recalling the definition of topological quasi *-algebra. Let $\mathcal{A}$ be a vector space and $\mathcal{A}_o$ a *-algebra contained in $\mathcal{A}$. We say that $\mathcal{A}$ is a *quasi *-algebra* with distinguished *-algebra $\mathcal{A}_o$ (or, simply, over $\mathcal{A}_o$) if (i) the right and left multiplications of an element of $\mathcal{A}$ by an element of $\mathcal{A}_o$ are always defined and linear; and (ii) an involution * (which extends the involution of $\mathcal{A}_o$) is defined in $\mathcal{A}$ with the property $(AB)^* = B^*A^*$ whenever the multiplication is defined. A quasi *-algebra $(\mathcal{A}, \mathcal{A}_o)$ is said to have a unit I if there exists an element $I \in \mathcal{A}_o$ such that $AI = IA = A$, $\forall\, A \in \mathcal{A}$. Finally, the quasi *-algebra $(\mathcal{A}, \mathcal{A}_o)$ is said to be *topological* if $\mathcal{A}$ carries a locally convex topology ξ such that (a) the involution is continuous and the multiplications are separately continuous; and (b) $\mathcal{A}_o$ is dense in $\mathcal{A}[\xi]$.

The most useful example of topological quasi *-algebra, as far as physical applications are concerned, is constructed in the following way. Let $\mathcal{H}$ be an Hilbert space and N an unbounded, self adjoint operator defined on a dense domain $D(N) \subset \mathcal{H}$. Let $D(N^k)$ be the domain of the operator N^k, $k \in \boldsymbol{N}$, and $\mathcal{D}$ the domain of all the powers of N: $\mathcal{D} := D^\infty(N) = \cap_{k\geq 0} D(N^k)$. The set $\mathcal{D}$ is dense in $\mathcal{H}$. On this set we introduce the topology t defined by the following seminorms: $\phi \in \mathcal{D} \to \|\phi\|_n := \|N^n\phi\|$, where n is a natural integer, and $\|\;\|$ is the norm of $\mathcal{H}$. Starting from $\mathcal{D}$ we can define, following Lassner, [5], the *-algebra $\mathcal{L}^+(\mathcal{D})$ of all the closable operators defined on $\mathcal{D}$ which, together with their adjoints, map $\mathcal{D}$ into itself. Moreover, calling $\mathcal{D}'$ the conjugate dual space of $\mathcal{D}$, endowed with the strong dual topology t', we define the set $\mathcal{L}(\mathcal{D}, \mathcal{D}')$ of all the continuous linear maps from $\mathcal{D}[t]$ into $\mathcal{D}'[t']$.

The topologies on $\mathcal{L}^+(\mathcal{D})$ and $\mathcal{L}(\mathcal{D}, \mathcal{D}')$ are introduced by means of the set $\mathcal{C}$ of all positive, bounded and continuous functions $f(x)$ on $\boldsymbol{R}_+$, which are decreasing faster than any inverse power of x. They satisfy the condition $\sup_{x\geq 0} f(x)x^k < \infty$, $\forall k \in \boldsymbol{N}$. The seminorms of the quasi-uniform topology τ on $\mathcal{L}^+(\mathcal{D})$ are

$$X \in \mathcal{L}^+(\mathcal{D}) \to \|X\|^{f,k} := \max\left\{\|f(N)XN^k\|, \|N^kXf(N)\|\right\}. \qquad (1)$$

where k is an integer and f belongs to $\mathcal{C}$. In [5] it has been proved that $\mathcal{L}^+(\mathcal{D})[\tau]$ is a complete locally convex topological *-algebra, with involution $X^\dagger := X^*|_{\mathcal{D}}$.

The seminorms of the uniform topology $\tau_{\mathcal{L}}$ on $\mathcal{L}(\mathcal{D}, \mathcal{D}')$ are defined by

$$X \in \mathcal{L}(\mathcal{D}, \mathcal{D}') \to \|X\|^f := \|f(N)Xf(N)\|. \qquad (2)$$

where, again, f belongs to $\mathcal{C}$. In [5] it has also been proved that $\mathcal{L}(\mathcal{D},\mathcal{D}')$, with the topology $\tau_{\mathcal{L}}$, is a topological quasi *-algebra over $\mathcal{L}^+(\mathcal{D})$.

3 Spin Models

The first application we want to discuss in this Section is the Almost Mean Field Ising Model we have introduced in 1991, [8]. This model is defined by the following finite volume hamiltonian

$$H_V = \frac{J}{|V|^\gamma}\sum_{i,j\in V}\sigma_i^3\sigma_j^3, \tag{3}$$

with $0<\gamma\leq 1$. Particularly relevant in the mathematical description of this model is the *almost magnetization* operator $S_V^3 := \frac{1}{|V|^\gamma}\sum_{p\in V}\sigma_p^3$. In fact, if A is a local observable, its regularized time evolution $\alpha_V^t(A) := e^{iH_Vt}Ae^{-iH_Vt}$ in general depends on t, A and S_V^3.

The model is defined on a lattice. To the lattice site p we attach the Hilbert space $\mathbb{C}^2$ and the algebra of the 2×2 matrices generated by the identity matrix and the Pauli matrices σ_i^α. The Hilbert space of the infinite lattice is $\mathcal{H}_{spin} = \otimes_{p\in\boldsymbol{Z}}\mathbb{C}_p^2$, and the algebra is the standard quasi local C*-spin algebra $\mathcal{A}_s$.

Let now $n=(n_1,n_2,n_3)$ be a unit vector in $\boldsymbol{R}^3$, and put $(\sigma\cdot n) = n_1\sigma^1+n_2\sigma^2+n_3\sigma^3$. The spectrum of $\sigma\cdot n$, is $\{1,-1\}$. We call $|n\rangle$ the unit eigenvector of $\mathbb{C}^2$ associated with 1.

Let (n,n^1,n^2) be an orthonormal basis of $\boldsymbol{R}^3$. We put $n^\pm = \frac{1}{2}(n^1\pm in^2)$ and, for $m=0,1$, $|m,n\rangle = (\sigma\cdot n^-)^m|n\rangle$. The vectors $|m,n\rangle$ are still eigenvectors of the operator $(\sigma\cdot n)$ with eigenvalue $(-1)^m$. Repeating this procedure for each lattice site we get, starting with a sequence of vectors in $\boldsymbol{R}^3$, $\{n\}:=\{n_p\}_{\{p\in\boldsymbol{Z}\}}$, a unit vector in $\mathcal{H}_{spin}$ defined as $|\{n\}\rangle = \otimes_p|n_p\rangle$. Furthermore, acting with the operators $(\sigma_p\cdot n_p^-)_p^m$, $p\in\boldsymbol{Z}$, on each $|n_p\rangle$, we get a new vector of $\mathcal{H}_{spin}$, $|\{m\},\{n\}\rangle := \otimes_p|m_p,n_p\rangle$. The set $\{|\{m\},\{n\}\rangle,\ m_p=0,1,\ \sum_p m_p<\infty\}$ forms an orthonormal basis in an Hilbert space $\{n\}$-depending, which we call $\mathcal{H}_{\{n\}}$, [3]. On this space we define the unbounded self-adjoint (number) operator $M_{\{n\}}$ by its action on the vectors of the basis

$$M_{\{n\}}|\{m\},\{n\}\rangle = \left(\mathbb{1}+\sum_p m_p\right)|\{m\},\{n\}\rangle.$$

Of course $M_{\{n\}}$ depends on $\{n\}$, that is on the Hilbert space where the operator acts.

As a matter of fact, the sequence $\{n\}$ cannot be chosen arbitrarily, but it has to be such that its 'almost' mean exists. In other words, we need to consider sequences such that $n_p^1 = n_p^2 = 0$ for almost all $p \in \boldsymbol{Z}$ and $\lim_{|V|\to\infty} \frac{1}{|V|^\gamma} \sum_{p\in V} n_p^3$ exists in $\boldsymbol{R}$.

The algebraic setting is fixed by the operator $M_{\{n\}}$: first we put $\mathcal{D}_{\{n\}} = \bigcap_k \mathcal{D}(M_{\{n\}}^k)$. Then we introduce the algebra $\mathcal{L}^\dagger(\mathcal{D}_{\{n\}})$, endowed with the quasi-uniform topology $\tau_{\{n\}}$, [5]. Finally we introduce a *-representation of the C*-spin algebra $\mathcal{A}_s$ on $\mathcal{L}^\dagger(\mathcal{D}_{\{n\}})$ and a topology τ_0 on $\mathcal{A}_s$ as follows:

$$\pi_{\{n\}}(\sigma_p^\alpha) \mid \{m\}, \{n\}\rangle = \sigma_p^\alpha \mid m_p, n_p\rangle \otimes (\prod_{p' \neq p} \otimes \mid m_{p'}, n_{p'}\rangle) \quad (\alpha = 1, 2, 3),$$

and

$$\|A\|_{\{n\}}^{f,k} := \max\left\{\|M_{\{n\}}^k \pi_{\{n\}}(A) f(M_{\{n\}})\|, \|f(M_{\{n\}})\pi_{\{n\}}(A) M_{\{n\}}^k\|\right\},$$

where f belongs to $\mathcal{C}$ and $k \geq 0$.

With these definitions, calling $\mathcal{A}$ the τ_0-completion of $\mathcal{A}_s$, it is possible to prove that:

- $(\mathcal{A}[\tau_0], \mathcal{A}_s)$ is a topological quasi *-algebra;
- all the powers of the almost magnetization S_3^V are τ_0-converging in $\mathcal{A}$;
- the finite volume dynamics α_V^t τ_0-converges to a one-parameter group of automorphisms α^t of $\mathcal{A}$
- α^t solves the τ_0-limit of the finite volume Heisenberg equation of motion.

Remark:– This is a first evidence of the relevance of topological quasi *-algebras in concrete applications. It is worthwhile to stress that not only the original algebraic approach, [2], but also the procedure proposed in references [6, 7] cannot be used here without making major changes. This means that C* and von Neumann algebras are really not the most convenient settings for the description of this model.

We now mention a second spin model, the almost mean field Heisenberg model, which can be conveniently described within the same algebraic framework, [9]. The finite volume hamiltonian is now

$$H_V = \frac{J}{|V|^\gamma} \sum_{i,j\in V} \sum_{\alpha=1}^{3} \sigma_i^\alpha \sigma_j^\alpha,$$

with $\frac{1}{2} < \gamma \leq 1$.

More spin models have been discussed using algebras and topologies strongly related to the ones above. For more details on these different models we refer to [10].

4 Free and Interacting Bosons

In this Section we show how the problem of describing the time evolution for free bosons, or for bosons interacting with the matter (via a simplified interaction), strongly suggests the use of a *natural* algebraic framework related to the model.

The formal hamiltonian H for the one mode free bosons is simply the number operator $N = a^\dagger a$, a and $a^\dagger$ being the annihilation and creation operators for the bosons. They satisfy the canonical commutation relation $[a, a^\dagger] = \mathbb{1}$. (More properly, N is the unique self-adjoint extension of the symmetric operator $a^\dagger a$.) Let now $\mathcal{D} := D^\infty(N) = \cap_{k\geq 0} D(N^k)$. This set is dense in the Fock-Hilbert space $\mathcal{H}$ constructed in the standard way. Starting from $\mathcal{D}$ we can define the *-algebra $\mathcal{L}^+(\mathcal{D})$. It is clear that all powers of a and $a^\dagger$ belong to this set. The topology in $\mathcal{L}^+(\mathcal{D})$ is the usual quasi-uniform topology defined by the seminorms

$$X \in \mathcal{L}^+(\mathcal{D}) \to \|X\|^{f,k} := \max\left\{\|f(N)XN^k\|, \|N^k X f(N)\|\right\}, \tag{4}$$

where $f \in \mathcal{C}$ and $k \geq 0$. We know that $\mathcal{L}^+(\mathcal{D})[\tau]$ is a complete locally convex topological *-algebra.

We further introduce some subspaces of $\mathcal{H}$ and some projection operators on these spaces: let $\mathcal{E}_l$ be the subspace generated by all the vectors which are proportional to $(a^\dagger)^l \Phi_0$. Then $\mathcal{F}_L$ be the direct sum $\mathcal{F}_L := \mathcal{E}_0 \oplus \mathcal{E}_1 \oplus \ldots\ldots \oplus \mathcal{E}_L$. Also, let $N = \sum_{l=0}^\infty l \Pi_l$ be the spectral decomposition of the number operator N. The operators Π_l are projection operators, as well as the operators $Q_L = \sum_{l=0}^L \Pi_l$. The following properties are obvious:

$$\Pi_k \Pi_l = \delta_{kl} \Pi_l, \quad \Pi_k^\dagger = \Pi_k; \qquad Q_L Q_M = Q_L, \quad \text{if } L \leq M, \quad Q_L^\dagger = Q_L.$$

It is clear that $\Pi_k : \mathcal{H} \to \mathcal{E}_k$, and $Q_L : \mathcal{H} \to \mathcal{F}_L$. The operator Q_L is used to cut-off the hamiltonian, by replacing a with $a_L := Q_L a Q_L$. The regularized hamiltonian is simply $H_L = Q_L N Q_L = N Q_L$ and the related time evolution is $\alpha_L^t(X) = e^{iH_L t} X e^{-iH_L t}$. This *occupation number* cut-off produces a self adjoint bounded operator H_L. In [11] we have proven that *the limits of* $\alpha_L^t(a^n)$ *and* $\alpha_L^t((a^\dagger)^n)$ *exist in* $\mathcal{L}^+(\mathcal{D})[\tau]$ *for all* $n \in \boldsymbol{N}$, which shows that $\mathcal{L}^+(\mathcal{D})[\tau]$ is a natural algebra for describing our model. The fact that the model itself suggests the construction of the algebra and the topology is, in our opinion, a strong indication of the validity of our approach: if we are interested in the description of a physical model, the model itself drives our choices!

The same algebraic framework turns out to be useful also in the analysis of the thermodynamical limit of the interacting model described by the following formal hamiltonian:

$$H_V = \frac{J}{|V|} \sum_{i,j \in V} \sigma_i^3 \sigma_j^3 + a^\dagger a + \gamma(a + a^\dagger)\sigma_V^3,$$

where $\sigma_V^3 = \frac{1}{|V|}\sum_{i \in V} \sigma_i^3$. Here, since the model is composite, the algebra $\mathcal{L}^+(\mathcal{D})$ must be replaced by $\mathcal{A} = B(\mathcal{H}_{spin}) \otimes \mathcal{L}^+(\mathcal{D})$. The topology on $\mathcal{A}$, τ_{comp}, is generated by seminorms which deal separately with the spin and with the boson variables:$\|XA\|^{f,k,\Psi} \equiv \|X\|^{f,k}\|A\Psi\|$, $X \in \mathcal{L}^+(\mathcal{D})$ and $A \in B(\mathcal{H}_{spin})$. It is worthwhile to remind also that Ψ cannot be a generic vector in $\mathcal{H}_{spin}$, see [3], but must belong to the set

$$\mathcal{F} = \left\{ \Psi \in \mathcal{H}_{spin} \; : \; \lim_{|V|,\infty} \frac{1}{|V|} \sum_{p \in V} \sigma_p^3 \Psi \sigma_\infty^3 \Psi, \;\; \|\sigma_\infty^3\| \leq 1 \right\}.$$

As before, the regularized hamiltonian is obtained by replacing a with $a_L := Q_L a Q_L$, so that the new hamiltonian $H_{V,L}$ depends on two, in principle, unrelated cutoffs. The existence of the limit of $\alpha_{V,L}^t(X) = e^{iH_{V,L}t} X e^{-iH_{V,L}t}$ is ensured by the following result, again contained in [11]: *the limit of* $\alpha_{V,L}^t(a)$ *for* $|V|$ *and* L *both diverging exists in* $\mathcal{A}[\tau]$. *Moreover, if the two cutoffs satisfy the relation* $|V| = L^r$, *for a certain integer* $r > 1$, *the same holds true also for* $\alpha_{V,L}^t(\sigma_\alpha^i)$.

Remark:– Other models, related to the above ones, can be conveniently analyzed in the algebraic setting described in [11]. It is clear that, again, some of the Haag and Kastler axioms are not satisfied, and that a different structure is to be required.

5 Conclusions

These and other models suggest a natural extension of the original Haag and Kastler framework, extension in which the original abstract quasi-local C*-algebra should be replaced by the algebra $\mathcal{L}^+(\mathcal{D})$ or with a topological quasi *-algebra of operators.

We want to conclude the paper simply by mentioning some new results in which the relevance of $\mathcal{L}^+(\mathcal{D})$ appears evident once again, [12].

The first result is related to the analysis of the thermodynamical limit for models containing modified quons. These are excitations whose annihilation and creation operators a and $a^\dagger$ satisfy the relation $aa^\dagger - f(q)a^\dagger a = \mathbb{1}$, $f(q)$ being a real function which assumes values ± 1 as well as values larger than 1.

Our most recent result is related to the extension of contraction mappings to $\mathcal{L}^+(\mathcal{D})$. Fixed point theorems have been proved and applications to physical systems are derived. In particular we have discussed the possibility of using this procedure for the analysis of the thermodynamical limit of quantum mechanical models.

Acknowledgments

It is a pleasure to thank Prof. A. Inoue, Prof. H. Kurose and all the staff at Fukuoka University for their warm hospitality.

References

[1] R.F. Streater, A.S. Wightman,*PCT, Spin and Statistics, and All That*, Benjamin Inc., New York, (1964)

[2] R. Haag, D. Kastler, *An algebraic approach to Quantum Field Theory*, J. Math. Phys., **5**, 848-861, (1964)

[3] W.Thirring and A.Wehrl, *On the Mathematical Structure of the B.C.S.-Model*, Commun.Math.Phys. **4**, 303-314 (1967)

[4] K. Hepp and E. H. Lieb: Helv. Phys. Acta **46**, 573, 1973

[5] G. Lassner, *Topological algebras and their applications in Quantum Statistics*, Wiss. Z. KMU-Leipzig, Math.-Naturwiss. R., **30**, 572-595, (1981), *Algebras of unbounded operators and quantum dynamics*, Physica, **124 A**, 471-480, (1984)

[6] G. Morchio, F. Strocchi,*Mathematical Structures for long range dynamics and symmetry breaking*, J. Math. Phys., **28**, 622-635 (1987)

[7] F. Bagarello, G. Morchio, *Dynamics of Mean-Field Spin Models from Basic Results in Abstract Differential Equations*, J. Stat. Phys. Vol.66, **3/4**, 849-866 (1992)

[8] F. Bagarello, C. Trapani, *"Almost" Mean Field Ising Model: An Algebraic Approach*, J. Stat. Phys. Vol.65, **3/4**, 469-482 (1991)

[9] F. Bagarello, C. Trapani, *A note on the algebraic approach to the "almost" mean field Heisenberg model*, Nuovo Cimento, **108B**, 779-784 (1993)

[10] F. Bagarello, C. Trapani, *The Heisenberg Dynamics of Spin Systems: A Quasi*-Algebras Approach*, J. Math. Phys., **37**, 4219-4234 (1996)

[11] F. Bagarello, *Applications of Topological *-Algebras of Unbounded Operators*, J. Math. Phys., **39**, 6091-6105, (1998)

[12] F. Bagarello, *Applications of Topological *-Algebras of Unbounded Operators to Modified Quons*, submitted to J. Stat. Phys.; F. Bagarello, *Fixed Point Results in Topological *-Algebras of Unbounded Operators*, in progress

76 ON CONTINUITY PROPERTIES OF VERTEX OPERATORS

Klaus-Detlef KÜRSTEN
Martin LÄUTER

Fakultät für Mathematik und
Informatik, Mathematisches Institut
Universität Leipzig
Augustusplatz 10, D04109
Leipzig, Germany

kuersten@mathematik.uni-leipzig.de
laeuter@mathematik.uni-leipzig.de

Abstract

Continuity properties of elements $Y(v, z)$ of complex realized vertex operator algebras in the sense of [2] are investigated by constructing a generalized scale $(H_{R,S})$ of Hilbert spaces depending on two parameters such that each $Y(v, z)$ may be considered in a natural way as a continuous linear operator between spaces belonging to this scale. As a consequence, products of vertex operators may be realized as factorization products as well as products of operators on a certain partial inner product space.

1 Introduction

We consider complex realized vertex operator algebras in the sense of [2]. Elements of such vertex operator algebras are linear operators $Y(v, z) : V_L \to \overline{V_L}$, where V_L is a certain graded vector space with the property that all its homogeneous subspaces are of finite dimension, whereas $\overline{V_L}$ denotes the algebraic completion of V_L. Moreover, $Y(v, z)$ depends on the parameter $v \in V_L$ and on the complex variable $z \in \mathbb{C}\backslash\{0\}$.

Continuity properties of such operators are described by constructing a generalized scale $(H_{R,S})$ of Hilbert spaces depending on two positive real parameters R, S, such that each operator $Y(v, z)$ may be extended to a continuous

H.G.W. Begehr et al. (eds.), Proceedings of the Second ISAAC Congress, Vol. 1, 661–669.

linear operator between certain spaces of this scale. More precisely, the scale is constructed in such a way that whenever there are given $v \in V_L$, $z \in \mathbb{C} \setminus \{0\}$, and positive real numbers R, R', S, S' such that $R < |z| < R'$ and $S < S'$, there are natural inclusions

$$V_L \subset H_{R,S} \subset H_{R',S'} \subset \overline{V_L}$$

and $Y(v, z)$ has a unique extension to a continuous linear operator

$$H_{R,S} \to H_{R',S'} .$$

On the basis of the scale $(H_{R,S})$ there will be defined a factorization product on the space of linear operators mapping V_L into $\overline{V_L}$, which is consistent with the known product of vertex operators. This way, the partial algebra of vertex operators is realized as a partial algebra of continuous operators. Alternatively, it is also possible to define on $\overline{V_L}$ a structure of a partial inner product space such that the partial algebra of vertex operators may be realized in a natural way as a partial algebra of operators on this partial inner product space.

2 Algebraic Constructions

For the definition of the spaces and operators we refer to [2], but to have them consistent with the Hilbert scales of spaces and the affiliated norms we give a short introduction of the notions here.

Let $\mathfrak{h}$ a $\mathbb{C}$-linear space of finite dimension with a base $\{e_1, \dots, e_{\dim \mathfrak{h}}\}$. Then let L be the lattice generated by that base of $\mathfrak{h}$. Fix an even non-degenerate symmetric bilinear form on $\mathfrak{h}$, i.e. $\langle a, a\rangle \in 2\mathbb{Z}$ for all $a \in L$.

From $\mathfrak{h}$ we get a Heisenberg algebra $\hat{\mathfrak{h}}_{\mathbb{Z}}$ by setting

$$\hat{\mathfrak{h}}_{\mathbb{Z}} = \mathbb{C}\mathfrak{c} \oplus \sum_{n \in \mathbb{Z}\setminus\{0\}} \mathfrak{h} \otimes t^n$$

and introducing a bracket $[\cdot, \cdot]$ by linear extension of

$$[\mathfrak{c}, \hat{\mathfrak{h}}_{\mathbb{Z}}] = 0,$$
$$[h \otimes t^m, g \otimes t^n] = \langle h, g\rangle \, m \, \delta_{n+m,0} \, \mathfrak{c}.$$

Then denote with $\hat{\mathfrak{h}}_{\mathbb{Z}}^-$ the negative part of $\hat{\mathfrak{h}}_{\mathbb{Z}}$: $\hat{\mathfrak{h}}_{\mathbb{Z}}^- = \sum_{n<0} \mathfrak{h} \otimes t^n$.

The next step is to construct the symmetric algebra V of $\hat{\mathfrak{h}}_{\mathbb{Z}}^-$. We may view at it as the polynomial algebra with the elements $\{(e_\kappa \otimes t^{-k}) : \kappa = 1, \dots, \dim \mathfrak{h},\ k \in \mathbb{N}\}$ of $\hat{\mathfrak{h}}_{\mathbb{Z}}^-$ as variables. These variables can be enumerated so that for the ith item $e_{\kappa_i} \otimes t^{-k_i}$ the equation $i = \kappa_i + (k_i - 1) \cdot \dim \mathfrak{h}$ holds. So each monomial in V can be identified with a sequence (l_i) in $\mathbb{N}_0 = \mathbb{N} \cup \{0\}$ with $\lim_{i\to\infty} l_i = 0$, and denoted by $|l_i\rangle$

$$(l_i) \sim \prod_{i=1}^{\infty} (e_{\kappa_i} \otimes t^{-k_i})^{l_i} =: |l_i\rangle .$$

A grading on V can be introduced by

$$\deg |l_i\rangle := \sum_{i=1}^{\infty} k_i l_i.$$

Denote with E_r the set $\{|l_i\rangle : \deg |l_i\rangle = r\}$ and $V_r = \operatorname{lin} E_r$, so that E_r is a fixed base of the homogeneous subspace V_r of V.

But V is only one of two tensor factors which constitute V_L. The second one depends on a certain extension of the additive group L. Recall that $\mathfrak{h}$ was a linear space with the base $\{e_1, \ldots, e_{\dim \mathfrak{h}}\}$, and L the lattice generated by that base. Let $\hat{L}$ denote a central extension of L by a finite cyclic group $< \kappa >$ of order u generated by κ, i.e.

$$0 \longrightarrow < \kappa > \longrightarrow \hat{L} \longrightarrow L \longrightarrow 0.$$

Further we will need the map $\hat{L} \to L$ in the above sequence and denote it by $\hat{L} \ni a \mapsto \overline{a} \in L$. Now let us fix a section $L \in \alpha \longrightarrow e_\alpha \in \hat{L}$ and chose $\omega \in \mathbb{C}$ so that $\omega^u = 1$.

The second tensor factor of V_L will be denoted by $\mathbb{C}\{L\}$. It is linearly isomorphic to the group algebra $\mathbb{C}[L]$ of L, but it has additionally a structure of an $\hat{L}$-module. We omit a complete definition, but outline the properties of $\mathbb{C}\{L\}$ needed here: There is a mapping $\iota : \hat{L} \to \mathbb{C}\{L\}$ such that $(\iota(e_\alpha))_{\alpha \in L}$ is a base of $\mathbb{C}\{L\}$ and

$$\begin{aligned} a \cdot \iota(b) &= \iota(ab), \\ \kappa \cdot \iota(b) &= \omega \iota(b) \end{aligned}$$

for all $a, b \in \hat{L}$.

Now we are ready to build the necessary spaces on which the complex realized vertex operators act:

$$V_L = V \otimes \mathbb{C}\{L\},$$

which has $(|l_i\rangle \otimes \iota(e_\alpha))_{|l_i\rangle \in \cup E_r, \alpha \in L}$ as a base. Its "algebraic completion" will be denoted by $\overline{V_L}$:

$$\overline{V_L} = \left\{ \sum \lambda_{|l_i\rangle,\alpha} \; |l_i\rangle \otimes \iota(e_\alpha) : \lambda_{|l_i\rangle,\alpha} \in \mathbb{C} \right\}.$$

Along the same lines we will get $\overline{V}$ and $\overline{\mathbb{C}\{L\}}$, respectively.

3 Construction of Hilbert spaces

Up to this point the constructions may be found in [2]. We can go further and introduce the topological aspects. Given $R > 0$ the space W_R is defined to be the Hilbert space with $(|l_i\rangle)_{|l_i\rangle \in \cup E_r}$ as an orthogonal base so that

$$\| \, |l_i\rangle \, \|_{W_R} = \frac{1}{R^{\deg |l_i\rangle}} \left(\prod_{i=1}^{\infty} \sqrt{l_i!} \right)$$

holds. All the W_R, $R > 0$, are linear subspaces of $\overline{V}$.

Furthermore let $\langle\cdot,\cdot\rangle_+$ denote a fixed scalar product on $\mathfrak{h} \times \mathfrak{h}$. Using this we can do something similar in the second component: Given $S > 0$, let G_S be the Hilbert space with $(\iota(e_\alpha)_{\alpha\in L})$ as an orthogonal base such that

$$\|\iota(e_\alpha)\|_{G_S} = \frac{1}{S^{\langle\alpha,\alpha\rangle_+}}.$$

Joining the efforts we define subspaces $H_{R,S}$ of $\overline{V_L}$, which are Hilbert spaces with orthogonal bases $(|l_i\rangle \otimes \iota(e_\alpha))$ and the defining condition for the norm

$$\|\, |l_i\rangle \otimes \iota(e_\alpha)\|_{H_{R,S}} = \|\, |l_i\rangle \,\|_{W_R} \cdot \|\iota(e_\alpha)\|_{G_S}.$$

4 Vertex Operators

Having the spaces, we can now move forward to inspecting some operators between them. Again we follow the lines of [2]. Let $h\in\mathfrak{h}$ and $n < 0$. Then the mapping

$$h(n) : x \otimes y \mapsto ((h \otimes t^n)x) \otimes y$$

extends to a linear operator $h(n)$ on V_L. It is just the multiplication operator by $h \otimes t^n$ in the first component.
For $n > 0$ and $h \in \mathfrak{h}$, let $\widetilde{h(n)}$ be the linear operator on V satisfying

$$\widetilde{h(n)}(g \otimes t^m) = [h \otimes t^n, g \otimes t^m]\mathbf{1}$$
$$\widetilde{h(n)}(p \cdot q) = (\widetilde{h(n)}p) \cdot q + p \cdot (\widetilde{h(n)}q),$$

i.e. the extension on products is by the product rule for derivations. Then $h(n) : V_L \to V_L$ is defined by

$$h(n) = \widetilde{h(n)} \otimes Id\,.$$

Note that two "differentiation" operators $h(n)$ and $g(m)$ $(n, m > 0)$ commute.

Operators $h(0)$, z^h, and a are defined as linear extensions of the mappings

$$h(0) : x \otimes \iota(b) \mapsto \langle h, \overline{b}\rangle\, x \otimes \iota(b) \qquad (h \in \mathfrak{h})$$
$$z^h : x \otimes \iota(b) \mapsto z^{\langle h,\overline{b}\rangle} x \otimes \iota(b) \qquad (h \in L, z \in \mathbb{C} \setminus \{0\})$$
$$a : x \otimes \iota(b) \mapsto x \otimes \iota(ab) \qquad (a \in \hat{L}).$$

These operators act essentially only on the second component $\mathbb{C}\{L\}$.

Using the above definitions one can define the following expressions which exist at least as formal series in the variable z with coefficients in $\mathrm{End}\,(V_L, \overline{V_L})$. Let $\alpha\in\mathfrak{h}$.

$$\alpha^\pm(z) = \sum_{n\in\pm\mathbb{N}} \alpha(n)\, z^{-n-1}$$
$$\alpha(z) = \alpha^-(z) + \alpha(0)z^{-1} + \alpha^+(z)$$
$$E^\pm(\alpha, z) = \exp\left(\sum_{n\in\pm\mathbb{N}} \frac{\alpha(n)}{n}\, z^{-n}\right),$$

where exp$(\cdot)$ stands for the exponential series expansion. One can see that, if $z \in \mathbb{C}\backslash\{0\}$ then $E^+(\alpha, z)$ and $\alpha^+(z)$ consist only of "differentiation" operators. In fact they belong to End (V_L). The operators $E^-(\alpha, z)$ and $\alpha^-(z)$ may be regarded as elements of End $(\overline{V_L})$ under the same condition.

Now we are ready to give definitions of vertex operators. For $\alpha\in\mathfrak{h}$ and $z\in\mathbb{C}$ set

$$X_0(\alpha, z) = E^-(\alpha, z)\, E^+(\alpha, z)$$

as an operator from V_L to $\overline{V_L}$.

For $a\in\hat{L}$ we set $v = \mathbf{1} \otimes \iota(a)$. Then

$$Y(v, z) = Y(a, z) := X_0(-\overline{a}, z)\, a\, z^{\overline{a}}.$$

Extending this formula for $v = \alpha_1(-n_1)\cdots\alpha_p(-n_p)1 \otimes \iota(a)$, where $\alpha_j\in\mathfrak{h}$ and $n_j\in\mathbb{N}$ the vertex operator $Y(v, z)$ is defined to be a "normal ordering" of the product

$$\left(\frac{1}{(n_1-1)!}\left(\frac{\mathrm{d}}{\mathrm{d}z}\right)^{n_1-1}\alpha_1(z)\right)\cdots\left(\frac{1}{(n_p-1)!}\left(\frac{\mathrm{d}}{\mathrm{d}z}\right)^{n_p-1}\alpha_p(z)\right)Y(a, z).$$

Note that normal ordering is a well-defined procedure which performs this product into a linear combination of products with the property that in each of these products each of the "multiplication" operators $\left(\frac{1}{(n-1)!}\left(\frac{\mathrm{d}}{\mathrm{d}z}\right)^{n-1}\alpha^-(z)\right)$, $E^-(-\overline{a}, z)$ stands to the left of each of the "differentiation" operators $\left(\frac{1}{(n-1)!}\left(\frac{\mathrm{d}}{\mathrm{d}z}\right)^{n-1}\alpha^+(z)\right)$, $E^+(-\overline{a}, z)$.

Finally, $Y(v, z)$ is extended linearly to arbitrary $v\in V_L$.

5 Continuity of the Operators

We switch again to consider the topological properties of the vertex operators. When examining the continuity of $Y(v, z)$ one sees that it is a linear combination concatenations of some operators of a much simpler structure and so one can examine their continuity properties first.

Proposition 1 *The operators* $\left(\frac{1}{(n-1)!}\left(\frac{\mathrm{d}}{\mathrm{d}z}\right)^{n-1}\alpha^\pm(z)\right)$, $\alpha(0)$, a *and* $z^{\overline{a}}$ *are continuous from* $H_{R,S}$ *to* $H_{R',S'}$ *whenever* $0 < R < R'$ *and* $0 < S < S'$.

The proofs are based on an almost straightforward estimation of norms.

But $X_0(\alpha, z)$ is different. The number of differentiation and/or multiplication operators coming together in the coefficients of the result is not bounded. Consequently, finer methods of estimating the coefficients have to be applied. Note that $X_0(\alpha, z)$ acts essentially on the first tensor component of V_L, i. e., that there exists $\widetilde{X_0}(\alpha, z) : V \to \overline{V}$ such that $X_0(\alpha, z) = \widetilde{X_0}(\alpha, z) \otimes Id$. Let now $|l_i\rangle \in E_r$ be given. Then $\widetilde{X_0}(\alpha, z)(|l_i\rangle)$ may be written as

$$z^{-r}\sum_{s=0}^{\infty} z^s \sum_{|l_i'\rangle\in E_s} \mu_{|l_i\rangle,|l_i'\rangle}\ |l_i'\rangle\,,$$

and the coefficients $\mu_{|l_i\rangle,|l'_i\rangle}$ may be calculated explicitely and estimated as follows (cf. [8]).

$$\mu_{|l_i\rangle,|l'_i\rangle} = \prod_{i=1}^{\infty} \left(\sum_{m_i=0}^{\min\{l_i,l'_i\}} \binom{l_i}{m_i} \frac{(-1)^{l'_i-m_i}}{(l'_i-m_i)!} \cdot \frac{\langle \alpha, e_{\kappa_i}\rangle^{l_i-m_i} \alpha_{\kappa_i}^{l'_i-m_i}}{k_i^{l'_i-m_i}} \right),$$

where α_{κ_i} is the coefficient of e_{κ_i} in the expansion of $\alpha = \sum_{j=1}^{\dim \mathfrak{h}} \alpha_j e_j$.

Proposition 2 *Given $\alpha \in \mathfrak{h}$ and $\varepsilon > 0$, there exists a constant C such that*

$$\left| \mu_{|l_i\rangle,|l'_i\rangle} \right| < C \cdot (1+\varepsilon)^{\deg|l_i\rangle + \deg|l'_i\rangle} \cdot \prod_{i=1}^{\infty} \sqrt{\frac{l_i!}{l'_i!}}$$

for all $|l_i\rangle, |l'_i\rangle \in \cup E_r$.

This is the main point to prove the continuity of the vertex operator $X_0(\alpha, z)$.

Proposition 3 *Suppose that $\alpha \in \mathfrak{h}$, $z \in \mathbb{C}\backslash\{0\}$, $0 < R < |z| < R'$ and $0 < S < S'$. Then the operator $X_0(\alpha, z)$ has a unique extension to a continuous linear operator from $H_{R,S}$ to $H_{R',S'}$.*

Finally we have all means to prove the following theorem, and since the continuity of $X_0(\alpha, z)$ poses the strongest restrictions on R, R', S and S', it resembles Proposition 3.

Theorem 1 *Suppose that $v \in V_L$, $z \in \mathbb{C}\backslash\{0\}$, $0 < R < |z| < R'$ and $0 < S < S'$. Then the operator $Y(v, z)$ has a unique extension to a continuous linear operator from $H_{R,S}$ to $H_{R',S'}$.*

6 Factorization Products

The theorem allows us to view products of vertex operators as factorization products. Let us start with a definition of factorization products adapted to the present context. Suppose that D, $\overline{D}$ are locally convex spaces and that D is a dense linear subspace of $\overline{D}$. Then factorization products on the space $\mathcal{L}(D, \overline{D})$ of continuous linear operators from D to $\overline{D}$ are defined as follows.

Definition 1 *Let $\mathfrak{K}$ be a fixed set of locally convex spaces so that for arbitrary $E, F \in \mathfrak{K}$ the following two conditions are satisfied:*
i) $D \subset E \subset \overline{D}$ and the embeddings $D \hookrightarrow E$ and $E \hookrightarrow \overline{D}$ are continuous.
ii) D is dense in $E \cap F$ w. r. t. the projective topology (i.e. the topology generated by the topologies induced from E and F).
The product $T_n \circ \cdots \circ T_1$ of elements of $\mathcal{L}(D, \overline{D})$ is said to be defined w. r. t. $\mathfrak{K}$, if there are spaces $E_0, \ldots, E_n \in \mathfrak{K}$ and continuous extensions $S_j \in \mathcal{L}(E_{j-1}, E_j)$ of T_j for every $j \in \{1, \ldots, n\}$. In this case

$$T_n \circ \cdots \circ T_1 \varphi := S_n(\cdots(S_1(\varphi))\cdots).$$

Clearly $T_n \circ \ldots \circ T_1 \in \mathcal{L}(D, \overline{D})$. An easy induction on n (using the second condition on $\mathfrak{K}$) shows that it does not depend on the special choice of $E_0, \ldots, E_n$ in $\mathfrak{K}$ (cf. [4]). But, in general, the factorization product depends on the choice of the set $\mathfrak{K}$. Factorization products were used also in other situations, e. g. in [3, 7, 1, 6].

V_L and $\overline{V_L}$ form a dual pair w. r. t. the pairing

$$\left(\sum_{|l_i\rangle,\alpha} \lambda_{|l_i\rangle,\alpha} \, |l_i\rangle \otimes \iota(e_\alpha), \sum_{|l_i\rangle,\alpha} \mu_{|l_i\rangle,\alpha} \, |l_i\rangle \otimes \iota(e_\alpha) \right) \mapsto \sum_{|l_i\rangle,\alpha} \lambda_{|l_i\rangle,\alpha} \cdot \mu_{|l_i\rangle,\alpha}.$$

So we endow them with their Mackey topologies. Since the vectors $|l_i\rangle \otimes \iota(e_\alpha)$ form an orthogonal base in each $H_{R,S}$, spaces of type $H_{R,S} \cap H_{R',S'}$ are isomorphic to Hilbert spaces with the same orthogonal base. This implies that the set $\mathfrak{K} = \{H_{R,S}\}_{R,S>0}$ satisfies the condition ii) in Definition 1.

Thus the product of vertex operators becomes the factorization product w. r. t. the class $\mathfrak{K}$:

Theorem 2 *Let elements $v_1, \ldots, v_n$ of V_L and complex numbers $z_1, \ldots, z_n$ be given so that $0 < |z_1| < \cdots < |z_n|$. Then the product*

$$Y(v_n, z_n) \circ \cdots \circ Y(v_1, z_1)$$

exists as the factorization product w. r. t. $\mathfrak{K} = \{H_{R,S}\}_{R,S>0}$.

7 Operators on PIP-spaces

Vertex operators and their products can be viewed in still another context. As a consequence of the above results they may be regarded as operators on a partial inner product spaces (PIP-spaces). The following definitions of PIP-spaces and operators on them are equivalent to the original definitions in [1].

A linear compatibility on a $\mathbb{C}$-vector space $\mathcal{V}$ is a symmetric binary relation $\sharp$ on $\mathcal{V}$ such that all sets of the type

$$M^\sharp \overset{\text{def}}{=} \{\varphi \in \mathcal{V} : \varphi \,\sharp\, \psi \text{ for all } \psi \in M\} \qquad (M \subset \mathcal{V})$$

(called assaying subspaces) are linear subspaces of $\mathcal{V}$. Let $\Gamma(\sharp)$ denote the graph of $\sharp$. A partial inner product on $(\mathcal{V}, \sharp)$ is a mapping

$$\Gamma(\sharp) \ni (\varphi, \psi) \mapsto \langle \varphi, \psi \rangle \in \mathbb{C}$$

which is linear in φ and satisfies $\langle \varphi, \psi \rangle = \overline{\langle \psi, \varphi \rangle}$, whenever $\varphi \,\sharp\, \psi$. A triple $(\mathcal{V}, \sharp, \langle \cdot, \cdot \rangle)$, where $\langle \cdot, \cdot \rangle$ is a partial inner product on $(\mathcal{V}, \sharp)$, is called a PIP-space. It is said to be nondegenerate if and only if $\langle \varphi, \psi \rangle = 0$ for all $\psi \in \mathcal{V}^\sharp$ implies that $\varphi = 0$.

Suppose now, that $(\mathcal{V}, \sharp, \langle \cdot, \cdot \rangle)$ is a nondegenerate PIP-space. In this case the restriction of $\langle \cdot, \cdot \rangle$ to $X \times \overline{X^\sharp}$, where X is an assaying subspace and $\overline{X^\sharp}$

denotes the associated conjugate linear space of $X^\sharp$, is a dual pairing. Each assaying subspace X of $\mathcal{V}$ will be endowed with the Mackey topology $\tau(X, \overline{X^\sharp})$.

Now the product of J.-P. Antoine and A. Grossmann [1] on

$$\mathrm{Op}(\mathcal{V}) = \mathcal{L}(\mathcal{V}^\sharp, \mathcal{V})$$

may be defined as factorization product.

Definition 2 *The product $T_n \circ \cdots \circ T_1$ of elements of* $\mathrm{Op}(\mathcal{V})$ *is said to be defined if there are assaying subspaces $E_0, \ldots, E_n$ of $\mathcal{V}$ and continuous extensions $S_j \in \mathcal{L}(E_{j-1}, E_j)$ of T_j. In this case*

$$T_n \circ \cdots \circ T_1 \varphi = S_n(\cdots(S_1\varphi)\cdots).$$

To apply this construction to our vertex operator situation we define the following compatibility relation on $\overline{V_L}$.

Definition 3 *Elements $\varphi = \sum \lambda_{|l_i\rangle,\alpha} |l_i\rangle \otimes \iota(e_\alpha)$ and $\psi = \sum \mu_{|l_i\rangle,\alpha} |l_i\rangle \otimes \iota(e_\alpha)$ of V_L are said to be compatible, denoted by $\varphi \sharp \psi$, if and only if*

$$\sum_{|l_i\rangle,\alpha} \left|\lambda_{|l_i\rangle,\alpha}\, \mu_{|l_i\rangle,\alpha}\right| < \infty.$$

In this case the partial scalar product is defined by

$$\langle \varphi, \psi \rangle = \sum \lambda_{|l_i\rangle,\alpha}\, \overline{\mu}_{|l_i\rangle,\alpha}.$$

Then it is not difficult to prove that the Hilbert spaces $H_{R,S}$ constructed above are assaying subspaces, i.e. they satisfy $H_{R,S} = (H_{R,S^\sharp})^\sharp$. Note that the space $\mathrm{Op}(\overline{V_L})$ of operators on $\overline{V_L}$ (viewed as operators in the PIP-space) coincides with the space $\mathcal{L}(V_L, \overline{V_L})$.

So we obtain the following result.

Theorem 3 *Each vertex operator $Y(v,z)$ with $v \in V_L$, $z \in \mathbb{C}\backslash\{0\}$ belongs to* $\mathrm{Op}(V_L)$. *If the conditions of Theorem 2 are satisfied, the product*

$$Y(v_n, z_n) \circ \cdots \circ Y(v_1, z_1)$$

exists as the product of operators on the PIP-space $\overline{V_L}$.

This structure seems to be more natural than the structure given by the scale of Hilbert spaces, since it does not depend on the special choice of the scalar product $\langle \cdot, \cdot \rangle_+$.

Acknowledgments

K.-D. Kürsten thanks Fukuoka University, Department of Applied Mathematics for the invitation to Fukuoka and the financial support.

References

[1] Antoine, J.-P., A. Grossmann: Partial inner product spaces I, II. *J. Funct. Anal.*, **23**:369–391, 1976.

[2] Frenkel, I., J. Lepowsky, A. Meurman: Vertex Operator Algebras and the Monster; Academic Press, 1988.

[3] Grossmann, A.: Homomorphisms and direct sums of nested Hilbert spaces. *Commun. Math. Phys.*, **4**:190–202, 1967.

[4] Kürsten, K.-D.: The completion of the maximal Op^*-algebra on a Fréchet domain. *Publ. RIMS Kyoto Univ.*, **22(1)**:151–175, 1986.

[5] Kürsten, K.-D., M. Läuter: Realization of vertex operators in generalized Hilbert scales (in preparation).

[6] Kürsten, K.-D.: On topological linear spaces of operators with a unitary domain of definition. *Wissenschaftliche Zeitschrift der Karl-Marx-Universität Leipzig, Mathematisch-Naturwissenschaftliche Reihe*, **39(6)**:623–655, 1990.

[7] Lassner, G., G. A. Lassner: Quasi-*-algebras and general Weyl quantization. In *Proc. 12. Intern. Conf. Differential Geometric Methods in Theoretical Physics*, pages 108–121, Berlin Heidelberg New York Tokyo, 1985. Springer-Verlag. Lecture Notes in Mathematics 1139.

[8] Läuter, M.: Realisierung von Kernoperatoren und Vertexoperatoren in Systemen stetiger linearer Operatoren mit partieller Multiplikation. Dissertation, Universität Leipzig, 1998.

77 A MULTIVARIABLE VON NEUMANN'S INEQUALITY

Takateru OKAYASU

Department of Mathematical Sciences, Faculty of Science, Yamagata University
Yamagata, 990-8560, Japan

1. It is a classical result due to von Neumann that any contraction T on a Hilbert space enjoys the so-called von Neumann inequality

$$||p(T)|| \leq \sup_{z \in \mathbf{T}} |p(z)|$$

for any polynomial p in one variable z, $\mathbf{T}$ the torus in the plane $\mathbf{C}$. It, however, is true further, that T enjoys the von Neumann inequality "of matricial type":

Theorem 1. *If T is a contraction on a Hilbert space, $\begin{pmatrix} p_{11} & \dots & p_{1m} \\ \vdots & & \vdots \\ p_{m1} & \dots & p_{mm} \end{pmatrix}$, a square matrix with entries p_{ij} $(1 \leq i, j \leq m)$ polynomials in one variable z, then the inequality*

$$\left\| \begin{pmatrix} p_{11}(T) & \dots & p_{1m}(T) \\ \vdots & & \vdots \\ p_{m1}(T) & \dots & p_{mm}(T) \end{pmatrix} \right\| \leq \sup_{z \in \mathbf{T}} \left\| \begin{pmatrix} p_{11}(z) & \dots & p_{1m}(z) \\ \vdots & & \vdots \\ p_{m1}(z) & \dots & p_{mm}(z) \end{pmatrix} \right\|$$

H.G.W. Begehr et al. (eds.), Proceedings of the Second ISAAC Congress, Vol. 1, 671–677.

holds, where the left hand side means the operator norm of the operator matrix $\begin{pmatrix} p_{11}(T) & \dots & p_{1m}(T) \\ \vdots & & \vdots \\ p_{m1}(T) & \dots & p_{mm}(T) \end{pmatrix}$ *acting on the ditect sum Hilbert space, the right hand side the supemum of the operator norm of the numerical matrix* $\begin{pmatrix} p_{11}(z) & \dots & p_{1m}(z) \\ \vdots & & \vdots \\ p_{m1}(z) & \dots & p_{mm}(z) \end{pmatrix}$, *z runs over the torus* $\mathbf{T}$, *respectively.*

In this paper, we will concern with that sort of inequality, namely,

$$\left\| \begin{pmatrix} p_{11}(T_1, T_2, \dots, T_n)) & \dots & p_{1m}(T_1, T_2, \dots, T_n)) \\ \vdots & & \vdots \\ p_{m1}(T_1, T_2, \dots, T_n)) & \dots & p_{mm}(T_1, T_2, \dots, T_n)) \end{pmatrix} \right\|$$

$$\leq \sup_{z_1, z_2, \dots, z_n \in \mathbf{T}} \left\| \begin{pmatrix} p_{11}(z_1, z_2, \dots, z_n) & \dots & p_{1m}(z_1, z_2, \dots, z_n) \\ \vdots & & \vdots \\ p_{m1}(z_1, z_2, \dots, z_n) & \dots & p_{mm}(z_1, z_2, \dots, z_n) \end{pmatrix} \right\|,$$

where T_1, T_2, $\dots$, T_n are commuting contractions on a Hilbert space, $\begin{pmatrix} p_{11} & \cdots & p_{1m} \\ \vdots & & \vdots \\ p_{m1} & \cdots & p_{mm} \end{pmatrix}$, a square matrix with entries $p_{ij}(1 \leq i, j \leq m)$ polynomials in n variables $z_1, z_2, \dots, z_n$. If $T_1, T_2, \dots, T_n$ enjoy this inequality for any square matrix $\begin{pmatrix} p_{11} & \cdots & p_{1m} \\ \vdots & & \vdots \\ p_{m1} & \cdots & p_{mm} \end{pmatrix}$, then we will say in short that they enjoy the von Neumann inequality.

We will state in Theorem 3, together with a brief proof of it, that if commuting contractions $S_1, S_2, \dots, S_m$ enjoy the von Neumann inequality, commuting contractions T_1, T_2, $\dots$, T_n also enjoy the von Neumann inequality, and

$$S_1, S_2, \dots, S_m, T_1, T_2, \dots, T_n$$

are commuting, then they enjoy the von Neumann inequality provided that each S_i double commutes with every T_j and T_1, T_2, $\dots$, T_n generate an injective von Neumann algebra.

2. For commuting contractions mutual dependency of enjoying the von Neumann inequality and admitting simultaneous unitary dilations is known:

Theorem 2 (See [8], Theorem 1). *Let $T_1, T_2, \ldots, T_n$ be commuting contractions on a Hilbert space $\mathcal{H}$. Then the following are equivalent:*

(i) $T_1, T_2, \ldots, T_n$ *enjoy the von Neumann inequality.*

(ii) $T_1, T_2, \ldots, T_n$ *admit simultaneous unitary dilations, in the sense that there exist a Hilbert space $\mathcal{K}$ which contains $\mathcal{H}$ and commuting unitaries $U_1, U_2, \ldots, U_n$ on $\mathcal{K}$ such that for any integers $m_1, m_2, \ldots, m_n \geq 0$ we have*

$$T_1^{m_1} T_2^{m_2} \cdots T_n^{m_n} = P U_1^{m_1} U_2^{m_2} \cdots U_n^{m_n}|_{\mathcal{H}},$$

where P is the projection on $\mathcal{K}$ onto $\mathcal{H}$, the right hand side the restriction on $\mathcal{H}$ of the operator $P\ U_1^{m_1}\ U_2^{m_2}\ \cdots\ U_n^{m_n}$.

So, the Sz.-Nagy theorem which asserts that any contraction on a Hilbert space admits a unitary dilation should imply Theorem 1. On the other hand, the Andô theorem [1] which asserts that any two commuting contractions admit simultaneous unitary dilations enables us to state that any two commuting contractions on a Hilbert space enjoy the von Neumann iequality.

But things appear remarkably different in the case of n commuting contractions, $n \geq 3$. We in fact can find three commuting contractions which do not enjoy the von Neumann inequality (Crabb and Davie [4], Parrott [9], Varopoulos [12], [14]), and so, do not admit simultaneous unitary dilations.

We say that $T_1, T_2, \ldots, T_n$ double commute if T_i, T_j commute and T_i^*, T_j commute for each pair i and j, distinct. If the given commuting contractions double commute, then the matter stands clear. That is, if commuting contractions $T_1, T_2, \ldots, T_n$ double commute, then they admit simultaneous unitary dilations (See Sz.-Nagy-Foiaş [10]), and so, they enjoy the von Neumann inequality.

It is known, on the other hand, that commuting isometries always admit simultaneous unitary dilations (See Sz.-Nagy-Foiaş [10]) and enjoy the von Neumann inequality.

3. A linear map ϕ of a C^*-algebra A into a C^*-algebra B is said to be completely positive if the square matrix $(\phi(a_{ij}))$ is positive provided so is a square matrix (a_{ij}), $a_{ij} \in A$. A C^*-algebra A is said to be injective if for any C^*-algebras B and C, $B \subset C$, any completely positive map of B into A extends to a completely positive map of C into A.

A von Neumann algebra R on a Hilbert space said to be semidiscrete if the inequality

$$\left\| \sum_{j=1}^{k} X_j Y_j \right\| \leq \left\| \sum_{j=1}^{k} X_j \otimes Y_j \right\|$$

holds for any $X_1, X_2, \ldots, X_k$ in the commutant R' of R and $Y_1, Y_2, \ldots, Y_k$ in R, where the right hand side means the operator norm of the operator $\sum_{j=1}^{k} X_j \otimes Y_j$ on te tensor product space $\mathcal{H} \otimes \mathcal{H}$. It is certainly a very significant matter that a von Neumann algebra R is injective if and only if it is semidiscrete (Connes [3], Effros-Lance [5], Choi-Effros [2], and Wassermann [13]).

A von Neumann algebras of type I is known to be injective. A von Neumann algebra of finite dimension, and even that of approximately finite dimension, is known to be injective.

Theorem 3 (Cf. [8], Theorem 2) **.** *Suppose that commuting contractions $S_1, S_2, \ldots, S_m$ on a Hilbert space $\mathcal{H}$ enjoy the von Neumann inequality, commuting contractions T_1, T_2, $\ldots$, T_n on $\mathcal{H}$, which generate an injective von Neumann algebra on $\mathcal{H}$ also enjoy the von Neumann inequality, and S_i double commutes with T_j for any pair i, j. Then the contractions*

$$S_1, S_2, \ldots, S_m, T_1, T_2, \ldots, T_n$$

enjoy the von Neumann inequality.

In general, a linear map ϕ of a subspace of a C^*-algebra into a C^*-algebra is said to be completely contractive if the norm of the square matrix $(\phi(a_{ij}))$ does not exceed one whenever so does the the norm of a square matrix (a_{ij}), $a_{ij} \in A$. Given commuting contractions $T_1, T_2, \ldots, T_n$, they enjoy the von Neumann inequality if and only if the map

$$q \longrightarrow q(T_1, T_2, \ldots, T_n)$$

of the space P^n of polynomials in n variables $z_1, z_2, \ldots, z_n$ in the torus $\mathbf{T}$ into $B(\mathcal{H})$ is completely contractive.

Since each of the maps

$$\phi : p \longrightarrow p(S_1, S_2, \ldots, S_m)$$

of P^m into $B(\mathcal{H})$ and

$$\psi : q \longrightarrow q(T_1, T_2, \ldots, T_n)$$

of P^n into $B(\mathcal{H})$ is completely contractive, the map $\phi \otimes \psi$ must be completely contractive. But by the assumption, the von Neumann algebra generated by $T_1, T_2, \ldots, T_n$ is injective, so, the map

$$\sum_{j=1}^{k} X_j \otimes Y_j \longrightarrow \sum_{j=1}^{k} X_j Y_j,$$

where $X_1, X_2, \ldots, X_k \in R'$ and $Y_1, Y_2, \ldots, Y_k \in R$, extends to a homomorphism π of the minimal tensor product $R' \otimes R$ into $B(\mathcal{H})$. It is completely contractive. So the map

$$\Phi : r \longrightarrow r(S_1, S_2, \ldots, S_m, T_1, T_2, \ldots, T_n)$$

of P^{m+n} into $B(\mathcal{H})$, which is nothing but the composition of π and $\phi \otimes \psi$, turns out to be completely contractive. Thus we have shown Theorem 3.

A GCR-operator means an operator which generate a GCR-algebra (or a postliminal algebra) ([6], [8]). Any GCR-operator is necessarily nuclear ([11]) and generates an injective von Neumann algebra.

Corollary 1. *Suppose that commuting contractions $S_1, S_2, \ldots, S_m$ on a Hilbert space $\mathcal{H}$ enjoy the von Neumann inequality, and that commuting contractions $T_1, T_2, \ldots, T_n$ on $\mathcal{H}$ are GCR-operators. If each S_i double commutes with every T_j and each T_k double commutes with every T_l ($k \neq l$), then the contractions*

$$S_1, S_2, \ldots, S_m, T_1, T_2, \ldots, T_n$$

enjoy the von Neumann inequality.

Any quasi-normal operator, especially any normal operator, and any isometry, is necessarily a GCR-operator ([7]).

Since commuting compact contractions generate a GCR-algebra, we have the following

Corollary 2. *Suppose that commuting contractions $S_1, S_2, \ldots, S_m$ on a Hilbert space $\mathcal{H}$ enjoy the von Neumann inequality, and that commuting compact contractions T_1, T_2, ..., T_n on $\mathcal{H}$ also enjoy the von Neumann*

inequality. If each S_i *double commutes with every* T_j*, then the contractions*

$$S_1, S_2, \ldots, S_m, T_1, T_2, \ldots, T_n$$

enjoy the von Neumann inequality.

References

[1] T. Andô, On a pair of commuting contractions, Acta Sci. Math. (Szeged) 24 (1963), 88-90.

[2] M.-D. Choi and E. G. Effros, Separable nuclear C^*-algebras and injectivity, Duke Math. J. 43 (1976), 309-322.

[3] A. Connes, Classification of injective factors, Ann. Math. 104 (1976), 73-116.

[4] M. J. Crabb and A. M. Davie, von Neumann inequality for Hilbert space operators, Bull. London Math. Soc. 7 (1975), 49-50.

[5] E. G. Effros and C. Lance, Tensor products of operator algebras, Adv. Math. 25 (1977), 1-34.

[6] I. Kaplansky, The structure of certain operator algebras, Trans. Amer. Math. Soc. 70 (1971), 219-225.

[7] T.Okayasu, On GCR-operators, Tôhoku Math. J. 21 (1969), 573-579.

[8] T. Okayasu, The von Neumann inequality and dilation theorems for contractions, "Op. Theory and Complex Analy.", Sapporo, 1991, Op. Theory Adv. Appl. 59 (1992), 285-291.

[9] S. Parrott, Unitary dilations for commuting contractions, Pacif. J. Math. 34 (1970), 481-490.

[10] B. Sz.-Nagy and C. Foiaş, Harmonic analysis of operators on Hilbert space, North-Holland, Amsterdam, 1970.

[11] M. Takesaki, On the cross-norm of the direct product of C^*-algebras, Tôhoku Math. J. 16 (1964), 111-122.

[12] N. Th. Varopoulos, On an inequality of von Neumann and application of the metric theory of tensor products to operator theory, J. Funct. Analy. 16 (1974), 83-100.

[13] S. Wassermann, Injective W^*-algebras, Math. Proc. Camb. Phil. Soc. 82 (1977), 39-47.

[14] N. Th. Varopoulos, On a commuting family of contraction operators, Rev. Roumaine Pures Appl. 21 (1976), 1283-1285.

78 CQ*-ALGEBRAS OF OPERATORS AND APPLICATION TO QUANTUM MODELS

Camillo TRAPANI

Dipartimento di Scienze Fisiche ed Astronomiche, Università di Palermo
I-90123 Palermo, Italy

1 Introduction and motivations

C^*-algebras constitute, as known, the cornerstones of the Haag-Kastler [1] algebraic approach to quantum systems, with infinitely many degrees of freedom: one usually starts, in fact, with associating to each bounded region V of the configuration space of the system the C^*-algebra $\mathcal{A}_V$ of local observables in V. The uniform completion $\mathcal{A}$ of the algebra $\mathcal{A}_0$ generated by the $\mathcal{A}_V$'s is then considered as the C^*-algebra of observables of the system.

There are, however, several physical models that do not fit into the Haag-Kastler framework: this is the case, for instance, of certain quantum statistical models for which the thermodynamical limit of some local observables fails to exist in the uniform topology; or, even, of Wightman quantum field theory, where point-like fields are not, in general, elements of any C^*-algebra.

Nevertheless, the Haag-Kastler approach is so elegant and so powerful to make reasonable any effort directed to preserve its main features. One way of doing this consists in assuming that there exists on $\mathcal{A}_0$ a weaker norm such that the completion of $\mathcal{A}_0$ with respect to this new norm contains all 'objects' of physical interest. This assumption calls on the stage partial algebraic structures since the completion of a locally convex algebra, where the multiplication is not jointly continuous, is, in general, only a *partial *-algebra* (indeed, the product of two elements exists if one of them lies in the original algebra).

Partial *-algebras introduced originally by Antoine and Karwowski [2] have been studied by several authors (see [3, 4] for complete references). Among them, *quasi *-algebras* and *CQ*-algebras* are of particular interest due to their

H.G.W. Begehr et al. (eds.), Proceedings of the Second ISAAC Congress, Vol. 1, 679–685.

reasonably simple algebraic structure and to their good topological behavior. In this note, we will shortly describe the class of CQ^*-algebras and discuss their possible applications to some quantum models. The study of CQ^*-algebras has been carried out in a series of papers by F.Bagarello and the author [5, 6, 7] and is still in progress.

The author whishes to thank Prof. A.Inoue for the invitation to present this discussion at the 2nd ISAAC Conference.

2 CQ*-algebras

We start with the following

Definition 2.1 *Let $\mathcal{A}$ be a right Banach module over the C^*-algebra $\mathcal{A}_\flat$, with isometric involution $*$ and such that $\mathcal{A}_\flat \subset \mathcal{A}$. Set $\mathcal{A}_\sharp = (\mathcal{A}_\flat)^*$. We say that $\{\mathcal{A}, *, \mathcal{A}_\flat, \flat\}$ is a CQ^*-algebra if*

(i) $\mathcal{A}_\flat$ *is dense in* $\mathcal{A}$ *with respect to its norm* $\| \, \|$

(ii) $\mathcal{A}_0 := \mathcal{A}_\flat \cap \mathcal{A}_\sharp$ *is dense in* $\mathcal{A}_\flat$ *with respect to its norm* $\| \, \|_\flat$

(iv) $(AB)^* = B^* A^*, \quad \forall A, B \in \mathcal{A}_0$

(iii) $\|B\|_\flat = \sup_{A \in \mathcal{A}} \|AB\|, \quad B \in \mathcal{A}_\flat$

Since $*$ is isometric, the space $\mathcal{A}_\sharp$ is itself a C*-algebra with respect to the involution $X^\sharp := (X^*)^{\flat *}$ and norm $\| X \|_\sharp := \| X^* \|_\flat$.
A CQ*-algebra is called *proper* if $\mathcal{A}_\sharp = \mathcal{A}_\flat$.

A CQ*-algebra is a partial *-algebra: indeed the multiplication of two elements $A, B \in \mathcal{A}$ is defined if either $A \in \mathcal{A}_\sharp$ or $B \in \mathcal{A}_\flat$.

A way to construct CQ*-algebras is provided by the following proposition [5]:

Proposition 2.2 *Let $\mathcal{A}_\flat$ be a C^*-algebra, with norm $\| \, \| \, \flat$ and involution $\flat$; let $\| \, \|$ be another norm on $\mathcal{A}$, weaker than $\| \, \|_\flat$ and such that*

(i) $\| AB \| \leq \| A \| \| B \|_\flat \qquad \forall A, B \in \mathcal{A}_\flat$

(ii) there exists a $\| \, \|_\flat$-dense subalgebra $\mathcal{A}_0$ of $\mathcal{A}_\flat$ where an involution * is defined with the property

$$\| A^* \| = \| A \|, \quad \forall A \in \mathcal{A}_0$$

then if $\mathcal{A}$ denotes the $\| \, \|$-completion of $\mathcal{A}_\flat$, $(\mathcal{A}, *, \mathcal{A}_\flat, \flat)$ is a CQ^*-algebra.

If $\mathcal{A}_\flat = \mathcal{A}_0$ and $\flat = *$, then the CQ*-algebra we obtain is proper [5].

Thus, for instance, if I is a compact interval of $\mathbb{R}$, the space $L^p(I, dx)$ (dx the Lebesgue measure on I) is a proper CQ*-algebra over the C*-algebra $C(I)$ of all continuous function on I [7].

The most typical example of a non-commutative CQ*-algebra is provided by certain spaces of linear operators acting on a *triplet* of Hilbert spaces.

Let $\mathcal{H}$ be a Hilbert space with scalar product $(.,.)$ and S a positive selfadjoint *unbounded* operator, with dense domain $D(S)$. We assume that $S \geq \mathbb{I}$. Then $D(S)$ becomes a Hilbert space, denoted by $\mathcal{H}_{+1}$, with the scalar product

$$(f,g)_{+1} = (Sf, Sg) \tag{1}$$

If $\mathcal{H}_{-1}$ denote the Hilbert space which is the conjugate dual of $\mathcal{H}_{+1}$, we get, in canonical way, a scale of Hilbert spaces

$$\mathcal{H}_{+1} \xrightarrow{i} \mathcal{H} \xrightarrow{j} \mathcal{H}_{-1} \tag{2}$$

where i is the identity map i of $\mathcal{H}_{+1}$ into $\mathcal{H}$ and j is the canonical embedding of $\mathcal{H}$ into $\mathcal{H}_{-1}$ (both are continuous and with dense range).
With obvious identifications, we can read (2) as a chain of topological inclusions.
Let $\mathcal{B}(\mathcal{H}_{+1}, \mathcal{H}_{-1})$ be the Banach space of bounded operators from $\mathcal{H}_{+1}$ into $\mathcal{H}_{-1}$ with its natural norm $\|\cdot\|_{+1,-1}$. In this space an *isometric* involution $A \mapsto A^*$ can be defined by:

$$(A^*f, g) = \overline{(Ag, f)} \quad \forall f, g \in \mathcal{H}_{+1}$$

Let $\mathcal{B}(\mathcal{H}_{+1})$ be the C^*-algebra of all bounded operators on $\mathcal{H}_{+1}$. Its natural involution is denoted here as $\flat$ and its C*-norm as $\|\cdot\|_\flat$.
Furthermore, let $\mathcal{B}(\mathcal{H}_{-1})$ be the C^*-algebra of all bounded operators on $\mathcal{H}_{-1}$ with involution $\sharp$ and C*-norm $\|\cdot\|_\sharp$.
Then $\mathcal{B}(\mathcal{H}_{+1})$ and $\mathcal{B}(\mathcal{H}_{-1})$ are (isomorphic to) subspaces of $\mathcal{B}(\mathcal{H}_{+1}, \mathcal{H}_{-1})$, and $A \in \mathcal{B}(\mathcal{H}_{+1})$ if, and only if, $A^* \in \mathcal{B}(\mathcal{H}_{-1})$.
The set

$$\mathcal{B}^+(\mathcal{H}_{+1}) = \{A \in \mathcal{B}(\mathcal{H}_{+1}, \mathcal{H}_{-1}) \ : A, A^* \in \mathcal{B}(\mathcal{H}_{+1})\}$$

is a *-algebra in $\mathcal{B}(\mathcal{H}_{+1}, \mathcal{H}_{-1})$.
If $A \in \mathcal{B}(\mathcal{H}_{+1}, \mathcal{H}_{-1})$ and $B \in \mathcal{B}(\mathcal{H}_{+1})$, then AB is well-defined and $AB \in \mathcal{B}(\mathcal{H}_{+1}, \mathcal{H}_{-1})$. Analogously, if $C \in \mathcal{B}(\mathcal{H}_{-1})$, CA is a well-defined element of $\mathcal{B}(\mathcal{H}_{+1}, \mathcal{H}_{-1})$.
Therefore, $\mathcal{B}(\mathcal{H}_{+1}, \mathcal{H}_{-1})$ is a *right Banach module* over the C*-algebra $\mathcal{B}(\mathcal{H}_{+1})$. But it is never a CQ*-algebra, because the density conditions of Definition 2.1 are not satisfied [8].
However, by Proposition 2.2 we can always construct a CQ*-algebra of operators acting in a scale of Hilbert spaces. We denote this CQ*-algebra with $(\mathcal{B}_c(\mathcal{H}_{+1}, \mathcal{H}_{-1}), *, \mathcal{B}_c(\mathcal{H}_{+1}), \flat)$. The problem of characterizing elements of $(\mathcal{B}_c(\mathcal{H}_{+1}, \mathcal{H}_{-1})$ has been considered in [8] where it has been proved that it consists of the class of operators *compact with respect to S*.

Once we have at our disposal CQ*-algebras of operators, the question arises as to whether an abstract CQ*-algebra $(\mathcal{A}, *, \mathcal{A}_\flat, \flat)$ can be realized as a CQ*-algebra of operators by a GNS-like construction, starting from a conveniently chosen *state*.

Let $(\mathcal{A}, *, \mathcal{A}_\flat, \flat)$ be a CQ*-algebra. A linear functional and ω on $\mathcal{A}$ is called an *admissible q-state* if

(i) $\omega(X^*X) \geq 0, \quad \forall X \in \mathcal{A}_\flat$

(ii) $\omega(X^*AY) = \overline{\omega(Y^*A^*X)}, \quad \forall A \in \mathcal{A}, \forall X, Y \in \mathcal{A}_\flat$

(iii) $\omega(\mathbb{I}) = 1$

(iv) $\forall A \in \mathcal{A}, \exists K_A > 0 : |\omega(X^*AY)|^2 \leq K_A \omega(X^\flat X)\omega(Y^\flat Y)$,
$\forall X, Y \in \mathcal{A}_\flat$

(v) $\{X_n\} \subset \mathcal{A}_\flat$ sequence s.t. $\lim_{n\to\infty} \omega(X_n^* X_n) = 0$ and
$\omega((X_n - X_m)^\flat (X_n - X_m)) \to 0 \Rightarrow \lim_{n\to\infty} \omega(X_n^\flat X_n) = 0$

Condition *(iv)* allows the definition of the representation on the whole $\mathcal{A}$ and we get the following:

Proposition 2.3 *For each admissible q-state ω on a CQ*-algebra $(\mathcal{A}, *, \mathcal{A}_\flat, \flat)$ there exists a triplet of Hilbert spaces*

$$\mathcal{H}_\flat \subset \mathcal{H}_0 \subset \mathcal{H}_\sharp,$$

a vector $\xi_\flat \in \mathcal{H}_\flat$; a linear map:

$$\pi : \mathcal{A} \to \mathcal{B}(\mathcal{H}_\flat, \mathcal{H}_\sharp)$$

and a $\flat$-representation $\pi_\flat : \mathcal{A}_\flat \to \mathcal{B}(\mathcal{H}_\flat)$ of $\mathcal{A}_\flat$ in $\mathcal{H}_\flat$ such that:
i) $\pi(A^*) = \pi(A)^*, \quad \forall A \in \mathcal{A}$
ii)$\pi(AX) = \pi(A)\pi_\flat(X), \quad \forall A \in \mathcal{A}, X \in \mathcal{A}_\flat$
iii) $\omega(A) =< \pi(A)\xi_\flat, \xi_\flat >, \quad \forall A \in \mathcal{A}$
where $< .,. >$ denotes the (extension of the) inner product of $\mathcal{H}_0$.

3 Applications to Quantum Models

Let us now discuss some application of the structures introduced above to certain spin systems (for details, see [9]).

Let V be a finite region of a d-dimensional lattice and $|V|$ the number of points in V. It is well known that in this case the local C^*-algebra $\mathcal{A}_V$ is isomorphic to the C^*-algebra of all $2^{|V|} \times 2^{|V|}$-matrices on the $2^{|V|}$-dimensional complex Hilbert space $\mathcal{H}_V = \bigotimes_{p \in V} C_p^2$ (where C_p^2 is the 2-dimensional complex Hilbert space at p); indeed $\mathcal{A}_V$ is generated by the Pauli operators $\sigma_p = (\sigma_p^1, \sigma_p^2, \sigma_p^3)$ and by the unit 2×2 matrix I_p at every point $p \in V$.

For each unit vector $n = (n_1, n_2, n_3)$ in $\mathbb{R}^3$, define

$$(\sigma \cdot n) = n_1\sigma^1 + n_2\sigma^2 + n_3\sigma^3.$$

The spectrum of $\sigma \cdot n$ is the set $\{1, -1\}$; let $|n\rangle$ be a unit eigenvector associated with 1.

To each infinite sequence of unit vectors in $\mathbb{R}^3$, $\{n\} = \{n_1, n_2, \cdots\}$ it corresponds a unit vector $|\{n\}\rangle = \bigotimes_p |n_p\rangle$, in the infinite tensor product $\mathcal{H}_\infty = \bigotimes_p C_p^2$. We put

$$\mathcal{A}_0 = \bigcup_V \mathcal{A}_V \quad \text{and} \quad \mathcal{D}^0_{\{n\}} = \mathcal{A}_0|\{n\}\rangle$$

and we denote the closure of $\mathcal{D}^0_{\{n\}}$ in $\mathcal{H}_\infty$ by $\mathcal{H}_{\{n\}}$. Then, to any sequence $\{n\}$ of 3-vectors it corresponds a state $|\{n\}\rangle$ of the system. Each of these states defines a faithful representation $\pi_{\{n\}}(\mathcal{A}_0)$ in the Hilbert space $\mathcal{H}_{\{n\}}$ in the canonical way [10]. Without giving the full details, we mention that this is done by constructing a special basis for $\mathcal{H}_{\{n\}}$ by *flipping* a finite number of spins in the *ground* state $|\{n\}\rangle$. This leads to introduce a new *quantum number* m (namely, a sequence of 0's and 1's that labels the flipped spins in $|\{n\}\rangle$). It turns out that the set $\{|\{m\},\{n\}\rangle = \underset{p}{\otimes}|m_p,n_p\rangle;\ m_p = 0,1,\ \sum_p m_p < \infty\}$, obtained in this way, is an orthonormal basis in $\mathcal{H}_{\{n\}}$. In this space we define the unbounded self-adjoint operator M by

$$M|\{m\},\{n\}\rangle = \left(\sum_p m_p\right)|\{m\},\{n\}\rangle.$$

M counts the number of the flipped spins in $|\{m\},\{n\}\rangle$ with respect to the ground state $|\{n\}\rangle$. The representation $\pi_{\{n\}}$ is then defined on the basis vectors $\{|\{m\},\{n\}\rangle\}$ by

$$\pi_{\{n\}}(\sigma^i_p)\mid\{m\},\{n\}\rangle = \sigma^i_p \mid m_p,n_p\rangle \otimes (\prod_{p'\neq p} \otimes \mid m_{p'},n_{p'}\rangle) \quad (i=1,2,3).$$

and extended in obvious way to the whole space $\mathcal{H}_{\{n\}}$.

In view of the construction of a topology where thermodynamical limits can be handled, we need to chose an appropriate domain for the representation $\pi_{\{n\}}$. Since M is a number operator, then e^M is a densely defined self-adjoint operator. Let $\mathcal{D}$ denote its domain. Then $\mathcal{D}$ can be made into a Hilbert space, denoted as $\mathcal{H}_M$, in canonical way. The norm in $\mathcal{H}_M$ is given by

$$\| f \|_M = \| e^M f \|, \quad f \in \mathcal{H}_M.$$

Taking the conjugate dual $\mathcal{H}_{\overline{M}}$ of $\mathcal{H}_M$, with respect to the scalar product of $\mathcal{H}_{\{n\}}$, we get the triplet of Hilbert spaces

$$\mathcal{H}_M \subset \mathcal{H}_{\{n\}} \subset \mathcal{H}_{\overline{M}}$$

Let $(\mathcal{B}_c(\mathcal{H}_M,\mathcal{H}_{\overline{M}}), *, \mathcal{B}_c(\mathcal{H}_M), \flat)$ be the CQ*-algebra of bounded operators acting in the triplet.
The norm of $\mathcal{B}_c(\mathcal{H}_M,\mathcal{H}_{\overline{M}})$ can be written in terms of the norm in $\mathcal{B}(\mathcal{H}_{\{n\}})$ by

$$\| X \|_{M,\overline{M}} = \| e^{-M} X e^{-M} \|, \quad X \in \mathcal{B}_c(\mathcal{H}_M,\mathcal{H}_{\overline{M}}).$$

Similarly the norm in $\mathcal{B}_c(\mathcal{H}_M)$ becomes

$$\| X \|_M = \| e^M X e^{-M} \|, \quad X \in \mathcal{B}_c(\mathcal{H}_M).$$

The CQ*-norm

$$\| \pi_{\{n\}}(A) \|_{M,\overline{M}} = \| e^{-M} \pi_{\{n\}}(A) e^{-M} \|$$

is, in general, weaker from $\| A \|$.
As seen above, to each $\{n\}$ it corresponds a CQ*-algebra; therefore a convenient topology $\tau_{\mathcal{F}}$ on $\mathcal{A}_0$ can be defined by taking into account the whole class $\mathcal{F}$ of *relevant* states of the system. The family $\mathcal{F}$ is the key of the construction, when dealing with a concrete model and it depends, in general, on the specific form of the local Hamiltonian.
The topology $\tau_{\mathcal{F}}$ is defined by the family of seminorms (we explicitly mention the dependence of M on $\{n\}$):

$$A \in \mathcal{A}_0 \longrightarrow \| e^{-M_{\{n\}}} \pi_{\{n\}}(A) e^{-M_{\{n\}}} \|, \quad \{n\} \in \mathcal{F}.$$

Clearly, $\tau_{\mathcal{F}}$ is a projective limit of CQ*-norms. The completion of $\mathcal{A}_0[\tau_{\mathcal{F}}]$ will be denoted with $\mathcal{A}$. The following proposition is now obvious.

Proposition 3.1 *Let $\mathcal{A}$ denote the completion of $\mathcal{A}_0[\tau_{\mathcal{F}}]$. Then $(\mathcal{A}[\tau_{\mathcal{F}}], \mathcal{A}_0)$ is a topological quasi *-algebra.*

If H_V is the finite volume Hamiltonian of the system, we put, as usual,

$$\delta_V(A) = i[H_V, A], \quad A \in \mathcal{A}_0.$$

Proposition 3.2 *Let the finite volume Hamiltonian H_V of the system be a polynomial $p_V(S_V^i, \sigma_k^j)$ in the variables S_V^i with $S_V^i \in \mathcal{A}_0$, $i = 1, 2, \ldots, N$ and a continuous function of the σ_k^j's . Assume that*

(i) $\forall A \in \mathcal{A}_0$, $[H_V, A]$ depends on V only through the S_V^i's;

(ii) $\lim_{|V| \to \infty} (S_V^i)^n$ exists in $\mathcal{A}[\tau_{\mathcal{F}}]$, $\forall n \in \mathbb{N}$, $i = 1, 2, \ldots, N$.

Then for each $A \in \mathcal{A}_0$ the limit

$$\lim_{|V| \to \infty} \delta_V(A)$$

*exist in $\mathcal{A}[\tau_{\mathcal{F}}]$. Setting $\delta(A) = \tau_{\mathcal{F}} - \lim_{|V| \to \infty} \delta_V(A)$, $A \in \mathcal{A}_0$, then δ is a *-derivation of $\mathcal{A}_0$.*

Of course, we expect δ to be the generator of a one-parameter group of automorphisms of the completion of $\mathcal{A}_0$ in a suitably chosen topology. In order to get this, additional conditions on H_V are needed.

Proposition 3.3 *Let the local Hamiltonian H_V of a spin model be of the form*

$$H_V = |V| p(S_V^1, S_V^2, S_V^3)$$

where p is a polynomial and

$$S_V^i = \frac{1}{|V|} \sum_{p \in V} \sigma_p^i, \quad i = 1, 2, 3.$$

Then, for each $A \in \mathcal{A}_0$, there exists a positive number r_A such that the local dynamics

$$\alpha_V^t(A) = e^{iH_V t} A e^{-iH_V t}$$

converges in the topology $\tau_{\mathcal{F}}$, when $|V| \to \infty$, to a limit which we denote by $\alpha^t(A)$, for any t with $|t| < r_A$.

The assumptions made on H_V are fulfilled by several interesting models such as the mean field spin models [?, BCS1] For the Heisenberg model, for instance, the local Hamiltonian has the form

$$H_V = \frac{J}{|V|} \sum_{p,q \in V} \sum_{i=1}^{3} \sigma_p^i \sigma_q^i, \tag{3}$$

and the operators S_V^i are defined as

$$S_V^i = \frac{1}{|V|} \sum_{p \in V} \sigma_p^i, \quad i = 1, 2, 3.$$

As a conclusion, we remark that many alternative ways of overcoming the problems posed by C*-approach, have been successfully proposed in the literature; these methods are, however mostly dependent on the specific model. This is not the case of the approach outlined here, which, for this reason, seem to be closer to the original Haag-Kastler set-up.

References

[1] R.Haag and D.Kastler, *An Algebraic Approach to Quantum Field Theory* J.Math.Phys. **5**, 848-861 (1964)

[2] J.-P.Antoine and W.Karwowski,*Partial *-Algebras of Closed Linear Operators in Hilbert Space*, Publ.RIMS, Kyoto Univ. **21**, 205-236 (1985)

[3] J.-P.Antoine, A.Inoue and C.Trapani, *Partial *-Algebras of Closable Operators: a review*, Rev. Math. Phys., Rev. Math. Phys. **8**, 1-42 (1996)

[4] J.-P.Antoine, F.Bagarello and C.Trapani, *Topological partial *-algebras: Basic properties and examples*,Rev. Math. Phys. **11** 267-302 (1999)

[5] F.Bagarello and C.Trapani,*States and representations of CQ* -algebras*, Ann. Inst. H. Poincaré **61**, 103-133(1994)

[6] F.Bagarello and C.Trapani, *CQ*-algebras: structure properties* Publ. RIMS, Kyoto Univ. **32**, 85-116 (1996)

[7] F.Bagarello and C.Trapani, *L^p-spaces as quasi *-algebras*, J.Math. Anal. Appl . **197**, 810-824 (1996)

[8] K.D.Kürsten and C.Trapani,*On certain Banach Modules of Operators in Hilbert spaces triplets*, in preparation

[9] F.Bagarello and C.Trapani, *The Heisenberg Dynamics of Spin Systems:A Quasi *-Algebras approach*, J. Math. Phys. **37**, 4219-4234 (1996)

[10] W.Thirring and A.Wehrl, *On the Mathematical Structure of the B.C.S.-Model*, Commun.Math.Phys. **4**, 303-314 (1967)

79 ON STRUCTURE OF C*-LIKE LOCALLY CONVEX *-ALGEBRAS

Atsushi INOUE

Department of Applied Mathematics
Fukuoka University
Nanakuma, Jonan-ku
Fukuoka 814-0180, Japan

a-inoue@fukuoka-u.ac.jp

Klaus-Detlef KÜRSTEN

Fakultät für Mathematik und
Informatik, Mathematisches Institut
Universität Leipzig
Augustusplatz 10, D04109
Leipzig, Germany

kuersten@mathematik.uni-leipzig.de

Abstract. The main purpose of this note is to study the structure of a class of locally convex $*$-algebras which behave like C*-algebras.

1. Introduction

In this note we define the notion of C*-like locally convex $*$-algebras and introduce some results obtained in [15] about structure of such a $*$-algebra.

A *locally convex $*$-algebra* means that it is a Hausdorff locally convex space and a $*$-algebra whose multiplication is separately continuous and the involution is continuous. In this note we assume that a locally convex $*$-algebra has identity *1*. The studies of locally convex $*$-algebras were begun with those of locally multiplicatively convex $*$-algebras by R. Arens [3] and E. Michael [17]. A locally convex $*$-algebra $\mathcal{A}$ is said to be *multiplicatively convex* (simply, m-convex) if there exists a family

H.G.W. Begehr et al. (eds.), Proceedings of the Second ISAAC Congress, Vol. 1, 687–696.

$\{p_\lambda\}_{\lambda \in \Lambda}$ of m^*-seminorms p_λ on $\mathcal{A}$ (that is, $p_\lambda(xy) \leq p_\lambda(x)p_\lambda(y)$ and $p_\lambda(x^*) = p_\lambda(x), {}^\forall x, y \in \mathcal{A}$) determining the topology, and if any p_λ is a C*-seminorm (that is, $p_\lambda(x^*x) = p_\lambda(x)^2, {}^\forall x \in \mathcal{A}$) then $\mathcal{A}$ is said to be a *locally C*-convex algebra.* A complete locally C*-convex algebra is also said to be a *pro-C*-algebra.* It is well-known that every complete locally m-convex $*$-algebra (pro-C*-algebra) is a projective limit of Banach $*$-algebras (C*-algebras). Pro C*-algebras have been studied from the mathematical interest and the physical applications [12, 18]. The studies of general locally convex $*$-algebras which are *not m-convex* were done by R. Arens [4], W. Zelazko [25], G. R. Allan [1, 2], P. G. Dixon [9] and etc. In particular, G. R. Allan practiced the systematic studies of such $*$-algebras. In [2] he defined the notion of GB*-algebras and investigated the structure of commutative GB*-algebras, and P. G. Dixon [9] investigated the structure of non-commutative GB*-algebras. Here we simply state the definition of GB*-algebras.

Let $\mathcal{A}$ be a locally convex $*$-algebra. We denote by $\mathcal{B}^*$ the collection of closed, bounded absolutely convex subsets $\mathbf{B}$ of $\mathcal{A}$ satisfying $1 \in \mathbf{B}, \mathbf{B}^* = \mathbf{B}$ and $\mathbf{B}^2 \subset \mathbf{B}$. For every $\mathbf{B} \in \mathcal{B}^*$, the linear span $\mathcal{A}[\mathbf{B}]$ of $\mathbf{B}$ forms a normed $*$-algebra equipped with the Minkowski functional $\| \; \|_\mathbf{B}$ of $\mathbf{B}$. If $\mathcal{A}[\mathbf{B}]$ is complete for every $\mathbf{B} \in \mathcal{B}^*$, then $\mathcal{A}$ is said to be *pseudo-complete.* If $\mathcal{A}$ is sequentially complete, then $\mathcal{A}$ is pseudo-complete. A pseudo-complete locally convex $*$-algebra $\mathcal{A}$ is said to be a *GB*-algebra* if $\mathcal{B}^*$ has the greatest member $\mathbf{B}_0$ and $(1 + x^*x)^{-1} \in \mathcal{A}[\mathbf{B}_0]$ for every $x \in \mathcal{A}$. If $\mathcal{A}$ is a GB*-algebra, then $\mathcal{A}[\mathbf{B}_0]$ is a C*-algebra. For more details refer to [2, 9]. It is difficult to study general locally convex $*$-algebras which are not m-convex even if the multiplication is jointly continuous, and so after the Allan and Dixon studies, it seems that the systematic studies of general locally convex $*$-algebras have not almost been developed.

Recently we have studied unbounded m^*-(or C*-) seminorms and tried to apply them to studies of locally convex $*$-algebras in [6, 7, 15]. Let $\mathcal{A}$ be a (locally convex) $*$-algebra. An m^*-(or C*-) seminorm p on a $*$-subalgebra $\mathcal{D}(p)$ of $\mathcal{A}$ is said to be an *unbounded m^*-(or C*-) seminorm* on $\mathcal{A}$. We give simple examples of unbounded C*-seminorms:

(1) Let $\mathcal{A}$ be a locally m-convex (or C*-convex) $*$-algebra and $\Gamma = \{p_\lambda\}_{\lambda \in \Lambda}$ a family of seminorms on $\mathcal{A}$ determining the topology. We put

$$\begin{cases} \mathcal{D}(p_\Gamma) = \{x \in \mathcal{A}; \; \sup_{\lambda \in \Lambda} p_\lambda(x) < \infty\}, \\ p_\Gamma(x) = \sup_{\lambda \in \Lambda} p_\lambda(x), \; x \in \mathcal{D}(p_\Gamma). \end{cases}$$

Then p_Γ is an unbounded m^*-(or C*-)norm on $\mathcal{A}$.

(2) Let $\mathcal{A}$ be a GB*-algebra over $\mathbf{B}_0$ ($\mathbf{B}_0$ is the greatest member in $\mathcal{B}^*$). We put

$$\begin{cases} \mathcal{D}(p_{\mathbf{B}_0}) = \mathcal{A}[\mathbf{B}_0], \\ p_{\mathbf{B}_0}(x) = \|x\|_{\mathbf{B}_0}, \; x \in \mathcal{D}(p_{\mathbf{B}_0}). \end{cases}$$

Then $p_{\mathbf{B}_0}$ is an unbounded C*-norm on $\mathcal{A}$.

(3) Let $\mathcal{D}$ be a dense subspace in a Hilbert space $\mathcal{H}$. We denote by $\mathcal{L}^\dagger(\mathcal{D})$ the set of all linear operators X from $\mathcal{D}$ to $\mathcal{D}$ such that $\mathcal{D}(X^*) \supset \mathcal{D}$ and $X^*\mathcal{D} \subset \mathcal{D}$. Then $\mathcal{L}^\dagger(\mathcal{D})$ is a $*$-algebra under the usual operations: $X+Y, \alpha X, XY$ and the involution $X^\dagger \equiv X^* \lceil \mathcal{D}$. A $*$-subalgebra $\mathcal{M}$ of $\mathcal{L}^\dagger(\mathcal{D})$ containing the identity operator I is said to be an *O^*-algebra* on $\mathcal{D}$ in $\mathcal{H}$ [16, 19, 22]. An O*-algebra $\mathcal{M}$ is said to be an *extended C*-algebra* if $\overline{\mathcal{M}_b}$ is a C*-algebra and $(I + X^*\overline{X})^{-1} \in \overline{\mathcal{M}_b}$ for each $X \in \mathcal{M}$, where $\mathcal{M}_b = \{X \in \mathcal{M}; X \text{ is bounded }\}$ [10, 14]. Let $\mathcal{M}$ be an O*-algebra on $\mathcal{D}$ in $\mathcal{H}$. We put

$$\begin{cases} \mathcal{D}(r_u) = \mathcal{M}_b, \\ r_u(X) = \|\overline{X}\| \;\; \text{(operator norm)} \;, X \in \mathcal{D}(r_u). \end{cases}$$

Then r_u is an unbounded C*-norm on $\mathcal{M}$.

(4) Let $\mathcal{A}$ be a $*$-algebra. A $*$-homomorphism π of $\mathcal{A}$ into an O*-algebra $\mathcal{L}^\dagger(\mathcal{D})$ satisfying $\pi(1) = I$ is said to be a *$*$-representation* of $\mathcal{A}$. Here we denote by $\mathcal{D}(\pi)$ and $\mathcal{H}_\pi$ the domain $\mathcal{D}$ and the Hilbert space $\mathcal{H}$, respectively. We put

$$\begin{cases} \mathcal{D}(r_\pi) = \mathcal{A}_b^\pi \equiv \{x \in \mathcal{A}; \; \pi(x) \text{ is bounded }\}, \\ r_\pi(x) = \|\overline{\pi(x)}\|, \; x \in \mathcal{D}(r_\pi). \end{cases}$$

Then r_π is an unbounded C*-seminorm on $\mathcal{A}$.

As seen in the above examples, the studies of unbounded C*-seminorms may be useful to those of the structure theory and the representation theory of locally convex $*$-algebras. In this note we introduce such one case.

We define the notion of M*-(or C*-)like locally convex $*$-algebras. Let $\mathcal{A}[\tau]$ be a locally convex $*$-algebra. A directed family $\Gamma = \{p_\lambda\}_{\lambda \in \Lambda}$ of seminorms on $\mathcal{A}$ determining the topology τ is said to be *M^*-like* if for any $\lambda \in \Lambda$ there exists $\lambda' \in \Lambda$ such that

$$p_\lambda(xy) \leq p_{\lambda'}(x) p_{\lambda'}(y) \text{ and } p_\lambda(x^*) \leq p_{\lambda'}(x), \quad {}^\forall x, y \in \mathcal{A}$$

and further, if

$$p_\lambda(x)^2 \leq p_{\lambda'}(x^*x), {}^\forall x \in \mathcal{A},$$

then Γ is said to be *C^*-like.* This means that each p_λ is not necessarily m-convex (or C*-convex), but $p_\Gamma = \sup_{\lambda \in \Lambda} p_\lambda$ is an m^*-(or C*-)norm on the $*$-subalgebra $\mathcal{D}(p_\Gamma) \equiv \{x \in \mathcal{A};\ \sup_{\lambda \in \Lambda} p_\lambda(x) < \infty\}$ of $\mathcal{A}$, that is, it is an unbounded m^*-(or C*-)norm on $\mathcal{A}$.

Definition 1.1. A locally convex $*$-algebra $\mathcal{A}[\tau]$ is said to be *M^*-like* (or *C^*-like*) if it is complete and there exists a M*-like (or C*-like) directed family $\Gamma = \{p_\lambda\}_{\lambda \in \Lambda}$ of seminorms on $\mathcal{A}$ determining the topology τ such that $\mathcal{D}(p_\Gamma)$ is τ-dense in $\mathcal{A}$.

Every Fréchet $*$-algebra $\mathcal{A}$ has an M*-like sequence of seminorms. In fact, $\mathcal{A}$ has a sequence $\{p_n\}_{n \in \mathbb{N}}$ of seminorms determining the topology satisfying $p_n(xy) \leq p_{n+1}(x)p_{n+1}(y)$ and $p_n(x^*) = p_n(x)$ for each $x, y \in \mathcal{A}$ and $n \in \mathbb{N}$ [25]. The Arens algebra $L^\omega[0,1] \equiv \bigcap_{1 \leq p < \infty} L^p[0,1]$ [4] is a C*-like locally convex $*$-algebra.

2. Structure of C*-like locally convex $*$-algebras

We study the structure C*-like locally convex $*$-algebras. Let $\mathcal{A}[\tau]$ be a complete locally convex $*$-algebra having a C*-like family $\Gamma = \{p_\lambda\}_{\lambda \in \Lambda}$ of seminorms on $\mathcal{A}$. Then we have the following

Lemma 2.1. (1) $p_\Gamma = \sup_{\lambda \in \Lambda} p_\lambda$ is an unbounded C*-norm on $\mathcal{A}$ such that the topology defined by p_Γ is finer than τ on $\mathcal{D}(p_\Gamma)$.

(2) We put

$$\mathcal{D}(p_\Gamma)_{\mathit{1}} = \{\alpha \mathit{1} + x;\ \alpha \in \mathbb{C}, x \in \mathcal{D}(p_\Gamma)\}.$$

Then $\mathcal{D}(p_\Gamma)_{\mathit{1}}$ is a unital C*-algebra with the C*-norm:

$$p_\Gamma^{\mathit{1}}(\alpha \mathit{1} + x) \equiv \sup\{p_\Gamma((\alpha \mathit{1} + x)y); y \in \mathcal{D}(p_\Gamma) \text{ s.t. } p_\Gamma(x) \leq 1\}.$$

(3) We denote by $\mathcal{U}(p_\Gamma)$ and $\mathcal{U}(p_\Gamma^{\mathit{1}})$ the unit balls of the C*-algebra $\mathcal{D}(p_\Gamma)$ and $\mathcal{D}(p_\Gamma)_{\mathit{1}}$, respectively. Then $\mathcal{U}(p_\Gamma^{\mathit{1}}) \in \mathcal{B}^*$ and $\mathcal{A}[\mathcal{U}(p_\Gamma^{\mathit{1}})] = \mathcal{D}(p_\Gamma)_{\mathit{1}}$.

We remark that $\mathcal{B}^*$ does not necessarily the greatest member, and if it has the greatest member $\mathbf{B}_0$, that is, $\mathcal{A}$ is a GB*-algebra, then $\mathcal{A}[\mathcal{U}(p_\Gamma^1)] = \mathcal{D}(p_\Gamma)_1 \subset \mathcal{A}[\mathbf{B}_0]$ and $p_\Gamma^1(x) = \|x\|_{\mathbf{B}_0}$ for each $x \in \mathcal{D}(p_\Gamma)_1$, but $\mathcal{D}(p_\Gamma)_1 \subsetneq \mathcal{A}[\mathbf{B}_0]$ in general. We consider when $\mathcal{A}$ is a GB*-algebra having $\mathbf{B}_0 = \mathcal{U}(p_\Gamma^1)$. For this question we have the following

Theorem 2.2. Let $\mathcal{A}$ be a complete locally convex $*$-algebra having a C*-like family $\Gamma = \{p_\lambda\}_{\lambda \in \Lambda}$ of the seminorms. Then the following statements are equivalent:

(i) $\mathcal{A}$ is a C*-like locally convex $*$-algebra.

(ii) $1 \in \mathcal{U}(p_\Gamma), (1+h^2)^{-1}$ and $h(1+h^2)^{-1}$ belong to $\mathcal{U}(p_\Gamma)$ for each $h^* = h \in \mathcal{A}$.

(iii) $\mathcal{A}$ is a GB*-algebra having $\mathbf{B}_0 = \mathcal{U}(p_\Gamma)$.

For this proof refer to Theorem 2.1 in [15]. By Theorem 2.2 we see that a C*-like locally convex $*$-algebra is independent for the method of taking the C*-like families. If $\mathcal{A}[\tau]$ is a locally convex $*$-algebra with a C*-like family $\Gamma = \{p_\lambda\}_{\lambda \in \Lambda}$ of seminorms such that $\mathcal{D}(p_\Gamma)$ is τ-dense in $\mathcal{A}$, then the completion $\overline{\mathcal{A}}[\tau]$ of $\mathcal{A}[\tau]$ is a C*-like locally convex $*$-algebra. Hence, the assumption of completeness of the locally convex $*$-algebra in Theorem 2.1 is not essential. By Theorem 2.2 and the Allan result ([2] Theorem 3.9) every commutative C*-like locally convex $*$-algebra is isomorphic to a $*$-algebra of continuous functions on a compact space which may be infinite on nowhere dense subset. In more detail we state this.

We denote by $\mathbb{C}^*$ the one point compactification of the complex plane $\mathbb{C}$, obtained by adjoining the point ∞. A partial algebraic structure is defined on $\mathbb{C}^*$ by : $\infty + \lambda = \infty$ $(\lambda \in \mathbb{C})$, $\infty \cdot \lambda = \infty$ $(\lambda \in \mathbb{C}^*,\ \lambda \neq 0)$ and $\overline{\infty} = \infty$.

Definition 2.3. A family $\mathcal{F}$ of continuous $\mathbb{C}^*$-valued functions on a topological space M is called a *$*$-algebra of $\mathbb{C}^*$-valued continuous functions* if the following conditions are satisfied:

(i) each $f \in \mathcal{F}$ takes the valued ∞ on at most a nowhere dense subset of M;

(ii) for any $f, g \in \mathcal{F}$ and $\alpha \in \mathbb{C}$, the funtions $f+g, \alpha f, fg, f^* (= \overline{f})$, defined pointwise on the dense subset of M where f and g are both finite, are extendible (necessarily uniquely) to continuous $\mathbb{C}^*$-valued functions on M which also belong to $\mathcal{F}$. A $*$-algebra of $\mathbb{C}^*$-valued continuous functions is said to be C*-*like* if it is a C*-like locally convex $*$-algebra.

By Theorem 2.2 and ([2] Theorem 3.9) we have the following

Theorem 2.4. Every commutative C*-like locally convex $*$-algebra is isomorphic to a C*-like algebra of $\mathbb{C}^*$-valued continuous functions on a compact space.

We next consider the noncommutative case.

Definition 2.5. An O*-algebra is said to be *C*-like* if it is a C*-like locally convex $*$-algebra.

For the relation of C*-like O*-algebras and extended C*-algebras we have the following

Proposition 2.6. Every C*-like O*-algebra is an extended C*-algebra.

For this proof refer to ([15] Proposition 2.6). By Theorem 2.2 and the P. G. Dixon result ([9] Theorem 7.13) we have the following

Theorem 2.7. Every C*-like locally convex $*$-algebra is isomorphic to a C*-like O*-algebra.

3. Examples

In this section we give some examples of C*-like locally convex $*$-algebras.

Example 3.1. Every pro-C*-algebra is a C*-like locally convex $*$-algebra.

Example 3.2. (1) The function algebra $C(\mathbb{R})$ of continuous $\mathbb{C}$-valued functions on $\mathbb{R}$ equipped with the topology defined by a family $\Gamma = \{p_n; n \in \mathbb{N}\}$ of C*-seminorms : $p_n(f) = \sup\{|f(t)|; -n \leq t \leq n\}$ is a pro-C*-algebra such that $\mathcal{D}(p_\Gamma)$ equals the C*-algebra $C_b(\mathbb{R}) \equiv \{f \in C(\mathbb{R}); f \text{ is bounded }\}$ and $p_\Gamma(f) = \sup\{|f(t)|; t \in \mathbb{R}\}$, $f \in C_b(\mathbb{R})$.

(2) Let $\{h_n\}$ be a sequence of continuous $\mathbb{C}^*$-valued functions on a compact space M such that $1 \leq h_1 \leq h_2 \leq \dots$ and $^\forall n \in \mathbb{N}, ^\exists n_1 \in \mathbb{N}$

s.t. $h_{n+1} \leq h_n^2 \leq h_{n_1}$. We denote by $\mathcal{F}_{h_n}$ the set of all continuous $\mathbb{C}^*$-valued functions f on M such that $\sup\{|f(t)|/h_n(t); t \in M\} < \infty$. Then $\mathcal{F}(\{h_n\}) \equiv \bigcup_{n=1}^{\infty} \mathcal{F}_{h_n}$ is a locally convex ∗-algebra with the C*-like sequence $\Gamma = \{p_n; n \in \mathbb{N}\}$ of seminorms on $\mathcal{F}(\{h_n\})$ defined by $p_n(f) = \sup\{e^{-\frac{h_1(t)}{n}}|f(t)|; t \in M\}$, $n \in \mathbb{N}$, and $\mathcal{D}(p_\Gamma) = C(M)$ and $p_\Gamma(f) = \sup\{|f(t)|; t \in M\}$, $f \in \mathcal{D}(p_\Gamma)$. Hence, the completion of $\overline{\mathcal{F}}(\{h_n\})$ of $\mathcal{F}(\{h_n\})$ is a C*-like locally convex ∗-algebra.

Example 3.3. (1) Let $(X, \mathcal{B}, \mu)$ be a finite measure space. Then the function algebra $L^\omega(X, \mathcal{B}, \mu) \equiv \bigcap_{1 \leq p < \infty} L^p(X, \mathcal{B}, \mu)$ is a C*-like locally convex ∗-algebra with a C*-like family $\{\mu(X)^{-1/p}\| \ \|_p; 1 \leq p < \infty\}$ of norms. In particular, the Arens algebra $L^\omega[0,1] \equiv \bigcap_{1 \leq p < \infty} L^p[0,1]$ is a C*-like locally convex ∗-algebra.

(2) We consider the noncommutative case. Let $\mathcal{M}$ be a von Neumann algebra with a faithful finite trace τ and $L^p(\tau)$ the Segal L^p-space with respect to τ [20]. Then $L^\omega(\tau) \equiv \bigcap_{1 \leq p < \infty} L^p(\tau)$ is a C*-like locally convex ∗-algebra with the C*-like family $\{\tau(I)^{-1/p}\| \ \|_p; 1 \leq p < \infty\}$ of norms [14].

Example 3.4. Here we consider when an O*-algebra is C*-like. Let $\mathcal{M}$ be an O*-algebra on $\mathcal{D}$ in $\mathcal{H}$. The topology $t_\mathcal{M}$ on $\mathcal{D}$ defined by a family $\{\| \ \|_X; X \in \mathcal{M}\}$ of seminorms : $\|\xi\|_X \equiv \|X\xi\|, \xi \in \mathcal{D}$ is called the *graph topology*. If $\mathcal{D}[t_\mathcal{M}]$ is complete, then $\mathcal{M}$ is said to be *closed*. Let $\mathcal{M}$ be a closed O*-algebra on $\mathcal{D}$ in $\mathcal{H}$. We denote by $\mathfrak{S}$ the set of all bounded subsets of $\mathcal{D}[t_\mathcal{M}]$. The topology on $\mathcal{M}$ determined by a family $\{p_\mathfrak{M}; \mathfrak{M} \in \mathfrak{S}\}$ is called the *uniform topology*, where $p_\mathfrak{M}$ is a seminorm on $\mathcal{M}$ defined by $p_\mathfrak{M}(X) = \sup_{\xi,\eta \in \mathfrak{M}} |(X\xi|\eta)|, X \in \mathcal{M}$. Since $\mathcal{M}$ is closed, it follows that $\mathfrak{M}$ is $t_\mathcal{M}$-bounded if and only if $t_{\mathcal{L}^\dagger(\mathcal{D})}$-bounded, so that the uniform topology is independent for closed O*-algebras on $\mathcal{D}$. Hence we denote it by $\tau_\mathcal{D}$.

(1) Suppose that there exists a direct set $\{\mathcal{H}_\lambda\}_{\lambda \in \Lambda}$ of closed subspaces of $\mathcal{H}$ such that

(i) $\mathcal{H}_\lambda \subset \mathcal{D}, \quad {}^\forall\lambda \in \Lambda$;

(ii) $\overline{\mathcal{M}\mathcal{H}_\lambda}$ (the norm closure) $\subset \mathcal{D}, \quad {}^\forall\lambda \in \Lambda$;

(iii) $^\forall\mathfrak{M} \in \mathfrak{S}, ^\exists\lambda \in \Lambda; \mathfrak{M} \subset \mathcal{H}_\lambda$.

Then, the completion $\overline{\mathcal{M}}[\tau_{\mathcal{D}}]$ of $\mathcal{M}[\tau_{\mathcal{D}}]$ is a C*-like O*-algebra.

(2) Let $\mathcal{A}[\tau]$ be a locally convex $*$-algebra with a family $\{p_\lambda\}_{\lambda\in\Lambda}$ of seminorms determining the topology satisfying $p_\lambda(x^*x) \leq p_\lambda(x)^2$ for each $\lambda \in \Lambda$ and $x \in \mathcal{A}$. Let π be a $*$-representation of $\mathcal{A}$ such that the O*-algebra $\pi(\mathcal{A})$ is closed. Suppose that π is uniformly continuous, that is, π is continuous from $\mathcal{A}[\tau]$ to $\pi(\mathcal{A})[\tau_{\mathcal{D}(\pi)}]$. Then the completion $\overline{\pi(\mathcal{A})}[\tau_{\mathcal{D}(\pi)}]$ of the O*-algebra $\pi(\mathcal{A})[\tau_{\mathcal{D}(\pi)}]$ is a C*-like O*-algebra.

(3) Let $\mathcal{M}$ be a closed O*-algebra on $\mathcal{D}$ in $\mathcal{H}$. Suppose that there exists a family $\mathfrak{S}_0$ of $t_{\mathcal{M}}$-bounded subsets of $\mathcal{D}$ such that

(i) $\Gamma \equiv \{p_{\mathfrak{M}}; \mathfrak{M} \in \mathfrak{S}_0\}$ determines the uniform topology $\tau_{\mathcal{D}}$;

(ii) for any $\mathfrak{M} \in \mathfrak{S}_0$ there exists an element $\mathfrak{M}'$ of $\mathfrak{S}_0$ such that

$$p^{\mathfrak{M}}(X) \equiv \sup_{\xi\in\mathfrak{M}} \|X\xi\| \leq p_{\mathfrak{M}'}(X) = \sup_{\xi,\eta\in\mathfrak{M}'} |(X\xi|\eta)|, \quad X \in \mathcal{M};$$

(iii) $\mathcal{D}(p_\Gamma) \equiv \{X \in \mathcal{M}; \sup_{\mathcal{M}\in\mathfrak{S}_0} p^{\mathfrak{M}}(X) < \infty\}$ is $\tau_{\mathcal{D}}$-dense in $\mathcal{M}$.

Then $\overline{\mathcal{M}}[\tau_{\mathcal{D}}]$ is a C*-like O*-algebra.

References

[1] G. R. Allan, A spectral theory for locally convex algebras, Proc. London Math. Soc. **15**(1965), 399-421.

[2] G. R. Allan, On a class of locally convex algebras, Proc. London Math. Soc. **17**(1967), 91-114.

[3] R. Arens, A generalization of normed rings, Pacific J. Math. **2**(1952), 455-471.

[4] R. Arens, The space L^w and convex topological rings, Bull. Amer. Math. Soc. **52**(1946), 931-935.

[5] S. J. Bhatt, A note on generalized B*-algebras I, II, J. Indian Math. Soc. **43**(1979), 253-257: ibid **44**(1980), 285-290.

[6] S. J. Bhatt, A. Inoue and H. Ogi, Unbounded C*-seminorms and unbounded C^*-spectral algebras, preprint(Fukuoka University), 1998.

[7] S. J. Bhatt, A. Inoue and H. Ogi, C*-spectrality and spectral invariance in locally convex $*$-algebras, preprint (Fukuoka Univ.), 1999.

[8] R. M.BROOKS, On Locally m-convex $*$-algebras, Pacific J. Math. **25**(1967), 5-23.

[9] P. G. DIXON, Generalized B*-algebras I, II, *Proc. London Math. Soc.* **21**(1970), 693-715: *J. London Math. Soc.* **5**(1972), 159-165.

[10] P. G. DIXON, Unbounded operator algebras, *Proc. London Math. Soc.* **23**(1971), 53-69.

[11] J. DIXMIER, C*-algebras. *North-Holland*, 1977.

[12] M. DUBOISE-VIOLLETTE, A generalization of the classical moment problem on $*$-algebras with applications to relativistic quantum filed theory I, II, Commun. Math. Phys. **43**(1975), 225-254.: ibid **54**(1977), 151-172.

[13] M. FRAGOULOPOULOU, An introduction to the representation theory of topological $*$-algebras, Schriftenreihe des Math. Instituts der Univ. Münster **48** (1988), 1-81.

[14] A. INOUE, On a class of unbounded operator algebras I, II, III, *Pacific J. Math.* **65** (1976), 77-95: ibid **66** (1976), 411-431: ibid **69** (1977), 105-115.

[15] A. INOUE and K. D. KÜRSTEN, On C*-like locally convex $*$-algebras, Preprint (Fukuoka Univ.), 1999.

[16] G. LASSNER, Topological algebras of operators, *Rep. Math. Phys.* **3**(1972), 279-293.

[17] E. MICHAEL, Locally multiplicatively convex topological algebras, Mem. Amer. Math. Soc. **11** (1952).

[18] N. C. PHILLIPS, Inverse limit of C*-algebras, J. Operator Theory **19** (1988), 159-195.

[19] R.T. POWERS, Self-adjoint algebras of unbounded operators, *Commun. Math. Phys.* **21**(1971), 85-124.

[20] I.E. SEGAL, A non-commutative extension of abstract integration, *Ann. Math.* **57**(1953), 401-457.

[21] K. SCHMÜDGEN, Lokal multiplikativ konvexe O_p^*-algebren, *Math. Nach.* **85**(1978), 161-170.

[22] K. Schmüdgen, Unbounded operator Algebras and Representation Theory. *Berlin: Akademie-Verlag and Basel: Birkhänser Verlag*, 1990.

[23] M. Takesaki, Theory of Operator Algebra I, *Springer-Verlag, New York, Heidelberg, Berlin*, 1979.

[24] A. W. Wood, Numerical ranges and generalized B*-algebras, *Proc. London Math. Soc.* **34**(1977), 245-268.

[25] W. Zelazko, Metric generalizations of Banach algebras, Rozprawy Math. **47** (1965).

80 INFINITE-DIMENSIONAL GROUPS APPLICATIONS TO INDUSTRIAL PROBLEMS

Boris S. KHOTS

Compressor Controls Corporation, 4308 NW 114 Street
Des Moines, Iowa 50322, U. S. A.

bkhots@cccglobal.com

Introduction This paper describes constructional mathematical models and methods, which permit to apply infinite-dimensional objects concepts to industrial problems. Sudden result here is the connection of this problem with Fifth Hilbert problem. David Hilbert formulated the following (fifth) problem in 1900: to develop a theory of continuous groups of transformations without supposition about differentiability of the functions defining the groups [1]. The process to solve this problem led to the development of the general theory of locally compact topological groups. In particular, answer for concrete question - whether any local locally Euclidean group is locally isomorphical to a certain Lie group? - is positive [2]. Turning to the infinite - dimensional case, the situation changes cardinally. In particular, the theory of unitary group representations, which plays a central role in the finite - dimensional case, now gives way to the theory of Lie algebras representations with higher weight. Currently, the role of the group $DiffS^1$ and its central extension - the Virassoro - Botta group - can be clearly seen in modern mathematical physics (stream theory, bi - dimensional conformal field theory, accurately solved problems of statistical physics). This role turned out to be different from that of finite - dimensional (and finite) groups in physical applications.

H.G.W. Begehr et al. (eds.), Proceedings of the Second ISAAC Congress, Vol. 1, 697–705.

Infinite-dimensional groups need strong and concrete models and definitions. First of all, we introduced the next new classes of the infinite - dimensional objects: P - spaces, P - groups, and P - algebras Lie [6].

Definition 1 Linear topological space R over field of real (complex) numbers is called P - space, if the following conditions are satisfied:

a) $R \supset \bigcup_{i=1}^{\infty} R_i, R_i$ are Banach spaces, R_i closed subspace in R_{i+1}, $i = 1, 2, \ldots$

b) There is the projector $P_i : R \to R_i$ for each $i = 1, 2, \ldots$

c) $x_n \to x$ $(x_n, x \in R, n \to \infty)$ if and only if $P_i(x_n) \to P_i(x)$ for each $i = 1, 2, \ldots (n \to \infty)$

In particular, Hilbert and Banach spaces are P-spaces.

Definition 2 Local topological group G is called P - group, if the following conditions are satisfied:

a) $G \subset R$ with $0 = e$, R is P - space , 0 is zero in R, e is unity element of G .

b) $G \cap R_i$ contains neighborhood of zero in R_i for each $i = 1, 2, \ldots$

c) $P_i(ab^{-1}) = P_i(P_i(a)P_i(b)^{-1})$ for each $i = 1, 2, \ldots$, $a, b \in G$

d) $|P_i(xy) - P_i(x) - P_i(y)||P_i(x)|^{-1} \to 0, |P_i(xy) - P_i(x) - P_i(y)||P_i(y)|^{-1} \to 0$, if
$|P_i(x)| \neq 0; |P_i(y)| \neq 0; |P_i(x)| + |P_i(y)| \to 0$; $|.|$ - norm in R_i for each $i = 1, 2, \ldots$,
$x, y \in G$

In particular, classical finite - dimensional groups Lie and Birkhoff groups are P - groups. Groups of homeomorphisms of some classes of the manifolds are P - groups. Some groups of the formal power series are P - groups.

Definition 3 Algebra Lie T is called P - algebra Lie, if the following conditions are satisfied:

a) T is P - space

b) $P_i([ab]) = P_i([P_i(a), P_i(b)])$ for each $i = 1, 2, \ldots$, $a, b \in G$, $[.,.]$ - Lie commutator

c) $|P_i([ab])| \leq |P_i(a)||P_i(b)|$ for each $i = 1, 2, \dots$, $a, b \in T$, $|.|$ - norm in R_i

In particular, classical finite - dimensional algebras Lie and normed algebras Lie are P - algebras Lie.

We have next the main properties of these objects [7, 8, 9, 10, 11, 12, 13, 14, 15] : There is the functor from the category of P - algebras Lie to the category of P - groups. Conversely, there is the functor from the category of P - groups to the category of P - algebras Lie . First statement, in particular, means, that each P - group (uniqely with isomorphism precision) corresponds with some P - algebra Lie; thus each subgroup corresponds with certain closed subalgebra; each normal subgroup corresponds with closed ideal; each continuous homomorphism of P - groups corresponds with the homomorphism of P - algebras. Conversely, each P - algebra Lie (uniqely with a local isomorphism precision) corresponds with some P - group; thus each closed subalgebra of P - algebra Lie corresponds with certain subgroup; each closed ideal corresponds with certain normal subgroup; each homomorphism of P - algebras corresponds with certain homomorphism of P - groups. P - groups have several common properties with classical finite - dimensional groups Lie. In particular, each P - group G has a "canonical coordinates" [7, 8, 12], i.e. there is some local embedding (local homeomorphism) j of G onto a neighborhood of zero of some P - space R with following conditions: $j(e) = 0; j(x^n) = nj(x)$. Here is e - the unity element in G; 0 - the zero in R; $n =$ ア $1,$ ア $2, \dots$; the last relation is correct if x is sufficiently nearly to e and x^n is defined. Proof of the last "canonical coordinates" statement for Birkhoff groups is given in [3].

A significant role in infinite-dimensional group concepts applications to industrial problems plays the group G of all homeomorphisms of the segment $[0, 1]$ onto itself. Let
$H = [h \in G : h(0) = 0]$,
L be a subset of H containing all elements without internal stationary points.

Theorem 1 For any $l \in L$ there exists a homeomorphism A_l of the interval $(0, 1)$ onto a real line such that the isomorphism AA_l of the group G onto the group GG of all homeomorphisms of a line, generated by this homeomorphism, transfers the element l to the element $p_1 \in GG$, where $p_1 = p_1(x) = x + 1$ is a parallel displacement by 1.
Proof:

Let
$l(x) > x$ for $x \in (0,1)$. If $x_0 \in (0,1)$, then let
$x_n = l^n(x_0), n = 0, -1, +1, -2, +2, \ldots$.
Let A_l^0 be a homeomorphism of the segment $[x_0, x_1]$ onto the segment $[0,1]$ such that $A_l^0(x_0) = 0, A_l^0(x_1) = 1$. Let now
$A_l(x) = A_l^0(x)$, if $x \in [x_0, x_1]$ and
$A_l(x) = n + A_l^0(l^{-n}(x))$, if $x \in [x_n, x_{n+1}], n = 0, -1, +1, -2, +2, \ldots$.
Now we define $AA_l : G \to GG$:
$AA_l(f) = A_l f A_l^{-1}, f \in G$, considering $f$first in $(0,1)$, then naturally extending it to $[0,1]$. We have
$AA_l(l) = p_1$.
In the same way, we can consider the case when for all $x \in (0,1) l(x) < x$.
Theorem 1 is proved.
Corollary 1.
Each element $h \in H$ belongs to a one-parameter subgroup of elements of H.
Corollary 2.
Each element $l \in L$ belongs to a one-parameter subgroup of elements of L.

Theorem 2 For any $l_1, l_2 \in L$ there is $g \in G$ such that
$l_1 = g l_2 g^{-1}$.
Proof:
Let
$AA_{l_1}(l_1) = p_1 = A_{l_1} l_1 A_{l_1}^{-1}$
$AA_{l_2}(l_2) = p_1 = A_{l_2} l_2 A_{l_2}^{-1}$.
Then:
$g = A_{l_1}^{-1} A_{l_2}$.
Theorem 2 is proved.

Theorem 3 Let
$L_1 = [l \in L : l(x) > x, x \in (0,1)]$ $L_2 = [l \in L : l(x) < x, x \in (0,1)]$.
If $h \in H$, then $hL_1h^{-1} = L_1, hL_2h^{-1} = L_2$.
If $h \in G - H$, then $hL_1h^{-1} = L_2, hL_2h^{-1} = L_1$.
Proof is obvious.

Theorem 4 $H = LL$.
Proof:
For $h \in H$ let
$h_1(x) = max(h(x), x), h_2(x) = min(h(x), x)$.

Then $h_1, h_2 \in H, h = h_1 h_2 = h_2 h_1$. Let $l \in L_2$. Then $l^{-1} \in L_1$. If $x \in (0,1)$, then $lh_2(x) < h_2(x) \leq x$, $h_1 l^{-1}(x) \geq l^{-1}(x) > x$, $lh_2 \in L_2$, $h_1 l^{-1} \in L_1$. Therefore $h = (h_1 l^{-1})(lh_2) \in LL$, i.e. $H \subset LL$.
Converse inclusion is obvious.
Theorem 4 is proved.
We introduce now a topology in H and consider some structural problems.

Theorem 5 The group H of all increase homeomorphisms of the segment [0,1] onto itself with compact-open topology is locally homeomorphic to some P-space, but isn't P-group. So, we have negative solution of Fifth Hilbert problem for infinite-dimensional case.
Proof:
The group operation $(. * .)$ in H is a superposition, i.e. for any $f, g \in H (f * g)(x) = f(g(x))$. The compact-open topology in this case is the same as induced by the metric $d(f,g) = sup[d(f(x), g(x)) : x \in [0,1]]$. We define s as the space of all sequences $(x_{n,i}), i = 1, 2, ..., 2^n, n = 0, 1, 2, \ldots, (x_{n,i}) \in (0,1)$, with the usual product topology. Thus $s = \prod_{n=0}^{\infty} \prod_{i=1}^{2^n} (0,1)_{n,i}$. A basis for this topology of s is the collection of sets of the form $\prod_{n=0}^{\infty} \prod_{i=1}^{2^n} U_{n,i}$, where $U_{n,i}$ is an open subset of $(0,1)$ and for all but a finite number of the (n,i)'s, $U_{n,i}$ is $(0,1)$. s is homeomorphic to the real Hilbert space l_2 [3, 4]. The real Hilbert space l_2 is a particular case of P - space. Let $(x_{n,i}) \subset s$. We will define $h \subset H$ associated with $(x_{n,i})$. Suppose that we have defined points $A_n = [0 = a_{n,0} < a_{n,1} < \ldots < a_{n,2^n} = 1]$ and $B_n = [0 = b_{n,0} < b_{n,1} < \ldots < b_{n,2^n} = 1]$ and defined h such that $h(a_{n,i}) = b_{n,i}$ for $i = 0, 1, \ldots, 2^n$. We extend the definition of h to a set of points $A_{n+1} \supset A_n$ onto $B_{n+1} \supset B_n$ with each of A_{n+1} and B_{n+1} having 2^{n+1} points. If n is odd, then let z_i be the midpoint of segment $[a_{n,i-1}, a_{n,i}]$ for $i = 1, 2, \ldots, 2^n$ and let $y_i = h(z_i) = x_{n,i}(b_{n,i} - b_{n,i-1}) + b_{n,i-1}$. If n is even, then let y_i be the midpoint of the segment $[b_{n,i-1}, b_{n,i}]$ and let $z_i = h^{-1}(y_i) = x_{n,i}(a_{n,i} - a_{n,i-1}) + a_{n,i-1}$. Then let $A_{n+1} = A_n \bigcup z_i, i = 1, 2, \ldots, 2^n$ and $B_{n+1} = B_n \bigcup y_i, i = 1, 2, \ldots, 2^n$. Then we will have $A_{n+1} = [0 = a_{n+1,0} < a_{n+1,1} < \ldots < a_{n+1,2^n} = 1]$ and $B_{n+1} = [0 = b_{n+1,0} < b_{n+1,1} < \ldots < b_{n+1,2^n} = 1]$ with $h(a_{n+1,i}) = b_{n+1,i}$. Proceeding in this way and letting $A = \bigcup_{n=1}^{\infty} A_n$ and $B = \bigcup_{n=1}^{\infty} B_n$, A and B will be dense in $[0,1]$ and $h[A] = B$ will be order preserving. Thus h will have a continuous extension to

an homeomorphism g_h from H. If we let $F : s \to H$ be defined by $F((x_{n,i})) = g_h$, then F is the homeomorphism of s onto H. So, H is the group, which homeomorphic (and locally homeomorphic in particular) to P-space s. Moreover, H is locally Banach group. Let $K_{g,d}$ is the set of increase homeomorphisms f of the segment $[0, 1]$ onto itself with condition $f(x) = x$ for $x \in [0, g] \bigcup [d, 1], 0 < g < d < 1$. $K_{g,d}$ is subgroup from H. If U is any open neighborhood of the unity element e in H (which is a $f(x) = x$ for all $x \in [0, 1]$), there are g and d with sufficiently small $d - g$ such that $K_{g,d} \subset U$. The other words, H contains "small" subgroups. The absence of these ("small") subgroups was the main part of the positive solution of the finite-dimensional Fifth Hilbert problem [2]. In our case we can conclude, that H is not locally isomorphic to some P - group, because statement "Each P - group has a "canonical coordinates" contradicts existing of "small" subgroups in H. Theorem 5 is proved. So, we can see that group H gives us negative solution of infinite-dimensional Fifth Hilbert problem.

Theorem 6 Now we can formulate the theorem, which gives necessary and sufficient conditions for some local topological group, locally homeomorphic to some P - space, to be P - group.

Local topological group G is local isomorphic to some P - group H if and only if there is the local topological "canonical" embedding of G onto neighborhood of zero of some P-space R. In other words, Let G be a local topological group and there is some local embedding (local homeomorphism) j of G onto a neighborhood of zero of some real P - space R with the following conditions:
$j(e) = 0; j(x^n) = nj(x); l^{-1}j(j^{-1}(lj(x)) * j^{-1}(lj(y)) \to j(x) + j(y)$, if $l \to 0, l \neq 0$. Here is e - the unity element in G; 0 - the zero in R; $n = 0,$ ア $1,$ ア $2, \ldots$; l - real (complex) number ; (. $*$.) - the group operation in G; $x, y \in G$; the relations are correct if x, y are sufficiently nearly to e and x^n is defined. Then G is locally isomorphic to some P - group H. So, with these conditions infinite-dimensional Fifth Hilbert problem has positive solution.
Theorem 6 is proved in [8, 9].

A significant role in infinite-dimensional group concepts applications to industrial problems play, besides group G of all homeomorphisms of the segment $[0, 1]$ onto itself, the groups T and Q. T is a group of all local analytical homeomorphisms t of the real line into itself with a stationary point 0:

$t = t(x) = \sum_{k=1}^{\infty} t_k x^k$ with condition $t_1 > 0$.
Group operation is superposition: $t_1 t_2 = (t_1 t_2)(x) = t_!(t_2(x))$.
Q is a group of all formal power series q with real coefficients:
$q = q(x) = \sum_{k=1}^{\infty} q_k x^k$ with condition $q_1 > 0$.
Group operation is superposition:
$q_1 q_2 = (q_1 q_2)(x) = q_!(q_2(x))$.
Both groups T and Q meet conditions of Theorem 6 and are P - groups.

Two examples of P - groups and P - algebras applications.

1. **Turbomachinery** Infinite-dimensional objects and connected with them mathematical models and algorithms permit to develop new generation of controllers in turbomachinery. In particular, it means that controller is:
   real time self-learning;
   real time self-organizing;
   real time self-orienting.
   Classical PI- and PID- control algorithms are not sufficient now, and controller has more powerful real time control algorithms.
   Compressor antisurge control may not require surge tests - anti-surge controller can see a surge limit line from its operation point without any a priori information.

2. **Genetics** Infinite-dimensional objects and connected with them mathematical models and algorithms permit to develop new methods for DNA (RNA) structure prediction; Gene-Gene and Gene-Trait dependences.
   Genetic Gene expression data, for example, is huge-dimensional data with noisy and discontinuous functions. Classical search methods - calculus-based, enumerative, random, and genetic algorithms - can't effective work in this situation.
   New methods can give us possibilities:
   in agriculture - to raise a productivity;
   for human - to defeat diseases;
   for nature - to check evolutionary theories.

I'd like to thank my colleagues and friends Dr. Ben Bowen, Ilya Markevich and Mei Guo for their participation in this work. And also I would like to thank Professor Masaru Nishihara and Professor Pavel Krutitskii for their help.

References

[1] D. Hilbert, Mathematical Problems, Speech presented at the II International Congress of Mathematicians, Paris, France, on August 8, 1900, Gessammelte Abhandlungen, v. III, 1935, 290-329.

[2] D. Montgomery and L. Zippin, Topological transformation groups, Interscience, New York, 1955.

[3] G. Birkhoff, Analytical Groups, Transactions of the American Mathematical Society, 43(1938), 61.

[4] R. D. Anderson and R. H. Bing, A complete elementary proof that Hilbert spase is homeomorphic to the countable infinite product of lines, Bulletin of the American Mathematical Society, 74(1968), 771-792.

[5] J. Keesling, Using flows to construct Hilbert space factors of function spaces, Transactions of the American Mathematical Society, 161(1971), 1-24.

[6] B. S. Khots, Groups of local analytical homeomorphisms of line and P-groups, Russian Math Surveys, v. XXXIII, # 3 (201), Moscow, London, 1978, 189 - 190.

[7] B. S. Khots, P-groups and P-algebras Lie, Soviet Math (Iz VUZ), v. 10(305), Kazan, Allerton, New York, 1987, 80.

[8] B. S. Khots, Development of P-groups and P-algebras Lie for Management Systems, Works Collection of the National Information Petrochemical Institute of Russia, v. 50, Moscow, 1987, 91-99.

[9] B. S. Khots, P-groups and P-algebras Lie, Soviet Math (Iz VUZ), dep # 2265-B87, Kazan, Allerton, New York, 1988, 16 pp.

[10] B. S. Khots, The groups of homeomorphisms of the segment, Abstracts of papers presented to the American Mathematical Society, v. 17, #3, issue 105, 1996.

[11] B. S. Khots, Criterion of P-groups,Abstracts of papers presented to the American Mathematical Society, v. 17, #3, issue 105, 1996.

[12] B. S. Khots, Campbell-Hausdorff series and P-algebras Lie, Abstracts of papers presented to the American Mathematical Society, v. 18, #2, issue 108, 1997.

[13] B. S. Khots, P-groups and canonical coordinates, Abstracts of papers presented to the American Mathematical Society, v. 18, #2, issue 108, 1997.

[14] B. S. Khots, Development of P-algebra Lie, Abstracts of papers presented to the American Mathematical Society, v. 18, #2, issue 108, 1997.

[15] B. S. Khots, Groups of analytical homeomorphisms of the low-dimensional manifolds are P-groups, Abstracts of papers presented to the American Mathematical Society, v. 19, #2, issue 112, 1998.

81 INTEGRAL OPERATORS RELATED TO BOUNDARY VALUE PROBLEMS IN 2-D DOMAINS WITH CUTS

KRUTISKII P.A.

Dept. of Mathematics, Faculty of Physics, Moscow State University
Moscow 117234, Russia

conf@math.phys.msu.su

In [12-21] we studied Dirichlet and Neuman problems for Laplace and Helmholtz equations in 2-D multiply connected domains bounded by closed curves and containing cuts. Cuts model screens, cracks, wings and other thin structures in applied sciences. Problems [12-21] were studied by a method of boundary integral equations. In the present note we discuss general properties of integral operators related to boundary value problems for elliptic PDEs in 2-D domains with cuts.

1. Integral operators.

On the Os axis we consider 2 sets of segments $\Gamma^1 = \bigcup\limits_{n=1}^{N_1} \Gamma_n^1$ and $\Gamma^2 = \bigcup\limits_{n=1}^{N_2} \Gamma_n^2$, where $\Gamma_n^k = [a_n^k, b_n^k]$, $N_1, N_2 \geq 0$. Suppose that all segments do not have common points. Set $\Gamma = \Gamma^1 \cup \Gamma^2$. We consider the system of integral equations on Γ with respect to the function $\mu(s)$:

$$(1.1a) \qquad \frac{1}{\pi}\int_{\Gamma^1} \frac{\mu(\sigma)}{\sigma - s} d\sigma + \int_{\Gamma} \mu(\sigma) v(s,\sigma) d\sigma = f(s), \qquad s \in \Gamma^1,$$

H.G.W. Begehr et al. (eds.), Proceedings of the Second ISAAC Congress, Vol. 1, 707–718.

$$\mu(s) + \int_{\Gamma} \mu(\sigma) w(s,\sigma) d\sigma = f(s), \qquad s \in \Gamma^2, \tag{1.1b}$$

$$\int_{\Gamma} \mu(\sigma) U_n(\sigma) d\sigma = \varphi_n, \qquad n = 1, ..., N_1. \tag{1.1c}$$

Here $v(s,\sigma)$, $w(s,\sigma)$, $U_1(s), ..., U_{N_1}(s)$, $f(s)$ are given functions, and $\varphi_1, ..., \varphi_{N_1}$ are given constants. Suppose that the given functions satisfy the following conditions

(W)

$$\begin{cases} w(s,\sigma) \in C^0(\Gamma^2 \times \Gamma) \text{ or } w(s,\sigma) \text{ is a polar kernel [11] on } \Gamma^2 \times \Gamma\,; \\ v(s,\sigma) \in C^{0,\lambda}(\Gamma^1 \times \Gamma)\,;\ f(s) \in C^{0,\lambda}(\Gamma^1) \oplus C^0(\Gamma^2) \text{ with } \lambda \in (0,1]; \\ \text{functions } U_1(s), ..., U_{N_1}(s) \text{ are continuous on } \Gamma^2 \text{ and in interior points} \\ \text{of } \Gamma^1, \text{ while they may have logarithmic singularities at the ends of } \Gamma^1. \end{cases}$$

For simplicity the Hölder exponent for $v(s,\sigma)$ and $f(s)$ is the same. If these exponents are different in practice, then as λ we can take the lesser exponent. We will seek the solution of the system of equations (1.1) in the space

$$C_q^{\omega}(\Gamma^1) \oplus C^0(\Gamma^2), \qquad \omega \in (0,1], \quad q \in [0,1) \tag{S}$$

with the norm $\|\cdot\|_{C_q^{\omega}(\Gamma^1) \cap C^0(\Gamma^2)} = \|\cdot\|_{C_q^{\omega}(\Gamma^1)} + \|\cdot\|_{C^0(\Gamma^2)}$.

Definition. Function $\mu(s)$ belongs to the space $C_q^{\omega}(\Gamma^1)$ with $\omega \in (0,1]$, $q \in [0,1)$, if function $\nu(s) = \mu(s) \prod_{n=1}^{N_1} |s - a_n^1|^q \, |s - b_n^1|^q$ belongs to the Hölder space $C^{0,\omega}(\Gamma^1)$ with exponent ω, and $\|\mu\|_{C_q^{\omega}(\Gamma^1)} = \|\nu\|_{C^{0,\omega}(\Gamma^1)}$.

System (1.1) consists of Cauchy singular integral equation of the 1-st kind (1.1a), equation of the 2-nd kind with continuous or polar kernel (1.1b) and a set of additional conditions (1.1c).

Lemma 1.1. *Let conditions* (**W**) *hold. If the periodical conditions* $f(a_n^2) = f(b_m^2)$, $w(a_n^2, \sigma) = w(b_n^2, \sigma)$ *hold on a segment* Γ_n^2 $(n = 1, ..., N_2)$ *for any* $\sigma \in \Gamma$, *then any solution of system (1.1) also meets periodical condition on* Γ_n^2: $\mu(a_n^2) = \mu(b_n^2)$.

The lemma can be proved by direct verification with the help of equation (1.1b).

Let us prove the Fredholm alternative for the system (1.1). To do this we reduce system (1.1) to the equivalent Fredholm equation of the 2-nd kind and index zero in the Banach space. Inverting Cauchy singular integral operator in (1.1a), we transform (1.1a) to the equivalent integral equation of the 2-nd kind [1]:

$$(1.2)\quad \mu(s)+\frac{1}{Q_1(s)}\int_\Gamma \mu(\sigma)A_1(s,\sigma)d\sigma+\frac{1}{Q_1(s)}\sum_{n=0}^{N_1-1} G_n s^n = \frac{1}{Q_1(s)}F_1(s),$$

$$s\in\Gamma^1,$$

where

$$A_1(s,\sigma) = -\frac{1}{\pi}\int_{\Gamma^1}\frac{v(\xi,\sigma)}{\xi-s}Q_1(\xi)d\xi, \quad F_1(s) = -\frac{1}{\pi}\int_{\Gamma^1}\frac{Q_1(\sigma)f(\sigma)}{\sigma-s}d\sigma,$$

$Q_1(s) = \prod_{n=1}^{N_1}\left|\sqrt{s-a_n}\sqrt{b_n-s}\right|\operatorname{sign}(s-a_n)$, and $G_0,...,G_{N_1-1}$ are arbitrary constants. Using properties of singular integrals [1], one can show that $F_1(s)$, $A_1(s,\sigma)$ are Hölder functions if $s\in\Gamma^1$, $\sigma\in\Gamma$. Consequently, any solution of (1.2) belongs to $C^\omega_{1/2}(\Gamma^1)$, and below we look for $\mu(s)$ on Γ^1 in this space.

Assuming that $\mathcal{L}\subset\Gamma$, we set $\delta(s,\mathcal{L})=0$ if $s\notin\mathcal{L}$, and $\delta(s,\mathcal{L})=1$ if $s\in\mathcal{L}$. Let $Q(s)=\delta(s,\Gamma^1)Q_1(s)+\delta(s,\Gamma^2)$, $s\in\Gamma$. Instead of $\mu(s)\in C^\omega_{1/2}(\Gamma^1)\oplus C^0(\Gamma^2)$ we introduce new unknown function $\mu_*(s)=\mu(s)Q(s)\in C^{0,\omega}(\Gamma^1)\oplus C^0(\Gamma^2)$ and rewrite (1.2), (1.1b) in the form of one equation of the 2-nd kind on Γ:

(1.3*a*)

$$\mu_*(s)+\int_\Gamma \mu_*(\sigma)Q^{-1}(\sigma)A(s,\sigma)d\sigma+\delta(s,\Gamma^1)\sum_{n=0}^{N_1-1}G_n s^n = F(s),\quad s\in\Gamma,$$

where $A(s,\sigma) = \delta(s,\Gamma^1)A_1(s,\sigma)+\delta(s,\Gamma^2)w(s,\sigma)$, $F(s) = \delta(s,\Gamma^1)F_1(s)+\delta(s,\Gamma^2)f(s)$.

We choose constants $G_0,...,G_{N_1-1}$ to satisfy conditions (1.1c). For this purpose we substitute $\mu(s)$ from (1.1b), (1.2) into (1.1c), then in terms of $\mu_*(s)$ we obtain

$$(1.3b)\quad \int_\Gamma Q^{-1}(\xi)\mu_*(\xi)l_n(\xi)d\xi+\sum_{k=0}^{N_1-1}B_{nk}G_k = H_n,\quad n=1,...,N_1\ ,$$

where

$$l_n(\xi) = -\int_\Gamma Q^{-1}(\sigma)A(\sigma,\xi)U_n(\sigma)d\sigma,$$

$$H_n = -\int_\Gamma Q^{-1}(\sigma)F(\sigma)U_n(\sigma)d\sigma+\varphi_n,$$

$$B_{nk} = -\int_D Q^{-1}(\sigma)\delta(\sigma,\Gamma^1)\sigma^k U_n(\sigma)d\sigma. \tag{1.4}$$

So, the system of equations (1.1) for $\mu(s)$ is reduced to system (1.3) with respect to the function $\mu_*(s)$ and constants $G_0, ..., G_{N_1-1}$. Clearly, any solution of system (1.3) yields a solution of system (1.1). As noted above, $F_1(s)$ and $A_1(s,\sigma)$ are Hölder functions, if $s \in \Gamma^1$, $\sigma \in \Gamma$. More precisely, [1], $F_1(s) \in C^{0,p}(\Gamma^1)$, $p = \min\{1/2, \lambda\}$, and $A_1(s,\sigma)$ belong to $C^{0,p}(\Gamma^1)$ in s uniformly with respect to $\sigma \in \Gamma$. Hence, the integral operator in (1.3a) acts from $C^0(\Gamma)$ into $C^{0,p}(\Gamma^1) \oplus C^0(\Gamma^2)$. We have proved

Lemma 1.2. *Let $F(s) \in C^{0,p}(\Gamma^1) \oplus C^0(\Gamma^2)$, $p = \min\{\lambda, 1/2\}$. If $\mu_*(s) \in C^0(\Gamma)$ satisfies equation (1.3a), then $\mu_*(s)$ belongs to $C^{0,p}(\Gamma^1) \oplus C^0(\Gamma^2)$.*

Condition $F(s) \in C^{0,p}(\Gamma^1) \oplus C^0(\Gamma^2)$ holds, if the conditions (**W**) hold. Therefore, below we look for $\mu_*(s)$ in $C^0(\Gamma)$. Integral operator with the kernel $w(s,\sigma)Q^{-1}(\sigma)$ is a compact operator mapping $C^0(\Gamma)$ into $C^0(\Gamma^2)$ thanks to Arzela theorem, or thanks to compactness of operators with a polar kernel. Since $A_1(s,\sigma) \in C^0(\Gamma^1 \times \Gamma)$, the integral operator in (1.3a):

$$\mathbf{A}\mu_* = \int_\Gamma \mu_*(\sigma)Q^{-1}(\sigma)A(s,\sigma)d\sigma$$

is a compact operator mapping $C^0(\Gamma)$ into itself. We rewrite (1.3a) in the operator form

$$(I + \mathbf{A})\mu_* + PG = F \ , \tag{1.5a}$$

where P is the operator multiplying the row $P = \delta(s,\Gamma^1)(s^0, ..., s^{N_1-1})$ by the column $G = (G_0, ..., G_{N_1-1})^T$. Operator P is a finite dimensional from E_{N_1} into $C^0(\Gamma)$, and so compact [2]. Now we write equation (1.3b) in the form

$$I_{N_1}G + L\mu_* + (B - I_{N_1})G = H \ , \tag{1.5b}$$

where $H = (H_1, ..., H_{N_1})^T$ is a column of N_1 elements, I_{N_1} is a unit operator in E_{N_1}, B is a matrix of dimension $(N_1 \times N_1)$, consisting of elements B_{nk} from (1.4). Operator L acts from $C^0(\Gamma)$ into E_{N_1}, so that $L\mu_* = (L_1\mu_*, ..., L_{N_1}\mu_*)^T$, where

$$L_n\mu_* = \int_\Gamma Q^{-1}(\xi)\mu_*(\xi)l_n(\xi)d\xi, \ n = 1, ..., N_1.$$

Operators $(B - I_{N_1})$, L are finite dimensional and so compact [2].

We consider columns $\tilde{\mu} = (\mu_*(s), G)^T$, $\tilde{F} = (F(s), H)^T$ in the Banach space $C^0(D) \times E_{N_1}$ with the norm $\|\tilde{\mu}\|_{C^0(\Gamma)\times E_{N_1}} = \|\mu_*\|_{C^0(\Gamma)} + \|G\|_{E_{N_1}}$. We rewrite system (1.5) in the form of one equation

$$(1.6) \qquad (\mathbf{I}+\mathbf{R})\tilde{\mu} = \tilde{F} \ , \quad \mathbf{R} = \begin{pmatrix} \mathbf{A} & P \\ L & B - I_{N_1} \end{pmatrix} ,$$

where $\mathbf{I}$ is a unit operator in the space $C^0(\Gamma) \times E_{N_1}$. Evidently, $\mathbf{R}$ is a copmact operator mapping $C^0(\Gamma) \times E_{N_1}$ into itself. Hence, (1.6) is a Fredholm equation of index zero in this space. Thus, system (1.1) is reduceed to the equivalent Fredholm equation of the 2-nd kind and index zero in the Banach space $C^0(\Gamma) \times E_{N_1}$ with respect to the function $\mu_*(s)$ and constants $G_1, ..., G_{N_1}$.

Theorem 1.1. *Let conditions* (**W**) *hold, then:*
1) system of equations (1.1) is Fredholm of index zero in the space (**S**)*;*
2) system (1.1) is equivalent to the Fredholm equation (1.6) of the 2-nd kind and index zero in the space $C^0(\Gamma) \times E_{N_1}$*;*
3) if $\tilde{\mu} = (\mu_*(s), G)^T$ *is a solution of equation (1.6) in* $C^0(\Gamma) \times E_{N_1}$*, then the function* $\mu_*(s)$ *automatically belongs to* $C^{0,p}(\Gamma^1) \oplus C^0(\Gamma^2)$ *with* $p = \min\{1/2, \lambda\}$ *and yields a solution* $\mu(s)$ *of equations (1.1) by the formula* $\mu(s) = Q^{-1}(s)\mu_*(s)$*;*
4) all solutions of system (1.1) in (**S**) *belong to* $C^p_{1/2}(\Gamma^1) \oplus C^0(\Gamma^2)$.

Corollary 1.1. *Let conditions* (**W**) *hold. If the homogeneous system (1.1) has only a trivial solution in the space* (**S**)*, then inhomogeneous system (1.1) has unique solution* $\mu(s) \in C^p_{1/2}(\Gamma^1) \oplus C^0(\Gamma^2)$ *in the space* (**S**) *for any right-hand side (i.e. for any function* $f(s)$*, satisfying conditions* (**W**) *and for any constants* $\varphi_1, ..., \varphi_{N_1}$*).*

Let $\Gamma^1 = \Upsilon^1 \cup \Upsilon^2$, $\Upsilon^1 = \bigcup_{n=1}^{N_0} \Gamma^1_n$, $\Upsilon^2 = \bigcup_{n=N_0+1}^{N_1} \Gamma^1_n$. We consider a system of integral equations, which includes equation with a logarithmic kernel:

$$(1.7a) \qquad \frac{1}{\pi}\int_{\Upsilon^1} \frac{\mu(\sigma)}{\sigma - s} d\sigma + \int_\Gamma \mu(\sigma) V(s,\sigma) d\sigma = f(s), \qquad s \in \Upsilon^1,$$

$$(1.7b) \quad -\frac{1}{\pi}\int_{\Upsilon^2} \mu(\sigma) \ln|s-\sigma| d\sigma + \int_\Gamma \mu(\sigma) u(s,\sigma) d\sigma = \mathcal{F}(s), \qquad s \in \Upsilon^2,$$

$$(1.7c) \qquad \mu(s) + \int_\Gamma \mu(\sigma) w(s,\sigma) d\sigma = f(s), \qquad s \in \Gamma^2,$$

$$\int_\Gamma \mu(\sigma)U_n(\sigma)d\sigma = \varphi_n, \qquad n = 1,...,N_0. \tag{1.7d}$$

Here $V(s,\sigma)$, $u(s,\sigma)$, $w(s,\sigma)$, $U_1(s),...,U_{N_0}(s)$, $\mathcal{F}(s)$, $f(s)$ are given functions, and $\varphi_1,...,\varphi_{N_0}$ are given constants. Suppose that the given functions satisfy the following smoothness conditions

$(\mathbf{W}_1)$

$$\begin{cases} V(s,\sigma) \in C^{0,\lambda}(\Upsilon^1 \times \Gamma);\ u_s'(s,\sigma) \in C^{0,\lambda}(\Upsilon^2 \times \Gamma);\ \mathcal{F}(s) \in C^{1,\lambda}(\Upsilon^2); \\ f(s) \in C^{0,\lambda}(\Upsilon^1) \oplus C^0(\Gamma^2) \text{ with } \lambda \in (0,1];\ \text{function } w(s,\sigma) \\ \text{is the same as in conditions } (\mathbf{W});\ \text{functions } U_1(s),...,U_{N_0}(s) \\ \text{are continuous on } \Gamma^2 \text{ and in interior points of } \Gamma^1, \\ \text{while at the ends of } \Gamma^1 \text{ they may have logarithmic singularities.} \end{cases}$$

The solution of system of equations (1.7) is looked for in the space $(\mathbf{S})$. For system (1.7) lemma 1.1 holds. By differentiating equation (1.7b) in s, we observe that system (1.7) is equivalent to system (1.1), where $v(s,\sigma) = \delta(s,\Upsilon^1)V(s,\sigma) + \delta(s,\Upsilon^2)u_s'(s,\sigma)$, function $f(s)$ on Υ^2 is equivalent to $\mathcal{F}'(s)$, and for $n = N_0+1,...,N_1$ we have $\varphi_n = \mathcal{F}(a_n)$, $U_n(\sigma) = -\pi^{-1}\delta(\sigma,\Upsilon^2)\ln|a_n - \sigma| + u(a_n,\sigma)$. Functions $w(s,\sigma)$, $v(s,\sigma)$, $f(s)$, $U_1(\sigma),...,U_{N_1}(\sigma)$ in obtained system (1.1) satisfy conditions $(\mathbf{W})$.

Theorem 1.2. *Let conditions* $(\mathbf{W}_1)$ *hold, then system (1.7) is equivalent to system (1.1), and all results outlined in theorem 1.1 for system (1.1) hold for system (1.7) also.*

Corollary 1.2. *Let conditions* $(\mathbf{W}_1)$ *hold. If homogeneous system (1.7) has only a trivial solution in the space* $(\mathbf{S})$*, then inhomogeneous system (1.7) has the unique solution* $\mu(s) \in C^p_{1/2}(\Gamma^1) \oplus C^0(\Gamma^2)$*, in the space* $(\mathbf{S})$ *for any right-hand side (i.e. for any functions* $f(s)$, $\mathcal{F}(s)$*, satisfying conditions* $(\mathbf{W}_1)$ *and for any constants* $\varphi_1,...,\varphi_{N_0}$*). Here* $p = \min\{1/2, \lambda\}$.

Let us discuss our results. The index of system (1.1a)–(1.1b) is N_1. If number of conditions (1.1c) is not equal N_1 exactly, then Fredholm alternative may not hold for system (1.1). Fredholm alternative may not hold for the system (1.1) in a space different from the space $(\mathbf{S})$. Indeed, suppose, that homogeneous system (1.1) has only a trivial solution in the space $(\mathbf{S})$. According to corollary 1.1, inhomogeneous system (1.1) has unique solution in this space. Note, that $C^{0,\omega}(\Gamma)$ is a subspace of $(\mathbf{S})$. Clearly, that homogeneous system (1.1) has only a trivial solution in $C^{0,\omega}(\Gamma)$, but inhomogeneous system (1.1) typically is not solvable in this space, since a solution of (1.1) typically has singularities at the ends of Γ^1. It follows from theorem 1.1, that the solution of (1.1) belongs to the space $(\mathbf{S})$, where $q = 1/2$.

Modern numerical methods for Cauchy singular integral equations are developed in [3-10].

2. Boundary value problems.

Theory of integral equations developed above is applicable to the analysis of boundary value problems for elliptic PDEs in 2-D domains with cuts. As an example, we discuss the Dirichlet problem for the propagative Helmholtz equation in an exterior multiply connected domain bounded by closed and open curves [14].

In the plane $x = (x_1, x_2) \in R^2$ we consider an exterior multiply connected domain bounded by simple open curves $\Xi_1^1, ..., \Xi_{N_1}^1 \in C^{2,\lambda}$ and simple closed curves $\Xi_1^2, ..., \Xi_{N_2}^2 \in C^{2,\lambda}$, $\lambda \in (0,1)$, so that the curves do not have points in common. Set $\Xi^1 = \bigcup_{n=1}^{N_1} \Xi_n^1$, $\Xi^2 = \bigcup_{n=1}^{N_2} \Xi_n^2$, $\Xi = \Xi^1 \cup \Xi^2$. The exterior connected domain bounded by Ξ^2 will be called $\mathcal{D}$. We assume that each curve Ξ_n^k is parametrized by the arc length s : $\Xi_n^k = \{x : \ x = x(s) = (x_1(s), x_2(s)),\ s \in \Gamma_n^k\}$, where $\Gamma_n^k \equiv [a_n^k, b_n^k]$, $n = 1, ..., N_k$, $k = 1, 2$, so that

$$a_1^1 < b_1^1 < ... < a_{N_1}^1 < b_{N_1}^1 < a_1^2 < b_1^2 < ... < a_{N_2}^2 < b_{N_2}^2.$$

Therefore points $x \in \Xi$ and values of the parameter s are in one-to-one correspondence except a_n^2, b_n^2, which correspond to the same point x for $n = 1, ..., N_2$. Suppose that the domain $\mathcal{D}$ is to the right when the parameter s increases on Ξ_n^2. Let

$$\Gamma^1 = \bigcup_{n=1}^{N_1} [a_n^1, b_n^1], \qquad \Gamma^2 = \bigcup_{n=1}^{N_2} [a_n^2, b_n^2], \qquad \Gamma = \bigcup_{k=1}^{2} \bigcup_{n=1}^{N_k} [a_n^k, b_n^k].$$

Since Ξ_n^2 $(n = 1, ..., N_2)$ are closed curves, in this section we set

(2.0)
$$C^{k,r}(\Gamma_n^2) = \{\mathcal{F}(s) : \ \mathcal{F}(s) \in C^{k,r}[a_n^2, b_n^2],\ \ \mathcal{F}^{(m)}(a_n^2) = \mathcal{F}^{(m)}(b_n^2),\ \ m = 0, k\ \},$$
$$k = 0, 1,\ \ r \in [0, 1]; \qquad C^{k,r}(\Gamma^2) = \bigoplus_{n=1}^{N_2} C^{k,r}(\Gamma_n^2).$$

In this section the space $C^0(\Gamma^2)$ in the condition **(S)** is understood according to this definition.

We say, that the function $u(x)$ belongs to the smoothness class **K** if

1) $u \in C^0(\overline{\mathcal{D}}) \cap C^2(\mathcal{D} \backslash \Xi^1)$,

2) $\nabla u \in C^0(\overline{\mathcal{D} \backslash \Xi^1} \backslash X)$, where X is a point-set, consisting of the end-points of Ξ^1, so that $X = \bigcup_{n=1}^{N_1} \left(x(a_n^1) \cup x(b_n^1)\right)$,

3) in the neighbourhood of any point $x(d) \in X$ for some constants $\mathcal{C} > 0$, $\epsilon > -1$ the inequality holds

$$|\nabla u| \leq \mathcal{C}\,|x - x(d)|^{\epsilon}, \tag{2.1}$$

where $x \to x(d)$ and $d = a_n^1$ or $d = b_n^1$, $n = 1, ..., N_1$.

In the definition of the class **K**, we consider Ξ^1 as a set of cuts. In particular, by $C^0(\overline{\mathcal{D}\backslash\Xi^1}\backslash X)$ we denote a class of functions, which are continuously extended on $\Xi^1\backslash X$ from the left and right, but their values on $\Xi^1\backslash X$ from the left and right can be different, so that these functions may have a jump across $\Xi^1\backslash X$.

Let us formulate the Dirichlet problem for the Helmholtz equation in the domain $\mathcal{D}\backslash\Xi^1$.

Problem **U**. To find a function $u(x)$ of the class **K** which satisfies the Helmholtz equation

$$u_{x_1x_1}(x) + u_{x_2x_2}(x) + \beta^2 u(x) = 0, \quad x \in \mathcal{D}\backslash\Xi^1, \quad \beta = \text{const} > 0, \tag{2.2}$$

the boundary condition

$$u(x(s))|_{\Xi} = f(s), \tag{2.3}$$

and the conditions at infinity

$$u = O(|x|^{-1/2}), \quad \frac{\partial u}{\partial |x|} - i\beta u = o(|x|^{-1/2}), \quad |x| = \sqrt{x_1^2 + x_2^2} \to \infty. \tag{2.4}$$

All conditions of the problem **U** must be satisfied in the classical sense.

On the basis of the Rellich lemma and energy equalities [11] we can easily prove the following assertion.

Theorem 2.1. *If* $\Xi \in C^{2,\lambda}$, $\lambda \in (0,1]$, *then the problem* **U** *has at most one solution.*

To construct the solution of the problem we assume that $f(s)$ from (2.3) is an arbitrary function from the Banach space $C^{1,\lambda}(\Gamma) = C^{1,\lambda}(\Gamma^1) \oplus C^{1,\lambda}(\Gamma^2)$, $\lambda \in (0,1]$. Here $C^{1,\lambda}(\Gamma^2)$ is understood according to definition (2.0). We seek the solution of the problem in the form of single layer potential on the open curves and in the form of a linear combination of double and single layer potential on the closed curves

$$u[\mu](x) = u_1[\mu](x) + u_2[\mu](x)\ , \tag{2.5}$$

$$u_1[\mu](x) = \int_{\Gamma^1} \mu(\sigma)\Phi(x, y(\sigma))d\sigma,$$

$$u_2[\mu](x) = \int_{\Gamma^2} \mu(\sigma) \left(\frac{\partial}{\partial \mathbf{n}_y} - i \right) \Phi(x, y(\sigma)) d\sigma,$$

where $\Phi(x, y) = (i/4)\mathcal{H}_0^{(1)}(\beta |x - y(\sigma)|)$ and $\mathcal{H}_0^{(1)}(z)$ is the Hankel function of the first kind [11]

$$\mathcal{H}_0^{(1)}(z) = \frac{\sqrt{2}\exp(iz - i\pi/4)}{\pi\sqrt{z}} \int_0^\infty \exp(-t)t^{-1/2} \left(1 + \frac{it}{2z}\right)^{-1/2} dt.$$

In addition,

$$y = y(\sigma) = (y_1(\sigma), y_2(\sigma)) \in \Xi,$$

$$|x - y(\sigma)| = \sqrt{(x_1 - y_1(\sigma))^2 + (x_2 - y_2(\sigma))^2},$$

and $\mu(\sigma)$ is an unknown density.

By $\int_{\Gamma^k} \ldots d\sigma$ we mean

$$\sum_{n=1}^{N_k} \int_{a_n^k}^{b_n^k} \ldots d\sigma.$$

We will seek $\mu(s)$ in the Banach space (**S**).

It can be checked directly [12], that for such $\mu(s)$ the function $u_1[\mu](x)$ belongs to the class **K** and satisfies all conditions of the problem **U** except the boundary condition (2.3). In particular, the inequality (2.1) holds with $\epsilon = -q$ if $q \in (0, 1)$. The potential $u_2[\mu](x)$ satisfies equation (2.2) and belongs to $C^0(\overline{\mathcal{D}}) \cap C^2(\mathcal{D})$. Besides, function (2.5) satisfies the conditions at infinity (2.3).

To satisfy the boundary condition we subsitute (2.5) in (2.3) and arrive at the integral equation for the density $\mu(s)$:

$$\int_{\Gamma^1} \mu(\sigma)\Phi(x, y(\sigma))d\sigma + \tag{2.6}$$

$$+\frac{1}{2}\delta(s)\mu(s) + \frac{i}{4} \int_{\Gamma^2} \mu(\sigma) \left(\frac{\partial}{\partial \mathbf{n}_y} - i \right) \Phi(x, y(\sigma)) d\sigma = f(s), \quad s \in \Gamma,$$

where $\delta(s) = 0$ if $s \in \Gamma^1$ and $\delta(s) = 1$ if $s \in \Gamma^2$.

One can show that any solution of equation (2.6) in the space (**S**) automatically belongs to $C_q^\omega(\Gamma^1) \oplus C^{1,\lambda/2}(\Gamma^2)$. It follows from the properties of potentials that function (2.5) with such $\mu(s)$ belongs to the class **K**. So, *if $\mu(s)$ is a solution of equation (2.6) in the space* (**S**), *then the solution of the problem* **U** *is given by (2.5).*

Equation (2.6) is a particular case of system (1.7), where $\Upsilon^1 = \emptyset$, $N_0 = 0$, $\Upsilon^2 = \Gamma^1$. It can be verified directly that kernels in the integral equation (2.6) meet conditions ($\mathbf{W}_1$). System (1.7) has been studied in theorem 1.2. Consequently, Fredholm alternative holds for equation (2.6). Using the uniqueness theorem 2.1 one can prove that the homogeneous equation (2.6) has only a trivial solution in the space (**S**). According to corollary 1.2 and theorem 1.2, the inhomogeneous equation (2.6) is uniquely solvable in the space (**S**) and its solution belongs to $C^p_{1/2}(\Gamma^1) \oplus C^0(\Gamma^2)$ with $p = \min\{\lambda, 1/2\}$. Thus, the solution of the problem **U** is given by expression (2.6), where $\mu(s)$ is a solution of equation (2.5) in the space specified above.

Theorem 2.2. *If* $\Xi \in C^{2,\lambda}$, $f(s) \in C^{1,\lambda}(\Gamma)$, $\lambda \in (0,1)$, *then the solution of the problem* **U** *exists and is given by (2.5), where* $\mu(s)$ *is a solution of the Fredholm equation (2.6) in the space* $C^p_{1/2}(\Gamma^1) \oplus C^0(\Gamma^2)$ *with* $p = \min\{\lambda, 1/2\}$. *Equation (2.6) is uniquely solvable in this space.*

It can be checked directly that the solution of the problem **U** satisfies condition (2.1) with $\epsilon = -1/2$. Explicit expressions for singularities of the solution gradient at the end-points of the open curves can be easily obtained with the help of formulas presented in [12].

The Neumann problem on wave scattering by 2-D obstacles and screens has been investigated in [13]. The Dirichlet and Neumann problems for the dissipative Helmholtz equation in both interior and exterior domains, bounded by closed and open curves have been studied in [15], [16]. Similar problems for harmonic functions were treated in exterior [17], [18] and interior [19], [20] domains. A nonlinear problem on stratified fluid flow over obstacles and wings has been studied in [21].

Acknowledgment.

The author is thankful to Professor K.Kajitani for the opportunity to carry out research in Japan. The support of the University of Tsukuba and the RFBR grant 99-01-01063 is greatfully acknowledged.

References.

1. Muskhelishvili N.I. *Singular integral equations.* Noordhoff, Groningen, 1972.

2. Krein S.G. (ed.) *Functional Analysis.* M.: Nauka, 1964.

3. Lifanov I.K. *Singular integral equations and discrete vortices.* VSP, Zeist, 1996.

4. Golberg M.A. *Numerical solution of Cauchy singular integral equations with constant coefficients.* J.Int. Eqs., 1985, v.9, p.127-151.

5. Golberg M.A. *The pertubed Galerkin method for Cauchy singular integral equations with constant coefficients.* Appl.Math.Comp., 1988, v.26, p.1-33.

6. Golberg M.A. *Discrete projection methods for Cauchy singular integral equations with constant coefficients.* Appl.Math.Comp., 1989, v.33, p.1-41.

7. Golberg M.A. (ed.) *Numerical solution of integral equations.* Plenum Press, N.Y., 1990.

8. Ioakimidis N.I. *On the weighted Galerkin method of numerical solution of Cauchy-type integral equations.* SIAM J.Num.Anal., 1981, v.18, p.1120-1127.

9. Ioakimidis N.I. *On the natural interpolation formula for Cauchy-type singular integral equations of the first kind.* Computing, 1981, v.26, p.73-77.

10. Ioakimidis N.I. *Further convergence results for the weighted Galerkin method of numerical solution of Cauchy-type singular integral equations.* Math. Comp., 1983, v.41, p.79-85.

11. Vladimiriov V.S. *Equations of mathematical physics*, Marcel Dekker, N.Y., 1971.

12. Krutitskii P.A. *Dirichlet problem for the Helmholtz equation outside cuts in a plane.* Comp. Maths. Math. Phys., 1994, v.34, No.8/9, pp.1073-1090.

13. Krutitskii P.A. *The Neumann problem on wave propagation in a 2-D external domain with cuts.* J.Math. Kyoto Univ., 1998, v.38, No.3, p.439-451.

14. Krutitskii P.A. *Wave propagation in a 2-D exterior domain with cuts.* Nonlinear Analysis, TMA, 1998, v.32, p.135-144.

15. Krutitskii P.A. *The Dirichlet problem for the dissipative Helmholtz equation in a plane domain bounded by closed and open curves.* Hiroshima Math. J., 1998, v.28, No.1, p.149-168.

16. Krutitskii P.A. *The Neumann problem for the 2-D Helmholtz equation in a domain, bounded by closed and open curves.* Int. J. Maths. Math. Sci., 1998, v.21, No.2, p.209-216.

17. Krutitskii P.A. *The 2-dimensional Dirichlet problem in an exterior domain with cuts.* Zeitschrift für Analysis und ihre Anwendungen, 1998, v.17, No.2, p.361-378.

18. Krutitskii P.A. *The 2-D Neumann problem in an exterior domain bounded by closed and open curves.* Math. Meth. Appl. Sci., 1997, v.20, p.1551-1562.

19. Krutitskii P.A. *The uniquely solvable integral equation for harmonic Dirichlet problem in a plane domain with cuts.* ZAMM, 1999, v.79, No.9, p.591-601.

20. Krutitskii P.A. *The Neumann problem in a plane domain bounded by closed and open curves.* Complex Variables, 1999, v.38, No.2, p.1-20.

21. Krutitskii P.A. *Fast stratified flow over several obstacles, including wings.* IMA Journal of Applied Mathematics, 1996, v.57, No.3, p.243-256.

82 ON COMPUTATION OF SOME ASYMPTOTIC CHARACTERISTICS OF VACUUM POLARIZATION IN SCHWARZSCHILD SPACE-TIME [1]

Juri M. RAPPOPORT

Department of Computer Science, Russian Academy of Sciences
Vlasov street, Building 27, Apt.8, Moscow 117335, Russian Federation

jmrap@landau.ac.ru

The evaluation of some asymptotic characteristics of the black hole quantum theory is carried out. The computations give the possibility of defining the vacuum polarization behavior of quantum scalar fields in Schwarzschild spacetime more precisely.

The computational analysis of the vacuum polarization behavior of quantum scalar fields in Schwarzschild spacetime [1] presents the significant physical interest. This work is devoted to the computation of some asymptotic characteristics of these physical phenomenons, in particular to the computation of the density of the virtual pairs of the massless scalar particles near the horizon of the events of the nonrotating black hole on orbital moment.

The following equation is obtained under the consideration of the problem of the vacuum polarization in the Schwarzschild spacetime on the horizon of the events [2]

$$\frac{d^2R}{d\xi^2} + \frac{(2s+1)}{\xi}\frac{dR}{d\xi} + \left[\frac{q(q-is)}{\xi^2} - 4l^2\right]R = 0, \tag{1}$$

[1] This work is supported partly by the U.S. Civilian Research and Development Foundation, grant RM1-361.

H.G.W. Begehr et al. (eds.), Proceedings of the Second ISAAC Congress, Vol. 1, 719–726.

for the determination of the function $R(\xi)$, where s - spin of the particle, l - orbital moment, q, s - dimensionless parameters.

The solution of (1) is expressed through modified Bessel functions $\xi^{-s}K_{s+iq}(2l\xi)$, $\xi^{-s}I_{-s-iq}(2l\xi)$. We have for the case $s=0$, representing the greatest interest, near the horizon of the events $R_l \sim a_l K_{iq}(2l\xi) + b_l I_{-iq}(2l\xi)$, where $a_l = e^{iq/2} l^{-iq}/(M\Gamma(-iq))$.

It follows from [2] that near the horizon of the events

$$\sum_{l=0}^{\infty}(2l+1)\,|R_l(w/r)|^2 \sim \frac{2}{M^2\Gamma(iq)\Gamma(-iq)}\int_0^\infty K_{iq}^2(2l\xi)\,dl = \frac{4w^2}{1-2M/r}. \tag{2}$$

Let us note, that the calculation of the sums, analogous (2), is realized on the basis of the numerical quadrature methods for arbitrary r in [3].

The following formula for the density function of the virtual particles near the horizon of the events of the black hole $\rho_l^{(r)}$ can be obtained on the basis of [2,3]

$$\rho_l^{(r)} = \frac{(2l+1)4(1-r_g/r)}{q^2\Gamma(iq)\Gamma(-iq)} K_{iq}^2(2l\xi), \tag{3}$$

where r - radius of the particle, r_g - gravitational radius, $\xi = r/r_g - 1$, ξ - dimensionless variable, l - orbital moment, w - frequency, q - particle's energy in dimensionless units, $q = 2wr_g$, $r_g \le r \le 3r_g$. Then the components of the matrix element of the square of the massless scalar field of the particles for the Unruh vacuum [2] have the form

$$f_l^{(r)} = \int_0^\infty q(\coth \pi q - 1)\rho_l^{(r)}(q)\,dq. \tag{4}$$

We obtain near the horizon, that $(u/\Phi^2/u) \sim (\sum_{l=0}^{\infty} f_l^{(r)} - 1/12)/(16\pi^2 r_g^2(1-r_g/r))$. The expression of the matrix element for the Unruh vacuum in the general case is adduced in [2]. Using the relation [4] $\Gamma(iq)\Gamma(-iq) = \pi/(q\sinh \pi q)$, it's possible to transform the expression (3) for the density of the virtual particles to the form

$$\rho_l^{(r)} = \frac{4(2l+1)(1-r_g/r)}{\pi}\frac{\sinh \pi q}{q} K_{iq}^2(2l\xi), \tag{5}$$

more convenient for calculations. Substituting the result of the (5) to (4), we simplify the expression for the components of the matrix element of the scalar field of the particles

$$f_l^{(r)} = \frac{4(2l+1)(1-r_g/r)}{\pi}\int_0^\infty e^{-\pi q} K_{iq}^2(2l\xi)\,dq. \tag{6}$$

So the computation of Macdonald functions of imaginary order is actual [5-7].

We perform the calculation of the density for $0.01 \leq q \leq 10$ on the basis of the usage of the programs [8] and carry out the investigation of its behavior depending on the distance from the centre of the black hole and orbital moment. The results of the computations permit to observe the law of the decreasing virtual particle's density as the radius function. It's possible to remark on the basis of the conducted estimations that the series giving the complete density of the pairs converges slow on the moments near the black hole but the convergence accelerates soon on the measure of the removal from it. The computation of the components is conducted also. The integrand in (6) decreases soon, therefore it's sufficient to perform the numerical integration on the finite interval, providing the execution of the calculations with the prescribed accuracy.

The results of the computations of the functions $\rho_l^{(r)}$ for $q = 1$, $l = 1$, $r/r_g = 1.5(0.5)3$ and $f_l^{(r)}$ for $l = 1$, $r/r_g = 1.5(0.5)3$ are collected in the special table.

Table

r/r_g	$\rho_l^{(r)}$	$f_l^{(r)}$
1.5	$.1236369 \cdot 10^{+01}$	$.6395799 \cdot 10^{-01}$
2.0	$.1889580 \cdot 10^{+00}$	$.7336420 \cdot 10^{-02}$
2.5	$.2392392 \cdot 10^{-01}$	$.8349492 \cdot 10^{-03}$
3.0	$.2933586 \cdot 10^{-02}$	$.1142627 \cdot 10^{-04}$

Thus the conducted computations make more precise the picture of the behavior of the vacuum polarization of the quantum scalar fields near the black holes.

Computation of some characteristics of dipole and quadrupole radiation

The effective radiation in the frames of classical electrodynamics is expressed by combination of the Macdonald function and its first derivative under the consideration of the problems of bremsstrahlung accompanying the scattering of the charged particle on the Coulomb centre [1,9].

The matrix elements of the operators of dipole and quadrupole moments are expressed through hypergeometric function $F(x)$ and its derivative [10,11]. We have for the case of Coulomb field of the attraction

$$F(x) = F(in_1, in_2, 1, x), \quad x = -4n_1 n_2 \sin^2(\Theta/2)/(n_1 - n_2)^2, \qquad (7)$$

where $n_1 = Ze^2m/p_1$, $n_2 = Ze^2m/p_2$ - Born parameters, P_1, P_2 - the impulses of the particles before and after the collision, $p_1 = |P_1|$, $p_2 = |P_2|$, m - its mass, Θ - the angle of the scattering, connected with the eccentricity of the corresponding classical orbit by the formula $\epsilon_{-1} = \sin(\Theta/2)$. The transition to the case of Coulomb fields of the repulsion is realized by the substitution $n_1 \to -n_1$, $n_2 \to -n_2$.

Let us denote $n = n_1 + n_2$, $r = n_2 - n_1$, $x_0 = (n_1 + n_2)/2\sqrt{n_1 n_2}$, $a = r\epsilon x_0$.

The following asymptotic decomposition, reducing the computation of $F(x)$ to the problem of the evaluation of the Macdonald function $K_{ir}(x)$ and its derivative $K'_{ir}(x)$, is deduced in [11]:

$$F(in_1, in_2, 1, x) =$$

$$\frac{(a/2e)^{in}}{\Gamma(in_1)\Gamma(in_2)} \frac{4}{\pi}(K_{ir}(a)[1 - \frac{i}{n}(\frac{2}{3} + a^2 - r^2)] + i\frac{a}{n}K'_{ir}(a) + O(n^{-2})). \quad (8)$$

Let us consider the bremsstrahlung arising from the scattering of the electron on the nucleus. The asymptotics of the matrix element of the operator of the dipole moment on the states of the continuous spectrum of the Coulomb field for large values of the order for Born parameter have the form [11]

$$d_{p_1p_2} = \frac{4\pi Ze^2}{(p_1 - p_2)^2} \frac{2e}{mw^2 n} e^{\pi r/2 + i\Phi}(ir(n_2p_2 + n_1p_1)(K_{ir}(a) + O(n^{-1}))$$
$$+a(n_2p_2 - n_1p_1)(K'_{ir}(a) + O(n^{-1}))), \quad (9)$$

where the phase multiplier

$$e^{i\Phi} = (\frac{mwa(1-x)}{(p_1+p_2)^2 e})^{in} \frac{\Gamma(-in_1)\Gamma(-in_2)}{\Gamma(in_1)\Gamma(in_2)},$$

and e - the charge of the electron; w - the frequency of the emitted photon.

The asymptotic expansions for the matrix elements of the quadrupole moments on the Coulomb states of the continuous spectrum are obtained and adduced also there. The properties and the computation of modified Bessel functions (Macdonald functions) are described in [8,12] in detail. The subsequent part of this section is devoted to the properties and computation of the derivative of this function $K'_{ir}(x)$.

Let us use the equality [4]

$$K_{\nu-1}(z) + K_{\nu+1}(z) = -2K'_\nu(z). \quad (10)$$

We obtain from here

$$K'_{ir}(x) = -\frac{K_{ir-1}(x) + K_{ir+1}(x)}{2} =$$

$$-\frac{K_{1-ir}(x) + K_{1+ir}(x)}{2} = -ReK_{1+ir}(x), \qquad (11)$$

where $ReK_{1+ir}(x) = \frac{K_{1+ir}(x)+K_{1-ir}(x)}{2}$; $ImK_{1+ir}(x) = \frac{K_{1+ir}(x)-K_{1-ir}(x)}{2i}$.

The functions $ReK_{1+ir}(x)$ and $ImK_{1+ir}(x)$ have the following integral representations:

$$ReK_{1+ir}(x) = \int_0^\infty e^{-x\cosh t}\cosh t\cos(rt)dt, \quad 0 < x < \infty, \quad 0 \le r < \infty; \qquad (12)$$

$$ImK_{1+ir}(x) = \int_0^\infty e^{-x\cosh t}\sinh t\sin(rt)dt, \quad 0 < x < \infty, \quad 0 \le r < \infty. \qquad (13)$$

The pair of functions $u_1(x) = ReK_{1+ir}(x)$ and $u_2(x) = ImK_{1+ir}(x)$ satisfy the system of the differential equations

$$u_1'' + \frac{1}{x}u_1' - (1 + \frac{1-r^2}{x^2})u_1 + \frac{2r}{x^2}u_2 = 0,$$
$$u_2'' + \frac{1}{x}u_2' - \frac{2r}{x^2}u_1 - (1 + \frac{1-r^2}{x^2})u_2 = 0. \qquad (14)$$

The functions $ReK_{1+ir}(x)$ and $ImK_{1+ir}(x)$ are connected with modified Bessel functions of the first kind by relations

$$ReK_{1+ir}(x) = \frac{\pi}{\sinh \pi r}\frac{ImI_{1+ir}(x) - ImI_{-1-ir}(x)}{2}, \qquad (15)$$

$$ImK_{1+ir}(x) = \frac{\pi}{\sinh \pi r}\frac{ReI_{-1-ir}(x) - ReI_{1+ir}(x)}{2}. \qquad (16)$$

The expansion of $I_{1+ir}(x)$ on positive degrees of x have the form

$$I_{1+ir}(x) = \frac{x}{2}(\cos(r\ln\frac{x}{2}) + i\sin(r\ln\frac{x}{2}))\sum_{k=0}^{\infty}\frac{(x^2/4)^k}{k!\Gamma(k+2+ir)} = \sum_{k=0}^{\infty}(a_k + ib_k), \quad (17)$$

where a_k and b_k are connected by the following recurrent relations:

$$a_0 + ib_0 = \frac{x}{2}\frac{\cos[r\ln(x/2)] + i\sin[r\ln(x/2)]}{\Gamma(2+ir)},$$
$$m_k = \frac{x^2(k+1)}{4k((k+1)^2+r^2)}, \quad n_k = \frac{x^2 r}{4k((k+1)^2+r^2)}, \qquad (18)$$
$$a_k = a_{k-1}m_k + b_{k-1}n_k, \quad b_k = b_{k-1}m_k - a_{k-1}n_k.$$

The expression of $I_{-1-ir}(x)$ on positive degrees of x have the form

$$I_{-1-ir}(x) = (\frac{x}{2})^{-1}(\cos(r\ln\frac{x}{2}) - i\sin(r\ln\frac{x}{2}))\sum_{k=0}^{\infty}\frac{(x^2/4)^k}{k!\Gamma(k-ir)} = \sum_{k=0}^{\infty}(c_k + id_k). \tag{19}$$

The following recurrent relations are convenient for practical application

$$\begin{gathered} c_0 + id_0 = (\frac{x}{2})^{-1}\frac{\cos[r\ln(x/2)] - i\sin[r\ln(x/2)]}{\Gamma(-ir)}, \\ p_k = \frac{x^2(k-1)}{4k((k-1)^2+r^2)}, \quad q_k = \frac{x^2 r}{4k((k-1)^2+r^2)}, \\ c_k = c_{k-1}p_k - d_{k-1}q_k, \quad d_k = d_{k-1}p_k + c_{k-1}q_k. \end{gathered} \tag{20}$$

The expansions (17), (19) converge for all $0 < x < \infty$ and $0 \le r < \infty$. It follows immediately from (15)-(17), (19) that for $x \to 0$

$$K_{1+ir}(x) \sim \frac{\pi}{\sinh(\pi r)}\frac{\sin[r\ln(x/2)] + i\cos[r\ln(x/2)]}{\Gamma(-ir)}\frac{1}{x}. \tag{21}$$

The program for the computation of the values of these functions in the domain $0.1 \le x \le 10$, $0 \le r \le 10$ with the accuracy on 8 significant digits was constructed by author. The formulas (15)-(20), giving the expansions of these functions into the series, their integral representations (12), (13) and the vector form of the Lanczos's tau-method [13] were used for the computations.

The formulas (15)-(20) are convenient for the implementation on computers , the calculations on them are realized fast, but they give the necessary precision (8 significant digits) not in all domain of the alteration of the values x and r. The Chebyshev polynomials of the 16-th degree were taken for the approximation by means of the tau-method. The approximating polynomials of more high degree are necessary for the increase of r, but too big computational error arises under its usage. In the remaining domain were used the integral representations (12), (13), describable for the convenience in the form

$$\begin{aligned} ReK_{1+ir}(x) &= e^{-x}\int_0^A e^{(1-\cosh t)x}\cosh t\cos(rt)dt, \\ ImK_{1+ir}(x) &= e^{-x}\int_0^A e^{(1-\cosh t)x}\sinh t\sin(rt)dt \end{aligned}$$

and computable on the basis of Filon's formulas.

The value of the upper limit of the integration A, ensuring the evaluation of the integrals with the error, no bigger then 5×10^{-9}, was taken from condition $e^{x(1-\cosh A)}\cosh A$

$< 10^{-8}$. The coincidence of the values, obtained on different formulas, on the boundaries of the domains of the calculation on them was used for the control. The results coincide with the results obtained on some universal programs but it's required the mean duration of the computation of one point of values by our programs with the accuracy in eight significant digits approximately on the order smaller.

The spectrum of the effective radiation accompanying the scattering of a charged particle on the Coulomb centre for the dipole radiation has the form [1]

$$\frac{dk_w^{(\pm)}}{dw} / \frac{16Z^3e^5}{3m^2v_0^2} = F_1^{(\pm)}(\nu),$$
$$F_1^{(\pm)}(\nu) = -\nu e^{\pm\pi\nu} K_{i\nu}(\nu) K'_{i\nu}(\nu), \tag{22}$$

where k_w the effective radiation, w is the frequency, e is the charge of particle, m is its mass, v_0 is the initial velocity, and $\nu = w/mv_0^3$. The signs plus and minus correspond to the attraction and repulsion, respectively.

It is possible to obtain the following expression for the spectrum of the bremsstrahlung for the quadrupole case on the basis of [10, 11]

$$\frac{d\Sigma^{(\pm)}}{dw} / \frac{64G}{15} = F_2^{(\pm)}(\nu),$$

$$F_2^{(\pm)}(\nu) = e^{\pm\pi\nu} \int_\nu^\infty dx (x^2 - \nu^2)\{[(x^2 - 2\nu^2) + 3(x^2 - \nu^2)^2] K_{i\nu}^2(x)$$
$$+ x^2[1 + (x^2 - \nu^2)] K_{i\nu}'^2(x) - x[3x^2 - 4\nu^2] K_{i\nu}(x) K'_{i\nu}(x)\}, \tag{23}$$
$$\nu = \frac{Za}{2\beta_0}\frac{hw}{E}, \quad a \approx \frac{1}{137}, \quad \beta_0 = \frac{v_0}{c}, \quad E = \frac{p_0^2}{2m} = \frac{mv_0^2}{2},$$
$$0 \le \nu \le \frac{Za}{2\beta_0} \approx \frac{Z}{274\beta_0} \sim 10^3 \div 10^4.$$

The programs of the computation of $F_1^{(\pm)}(\nu)$ and $F_2^{(\pm)}(\nu)$ were elaborated on the basis of the programs of the evaluation of $K_{ir}(x)$ and $K'_{ir}(x)$ for the values of ν in the limits $0 \le \nu \le 10$. The numerical calculations gave the coincidence in the behavior of $F_1^{(\pm)}(\nu)$ and $F_2^{(\pm)}(\nu)$ with the results of analytical considerations [1] for the limiting cases of the large and small frequencies and allowed to make more precise their behavior for the intermediate values of the frequencies.

References

[1] Landau L.D., Lifschitz E.M.: *The theory of the field*; Nauka, Moscow, 1973. -304p.

[2] Candelas P.: Vacuum polarization in Schwarzschild spacetime; *Phys. Rev.* **21**, No. 8 (1980), 2185–2202.

[3] Fawcett M.S.: The energy-momentum tensor near black hole; *Commun. Math. Phys.* **89**, No.1 (1983), 103–115.

[4] Lebedev N.N.: *Special functions and their applications*; Fizmatgiz, M.-L., 1963. -358p.

[5] Lozier D.W., Olver F.W.J.: Numerical evaluation of special functions.-NISTIR 5383, Gaithersburg, MD, 1994.-47p.

[6] Lozier D.W., Olver F.W.J.: Airy and Bessel functions by parallel integration of ODEs; in Proceedings 6-th SIAM Conf. Scientific Computing, 1994, v.2, 531–538.

[7] Ehrenmark U.T.: The numerical inversion of two classes of Kontorovitch-Lebedev transform by direct quadrature; *Journ. Comp. Appl. Phys.*, **61**, (1995), 43–72.

[8] Rappoport J.M.: The programs and some methods of the Macdonald's function computation. - M., Akad.Nauk SSSR, OVM, 1991.-16p.

[9] Berestetski V.B., Lifschitz E.M., Pitaevski L.P.: *The relativistic quantum theory*; Nauka, Moscow, 1968, part 1. - 480p.

[10] Galtsov D.V., Grats Y.V.: The gravitational radiation for Coulomb collisions; *Vestnik Vuzov, Physics*, No. 2 (1974), 94–99.

[11] Galtsov D.V., Grats Y.V.: The semiclassical limit in the problem of the brems-strahlung; *Theor. and Math. Phys.*, **28**, No. 2 (1976), 201–210.

[12] Zurina M.I., Karmazina L.N.: *The tables of the modified Bessel function of imaginary order*; M., VC AN SSSR, 1967.-342p.

[13] Rappoport J.M.: The Tau-method and the computation of the Bessel functions of the complex order; in Proceedings of the Cornelius Lanczos International Centenary Conference, SIAM, Raleigh, 1994, 353–355.

83 COMPUTATION OF ONE-PARAMETRIC FAMILIES OF SOLUTIONS OF NONLINEAR EQUATIONS

Vladilen A. TRENOGIN

Department of Mathematics, Moscow State Steel and Alloys Institute
Leninsky prospect 4, Moscow, 117936, Russia

vladilen@trenogin.misa.ac.ru

1991 Mathematics Subject Classification: 58 E 07

Abstract. *Nonlinear analytical equation with the operator which maps one real Banach space to another one is considered. General Lyapunov-Schmidt bifurcation equation is deduced. On its basis the partial bifurcation equation describing the set of all small solutions with given tangent vector is extracted. The proposed approach leads to natural parametrizations of the curves of solutions and uses the abstract variant of Newton diagram method.*

Introduction.

The aim of this paper is to consider the different possible cases of one-parametric analytical manifolds defined by nonlinear equations in the Banach spaces near nonregular point and to investigate their bifurcations (see [1-2]).
Let X, Y be real Banach spases and let $F(x)$ be the nonlinear operator which maps the neighborhood S of zero of X in a neighborhood of zero of Y and also $F(0) = 0$. Below we investigate the situations where the equation

$$F(x) = 0 \tag{1}$$

has near the point 0 several smooth one-parametric families of solutions (the curves of solutions). Moreover we use the natural parametrizations

H.G.W. Begehr et al. (eds.), Proceedings of the Second ISAAC Congress, Vol. 1, 727–735.

of these curves. The simplest such problem was considered in the paper [3-4]. Our approach is a some combination of the methodic from [3-7] and [1]. The method of [1] gives us the possibility to investigate completely all different one-parametric famililies of small solutions. For the case of equation (1) our approuch consists in the construction and analysis of different corresponding bifurcation equations. As in [3-4] the equation (1) does not contains free parameters. The case of small real parameter was considered in [8]. The questions of optimal parametrization of multiparameter family of solutions were studied in [9].

We represent the equation (1) in the form

$$Bx = \sum_{k=2}^{\infty} R_k x^k, \tag{2}$$

where the operator $B \in L(X, Y)$ and $R_k x^k$ are k-power operators. We require that B satisfy the following conditions (the designations from [1-2] are used: $N = N(B)$ is the kernel of B, $W = R(B)$ is the range of B).

I. *There exists the projector $P \in L(X)$ which is not equal to 0 or to I such that $PX = N$. Set $U = (I - P)X$, then the direct decomposition $X = N \bigoplus U$ is valid.*

II. *There exists the projector $Q \in L(Y)$ which is not equal to 0 or to I such that $(I - Q)Y = W$. Set $QY = Z$, then the direct decomposition $Y = Z \bigoplus W$ is valid.*

Note that we don't suppose the finite dimensionality of N or Z. The conditions I and II mean the topological completability of subspaces N and W in the spaces X and Y respectively (see [5]).
For the simplicity we assumed that the right hand side of equation (2) is the analytical operator at the point $x = 0$. Otherwise we shall require realization of the following condition.

III. *There exists $r > 0$ such that the following numerical serie $\sum_{k=2}^{\infty} \| R_k \| r^k$ converges.*

¿From III it follows that for all x such that $\| x \| \leq r$ the abstract power serie $\sum_{k=2}^{\infty} R_k x^k$ converges absolutely and uniformely.

General bifurcation equation.

At first we shall give short deduction of general Lyapunov-Shmidt bifurcation equation (BE) which describes all zero set of equation (2) near the point 0. In contrast to the initial equation (2) BE doesn't contain the linear term, but BE in more narrow spaces is given. The analogues construction for the equations with parameters we offered in [6]. Let $B_U \in L(U, W)$ be the operator defined by $u \in U$ by formula $B_U u = Bu$. Note that B_U has inverse operator $\Gamma \in L(W, U)$. Set in (2)

$$x = \varphi + u, \quad \varphi \in N, \quad u \in U \tag{3}$$

and project (2) on W and on Z. Thus we obtain the system equivalent to (2)

$$B_U u = \sum_{k=2}^{\infty}(I - Q)R_k(\varphi + u)^k, \tag{4}$$

$$\sum_{k=2}^{\infty} QR_k(\varphi + u)^k = 0. \tag{5}$$

Thanks of the analytical variant of implicit operator theorem (see, for example, [1-2]) the equation (4) has the unique small analytical solution represented by the power serie

$$u = \sum_{s=2}^{\infty} u_s\varphi^s. \tag{6}$$

The terms of expressions (5) may be consecutively computed:
$u_2\varphi^2 = \Gamma R_2\varphi^2, \quad u_3\varphi^3 = \Gamma(2R_2\varphi u_2\varphi^2) + R_3\varphi^3, \ldots$
Setting it in (5) we obtain the Lyapunov-Schmidt bifurcation equation (BE) which also is analytical:

$$\sum_{r=2}^{\infty} L_r\varphi^r = 0. \tag{7}$$

For every solution φ of BE (7) we get the corresponding solution of (2) by formula (3) with $u = u(\varphi)$. If Θ is a set of all small solutions of BE, then the expression

$$x = \varphi + u(\varphi), \quad \varphi \in \Theta \tag{8}$$

gives us the complete representation of solutions of equation (2). Note that the terms of BE are also consecutively computed
$L_2\varphi^2 = QR_2\varphi^2, \quad L_3\varphi^3 = Q(2R_2\varphi u_2\varphi^2 + R_3\varphi^3),$

$L_4\varphi^4 = Q(2R_2\varphi u_3\varphi^3 + 3R_3\varphi^2 u_2\varphi^2 + R_4 q\varphi^4), \ldots$

In conclusion of this section we shall consider the case when the first term of BE is the term more high degree than 2. Suppose that the operator $R(x)$ satisfies the following special condition which generalize condition introduced in [4].

IV. *There exists the natural number* $l \geq 2$, *such that if* $l = 2$, *then* $QR_2\varphi^2 \neq 0$ *identically on* N, *and if* $l > 2$, *then* $R_k\varphi^k = 0, k = 2, ..l-1$ *on* N *but* $QR_l\varphi^l \neq 0$ *identically on* N.

For $l > 2$ the system (4)-(5) may be written in the form

$$B_U u = \sum_{k=2}^{l-1}(I-Q)[R_k(\varphi+u)^k - R_k(\varphi)^k] + \sum_{k=l}^{\infty}(I-Q)R_l(\varphi+u)^l, \quad (9)$$

$$\sum_{k=2}^{\infty} QR_k(\varphi+u)^k = 0. \quad (10)$$

Now from the first equation we get

$$u = \sum_{k=l}^{\infty} u_k\varphi^k,$$

where

$u_l = \Gamma(I-Q)R_l\varphi^l, \quad u_{l+1} = \Gamma(I-Q)R_{l+1}\varphi^{l+1} + \Gamma(I-Q)R_2 u_l\varphi$ and so on.

Here BE also has the special form

$$\sum_{r=l}^{\infty} L_r\varphi^r = 0. \quad (11)$$

It can be shown that

$L_l\varphi^l = QR_l\varphi^l, \quad L_{l+1}\varphi^{l+1} = QR_{l+1}\varphi^{l+1} + 2QR_2\varphi\Gamma(I-Q)R_l\varphi^l$ and so on.

These two variants of BE give us the formal possibility to investigate completely the set of all solutions of the equation (2). Below we consider the case of one-parametric solutions which is most often occured in practice.

Bifurcation equation in direction of given vector.

Here we shall deduce the BE in the direction of given tangent vector.

Lemma 1. *Suppose the conditions I,II,III are valid. If there exists the vector* $\varphi_* \neq 0$ *such that the equation (2) has the one-parametric family of solutions with small parameter* s *of the form*

$$x = s\varphi_* + sz(s), \quad z(s) \to 0, \quad s \to 0, \tag{12}$$

then $\varphi_* \in N$.

Lemma 2. *Let* $\varphi_* \in N$. *Fix the functional* $\gamma_* \in N^*$ *such that* $< \varphi_*, \gamma_* >= 1$.
The operator $P_* \in L(N)$ *defined by* $\varphi \in N$ *as* $P_* = I- < .,\gamma_* > \varphi_*$ *is a projector in* N. *Besides* $P_*\varphi_* = 0$ *and* $P_*U = 0$. *If* $\varphi \neq k\varphi*$ *for all* $k \in R$, *then* $P_*\varphi \neq 0$.

Introduce one-dimensional subspace $N_* = sp\{\varphi_*\}$ with basic vector φ_* and the subspace $V = P_*N$. Now we have the direct decomposition

$$N = N_* \check{+} V. \tag{13}$$

Lemma 3. *Let the conditions I.II.IV are fulfilled. If the equation (2) has the solution in the form (12) then* φ_* *is a solution of the following equation*

$$QR_l\varphi^l = 0. \tag{14}$$

Corollary. *The equation (14) we shall call determining equation. Only its solutions may serve the possible tangent vectors to one-parametric families of solutions of the equation (2).*

Corollary. *Let* l *even. Consider the expression* $< R_l\varphi^l, \psi >$ *defined at all* $\varphi \in N$. *If this expression by all* $\varphi \neq 0$ *has the fixed sign at least for one* $\psi \in N(B^*)$, *then determining equation hasn't nontrivial solutions. In this case the initial problem (2) hasn't nontrivial small solution with the tangent vector* φ_*.

By using the representation of N from lemma 2 we will look for the solutions of (11) in the form

$$\varphi = s(\varphi_* + v), \tag{15}$$

where $v = v(s) \in V$, $v = v(s) \to 0$. Under the necessary condition (14) we have the following operator equation with real parameter s for $v = v(s)$

$$\sum_{k=1}^{l} C_l^k L_l \varphi_*^{l-k} v^k + \sum_{r=l+1}^{\infty} L_r(\varphi_* + v)^r s^{r-l} = 0 \tag{16}$$

where C_l^k are binomial coefficients. This equation we will call φ_*-BE or bifurcation equation in direction of the tangent vector φ_*. Note that for every such tangent vector we can construct the corresponding BE. For every solution $v(s)$ of φ_*-BE we obtain the one-parametric family of solutions of the initial equation (2) by means of formula (15). At first we consider the more simple case. Let M_{10} is the restriction of operator $lL_l\varphi_*^{l-1}$ on V.

Definition. *Consider the nonlinear operator $M(\varphi) = 0$ which maps a neighborhood S of the point $\varphi_* \in V$ into a neighborhood of the point $0 \in Z$ where V and Z are Banach spaces. Suppose M is differentiable (in Freshe sense) in φ_*. We call φ_* the simple solution of equation $M(\varphi) = 0$ iff $M'(\varphi_*)$ is continuously invertible operator that is the operator $M'(\varphi_*)$ maps V on Z bijectively.*

Theorem. *If φ_* is the simple solution of the determining equation then BE (16) has unique small analytic solution $v(s)$ at $s = 0$. This small solution $v{=}v(s)$ generates the analytisal branch (15) and (8) of solutions of initial equation (2) with tangent vector φ_* by formula $x = s(\varphi_* + v(s) + u(s, v(s))$.*

Proof follows from analytical variant of implicit operator theorem. Note that the result of theorem is contained in [3] and for $l = 2$ in [4]. Let (14) has more than one simple solution. According to theorem for every simple solution of the determining equation there exists the corresponding one-parametric family of solutions of initial equation (2). So in the point $x = 0$ several analytical branches of solution of (2) are intersected. That is the bifurcation portrait. The situations where the solution φ_* isn't simple solution of (14) need in the additional investigation which will be done below.

Newton diagram and some essentially nonlinear equations in the Banach spaces.

This section is devoted to some common cases of computations of one-parametric solutions families with given tangent vector φ_*. Now we can rewrite φ_*−BE (16) in the form

$$M_{10}v + M_{01}s + \sum_{i+j\leq 2} M_{ij}v^i s^j = 0. \tag{17}$$

That is standard nonlinear equation wich was analyzed in [10]. Further we assume that the operator M_{10} is generalized Noeter operator in sense [5]: its kernel $N(M_{10})$ and its range $R(M_{10})$ are closed and have the closed topological supplemmments (see conditions I,II). In Theorem we studied the case if $N(M_{10}) = 0$ and $R(M_{10}) = Y$. Consider the additional different possible cases.
1) If $N(M_{10}) \neq \{0\}$ and $R(M_{10}) = Z$, then we have many-parametric situation which now is not studied.
2) If $N(M_{10}) = \{0\}$ and $R(M_{10}) \neq Z$, then we have one-parametric family of small solutions by the additional condition fulfilment.
3) If $N(M_{10}) \neq \{0\}$ and $R(M_{10}) \neq Z$, then we have the situation where the linear part can be droped by means of Lyapunov-Schmidt procedure (see [1]) as in the section 1.
4) If $M_{10} = 0$ we obtain the little-studied situation. Small solutions of this BE are connected with the nonstandard algebraic equation in the Banach spaces.
At first we consider cases, where the increasing part of Newton diagram consists of unique segment. Another possibilities may be considered analogously to the classical Newton diagram metod applications (see [1]). Here we give some conclusions for BE (16). Let $L_l\varphi_*^{l-1} = 0$ but $L_l\varphi_*^{l-2}v^2 \neq 0$ on V identically and $L_{l+1}\varphi_*^{l+1} \neq 0$. Then principal term of v has the form $v_{\frac{1}{2}}\sqrt{|s|}$. This term from the following determinig "algebraic" equation $\frac{1}{2}l(l-1)L_l\varphi_*^{l-2}v_{\frac{1}{2}}^2 = -\chi L_{l+1}\varphi_*^{l+1}$ with $\chi = signs$ must be defined. This is quadratic equation in Banach spaces which one must investigate. Similar equations are studied, for example, in [10] (see references there). More general case appears where $L_{l+1}\varphi_*^{l+1} \neq 0$ but $L_l\varphi_*^{l-j}v^j = 0$ on V at $j = 2, ...r-1$ and $L_l\varphi_*^{l-r}v^r \neq 0$ on V identically. Now the principal term of v has the form $v_{\frac{1}{r}}|s|^{\frac{1}{r}}$ and must be defined from the equation $C_l^r L_l\varphi_*^{l-r}v^{1/r} = -\chi L_{l+1}\varphi_*^{l+1}$. If by these cases the corresponding principal term is a simple solution of the corresponding equation (see Definition) then analogously theorem we obtain by using implicit operator theorem the unique correction term of small solution to its leading term. Otherwise, if this term is not simple solution, we

have to use Lyapunov-Schmidt procedure for annual the linear part as we make by BE obtaining. The corresponding precise statements of these propositions can be made as by Theorem formulation.

In conclusion we consider some addition to theorem, where two branches of solutions with the given tangent vector may exist. This example shows an important role of continuously invertibility of operator $M_{10} = L_l \varphi_*^{l-1}$ in the Theorem. Here the bifurcation of one-parametric family of solutions is observed.

Example. Consider the case $l = 2$. Suppose $n = 2$ and fix the basis φ_1, φ_2 in N. Suppose $\dim N(B^*) = 1$ and fix the basic vector ψ in it. We shall look for the tangent vector φ_* as $\varphi_* = c_1\varphi_1 + c_2\varphi_2$. The condition (14) is the equality to zero of quadratic function $< R_2\varphi_1^2, \psi > c_1^2 + 2 < R_2\varphi_1\varphi_2, \psi > c_1c_2 + < R_2\varphi_2^2, \psi > c_2^2 = 0$. If this function is positive or negative that is its discriminant $\Delta < 0$, then the equation (2) hasn't the small real family of solutions with the tangent vector φ_*. If this discriminant is positive then there exist two linearly independant tangent vectors and two real branches of solutions differentiable in zero. If $\Delta = 0$, then there exists one double tangent vector φ_*. Without loss of generality one can suppose that it is φ_1. In this case $< R_2\varphi_1^2, \psi >= 0$, c_1 is an arbitrary real number and ▪ $2 = 0$. We can set $< R_2\varphi_1\varphi_2, \psi >= 0$, $\alpha = R_2\varphi_2^2, \psi >\neq 0$ and $\beta =< R_2\varphi_1$束 $_2\varphi_1^2, \psi >\neq 0$. One can show that by $s \to 0$ the following asymptotic representation is valid $x(s) = s\varphi_1 \pm s\sqrt{s\beta/\alpha}\varphi_2 + o(| s |^{3/2})$. Thus under this conditions the equation (2) has 2 small real branches of solutions defined in that half-neighborhood of the point $s = 0$ where the subradical expression is positive. In this case the curve of solution has the edge. By the general situation for every segment of Newton diagram we develop the analogues technique. This propositions may be written as definite rigorous mathematical assertions analogously to mentioned theorem.

References.

[1]. *Vainberg N.M.,Trenogin V.A.* Theory of Branching of Solutions of Noordhoff Int. Publ., Leyden, Netherlands, 1974 (Russian original: Nauka, Moscow, 1969; German transl. Acad. Verlag, Berlin, 1973).

[2]. *Trenogin V.A.* Functional Analysis. Nauka, Moscow, 1980, 440 p (in Russian). (2d Ed. 1993, French.transl. Mir Publ., 1985) .

[3]. *Crandall M.G.,Rabinowitz P.H.* Bifurcation from simple eigenvalues. J.Fuct.Anal.8, p. 321-340 (1971).

[4]. *Magnus R.J.* On the local structure of the zero set of a Banach space valued mappings. J.Fuct.Anal. 22, p. 58-72 (1976).

[5]. *Trenogin V.A.* Generalized Noeter and Fredholm Operators. Abstr.of Int. Conf. to 90 years of acad. Nicolsky S.M.,Moscow, 1995 (in Russian).
[6]. *Trenogin V.A.* Improvement of invertibility of linear operators and Lyapunov-Schmidt branching equation. Vestnik RUDN, Math.Serie 3(1), 1996, p. 127-142 (in Russian).
[7]. *Trenogin V.A.* One-parametric solutions families of nonlinear equations and DE in Banach spaces. Works of Int. Conf. to 75 years of Kudryavzev L.D., v.2, p. 161-165, RUDN Publ., 1998 (in Russian).
[8]. *Sidorov N.A.* Explicite and implicite parametrization by construction of branching solutions by iterational methods. Math.Sbornik, .186(2), 1995, p. 129-144.
[9]. *Trenogin V.A.* Lyusternik's Theorem and the Best Parametrization of Solutions to Nonlinear Equations, Funct.Anal. and Its Appl. vol. 32(1), p. 1-4.
[10]. *Argyros I.K.* On quadratic equation. Rev.Acad.Cienc.Zaratoga Ser.2, 44 (1989). p. 23-32.

84 PARAMETER CONTINUATION METHOD FOR ORDINARY DIFFERENTIAL EQUATIONS

VETEKHA V. G.

Mathematical Analysis Chair
The Moscow State Institute of Electronics and Mathematics
Bolshoi Trekhsvyatitelsky pereulok, 3/12, Moscow, 109028, Russia

Abstract

The article is devoted to the application of parameter continuation method to solving the boundary value problem for the ordinary differential equations (DE) of second order. For the investigated DE the *a priori* estimate of the solution and the approximate solution are received on the basis of parameter continuation method. The error of the approximate solution is based on the coerciveness of the integral equation corresponding to the investigated boundary value problem.

Introduction

Parameter continuation method is inclusion of the equation $P(x) = 0$ into the one-parametric family of equations $F(x, \lambda) = 0, \quad \lambda \in [0; 1]$ connecting the given equation $(\lambda = 1)$ with solvable equation $(\lambda = 0)$ and study the dependence of the solutions from the parameter. Parameter continuation method was suggested and developed by Bernstein and Schauder [1], [2] for the proof the theorem of the existence of solution of the nonlinear elliptic boundary value problems. The abstract case of continuation method for linear operators was considered in [3]. Other situations were considered in [4-11].

H.G.W. Begehr et al. (eds.), Proceedings of the Second ISAAC Congress, Vol. 1, 737–742.

For example, we consider the result of the work [11]. Let us consider the following equation in Banach space X

$$x + Dx = y, \quad D(0) = 0$$

where D is nonlinear operator, and let D be Lipshitz continuous operator:

$$||Dx_1 - Dx_2|| \leq L \cdot ||x_1 - x_2||, \quad L = const > 0.$$

Suppose the mapping $A(\lambda)x = x + \lambda Dx$ satisfies the following condition $\exists \gamma > 0 : \forall x_1, x_2 \in X$

$$||A(\lambda)x_1 - A(\lambda)x_2|| \geq \gamma ||x_1 - x_2||, \quad \lambda \in [0; 1].$$

It is easy to verify that $\gamma \leq 1$. From [7] it follows that for every $\lambda \in [0; 1]$ the equation $A(\lambda)x = y$ has the unique solution and that from 0 to 1 we can go by $N \geq [L/\gamma] + 1$ steps by parameter λ (step by parameter λ is $\Delta\lambda = 1/N$). Let us x_n is the solution of the equation

$$A(\lambda_n)x = y, \quad \lambda_n = n \cdot \Delta\lambda,$$

Let $x_{n,s}$ be the s-th iteration of method of serial approximations (with contractive coefficient $q = L/(N\gamma) < 1$)

$$x_{n,k} = A^{-1}(\lambda_{n-1})(y - \Delta\lambda D x_{n,k-1}), \quad x_{n,0} = x_{n-1,k}$$

in the n-th step by the parameter λ. On each step by the parameter λ the s iterations of method of serial approximations are made. In the case, considered above, we proved the following proposition:

THEOREM 1. The estimations

$$||x_n - x_{n,s}|| \leq \frac{1 - b^n}{\gamma(1 - b)} \cdot \frac{q^s(1 + q)}{1 - q} \cdot ||y||,$$

$$b = \frac{q^s(1 + q)}{1 - q}, \quad q = \frac{L}{N\gamma} < 1,$$

are valid. If $n = N$ we have the estimation of approximate solution for the equation (0.1).

Therefore we prove the convergence of the net method [3]

$$\{x_{n,s}\} \xrightarrow{T} x_n.$$

The Gaponenko's [5] estimation ($\gamma = 1$) is

$$||x_n - x_{n.s}|| \leq q^{n+1} \frac{e^L - 1}{(1 - q)(e^q - 1)} \cdot ||y||, \quad q = \frac{L}{N} < 1.$$

THE APPLICATION OF THE PARAMETER CONTINUATION METHOD TO THE BOUNDARY VALUE PROBLEM

Consider the differential equation (DE) on $(0;\ \pi)$

$$z'' + g(t, z) = f(t), \quad g(t, 0) = 0 \tag{1}$$

with the boundary conditions

$$z(0) = z(\pi) = 0. \tag{2}$$

Suppose the function $g(t, z)$ is continuous on t and z, $f(t)$ is continuous on $(0;\ \pi)$,

$$M_\pi = ||f||_C = \max_{t \in [0;\,\pi]} |f(t)|$$

and inequality holds

$$\alpha_1^2 \le \frac{g(t, z_1) - g(t, z_2)}{z_1 - z_2} \le \alpha_2^2, \tag{3}$$

where $1 < \alpha_1 < \alpha_2 < 2$.

We proved the following theorem:

THEOREM 2. There exists the unique solution of the boundary value problems (1), (2) and the estimate holds

$$||z(t)||_C \le M_\pi E, \quad \text{where} \tag{4}$$

$$E = \max \left\{ 1 + \sqrt{1 + \left(\frac{4\pi}{\alpha_1^2} - \frac{4}{\alpha_1^3} \tan \frac{\pi \alpha_1}{2} \right)^2};\right.$$

$$\left. 1 + \sqrt{1 + \left(\frac{\pi}{4} + \frac{4}{\alpha_2^2} \left| \frac{\pi}{2} - \frac{1}{\alpha_2} \tan \frac{\pi \alpha_2}{4} \right| \right)^2} \right.$$

REMARK 1.

The solution of the boundary value problem $(\alpha l \ne \pi n, n = 0, 1, ...)$

$$x'' + \alpha^2 x = f(t), \quad x(0) = x(l) = 0 \tag{5}$$

may be written in the form

$$x(t) = -G_\alpha f \tag{6}$$

where the Green operator G_α is expressed by formula

$$(G_\alpha \varphi)(t) = \int_0^l K_\alpha(t, s) \varphi(s) ds \tag{7}$$

with the corresponding Green function

$$K_\alpha(t,s) = \begin{cases} \dfrac{\sin\alpha s \sin\alpha(l-t)}{\alpha\sin l\alpha}, & s \le t \\ \dfrac{\sin\alpha t \sin\alpha(l-s)}{\alpha\sin l\alpha}, & s > t \end{cases}$$

The following estimation is valid $||G_\alpha\varphi||_{L_2} \le \omega_\alpha ||\varphi||_{L_2}$

$$\text{where} \quad \omega_\alpha = \frac{1}{|\alpha\sin l\alpha|}\left(\frac{l\sin 2\alpha l}{8\alpha} - \frac{\sin^2\alpha l}{2\alpha^2} + \frac{l^2}{4}\right)^{1/2}.$$

The boundary value problem (1), (2) we can rewrite as integral equation

$$z(t) + Dz(t) = y(t), \text{ where} \tag{8}$$

$$Dz = G_\alpha\left[\alpha^2 z(s) - g(s, z(s))\right], \quad y = -G_\alpha f, \quad \alpha \in [\alpha_1, \alpha_2]. \tag{9}$$

We prove the theorem about the existence and uniqueness of the solution of boundary value problem by using parameter continuation method. For this purpose we consider the integral equation with parameter

$$Q(\lambda)z = z + \lambda Dz = y, \quad \lambda \in [0;1]. \tag{10}$$

LEMMA 1. If

$$Q(\lambda)z_1 - Q(\lambda)z_2 = -G_\alpha\varphi,$$

where $\varphi(t)$ is continuous function, then the estimate $||z_1 - z_2|| \le ||\varphi(t)||_C \cdot E$ holds, where E is defined by (4).

If we consider the operator

$$\Phi = \frac{d^2}{dt^2} + \alpha^2, \quad 1 < \alpha < 2 \quad \text{and} \quad y_i(0) = y_i(\pi) = 0, \quad i = 1;2,$$

then on the base of the Lemma 1 we receive $Q^{-1}(\lambda)$ - coerciveness of operator Φ [6]:

$$||\Phi y_1 - \Phi y_2||_C \ge \frac{1}{E}||Q^{-1}(\lambda)y_1 - Q^{-1}(\lambda)y_2||_C, \ \forall\lambda \in [0;1]. \tag{11}$$

Suppose we know the approximate solution $\hat{z}$ of the equation

$$Q(\lambda)z = y.$$

Let z_T be the exact solution of this equation, i.e. $Q(\lambda)z_T = y$, while $\tilde{z}$ the exact solution of the equation

$$Q(\lambda + \Delta\lambda)z = y, \quad \Delta\lambda \leq \frac{1}{3E}. \tag{12}$$

Suppose $||z_T - \hat{z}||_C \leq \delta$. We find the approximate solution of the equation (12) by using the iterative procedures:

$$z_{n+1} = Q^{-1}(\lambda)(y - \Delta\lambda D z_n), \quad z_0 = \hat{z}. \tag{13}$$

This procedures is convergent (we use here inequalities (11) and (3)):

$$||z_{n+2} - z_{n+1}||_C \leq q||z_{n_{n+1}} - z_n||_C, \text{ where } \quad q = 3E\Delta\lambda < 1.$$

We have

$$||\tilde{z} - z_{n+1}||_C \leq ||z_{n+1} - z_n||_C + ||z_{n+2} - z_{n+1}||_C + \ldots <$$
$$< (q^n + q^{n+1} + \ldots)||z_1 - \hat{z}||_C = \frac{q^n}{1-q}||z_1 - \hat{z}||_C.$$

and

$$||z_1 - z_T||_C \leq 3E\Delta\lambda||z_0||_C \leq 3E^2 M_\pi \Delta\lambda = B.$$

We obtain the following estimation

$$||\tilde{z} - z_n||_C \leq \frac{q^n}{1-q}(B + \delta), \text{ where } \quad B = 3\Delta\lambda M_\pi E^2.$$

Suppose z_n is the exact solution of the equation $Q(n\Delta\lambda)z = y$; while $z_{n,s}$ is the s-th iteration on the n-th step by parameter. It is easy to show that following estimations

$$||z_n - z_{n,s}||_C < Bb\frac{1-b^n}{1-b}, \quad b = \frac{q^s}{1-q} \tag{14}$$

are valid (see [11]) (on each step by parameter λ the s iterations of method of serial approximations are made).

We obtain the estimation (14) by using the method of mathematical induction. One can show that the parameter λ changes from 0 to 1 in $N > [3E] + 1$ steps. In the case $n = N$ we receive the approximate solution and corresponding error for the equation (8).

References

[1] Bernstein S.N., Math. An., 1904, Bd.59.

[2] Schauder J., Math. Z., Bd.38, 1934.

[3] Trenogin V.A., Functional Analysis (in Russian), Nauka, Moscow, 1980 and 1993 (Nauka, Novosibirsk).

[4] Donden A., Rev. Roum math. Pures et appl., v.25, No.10, p.1459-1475, 1980.

[5] Gaponenko J.L., Journ. Vych. Math. i Math. Phys., v.26, No.8, p.1123-1131, 1986.

[6] Trenogin V.A., Locally invertible operators and parameter continuation method, Funct. Anal. Appl., v.30, no 2, p.147-148, 1996.

[7] Trenogin V.A., Global invertibility of nonlinear operators and the method of continuation with respect to a parameter, Doklady Math., v.54, no 2, p.730-732, 1995.

[8] Trenogin V.A., Invertibility of nonlinear operators and parameter continuation method (English summary), Spectral and scattering theory, Editor A.Ramm (Newark, DE, 1997), p.189-197, Plenum, New York, 1998.

[9] Trenogin V.A., Properties of resolvent sets and estimations of resolvent of nonlinear operators, Doklady Math., v.57, no 2, p.190-192, 1998.

[10] Trenogin V.A., One-parametric solutions families of nonlinear equations and DE in Banach spaces, Works of International Conference devoted to 75 years of Kudryavzev L.D., v.2, p.161-165, RUDN Publ., 1998 (in Russian).

[11] Vetecha V.G., Estimations of the Approximate Solutions in Parameter Continuation Method (in Russian), Vestnik rossijskogo universiteta druzhby Narodov, Ser. Matematika, no 4-5(1), p.29-36, 1997-98.

85 SCHWARTZ' KERNEL THEOREM FOR FOURIER HYPERFUNCTIONS

E. BRÜNING

Department of Mathematics & Applied Mathematics, University of Durban-Westville
Private Bag X54001, Durban 4000, South Africa

ebruning@pixie.udw.ac.za

Abstract

An appropriate general version of the kernel theorem of L. Schwartz is formulated for Fourier hyperfunctions and a direct functional analytic proof is presented. This talk is based on a joint paper [2] with S. Nagamachi.

1 Introduction

L. Schwartz' kernel theorem has found a wide range of applications in Mathematics, Applied Mathematics and Mathematical Physics. Though it is a relatively abstract result in the theory of generalized functions it seems to have been conceived from the beginning with concrete applications in mind (see 20, 21 and in particular Schwartz' talk at the ICM 1950 [20]). Naturally this result has been extended in various directions (for instance [7, 1, 5, 4]). In his comprehensive studies of ultra-distributions [13, 14] Komatsu presents various versions of kernel theorems for this class of generalized functions.

H.G.W. Begehr et al. (eds.), Proceedings of the Second ISAAC Congress, Vol. 1, 743–754.

Soon, the abstract core of this result has been formulated as the so-called *abstract kernel theorem* in the context of nuclear spaces (see [19] and references there). Various other 'abstract kernel theorems' can be found in [13, 14].

The kernel theorem has played an important role in (standard) relativistic quantum field theory (see [24, 22, 11, 9]) which is formulated in terms of tempered distributions. In our program in hyperfunction quantum field theory (i.e., relativistic quantum field theory in terms of Fourier hyperfunctions) we encountered a problem which required the analogue of Schwartz' kernel theorem in the setting of Fourier hyperfunctions. The special cases for Fourier hyperfunctions which we found in the literature [3, 17, 10] did not solve our problem.

There are several equivalent versions of L. Schwartz' kernel theorem. Our reference point for this result is

Theorem 1.1. *Let G_i be an open nonempty subset of $\mathbb{R}^{n_i}$ for $i = 1, 2$ and denote by $\mathcal{D}(G_i)$ the standard Schwartz space of C^∞-functions with compact support in G_i. Then every separately continuous bilinear form b on $\mathcal{D}(G_1) \times \mathcal{D}(G_2)$ comes from a unique distribution T on $G_1 \times G_2$, i.e., there is a unique $T \in \mathcal{D}(G_1 \times G_2)'$ such that*

$$b(f_1, f_2) = T(f_1 \otimes f_2) \ \ \forall f_i \in \mathcal{D}(G_i), \ i = 1, 2.$$

Note that this result implies the following canonical isomorphisms which have many applications in distribution theory:

$$\mathcal{D}'(G_1 \times G_2) \cong \mathcal{D}'(G_1)\hat{\otimes}\mathcal{D}'(G_2) \cong L_b(\mathcal{D}(G_2), \mathcal{D}'(G_1)).$$

There are two different strategies of proving this result:

1. Functional analytic proof, i.e., properties of the underlying function spaces (nuclearity, barreledness, density arguments, properties of inductive limits).

2. Properties of solutions of special linear partial differential operators (heat kernel) and detailed direct estimates [8, 3].

We will use the first strategy.

Apparently the first version of the kernel theorem for Fourier hyperfunctions is due to S. Nagamachi and N. Mugibayashi, [17]. They give a rather indirect functional analytic proof relying on results from [6] and [15].

If we denote by $\mathcal{B}(\underset{\sim}{\mathcal{Q}}(\boldsymbol{D}^k), \underset{\sim}{\mathcal{Q}}(\boldsymbol{D}^m))$ the space of separately continuous bilinear forms on $\underset{\sim}{\mathcal{Q}}(\boldsymbol{D}^k) \times \underset{\sim}{\mathcal{Q}}(\boldsymbol{D}^m)$, the result of Nagamachi and Mugibayashi can be formulated as

$$\mathcal{B}(\underset{\sim}{\mathcal{Q}}(\boldsymbol{D}^k), \underset{\sim}{\mathcal{Q}}(\boldsymbol{D}^m)) \cong \underset{\sim}{\mathcal{Q}}(\boldsymbol{D}^{k+m})'.$$

However, in our recent investigations in hyperfunction quantum field theory we need a result of the form

$$\mathcal{B}(\underset{\sim}{\mathcal{Q}}(K), \underset{\sim}{\mathcal{Q}}(L)) \cong \underset{\sim}{\mathcal{Q}}(K \times L)'$$

for general closed subsets $K \subset \boldsymbol{D}^k$, respectively $L \subset \boldsymbol{D}^m$. And it is this version of the kernel theorem which we intend to prove.

Our functional analytic strategy of proof is straight forward and direct. First we collect the necessary properties of the relevant test function spaces of Fourier hyperfunctions for a simple version of the abstract kernel theorems. Then we identify the completion of their projective tensor product in the correct way.

2 Some facts from functional analysis

We assume some familiarity with Schwartz' distribution theory and the basic facts about Fourier hyperfunctions. Now we recall a few concepts and results from functional analysis which will be needed later.

Denote by $\mathcal{P}$ the system of seminorms which defines the topology of a HLCTVS F. Such a space is called *nuclear* if, and only if, for every $p \in \mathcal{P}$ there exists $q \in \mathcal{P}$, $p \leq q$, such that the canonical map $F_q \to F_p$ is nuclear. Here F_q is the Banach space canonically induced by the semi-norm q.

A linear map $T : F_q \to F_p$ is *nuclear* iff there are a) an equicontiuous sequence t_j of linear functionals on F_q, b) a bounded sequence of vectors e_j in F_p, and c) an absolutely summable sequence of numbers λ_j, such that for all $x \in F_q$

$$T(x) = \sum_{j=1}^{\infty} \lambda_j t_j(x) e_j.$$

With this in mind one can formulate a simple version of an abstract kernel theorem (Theorem 7.4.3 of [19]) the proof of which is straight forward.

Theorem 2.1 (An abstract kernel theorem). *Let E and F be two HLCTVS with F nuclear. Then every continuous bilinear form b on $E \times F$ is nuclear, i.e., $\forall x \in E$ and $\forall y \in F$,*

$$b(x,y) = \sum_{j=1}^{\infty} \lambda_j t_j(x) h_j(y)$$

with the following specification: t_j is an equicontinuous sequence of linear forms on E, h_j is an equicontinuous sequence of linear forms on F, and the sequence of numbers λ_j satisfies $\sum_{j=1}^{\infty} |\lambda_j| < \infty$.

3 Fourier hyperfunctions

We collect here some facts about Fourier hyperfunctions which are used later. Good references for these results are [12] and [16].

Denote by $\boldsymbol{D}^k$ the radial compactification of $\boldsymbol{R}^k$. For a closed subset $K \subset \boldsymbol{D}^k$ choose a fundamental sequence of neighborhoods $\tilde{U}_j$ of K in $\boldsymbol{D}^k + i\boldsymbol{R}^k$. For each $j \in \mathbb{N}$ introduce the Banach space $\mathcal{O}_j^c(\tilde{U}_j)$ of functions which are analytic on $\tilde{U}_j$ and continuous on the closure of this set, equipped with the norm

$$p_j(g) = \sup_{\tilde{U}_j \cap \boldsymbol{C}^k} e^{\frac{1}{j}|z|} |g(z)|.$$

Finally form the inductive limit of these Banach spaces

$$\underset{\sim}{\mathcal{O}}(K) = \text{ind} \lim_{j \to \infty} \mathcal{O}_j^c(\tilde{U}_j).$$

Note that this inductive limit is not strict so that a number of powerful results from functional analysis do not apply. This construction includes the case $K = \boldsymbol{D}^k$ as well. Continuous linear functionals on $\underset{\sim}{\mathcal{O}}(\boldsymbol{D}^k)$ are called *F*ourier hyperfunctions on $\boldsymbol{R}^k$. Fourier hyperfunctions form a flabby sheaf on $\boldsymbol{R}^k$. The sheaf of Schwartz distributions is not flabby. Every (tempered) distribution is a Fourier hyperfunction. However the space of Fourier hyperfunctions is much larger: While distributions are always of finite order locally, hyperfunctions can be of infinite order.

Note furthermore that $\underset{\sim}{\mathcal{O}}(\boldsymbol{D}^k)$ is (sequentially) dense in the space $\underset{\sim}{\mathcal{O}}(K)$, another property which distinguishes Fourier hyperfunctions and distributions in an important way.

The spaces $\underset{\sim}{\mathcal{O}}(K)$, $K \subset \boldsymbol{D}^k$ closed, are DFS-spaces (dual Frechet Schwartz spaces) and as such barreled. They are also nuclear.

4 Topological preliminaries

The detailed definitions of the various spaces used here can be found in [12] and [17, 18]. Sufficient details about the topologies of these spaces are given below in the proof of Lemma 4.2.

Lemma 4.1. *Let K be a closed subset of $\boldsymbol{D}^k$ and L a closed subset of $\boldsymbol{D}^m$. Then $E = \underset{\sim}{\mathcal{O}}(K) \otimes \underset{\sim}{\mathcal{O}}(L)$ is dense in $\underset{\sim}{\mathcal{O}}(K \times L)$.*

Proof. According to Theorem 2.7 of [18] the space $\underset{\sim}{\mathcal{O}}(\boldsymbol{D}^k)$ is (sequentially) dense in the space $\underset{\sim}{\mathcal{O}}(K)$ and according to Proposition 3.5 of [17] the tensor product $\underset{\sim}{\mathcal{O}}(\boldsymbol{D}^k) \otimes \underset{\sim}{\mathcal{O}}(\boldsymbol{D}^m)$ is dense in the space $\underset{\sim}{\mathcal{O}}(\boldsymbol{D}^{k+m})$. Thus we have the following dense inclusions

$$\underset{\sim}{\mathcal{O}}(\boldsymbol{D}^k) \otimes \underset{\sim}{\mathcal{O}}(\boldsymbol{D}^m) \subset \underset{\sim}{\mathcal{O}}(\boldsymbol{D}^{k+m}) \subset \underset{\sim}{\mathcal{O}}(K \times L)$$

$$\underset{\sim}{\mathcal{O}}(\boldsymbol{D}^k) \otimes \underset{\sim}{\mathcal{O}}(\boldsymbol{D}^m) \subset \underset{\sim}{\mathcal{O}}(K) \otimes \underset{\sim}{\mathcal{O}}(L).$$

And the statement of the lemma follows. □

Lemma 4.2. *Let K be a closed subset of $\boldsymbol{D}^k$ and L a closed subset of $\boldsymbol{D}^m$. Then $\underset{\sim}{\mathcal{O}}(K \times L)$ induces on $E = \underset{\sim}{\mathcal{O}}(K) \otimes \underset{\sim}{\mathcal{O}}(L)$ a topology τ which is finer than the ε-topology.*

Proof. Recall

$$\underset{\sim}{\mathcal{O}}(K) = \operatorname{ind} \lim_{j \to \infty} \mathcal{O}_j^c(\tilde{U}_j)$$

with a fundamental sequence of neighborhoods $\tilde{U}_j$ of $K \subset \boldsymbol{D}^k$ and similarly

$$\underset{\sim}{\mathcal{O}}(L) = \operatorname{ind} \lim_{j \to \infty} \mathcal{O}_j^c(\tilde{V}_j)$$

with a fundamental sequence of neighborhoods $\tilde{V}_j$ of $L \subset \boldsymbol{D}^m$. Hence a neighborhood of zero in $\underset{\sim}{\mathcal{O}}(K)$ is of the form $U = \operatorname{ach}(\bigcup_{j=1}^{\infty} U_j)$ with neighborhoods of zero U_j in the Banach space $\mathcal{O}_j^c(\tilde{U}_j)$ which can be assumed to be given as follows:

$$U_j = \{g \in \mathcal{O}_j^c(\tilde{U}_j);\ p_j(g) \le r_j\}$$

for some $r_j > 0$ where the norms p_j are defined by

$$p_j(g) = \sup_{\tilde{U}_j \cap \boldsymbol{C}^k} e^{\frac{1}{j}|z|} |g(z)|.$$

The absolutely convex hull of a set M is denote by $\mathrm{ach}(M)$.

Similarly a neighborhood of zero V in $\underset{\sim}{Q}(L)$ is of the form $V = \mathrm{ach}(\bigcup_{j=1}^{\infty} V_j)$ with neighborhoods of zero V_j in the Banach space $\mathcal{O}_j^c(\tilde{V}_j)$ given by

$$V_j = \{h \in \mathcal{O}_j^c(\tilde{V}_j);\ q_j(h) \leq \rho_j\}$$

for some $\rho_j > 0$ where the norms q_j are defined by

$$q_j(h) = \sup_{\tilde{V}_j \cap \boldsymbol{C}^m} e^{\frac{1}{j}|\zeta|} |h(\zeta)|.$$

Then a neighborhood of zero in E for the topology ε is of the form

$$H = \{f \in E;\ q_{U,V}(f) < \delta\} \tag{4.1}$$

for some $\delta > 0$, some neighborhoods of zero U in $\underset{\sim}{Q}(K)$, respectively V in $\underset{\sim}{Q}(L)$ where

$$q_{U,V}(f) = \sup_{T \in U^o} \sup_{S \in V^o} |\langle T \otimes S, f \rangle|. \tag{4.2}$$

U^o, respectively V^o denote the absolute polar of U in $\underset{\sim}{Q}(K)'$, respectively of V in $\underset{\sim}{Q}(L)'$.

Neighborhoods of zero W in

$$\underset{\sim}{Q}(K \times L) = \mathrm{ind} \lim_{j \to \infty} \mathcal{O}_j^c(\tilde{U}_j \times \tilde{V}_j)$$

are specified similarly

$$W = \mathrm{ach}(\bigcup_{j=1}^{\infty} W_j) \tag{4.3}$$

$$W_j = \{f \in \mathcal{O}_j^c(\tilde{U}_j \times \tilde{V}_j);\ \|f\|_j < \sigma_j\} \tag{4.4}$$

for some $\sigma_j > 0$ and norms $\|\cdot\|_j$ given by

$$\|f\|_j = \sup_{z \in \tilde{U}_j \cap \boldsymbol{C}^k} \sup_{\zeta \in \tilde{V}_j \cap \boldsymbol{C}^m} e^{\frac{1}{j}(|z|+|\zeta|)} |f(z,\zeta)|. \tag{4.5}$$

The topology τ on E is finer than the topology ε if for every neighborhood of zero H for ε there is a neighborhood of zero W as above such that $W \cap E \subset H$ or, since H is absolutely convex,

$$W_j \cap E \subset H, \quad j = 1, 2, \cdots. \tag{4.6}$$

For $f \in E \cap W_j$, $T \in U^o$, and $S \in V^o$ an elementary calculation shows

$$\langle T \otimes S, f \rangle = \langle T_x, \langle S_y, f(x,y) \rangle \rangle. \tag{4.7}$$

In order to estimate the right hand side we note first that $\langle S_y, f(\cdot, y) \rangle$ belongs to the space $\underset{\sim}{\mathcal{O}}(K)$ and therefore

$$|\langle T_x, \langle S_y, f(x,y) \rangle \rangle| \le p_j'(T) p_j(\langle S_y, f(\cdot, y) \rangle).$$

According to the definitions of the norms involved we can continue this estimate as follows:

$$\begin{aligned} p_j(\langle S_y, f(\cdot, y) \rangle) &= \sup_{\tilde{U}_j \cap \boldsymbol{C}^k} e^{\frac{1}{j}|z|} |(\langle S_y, f(z,y) \rangle| \\ &\le \sup_{\tilde{U}_j \cap \boldsymbol{C}^k} e^{\frac{1}{j}|z|} q_j'(S) \sup_{\tilde{V}_j \cap \boldsymbol{C}^m} e^{\frac{1}{j}|\zeta|} |f(z,\zeta)| \\ &= q_j'(S) \|f\|_j \end{aligned}$$

and thus, for $j = 1, 2, \dots$,

$$|\langle T \otimes S, f \rangle| \le p_j'(T) q_j'(S) \, \|f\|_j \tag{4.8}$$

where the prime indicates the dual norm. Using standard properties of polar sets and the explicit definitions of the neighborhoods U and U_j respectively V and V_j one deduces that $T \in U^o$, respectively $S \in V^o$, implies $T \in U_j^o$ respectively $S \in V_j^o$ for all $j = 1, 2, \dots$. According to the definitions of the dual norm and the absolute polar set we get for $T \in U_j^0$

$$p_j'(T) = \frac{1}{r_j} \sup \{ |\langle T, g \rangle| \; g \in V_j \} \le \frac{1}{r_j}$$

and similarly for $S \in V_j^o$: $q_j'(S) \le \frac{1}{\rho_j}$. It follows

$$q_{U,V}(f) \le \frac{1}{r_j \rho_j} \|f\|_j. \tag{4.9}$$

Now, given the $\delta > 0$ in the definition of the neighborhood H, choose $\sigma_j > 0$ such that $\sigma_j \le r_j \rho_j \delta$. Then, for all $f \in W_j \cap E$ we know $q_{U,V}(f) < \delta$, hence $W_j \cap E \subset H$ and we conclude. □

Proposition 4.3. *Let K be a closed subset of $\boldsymbol{D}^k$ and L a closed subset of $\boldsymbol{D}^m$. Then*

$$\underset{\sim}{\mathcal{O}}(K) \tilde{\otimes}_{\pi,\varepsilon} \underset{\sim}{\mathcal{O}}(L) = \underset{\sim}{\mathcal{O}}(K \times L)$$

where $\cdots \tilde{\otimes}_{\pi,\varepsilon} \cdots$ denotes the completion of the tensor product $E = \underset{\sim}{\mathcal{O}}(K) \otimes \underset{\sim}{\mathcal{O}}(L)$ with respect to the projective tensor product topology π, respectively with respect to the tensor product topology ε.

Proof. Since the projective topology π is the finest locally convex topology on E we know $\varepsilon \leq \pi$ and $\tau \leq \pi$ where τ denotes the relative topology on E induced by the topology of $\underset{\sim}{\mathcal{Q}}(K \times L)$. Lemma 2.2 shows that $\varepsilon \leq \tau$. This then implies for the completions of the space E with respect to these three topologies the following relation

$$\tilde{E}[\pi] \subset \tilde{E}[\tau] \subset \tilde{E}[\varepsilon].$$

By Proposition 2.12 of [16] the spaces $\underset{\sim}{\mathcal{Q}}(K)$ and $\underset{\sim}{\mathcal{Q}}(L)$ are nuclear, hence Theorem 50.1 of [23] implies that the completions of the tensor product E with respect to the topologies π and ε are equal, i.e., in our notation $\tilde{E}[\pi] = \tilde{E}[\varepsilon]$. Now Lemma 2.1 implies $\tilde{E}[\tau] = \underset{\sim}{\mathcal{Q}}(K \times L)$ and thus we conclude. □

5 The kernel theorem for Fourier hyperfunctions

Theorem 5.1. *Let K be a closed subset of $\boldsymbol{D}^k$ and L a closed subset of $\boldsymbol{D}^m$. Then, for every separately continuous bilinear form B on $G = \underset{\sim}{\mathcal{Q}}(K) \times \underset{\sim}{\mathcal{Q}}(L)$ there is a unique Fourier hyperfunction F_B on $K \times L$, i.e. $F_B \in \underset{\sim}{\mathcal{Q}}(K \times L)'$ such that for all $(g, h) \in G$*

$$B(g,h) = F_B(g \otimes h).$$

Proof. Since $\underset{\sim}{\mathcal{Q}}(K)$ and $\underset{\sim}{\mathcal{Q}}(L)$ are DFS-spaces (see [17]) they are barreled and thus Theorem 41.1 of [23] implies that every separately continuous bilinear form B on G is continuous. Furthermore, the spaces $\underset{\sim}{\mathcal{Q}}(K)$ and $\underset{\sim}{\mathcal{Q}}(L)$ are nuclear according to Proposition 2.12 of [16]. Hence Theorem 2.1 applies and proves that every such bilinear form B is actually nuclear, i.e., there is an equicontinuous sequence (T_j) in $\underset{\sim}{\mathcal{Q}}(K)'$, an equicontinuous sequence (S_j) in $\underset{\sim}{\mathcal{Q}}(L)'$, and a sequence (λ_j) of real numbers with $\sum_{j=1}^{\infty} |\lambda_j| < \infty$ such that for all $(g, h) \in G$

$$B(g,h) = \sum_{j=1}^{\infty} \lambda_j \langle T_j, g\rangle \langle S_j, h\rangle.$$

When we write the right hand side of this equation in the form

$$\sum_{j=1}^{\infty} \lambda_j (T_j \otimes S_j)(g \otimes h)$$

we see immediately that B is actually given by the continuous linear form

$$F_B = \sum_{j=1}^{\infty} \lambda_j T_j \otimes S_j$$

on $\underset{\sim}{\mathcal{O}}(K) \otimes_\pi \underset{\sim}{\mathcal{O}}(L)$ (compare Theorem 43.4 of [23]). Clearly F_B extends to a continuous linear form on the completion of this space which is equal to $\underset{\sim}{\mathcal{O}}(K \times L)$ according to Proposition 4.3. Thus we conclude. □

The general form of the kernel theorem for Fourier hyperfunctions is now a straight forward consequence of Theorem 5.1.

Theorem 5.2. *Let K_i be a closed subset of $\boldsymbol{D}^{k_i}$ for $i = 1, 2, \dots, N$. Then, for every separately continuous N-linear form B on $G = \underset{\sim}{\mathcal{O}}(K_1) \times \cdots \times \underset{\sim}{\mathcal{O}}(K_N)$ there is a unique Fourier hyperfunction F_B on $K_1 \times \cdots \times K_N$, i.e., $F_B \in \underset{\sim}{\mathcal{O}}(K_1 \times \cdots \times K_N)'$ such that for all $(g_1, \dots, g_N) \in G$*

$$B(g_1, \dots, g_N) = F_B(g_1 \otimes \cdots \otimes g_N).$$

Proof. On the basis of Theorem 5.1 a straight forward proof of induction with respect to the number N of arguments establishes this general form. □

We have to comment on this result. $K_1 \times \cdots \times K_N$ is a subset of $\boldsymbol{D}^{k_1} \times \cdots \times \boldsymbol{D}^{k_N}$, but not of $\boldsymbol{D}^n$, $n = \sum_{i=1}^N k_i$, in general. Accordingly elements $\mu \in \underset{\sim}{\mathcal{O}}(K_1 \times \cdots \times K_N)'$ have to be considered as elements of $\underset{\sim}{\mathcal{O}}(\prod_{i=1}^N \boldsymbol{D}^{k_i})'$ with support in $K_1 \times \cdots \times K_N$. But hyperfunctions in n variables are defined on $\boldsymbol{D}^n$, the radial compactification of $\boldsymbol{R}^n$; and in this sense we need the above result. However it is not difficult to show

$$\underset{\sim}{\mathcal{O}}(\boldsymbol{D}^n)' = \underset{\sim}{\mathcal{O}}(\prod_{i=1}^N \boldsymbol{D}^{k_i})'$$

though $\boldsymbol{D}^n \neq \prod_{i=1}^N \boldsymbol{D}^{k_i}$.

6 Conclusion

In conclusion I want to comment on the relation of our kernel theorem to some earlier related results. It is fairly straight forward to show that the above strategy of proof applies equally well to Schwartz distributions (general and tempered). Thus, compared to the proofs given in the literature [21, 23, 13, 14] we get a more direct (and simpler as we think) proof of this well known result. The results in [13, 14] amount to a comprehensive extension of Schwartz' theory to the case of various classes of ultradistributions. The new abstract kernel theorem in [14], i.e., Theorem 2.3 unfortunately does not cover the case of Fourier hyperfunctions since in the case of Fourier hyperfunctions the inductive limits representing the test-functions spaces are not strict.

In [10] a kernel theorem is stated for Fourier hyperfunctions which would cover the case we are interested in. Unfortunately, the proof seems to contain some gaps (the core of the proof is just a nearly literal repetition of the proof of Schwartz' kernel theorem as given in Trèves [23] and some argument seems to be necessary).

References

[1] W. Bogdanowicz, *A proof of Schwartz's theorem on kernels*, Studia Math. **20** (1961), 77-85.

[2] E. Brüning and S. Nagamachi, *Kernel theorem for Fourier hyperfunctions*, submitted for publication, September 1998.

[3] S-Y. Chung, D. Kim, and E. G. Lee, *Schwartz kernel theorem for the Fourier hyperfunctions*, Tsukuba J. Math. **19** (1995), no. 2, 377-385.

[4] K. Ehrenpreis, *On the theorem of kernels of Schwartz*, Proc. Amer. Math. Soc. **7** (1956), 713-718.

[5] H. Gask, *A proof of Schwartz's kernel theorem*, Math. Scand. **8** (1960), 327-332.

[6] I.M. Gelfand and G.E. Shilov, *Generalized Functions*, vol. 2, Academic Press, New York London, 1968.

[7] I.M. Gelfand and N.Ya. Vilenkin, *Generalized Functions*, vol. 4, Academic Press, New York London, 1964.

[8] L. Hörmander, *The analysis of linear partial differential operators I*, Grundlehren der mathematischen Wissenschaften, vol. 256, Springer-Verlag, Berlin Heidelberg New York Tokyo, 1983.

[9] S.S. Horuzhy, *Introduction to algebraic quantum field theory*, Kluwer Academic Publishers, Dordrecht Boston London, 1990.

[10] Y. Ito, *Fourier hyperfunctions of general type*, Journal of Mathematics of Kyoto University **28** (1988), no. 2, 213-265.

[11] R. Jost, *The general theory of quantized fields*, American Mathematical Society, Providence, Rhode Island, 1965.

[12] A. Kaneko, *Introduction to hyperfunctions*, Mathematics and its Applications (Japanese Series), Kluwer Academic Publishers, Dordrecht Boston London, 1988.

[13] H. Komatsu, *Ultradistributions II. The kernel theorem and ultradistributions with support in a manifold*, J. Fac. Sci. Univ. Tokyo Sect. IA **24** (1977), 607-628.

[14] H. Komatsu, *Ultradistributions III. Vector valued ultradistributions and the theory of kernels*, J. Fac. Sci. Univ. Tokyo Sect. IA **29** (1982), 653-717.

[15] B.S. Mityagin, *Nuclearity and other properties of spaces of type S (in Russian)*, Trudy Moskov. Mat. Obsc. **9** (1960), 317-328.

[16] S. Nagamachi, *The theory of vector valued Fourier hyperfunctions of mixed type I.*, Publ. RIMS, Kyoto Univ. **17** (1981), 25-63.

[17] S. Nagamachi and N. Mugibayashi, *Hyperfunction quantum field theory*, Commun. Math. Phys. **46** (1976), 119-134.

[18] T. Nishimura and S. Nagamachi, *On supports of Fourier hyperfunctions*, Math. Japonica **35** (1990), 293-313.

[19] A. Pietsch, *Nuclear locally convex spaces*, Ergebnisse der Mathematik und ihrer Grenzgebiete, Springer-Verlag, Berlin Heidelberg New York, 1972.

[20] L. Schwartz, *Théorie des noyaux*, Proceedings of the International Congress of Mathematicians: Cambridge Massachusetts, U.S.A., August 30 - September 6, 1950, vol. 1, IMU, American Mathematical Society, 1952, pp. 220-230.

[21] L. Schwartz, *Espaces de fonctions différentiables à valeurs vectorielles*, J. Analyse Math. **4** (1955), 88-148.

[22] R.F. Streater and A.S. Wightman, *PCT, spin and statistics, and all that*, Benjamin, New York, 1964.

[23] F. Treves, *Topological vector spaces, distributions and kernels*, Pure and Applied Mathematics, Academic Press, New York London, 1967.

[24] A.S. Wightman and L. Gårding, *Fields as operator-valued distributions in relativistic quantum field theory*, Arkiv för Fysik **28** (1964), 129-184.

86 EVERY STIELTJES MOMENT PROBLEM HAS SOLUTIONS IN GELFAND–SHILOV SPACES

Jaeyoung CHUNG

Department of Mathematics, Kunsan National University
Kunsan, 573–360 Korea

jychung@ks.kunsan.ac.kr

Soon–Yeong CHUNG

Department of Mathematics, Sogang University
Seoul 121–742 Korea

sychung@ccs.sogang.ac.kr

Dohan KIM

Department of Mathematics, Seoul National University
Seoul 151–742 Korea

dhkim@math.snu.ac.kr

1991 Mathematics Subject Classification: Primary 30E05, 46F15; Secondary 35K05
Key words: Stieltjes moment problem, Gel'fand–Shilov space, uniqueness

H.G.W. Begehr et al. (eds.), Proceedings of the Second ISAAC Congress, Vol. 1, 755–759.

Abstract

In 1989 A. J. Duran proved that for an arbitrary sequence $\{\mu_n\}$ there exists a function in the Schwartz class with support in the positive real line, whose moment is $\{\mu_n\}$. In this paper we prove that the moment function of an arbitrary sequence $\{\mu_n\}$ can be found in Gelfand-Shilov spaces $S^\beta(0,\infty)$ which is a subspace of Schwartz space.

1 Introduction

The moment problem originated by Stieltjes in 1894 has played an important role in analysis. If a sequence $\{\mu_n\}$ and a function ϕ can be written as

$$\int_I x^n \phi(x)\, dx = \mu_n, \quad \text{for } n = 0, 1, 2, \dots ,$$

then ϕ and μ_n are called the moment function and the n–th moment, respectively. The moment problem asks mainly when a sequence $\{\mu_n\}$ is the moment sequence of a function ϕ defined on I. The Stieltjes moment problem which treats the case of $I = [0,\infty)$ was considered by Stieltjes who dealt with the case of positive measures $d\mu(x)$ instead of moment function ϕ. In 1939, Boas proved that the moment function α of bounded variation can be found for any given sequence. Recently, Duran [D] showed an improved result that for any sequence there exists the moment function in the Schwartz space of rapidly decreasing functions. In 1994, Durán and Estrada [DE] made, as a part of their paper, a different proof based on the asymptotic analysis. It is natural to ask what is the smallest function space which ensures the existence of the moment function for an arbitrarily given sequence. In this paper we'll see that the moment function of an any given sequence can be found in the Gel'fand–Shilov space $S^\beta(0,\infty)$ which is a subspace of the Schwartz space.

2 Preliminaries

We introduce some function spaces and properties of asymptotic power series of analytic functions.

The following lemma based on the Ritt's theorem describes the existence of a bounded analytic function which has an arbitrarily given asymptotic power series.

Lemma 2.1. *Let $\{a_n\}$ be an arbitrary sequence of complex numbers. Then there exists a bounded analytic function F defined in the sector $S = \{re^{i\theta} | r > 0, \alpha < \theta < \beta\}$ such that*

$$F(z) \sim a_0 + a_1 z + a_2 z^2 + \cdots$$

as $z \to 0$ in S.

For $\alpha \geq 0$ and $\beta \geq 0$ we define function spaces S_α, S^β, and S_α^β by

$$\begin{aligned}
S_\alpha &= \left\{\phi \in C^\infty(\mathbb{R}) \Big| \sup_{x\in\mathbb{R}} |x^p \phi^{(q)}(x)| \leq C_q h^p p!^\alpha \quad \exists C_q, h > 0\right\}, \\
S^\beta &= \left\{\phi \in C^\infty(\mathbb{R}) \Big| \sup_{x\in\mathbb{R}} |x^p \phi^{(q)}(x)| \leq C_p h^q q!^\beta \quad \exists C_p, h > 0\right\}, \\
S_\alpha^\beta &= \left\{\phi \in C^\infty(\mathbb{R}) \Big| \sup_{x\in\mathbb{R}} |x^p \phi^{(q)}(x)| \leq C h^{p+q} p!^\alpha q!^\beta \quad \exists C, h > 0\right\}.
\end{aligned}$$

Modifying above fuction spaces, we introduce $S_\alpha(0,\infty)$, $S^\beta(0,\infty)$, and $S_\alpha^\beta(0,\infty)$ by

$$\begin{aligned}
S_\alpha(0,\infty) &= \{\phi \in S_\alpha | \phi(x) = 0, \quad \text{for } x \leq 0\}, \\
S^\beta(0,\infty) &= \{\phi \in S^\beta | \phi(x) = 0, \quad \text{for } x \leq 0\}, \\
S_\alpha^\beta(0,\infty) &= \left\{\phi \in S_\alpha^\beta | \phi(x) = 0, \quad \text{for } x \leq 0\right\}.
\end{aligned}$$

Throughout this paper, the Fourier–Laplace transform $\hat{\phi}$ of a function ϕ is defined by

$$\hat{\phi}(\xi) = \int_{-\infty}^{\infty} \phi(x) e^{ix\xi} \, dx.$$

The Fourier–Laplace transform of a function which belongs to $S_\alpha(0,\infty)$, $S^\beta(0,\infty)$, and $S_\alpha^\beta(0,\infty)$ can be described in the following lemma.

Lemma 2.2. *A function ψ defined on $\mathbb{R}$ is the Fourier–Laplace transform of a function ϕ of class $S^\beta(0,\infty)$ if and only if*

(i) *ψ is of class S_β.*

(ii) *ψ can be extended to a continuous function Ψ in the closed upper half plane $Im\, z \geq 0$ and analytic in $Im\, z > 0$.*

(iii) *Ψ vanishes as $z \to \infty$ in $Im\, z \geq 0$.*

Let G_α be an analytic function defined by

$$G_\alpha(z) = \exp\left(e^{\pi i(1-\frac{1}{2\alpha})}\sqrt[\alpha]{z+i}\right)$$

in the region $D := \left\{z \Big| -\frac{\pi}{2} < \arg z < \frac{3\pi}{2}\right\}$. Then $G_\alpha(z) \neq 0$ implies that G_α and $1/G_\alpha$ are analytic in the open unit disk $\{z : |z| < 1\}$. Furthermore, the image of $h_\alpha(z) := e^{\pi i(1-\frac{1}{2\alpha})}\sqrt[\alpha]{z+i}$ is contained in the sector $\pi - \frac{\pi}{2\alpha} < \arg z < \pi + \frac{\pi}{2\alpha}$ and $\cos(\arg h_\alpha(z)) \leq -k_\alpha$ for some constant $k_\alpha > 0$.

From these properties we may decide the behavior of G_α at infinity. Let $z + i = re^{i\theta}$. Then we have

$$\begin{aligned} |G_\alpha(z)| &= \exp \operatorname{Re}\left(e^{\pi i(1-\frac{1}{2\alpha})}(re^{i\theta})^{1/\alpha}\right) \\ &= \exp\left(-k_\alpha r^{1/\alpha}\right). \end{aligned}$$

This implies $G_\alpha(z) \to 0$ as $z \to \infty$ in D.

3 Main Theorem

We are ready to state prove the main theorems.

Theorem 3.1. *For any sequence $\{\mu_n\}$ of complex numbers there exists*

$$\phi \in S'^\beta(0, \infty)$$

such that

$$\mu_n = \int_0^\infty x^n \phi(x)\, dx \quad \textit{for } n = 0, 1, 2, \ldots$$

The complete proof will be published elsewhere.

References

[D] A. J. Duran, *The Stieltjes moment problem for rapidly decreasing functions*, Proc. Amer. Math. Soc. **107** (1989), 731–741.

[DE] A. L. Durán and R. Estrada, *Strong moment problems for rapidly decreasing smooth functions*, Proc. Amer. Math. Soc. **120** (1994), 529–534.

[EK] R. Estrada and R. P. Kanwal, *Asymptotic Analysis*, Birkhäuser, Boston 1994.

[GS] I. M. Gel'fand and G. E. Shilov, *Generalized functions* II, Academic Press, New York 1968

[P] D. H . Pahk, *Hypoelliptic convolution equations in the space* $\mathcal{K}_e'$, Trans. Amer. Math. Soc., **298** (1986), 485–495.

87 ROUMIEU TYPE TEMPERED ULTRADISTRIBUTIONS AND FOURIER HYPERFUNCTIONS

Atsuhiko EIDA

Faculty of Engineering Science, Tokyo University of Technology
1404-1, Katakura, Hachioji
Tokyo 192-8580, Japan

eida@cc.teu.ac.jp

Stevan PILIPOVIĆ

Institute of Mathematics, University of Novi Sad
Trg D.Obradovića 4
21000 Novi Sad, Yugoslavia

pilipovic@unsim.ns.ac.yu

0. Introduction

In [7] we studied spaces of tempered ultradifferential functions and tempered ultradistributions which are globally expressed for the usage of Fourier transforms on $\mathbf{R}^n$. These spaces $\mathcal{S}^*, \mathcal{S}'^*$ are topologically automorphic to themselves through the Fourier transformation. On the other hand, Kawai and Kaneko have introduced the spaces of Fourier hyperfunctions which are homologically constructed on $\mathbf{D}^n$, the compactification of $\mathbf{R}^n$ to the infinity (cf. [3] and [9]). These spaces are also

H.G.W. Begehr et al. (eds.), Proceedings of the Second ISAAC Congress, Vol. 1, 761–769.

topologically automorphic to themselves through the Fourier transformation.

But the difference is that the spaces of Fourier hyperfunctions can be locally expressed from the viewpoint of sheaves. So our aim is to proceed this idea to the microlocalization as Kaneko did in [3] and [4]. We introduce the sheaf $\mathcal{S}b^s$, $s \in (0,1)$, which is used for the definition of the wave front at $(x_0\infty, \xi) \in \mathbf{D}^n \times \mathbf{R}^n \setminus \{0\}$.

The main idea of our construction of the new sheaf is reflected through appropriate bounds which sections have to satisfy as imaginary part of the argument tends to zero.

The paper presents a part of our investigations of the microlocal decomposition in certain classes of hyperfunctions and microfunctions.

1. Notation and notions

We refer to [5] and [6] for the definitions of spaces of ultradifferentiable functions and ultradistributions of Beurling type and Roumieau type.

Let $t > 1$ and $h > 0$. The space of smooth functions φ on $\mathbf{R}^n$ equipped with the topology induced by the norm $\sigma_{h,2}$,

$$\sigma_{h,2}(\varphi) := (\sum_{\alpha,\beta\in\mathbf{N}_0^n} \int_{\mathbf{R}^n} \mid \frac{h^{|\alpha+\beta|}}{\alpha!^t\beta!^t}(1+\mid x\mid^2)^{|\beta|/2}\varphi^{(\alpha)}(x)\mid^2 dx)^{1/2} < \infty,$$

is denoted by $\mathcal{S}_2^h$. The strong dual of $\mathcal{S}^{\{t\}} = \operatorname{ind}\lim_{h\to 0}\mathcal{S}_2^h$ is called the space of tempered ultradistributions of Roumieau type. For the properties of these spaces as well as of Beurling type spaces, we refer the reader to [7] and [8]. Note, $\mathcal{S}^{\{t\}}$ is an $(LS)-$ space and

$$\mathcal{D}^{\{t\}} \hookrightarrow \mathcal{S}^{\{t\}} \hookrightarrow \mathcal{E}^{\{t\}}, \quad \mathcal{S}^{\{t\}} \hookrightarrow \mathcal{S},$$

where "$A \hookrightarrow B$" means that A is dense in B and the inclusion mapping is continuous.

Recall [8], an $f \in \mathcal{D}'^{\{t\}}$ is in $\mathcal{S}'^{\{t\}}$ if and only if $f = P(\partial)F$, where F is a continuous function such that for every $k > 0$ there exists $C_k > 0$,

$$\mid F(x)\mid \leq C_k e^{k|x|^{1/t}}, \quad x \in \mathbf{R}^n$$

where P is an ultradifferential operator of $\{t\}-$ class ([5]).

The Fourier transformation is an isomorphism of $\mathcal{S}^{\{t\}}$ onto itself.

2. Sheaves of Fourier hyperfunctions of type $[s, 0]$ ([4])

Let $\mathbf{D}^n = \mathbf{R}^n \sqcup S_\infty^{n-1}$ be the directional compactification of $\mathbf{R}^n$ with the natural topology: A fundamental system of neighbourhood of $a\infty \in S_\infty^{n-1}$ is given by sets of the form

$$(\Gamma \cap \{|x| > R\}) \sqcup (\Gamma \cap S^{n-1})_\infty,$$

where Γ runs over open cones containing a and $R \gg 0$. Let $\mathcal{O}$ be the sheaf of holomorphic functions on $\mathbf{C}^n$. Kaneko introduced in [4] the sheaves $\tilde{\mathcal{O}}^{s,\delta}$, $s > 0$, $\delta \in \mathbf{R}$. We need such sheaves with $s \in (0,1)$ and $\delta = 0$. Recall, for open $U \subset \mathbf{D}^n + i\mathbf{R}^n$,

$$\tilde{\mathcal{O}}^{s,0}(U) = \tilde{\mathcal{O}}^s(U) = \{F(z) \in \mathcal{O}(U \cap \mathbf{C}^n);\ \forall K \subset\subset U\ \forall \varepsilon > 0,$$

$$\sup_{z \in K \cap \mathbf{C}^n} |F(z)| e^{-\varepsilon |z|^s} < \infty\},$$

where $K \subset\subset U$ means that the closure of K in $\mathbf{D}^n + i\mathbf{R}^n$ is a compact set contained in U.

Let Ω be an open subset of $\mathbf{D}^n$ and Γ be an open cone in $\mathbf{R}^n$. Then $\tilde{\mathcal{O}}^s(\Omega + i\Gamma 0) := \varinjlim \tilde{\mathcal{O}}^s(U)$, U running over infinitesimal wedges with edge Ω and opening Γ.

The sheaf $\mathcal{P}^s$ on $\mathbf{D}^n$ is defined by $\tilde{\mathcal{O}}^s|_{\mathbf{D}^n}$ and the sheaf of Fourier hyperfunctions on $\mathbf{D}^n$ is defined by $\mathcal{Q}^s = \mathcal{H}^n_{\mathbf{D}^n}(\tilde{\mathcal{O}}^s)$ ([4]).

For every open set $\Omega \subset \mathbf{D}^n$, $\mathcal{Q}^s(\Omega) = H^n_\Omega(\mathbf{C}^n, \tilde{\mathcal{O}}^s)$. This cohomological expression can be translated in the following way. $\mathcal{Q}^s(\Omega) = X/Y$, where

$$X = \sum_\Gamma \tilde{\mathcal{O}}^s(\Omega + i\Gamma 0), \Gamma \text{ runs over all open convex cones}$$

and Y is a $\mathbf{C}$-linear subspace of X generated by all the elements of the form

$$F_1(z) + F_2(z) - F_3(z),\ F_j(z) \in \tilde{\mathcal{O}}^s(\Omega + i\Gamma_j 0), j = 1, 2, 3,$$

such that $\Gamma_3 \subset \Gamma_1 \cap \Gamma_2$ and that $F_1(z) + F_2(z) = F_3(z)$ on a common domain of definition.

We identify $\sum_{j=1}^N F_j(x + i\Gamma_j 0)$ with the corresponding element in X. Note,

$$\sum_{j=1}^N F_j(x + i\Gamma_j 0) \sim 0 \quad \text{if and only if there exists}$$

$$H_{jk}(z) \in \tilde{\mathcal{O}}^s(\Omega + i(\Gamma_j + \Gamma_k)0),\ H_{jk}(z) = -H_{kj}(z),\ j, k = 1, ..., N$$

$$F_j(z) = \sum_{k=1}^{N} H_{jk}(z), \quad j = 1, ..., N, \quad \text{on a common domain.}$$

The singular spectrum of an $f \in \mathcal{Q}^s(\Omega)$ is defined as follows: f is microanalytic at $(a, \xi) \in \Omega \times S^{n-1}$ if there exists a local boundary value expression $f = \sum_{j=1}^{r} F_j(x + i\Gamma_j 0)$ on a neighbourhood of a such that $\Gamma_j \cap \{\langle y, \xi \rangle < 0\} \neq 0$ for all j. The complement of the set of points where f is microlocally analytic is called the singular spectrum of f and denoted by $SS^s f$. (Refer to [3] and [4] for details.)

If $s = 1$, we omit the super-index. In this case $\mathcal{Q}$ is the well known space of Fourier hyperfunctions and we have the topological automorphism through the Fourier transformation: $\mathcal{F} : \mathcal{Q}(\mathbf{D}^n) \xrightarrow{\sim} \mathcal{Q}(\mathbf{D}^n)$.

3.Tempered hyperfunctions

In the sequel, we use the notation $s \in (0, 1)$ and $1/s = t$. For a cone γ, γ^0 denotes the polar cone; its interior is denoted by $int\gamma^o$ and $\bar{\gamma}$ denotes the closure of γ.

Recall, a cone is acute if it does not contain stright lines. In this case $int\gamma^o \neq \emptyset$.

Definition 1. *Let γ be an open cone such that $\bar{\gamma}$ is acute; denote $\Gamma = int\gamma^0$. $\mathcal{O}'^s_\tau(\mathbf{R}^n + i\Gamma) = \mathcal{O}'^s_\tau$ is the space functions $F \in \mathcal{O}(\mathbf{R}^n + i\Gamma)$ which satisfy*

$$(\forall a > 0)(\forall b > 0)(\forall \varepsilon > 0)(\exists M > 0).$$

$$|F(x + iy)| \leq M e^{a|x|^s + b/|y|^{s/(1-s)} + \varepsilon|y|}, \; x + iy \in \mathbf{R}^n + i\Gamma. \tag{1}$$

This space corresponds to the Roumieau space of tempered ultradistributions supported by $\bar{\gamma}$. This is a consequence of following theorems (with the given notation).

Theorem 1 ([1]) *The Laplace transformation is an algebraic isomorphism of*
$\mathcal{S}'^{\{t\}}(\bar{\gamma})$ onto $\mathcal{O}'^s_\tau(\mathbf{R}^n + i\Gamma)$:

$$f \mapsto \mathcal{L}(f) = F(z) = \langle f(u), \eta(u)e^{izu} \rangle, \quad z \in \mathbf{R}^n + iC,$$

where $\eta \in \mathcal{E}^{\{t\}}$ *and* $\eta = \left(\begin{cases} 1 & \text{in a neighbourhood of } \bar{\gamma} \\ 0 & \text{out of a neighbourhood of } \bar{\gamma} \end{cases} \right).$

In particular, $F(x + i0) = \; bvF = \mathcal{F}(f)(x) \in \mathcal{S}'^{\{t\}}$.

Theorem 2 ([8]) *Let Γ be an open convex cone in $\mathbf{R}^n$ and F be an analytic function in*

$$Z = \{z \in \mathbf{C}^n; \;\; Imz \in \Gamma, \; | \; Imz \; |< h\}$$

for some $h > 0$. Moreover, assume that

$$(\forall a > 0) \; (\forall b > 0) \; (\exists C > 0)$$

$$| \; F(x + \sqrt{-1}y) \; |\leq Ce^{a|x|^s + b/|y|^{s/(1-s)}}, \;\; x + \sqrt{-1}y \in Z.$$

Then,

$$F(x + \sqrt{-1}y) \overset{\mathcal{S}'\{t\}}{\longrightarrow} F(x + \sqrt{-1}0), \;\; y \to 0, \;\; y \in \Gamma.$$

Definition 2. *Let Γ be an open convex cone. $\tilde{\mathcal{O}}'^{s}_{\tau}(\Omega + i\Gamma 0)$ is the space of functions $F \in \mathcal{O}((\mathbf{R}^n \cap \Omega) + i\Gamma 0)$ such that*

$$(\forall K \subset\subset \Omega) \;\; (\forall \Delta \subset\subset \Gamma)(\forall a > 0)(\forall b > 0)(\exists c > 0) \tag{2}$$

$$|F(x + iy)| < ce^{a|x|^s + b/|y|^{s/(1-s)}}, \; x + iy \in (\mathbf{R}^n \cap K) + i\Delta, |y| < 1/c.$$

Then, $\mathcal{S}b^s(\Omega)$ is a $\mathbf{C}$-vector space defined as X'/Y', where X' and Y' are defined by

$$X' = \sum_{\Gamma} \mathcal{O}'^{s}_{\tau}(\Omega + i\Gamma 0), \; \Gamma \textit{ runs over all open convex cones}$$

and Y' is a $\mathbf{C}$-linear subspace of X' generated by all the elements of the form

$$F_1(z) + F_2(z) - F_3(z), \; F_j(z) \in \mathcal{O}'^{s}_{t}(\Omega + i\Gamma_j 0), j = 1, 2, 3,$$

such that $\Gamma_3 \subset \Gamma_1 \cap \Gamma_2$ and that $F_1(z) + F_2(z) = F_3(z)$ on a common domain of definition.

It consists of sums

$$\sum_{j=1}^{N} F_j(x + i\Gamma_j 0), \;\; F_j \in \mathcal{O}'^{s}_{\tau}(\Omega + i\Gamma_j 0). \tag{3}$$

Clearly, $\Omega \mapsto \mathcal{S}b^s(\Omega)$ is a sheaf. It is denoted by $\mathcal{S}b^s$.

Theorem 3. *1) The natural mapping $\mathcal{S}b^s \to \mathcal{Q}^s$ defined by*

$$\sum_{j=1}^{N} F_j(x + i\Gamma_j) \to \sum_{j=1}^{N} F_j(x + i\Gamma_j)$$

is injective. In particular, $\mathcal{S}b^s$ is a subsheaf of $\mathcal{Q}^s$.

2) $\mathcal{S}b^s$ is a soft sheaf.

Proof. The first assertion is trivial and to prove the second one we have to use the partition of unity with elements in $\mathcal{D}'^{\{t\}}$.

Now, we will define the Laplace transform $\mathcal{L} : \mathcal{S}'^{\{t\}} \mapsto \mathcal{S}b^s$. Let $K \subset \mathbf{R}^n$ be a compact set such that $intK \ni 0$ and $\gamma_i, i = 1, ..., p$, be open convex cones such that $\cup_{i=1}^{p} \gamma_i = \mathbf{R}^n \setminus \{0\}$. Assume that the cones $\bar{\gamma}_i, i = 1, .., p$, are acute. Let $\eta_0, \eta_1, ..., \eta_p$ be a partition of unity such that supp $\eta_0 \subset K$, supp $\eta_i \subset \gamma_i, i = 1, ..., p$. If $f \in \mathcal{S}'^{\{t\}}$, then

$$f = f_0 + \sum_{1}^{p} f_i, \; f_0 = f\eta_0 \in \mathcal{E}'^{\{t\}}, \; f_i = f\eta_i \in \mathcal{S}'^{\{t\}}(\bar{\gamma}_i), \; i = 1, ..., p.$$

We define the Laplace transform by

$$\mathcal{L}f = F_0 + \sum_{1}^{p} F_i,$$

where

$$F_0(z) = \mathcal{L}f_0(z), z \in \mathbf{C}^n, F_i(z) = \mathcal{L}f_i(z), Imz \in \Gamma_i = int\gamma_i^o, i = 1, ..., p.$$

In the first part of next assertion we prove that $\mathcal{L} : \mathcal{S}'^t \to \mathcal{S}b^s(\mathbf{D}^n)$ does not depend on the given decomposition.

Let $\tilde{K}, \tilde{\gamma}_j, j = 1, ..., r$, be another covering of $\mathbf{R}^n$ and $\tilde{\eta}_j, j = 0, ..., r$, be the corresponding partition of unity with the same properties as above.

Theorem 4. *(i) Let $f \in \mathcal{S}'^{\{t\}}$ and $f = \tilde{f}_0 + \sum_1^r \tilde{f}_j$, where*

$$\tilde{f}_0 = \tilde{f}\eta_0 \in \mathcal{E}'^{\{t\}}, \; \tilde{f}_j = \tilde{f}\eta_j \in \mathcal{S}'^{\{t\}}(\bar{\tilde{\gamma}}_j), \tilde{F}_j = \mathcal{L}(\tilde{f}_j), j = 0, ..., r.$$

Then,

$$F_0 + \sum_{1}^{p} F_i - (\tilde{F}_0 + \sum_{1}^{r} \tilde{F}_j) \in Y'.$$

(ii) Let

$$j(f) = \mathcal{L}\mathcal{F}^{-1}(f).$$

Then, j is an algebraic isomorphism of $\mathcal{S}'^{\{t\}}$ onto $\mathcal{S}b^s(\mathbf{D}^n)$ such that

$$bv(j(f)) = f.$$

Proof. (i) For every $i = 1, ..., p$,

$$F_i(z) = \mathcal{L}(\sum_{j=0}^{r} f_i\tilde{\eta}_j)(z), Im z \in \Gamma_i = \text{ int } \bar{\gamma}_i.$$

This implies that

$$F(x + i\Gamma_i 0) - \mathcal{L}(\tilde{\eta}_0 f_i)(z) - \sum_{j=1}^{r} \mathcal{L}(\tilde{\eta}_j f_i)(x + i\text{int } \tilde{\gamma}_j) \in Y',$$

where $\tilde{\Gamma}_j, \text{int}\tilde{\gamma}_j,\ j = 1, ..., r$. Clearly, $F_0 - \sum_{j=0}^{r} \mathcal{L}(\tilde{\eta}_j f_0)(z) \in Y'$. This implies assertion in (i).

(ii) Let $f \in \mathcal{S}b^s(\mathbf{D}^n)$ be given by a single element $F(x + i\Gamma 0)$ which satisfies (2) (where $\Gamma = int\bar{\gamma}$). Then Theorems 1 and 2 imply that

$$\text{bv}F(x + i\Gamma 0) \in \mathcal{S}'^{\{t\}}$$

and $\mathcal{L}(\mathcal{F}^{-1}(\text{ bv}F(x + iC0))) = f$.

Conversely, let $f \in \mathcal{S}'^{\{t\}}$. Then, $\hat{f} \in \mathcal{S}'^{\{t\}}$. We decompose $\hat{f}$ by using the partition of unity $\eta_i, i = 1, ..., p$, and the notation given before Theorem 4 as $\hat{f} = \hat{f}_0 + \sum_{i=1}^{p} \hat{f}_i$. We have $\mathcal{L}\hat{f}_i \in \mathcal{O}'^s_t(\mathbf{R}^n + i\Gamma_i)$ and

$$\sum_{i=1}^{p} \text{bv}((\mathcal{L}\hat{f}_i)(x + i\Gamma_i 0)) = f.$$

Now the proof follows from Theorem 1.

We define quotient sheaves $\mathcal{S}b^s/\mathcal{P}^s$ and $\mathcal{Q}^s/\mathcal{S}b^s$ by

$$0 \to \mathcal{P}^s \to \mathcal{S}b^s \to \mathcal{S}b^s/\mathcal{P}^s \to 0,$$

$$0 \to \mathcal{S}b^s \to \mathcal{Q}^s \to \mathcal{Q}^s/\mathcal{S}b^s \to 0.$$

Theorem 5. *For every open $\Omega \subset D^n$,*

$$\mathcal{S}b^s/\mathcal{P}^s(\Omega) = \mathcal{S}b^s(\Omega)/\mathcal{P}^s(\Omega),$$

$$\mathcal{Q}^s/\mathcal{S}b^s(\Omega) = \mathcal{Q}^s(\Omega)/\mathcal{S}b^s(\Omega).$$

Proof The first assertion follows from $H^k(\Omega, \mathcal{P}^s) = 0,\ k > 1,\ \Omega \subset \mathbf{D}^n$. The softness of $\tilde{\mathcal{S}}b^s$ implies $H^k(\Omega, \tilde{\mathcal{S}}b^s) = 0,\ k > 1,\ \Omega \subset D^n$.

Denote by i a natural mapping

$$\mathcal{S}b^s(\Omega) \to \mathcal{S}b^s(\Omega)/\mathcal{P}^s(\Omega).$$

We define

$$\mathrm{singsupp}^s f = \mathrm{supp} i(f), \quad f \in \mathcal{S}b^s(\mathbf{D}^n).$$

Theorem 6. *Let $f \in \mathcal{S}'^{\{t\}}$. Then*

$$\mathbf{R}^n \cap \mathrm{singsupp}^s j(f) = \mathrm{singsupp}_{\mathrm{a}} f$$

and

$$\mathrm{singsupp}^s j(f) = \mathrm{supp} i(j(f)).$$

($\mathrm{singsupp}_{\mathrm{a}} f$ *denotes the analytic singular support of* f).

Proof. It follows from the edge of the wedge theorem (cf. [2]).

The isomorphism j enables us to define the analytic singular support and the analytic wave front set of an $f \in \mathcal{S}'^{\{t\}}$ at $x_0\infty$ and $(x_0\infty, \xi_0)$ respectively, where $(x_0\infty, \xi_0)$ denotes the point on S^{n-1}_∞ in the direction $\xi_0 \in \mathbf{R}^n \setminus \{0\}$.

Definition 3. *Let $f \in \mathcal{S}'^{\{t\}}$. Then*

$x_0\infty \notin \mathrm{singsupp}^s f$ *if* $x_0\infty \notin \mathrm{singsupp}^s jf$.

$(x_0\infty, \xi_0) \notin SS^s f$ *(f is s-micro-analytic at this point) if there exists a local boundary value expression* (3) *of f in a neighbourhood of $x_0\infty$ such that*

$$\Gamma_j \cap \{\langle y, \xi_0 \rangle < 0\} \neq \emptyset,\ j = 1, \ldots, N.$$

Theorem 7. *Let $f \in \mathcal{S}'^{\{t\}}$. Then*

$$x_0\infty \notin \mathrm{singsupp}^s f \Longleftrightarrow \forall \xi \in \mathbf{R}^n \setminus \{0\},\ (x_0\infty, \xi) \notin SS^s f.$$

References

[1] R. Carmichael and S. Pilipović, On the convolution and Laplace transformation in the space of Beurling-Gevrey tempered ultradistributions, Math. Nachr., 158 (1992), 119-132.

[2] A. Eida and S. Pilipović, On the microlocal decomposition of some classes of ultradistributions, Math. Proc. Camb. Phil. Soc., 125(1999), 455-461.

[3] A.Kaneko, Introduction à la Théorie des Hyperfonctions I, II, Cours de DEA, Université de Grenoble (1978).

[4] A. Kaneko, On the global existence of real analytic solutions of linear partial differential equations on unbounded domain, J. Fac. Sci., Univ. of Tokyo, Sect. IA 32 (1985), 319-372.

[5] H.Komatsu, Microlocal Analysis in Gevrey Classes and in Convex Domain, Springer, Lect. Not. Math. 1726 (1989), 426-493.

[6] H.Komatsu, Structure theorems and a characterization, J. Fac. Sci., Univ. of Tokyo, Sect. IA 20 (1973), 25-105.

[7] D. Kovačević and S. Pilipović, Structural properties of the space of tempered ultradistributions, Proc. Conf. "Complex Analysis and Generalized Functions", Varna 1991, 169-184.

[8] S. Pilipović, Characterization of bounded sets in spaces of ultradistributions, Proc. AMS, 120(1994), 1191-1206.

[9] M.Sato, T.Kawai and M.Kashiwara, Microfunctions and pseudo - differential equations, Lect. Notes in Math. No 287 (1973), 265-529, Springer.

88 HOLOMORPHIC FUNCTIONS ON THE DUAL LIE BALL

Keiko FUJITA

Faculty of Culture and Education
Saga University
Saga, 840-8502 Japan

keiko@cc.saga-u.ac.jp

Mitsuo MORIMOTO

Department of Mathematics
International Christian University
Tokyo, 181-8585 Japan

mitsuo@icu.ac.jp

1 Introduction

Let

$$F(z,w) = \sum_{n,m=0}^{\infty} a_{n,m} z^n w^m$$

be a Taylor expansion of a germ of holomorphic functions at $(0,0) \in \mathbb{C}^2$. Then we know the following two theorems:

H.G.W. Begehr et al. (eds.), Proceedings of the Second ISAAC Congress, Vol. 1, 771–780.

Theorem 1. *Let* $r > 0$.

$$\limsup_{n+m\to\infty} |a_{n,m}|^{1/(n+m)} \le 1/r$$

if and only if $F(z,w)$ *is a holomorphic function on the polydisk*

$$\{(z,w) \in \mathbb{C}^2;\ |z| < r \text{ and } |w| < r\}.$$

Theorem 2. *Let* $r > 0$.

$$\limsup_{n+m\to\infty} \left|\frac{n!m!}{(n+m)!} a_{n,m}\right|^{1/(n+m)} \le 1/(2r)$$

if and only if $F(z,w)$ *is a holomorphic function on the domain*

$$\{(z,w) \in \mathbb{C}^2;\ |z| + |w| < 2r\}.$$

Theorem 1 is a theorem of the Cauchy-Hadamard type and was generalized in [5] for the double series expansion in higher dimensions (Theorem 4). A generalization of Theorem 2 was found in [3] for the double series expansion in higher dimensions (Theorem 7). The original proof of Theorem 7 relied on some results in [3] and the Fourier-Borel transformation for entire functionals on $\mathbb{C}^{n+1}$. (See Fujita [2] and Iwahara [4] for related topics.)

In this paper we shall give a direct proof of Theorem 7. The same idea can be applied to characterize holomorphic functions on the Euclidean ball by the double series expansion (Theorem 9).

2 Double series expansion

Let $N(z)$ be a norm on $\mathbb{C}^{n+1}$. Let f be a germ of holomorphic functions at $0 \in \mathbb{C}^{n+1}$ and

$$f(z) = \sum_{k=0}^{\infty} f_k(z) \tag{1}$$

be its expansion into homogeneous polynomials f_k of degree k. We put

$$\|f_k\|_{1,N} = \sup\{|f_k(z)|;\ N(z) \le 1\}.$$

The ball $\tilde{B}(r,N)$ of radius r with respect to $N(z)$ is defined by

$$\tilde{B}(r,N) = \{z \in \mathbb{C}^{n+1};\ N(z) < r\}.$$

The following is also a theorem of the Cauchy-Hadamard theorem:

Theorem 3. *Let $r > 0$ and let $N(z)$ be a norm on $\mathbb{C}^{n+1}$. Let f be given by (1). Then $f(z)$ is a holomorphic function on the ball $\tilde{B}(r, N)$ if and only if*

$$\limsup_{k\to\infty}(\|f_k\|_{1,N})^{1/k} \leq 1/r.$$

Let $f_k(z)$ be a homogeneous polynomial of degree k. Then f_k can be uniquely expanded as follows:

$$f_k(z) = f_{k,k}(z) + z^2 f_{k,k-2}(z) + (z^2)^2 f_{k,k-4}(z) + \cdots$$

$$+(z^2)^{[k/2]} f_{k,k-2[k/2]}(z),$$

where $z^2 = z_1^2 + z_2^2 + \cdots + z_{n+1}^2$ and $f_{k,k-2l}(z)$ is a homogeneous *harmonic* polynomial of degree $k - 2l$;

$$\Delta f_{k,k-2l}(z) = \left(\frac{\partial^2}{\partial z_1^2} + \frac{\partial^2}{\partial z_2^2} + \cdots + \frac{\partial^2}{\partial z_{n+1}^2}\right) f_{k,k-2l}(z) = 0.$$

Therefore, if a germ f of holomorphic functions at 0 is expanded as in (1), then it can be expanded further into the *double series*:

$$f(z) = \sum_{k=0}^{\infty}\sum_{l=0}^{[k/2]} (z^2)^l f_{k,k-2l}(z). \tag{2}$$

3 Lie norm and Lie ball

Let $\|x\|$ be the Euclidian norm on $\mathbb{R}^{n+1}$. The real unit sphere of dimension n is denoted by $\mathbb{S} = \mathbb{S}^n = \{x \in \mathbb{R}^{n+1};\ \|x\| = 1\}$. The *Lie norm* $L(z)$ on $\mathbb{C}^{n+1}$ is the cross norm of $\|x\|$:

$$L(z) = \inf\left\{\sum_{j=1}^{M} |\lambda_j|\|x_j\|;\ z = \sum_{j=1}^{M} \lambda_j x_j, \lambda_j \in \mathbb{C}, x_j \in \mathbb{R}^{n+1}\right\}.$$

We know (see, for example, [6]) that

$$L(z) = \sqrt{\|z\|^2 + \sqrt{\|z\|^4 - |z^2|^2}}, \tag{3}$$

where $\|z\|^2 = |z_1|^2 + |z_2|^2 + \cdots + |z_{n+1}|^2$. Note that $L(x) = \|x\|$ for $x \in \mathbb{R}^{n+1}$ and that

$$L(z) = \max\{|z_1 + iz_2|, |z_1 - iz_2|\} \text{ for } z = (z_1, z_2, 0, \cdots, 0) \in \mathbb{C}^{n+1}. \tag{4}$$

The open ball of radius r with respect to $L(z)$ is called the *Lie ball* and is denoted by $\tilde{B}(r) = \{z \in \mathbb{C}^{n+1};\ L(z) < r\}$. The Shilov boundary of $\tilde{B}(r)$ is the Lie sphere $\Sigma_r = \{re^{i\theta}x;\ \theta \in \mathbb{R},\ x \in \mathbb{S}^n\}$. For a homogeneous polynomial f_k we have

$$\|f_k\|_{1,L} = \sup\{|f_k(x)|;\ x \in \mathbb{S}\} \equiv \|f_k\|_{C(\mathbb{S})}. \tag{5}$$

Because of (4), the following is a generalization of Theorem 1:

Theorem 4 (Morimoto [5]). *Let $r > 0$ and let f be given by (2). Then f is holomorphic on the Lie ball $\tilde{B}(r)$ if and only if*

$$\limsup_{k\to\infty} \left(\|f_{k,k-2l}\|_{C(\mathbb{S})}\right)^{1/k} \leq 1/r. \tag{6}$$

In the double series expansion (2) the parameter l runs between 0 and $[k/2]$. Because $k \leq k + (k-2l) = 2k - 2l \leq 2k$, the condition (6) is equivalent to

$$\limsup_{2k-2l\to\infty} (r^k\|f_{k,k-2l}\|_{C(\mathbb{S})})^{1/(2k-2l)} \leq 1.$$

For $0 < \lambda < r$, we put $\tilde{B}(\lambda, r) = \{z \in \mathbb{C}^{n+1};\ |z^2| < \lambda^2,\ L(z) < r\}$ and $\Sigma_{\lambda,r} = \{z \in \mathbb{C}^{n+1};\ |z^2| = \lambda^2,\ L(z) = r\}$. If a holomorphic function f on $\tilde{B}(\lambda, r)$ is given by (2), then $f_{k,k-2l}$ is represented by an integral over $\Sigma_{\mu,R}$ with respect to the normalized invariant measure (see Morimoto [7]):

$$f_{k,k-2l}(z) = \mu^{-4l} N(k-2l, n)/L_{k-2l,\mu,R} \int_{\Sigma_{\mu,R}} (\bar{w}^2)^l \tilde{P}_{k-2l,n}(z, \bar{w}) f(w) d\dot{w}, \tag{7}$$

where $\mu = \lambda/(1+\epsilon)$ and $R = r/(1+\epsilon)$ for $\epsilon > 0$,

$$\begin{aligned} N(k,n) &= (2k+n-1)(k+n-2)!/(k!(n-1)!),\\ L_{k,\mu,R} &= \mu^{2k} P_{k,n}((\mu^2/R^2 + R^2/\mu^2)/2),\\ \tilde{P}_{k,n}(z,w) &= (\sqrt{z^2})^k (\sqrt{w^2})^k P_{k,n}((z/\sqrt{z^2}) \cdot (w/\sqrt{w^2})) \end{aligned} \tag{8}$$

and $P_{k,n}(t)$ is the Legendre polynomial of degree k and of dimension $n+1$. Note that

$$\begin{aligned} &2^{-k}\gamma_{k,n} R^{2k} \leq L_{k,\mu,R} \leq R^{2k},\\ &\gamma_{k,n} = \Gamma(k + (n+1)/2) 2^k / (N(k,n)\Gamma((n+1)/2)k!) \end{aligned} \tag{9}$$

(see Fujita [1]). We can generalize Theorem 4 as follows:

Theorem 5. *Let $r > \lambda > 0$ and let f be given by (2). Then f is holomorphic on the domain $\tilde{B}(\lambda, r)$ if and only if*

$$\limsup_{k\to\infty} \left(\lambda^{2l} r^{k-2l} \|f_{k,k-2l}\|_{C(\mathbb{S})}\right)^{1/k} \le 1. \tag{10}$$

Proof. First suppose the inequality (10) holds. Then for any $\epsilon > 0$ there is $C \ge 0$ such that

$$\|f_{k,k-2l}\|_{C(\mathbb{S})} \le C(1+\epsilon)^k / (\lambda^{2l} r^{k-2l}). \tag{11}$$

Therefore, by (5) and (11) we have

$$\begin{aligned}
&\sum_{k=0}^{\infty} \sum_{l=0}^{[k/2]} (z^2)^l |f_{k,k-2l}(z)| \\
&\le C \sum_{k=0}^{\infty} \sum_{l=0}^{[k/2]} |z^2|^l \|f_{k,k-2l}\|_{C(\mathbb{S})} L(z)^{k-2l} \\
&\le C \sum_{k=0}^{\infty} \sum_{l=0}^{[k/2]} ((1+\epsilon)L(z)/r)^k (r^2|z^2|/(\lambda^2 L(z)^2))^l \\
&= C \sum_{l=0}^{\infty} \sum_{k=2l}^{\infty} ((1+\epsilon)L(z)/r)^k (r^2|z^2|/(\lambda^2 L(z)^2))^l \\
&= Cr/(r-(1+\epsilon)L(z)) \sum_{l=0}^{\infty} ((1+\epsilon)^2|z^2|/\lambda^2)^l.
\end{aligned}$$

This converges uniformly on compact sets in $\tilde{B}(\lambda/(1+\epsilon), r/(1+\epsilon))$. Because $\epsilon > 0$ is arbitrary, (2) defines a holomorphic function f on $\tilde{B}(\lambda, r)$.

Conversely, if $f \in \mathcal{O}(\tilde{B}(\lambda, r))$, then the integral representation (7) implies

$$\|f_{k,k-2l}\|_{C(\mathbb{S})} \le \mu^{-4l} N(k-2l, n)/L_{k-2l,\mu,R}\mu^{2l} R^{k-2l} \|f\|_{C(\Sigma_{\mu,R})},$$

where $\mu = \lambda/(1+\epsilon)$, $R = r/(1+\epsilon)$, and $\|f\|_{\Sigma_{\mu,R}} = \sup\{|f(z)|;\ z \in \Sigma_{\mu,R}\}$. Then we have

$$\lambda^{2l} r^{k-2l} \|f_{k,k-2l}\|_{C(\mathbb{S})} \le N(k-2l, n)(1+\epsilon)^k (R^{2(k-2l)}/L_{k-2l,\mu,R}) \|f\|_{C(\Sigma_{\mu,R})}.$$

Therefore, (8) and (9) imply

$$\limsup_{k\to\infty} \left(\lambda^{2l} r^{k-2l} \|f_{k,k-2l}\|_{C(\mathbb{S})}\right)^{1/k} \le 1+\epsilon,$$

from which results (10) because $\epsilon > 0$ is arbitrary. □

4 Dual Lie norm and dual Lie ball

Let $N(z)$ be a norm on $\mathbb{C}^{n+1}$. The norm $N^*(z)$ dual to $N(z)$ is defined by

$$N^*(z) = \sup\{|z \cdot \zeta|;\ N(\zeta) \le 1\},$$

where $z \cdot \zeta = z_1\zeta_1 + z_2\zeta_2 + \cdots + z_{n+1}\zeta_{n+1}$. The norm $L^*(z)$ dual to $L(z)$ is called the *dual Lie norm*. We know

$$L^*(z) = \sqrt{(\|z\|^2 + |z^2|)/2}. \tag{12}$$

We have $L^*(z) \le L(z) \le 2L^*(z)$. Note that $L^*(x) = \|x\| = L(x)$ for $x \in \mathbb{R}^{n+1}$ and that

$$L^*(z) = \frac{1}{2}(|z_1 + iz_2| + |z_1 - iz_2|) \text{ for } z = (z_1, z_2, 0, \cdots, 0) \in \mathbb{C}^{n+1}. \tag{13}$$

The ball of radius r with respect to $L^*(z)$ is called the *dual Lie ball* and is denoted by $\tilde{B}^*(r) = \{z \in \mathbb{C}^{n+1};\ L^*(z) < r\}$.

Lemma 6. (i) $L^*(z) = \frac{1}{2}\left(L(z) + |z^2|/L(z)\right)$.

(ii) $\tilde{B}^*(r) = \bigcup\left\{\tilde{B}(\lambda, R);\ 0 < \lambda < r,\ R + \lambda^2/R < 2r\right\}$.

Proof. (3) and (12) imply (i), from which results (ii). □

Because of (13), the following is a generalization of Theorem 2:

Theorem 7. *Let $r > 0$ and let f be given by (2). Then f is holomorphic on the dual Lie ball $\tilde{B}^*(r)$ if and only if*

$$\limsup_{k\to\infty} \left((l!(k-l)!/k!)\|f_{k,k-2l}\|_{C(\mathbb{S})}\right)^{1/k} \le 1/(2r). \tag{14}$$

Proof. Suppose first the inequality (14) holds. Then for any $\epsilon > 0$ there is $C \ge 0$ such that

$$\|f_{k,k-2l}\|_{C(\mathbb{S})} \le Ck!/(l!(k-l)!)((1+\epsilon)/(2r))^k.$$

Therefore, we have

$$\begin{aligned}
&\sum_{k=0}^{\infty}\sum_{l=0}^{[k/2]} |(z^2)^l f_{k,k-2l}(z)| \\
&\le \sum_{k=0}^{\infty}\sum_{l=0}^{[k/2]} |z^2|^l \|f_{k,k-2l}\|_{C(\mathbb{S})} L(z)^{k-2l} \\
&\le C\sum_{k=0}^{\infty}\sum_{l=0}^{[k/2]} k!/(l!(k-l)!)((1+\epsilon)/(2r))^k L(z)^{k-l}(|z^2|/L(z))^l \\
&\le C\sum_{k=0}^{\infty}((1+\epsilon)/(2r))^k \sum_{l=0}^{k} k!/(l!(k-l)!)L(z)^{k-l}(|z^2|/L(z))^l \\
&= C\sum_{k=0}^{\infty}((1+\epsilon)/(2r))^k (L(z)+|z^2|/L(z))^k \\
&= C\sum_{k=0}^{\infty}((1+\epsilon)/r)^k L^*(z)^k
\end{aligned}$$

by Lemma 6. Therefore, the double series (2) converges uniformly on compact sets in $\tilde{B}^*(r/(1+\epsilon))$. As $\epsilon > 0$ is arbitrary, (2) defines a holomorphic function f on $\tilde{B}^*(r)$.

Conversely, let f be a holomorphic function on $\tilde{B}^*(r)$. Because of Lemma 6, f is holomorphic on $\tilde{B}(\lambda, R)$ if $R+\lambda^2/R < 2r$. By the integral representation (7) (see the proof of Theorem 5), for any $\epsilon > 0$ there is $C_{k,l} \ge 0$ such that $\limsup_{k\to\infty} \sqrt[k]{C_{k,l}} = 1$ and

$$\|f_{k,k-2l}\|_{C(\mathbb{S})} \le C_{k,l}/(\lambda^{2l}R^{k-2l}) \tag{15}$$

for any $\lambda > 0$ and $R > 0$ with $(R+\lambda^2/R) \le 2r/(1+\epsilon)$, $k = 1, 2, \cdots$, and $l = 0, 1, 2, \cdots, [k/2]$. Taking $\lambda = 2r\sqrt{l(k-l)}/(k(1+\epsilon))$ and $R = 2r(k-l)/(k(1+\epsilon))$ in (15) we get

$$\|f_{k,k-2l}\|_{C(\mathbb{S})} \le C_{k,l}((1+\epsilon)/(2r))^k \cdot k^k/(l^l(k-l)^{k-l})$$

for $k = 1, 2, \cdots$. Therefore, we have

$$\limsup_{k\to\infty}\left(l^l(k-l)^{k-l}/k^k\|f_{k,k-2l}\|_{C(\mathbb{S})}\right)^{1/k} \le (1+\epsilon)/(2r).$$

As $\epsilon > 0$ can be arbitrarily small, we have

$$\limsup_{k\to\infty}\left(l^l(k-l)^{k-l}/k^k\|f_{k,k-2l}\|_{C(\mathbb{S})}\right)^{1/k} \le 1/(2r).$$

By the Stirling formula, we get (6). □

5 Holomorphic function on the Euclidean ball

Let $\tilde{U}(r) = \{z \in \mathbb{C}^{n+1}; \|z\| < r\}$ be the open Euclidean ball.

Lemma 8. (i) $\|z\|^2 = \left(L(z)^2 + (|z^2|/L(z))^2\right)/2$.

(ii) $\tilde{U}(r) = \bigcup\left\{\tilde{B}(\lambda, R);\ 0 < \lambda < r,\ R^2 + \lambda^4/R^2 < 2r^2\right\}$.

Proof. (3) implies (i), from which results (ii). □

Theorem 9. *Let $r > 0$ and let f be given by (2). Then f is holomorphic on the Euclidean ball $\tilde{U}(r)$ if and only if*

$$\limsup_{k\to\infty}((l!(k-l)!/k!)^{1/2}\|f_{k,k-2l}\|_{C(\mathbb{S})})^{1/k} \le 1/(\sqrt{2}r). \tag{16}$$

Proof. Suppose first the inequality (16) holds. For any $\epsilon > 0$ there is $C \geq 0$ such that

$$\|f_{k,k-2l}\|_{C(\mathbb{S})} \le C(k!/(l!(k-l)!))^{1/2}((1+\epsilon)/(\sqrt{2}r))^k$$

for $k = 0, 1, 2, \cdots$, $l = 0, 1, \cdots, [k/2]$. Therefore, by Lemma 8 we have

$$\begin{aligned}
&\sum_{k=0}^{\infty}\sum_{l=0}^{[k/2]}|(z^2)^l f_{k,k-2l}(z)| \\
&\le \sum_{k=0}^{\infty}\sum_{l=0}^{[k/2]}|z^2|^l\|f_{k,k-2l}\|_{C(\mathbb{S})}L(z)^{k-2l} \\
&\le C\sum_{k=0}^{\infty}\sum_{l=0}^{[k/2]}(k!/(l!(k-l)!))^{1/2}((1+\epsilon)/(\sqrt{2}r))^k L(z)^{k-l}(|z^2|/L(z))^l \\
&\le C\sum_{k=0}^{\infty}((1+\epsilon)/(\sqrt{2}r))^k\sum_{l=0}^{k}(k!/(l!(k-l)!))^{1/2}L(z)^{k-l}(|z^2|/L(z))^l \\
&\le C\sum_{k=0}^{\infty}((1+\epsilon)/(\sqrt{2}r))^k\sqrt{k+1}\,\times \\
&\left(\sum_{l=0}^{k}k!/(l!(k-l)!)L(z)^{2(k-l)}(|z^2|/L(z))^{2l}\right)^{1/2} \\
&\le C\sum_{k=0}^{\infty}((1+\epsilon)/(\sqrt{2}r))^k\sqrt{k+1}(L(z)^2+(|z^2|/L(z))^2)^{k/2} \\
&= C\sum_{k=0}^{\infty}((1+\epsilon)/r)^k\sqrt{k+1}\|z\|^k.
\end{aligned}$$

This calculation shows that the double series (2) converges uniformly on compact sets in $\tilde{U}(r/(1+\epsilon))$. As $\epsilon > 0$ is arbitrary, (2) defines a holomorphic function f on $\tilde{U}(r)$.

Conversely, let f be a holomorphic function on $\tilde{U}(r)$. Then f is holomorphic on $\tilde{B}(\lambda, R)$ if $R^2 + \lambda^4/R^2 \leq 2r^2$. By the integral representation (7) (see the proof of Theorem 5), for any $\epsilon > 0$ there exists $C_{k,l} \geq 0$ such that $\limsup_{k\to\infty} \sqrt[k]{C_{k,l}} = 1$ and

$$\|f_{k,k-2l}\|_{C(\mathbb{S})} \leq C_{k,l}/(\lambda^{2l}R^{k-2l}).$$

for any $\lambda > 0$ and $R > 0$ with $R^2 + \lambda^4/R^2 \leq (2r/(1+\epsilon))^2$, $k = 1, 2, \cdots$, and $l = 0, 1, 2, \cdots, [k/2]$. Taking $\lambda^2 = 2r^2\sqrt{(k-l)l}/(k(1+\epsilon))$ and $R^2 = 2r^2(k-l)/(k(1+\epsilon))$, we get

$$\|f_{k,k-2l}\|_{C(\mathbb{S})} \leq C((1+\epsilon)/(\sqrt{2}r))^k(k^k/(l^l(k-l)^{(k-l)})^{1/2}$$

for $k = 1, 2, \cdots$. Therefore, we get

$$\limsup_{k\to\infty}((l^l(k-l)^{k-l}/k^k)^{1/2}\|f_{k,k-2l}\|_{C(\mathbb{S})})^{1/k} \leq (1+\epsilon)/(\sqrt{2}r).$$

Because $\epsilon > 0$ is arbitrary, the Stirling formula implies (16). □

References

[1] Fujita, K., Hilbert spaces of eigen-functions of the Laplacian, "Reproducing Kernels and their Applications", Kluwer Academic Publishers, 1999, pp. 65 – 80.

[2] Fujita, K., Characterization of entire functions of exponential type with respect to the Lie norm, submitted to RIMS Kokyuroku.

[3] Fujita, K., On the double series expansion with harmonic components, to appear in "Finite or Infinite Dimensional Complex Analysis: Seventh International Colloquium", Lecture Notes in Pure and Applied Mathematics, Marcel Dekker.

[4] Iwahara, Y., Entire functions of exponential type related to the Lie ball, Master's thesis, Sophia University, 1998.

[5] Morimoto, M., Analytic functionals on the Lie sphere, Tokyo J. Math., 3(1980), 1 – 35.

[6] Morimoto, M., Analytic Functionals on the Sphere, Translations of Mathematical Monographs, Volume 178, AMS, 1998.

[7] Morimoto, M., A generalization of the Cauchy-Hua integral formula on the Lie ball, Tokyo J. Math., 22(1999), 177 – 192.

89 SATO HYPERFUNCTIONS VALUED IN A LOCALLY CONVEX SPACE

Yoshifumi ITO

Faculty of Integrated Arts and Sciences, the University of Tokushima
Tokushima, 770-8502 JAPAN

y-ito@ias.tokushima-u.ac.jp

Introduction

This is the report of the results of Y. Ito [3]-[5], [12], [22]. As for Sato-Fourier hyperfunctions, we refer to the works [1], [6]-[11], [13]-[21], [23]-[26].

At first, we look back briefly on the history of the discovery of the concepts of generalized functions.

The concepts of functions and derivatives were generalized for the requirement of researches on analysis, especially on partial differential equations. Physicists and engineers had already used singular functions such as Dirac's δ-function and Heaviside's function as convenient tools, but it was not clear what they meant mathematically.

Hadamard had found out the concept of finite part of divergent integral in the construction of a fundamental solution of the wave equation. In 1945, Schwartz established the theory of distributions and generalized systematically pioneering works. Thereby he gave the mathematical

H.G.W. Begehr et al. (eds.), Proceedings of the Second ISAAC Congress, Vol. 1, 781–788.

foundation to the methods used in the research on physics and on partial differential equation.

The purpose of introducing the concept of generalized functions was to make it possible to carry out the fundamental operations "differentiation and integration" of analysis in a generalized sense for singular "functions" for which we cannot differentiate and integrate in the sense of classical analysis. There are many singular "functions" of this kind which describe physical quantities.

In 1958, Mikio Sato found out the theory of Sato hyperfunctions. We consider hyperfunctions as generalized functions, among which are the following: Radon measures, Schwartz distributions, Roumieu-Beurling ultradistributions, infrahyperfunctions (Petzsche hyperfunctions), analytic functionals, Sato hyperfunctions, Fourier hyperfunctions, modified Fourier hyperfunctions, mixed Fourier hyperfunctions, partial Fourier hyperfunctions, partial modified Fourier hyperfunctions, partial mixed Fourier hyperfunctions, Laplace hyperfunctions and their vector-valued versions.

The concept of hyperfunctions is a generalization of that of functions or measures. We must define them independently on the choice of coordinate system. But many examples of Schwartz distributions and Sato hyperfunctions depend on the choice of coordinate system, because they are defined as "special hyperfunctions".

Those which do not depend on the special choice of the coordinate system are the mathematical structure such as the space of hyperfunctions and the operation on them. If those properties change accordingly to the choice of coordinate system, we are in trouble when we understand the mathematical substances such as hyperfunctions.

The theory of Schwartz distributions is a theory in the C^∞-category. In this respect, we carry out the analysis in a real domain on the basis of informations on the analysis in a real domain. Contrary to this, the theory of Sato-Fourier-Laplace hyperfunctions is a theory in the real-analytic category. In this respect, we carry out the analysis in a real domain on the basis of informations on the analysis in a complex domain.

For the time being we can understand the difference of two theories shematically as above. But, if we consider carefully the situation, we can use the information of the analysis in a complex domain when we treat the Fourier transformation in the theory of Schwartz distributions. On the other hand, even considering microlocal analysis, we can obtain real-analytic informations by using the concept of analytic wave front sets. This means that real-analytic functions, being C^∞-functions, can be treated in the framework of the theory of C^∞-category.

Then, in the theory of Sato-Fourier-Laplace hyperfunctions, can we obtain the information of C^∞-category? In fact, when we represent Sato-Fourier-Laplace hyperfunctions as boundary values of holomorphic functions, we can extract the informations of C^∞-category such as being Schwartz distributions, by restricting the growth condition of defining functions near the boundary [2], [27].

Thus considering the situations, we find that the standard of these classification becomes fairly mental. But, in the point that it does or does not become a natural framework to each theory, this classification surely has a meaning.

As far as known up to the present, almost all hyperfunctions have the twofold realization as follows.

(1) the realization as a continuous linear functional on a space of functions.

(2) the realization as a boundary value of holomorphic functions.

Considering the case of Sato hyperfunctions, in the realization (2), we can treat the case up to the Fréchet space valued ones. But in the realization (1), we can treat the case of the general locally convex space valued ones.

In this report, we introduce the concept of Sato hyperfunctions valued in a general locally convex space in the realization (1).

For the limit of the method of tensor product, this generalization is considered as the ultimate case.

1. Sato hyperfunctions valued in a locally convex space

Let M be an n-dimensional and real-analytic manifold countable at infinity and X a complexification of M.

Let E be an arbitrary, locally convex, Hausdorff and topological vector space over the complex number field as far as the contrary is not mentioned.

For a compact subset K of M, we denote by $\mathcal{A}(K)$ the space of functions which are holomorphic in a complex neighborhood of K. Then analytic-linear mappings are defined to be continuous linear mappings from $\mathcal{A}(K)$ to E as vector-valued analytic functionals.

Theorem 1. *For every compact subset K of M, we denote by*

$$\mathcal{A}'(K;\, E) = L(\mathcal{A}(K),\, E)$$

the space of analytic-linear mappings on K. Then the following are valid.

(1) *For any two compact sets K_1 and K_2 in M with $K_1 \subset K_2$, there exists a continuous injection $i_{K_1,K_2} : \mathcal{A}'(K_1;\, E) \longrightarrow \mathcal{A}'(K_2;\, E)$.*

(2) *For two compact sets K_1, K_2 in M with $K_1 \subset K_2$, such that every connected component of K_2 intersects K_1, i_{K_1,K_2} has the dense range in $\mathcal{A}'(K_2;\, E)$.*

(3) *For an arbitrary family $\{K_i\}_{i \in I}$ of at most countable compact sets in M, we have $\cap_{i \in I} \mathcal{A}'(K_i;\, E) \cong \mathcal{A}'(\, \cap_{i \in I} K_i;\, E)$.*

(4) $\mathcal{A}'(\emptyset;\, E) = 0$.

Let Ω be an open subset of M and K a compact subset of Ω. We call the "envelope" of K (in Ω) and denote by $\tilde{K}$, the union of K and all the relatively compact connected components (in Ω) of $\Omega \setminus K$. It is again a compact set.

Theorem 2. *Let Ω be a relatively compact subset of M and K_1, $K_2\, (K_1 \subset K_2)$ two compact subsets of Ω such that $K_i = \tilde{K}_i$ holds $(i = 1, 2)$. Then $\mathcal{A}'((\Omega \setminus K_2)^{\mathrm{cl}};\, E)$ is dense in $\mathcal{A}'((\Omega \setminus K_1)^{\mathrm{cl}};\, E)$.*

Now we consider vector-valued Sato hyperfunctions on a relatively compact open set in M.

Let Ω be a relatively compact open subset of M. We put

$$\mathcal{B}(\Omega;\, E) = \mathcal{A}'(\Omega^{\mathrm{cl}};\, E)/\mathcal{A}'(\partial\Omega;\, E).$$

Since $\mathcal{A}'(\partial\Omega;\, E)$ is dense in $\mathcal{A}'(\Omega^{\mathrm{cl}};\, E)$, $\mathcal{B}(\Omega;\, E)$ has no nontrivial topology.

Definition 1. An element of $\mathcal{B}(\Omega;\, E)$ is called a Sato hyperfunction on Ω valued in E or an E-valued Sato hyperfunction on Ω.

Proposition 1. *Let Ω be a relatively compact open set in M. Then the collection $\{B(\omega;\, E); \omega$ is an open subset of $\Omega\}$ becomes a presheaf (of vector spaces) over Ω.*

Proposition 2. *We use the notation in Proposition 1. Let $\omega = \cup_{i \in I} \omega_i$ be a union of open subsets ω_i of $\Omega (i \in I)$ and $T \in \mathcal{B}(\omega;\, E)$ with $T|_{\omega_i} = 0$ for all $i \in I$. Then we have $T = 0$.*

Thus we have seen that the presheaf $\{\mathcal{B}(\omega :\, E); \omega$ is an open subset of $\Omega\}$ satisfies the null condition. But if E is not a Fréchet space, this presheaf does not satisfy the glueing condition. Thus, in general, we can not prove this presheaf is a sheaf unless E is a Fréchet space.

By virtue of Proposition 2, we can define the support of $T \in \mathcal{B}(\Omega;\, E)$. Namely, a compact set K in Ω is said to be the support of T if $\Omega \setminus K$ is the largest open set in Ω for which $T|_{\Omega\setminus K} = 0$ holds. Then we denote $K = \mathrm{supp}(T)$.

Proposition 3. *We use the notation in Proposition 1. Let K be a compact subset of Ω and put*

$$\mathcal{B}_K(\Omega;\, E) = \{T \in \mathcal{B}(\Omega;\, E); \mathrm{supp}(T) \subset K\}.$$

Then we have the inclusion

$$\mathcal{A}'(K;\, E) \subset \mathcal{B}_K(\Omega;\, E).$$

In order to prove the inclusion $\mathcal{B}_K(\Omega;\, E) \subset \mathcal{A}'(K;\, E)$, it is sufficient to know the following.

Problem A. For two compact sets K_1 and K_2 in M and put $K = K_1 \cup K_2$. Then is the sequence

$$\mathcal{A}'(K_1;\, E) \oplus \mathcal{A}'(K_2;\, E) \longrightarrow \mathcal{A}'(K;\, E) \longrightarrow 0$$

exact?

If E is a Fréchet space, the answer to Problem A is affirmative. But in general we do not know any answer.

Next we consider E-valued Sato hyperfunctions on M.

Let $\{\mathcal{B}_1(\Omega;\, E); \Omega \text{ is an open set in } M\}$ be the presheaf over M defined as follows:

$$\mathcal{B}_1(\Omega;\, E) = \{0\}, \text{ if } \Omega \text{ is not relatively compact,}$$
$$\mathcal{B}_1(\Omega;\, E) = \mathcal{B}(\Omega;\, E), \text{ if } \Omega \text{ is relatively compact.}$$

The restriction mappings are defined as follows:

$$\mathcal{B}_1(\Omega;\, E) \longrightarrow \mathcal{B}_1(\omega;\, E)$$

$$0 \longrightarrow 0 \text{ if } \Omega \text{ is not relatively compact with } \Omega \supset \omega,$$

$$T \longrightarrow T|_\omega \text{ if } \Omega \text{ is relatively compact with } \Omega \supset \omega.$$

This presheaf satisfies the null condition of sheaves but not the glueing condition.

We denote by ${}^E\mathcal{B}$ the sheaf associated to this presheaf $\{\mathcal{B}_1(\Omega;\, E); \Omega$ is an open set in $M\}$. It is a sheaf of vector spaces over M.

Definition 2. The sheaf ${}^{E}\mathcal{B}$ is called the sheaf of E-valued Sato hyperfunctions over M.

Here we present the following problem.

Problem B. Is the sheaf ${}^{E}\mathcal{B}$ flabby?

If E is a Fréchet space, the answer to Problem B is affirmative. But in general we do not know any answer. If the presheaf $\{\mathcal{B}(\omega, E) = \mathcal{A}'(\omega^{\text{cl}}; E)/\mathcal{A}'(\partial\omega; E); \omega$ is an open subset of $\Omega\}$ over a relatively compact open set Ω in M becomes a flabby sheaf, then we can show that the answer to the Problem B is affirmative.

We can define some operations on analytic-linear mappings and E-valued Sato hyperfunctions such as tensor products, multiplication by a holomorphic or real-analytic functions, differentiation and pull back by an analytic diffeomorphism. Thereby we can prove Kernel Theorems for analytic-linear mappings and E-valued Sato hyperfunctions.

Acknowledgements. I wish to thank Professor Morimoto for his invitation to this Congress and for useful remarks in the preparation of this work.

References

[1] P. D. F. Ion and T. Kawai, Theory of vector-valued hyperfunctions, Publ. RIMS, Kyoto Univ., **11**(1975), 1-19.

[2] M. Itano, On the distributional boundary values of vector valued holomorphic functions, J. Sci. Hiroshima Ser.A **32**(1968), 397-440.

[3] Y. Ito, On the theory of vector valued hyperfunctions, J. Math. Tokushima Univ., **13**(1979), 29-51.

[4] Y. Ito, Theory of analytic linear mappings, I. General theory, J. Math. Tokushima Univ., **14**(1980), 25-74.

[5] Y. Ito, Analytic linear mappings and vector valued hyperfunctions, J. Math. Tokushima Univ., **15**(1981), 1-51.

[6] Y. Ito, On the abstract Cauchy problems in the sense of Fourier hyperfunctions, J. Math. Tokushima Univ., **16**(1982), 25-31.

[7] Y. Ito, Fourier hyperfunction semi-groups, J. Math. Tokushima Univ., **16**(1982), 33-53.

[8] Y. Ito, Theory of (vector valued) Fourier hyperfunctions. Their realization as boundary values of (vector valued) slowly increasing holomorphic functions, (I), J. Math. Tokushima Univ., **18**(1984), 57-101.

[9] Y. Ito, Theory of (vector valued) Fourier hyperfunctions. Their realization as boundary values of (vector valued) slowly increasing holomorphic functions, (II), J. Math. Tokushima Univ., **19**(1985), 25-61.

[10] Y. Ito, Stationary Fourier hyperprocesses, Publ. RIMS, Kyoto Univ., **22-1**(1986), 31-42.

[11] Y. Ito, Theory of (vector valued) Sato hyperfunctions. Their realization as boundary values of (vector valued) holomorphic functions, J. Math. Tokushima Univ., **20**(1986), 27-39.

[12] Y. Ito, Fourier hyperfunctions of general type, J. Math. Kyoto Univ., **28-2**(1988), 213-265.

[13] Y. Ito, Theory of (vector valued) Fourier hyperfunctions. Their realization as boundary values of (vector valued) slowly increasing holomorphic functions, (III), J. Math. Tokushima Univ., **23**(1989), 23-38.

[14] Y. Ito, Theory of (vector valued) Fourier hyperfunctions. Their realization as boundary values of (vector valued) slowly increasing holomorphic functions, (IV), J. Math. Tokushima Univ., **24**(1990), 13-21.

[15] Y. Ito, Vector valued Fourier hyperfunctions, J. Math. Kyoto Univ., **32-2**(1992), 259-285.

[16] Y. Ito, Theory of (vector valued) Fourier hyperfunctions. Their realization as boundary values of (vector valued) slowly increasing holomorphic functions, (V), J. Math. Tokushima Univ., **26**(1992), 39-50.

[17] Y. Ito, Boundary values of (slowly increasing) holomorphic functions, J. Math. Tokushima Univ., **27**(1993), 23-35.

[18] Y. Ito, Stationary Fourier hyperfields, J. Math. Tokushima Univ., **28**(1994), 39-54.

[19] Y. Ito, Theory of Fourier microfunctions of several types (I), J. Math. Tokushima Univ., **29**(1995), 55-75.

[20] Y. Ito, Theory of Fourier microfunctions of several types (II), J. Math. Tokushima Univ., **30**(1996), 45-69.

[21] Y. Ito, Theory of Fourier microfunctions of several types (III), J. Math. Tokushima Univ., **31**(1997), 29-61.

[22] Y. Ito, Sato hyperfunctions valued in a locally convex space, J. Math. Tokushima Univ., **32**(1998), 27-41.

[23] Y. Ito, Theory of (vector valued) Sato hyperfunctions on a real analytic manifold, to appear.

[24] Y. Ito, Theory of general Fourier hyperfunctions, to appear.

[25] Y. Ito and S. Nagamachi, Theory of H-valued Fourier hyperfunctions, Proc. Japan Acad., **51-7**(1975), 558-561.

[26] Y. Ito and S. Nagamachi, On the theory of vector valued Fourier hyperfunctions, J. Math. Tokushima Univ., **9**(1975), 1-33.

[27] D.Vogt, Vektorwertige Distributionen als Randverteilungen holomorpher Funktionen, Manuscripta Math. **17**(1975), 267-290.

90 DENSITY THEOREMS FOR EXPONENTIAL SYSTEMS IN THE BERGMAN SPACE OF HOLOMORPHIC FUNCTIONS

LÊ Hai Khôi [1]

Hanoi Institute of Information Technology
Hoang Quoc Viet Rd., Cau Giay Dist.
10000 Hanoi, Vietnam

Lhkhoi@ ioit.ncst.ac.vn

Abstract

The aim of this note is to study the density of exponential systems in the Bergman space of holomorphic functions in a bounded convex domain of $\mathbf{C}^n$. The weighted space of entire functions is involved in this study.

1 Introduction

Let H be a Hilbert space of holomorphic functions in $\mathbf{C}^n$ with the inner product $(. \mid .)_H$. Suppose that for every $\zeta \in \mathbf{C}^n$ the function $e_\zeta(z) = e^{\zeta \cdot z}$ belongs to H. We shall be concerned with a question whether there exists a sequence $\Lambda = \left(\lambda^k\right)_{k=1}^{\infty}$ of complex vectors in $\mathbf{C}^n$ such that the

[1]Supported in part by Grant-in-Aid for Scientific Research (No. 11640181), Ministry of Education, Science and Culture (Japan) and the National Basic Research Program in Natural Sciences (Vietnam).

H.G.W. Begehr et al. (eds.), Proceedings of the Second ISAAC Congress, Vol. 1, 789–797.

corresponding exponential system $\mathcal{E}_\Lambda = \left(e^{\lambda_k \cdot z}\right)_{k=1}^{\infty}$ is dense in H for the Bergman space of holomorphic functions in a bounded convex domain of $\mathbf{C}^n$. Related topics are also treated.

We consider the following transformation on the space H:

$$\tilde{f}(\zeta) = (f \mid e_\zeta)_H, \ f \in H, \ \zeta \in \mathbf{C}^n.$$

This is an entire function in $\mathbf{C}^n$.

The density of the exponential system $\mathcal{E}_\Lambda$ in the space H is equivalent to the implication

$$\tilde{f}(\lambda^k) = 0, \ \forall k \geq 1 \implies \tilde{f} \equiv 0.$$

The ideas of our study is to find conditions for the transformation $\tilde{f}$ to be in some weighted space of entire functions for which there exists a discrete set of uniqueness. The last question, as is well known, can be derived from the existence of discrete weakly sufficient sets.

The author thanks Professors Joji Kajiwara and Mitsuo Morimoto for attention to this work.

2 The weighted space of entire functions

In this section we introduce a weighted space of entire functions in a general setting which will play a crucial role in our discussions.

Let $\Phi = (\varphi_k)_{k=1}^{\infty}$ be an increasing sequence of real functions defined in $\mathbf{C}^n$. To each φ_k we associate the following Banach space

$$E_k = \left\{ f \text{ entire in } \mathbf{C}^n; \ |f|_k = \sup_{\mathbf{C}^n} \frac{|f(\zeta)|}{e^{\varphi_k(\zeta)}} < +\infty \right\}, \ k = 1, 2, \ldots$$

Then the vector space $E_\Phi = E_\Phi(\mathbf{C}^n) = \cup_{k=1}^{\infty} E_k$ can be endowed with a topology of internal inductive limit of the spaces E_k and is called a *weighted space of entire functions.*

For an arbitrary subset S of $\mathbf{C}^n$ we can produce in the space E_Φ another weaker internal inductive topology with respect to the seminormed spaces

$$E_k^S = \left\{ f \in E_\Phi; \ |f|_k^S = \sup_{S} \frac{|f(\zeta)|}{e^{\varphi_k(\zeta)}} < +\infty \right\}, \ k = 1, 2, \ldots$$

In the case when these two topologies are equivalent, the set S is called a *weakly sufficient set* for the space E_Φ (see, e.g., [7]).

It can be verified that a set S is weakly sufficient for the space E_Φ if and only if $\forall k\ \exists m : E_k^S \hookrightarrow E_m$, here and in the sequel the symbol $\hookrightarrow$ means the continuous imbedding. From this fact it follows that any weakly sufficient set for the space E_Φ is a set of uniqueness for this space. Furthermore, it was proved in [1] that under certain conditions if S is a set of uniqueness and $\forall k\ \exists m : E_k^S \subset E_m$, then S is a weakly sufficient set.

There are several papers that dealt with the existence of a discrete weakly sufficient set in the weighted space E_Φ under rather general assumptions imposed on the weight system $\Phi = (\varphi_k)$ (see, e.g; [2, 3] and references therein).

We refer the reader to [1 - 3] for having information in details. It should be noted that in the paper [3] the sequence $\Lambda = (\lambda^k)$ is constructed explicitly, while in [2] it was proved only the existence of such a sequence. Also we mention that the weak sufficiency of the sequence Λ is related to the possibility of representation of holomorphic functions by Dirichlet series with frequencies (λ^k).

3 The Bergman space of holomorphic functions

Let Ω be a bounded convex domain in $\mathbf{C}^n$ with the supporting function $H_\Omega(\zeta)$ $(0 \in \Omega)$. Denote by $\mathcal{O}(\Omega)$ the space of holomorphic functions on Ω with the topology of uniform convergence on compact subsets of Ω.

The Bergman space on the domain Ω is defined as follows

$$A^2(\Omega) = L^2(\Omega) \cap \mathcal{O}(\Omega) = \left\{ f \in \mathcal{O}(\Omega);\ \|f\|_2 = \left[\int_\Omega |f(z)|^2 dV(z)\right]^{1/2} < +\infty \right\},$$

where $V(\zeta)$ is the Lebesgue measure. This is a Hilbert space with the inner product

$$(f \mid g) = \int_\Omega \overline{f(z)} g(z) dV(z).$$

It is obvious, due to the boundedness of the domain Ω, that $e_\zeta(z) \in A^2(\Omega)$, $\forall \zeta \in \mathbf{C}^n$. The considered transformation of functions from the Bergman space $A^2(\Omega)$ has the form

$$\tilde{f}(\zeta) = \int_\Omega \overline{f(z)} e^{\zeta \cdot z} dV(z),\ f \in A^2(\Omega),\ \zeta \in \mathbf{C}^n. \tag{3.1}$$

Furthermore, as is well known, the dual space $\mathcal{O}(\Omega)^*$ can be identified with the space $[1, H_\Omega)$ of entire functions F on $\mathbf{C}^n$ with the rate of growth

$$|F(\zeta)| = O\left(e^{(1-\varepsilon)H_\Omega(\zeta)}\right).$$

Moreover, this space $[1, H_\Omega)$ can be expressed in the form of weighted space of entire functions E_Φ introduced above with $\varphi_k(\zeta) = q_k H_\Omega(\zeta)$, where $0 < (q_k) \uparrow 1$.

We prove the following result.

Theorem 3.1. *Let $\Lambda = (\lambda^k)_{k=1}^\infty$ be a sequence of complex vectors in $\mathbf{C}^n$ that is the set of uniqueness for the space $[1, aH_\Omega)$ for some $a > 1$. Then the corresponding exponential system $\mathcal{E}_\Lambda$ is dense in the space $A^2(\Omega)$.*

Proof. Since the space $H^\infty(\Omega)$ of bounded holomorphic functions in Ω is dense in the space $A^2(\Omega)$, it suffices to prove that the system $\mathcal{E}_\Lambda$ is dense in $H^\infty(\Omega)$.

We shall show that if $f \in H^\infty(\Omega)$, then $\tilde{f}$ belongs to the space $[1, aH_\Omega)$ for some $a > 1$. Indeed, let $f \in H^\infty(\Omega)$. Due to its boundedness in Ω, from (3.1) it follows that for some (for any) $a > 1$

$$|\tilde{f}(\zeta)| \le C_\Omega e^{H_\Omega(\zeta)} < C_\Omega e^{aH_\Omega(\zeta)}, \ \forall \zeta \in \mathbf{C}^n \setminus \{0\},$$

which means that $\tilde{f}$ belongs to the space $[1, aH_\Omega)$.

Consequently, if $\tilde{f}(\lambda^k) = 0, \ \forall k \ge 1$ then $\tilde{f}$ should be zero identically.

Corollary 3.2. *There exists a sequence $\Lambda = (\lambda^k)$ that is a set of uniqueness for the space $[1, aH_\Omega)$. Moreover, this sequence can be constructed explicitly.*

Proof. Such a discrete set of uniqueness in Theorem 3.1 always exists, for from [3] it follows, in particular, that it can be constructed explicitly a sequence $\Lambda = (\lambda^k) \subset \mathbf{C}^n$ that is weakly sufficient for all spaces $[1, aH_\Omega)$, $a > 0$. We can also see by the method of [3] that for every $a > 0$ there is a discrete weakly sufficient set Λ for the space $[1, aH_\Omega)$. In any case Λ is a set of uniqueness for $[1, aH_\Omega)$.

4 The Hilbert space of holomorphic Dirichlet series

Throughout this section the following are considered to be given: a bounded convex domain Ω in $\mathbf{C}^n$ with supporting function $H_\Omega(\zeta)$, a sequence $\Lambda = (\lambda^k)_{k=1}^\infty$, $\lambda^k = (\lambda_1^k, \dots, \lambda_n^k)$, satisfying condition

$$\lim_{k\to\infty} \frac{\log k}{|\lambda^k|} = 0.$$

As is well known, a Dirichlet series

$$\sum_{k=1}^{\infty} c_k e^{\lambda^k . z}, \quad z \in \Omega,$$

converges for the compact-open topology of the space $\mathcal{O}(\Omega)$ if and only if it converges absolutely for this topology. Moreover, the coefficients of such series are characterized completely by the following formula

$$\limsup_{k\to\infty} \frac{\log|c_k| + H_\Omega(\lambda^k)}{|\lambda^k|} \leq 0. \tag{4.1}$$

In connection with this we denote

$$E_\Omega = \{(c_k); (4.1) \text{ is satisfied}\}.$$

This sequence space E_Ω, together with the other sequence spaces, which are closely related to the class of sequences of coefficients of Dirichlet series, as well as associated to them fucntions spaces, were introduced and studied systematically in a series of our works (see, e.g., [4 - 6]).

Further, to each sequence of real numbers (α_k) we associate the following sequence space

$$X_\alpha = \left\{(c_k);\ \|c\|_\alpha^2 = \sum_{k=1}^{\infty} |c_k|^2 e^{2\alpha_k} < +\infty\right\},$$

which is a Hilbert space with the scalar product

$$(c \mid d) = \sum_{k=1}^{\infty} c_k \overline{d}_k e^{2\alpha_k}, \ \forall c = (c_k),\ d = (d_k) \in X_\alpha.$$

Also denote

$$\alpha_* = \alpha_*(\Omega) = \liminf_{k\to\infty} \frac{\alpha_k - H_\Omega(\lambda^k)}{|\lambda^k|}.$$

It was proved in [6] that $X_\alpha \subset E_\Omega$ if and only if $\alpha_* \geq 0$ and moreover, these spaces never coincide.

In the sequel we suppose that for a sequence of real numbers (α_k) the condition $\alpha_* \geq 0$ is always satisfied. Let $(c_k) \in X_\alpha$. Then the function

$$f(z) = \sum_{k=1}^{\infty} c_k e^{\lambda^k . z}, \quad z \in \Omega, \tag{4.2}$$

belongs to the space $\mathcal{O}(\Omega)$, the series in the right-hand side converges (absolutely) in the compact-open topology of $\mathcal{O}(\Omega)$. Therefore, the function space associated to it

$$X_\alpha(\Lambda,\Omega)=\left\{f(z)=\sum_{k=1}^{\infty}c_ke^{\lambda^k.z};\quad (c_k)\in X_\alpha,\ z\in\Omega\right\}$$

is a subspace of the space $\mathcal{O}(\Omega)$.

Note that among all representations of a function $f\in X_\alpha(\Lambda,\Omega)$ in the form (4.2) there exists a unique one with the minimum Hilbert norm in the space X_α [6], denoted by

$$f(z)=\sum_{k=1}^{\infty}f_k^*e^{\lambda^k.z}.$$

We call $\left(f_k^*\right)$ a minimum coefficients sequence of f.

Then $X_\alpha(\Lambda,\Omega)$ is a Hilbert space with the inner product

$$(f\mid g)=\sum_{k=1}^{\infty}\overline{f_k^*}g_k^*e^{2\alpha_k},\ \forall f,\ g\in X_\alpha(\Lambda,\Omega).$$

Since $e^{\lambda^k.z}\in X_\alpha(\Lambda,\Omega)$, $\forall k\geq 1$, a question arises: when does the inclusion $X_\alpha(\Lambda,\Omega)\subset A^2(\Omega)$ hold?

Proposition 4.1. *If $\alpha_*>0$ then $X_\alpha(\Lambda,\Omega)\hookrightarrow A^2(\Omega)$.*

Proof. Let $c=\left(c_k\right)$ be an element of X_α. Consider its associated Dirichlet series

$$f(z)=\sum_{k=1}^{\infty}c_ke^{\langle\lambda^k,z\rangle},\quad z\in\Omega.$$

By Cauchy-Schwarz inequality, for every $z\in\Omega$, we have

$$|f(z)|=\left|\sum_{k=1}^{\infty}c_ke^{\langle\lambda^k,z\rangle}\right|\leq\sqrt{\sum_{k=1}^{\infty}|c_k|^2e^{2\alpha_k}}\cdot\sqrt{\sum_{k=1}^{\infty}e^{2(\Re\langle\lambda^k,z\rangle-\alpha_k)}}$$

$$=\|c\|_\alpha\cdot\sqrt{\sum_{k=1}^{\infty}e^{2(\Re\langle\lambda^k,z\rangle-\alpha_k)}}\leq\|c\|_\alpha\cdot\sqrt{\sum_{k=1}^{\infty}e^{2[H_\Omega(\lambda^k)-\alpha_k]}}.$$

We suffice to prove that the last series converges. From the definition of α_* it follows that for $\varepsilon>0$ there exists N such that $\forall k\geq N$

$$\alpha_k\geq(\alpha_*-\varepsilon)|\lambda^k|+H_\Omega(\lambda^k),$$

or equivalently,

$$H_\Omega(\lambda^k) - \alpha_k \le (\varepsilon - \alpha_*)|\lambda^k|.$$

As $\alpha_* > 0$ we can choose from the begining $\varepsilon < \alpha_*$. Then the desired convergence follows.

Thus, there exists a positive constant C such that

$$\|f\|_{A^2(\Omega)} \le C\|c\|_\alpha, \ \forall c = \big(c_k\big) \in X_\alpha,$$

which shows that $f(z) \in A^2(\Omega)$. The proposition is proved.

We can ask whether the condition $\alpha_* > 0$ is necessary for such an inclusion. Obviously, this is so if we have

$$\sum_{k=1}^{\infty} e^{2[H_\Omega(\lambda^k)-\alpha_k]} < +\infty. \tag{4.3}$$

The consition $\alpha_* > 0$ is sufficient for (4.3) to hold, but it is not necessary at all. We can show that there exist sequences $\big(\alpha_k\big)$ with $\alpha_* = 0$ for which (4.3) holds. Indeed, consider the following $\big(\alpha_k\big)$:

$$\alpha_k = H_\Omega(\lambda^k) + \tau \log k, \ \ \tau > 1/2.$$

It is clear that $\alpha_* = 0$. Furthermore, $H_\Omega(\lambda^k) - \alpha_k = -\tau \log k$ and (4.3) is satisfied. This means that $X_\alpha(\Lambda, \Omega) \hookrightarrow A^2(\Omega)$.

Combining Theorem 3.1 and Proposition 4.1 yields

Theorem 4.2. *There exist* $\Lambda = \big(\lambda^k\big)$ *and* $\big(\alpha_k\big)$ *such that* $X_\alpha(\Lambda, \Omega)$ *is dense in the Bergman space* $A^2(\Omega)$.

Another question is whether $X_\alpha(\Lambda, \Omega)$ is closed in $A^2(\Omega)$? Note that Proposition 4.1 actually shows that if $\alpha_* > 0$ then $X_\alpha(\Lambda, \Omega)$ is contained not only in $A^2(\Omega)$ but also in $H^\infty(\Omega)$.

The following result, which shows that no infinite-dimensional closed subspace of the Bergman space can contain only bounded functions, was kindly informed to the author by S. Axler.

Proposition 4.3. *Let* Ω *be a bounded open set of* $\mathbf{C}^n$. *Let* X *be an infinite-dimensional closed subspace of the space* $A^2(\Omega)$. *Then* X *contains an unbounded function.*

Proof. Suppose that $X \subset H^\infty(\Omega)$. The closed graph theorem implies that the inclusion map from X (with the L^2-norm) to H^∞ is continuous. Thus there is a constant C such that

$$\|f\|_\infty < C,$$

for all f in X with $\|f\|_2 = 1$.

Because X is an infinite-dimensional subspace of $A^2(\Omega)$, X has an orthonormal basis $\big(f_m\big)$. Since each $\|f_m\|_2 = 1$, we have $\|f_m\|_\infty < C$. For each point $z \in \Omega$, there exists $k_z \in A^2(\Omega)$ such that

$$h(z) = \langle h, k_z \rangle,$$

for all $h \in A^2(\Omega)$. Thus $f_m(z) = \langle f_m, k_z \rangle$. As $\big(f_m\big)$ is an orthonormal basis, we have

$$\sum_m |f_m(z)|^2 = \sum_m |\langle f_m, k_z \rangle|^2 < \infty,$$

for each $z \in \Omega$. In particular,

$$f_m(z) \to 0 \text{ as } n \to \infty,$$

for every $z \in \Omega$. This condition, combined with the uniform boundedness of $\big(f_m\big)$ on Ω, implies that $\|f_m\|_2$ tends to 0 (here we use Lebesgue dominated convergence theorem), contradicting $\|f_m\|_2 = 1$. This contradiction completes the proof.

References

[1] Abanin, A. V., *Some criteria for weak sufficiency* (Russian), Mat. Zametki **40** (1986), 442–454. English transl. in Math. Notes **40** (1986), 757–764.

[2] Epifanov, O. V., *Variations of weakly sufficient sets in spaces of analytic functions* (Russian), Izv. Vyssh. Uchebn. Zaved. Mat. (1986), N7, 50–56. English transl. in Soviet Math. (Iz.VUZ) **30** (1986), 67–74.

[3] Lê Hai Khôi, *Espaces conjugués, ensembles faiblement suffisants discrets et systèmes de représentation exponentielle*, Bull. Sci. Math. (2) **113** (1989), 309–347.

[4] Lê Hai Khôi, *Holomorphic Dirichlet series in several variables*, Math. Scand. **77** (1995), 85–107.

[5] Lê Hai Khôi, *Convolution operators on holomorphic Dirichlet series*, Tokyo J. Math. **20** (1997), 389–402.

[6] Lê Hai Khôi, *Hilbert spaces of holomorphic Dirichlet series and applications to convolution equations*, J. Math. Anal. Appl. **206** (1997), 10–24.

[7] Schneider, D. M., *Sufficient sets for some spaces of entire functions*, Trans. Amer. Math. Soc. **197** (1974), 161–180.

International Society for Analysis, Applications and Computation

1. H. Florian et al. (eds.): *Generalized Analytic Functions*. Theory and Applications to Mechanics. 1998 ISBN 0-7923-5043-X
2. H.G.W. Begehr et al. (eds.): *Partial Differential and Integral Equations*. 1999 ISBN 0-7923-5482-6
3. S. Saitoh, D. Alpay, J.A. Ball and T. Ohsawa (eds.): *Reproducing Kernels and their Applications*. 1999 ISBN 0-7923-5618-7
4. R.P. Gilbert, J. Kajiwara and Y.S. Xu (eds.): *Recent Developments in Complex Analysis and Computer Algebra*. 1999 ISBN 0-7923-5999-2
5. R.P. Gilbert, J. Kajiwara and Y.S. Xu (eds.): *Direct and Inverse Problems of Mathematical Physics*. 2000 ISBN 0-7923-6005-2
6. H.G.W. Begehr, A. Okay Celebi and W. Tutschke (eds.): *Complex Methods for Partial Differential Equations*. 1999 ISBN 0-7923-6000-1
7. H.G.W. Begehr, R.P. Gilbert and J. Kajiwara (eds.): *Proceedings of the Second ISAAC Congress*. Volume 1. 2000 ISBN 0-7923-6597-6
8. H.G.W. Begehr, R.P. Gilbert and J. Kajiwara (eds.): *Proceedings of the Second ISAAC Congress*. Volume 2. 2000 ISBN 0-7923-6598-4

KLUWER ACADEMIC PUBLISHERS – DORDRECHT / BOSTON / LONDON